Quantum Information with Continuous Variables of Atoms and Light

Quantum Information with Continuous Variables of Atoms and Light

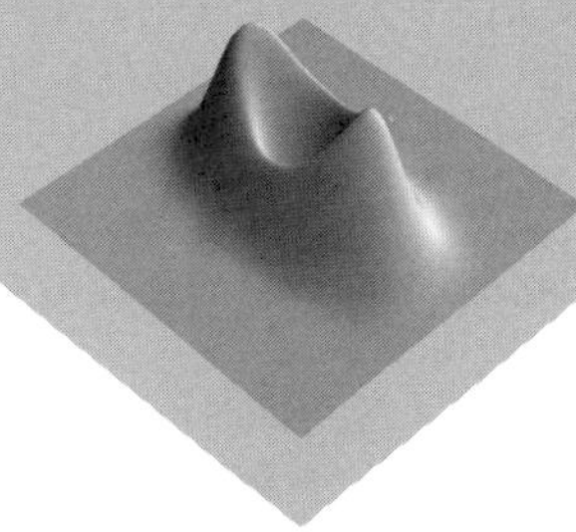

Editors

N. J. CERF
Université Libre de Bruxelles, Belgium

G. LEUCHS
Universität Erlangen-Nürnberg, Germany

E. S. POLZIK
Niels Bohr Institute, Denmark

ICP

Imperial College Press

Published by

Imperial College Press
57 Shelton Street
Covent Garden
London WC2H 9HE

Distributed by

World Scientific Publishing Co. Pte. Ltd.
5 Toh Tuck Link, Singapore 596224
USA office: 27 Warren Street, Suite 401-402, Hackensack, NJ 07601
UK office: 57 Shelton Street, Covent Garden, London WC2H 9HE

British Library Cataloguing-in-Publication Data
A catalogue record for this book is available from the British Library.

QUANTUM INFORMATION WITH CONTINUOUS VARIABLES OF ATOMS AND LIGHT

ISBN-13 978-1-86094-760-5
ISBN-10 1-86094-760-3
ISBN-13 978-1-86094-776-6 (pbk)
ISBN-10 1-86094-776-X (pbk)

Printed in Singapore by Mainland Press

... continuous quantum variables are the language used in the original formulation of the EPR gedankenexperiment:

Thus, by measuring either P [the momentum] or Q [the coordinate of the first system] we are in a position to predict with certainty, and without in any way disturbing the second system, either the value of the quantity P [...] or the value of the quantity Q [...]. In accordance with our criterion of reality, in the first case we must consider the quantity P as being an element of reality, in the second case the quantity Q is an element of reality. But, as we have seen, both wave functions [the eigenfunctions of P and Q] belong to the same reality.

A. Einstein, B. Podolsky and N. Rosen (1935)

Preface

This book is a joint effort of a number of leading research groups actively developing the field of quantum information processing and communication (QIPC) with continuous variables. The term "continuous" refers to the fact that the description of quantum states within this approach is carried out in the phase space of canonical variables, x and p, which are indeed continuous variables over an infinite dimensional Hilbert space. Historically, the field of QIPC with continuous variables has dealt mostly with Gaussian states, such as coherent states, squeezed states, or Einstein-Podolsky-Rosen (EPR) two-mode entangled states. A powerful mathematical formalism for Gaussian states, which are completely described by only first and second order momenta, is presented in the first part of this book in the chapters by G. Adesso and F. Illuminati (entanglement properties of Gaussian states) and by J. Eisert and M. M. Wolf (Gaussian quantum channels). This is a useful tool in the study of entanglement properties of harmonic chains (see chapter by K. M. R. Audenaert *et al.*), as well as in the description of quantum key distribution based on coherent states (see chapter by F. Grosshans *et al.*). A more exotic topic involving Gaussian states is covered in the chapter by O. Krüger and R. F. Werner (Gaussian quantum cellular automata).

Gaussian operations on Gaussian states alone do not allow for the purification and distillation of continuous-variable entanglement, features which are critical for error corrections in QIPC, so that the recourse to non-Gaussian operations is necessary (see chapter by J. Fiurášek *et al.*). Non-Gaussian operations are also crucial in order to build loophole-free Bell tests that rely on homodyne detection (see chapter by R. García-Patrón). Interestingly, the continuous-variable formalism is also appropriate for the analysis of non-Gaussian states, such as Fock states, qubit (quantum bit) states, and coherent superposition (Schrödinger cat) states. Indeed, the Wigner function over an infinite dimensional Hilbert space provides the most complete description of any state, including a discrete variable, qubit state. The Hilbert space may be spanned by the Fock state basis in the case of a single field mode, or, in the case of single photons, by the spectral mode functions. The characterization of such non-Gaussian states by homodyne tomography is reviewed in the chapter by G. M. D'Ariano *et al.* Then, recent theoretical developments in the generation of particular non-Gaussian states (Schrödinger cat states) are presented in the chapter by H. Jeong and T. C. Ralph.

Continuous variables have played a particularly important role in QIPC with light, due to the highly efficient and well experimentally developed method of "homodyne detection", which provides a direct access to the canonical variables of light. This area of "optical continuous variables" is covered in the second part of this book. Here, the variables x and p are the two quadrature phase operators associated with the sine and cosine components of the electromagnetic field. By mixing the quantum light field under investigation with a strong classical "local oscillator" light on a beam splitter, the variables x and p can readily be observed, and hence a complete description of the quantum field is obtained. If one takes into account the polarization of light as an additional degree of freedom, the Stokes operators have to be introduced and the notions of polarization squeezing and polarization entanglement arise, as described in the chapter by N. Korolkova.

Several recent experiments with continuous variables of light are presented in this part of the book. For example, the chapters by J. Laurat *et al.*, O. Glöckl *et al.*, and V. Josse *et al.* present the generation of EPR entangled light via the optical nonlinearities provided by solid state materials and cold atoms. Some other chapters present several applications of optical continuous variables to QIPC protocols, such as quantum teleportation by N. Takei *et al.*, quantum state sharing by T. Tyc *et al.*, and quantum cloning by U. L. Andersen *et al.* Applications of continuous-variable squeezing to ultra-precise measurements are covered in the chapters by C. Fabre *et al.* (quantum imaging) and by R. Schnabel (towards squeezing-enhanced gravitational wave interferometers). For single-photon states, the concept of canonical continuous variables can be transferred to other observables, e.g. the position x and wave vector k, as shown in the chapter by L. Zhang *et al.*

The non-Gaussian operations such as photon counting combined with the continuous-variable homodyne-based analysis of the light conditioned on photon counting take QIPC with optical continuous variables into a new domain. This domain, where the purification of entanglement and error correction is, in principle, possible, is explored experimentally in the chapters by J. Wenger *et al.* (photon subtracted squeezed states) and by

A. I. Lvovsky and M. G. Raymer (single-photon Fock states). The latter chapter reports on the progress in experimental quantum tomography and state reconstruction.

Another avenue in QIPC with continuous variables has opened up when it was realized that multi-atomic ensembles can well serve as efficient storage and processing units for quantum information. The third part of this book is devoted to the development and application of this approach based on "atomic continuous variables". The quantum interface between light pulses carrying quantum information and atomic processors has become an important ingredient in QIPC, as some of the most spectacular recent developments of the light-atoms quantum interface have been achieved with atomic ensembles. The continuous-variable approach to atomic states has proven to be very competitive compared to the historically first single atom and cavity QED approach.

The theory of quantum non-demolition measurement on light transmitted through atoms, quantum feedback, and multi-pass interaction of light with atoms, is presented in the chapters by L. B. Madsen and K. Mølmer and by R. van Handel *et al.* Experiments on spin squeezing of atoms are described in the chapter by J. M. Geremia, while the theory and experiments of EPR entanglement of distant atomic objects and quantum memory for light are presented in the chapter by K. Hammerer *et al.* Atomic ensembles can also serve as sources of qubit-type entanglement. In this case, a single qubit state is distributed over the entire multi-atomic ensemble, providing thus a conceptual bridge between a discrete computational variable and a continuous (or collective) variable used as its physical implementation. The work towards the implementation of a promising proposal for the generation of such type of entanglement conditioned on photon detection (the Duan-Lukin-Cirac-Zoller protocol) is presented in the chapter by C. W. Chou *et al.* Interestingly, such an analysis of qubits in the continuous-variable language makes the old sharp boundary between continuous and discrete variables softer. Finally, the theory of decoherence suppression in quantum memories for photons is discussed in the chapter by M. Fleischhauer and C. Mewes.

In summary, this book is aimed at providing a comprehensive review of the main recent progresses in continuous-variable quantum information processing and communication, a field which has been rapidly developing both theoretically and experimentally over the last five years. It was originally intended to review the main advances that had resulted from the project "Quantum Information with Continuous Variables" (QUICOV) funded by the European Commission from 2000 to 2003. However, given the unexpected pace at which new paradigms and applications continued to appear, it soon became clear that this objective had become too restrictive. Instead, this book evolved into a compilation of the even more recent achievements that were reported in the series of workshops especially devoted to continuous-variable QIPC that took place in Brussels (2002), Aix-en-Provence (2003), Veilbronn (2004), and Prague (2005). Yet, the picture would not have been complete without the contributions of several additional world experts, which have rendered this book fairly exhaustive. We are confident that the various directions explored in the 27 chapters of this book will form a useful basis in order to approach continuous-variable QIPC. This is, however, probably not the end of the story, and we expect that future developments in this field will open new horizons in quantum state engineering, quantum computing and communication.

We warmly thank Gerlinde Gardavsky for her careful work on preparing the lay-out, correcting and proof-reading this book.

Nicolas J. Cerf
Gerd Leuchs
Eugene S. Polzik
Editors

Contents

Part III: Atomic Continuous Variables

Part I: Fundamental Concepts

Chapter 1

Bipartite and Multipartite Entanglement of Gaussian States

Gerardo Adesso and Fabrizio Illuminati
Dipartimento di Fisica "E. R. Caianiello", Università di Salerno; CNR-Coherentia, Gruppo di Salerno; and INFN Sezione di Napoli-Gruppo Collegato di Salerno, Via S. Allende, 84081 Baronissi (SA), Italy
email: gerardo@sa.infn.it, illuminati@sa.infn.it

In this chapter we review the characterization of entanglement in Gaussian states of continuous variable systems. For two-mode Gaussian states, we discuss how their bipartite entanglement can be accurately quantified in terms of the global and local amounts of mixedness, and efficiently estimated by direct measurements of the associated purities. For multimode Gaussian states endowed with local symmetry with respect to a given bipartition, we show how the multimode block entanglement can be completely and reversibly localized onto a single pair of modes by local, unitary operations. We then analyze the distribution of entanglement among multiple parties in multimode Gaussian states. We introduce the continuous-variable tangle to quantify entanglement sharing in Gaussian states and we prove that it satisfies the Coffman-Kundu-Wootters monogamy inequality. Nevertheless, we show that pure, symmetric three–mode Gaussian states, at variance with their discrete-variable counterparts, allow a promiscuous sharing of quantum correlations, exhibiting both maximum tripartite residual entanglement and maximum couplewise entanglement between any pair of modes. Finally, we investigate the connection between multipartite entanglement and the optimal fidelity in a continuous-variable quantum teleportation network. We show how the fidelity can be maximized in terms of the best preparation of the shared entangled resources and, viceversa, that this optimal fidelity provides a clearcut operational interpretation of several measures of bipartite and multipartite entanglement, including the entanglement of formation, the localizable entanglement, and the continuous-variable tangle.

1. Introduction

One of the main challenges in fundamental quantum theory as well as in quantum information and computation sciences lies in the characterization and quantification of bipartite entanglement for mixed states, and in the definition and interpretation of multipartite entanglement both for pure states and in the presence of mixedness. While important insights have been gained on these issues in the context of qubit systems, a less satisfactory understanding has been achieved until recent times on higher-dimensional systems, as the structure of entangled states in Hilbert spaces of high dimensionality exhibits a formidable degree of complexity. However, and quite remarkably, in infinite-dimensional Hilbert spaces of continuous-variable systems, ongoing and coordinated efforts by different research groups have led to important progresses in the understanding of the entanglement properties of a restricted class of states, the so-called Gaussian states. These states, besides being of great importance both from a fundamental point of view and in practical applications, share peculiar features that make their structural properties amenable to accurate and detailed theoretical analysis. It is the aim of this chapter to review some of the most recent results on the characterization and quantification of bipartite and multipartite entanglement in Gaussian states of continuous variable systems, their relationships with standard measures of purity and mixedness, and their operational interpretations in practical applications such as quantum communication, information transfer, and quantum teleportation.

2. Gaussian States of Continuous Variable Systems

We consider a continuous variable (CV) system consisting of N canonical bosonic modes, associated to an infinite-dimensional Hilbert space $\mathcal{H}$ and described by the vector $\hat{X} = \{\hat{x}_1, \hat{p}_1, \ldots, \hat{x}_N, \hat{p}_N\}$ of the field quadrature ("position" and "momentum") operators. The quadrature phase operators are connected to the annihilation $\hat{a}_i$ and creation $\hat{a}_i^\dagger$ operators of each mode, by the relations $\hat{x}_i = (\hat{a}_i + \hat{a}_i^\dagger)$ and $\hat{p}_i = (\hat{a}_i - \hat{a}_i^\dagger)/i$. The canonical commutation relations for the $\hat{X}_i$'s can be expressed in matrix form: $[\hat{X}_i, \hat{X}_j] = 2i\Omega_{ij}$, with the symplectic form $\Omega = \oplus_{i=1}^{n} \omega$ and $\omega = \delta_{ij-1} - \delta_{ij+1}$, $i, j = 1, 2$.

Quantum states of paramount importance in CV systems are the so-called Gaussian states, *i.e.* states with Gaussian characteristic functions and quasi–probability distributions.[1] The interest in this special class of states (important examples include vacua, coherent, squeezed, thermal, and squeezed-thermal states of the electromagnetic field) stems from the feasibility to produce and control them with linear optical elements, and from the increasing number of efficient proposals and successful experimental implementations of CV quantum information and communication processes involving multimode Gaussian states (see Ref. 2 for recent reviews). By definition, a Gaussian state is completely characterized by first and second moments of the canonical operators. When addressing physical properties invariant under local unitary transformations, such as mixedness and entanglement, one can neglect first moments and completely characterize Gaussian states by the $2N \times 2N$ real covariance matrix (CM) $\boldsymbol{\sigma}$, whose entries are $\sigma_{ij} = 1/2\langle\{\hat{X}_i, \hat{X}_j\}\rangle - \langle\hat{X}_i\rangle\langle\hat{X}_j\rangle$. Throughout this chapter, $\boldsymbol{\sigma}$ will be used indifferently to indicate the CM of a Gaussian state or the state itself. A real, symmetric matrix $\boldsymbol{\sigma}$ must fulfill the Robertson-Schrödinger uncertainty relation[3]

$$\boldsymbol{\sigma} + i\Omega \geq 0, \tag{1}$$

to be a *bona fide* CM of a physical state. Symplectic operations (*i.e.* belonging to the group $Sp_{(2N,\mathcal{R})} = \{S \in SL(2N, \mathcal{R}) \,:\, S^T\Omega S = \Omega\}$) acting by congruence on CMs in phase space, amount to unitary operations on density matrices in Hilbert space. In phase space, any N-mode Gaussian state can be transformed by symplectic operations in its Williamson diagonal form[4] $\boldsymbol{\nu}$, such that $\boldsymbol{\sigma} = S^T\boldsymbol{\nu}S$, with $\boldsymbol{\nu} = \operatorname{diag}\{\nu_1, \nu_1, \ldots \nu_N, \nu_N\}$. The set $\Sigma = \{\nu_i\}$ of the positive-defined eigenvalues of $|i\Omega\boldsymbol{\sigma}|$ constitutes the symplectic spectrum of $\boldsymbol{\sigma}$ and its elements, the so-called symplectic eigenvalues, must fulfill the conditions $\nu_i \geq 1$, following from Eq. (1) and ensuring positivity of the density matrix associated to $\boldsymbol{\sigma}$. We remark that the full saturation of the uncertainty principle can only be achieved by pure N-mode Gaussian states, for which $\nu_i = 1 \; \forall i = 1, \ldots, N$. Instead, those mixed states such that $\nu_{i \leq k} = 1$ and $\nu_{i > k} > 1$, with $1 \leq k \leq N$, partially saturate the uncertainty principle, with partial saturation becoming weaker with decreasing k. The symplectic eigenvalues ν_i are determined by N symplectic invariants associated to the characteristic polynomial of the matrix $|i\Omega\boldsymbol{\sigma}|$. Global invariants include the determinant $\operatorname{Det}\boldsymbol{\sigma} = \prod_i \nu_i^2$ and the quantity $\Delta = \sum_i \nu_i^2$, which is the sum of the determinants of all the 2×2 submatrices of $\boldsymbol{\sigma}$ related to each mode.[5]

The degree of information about the preparation of a quantum state ϱ can be characterized by its *purity* $\mu \equiv \operatorname{Tr}\varrho^2$, ranging from 0 (completely mixed states) to 1 (pure states). For a Gaussian state with CM $\boldsymbol{\sigma}$ one has[6]

$$\mu = 1/\sqrt{\operatorname{Det}\boldsymbol{\sigma}}\,. \tag{2}$$

As for the entanglement, we recall that positivity of the CM's partial transpose (PPT)[7] is a necessary and sufficient condition of separability for $(M+N)$-mode bisymmetric Gaussian states (see Sec. 4) with respect to the $M|N$ bipartition of the modes,[8] as well as for $(M+N)$-mode Gaussian states with fully degenerate symplectic spectrum.[9] In the special, but important case $M = 1$, PPT is a necessary and sufficient condition for separability of all Gaussian states.[11,10] For a general Gaussian state of any $M|N$ bipartition, the PPT criterion is replaced by another necessary and sufficient condition stating that a CM $\boldsymbol{\sigma}$ corresponds to a separable state if and only if there exists a pair of CMs $\boldsymbol{\sigma}_A$ and $\boldsymbol{\sigma}_B$, relative to the subsystems A and B respectively, such that the following inequality holds:[11] $\boldsymbol{\sigma} \geq \boldsymbol{\sigma}_A \oplus \boldsymbol{\sigma}_B$. This criterion is not very useful in practice. Alternatively, one can introduce an operational criterion based on a nonlinear map, that is independent of (and strictly stronger than) the PPT condition.[12]

In phase space, partial transposition amounts to a mirror reflection of one quadrature in the reduced CM of one of the parties. If $\{\tilde{\nu}_i\}$ is the symplectic spectrum of the partially transposed CM $\tilde{\boldsymbol{\sigma}}$, then a $(1+N)$-mode (or bisymmetric $(M+N)$-mode) Gaussian state with CM $\boldsymbol{\sigma}$ is separable if and only if $\tilde{\nu}_i \geq 1\ \forall i$. A proper measure of CV entanglement is the *logarithmic negativity*[13] $E_{\mathcal{N}} \equiv \log \|\tilde{\varrho}\|_1$, where $\|\cdot\|_1$ denotes the trace norm, which constitutes an upper bound to the *distillable entanglement* of the state ϱ. It can be computed in terms of the symplectic spectrum $\tilde{\nu}_i$ of $\tilde{\boldsymbol{\sigma}}$:

$$E_{\mathcal{N}} = \max\left\{0, -\sum\nolimits_{i:\tilde{\nu}_i<1} \log \tilde{\nu}_i\right\}. \tag{3}$$

$E_{\mathcal{N}}$ quantifies the extent to which the PPT condition $\tilde{\nu}_i \geq 1$ is violated.

3. Two–Mode Gaussian States: Entanglement and Mixedness

Two–mode Gaussian states represent the prototypical quantum states of CV systems, and constitute an ideal test-ground for the theoretical and experimental investigation of CV entanglement.[14] Their CM can be written is the following block form

$$\boldsymbol{\sigma} \equiv \begin{pmatrix} \boldsymbol{\alpha} & \boldsymbol{\gamma} \\ \boldsymbol{\gamma}^T & \boldsymbol{\beta} \end{pmatrix}, \tag{4}$$

where the three 2×2 matrices $\boldsymbol{\alpha}$, $\boldsymbol{\beta}$, $\boldsymbol{\gamma}$ are, respectively, the CMs of the two reduced modes and the correlation matrix between them. It is well known[10] that for any two–mode CM $\boldsymbol{\sigma}$ there exists a local symplectic operation

$S_l = S_1 \oplus S_2$ which takes $\boldsymbol{\sigma}$ to its standard form $\boldsymbol{\sigma}_{sf}$, characterized by

$$\boldsymbol{\alpha} = \mathrm{diag}\{a, a\}, \quad \boldsymbol{\beta} = \mathrm{diag}\{b, b\}, \quad \boldsymbol{\gamma} = \mathrm{diag}\{c_+, c_-\}. \tag{5}$$

States whose standard form fulfills $a = b$ are said to be symmetric. Any pure state is symmetric and fulfills $c_+ = -c_- = \sqrt{a^2-1}$. The uncertainty principle Ineq. (1) can be recast as a constraint on the $Sp_{(4,\mathbb{R})}$ invariants $\mathrm{Det}\boldsymbol{\sigma}$ and $\Delta(\boldsymbol{\sigma}) = \mathrm{Det}\boldsymbol{\alpha} + \mathrm{Det}\boldsymbol{\beta} + 2\,\mathrm{Det}\boldsymbol{\gamma}$, yielding $\Delta(\boldsymbol{\sigma}) \leq 1 + \mathrm{Det}\boldsymbol{\sigma}$. The standard form covariances a, b, c_+, and c_- can be determined in terms of the two local symplectic invariants

$$\mu_1 = (\mathrm{Det}\,\boldsymbol{\alpha})^{-1/2} = 1/a\,, \quad \mu_2 = (\mathrm{Det}\,\boldsymbol{\beta})^{-1/2} = 1/b, \tag{6}$$

which are the marginal purities of the reduced single–mode states, and of the two global symplectic invariants

$$\mu = (\mathrm{Det}\,\boldsymbol{\sigma})^{-1/2} = [(ab - c_+^2)(ab - c_-^2)]^{-1/2}\,, \quad \Delta = a^2 + b^2 + 2c_+c_-, \tag{7}$$

where μ is the global purity of the state. Eqs. (6-7) can be inverted to provide the following physical parametrization of two–mode states in terms of the four independent parameters μ_1, μ_2, μ, and Δ:[15]

$$a = \frac{1}{\mu_1}, \quad b = \frac{1}{\mu_2}, \quad c_\pm = \frac{\sqrt{\mu_1\mu_2}}{4}\left(\epsilon_- \pm \epsilon_+\right), \tag{8}$$

with $\epsilon_\mp \equiv \sqrt{\left[\Delta - (\mu_1 \mp \mu_2)^2/(\mu_1^2\mu_2^2)\right]^2 - 4/\mu^2}$. The uncertainty principle and the existence of the radicals appearing in Eq. (8) impose the following constraints on the four invariants in order to describe a physical state

$$\mu_1\mu_2 \leq \mu \leq \frac{\mu_1\mu_2}{\mu_1\mu_2 + |\mu_1 - \mu_2|}\,, \tag{9}$$

$$\frac{2}{\mu} + \frac{(\mu_1 - \mu_2)^2}{\mu_1^2\mu_2^2} \leq \Delta \leq 1 + \frac{1}{\mu^2}\,. \tag{10}$$

The physical meaning of these constraints, and the role of the extremal states (*i.e.* states whose invariants saturate the upper or lower bounds of Eqs. (9-10)) in relation to the entanglement, will be investigated soon.

In terms of symplectic invariants, partial transposition corresponds to flipping the sign of $\mathrm{Det}\,\boldsymbol{\gamma}$, so that Δ turns into $\tilde{\Delta} = \Delta - 4\,\mathrm{Det}\,\boldsymbol{\gamma} = -\Delta + 2/\mu_1^2 + 2/\mu_2^2$. The symplectic eigenvalues of the CM $\boldsymbol{\sigma}$ and of its partial transpose $\tilde{\boldsymbol{\sigma}}$ are promptly determined in terms of symplectic invariants

$$2\nu_\mp^2 = \Delta \mp \sqrt{\Delta^2 - 4/\mu^2}, \quad 2\tilde{\nu}_\mp^2 = \tilde{\Delta} \mp \sqrt{\tilde{\Delta}^2 - 4/\mu^2}\,, \tag{11}$$

where in our naming convention $\nu_- \leq \nu_+$ in general, and similarly for the $\tilde{\nu}_\mp$. The PPT criterion yields a state $\boldsymbol{\sigma}$ separable if and only if $\tilde{\nu}_- \geq 1$.

Since $\tilde{\nu}_+ > 1$ for all two–mode Gaussian states, the quantity $\tilde{\nu}_-$ also completely quantifies the entanglement, in fact the logarithmic negativity Eq. (3) is a monotonically decreasing and convex function of $\tilde{\nu}_-$, $E_{\mathcal{N}} = \max\{0, -\log \tilde{\nu}_-\}$. In the special instance of symmetric Gaussian states, the *entanglement of formation*[16] E_F is also computable[17] but, being again a decreasing function of $\tilde{\nu}_-$, it provides the same characterization of entanglement and is thus fully equivalent to $E_{\mathcal{N}}$ in this subcase.

A first natural question that arises is whether there can exist two-mode Gaussian states of finite maximal entanglement at a given amount of mixedness of the global state. These states would be the analog of the maximally entangled mixed states (MEMS) that are known to exist for two-qubit systems.[18] Unfortunately, it is easy to show that a similar question in the CV scenario is meaningless. Indeed, for any fixed, finite global purity μ there exist infinitely many Gaussian states which are infinitely entangled. However, we can ask whether there exist maximally entangled states at fixed global *and* local purities. While this question does not yet have a satisfactory answer for two-qubit systems, in the CV scenario it turns out to be quite interesting and nontrivial. In this respect, a crucial observation is that, at fixed μ, μ_1 and μ_2, the lowest symplectic eigenvalue $\tilde{\nu}_-$ of the partially transposed CM is a monotonically increasing function of the global invariant Δ. Due to the existence of exact *a priori* lower and upper bounds on Δ at fixed purities (see Ineq. 10), this entails the existence of both maximally *and* minimally entangled Gaussian states. These classes of extremal states have been introduced in Ref. 19, and completely characterized (providing also schemes for their experimental production) in Ref. 15, where the relationship between entanglement and information has been extended considering generalized entropic measures to quantify the degrees of mixedness. In particular, there exist maximally and minimally entangled states also at fixed global and local generalized Tsallis p-entropies.[15] In this short review chapter, we will discuss only the case in which the purities (or, equivalently, the linear entropies) are used to measure the degree of mixedness of a quantum state. In this instance, the Gaussian maximally entangled mixed states (GMEMS) are two–mode squeezed thermal states, characterized by a fully degenerate symplectic spectrum; on the other hand, the Gaussian least entangled mixed states (GLEMS) are states of partial minimum uncertainty (*i.e.* with the lowest symplectic eigenvalue of their CM being equal to 1). Studying the separability of the extremal states (via the PPT criterion), it is possible to classify the entanglement properties of

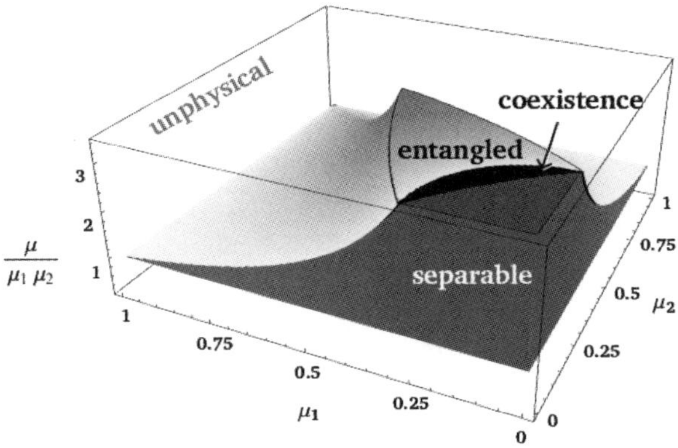

Fig. 1. Classification of the entanglement for two–mode Gaussian states in the space of marginal purities $\mu_{1,2}$ and normalized global purity $\mu/\mu_1\mu_2$. All physical states lie between the horizontal plane of product states $\mu = \mu_1\mu_2$, and the upper limiting surface representing GMEMMS. Separable states (dark grey area) and entangled states are well distinguished except for a narrow coexistence region (depicted in black). In the entangled region the average logarithmic negativity (see text) grows from white to medium grey. The expressions of the boundaries between all these regions are collected in Eq. (12).

all two–mode Gaussian states in the manifold spanned by the purities:

$$\begin{aligned} &\mu_1\mu_2 \leq \mu \leq \tfrac{\mu_1\mu_2}{\mu_1+\mu_2-\mu_1\mu_2}, && \Rightarrow \text{separable;} \\ &\tfrac{\mu_1\mu_2}{\mu_1+\mu_2-\mu_1\mu_2} < \mu \leq \tfrac{\mu_1\mu_2}{\sqrt{\mu_1^2+\mu_2^2-\mu_1^2\mu_2^2}}, && \Rightarrow \text{coexistence;} \\ &\tfrac{\mu_1\mu_2}{\sqrt{\mu_1^2+\mu_2^2-\mu_1^2\mu_2^2}} < \mu \leq \tfrac{\mu_1\mu_2}{\mu_1\mu_2+|\mu_1-\mu_2|}, && \Rightarrow \text{entangled.} \end{aligned} \tag{12}$$

In particular, apart from a narrow "coexistence" region where both separable and entangled Gaussian states can be found, the separability of two–mode states at given values of the purities is completely characterized. For purities that saturate the upper bound in Ineq. (9), GMEMS and GLEMS coincide and we have a unique class of states whose entanglement depends *only* on the marginal purities $\mu_{1,2}$. They are Gaussian maximally entangled states for fixed marginals (GMEMMS). The maximal entanglement of a Gaussian state decreases rapidly with increasing difference of marginal purities, in analogy with finite-dimensional systems.[20] For symmetric states ($\mu_1 = \mu_2$) the upper bound of Ineq. (9) reduces to the trivial bound $\mu \leq 1$ and GMEMMS reduce to pure two–mode states. Knowledge of the global and marginal purities thus accurately characterizes the entanglement of two-mode Gaussian states, providing strong sufficient conditions and exact, analytical lower and upper bounds. As we will now show, marginal and global purities allow as well an accurate quantification

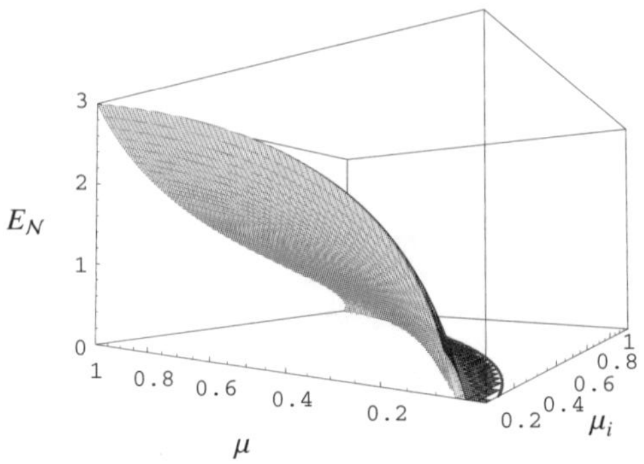

Fig. 2. Maximal and minimal logarithmic negativities as functions of the global and marginal purities of symmetric two-mode Gaussian states. The darker (lighter) surface represents GMEMS (GLEMS). In this space, a generic two–mode mixed symmetric state is represented by a dot lying inside the narrow gap between the two extremal surfaces.

of the entanglement. Outside the region of separability, GMEMS attain maximum logarithmic negativity $E_{\mathcal{N}\,\mathrm{max}}$ while, in the region of nonvanishing entanglement (see Eq. (12)), GLEMS acquire minimum logarithmic negativity $E_{\mathcal{N}\,\mathrm{min}}$. Knowledge of the global purity, of the two local purities, and of the global invariant Δ (*i.e.*, knowledge of the full covariance matrix) would allow for an exact quantification of the entanglement. However, we will now show that an estimate based only on the knowledge of the experimentally measurable global and marginal purities turns out to be quite accurate. We can in fact quantify the entanglement of Gaussian states with given global and marginal purities by the *average logarithmic negativity* $\bar{E}_\mathcal{N} \equiv (E_{\mathcal{N}\,\mathrm{max}} + E_{\mathcal{N}\,\mathrm{min}})/2$ We can then also define the relative error $\delta\bar{E}_\mathcal{N}$ on $\bar{E}_\mathcal{N}$ as $\delta\bar{E}_\mathcal{N}(\mu_{1,2},\mu) \equiv (E_{\mathcal{N}\,\mathrm{max}} - E_{\mathcal{N}\,\mathrm{min}})/(E_{\mathcal{N}\,\mathrm{max}} + E_{\mathcal{N}\,\mathrm{min}})$. It is easy to see that this error decreases *exponentially* both with increasing global purity and decreasing marginal purities, *i.e.* with increasing entanglement, falling for instance below 5% for symmetric states ($\mu_1 = \mu_2 \equiv \mu_i$) and $\mu > \mu_i$. The reliable quantification of quantum correlations in genuinely entangled two-mode Gaussian states is thus always assured by the experimental determination of the purities, except at most for a small set of states with very weak entanglement (states with $E_\mathcal{N} \lesssim 1$). Moreover, the accuracy is even greater in the general non-symmetric case $\mu_1 \neq \mu_2$, because the maximal achievable entanglement decreases in such an instance. In Fig. 2, the surfaces of extremal logarithmic negativities are plotted versus μ_i and μ for symmetric states. In the case $\mu = 1$ the upper and lower bounds coincide, since for pure states the entanglement is completely quantified by the marginal purity. For mixed states this is not the case, but, as the plot shows, knowledge of the global and marginal purities strictly bounds the entanglement both from above and from below. This analysis shows that the average logarithmic negativity $\bar{E}_{\mathcal{N}}$ is a reliable estimate of the logarithmic negativity $E_{\mathcal{N}}$, improving as the entanglement increases. We remark that the purities may be directly measured experimentally, without the need for a full tomographic reconstruction of the whole CM, by exploiting quantum networks techniques[21] or single–photon detections without homodyning.[22]

Finally, it is worth remarking that most of the results presented here (including the sufficient conditions for entanglement based on knowledge of the purities), being derived for CMs using the symplectic formalism in phase space, retain their validity for generic non Gaussian states of CV systems. For instance, any two-mode state with a CM equal to that of an entangled two-mode Gaussian state is entangled as well.[23] Our methods may thus serve to detect entanglement for a broader class of states in infinite-dimensional Hilbert spaces. The analysis briefly reviewed in this paragraph on the relationships between entanglement and mixedness, can be generalized to multimode Gaussian states endowed with special symmetry under mode permutations, as we will show in the next section.

4. Multimode Gaussian States: Unitarily Localizable Entanglement

We will now consider Gaussian states of CV systems with an arbitrary number of modes, and briefly discuss the simplest instances in which the techniques introduced for two–mode Gaussian states can be generalized and turn out to be useful for the quantification and the scaling analysis of CV multimode entanglement. We introduce the notion of *bisymmetric* states, defined as those $(M + N)$-mode Gaussian states, of a generic bipartition $M|N$, that are invariant under local mode permutations on the M-mode and N-mode subsystems. The CM $\boldsymbol{\sigma}$ of a $(M + N)$-mode bisymmetric Gaussian state results from a correlated combination of the fully symmetric blocks $\boldsymbol{\sigma}_{\alpha^M}$ and $\boldsymbol{\sigma}_{\beta^N}$:

$$\boldsymbol{\sigma} = \begin{pmatrix} \boldsymbol{\sigma}_{\alpha^M} & \boldsymbol{\Gamma} \\ \boldsymbol{\Gamma}^{\mathsf{T}} & \boldsymbol{\sigma}_{\beta^N} \end{pmatrix}, \tag{13}$$

where $\boldsymbol{\sigma}_{\alpha^M}$ ($\boldsymbol{\sigma}_{\beta^N}$) describes a M-mode (N-mode) reduced Gaussian state completely invariant under mode permutations, and $\boldsymbol{\Gamma}$ is a $2M \times 2N$ real matrix formed by identical 2×2 blocks $\boldsymbol{\gamma}$. Clearly, $\boldsymbol{\Gamma}$ is responsible for the correlations existing between the M-mode and the N-mode parties. The identity of the submatrices $\boldsymbol{\gamma}$ is a consequence of the local invariance under mode exchange, internal to the M-mode and N-mode parties. A first observation is that the symplectic spectrum of the CM $\boldsymbol{\sigma}$ Eq. (13) of a bisymmetric $(M+N)$-mode Gaussian state includes two degenerate eigenvalues, with multiplicities $M-1$ and $N-1$. Such eigenvalues coincide, respectively, with the degenerate eigenvalue ν_{α}^{-} of the reduced CM $\boldsymbol{\sigma}_{\alpha^M}$ and the degenerate eigenvalue ν_{β}^{-} of the reduced CM $\boldsymbol{\sigma}_{\beta^N}$, with the same respective multiplicities. Equipped with this result, one can prove[8] that $\boldsymbol{\sigma}$ can be brought, by means of a local unitary operation, with respect to the $M|N$ bipartition, to a tensor product of single-mode uncorrelated states and of a two-mode Gaussian state with CM $\boldsymbol{\sigma}^{eq}$. Here we give an intuitive sketch of the proof (the detailed proof is given in Ref. 8). Let us focus on the N-mode block $\boldsymbol{\sigma}_{\beta^N}$. The matrices $i\boldsymbol{\Omega}\boldsymbol{\sigma}_{\beta^N}$ and $i\boldsymbol{\Omega}\boldsymbol{\sigma}$ possess a set of $N-1$ simultaneous eigenvectors, corresponding to the same (degenerate) eigenvalue. This fact suggests that the phase-space modes corresponding to such eigenvectors are the same for $\boldsymbol{\sigma}$ and for $\boldsymbol{\sigma}_{\beta^N}$. Then, bringing by means of a local symplectic operation the CM $\boldsymbol{\sigma}_{\beta^N}$ in Williamson form, any $(2N-2)\times(2N-2)$ submatrix of $\boldsymbol{\sigma}$ will be diagonalized because the normal modes are common to the global and local CMs. In other words, no correlations between the M-mode party with reduced CM $\boldsymbol{\sigma}_{\alpha^M}$ and such modes will be left: all the correlations between the M-mode and N-mode parties will be concentrated in the two conjugate quadratures of a single mode of the N-mode block. Going through the same argument for the M-mode block with CM $\boldsymbol{\sigma}_{\alpha^M}$ will prove the proposition and show that the whole entanglement between the two multimode blocks can always be concentrated in only two modes, one for each of the two multimode parties.

While, as mentioned, the detailed proof of this result can be found in Ref. 8 (extending the findings obtained in Ref. 24 for the case $M=1$), here we will focus on its relevant physical consequences, the main one being that the bipartite $M\times N$ entanglement of bisymmetric $(M+N)$-mode Gaussian states is *unitarily localizable*, *i.e.*, through local unitary operations, it can be fully concentrated on a single pair of modes, one belonging to party (block) M, the other belonging to party (block) N. The notion of "unitarily localizable entanglement" is different from that of "localizable entanglement" originally introduced by Verstraete, Popp, and Cirac for spin systems.[25] There, it was defined as the maximal entanglement concentrable on two chosen spins through local *measurements* on all the other spins. Here, the local operations that concentrate all the multimode entanglement on two modes are *unitary* and involve the two chosen modes as well, as parts of the respective blocks. Furthermore, the unitarily localizable entanglement (when computable) is always stronger than the localizable entanglement.

In fact, if we consider a generic bisymmetric multimode state of a $M|N$ bipartition, with each of the two target modes owned respectively by one of the two parties (blocks), then the ensemble of optimal local measurements on the remaining ("assisting") $M+N-2$ modes belongs to the set of local operations and classical communication (LOCC) with respect to the considered bipartition. By definition the entanglement cannot increase under LOCC, which implies that the localized entanglement (à la Verstraete, Popp, and Cirac) is always less or equal than the original $M \times N$ block entanglement. On the contrary, *all* of the same $M \times N$ original bipartite entanglement can be unitarily localized onto the two target modes, resulting in a reversible, of maximal efficiency, multimode/two-mode entanglement switch. This fact can have a remarkable impact in the context of quantum repeaters[26] for communications with continuous variables. The consequences of the unitary localizability are manifold. In particular, as already previously mentioned, one can prove that the PPT (positivity of the partial transpose) criterion is a necessary and sufficient condition for the separability of $(M+N)$-mode bisymmetric Gaussian states.[8] Therefore, the multimode block entanglement of bisymmetric (generally mixed) Gaussian states with CM $\boldsymbol{\sigma}$, being equal to the bipartite entanglement of the equivalent two-mode localized state with CM $\boldsymbol{\sigma}^{eq}$, can be determined and quantified by the logarithmic negativity in the general instance and, for all multimode states whose two–mode equivalent Gaussian state is symmetric, the entanglement of formation between the M-mode party and the N-mode party can be computed exactly as well.

For the sake of illustration, let us consider fully symmetric $2N$-mode Gaussian states described by a $2N \times 2N$ CM $\boldsymbol{\sigma}_{\beta^{2N}}$. These states are trivially bisymmetric under any bipartition of the modes, so that their block entanglement is always localizable by means of local symplectic operations. This class of states includes the pure, CV GHZ–type states (discussed in Refs. 27, 24) that, in the limit of infinite squeezing, reduce to the simultaneous eigenstates of the relative positions and the total momentum and coincide with the proper Greenberger-Horne-Zeilinger[28] (GHZ) states of CV systems.[27] The standard form CM $\boldsymbol{\sigma}^{p}_{\beta^{2N}}$ of this particular class of pure, symmetric multimode Gaussian states depends only on the local mixedness parameter $b \equiv 1/\mu_\beta$, which is the inverse of the purity of any single-mode reduced block, and it is proportional to the single-mode squeezing. Exploiting our previous analysis, we can compute the entanglement between a block of K modes and the remaining $2N-K$ modes for pure states (in this

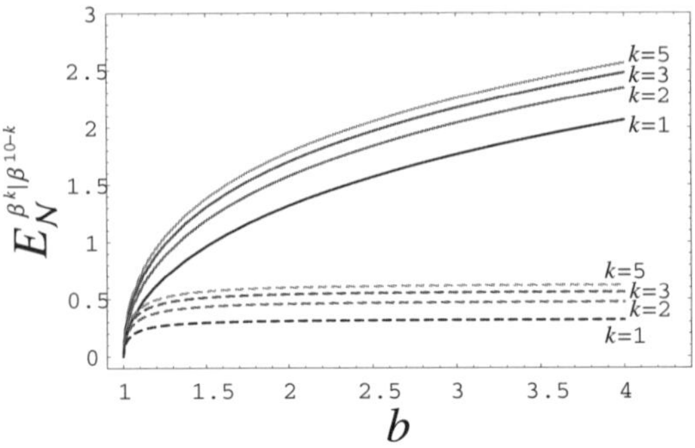

Fig. 3. Hierarchy of block entanglements of fully symmetric $2N$-mode Gaussian states of $K \times (2N-K)$ bipartitions ($N = 10$) as a function of the single-mode mixedness b, for pure states (solid lines) and for mixed states obtained from $(2N+4)$-mode pure states by tracing out 4 modes (dashed lines).

case the block entanglement is simply the Von Neumann entropy of any of the reduced blocks) and, remarkably, for mixed states as well.

We can in fact consider a generic $2N$-mode fully symmetric mixed state with CM $\boldsymbol{\sigma}^{p\backslash Q}_{\beta^{2N}}$, obtained from a pure fully symmetric $(2N+Q)$-mode state by tracing out Q modes. For any Q and any dimension N of the block ($K \leq N$), and for any nonzero squeezing (*i.e.* for any $b > 1$) one has that the state exhibits genuine multipartite entanglement, as first remarked in Ref. 27 for pure states: each K-mode party is entangled with the remaining $(2N-K)$-mode block. Furthermore, the genuine multipartite nature of the entanglement can be precisely quantified by observing that the logarithmic negativity between the K-mode and the remaining $(2N-K)$-mode block is an increasing function of the integer $K \leq N$, as shown in Fig. 3. The optimal splitting of the modes, which yields the maximal, unitarily localizable entanglement, corresponds to $K = N/2$ if N is even, and $K = (N-1)/2$ if N is odd. The multimode entanglement of mixed states remains finite also in the limit of infinite squeezing, while the multimode entanglement of pure states diverges with respect to any bipartition, as shown in Fig. 3. For a fixed amount of local mixedness, the scaling structure of the multimode entanglement with the number of modes can be analyzed as well, giving rise to an interesting result.[8] Let us consider, again for the sake of illustration, the class of fully symmetric $2N$-mode Gaussian states, but now at fixed single-mode purity. It is immediate to see that the entanglement between any two modes decreases with N, while the $N|N$ entanglement increases (and diverges for pure states as $N \to \infty$): the quantum correlations become distributed among all the modes. This is a clear signature of genuine multipartite entanglement and suggests a detailed analysis of its sharing properties, that will be discussed in the next section. The scaling structure of multimode entanglement also elucidates the power of the unitary localizability as a strategy for entanglement purification, with its efficiency improving with increasing number of modes. Finally, let us remark that the local symplectic operations needed for the unitary localization can be implemented by only using passive[29] and active linear optical elements such as beam splitters, phase shifters and squeezers, and that the original multimode entanglement can be estimated by the knowledge of the global and local purities of the equivalent, localized two–mode state (see Refs. 8, 24 for a thorough discussion), along the lines presented in Sec. 3 above.

5. Entanglement Sharing of Gaussian States

Here we address the problem of entanglement sharing among multiple parties, investigating the structure of multipartite entanglement.[30,31] Our aim is to analyze the distribution of entanglement between different (partitions of) modes in CV systems. In Ref. 32 Coffman, Kundu and Wootters (CKW) proved for a three-qubit system ABC, and conjectured for N qubits (this conjecture has now been proven by Osborne and Verstraete[33]) that the entanglement between, say, qubit A and the remaining two–qubits partition (BC) is never smaller than the sum of the A|B and A|C bipartite entanglements in the reduced states. This statement quantifies the so-called *monogamy* of quantum entanglement,[34] in opposition to the classical correlations which can be freely shared. One would expect a similar inequality to hold for three–mode Gaussian states, namely

$$E^{i|(jk)} - E^{i|j} - E^{i|k} \geq 0, \tag{14}$$

where E is a proper measure of CV entanglement and the indices $\{i, j, k\}$ label the three modes. However, an immediate computation on symmetric states shows that Ineq. (14) can be violated for small values of the single-mode mixedness b using either the logarithmic negativity $E_{\mathcal{N}}$ or the entanglement of formation E_F to quantify the bipartite entanglement. This is not a paradox;[31] rather, it implies that none of these two measures is the proper candidate for approaching the task of quantifying entanglement sharing in CV systems. This situation is reminiscent of the case of qubit systems, for which the CKW inequality holds using the tangle τ,[32] but fails if one chooses equivalent measures of bipartite entanglement such as the concurrence[35] (*i.e.* the square root of the tangle) or the entanglement of formation itself. Related problems on inequivalent entanglement measures for the ordering of Gaussian states are discussed in Ref. 36.

We then wish to define a new measure of CV entanglement able to capture the entanglement distribution trade-off via the monogamy inequality (14). A rigorous treatment of this problem is presented in Ref. 30. Here we briefly review the definition and main properties of the desired measure that quantifies entanglement sharing in CV systems. Because it can be regarded as the continuous-variable analogue of the tangle, we will name it, in short, the *contangle.*

For a pure state $|\psi\rangle$ of a $(1+N)$-mode CV system, we can formally define the contangle as

$$E_\tau(\psi) \equiv \log^2 \|\tilde{\varrho}\|_1 , \quad \varrho = |\psi\rangle\langle\psi| . \tag{15}$$

$E_\tau(\psi)$ is a proper measure of bipartite entanglement, being a convex, increasing function of the logarithmic negativity $E_{\mathcal{N}}$, which is equivalent to the entropy of entanglement in all pure states. For pure Gaussian states $|\psi\rangle$ with CM $\boldsymbol{\sigma}^p$, one has $E_\tau(\boldsymbol{\sigma}^p) = \log^2(1/\mu_1 - \sqrt{1/\mu_1^2 - 1})$, where $\mu_1 = 1/\sqrt{\mathrm{Det}\,\boldsymbol{\sigma}_1}$ is the local purity of the reduced state of mode 1, described by a CM $\boldsymbol{\sigma}_1$ (considering $1 \times N$ bipartitions). Definition (15) is extended to generic mixed states ϱ of $(N+1)$-mode CV systems through the convex-roof formalism, namely:

$$E_\tau(\varrho) \equiv \inf_{\{p_i,\psi_i\}} \sum_i p_i E_\tau(\psi_i), \tag{16}$$

where the infimum is taken over the decompositions of ϱ in terms of pure states $\{|\psi_i\rangle\}$. For infinite-dimensional Hilbert spaces the index i is continuous, the sum in Eq. (16) is replaced by an integral, and the probabilities $\{p_i\}$ by a distribution $\pi(\psi)$. All multimode mixed Gaussian states $\boldsymbol{\sigma}$ admit a decomposition in terms of an ensemble of pure Gaussian states. The infimum of the average contangle, taken over all pure Gaussian decompositions only, defines the *Gaussian contangle* G_τ, which is an upper bound to the true contangle E_τ, and an entanglement monotone under Gaussian local operations and classical communications (GLOCC).[36,37] The Gaussian contangle, similarly to the Gaussian entanglement of formation,[37] acquires the simple form $G_\tau(\boldsymbol{\sigma}) \equiv \inf_{\boldsymbol{\sigma}^p \leq \boldsymbol{\sigma}} E_\tau(\boldsymbol{\sigma}^p)$, where the infimum runs over all pure Gaussian states with CM $\boldsymbol{\sigma}^p \leq \boldsymbol{\sigma}$.

Equipped with these properties and definitions, one can prove several results.[30] In particular, the general (multimode) monogamy inequality $E^{i_m|(i_1\ldots i_{m-1}i_{m+1}\ldots i_N)} - \sum_{l\neq m} E^{i_m|i_l} \geq 0$ is satisfied by all pure three-mode and all pure *symmetric* N-mode Gaussian states, using either E_τ or G_τ to quantify bipartite entanglement, and by all the corresponding mixed states using G_τ. Furthermore, there is numerical evidence supporting the conjecture that the general CKW inequality should hold for all *nonsymmet-*

ric N-mode Gaussian states as well.[a] The sharing constraint (14) leads to the definition of the *residual contangle* as a tripartite entanglement quantifier. For nonsymmetric three-mode Gaussian states the residual contangle is partition-dependent. In this respect, a proper quantification of tripartite entanglement is provided by the *minimum* residual contangle

$$E_\tau^{i|j|k} \equiv \min_{(i,j,k)} \left[E_\tau^{i|(jk)} - E_\tau^{i|j} - E_\tau^{i|k} \right], \tag{17}$$

where (i,j,k) denotes all the permutations of the three mode indexes. This definition ensures that $E_\tau^{i|j|k}$ is invariant under mode permutations and is thus a genuine three-way property of any three-mode Gaussian state. We can adopt an analogous definition for the minimum residual Gaussian contangle $G_\tau^{i|j|k}$. One finds that the latter is a proper measure of genuine tripartite CV entanglement, because it is an entanglement monotone under tripartite GLOCC for pure three-mode Gaussian states.[30]

Let us now analyze the sharing structure of multipartite CV entanglement, by taking the residual contangle as a measure of tripartite entanglement. We pose the problem of identifying the three–mode analogues of the two inequivalent classes of fully inseparable three–qubit states, the GHZ state[28] $|\psi_{\rm GHZ}\rangle = (1/\sqrt{2})\,[|000\rangle + |111\rangle]$, and the W state[38] $|\psi_W\rangle = (1/\sqrt{3})\,[|001\rangle + |010\rangle + |100\rangle]$. These states are both pure and fully symmetric, but the GHZ state possesses maximal three-party tangle with no two-party quantum correlations, while the W state contains the maximal two-party entanglement between any pair of qubits and its tripartite residual tangle is consequently zero.

Surprisingly enough, in symmetric three–mode Gaussian states, if one aims at maximizing (at given single–mode squeezing b) either the two–mode contangle $E_\tau^{i|l}$ in any reduced state (*i.e.* aiming at the CV W-like state), or the genuine tripartite contangle (*i.e.* aiming at the CV GHZ-like state), one finds the same, unique family of pure symmetric three–mode squeezed states. These states, previously named "GHZ-type" states,[27] have been introduced for generic N–mode CV systems in the previous section, where their multimode entanglement scaling has been studied.[8,24] The peculiar nature of entanglement sharing in this class of states, now baptized

[a]Very recently, the conjectured monogamy inequality for all (pure or mixed) N-mode Gaussian states has been indeed proven by considering a slightly different version of the continuous-variable tangle, defined in terms of the (convex-roof extended) squared negativity instead of the squared logarithmic negativity. See T.Hiroshima, G. Adesso, and F. Illuminati, quant-ph/0605021.

CV GHZ/W states, is further confirmed by the following observation. If one requires maximization of the 1×2 bipartite contangle $E_\tau^{i|(jk)}$ under the constraint of separability of all two–mode reductions, one finds a class of symmetric mixed states whose tripartite residual contangle is strictly smaller than the one of the GHZ/W states, at fixed local squeezing.[39] Therefore, in symmetric three–mode Gaussian states, when there is no two–mode entanglement, the three-party one is not enhanced, but frustrated.

These results, unveiling a major difference between discrete-variable and CV systems, establish the *promiscuous* structure of entanglement sharing in symmetric Gaussian states. Being associated with degrees of freedom with continuous spectra, states of CV systems need not saturate the CKW inequality to achieve maximum couplewise correlations. In fact, without violating the monogamy inequality (14), pure symmetric three–mode Gaussian states are maximally three-way entangled and, at the same time, maximally robust against the loss of one of the modes due, for instance, to decoherence, as demonstrated in full detail in Ref. 39. This fact may promote these states, experimentally realizable with the current technology,[40] as candidates for reliable CV quantum communication. Exploiting a three–mode CV GHZ/W state as a quantum channel can ensure for instance a tripartite quantum information protocol like a teleportation network or quantum secret sharing; or a standard, highly entangled two–mode channel, after a unitary (reversible) localization has been performed through a single beam splitter; or, as well, a two–party quantum protocol with better-than-classical efficiency, even if one of the modes is lost due to decoherence. We will next focus on a relevant applicative setting of CV multipartite entanglement, in which various of its properties discussed so far will come in a natural relation.

6. Exploiting Multipartite Entanglement: Optimal Fidelity of Continuous Variable Teleportation

In this section we analyze an interesting application of multipartite CV entanglement: a quantum teleportation-network protocol, involving N users who share a genuine N-partite entangled Gaussian resource, completely symmetric under permutations of the modes. In the standard multiuser protocol, proposed by Van Loock and Braunstein,[41] two parties are randomly chosen as sender (Alice) and receiver (Bob), but, in order to accomplish teleportation of an unknown coherent state, Bob needs the results of $N-2$ momentum detections performed by the other cooperating parties.

A nonclassical teleportation fidelity (i.e. $\mathcal{F} > \mathcal{F}^{cl}$) between *any* pair of parties is sufficient for the presence of genuine N-partite entanglement in the shared resource, while in general the converse is false (see *e.g.* Fig. 1 of Ref. 41). The *fidelity*, which quantifies the success of a teleportation experiment, is defined as $\mathcal{F} \equiv \langle \psi^{in} | \varrho^{out} | \psi^{in} \rangle$, where "in" and "out" denote the input and the output state. $\mathcal{F}$ reaches unity only for a perfect state transfer, $\varrho^{out} = |\psi^{in}\rangle\langle\psi^{in}|$, while without entanglement in the resource, by purely classical communication, an average fidelity of $\mathcal{F}_{cl} = 1/2$ is the best that can be achieved if the alphabet of input states includes all coherent states with even weight.[42] This teleportation network has been recently demonstrated experimentally[43] by exploiting three-mode squeezed Gaussian states,[40] yielding a best fidelity of $\mathcal{F} = 0.64 \pm 0.02$, an index of genuine tripartite entanglement. Our aim is to determine the optimal multi-user teleportation fidelity, and to extract from it a quantitative information on the multipartite entanglement in the shared resources. By "optimal" here we mean maximization of the fidelity over all local single-mode unitary operations, at fixed amounts of noise and entanglement in the shared resource. We consider realistically mixed N-mode Gaussian resources, obtained by combining a mixed momentum-squeezed state (with squeezing parameter r_1) and $N-1$ mixed position-squeezed states (with squeezing parameter $r_2 \neq r_1$ and in principle a different noise factor) into an N-splitter[41] (a sequence of $N-1$ suitably tuned beam splitters). The resulting state is a completely symmetric mixed Gaussian state of a N-mode CV system. For a given thermal noise in the individual modes (comprising the unavoidable experimental imperfections), all the states with equal average squeezing $\bar{r} \equiv (r_1 + r_2)/2$ are equivalent up to local single–mode unitary operations and possess, by definition, the same amount of multipartite entanglement with respect to any partition. The teleportation efficiency, instead, depends separately on the different single–mode squeezings. We have then the freedom of unbalancing the local squeezings r_1 and r_2 without changing the total entanglement in the resource, in order to single out the optimal form of the resource state, which enables a teleportation network with maximal fidelity. This analysis is straightforward (see Ref. 44 for details), but it yields several surprising side results. In particular, one finds that the optimal form of the shared N-mode symmetric Gaussian states, for $N > 2$, is neither unbiased in the x_i and p_i quadratures (like the states discussed in Ref. 45 for $N = 3$), nor constructed by N equal squeezers ($r_1 = r_2 = \bar{r}$). This latter case, which has been implemented experimentally[43] for $N = 3$, is clearly not optimal, yielding fidelities lower than $1/2$ for $N \geq 30$ and $\bar{r}$ falling in

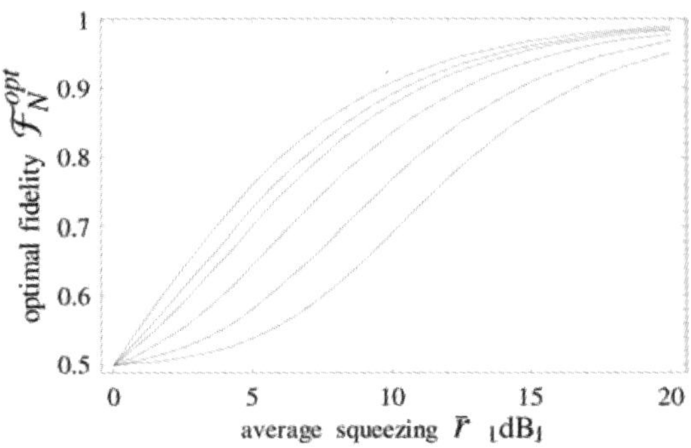

Fig. 4. Plot of the optimal fidelity for teleporting an arbitrary coherent state from any sender to any receiver chosen from N ($N = 2, \ldots, 50$) parties, exploiting N-party entangled, pure symmetric Gaussian states as resources. A nonclassical fidelity $\mathcal{F}_N^{opt} > 0.5$ is always assured for any N, if the shared entangled resources are prepared in their optimal form.

a certain interval.[41] According to the authors of Ref. 41, the explanation of this paradoxical behavior should lie in the fact that their teleportation scheme might not be optimal. However, a closer analysis shows that the problem does not lie in the choice of the protocol, but rather in the choice of the resource states. If the shared N-mode squeezed states are prepared, by local unitary operations, in the optimal form (described in detail in Ref. 44), the teleportation fidelity $\mathcal{F}^{opt}$ is guaranteed to be nonclassical (see Fig. 4) as soon as $\bar{r} > 0$ for any N, in which case the considered class of pure states is genuinely multiparty entangled, as we have shown in the previous sections. In fact, one can show[44] that this nonclassical optimal fidelity is *necessary and sufficient* for the presence of multipartite entanglement in any multimode symmetric Gaussian state used as a shared resource for CV teleportation. These findings yield quite naturally a direct operative way to quantify multipartite entanglement in N-mode (mixed) symmetric Gaussian states, in terms of the so-called *Entanglement of Teleportation*,[44] defined as the normalized optimal fidelity

$$E_T \equiv \max\left\{0, \left(\mathcal{F}_N^{opt} - \mathcal{F}_{cl}\right)/\left(1 - \mathcal{F}_{cl}\right)\right\}, \tag{18}$$

going from 0 (separable states) to 1 (CV GHZ/W state). Moreover, one finds that the optimal shared entanglement that allows for the maximal fidelity is *exactly* the CV counterpart of the localizable entanglement, originally introduced for spin systems by Verstraete, Popp, and Cirac.[25] The CV localizable entanglement (not to be confused with the unitarily localizable entanglement introduced in Sec. 4) thus acquires a suggestive operational meaning in terms of teleportation processes. In fact, the localizable entanglement of formation (computed by finding the optimal set of local measurements — unitary transformations and nonunitary momentum detections — performed on the assisting modes to concentrate the highest possible entanglement onto Alice and Bob pair of modes) is a monotonically increasing function of E_T: $E_F^{loc} = f[(1-E_T)/(1+E_T)]$, with $f(x) \equiv \frac{(1+x)^2}{4x}\log\frac{(1+x)^2}{4x} - \frac{(1-x)^2}{4x}\log\frac{(1-x)^2}{4x}$. For $N=2$ (standard two-user teleportation[46]) the state is already localized and $E_F^{loc} = E_F$, so that E_T is equivalent to the entanglement of formation E_F of two-mode Gaussian states. Remarkably, for $N=3$, *i.e.* for three-mode pure Gaussian resource states, the residual contangle $E_\tau^{i|j|k}$ introduced in Sec. 5 (see Eq. (17)) turns out to be itself a monotonically increasing function of E_T:

$$E_\tau^{i|j|k} = \log^2 \frac{2\sqrt{2}E_T - (E_T+1)\sqrt{E_T^2+1}}{(E_T-1)\sqrt{E_T(E_T+4)+1}} - \frac{1}{2}\log^2 \frac{E_T^2+1}{E_T(E_T+4)+1}. \quad (19)$$

The quantity E_T thus represents another *equivalent* quantification of genuine tripartite CV entanglement and provides the latter with an operational interpretation associated to the success of a three-party teleportation network. This suggests a possible experimental test of the promiscuous sharing of CV entanglement, consisting in the successful (with nonclassical optimal fidelity) implementation of both a three-user teleportation network exploiting pure symmetric Gaussian resources, and of two-user standard teleportation exploiting any reduced two-mode channel obtained discarding a mode from the original resource.

Besides their theoretical aspects, the results reviewed in this section are of direct practical interest, as they answer the experimental need for the best preparation recipe of an entangled squeezed resource, in order to implement quantum teleportation and in general CV communication schemes with the highest possible efficiency.

7. Conclusions and Outlook

We have reviewed some recent results on the entanglement of Gaussian states of CV systems. For two-mode Gaussian states we have shown how bipartite entanglement can be qualified and quantified via the global and local degrees of purity. Suitable generalizations of the techniques introduced for two-mode Gaussian states allow to analyze various aspects of entanglement in multimode CV systems, and we have discussed recent findings on the scaling, localization, and sharing properties of multipartite entanglement in symmetric, bisymmetric, and generic multimode Gaussian states. Finally, we have shown that many of these properties acquire a clear and simple operational meaning in the context of CV quantum communication and teleportation networks. Generalizations and extensions of these results appear at hand, and we may expect further progress along these lines in the near future, both for Gaussian and non Gaussian states.

A good portion of the material reported in this chapter originates from joint work with our friend and colleague Alessio Serafini, whom we warmly thank for the joy of collaborating together. It is as well a pleasure to acknowledge stimulating exchanges over the last two years with Nicolas Cerf, Ignacio Cirac, Silvio De Siena, Jens Eisert, Marie Ericsson, Rosario Fazio, Jaromír Fiurášek, Tohya Hiroshima, Ole Krüger, Gerd Leuchs, Norbert Lütkenhaus, Chiara Macchiavello, Klaus Mølmer, Tobias Osborne, Matteo Paris, Martin Plenio, Eugene Polzik, Mario Rasetti, Gustavo Rigolin, Anna Sanpera, Peter van Loock, Frank Verstraete, David Vitali, Reinhard Werner, Michael Wolf, Bill Wootters, and Paolo Zanardi.

References

1. *Quantum Information Theory with Continuous Variables*, S. L. Braunstein and A. K. Pati Eds. (Kluwer, Dordrecht, 2002).
2. S. L. Braunstein and P. van Loock, Rev. Mod. Phys. **77**, 513 (2005); J. Eisert and M. B. Plenio, Int. J. Quant. Inf. **1**, 479 (2003).
3. R. Simon, E. C. G. Sudarshan, and N. Mukunda, Phys. Rev. A **36**, 3868 (1987).
4. J. Williamson, Am. J. Math. **58**, 141 (1936); see also V. I. Arnold, *Mathematical Methods of Classical Mechanics* (Springer-Verlag, New York, 1978).
5. A. Serafini, F. Illuminati, and S. De Siena, J. Phys. B **37**, L21 (2004).
6. M. G. A. Paris, F. Illuminati, A. Serafini, and S. De Siena, Phys. Rev. A **68**, 012314 (2003).
7. A. Peres, Phys. Rev. Lett. **77**, 1413 (1996); R. Horodecki, P. Horodecki, and M. Horodecki, Phys. Lett. A **210**, 377 (1996).
8. A. Serafini, G. Adesso, and F. Illuminati, Phys. Rev. A **71**, 032349 (2005).
9. A. Botero and B. Reznik, Phys. Rev. A **67**, 052311 (2003); G. Giedke, J. Eisert, J. I. Cirac, and M. B. Plenio, Quant. Inf. Comp. **3**, 211 (2003).
10. R. Simon, Phys. Rev. Lett. **84**, 2726 (2000); L.-M. Duan, G. Giedke, I. Cirac, and P. Zoller, *ibid.* **84**, 2722 (2000).
11. R. F. Werner and M. M. Wolf, Phys. Rev. Lett. **86**, 3658 (2001).
12. G. Giedke, B. Kraus, M. Lewenstein, and J. I. Cirac, Phys. Rev. Lett. **87**, 167904 (2001).
13. G. Vidal and R. F. Werner, Phys. Rev. A **65**, 032314 (2002); K. Życzkowski, P. Horodecki, A. Sanpera, and M. Lewenstein, Phys. Rev. A **58**, 883 (1998); J. Eisert, *PhD Thesis* (University of Potsdam, Potsdam, 2001); M. B. Plenio, Phys. Rev. Lett. **95**, 090503 (2005).
14. J. Laurat, G. Keller, J.-A. Oliveira-Huguenin, C. Fabre, T. Coudreau, A. Serafini, G. Adesso, and F. Illuminati, J. Opt. B **7**, S577 (2005).
15. G. Adesso, A. Serafini, and F. Illuminati, Phys. Rev. A **70**, 022318 (2004).

16. C. H. Bennett, D. P. DiVincenzo, J. A. Smolin, and W. K. Wootters, Phys. Rev. A **54**, 3824 (1996).
17. G. Giedke, M. M. Wolf, O. Krüger, R. F. Werner, and J. I. Cirac, Phys. Rev. Lett. **91**, 107901 (2003).
18. S. Ishizaka and T. Hiroshima, Phys. Rev. A **62**, 022310 (2000); F. Verstraete, K. Audenaert, and B. De Moor, *ibid.* **64**, 012316 (2001); W. J. Munro, D. F. V. James, A. G. White, and P. G. Kwiat, *ibid.* **64**, 030302 (2001).
19. G. Adesso, A. Serafini, and F. Illuminati, Phys. Rev. Lett. **92**, 087901 (2004).
20. G. Adesso, F. Illuminati, and S. De Siena, Phys. Rev. A **68**, 062318 (2003).
21. A. K. Ekert, C. M. Alves, D. K. L. Oi, M. Horodecki, P. Horodecki, and L. C. Kwek, Phys. Rev. Lett. **88**, 217901 (2002); R. Filip, Phys. Rev. A **65**, 062320 (2002).
22. J. Fiurášek and N. J. Cerf, Phys. Rev. Lett. **93**, 063601 (2004); J. Wenger, J. Fiurášek, R. Tualle-Brouri, N. J. Cerf, and P. Grangier, Phys. Rev. A **70**, 053812 (2004).
23. P. van Loock, Fortschr. Phys. **50**, 12 1177 (2002).
24. G. Adesso, A. Serafini, and F. Illuminati, Phys. Rev. Lett. **93**, 220504 (2004).
25. F. Verstraete, M. Popp, and J. I. Cirac, Phys. Rev. Lett. **92**, 027901 (2004).
26. H.-J. Briegel, W. Dür, J. I. Cirac, and P. Zoller, Phys. Rev. Lett. **81**, 5932 (1998).
27. P. van Loock and A. Furusawa, Phys. Rev. A **67**, 052315 (2003).
28. D. M. Greenberger, M. A. Horne, A. Shimony, and A. Zeilinger, Am. J. Phys. **58**, 1131 (1990).
29. M. M. Wolf, J. Eisert, and M. B. Plenio, Phys. Rev. Lett. **90**, 047904 (2003).
30. G. Adesso and F. Illuminati, New J. Phys. **8**, 15 (2006).
31. G. Adesso and F. Illuminati, quant-ph/0506213, Int. J. Quant. Inf. (2006), in press.
32. V. Coffman, J. Kundu, and W. K. Wootters, Phys. Rev. A **61**, 052306 (2000).
33. T. J. Osborne and F. Verstraete, Phys. Rev. Lett. **96**, 220503 (2006).
34. B. M. Terhal, IBM J. Res. & Dev. **48**, 71 (2004), and quant-ph/0307120.
35. W. K. Wootters, Phys. Rev. Lett. **80**, 2245 (1998).
36. G. Adesso and F. Illuminati, Phys. Rev. A **72**, 032334 (2005).
37. M. M. Wolf, G. Giedke, O. Krüger, R. F. Werner, and J. I. Cirac, Phys. Rev. A **69**, 052320 (2004).
38. W. Dür, G. Vidal, and J. I. Cirac, Phys. Rev. A **62**, 062314 (2000).
39. G. Adesso, A. Serafini, and F. Illuminati, Phys. Rev. A **73**, 032345 (2006).
40. T. Aoki, N. Takei, H. Yonezawa, K. Wakui, T. Hiraoka, A. Furusawa, and P. van Loock, Phys. Rev. Lett. **91**, 080404 (2003).
41. P. van Loock and S. L. Braunstein, Phys. Rev. Lett. **84**, 3482 (2000).
42. S. L. Braunstein, C. A. Fuchs, and H. J. Kimble, J. Mod. Opt. **47**, 267 (2000); K. Hammerer, M. M. Wolf, E. S. Polzik, and J. I. Cirac, Phys. Rev. Lett. **94**, 150503 (2005).
43. H. Yonezawa, T. Aoki, and A. Furusawa, Nature **431**, 430 (2004).
44. G. Adesso and F. Illuminati, Phys. Rev. Lett. **95**, 150503 (2005).
45. W. P. Bowen, P. K. Lam, and T. C. Ralph, J. Mod. Opt. **50**, 801 (2003).
46. S. L. Braunstein and H. J. Kimble, Phys. Rev. Lett. **80**, 869 (1998).

Chapter 2

Gaussian Quantum Channels

J. Eisert[1,2] and M. M. Wolf[3]
[1] *Blackett Laboratory, Imperial College London, London SW7 2BW, UK*
[2] *Institute for Mathematical Sciences, Imperial College London, London SW7 2PW, UK*
[3] *Max-Planck-Institut für Quantenoptik, 85748 Garching, Germany*

1. Introduction

Any physical operation that reflects the time evolution of the state of a quantum system can be regarded as a channel. In particular, quantum channels grasp the way how quantum states are modified when subjected to noisy quantum communication lines. Couplings to other external degrees of freedom, often beyond detailed control, will typically lead to losses and decoherence, effects that are modeled by appropriate non-unitary quantum channels.

Gaussian quantum channels play a quite central role indeed, not only in the context of the present book. After all, good models for the transmission of light through fibers are provided by Gaussian channels. This is no accident: linear couplings of bosonic systems to other bosonic systems with quadratic Hamiltonians can in fact appropriately be said to be ubiquitous in physics. In this optical context then, the time evolution of the modes of interest, disregarding the modes beyond control, is then reflected by a Gaussian bosonic channel. Random classical noise, introduced by Gaussian random displacements in phase space, gives also rise to a Gaussian quantum channel, as well as losses that can be modelled as a beam splitter like interaction with the vacuum or a thermal mode.

This chapter provides a brief introduction into the theory of Gaussian quantum channels. After setting the notation and introducing to the elementary concepts, we provide a number of practically relevant examples.

Emphasis will later be put on questions concerning capacities: Capacities come in several flavours, and essentially quantify the usefulness of a quantum channel for the transmission of classical or quantum information. We will briefly highlight several major results that have been achieved in this field. Finally, we discuss a number of open questions, notably related to the intriguing but interesting and fundamental questions of additivities of quantum channel capacities.

2. Gaussian Channels

In mathematical terms a *quantum channel* is a completely positive trace-preserving map $\rho \longmapsto T(\rho)$ that takes states, i.e., density operators ρ acting on some Hilbert space $\mathcal{H}$, into states.[a] For simplicity we will always assume that output and input Hilbert spaces are identical. Every channel can be conceived as reduction of a unitary evolution in a larger quantum system. So for any channel T there exists a state ρ_E on a Hilbert space $\mathcal{H}_E$, and a unitary U such that

$$T(\rho) = \text{tr}_E[U(\rho \otimes \rho_E)U^\dagger]. \tag{1}$$

The system labeled E serves as an environment, embodying degrees of freedom of which elude the actual observation, inducing a decoherence process. The channel is then a local manifestation of the unitary evolution of the joint system. A *Gaussian channel*[2–5,1] is now a channel of the form as in Eq. (1), where U is a Gaussian unitary, determined by a quadratic bosonic Hamiltonian, and ρ_E is a Gaussian state.[6] In many cases, of which the lossy optical fiber is the most prominent one, this restriction to quadratic Hamiltonians gives a pretty good description of the physical system. Note that although the channel is assumed to be Gausssian in the entire chapter, the input states are not necessarily taken to be Gaussian.

2.1. *Preliminaries*

It seems appropriate for the following purposes to briefly fix the notation concerning Gaussian states and their transformations.[5–8] For a quantum system with n modes, i.e., n canonical degrees of freedom, the *canonical coordinates* will be denoted as $R = (x_1, p_1, ..., x_n, p_n)$. Most naturally,

---

[a]This expression refers to the Schrödinger picture of quantum channels. Equivalently, one can define the dual linear map T^* in the Heisenberg picture via $\text{tr}[\rho T^*(A)] = \text{tr}[T(\rho)A]$, which in turn is then completely positive and unital.

these operators can be conceived as corresponding to field quadratures. Although all statements in this chapter hold true for any physical system having canonical coordinates, we will often refer to the optical context when intuitively describing the action of a channel. The creation and annihilation operators are related to these canonical coordinates according to $x_i = (a_i + a_i^\dagger)/\sqrt{2}$ and $p_i = -i(a_i - a_i^\dagger)/\sqrt{2}$. The coordinates satisfy the canonical commutation relations, which can be expressed in terms of the *Weyl operators* or *displacement operators* $W_\xi = e^{i\xi^T\sigma R}$ with $\xi \in \mathbb{R}^{2n}$:

$$W_\xi^\dagger W_{\xi'} = W_{\xi'} W_\xi^\dagger e^{i\xi^T\sigma\xi'}, \quad \sigma = \bigoplus_{i=1}^{n} \begin{pmatrix} 0 & 1 \\ -1 & 0 \end{pmatrix}, \tag{2}$$

where we have set $\hbar = 1$. The matrix σ defines the symplectic scalar product, simply indicating that position and momentum of the same mode do not commute.

The Fourier transform of the ordinary Wigner function in phase space $\mathbb{R}^{2n}$ is the *characteristic function*

$$\chi_\rho(\xi) = \mathrm{tr}[\rho W_\xi], \tag{3}$$

from which the state can be reobtained as $\rho = \int d^{2n}\xi\ \chi_\rho(\xi) W_\xi^\dagger/(2\pi)^n$. Gaussian states are exactly those having a Gaussian characteristic function, and therefore a Gaussian Wigner function in phase space:

$$\chi_\rho(\xi) = e^{-\xi^T\Gamma\xi/4 + D^T\xi}. \tag{4}$$

Here, the $2n \times 2n$-matrix Γ and the vector $D \in \mathbb{R}^{2n}$ are essentially the first and second moments: they are related to the covariance matrix γ and the displacements d as $\Gamma = \sigma^T\gamma\sigma$ and $D = \sigma d$. This choice is then consistent with the definition of the *covariance matrix* as having entries $\gamma_{j,k} = 2\mathrm{Re}\left\langle (R_j - d_j)(R_k - d_k)\right\rangle_\rho$, $j, k = 1, ..., 2n$, with $d_j = \mathrm{tr}[R_j\rho]$. In general, states satisfy the Heisenberg uncertainly principle, which can be expressed as $\gamma + i\sigma \geq 0$. This is a simple semi-definite constraint onto any matrix of second moments, also obeyed by every non-Gaussian state.

2.2. *General Gaussian channels*

The simplest Gaussian channel is a lossless unitary evolution, governed by a quadratic bosonic Hamiltonian:

$$\rho \longmapsto U\rho U^\dagger, \quad U = e^{\frac{i}{2}\sum_{k,l} H_{kl} R_k R_l}, \tag{5}$$

with H being a real and symmetric $2n \times 2n$ matrix. Such unitaries correspond to a representation of the real symplectic group $Sp(2n, \mathbb{R})$, formed by those real matrices for which $S\sigma S^T = \sigma$.[7-9] These are exactly the linear transformations which preserve the commutation relations. The relation between such a *canonical* transformation in phase space and the corresponding unitary in Hilbert space is given by $S = e^{H\sigma}$. Needless to say, Gaussian unitaries are ubiquitous in physics, in particular in optics, and this is the reason why Gaussian channels play such an important role. Notably, the action of ideal beam splitters, phase shifters, and squeezers correspond to symplectic transformations.[b]

It is often instructive to consider transformations on the level of Weyl operators in the Heisenberg picture. For a symplectic transformation we have $W_\xi \longmapsto W_{S^{-1}\xi}$. The action of a *general Gaussian channel* $\rho \longmapsto T(\rho)$ can be phrased as

$$W_\xi \longmapsto W_{X\xi}\, e^{-\frac{1}{2}\xi^T Y \xi}, \tag{6}$$

where X, Y are real $2n \times 2n$-matrices.[5,7,1] Additional linear terms in the quadratic form are omitted since they merely result in displacements in phase space, which are not interesting for later purpose. Not any transformation of the above form is possible: complete positivity of the channel dictates that[c]

$$Y + i\sigma - iX^T\sigma X \geq 0. \tag{7}$$

Depending on the context it may be more appropriate or transparent to formulate a Gaussian channel in the Schrödinger picture $\rho \longmapsto T_{X,Y}(\rho)$ or to define it as a transformation of covariance matrices

$$\gamma \longmapsto X^T\gamma X + Y. \tag{8}$$

This is the most general form of a Gaussian channel. Roughly speaking X serves the purpose of amplification or attenuation and rotation in phase space, whereas the Y contribution is a noise term which may consist of quantum (required to make the map physical) as well as classical noise. Interestingly, X may be any real matrix, and hence, any map $\gamma \longmapsto X^T\gamma X$ can be approximately realized, as long as 'sufficient noise' is added. In this

---

[b]Any such S can be decomposed into a *squeezing component*, and a *passive operation*.[9] So one may write $S = VZW$, with $V, W \in K(n) = Sp(2n, \mathcal{R}) \cap SO(2n)$ are orthogonal symplectic transformations, forming the subgroup of passive, i.e., number-preserving, operations. In turn, $Z = \text{diag}\,(z_1, 1/z_1, \ldots, z_n, 1/z_n)$ with $z_1, ..., z_n \in \mathcal{R}\backslash\{0\}$ are local single-mode squeezings.

[c]The case of a single mode is particularly transparent. Then, mixedness can be expressed entirely in terms of determinants, and hence, the above requirement can be cast into the form $Y \geq 0$, and $\text{Det}\,[Y] \geq (\text{Det}\,[X] - 1)^2$.

language, it also becomes immediately apparent how much noise will be introduced by any physical device approximating amplification or time reversal, meaning phase conjugation in an optical context. For second moments far away from minimal uncertainty, this additional noise may hardly have an impact (so classical fields can be phase conjugated after all), whereas close to minimal uncertainty this is not so any longer.

2.3. *Important examples of Gaussian channels*

The practically most important Gaussian channel is probably an idealized action of a fiber. Moreover, as mentioned earlier, any situation where a quadratic coupling to a Gaussian environment provides a good description can be cast into the form of a Gaussian channel. We will in the following consider a number of important special cases of Gaussian channels for single modes:

1. The *classical noise channel* merely adds classical Gaussian noise to a quantum state, i.e., $X = \mathbb{1}$, $Y \geq 0$.[3,10–12] In Schrödinger picture this channel can be represented by a random displacement according to a classical Gaussian probability distribution:

$$T(\rho) = \frac{1}{4\pi\sqrt{\operatorname{Det} Y}} \int d^2\xi \; W_\xi \rho W_\xi^\dagger \; e^{-\frac{1}{4}\xi^T Y^{-1}\xi}. \tag{9}$$

2. In the *thermal noise channel*[3,11] a mode passively interacts with another mode in a thermal state, $\rho \longmapsto T(\rho) = \operatorname{tr}_E[U_\eta(\rho \otimes \rho_E)U_\eta^\dagger]$. The result is as if the mode had been coupled in with a beam splitter of some transmitivity η.[d] For the second moments, we have that

$$\gamma \longmapsto [S_\eta(\gamma \oplus c\mathbb{1}_2)S_\eta^T]_E, \tag{10}$$

where $c\mathbb{1}_2$, $c \geq 1$, is the covariance matrix of a thermal *Gibbs state*

$$\rho_E = \frac{2}{c+1}\sum_{n=0}^{\infty}\left(\frac{c-1}{c+1}\right)^n |n\rangle\langle n| \tag{11}$$

with mean photon number $(c-1)/2$. $[.]_E$ denotes the leading 2×2 submatrix. The passive symplectic transformation S_η is given by

$$S_\eta = \begin{bmatrix} \sqrt{\eta}\,\mathbb{1}_2 & \sqrt{1-\eta}\,\mathbb{1}_2 \\ -\sqrt{1-\eta}\,\mathbb{1}_2 & \sqrt{\eta}\,\mathbb{1}_2 \end{bmatrix}, \quad \eta \in [0,1]. \tag{12}$$

[d] In the Heisenberg picture this means that the annihilation operator transforms as $a \mapsto \sqrt{\eta}\, a + \sqrt{1-\eta}\, b$, where b is the annihilation operator of the ancillary mode.

So we obtain

$$\gamma \longmapsto \eta\gamma + (1-\eta)c\mathbb{1}_2. \tag{13}$$

3. The *lossy channel* is obtained by setting $c = 1$ in Eq. (13). It reflects photon loss with probability $1-\eta$. This channel is the prototype for optical communication through a lossy fiber, since thermal photons (leading to a contribution $c > 1$) are negligible at room temperature. When using an optical fiber of length l and *absorbtion length* l_A we may set $\eta = e^{-l/l_A}$. The lossy channel with $X = \sqrt{\eta}\mathbb{1}_2$, $Y = (1-\eta)\mathbb{1}_2$ is also called *attenuation channel*.[3]

4. The *amplification channel* [3] is of the form

$$X = \sqrt{\eta}\mathbb{1}_2, \;\; Y = (\eta - 1)\mathbb{1}_2, \quad \eta \in (1, \infty). \tag{14}$$

Here, the term Y is a consequence of the noise that is added due to Heisenberg uncertainty. Note that a classical noise channel can be recovered as a concatenation of a lossy channel, followed by an amplification.

All these examples correspond to a single mode characterized by a fixed frequency ω. This is often referred to as the narrowband case as opposed to *broadband channels*,[13–15] which consist out of many uncoupled single-mode channels, each of which corresponds to a certain frequency ω_i, $i = 1, 2, \ldots$. Best studied is the simple homogeneous case of a lossy broadband channel (equally spaced frequencies ω_i, with equal transmittivity η in all the modes).

It shall finally be mentioned that the very extensive literature on harmonic open quantum systems is essentially concerned with Gaussian channels of a specific kind, yet one where the environment consists of infinitely many modes, where the linear coupling is characterized by some spectral density.

3. Entropies and Quantum Mutual Information

3.1. *Output entropies*

Channels describing the physical transmission of quantum states typically introduce noise to the states as a consequence of a decoherence process. Pure inputs are generally transformed into mixed outputs, so into states ρ having a positive von-Neumann entropy

$$S(\rho) = -\mathrm{tr}[\rho \log \rho]. \tag{15}$$

The entropy of the output will clearly depend on the input and the channel itself, and the minimal such entropy can be taken as a characteristic feature of the quantum channel. Introducing more generally the α *Renyi entropies* for $\alpha \geq 0$ as

$$S_\alpha(\rho) = \frac{1}{1-\alpha} \log \mathrm{tr}[\rho^\alpha] \tag{16}$$

this *minimal output entropy*[16] is then defined as[e]

$$\nu_\alpha(T) = \inf_\rho (S_\alpha \circ T)(\rho). \tag{17}$$

The Renyi entropies[17] are derived from the α-norms of the state, $\|\rho\|_\alpha = \mathrm{tr}[\rho^\alpha]^{1/\alpha}$. In case of the limit $\lim_{\alpha \searrow 1}$ one retains the von-Neumann entropy, i.e., $\lim_{\alpha \searrow 1} S_\alpha(\rho) = S(\rho)$; for $\alpha = 2$, this is the *purity* in the closer sense. Roughly speaking, the smaller the minimal output entropy, the less decohering is the channel (see, e.g., Ref. 18). The actual significance of this quantity yet originates from its intimate relationship concerning questions of capacities. This will be elaborated on in the subsequent section.

3.2. *Mutual information and coherent information*

In Shannon's seminal channel coding theorem the capacity of a classical channel is expressed in terms of the classical mutual information.[20] In fact, as we will see below, the quantum analogue of this quantity plays a similar role in quantum information theory. For any quantum channel T and any quantum state ρ acting on a Hilbert space $\mathcal{H}$, one defines the *quantum mutual information* $I(\rho, T)$ as

$$I(\rho, T) = S(\rho) + (S \circ T)(\rho) - S(\rho, T), \tag{18}$$

where $S(\rho, T) = (\mathbb{1} \otimes T)(|\psi\rangle\langle\psi|)$ and $|\psi\rangle \in \mathcal{H}_D \otimes \mathcal{H}$ is any purification of the state $\rho = \mathrm{tr}_D[|\psi\rangle\langle\psi|]$.[3,19] It is not difficult to see that $I(\rho, T)$ does not depend on the chosen purification. The quantum mutual information has many desirable properties: it is positive, concave with respect to ρ, and additive with respect to quantum channels of the form $T^{\otimes n}$. The latter property comes in very handy when relating this quantity to the entanglement-assisted classical capacity. An important part of the quantum mutual information is the *coherent information* given by

$$J(\rho, T) = (S \circ T)(\rho) - S(\rho, T). \tag{19}$$

$J(\rho, T)$ can be positive as well as negative, it is convex with respect to T but its convexity properties with respect to ρ are unclear.

---

[e] We use the notation $(S \circ T)(\rho) = S(T(\rho))$.

3.3. *Entropies of Gaussian states and extremal properties*

When maximizing the rate at which information can be sent through a Gaussian channel, Gaussian states play an important role. In fact, in many cases it turns out that encoding the information into Gaussian states leads to the highest transmission rates. This is mainly due to the fact that for a given covariance matrix many entropic quantities take on their extremal values for Gaussian states. These entropic quantities, and in fact any unitarily invariant functional, can for Gaussian states immediately be read off the symplectic spectrum of the covariance matrix: any covariance matrix γ of n modes can be brought to the *Williamson normal form*[21], $\gamma \longmapsto S\gamma S^T = \mathrm{diag}(c_1, c_1, c_2, c_2, ..., c_n, c_n)$ with an appropriate $S \in Sp(2n, \mathbb{R})$, and $\{c_i : i = 1, ..., n\}$ being the positive part of the spectrum of $i\sigma\gamma$. This is nothing but the familiar normal mode decomposition with the c_i corresponding to the normal mode frequencies. Then, the problem of evaluating any of the above quantities is reduced to a single-mode problem. For example, the von-Neumann entropy is given by[2]

$$S(\rho) = \sum_{i=1}^{n} g\Big(\frac{c_i - 1}{2}\Big), \tag{20}$$

where $g(N) = (N+1)\log(N+1) - N\log N$ is the entropy of a thermal Gaussian state with average photon number N. Similar expressions can be found for the other entropic quantities.

Consider now any state $\tilde{\rho}$ which has the same first and second moments as its Gaussian counterpart ρ. Then

$$S(\rho) - S(\tilde{\rho}) = S(\tilde{\rho}, \rho) + \mathrm{tr}\left[(\tilde{\rho} - \rho)\log\rho\right], \tag{21}$$

where the first term is the nonnegative relative entropy, and the second term vanishes since the expectation value of the operator $\ln\rho$ depends only on the first and second moments. Hence, the Gaussian state has the largest entropy among all states with a given covariance matrix.[2] A more sophisticated argument shows that the same holds true for the quantum mutual information:[3] For any Gaussian channel T and fixed first and second moments of ρ, the respective Gaussian state maximizes $I(\rho, T)$. Whether a similar statement also holds for the coherent information is not known.

3.4. *Constrained quantities*

There are essentially two subtleties[22,23] that arise in the infinite-dimensional context as we encounter it here for Gaussian quantum channels: on the one hand, there is the necessity of natural input constraints, such as one of finite mean energy. Otherwise, the capacities diverge. On the other hand, there is the possibility of continuous state ensembles.[f] The need for a constraint is already obvious when considering the von-Neumann entropy: On a state space over an infinite dimensional Hilbert space, the von-Neumann entropy is not (trace-norm) continuous, but only lower semi-continuous,[g] and almost everywhere infinite.

This problem can be tamed by introducing an appropriate constraint. For our purposes, we may take the Hamiltonian $H = \sum_{i=1}^{n}(x_i^2 + p_i^2)/2$. Then, instead of taking all states into account, one may consider the subset

$$\mathcal{K} = \{\rho : \mathrm{tr}[\rho H] < h\}. \tag{22}$$

introducing for some $h > 0$ a *constraint on the mean energy*[h] or mean photon number $N = h - 1/2$. Similarly, for tensor products we consider $\mathcal{K}^{\otimes n} = \{\rho : \mathrm{tr}[\rho H^{\otimes n}] < nh\}$. On this very natural subset $\mathcal{K}$ the von-Neumann entropy and the classical information capacity retain their continuity. In fact, many entanglement measures also retain the continuity properties familiar in the finite-dimensional context, such that, e.g., the entropy of a subsystem for pure states can indeed be interpreted as the distillable entanglement.[24]

4. Capacities

In classical information theory a single number describes how much information can reliably be sent through a channel: its *capacity*. In quantum information theory the situation is more complicated and each channel is characterized by a number of different capacities.[25] More precisely, which capacity is the relevant one depends on whether we want to transmit classical or quantum information, and on the resources and protocols we allow for. An important resource that we must consider is entanglement shared between sender and receiver. The presence or absence of this resource together with the question about sending classical or quantum information leads to four basic capacities, which we will discuss in the following.

---

[f]This is understood as taking into account probability measures on the set of quantum states. For an approach in the language of probability and operator theory, see Ref. 22.
[g]This means that if, for a state ρ, $\{\rho_n\}$ is a sequence of states for which $\rho_n \to \rho$ in trace-norm as $n \to \infty$, then $S(\rho) \leq \liminf_{n\to\infty} S(\rho_n)$.
[h]More general constraints than this one can be considered, leading to *compact subsets of state space* on which one retains continuity properties in particular for the von-Neumann entropy and the classical information capacity.[17,22–24] Essentially, any unbounded positive operator H with a spectrum without limiting points would also be appropriate, such that $\mathrm{tr}\exp[-\beta H] < \infty$ for all $\beta > 0$.

4.1. *Classical information capacity*

The *classical information capacity* is the asymptotically achievable number of classical bits that can be reliably transmitted from a sender to a receiver per use of the channel. Here, it is assumed that the parties may coherently encode and decode the information in the sense that they may use entangled states as codewords at the input and joint measurements over arbitrary channel uses at the output. This answers essentially the question of how useful a quantum channel is for the transmission of classical information.

This capacity is derived from the single-shot expression,[26,27] appropriately constrained as above,

$$C_1(T,\mathcal{K}) = \sup \left[S\Big(\sum_i p_i T(\rho_i)\Big) - \sum_i p_i (S\circ T)(\rho_i) \right], \tag{23}$$

where the supremum is taken over all probability distributions and sets of states satisfying $\rho = \sum_i p_i \rho_i$ under the constraint $\rho \in \mathcal{K}$.[22,23,i] By the Holevo-Schumacher-Westmoreland (HSW) theorem,[26,27] this single-shot expression gives the capacity if the encoding is restricted to product states. Hence, the full classical information capacity can formally be expressed as the regularization of C_1,

$$C(T,\mathcal{K}) = \lim_{n\to\infty} \frac{1}{n} C_1(T^{\otimes n}, \mathcal{K}^{\otimes n}). \tag{24}$$

Clearly, $C(T,\mathcal{K}) \geq C_1(T,\mathcal{K})$ since the latter does not allow for inputs which are entangled over several instances of the channel. Yet, it is in general not known whether this possibility comes along with any advantage at all, so whether entangled inputs facilitate a better information transfer. This will be remarked on later.

Note that in this infinite-dimensional setting, the constraint is required to obtain a meaningful expression for the capacity: for all non-trivial Gaussian channels the optimization over all input ensembles in Eq. (23) would lead to an infinite capacity. This can simply be achieved by encoding the information into phase space translates of any signal state. Then no matter how much noise is induced by the channel, we can always choose the spacing between the different signal states sufficiently large such that they can be distinguished nearly perfectly at the output.

---

[i] The above constraint also ensures that $(S\circ T)(\rho) < \infty$. The convex hull function of $S\circ T$, given by $\rho \longmapsto \hat{S}(\rho,T) = \inf \sum_i p_i (S\circ T)(\rho_i)$ in Eq. (23), with the infimum being taken over all ensembles with $\sum_i p_i\rho_i = \rho$, is still convex in the unconstrained case, but no longer continuous, however, lower semi-continuous in the above sense.

Let us now follow the lines of Ref. 28 and sketch the derivation of the classical capacity for lossy channels. First of all, a lower bound on $C(T,\mathcal{K})$ can be obtained by choosing an explicit input ensemble for Eq. (23). Random coding over coherent states according to a classical Gaussian probability distribution leads to an average input state of the form

$$\rho \propto \int d^2\xi\, W_\xi|0\rangle\langle 0|W_\xi^\dagger\, e^{-\frac{1}{4}\xi^T V^{-1}\xi}\,, \tag{25}$$

with covariance matrix $\gamma = \mathbb{1}+V$. Hence, if we choose $V = 2N\mathbb{1}$, the average number of photons in the input state will be $\mathrm{tr}\,[\rho a^\dagger a] = N$. The constraint set $\mathcal{K}$ hence corresponds to the choice of $h = N + 1/2$. After passing a lossy channel with transmittivity η this changes to $\mathrm{tr}\,[T(\rho)a^\dagger a] = \eta N$, and since $T(\rho)$ is a thermal state, its entropy is given by $(S \circ T)(\rho) = g(\eta N)$. The action of a lossy channel on a coherent input state is to shift the state by a factor η towards the origin in phase space. In other words, the channel maps coherent states onto coherent states and since the latter have zero entropy, we have[3]

$$C_1(T,\mathcal{K}) \geq (S \circ T)(\rho) = g(\eta N). \tag{26}$$

Assume now that $\tilde{\rho}$ is the average input state optimizing $C_1(T^{\otimes n},\mathcal{K}^{\otimes n})$ under a given constraint for the mean energy as described above. Then

$$C_1(T^{\otimes n},\mathcal{K}^{\otimes n}) \leq (S \circ T^{\otimes n})(\tilde{\rho}) \leq \sum_{i=1}^{n} (S \circ T)(\tilde{\rho}_i), \tag{27}$$

where $\tilde{\rho}_i$ is the reduction of $\tilde{\rho}$ to the i-th mode and the second inequality is due to the subadditivity of the von-Neumann entropy. Since for a fixed average photon number $\mathrm{tr}\,[\tilde{\rho}_i a^\dagger a] = N_i$ the entropy is maximized by a Gaussian state, we have in addition that $(S \circ T)(\tilde{\rho}_i) \leq g(\eta N_i)$.

Together with the lower bound this implies that the classical capacity of a lossy channel is indeed given by $C(T,\mathcal{K}) = g(\eta N)$,[28] if the average number of input photons per channel use is restricted to be not larger than N, corresponding to the constrained associated with $\mathcal{K}$. Hence random coding over coherent states turns out to be optimal and neither non-classical signal states nor entanglement is required in the encoding step.[j]

[j] Of course, there might also be optimal encodings which do exploit a number state alphabet or entanglement between successive channel uses.

An immediate consequence of this result is that the classical capacity of the homogeneous broadband channel T is given by

$$C(T,\mathcal{K}) = t\frac{\sqrt{\eta}}{\ln 2}\sqrt{\frac{\pi P}{3}} + \mathcal{O}(1/t) , \tag{28}$$

where P is the average input power and t is the transmission time related to the frequency spacing $\delta\omega = 2\pi/t$. For the lossless case $\eta = 1$ this capacity was derived in Ref. 15, 29.

4.2. *Quantum capacities and coherent information*

The *quantum capacity* is the rate at which qubits can be reliably transmitted through the channel from a sender to a receiver. This transmission is done again employing appropriate encodings and decodings before envoking instances of the quantum channel.[3] This capacity can be made precise using the *norm of complete boundedness*.[k] The question is how well the identity channel can be approximated in this norm. More specifically,[30] the quantum capacity Q is the supremum of $c \geq 0$ such that for all $\varepsilon, \delta > 0$ there exist $n, m \in$, decodings T_{D} and encodings T_{E} with

$$\left|\frac{n}{m} - c\right| < \delta, \quad \left\|\mathrm{Id}_2^{\otimes n} - T_{\mathrm{D}} T^{\otimes m} T_{\mathrm{E}}\right\|_{\mathrm{cb}} < \varepsilon. \tag{29}$$

One may also consider a weaker instance, allowing for ε-errors, and then look at a Q_ε-capacity.[3] It is known that the quantum capacity does not increase if we allow for additional classical forward communication.[31]

In Ref. 32 it was proven that the quantum capacity $Q(T)$ can be expressed in terms of the coherent information as

$$Q(T) = \lim_{n\to\infty} \frac{1}{n} \sup_{\rho} J\big(\rho, T^{\otimes n}\big) . \tag{30}$$

Unfortunately, the asymptotic regularization is required in general, since the supremum over the coherent information is known to be not additive.[l] However, the single-shot quantity $\sup_\rho J(\rho, T)$ already gives a useful lower bound on $Q(T)$. For the classical noise Gaussian channel and Gaussian ρ this was first shown to be attainable in Ref. 10, based on earlier work,[34] using methods of quantum stabilizer codes that embed a finite-dimensional

---

[k] This is defined as $||T||_{\mathrm{cb}} = \sup_n ||\mathrm{Id}_n \otimes T||$, where $||T|| = \sup_X ||T(X)||_1/||X||_1$.

[l] Note also that while the subtleties in the infinite-dimensional context have been fleshed out and precisely clarified for the classical information capacity,[22,23] the entanglement-assisted capacity,[33] and measures of entanglement,[24,23] questions of continuity related to the quantum capacity are to our knowledge still awaiting a rigorous formulation.

protected code space in an infinite-dimensional one. For more general thermal noise channels, this is given by[3]

$$J(\rho, T) = g(N') - g\left(\frac{D + N' - N - 1}{2}\right) - g\left(\frac{D - N' + N - 1}{2}\right), \quad (31)$$

$$D = \sqrt{(N + N' + 1)^2 - 4\eta N(N+1)}\,, \quad (32)$$

where $N' = \eta N + (1 - \eta)(c - 1)/2$ is the average photon number at the channel output. In fact, the same bound holds for the amplification channel, for which $\eta > 1$ and $N' = \eta N + (\eta - 1)(c + 1)/2$. For broadband channels, lower bounds of this kind on the quantum capacity were discussed in Ref. 13.

A computable upper bound on the quantum capacity of any channel is given by $Q(T) \leq \log ||T\theta||_{\text{cb}}$.[3] For finite-dimensional systems θ is the matrix transposition, which corresponds to the momentum-reversal operation in the continuous variables case. This bound is zero for *entanglement breaking* channels[m] and additive for tensor products of channels. For attenuation and amplification channels with classical noise, i.e., channels acting as $\gamma \mapsto \eta\gamma + |1 - \eta|c$, this leads to[3]

$$Q(T) \leq \log(1 + \eta) - \log|1 - \eta| - \log c. \quad (33)$$

Note that this bound is finite for all $\eta \neq 1$. This is remarkable since it does not depend on the input energy. That is, unlike the classical capacity, the unconstrained quantum capacity does typically not diverge. Moreover, it is even zero in the case $\eta \leq 1/2$, since then the no-cloning theorem forbids an asymptotic error-free transmission of quantum information.

4.3. *Entanglement-assisted capacities*

Needless to say, in a quantum information context, it is meaningful to see what rates can be achieved for the transfer of classical information when entanglement is present. This is the kind of information transfer considered in the *entanglement-assisted classical capacity* C_E.[33,35] It is defined as the rate at which bits that can be transmitted in a reliable manner in the presence of an unlimited amount of prior entanglement shared between the sender and the receiver. In just the same manner, the *entanglement-assisted quantum capacity* Q_E may be defined.[13,14,33] Similarly, this quantifies the rate at which qubits can asymptotically be reliably transmitted per channel

[m] A channel is called entanglement breaking if it corresponds to a measure and repreparation scheme.

use, again in the presence of unlimited entanglement. Exploiting teleportation and dense coding is not difficult to see that $2Q_E = C_E$. Now, the entanglement-assisted capacity C_E is intimately related to the quantum mutual information, as just the supremum of this quantity with respect to all states $\rho \in \mathcal{K}$ as in Eq. (22)

$$C_E(T, \mathcal{K}) = \sup_\rho I(\rho, T). \tag{34}$$

Again, with this constraint,[33] the quantity regains the appropriate continuity properties.[n] Note that in this case, no asymptotic version has to be considered, and due to the additivity of the quantum mutual information the single-shot expression already provides the capacity.

In a sense Eq. (34) is the direct analogue of Shannon's classical coding theorem. The latter states that the classical capacity of a classical channel is given by the maximum mutual information. The main difference is, however, that in the classical case shared randomness does not increase the capacity, whereas for quantum channels shared entanglement typically increases the capacity,

$$C(T, \mathcal{K}) \leq C_E(T, \mathcal{K}). \tag{35}$$

Again, similar to the classical case C_E is conjectured to characterize equivalence classes of channels within which all channels can efficiently simulate one another.[35]

For Gaussian channels the extremal property of Gaussian states with respect to the quantum mutual information allows us to calculate $C_E(T, \mathcal{K})$ by only maximizing over constrained Gaussian states ρ. For attenuation channels with classical noise, i.e., $\gamma \mapsto \eta\gamma + (1-\eta)c$ with $0 \leq \eta \leq 1$, it was shown in Ref. 3 that

$$C_E(T, \mathcal{K}) = g(N) + J(\rho, T)\ , \tag{36}$$

with the coherent information $J(\rho, T)$ taken from Eq. (31). For the homogeneous broadband lossy channel, extensively discussed in Refs. 13, 14, it holds again that $C_E(T, \mathcal{K}) \propto t\sqrt{P}$.

[n] In a more general formulation – i.e., for non-Gaussian constrained channels, or for Gaussian channels with different constraints – one has to require that $\sup_{\rho\in\mathcal{K}}(S \circ T)(\rho) < \infty$.[33]

5. Additivity Issues

In the previous sections, we have encountered additivity problems of several quantities related to quantum channels. Such questions are at the core of quantum information theory: essentially, the questions is whether for products channels one can potentially gain from utilizing entangled inputs. This applies in particular to the additivity of the single-shot expression C_1 and the minimal output entropy.[o,16] A number of partial results on additivity problems have been found. Yet, a conclusive answer to the most central additivity questions is still lacking. In particular, it is one of the indeed intriguing open questions of quantum information science whether the single-shot expression C_1 in Eq. (23) is already identical to the classical information capacity as it is true for the case of the lossy channel.[28]

5.1. *Equivalence of additivity problems*

Interestingly, a number of additivity questions are related in the sense that they are either all true or all false. This connection is particularly well-established in the finite-dimensional context:[36–38] then, the equivalence of the (i) additivity of the minimum output 1-entropy, the von-Neumann entropy, (ii) the additivity of the single-shot expression C_1, (iii) the additivity of the entanglement of formation, and (iv) the strong superadditivity of the entanglement of formation have been shown to be equivalent.[36–38] This equivalence, besides being an interesting result in its own right, provides convenient starting points for general studies on additivity, as in particular the minimal output entropies appear much more accessible than the classical information capacity.

In the infinite-dimensional context, the argument concerning equivalence is somewhat burdened with technicalities. We will here state the main part of an equivalence theorem of additivity questions concerning any pair T_1, T_2 of Gaussian channels.[23] The following properties (1) and (2) are equivalent and imply (3):

(1) For any state ρ on the product Hilbert space and for all appropriately

[o]In the context of entanglement measures, additivity refers to the property that for a number of uncorrelated bi-partite systems, the degrees of entanglement simply add up to the total entanglement.

constraint sets $\mathcal{K}_1$ and $\mathcal{K}_2$ we have that[p]

$$C_1(T_1 \otimes T_2, \mathcal{K}_1 \otimes \mathcal{K}_2) = C_1(T_1, \mathcal{K}_1) + C_1(T_2, \mathcal{K}_2). \tag{37}$$

(2) For any state ρ with $(S \circ T_1)(\text{tr}_2[\rho]) < \infty$ and $(S \circ T_2)(\text{tr}_1[\rho]) < \infty$

$$\hat{S}(\rho, T_1 \otimes T_2) \geq \hat{S}(\text{tr}_2[\rho], T_1) + \hat{S}(\text{tr}_1[\rho], T_2), \tag{38}$$

where for a channel T and $\sum_i p_i \rho_i = \rho$

$$\hat{S}(\rho, T) = \inf \sum_i p_i (S \circ T)(\rho_i). \tag{39}$$

(3) For the minimal output entropies

$$\bar{\nu}_1(T_1 \otimes T_2) = \nu_1(T_1) + \nu_1(T_2) \tag{40}$$

where the bar indicates that in order to evaluate the minimal output entropy of $T_1 \otimes T_2$, the infimum is taken only over all pure states ρ such that $S(\text{tr}_2[\rho]) = S(\text{tr}_1[\rho]) < \infty$ and $(S \circ (T_1 \otimes T_2))(\rho) < \infty$.

In particular, this means that once a general answer to (1) or (2) was known for Gaussian channels, a general single-shot expression for the classical information capacity of such channels would be available, solving a long-standing open question. Moreover, it was proven that the above implications hold true if one of the additivity conjectures is proven for the general finite dimensional case.[23]

5.2. *Gaussian inputs to Gaussian channels*

In restricted settings, quite far-reaching statements concerning additivity can yet be made. For example, if one not only considers Gaussian channels, but requires that the encoding is done entirely in Gaussian terms, the additivity for minimal output entropies can be proven in quite some generality.[39] The *Gaussian minimal output entropy* is defined as

$$\nu'_\alpha(T) = \inf_\rho (S_\alpha \circ T)(\rho), \tag{41}$$

where the infimum is taken over all Gaussian states. Then one finds that the minimal output α-entropy for single-mode Gaussian channels $T_1, ..., T_n$, as

[p] This has to hold for all compact subsets $\mathcal{K}_1$ and $\mathcal{K}_2$ of state space for which $(S \circ T_i)(\rho) < \infty$ for all states $\rho \in \mathcal{K}_i$, $i = 1, 2$, and such that $C_1(T_1, \mathcal{K}_1), C_1(T_2, \mathcal{K}_2) < \infty$. Note that these assumptions are in particular satisfied if $\mathcal{K}_1$ and $\mathcal{K}_2$ are defined by an energy constraint.

in Eq. (8) characterized by $X_1, ..., X_n$ and $Y_1, ..., Y_n$, $Y_i \geq 0$, and $\det[X_i] = \det[X_j]$ for all i, j is additive for all $\alpha \in (1, \infty)$. This includes the important case of identical Gaussian channels T,

$$\nu'_\alpha(T^{\otimes n}) = n\nu'_\alpha(T) \tag{42}$$

for all n and all $\alpha \in (1, \infty)$. Moreover, for $\alpha = 2$ this kind of additivity was proven for arbitrary multi-mode Gaussian channels for which Det $[X_i] \neq 0$. It is after all not known whether this requirement constitutes a restriction to generality. It is an interesting open question whether non-Gaussian encodings can be advantageous when transmitting classical information through Gaussian channels. Frequently, it is taken as a reasonable conjecture that Gaussian encodings to achieve the classical information capacity are always appropriate. And, in fact, as we saw, this is true for the important case of the lossy channel.[28]

5.3. *Integer output entropies and Gaussian inputs*

In turn, for specific channels, the unconstrained minimal output α-entropies for tensor products can be identified for integer α. These *integer instances of output purities* are not immediately related to the question of the classical information capacity, for which the limit $\alpha \searrow 1$ is needed. However, they provide a strong indication of additivity also in the general case. Notably, for the single-mode classical and thermal noise channels T,

$$\nu_p(T^{\otimes n}) = n\nu_p(T) \tag{43}$$

has been established for integer p.[11] The concept of entrywise positive maps also provides a general framework for assessing integer minimal output entropies for Gaussian channels,[40] generalizing previous results.

6. Outlook

This chapter was concerned with the theory of Gaussian quantum communication channels. Such channels arise in several practical contexts, most importantly as models for lossy fibers. Emphasis was put on questions related to capacities, which give the best possible bounds on the rates that can be achieved when using channels for the communication of quantum or classical information.

Though many basic questions have been solved over the last few years, many interesting questions in the theory of bosonic Gaussian channels are essentially open. This applies in particular to additivity issues: general formulae for the classical information capacity are simply not available before a resolution of these issues. For specific channels, a number of methods can yet be applied to find additivity of output purities. It may be interesting to see how far the idea of relating minimal 1-entropies to 2-entropies as in Ref. 41 could be extended in the infinite-dimensional context.

Then, there is the old conjecture that to take Gaussian ensembles does not constitute a restriction of generality anyway when transmitting information through a Gaussian quantum channel. In the light of this conjecture, it would be interesting whether a complete theory of quantum communication can be formulated, restricting both to Gaussian ensembles and Gaussian channels.

Finally, all what has been stated on capacities in this chapter refers to the case of memoryless channels. For Gaussian channels with memory, the situation can be quite different. For example, notably, the classical information capacity can be enhanced using entangled instead of product inputs.[42,43] It would in this context also be interesting to see the program of Ref. 44 implemented in the practically important case of Gaussian quantum channels.

Acknowledgments

This work has been supported by the EPSRC (GR/S82176/0, QIP-IRC), the European Union (QUPRODIS, IST-2001-38877), the DFG (Schwerpunktprogramm QIV), and the European Research Councils (EURYI).

References

1. B. Demoen, P. Vanheuswijn, and A. Verbeure, Lett. Math. Phys. **2**, 161 (1977).
2. A.S. Holevo, M. Sohma, and O. Hirota, Phys. Rev. A **59**, 1820 (1999).
3. A.S. Holevo and R.F. Werner, Phys. Rev. A **63**, 032312 (2001).
4. J. Eisert and M.B. Plenio, Phys. Rev. Lett. **89**, 097901 (2002).
5. G. Lindblad, J. Phys. A **33**, 5059 (2000).
6. A.S. Holevo, *Probabilistic Aspects of Quantum Theory* (North-Holland, Amsterdam, 1982), Chapter 5.
7. J.I. Cirac, J. Eisert, G. Giedke, M.B. Plenio, M. Lewenstein, M.M. Wolf, and R.F. Werner, textbook in preparation (2005).
8. J. Eisert and M.B. Plenio, Int. J. Quant. Inf. **1**, 479 (2003).
9. Arvind, B. Dutta, N. Mukunda, and R. Simon, Pramana **45**, 471 (1995); quant-ph/9509002.

10. J. Harrington and J. Preskill, Phys. Rev. A **64**, 062301 (2001).
11. V. Giovannetti, S. Lloyd, L. Maccone, J.H. Shapiro, and B.J. Yen, Phys. Rev. A **70**, 022328 (2004).
12. C.M. Caves and K. Wodkiewicz, quant-ph/0409063.
13. V. Giovannetti, S. Lloyd, L. Maccone, and P.W. Shor, Phys. Rev. A **68**, 062323 (2003).
14. V. Giovannetti, S. Lloyd, L. Maccone, and P.W. Shor, Phys. Rev. Lett. **91**, 047901 (2003).
15. C.M. Caves and P.D. Drummond, Rev. Mod. Phys. **66**, 481 (1994).
16. G.G. Amosov, A.S. Holevo, and R.F. Werner, Problems in Information Transmission **36**, 25 (2000).
17. A. Wehrl, Rev. Mod. Phys. **50**, 221 (1978).
18. A. Serafini, F. Illuminati, M.G.A. Paris, and S. De Siena, Phys. Rev. A **69**, 022318 (2004).
19. C. Adami and N.J. Cerf, Phys. Rev. A **57**, 4153 (1998).
20. C.E. Shannon, The Bell System Tech. J. **27**, 379, (1948); ibid. **27**, 623, (1948).
21. J. Williamson, Am. J. Math. **58**, 141 (1936); see also V.I. Arnold, *Mathematical Methods of Classical Mechanics*, (Springer-Verlag, New York, 1978).
22. A.S. Holevo and M.E. Shirokov, Theory Prob. Appl. **50**, 86 (2005).
23. M.E. Shirokov, quant-ph/0411091 (2004).
24. J. Eisert, C. Simon, and M.B. Plenio, J. Phys. A **35**, 3911 (2002).
25. P.W. Shor, Math. Prog. **97**, 311 (2003).
26. A.S. Holevo, IEEE Trans. Inf. Theory **44**, 269 (1998).
27. B. Schumacher and M.D. Westmoreland, Phys. Rev. A **56**, 131 (1997).
28. V. Giovannetti, S. Guha, S. Lloyd, L. Maccone, J.H. Shapiro, and H.P. Yuen, Phys. Rev. Lett. **92**, 027902 (2004).
29. H.P. Yuen and M. Ozawa, Phys. Rev. Lett. **70**, 363 (1992).
30. D. Kretschmann and R.F. Werner, New J. Phys. **6**, 26 (2004).
31. C.H. Bennett, D.P. DiVincenzo, J.A. Smolin, and W.K. Wootters, Phys. Rev. A **54**, 3824 (1996); H. Barnum, E. Knill, and M.A. Nielsen, IEEE Trans. Inf. Th. **46**, 1317 (2000).
32. P.W. Shor, *The quantum channel capacity and coherent information*, lecture notes, MSRI Workshop on Quantum Computation (2002); I. Devetak, IEEE Trans. Inf. Th. **51**, 44 (2005); S. Lloyd, Phys. Rev. A **55**, 1613 (1997).
33. A.S. Holevo, quant-ph/0211170 (2002).
34. D. Gottesman, A. Kitaev, and J. Preskill, Phys. Rev. A **64**, 012310 (2001).
35. C.H. Bennett, P.W. Shor, J.A. Smolin, A.V. Thapliyal, IEEE Trans Inf. Th. **48**, 2637 (2002).
36. P.W. Shor, Comm. Math. Phys. **246**, 453 (2004).
37. K. Matsumoto, T. Shimono, and A. Winter, Commun. Math. Phys. **246**, 427 (2004).
38. K.M.R. Audenaert and S.L. Braunstein, Commun. Math. Phys. **246**, 443 (2004).
39. A. Serafini, J. Eisert, and M.M. Wolf, Phys. Rev. A **71**, 012320 (2005).
40. C. King, M. Nathanson, and M.B. Ruskai, Lin. Alg. Appl. **404**, 367 (2005).
41. M.M. Wolf and J. Eisert, New J. Phys. **7**, 93 (2005).

42. N.J. Cerf, J. Clavareau, C. Macchiavello, and J. Roland, Phys. Rev. A **72**, 042330 (2005).
43. G. Ruggeri, G. Soliani, V. Giovannetti, and S. Mancini, Europhys. Lett. **70**, 719 (2005).
44. D. Kretschmann and R.F. Werner, quant-ph/0502106 (2005).

Chapter 3

Entanglement in Systems of Interacting Harmonic Oscillators

K. M. R. Audenaert, J. Eisert and M. B. Plenio
QOLS, Blackett Laboratory, Imperial College London
Prince Consort Road, London SW7 2BW, UK and
Institute for Mathematical Sciences, Imperial College London
53 Prince's Gate, London SW7 2PG, UK

emails: k.audenaert@imperial.ac.uk, j.eisert@imperial.ac.uk, m.plenio@imperial.ac.uk

1. Introduction

Recently, it has become clear that methods and ideas from quantum information science provide new tools to assess quantum correlations and entanglement in strongly correlated many-body systems. This applies both to ground state as well as to non-equilibrium properties. The study of entanglement in coupled harmonic systems with many degrees of freedom, specifically, is considerably simplified if one focuses attention onto a class of quantum states which plays a central role in this book: this is the class of Gaussian states. Then, a reduced description is available in terms of first and second moments of the states, and one may refrain from explicitly referring to the quantum states at hand. Such reduced descriptions in terms of moments have a long tradition in the study of many-body systems in fermionic systems: the classic works on exact solutions of spin systems utilizing a mapping onto fermionic Gaussian systems provide examples. Here we will be concerned with the entanglement properties of bosonic harmonic systems with many degrees of freedom. In recent years a detailed theory of entanglement of Gaussian states has been developed, much of which has been outlined in previous chapters of this book.

To set the notation for this chapter, we will first provide a brief review of some concepts and formulae that will be employed in the remainder of this chapter. For more details consult tutorial reviews[1,2] or the preceding sections of this book. We will then employ this to study the static (Sec. 3) and dynamic (Sec. 4) properties of entanglement in such systems.

2. Systems of Harmonic Oscillators

Throughout this chapter, we will use units such that Planck's constant $\hbar$ and Boltzmann's constant k are set to unity. In these units, both temperature and oscillator frequency are commensurate with energy. Harmonic oscillators are quantum systems having canonical coordinates associated with position and momentum. In a system with n canonical degrees of freedom, it is convenient to arrange these in the form of a vector $R^T = (q_1, \ldots, q_n,\ p_1, \ldots, p_n)$. The characteristic feature distinguishing the quantum harmonic oscillator from its classical counterpart is the canonical commutation relation (CCR) between position and momentum. They can be expressed as $[R_j, R_k] = i\sigma_{j,k}$ where σ is the so-called *symplectic matrix*, the real skew-symmetric block diagonal $2n \times 2n$-matrix

$$\sigma = \begin{bmatrix} 0 & \mathbb{I}_n \\ -\mathbb{I}_n & 0 \end{bmatrix}. \tag{1}$$

On the level of Weyl displacement operators, defined as $W_\xi = \exp(i\xi^T \sigma R)$, the CCR become $W_\xi^\dagger W_{\xi'} = W_{\xi'}^\dagger W_\xi e^{i\xi^T\sigma\xi'}$ for $\xi, \xi' \in \mathbf{R}^{2n}$.

As mentioned before, Gaussian states can be fully characterized by their first and second moments. Gaussian states are those for which the *characteristic function* χ_ρ, defined as the expectation value of the Weyl displacement operator $\chi_\rho(\xi) = \mathrm{Tr}[\rho W_\xi]$, is a Gaussian in phase space, so

$$\chi_\rho(\xi) = \chi_\rho(0) \exp\left(-\frac{1}{4}(\sigma\xi)^T \gamma(\sigma\xi) - d^T(\sigma\xi)\right). \tag{2}$$

The covariance matrix γ – embodying the second moments – is in turn the real symmetric $2n \times 2n$ matrix

$$\gamma_{j,k} = 2\mathrm{Re}\ \mathrm{Tr}\left[\rho\left(R_j - \langle R_j\rangle_\rho\right)\left(R_k - \langle R_k\rangle_\rho\right)\right]. \tag{3}$$

This definition holds true also for non-Gaussian states, and is consistent with the convention made in Eq. (2). With this convention, the covariance matrix of the n-mode vacuum is $\gamma = \mathbb{I}_{2n}$.

Covariance matrices can clearly be diagonalized, not only with orthogonal matrices, but also with matrices corresponding to transformations preserving the CCR. Such transformations are called *symplectic transformations.* The elements of the real *symplectic group* are formed by those matrices S satisfying $S\sigma S^T = \sigma$. When diagonalizing a covariance matrix under congruence with such a matrix, the resulting diagonal elements are up to double degeneracy the *symplectic eigenvalues* γ_j. They can be obtained directly as the ordinary eigenvalues of the matrix $\gamma^{1/2} i\sigma\gamma^{1/2}$. In its diagonal form, the covariance matrix formally describes a set of uncorrelated modes. This is the normal mode decomposition: if one is allowed for global transformations, operating on all degrees of freedom, a system of interacting oscillators can be formally decoupled via this symplectic diagonalisation. This procedure corresponds to the familiar one of finding the dispersion relation of phonons in a crystal lattice.

The covariance matrix entails the complete information about the entanglement properties of Gaussian states. So any entanglement measure of bipartite systems can be expressed as properties of this matrix. For pure states ρ, entanglement can most reasonably be quantified in terms of the von-Neumann entropy of a reduction to one part A of a bipartite system AB, $E(\rho) = S(\rho_A) = -\mathrm{tr}[\rho_A \log_2 \rho_A]$. This measure of entanglement is called *entropy of entanglement.*[3] It provides the rate at which one can distill maximally entangled qubit pairs from a given state, so its *distillable entanglement* (as well as the rate that is necessary in a preparation procedure, the *entanglement cost*).

In turn, there is no unique way to quantify entanglement for mixed states, in the sense that a certain number of reasonable requirements already uniquely define a single quantity grasping the degree of entanglement. There are several different measures capturing entanglement with respect to different operational interpretations. In particular, the distillable entanglement and the entanglement cost are generally different from each other. For the purposes of this work we settle for the *logarithmic negativity*, which is comparatively easy to compute and has an interpretation as a cost function.[4–8] Moreover, and importantly, it is a general upper bound to the distillable entanglement also for mixed quantum states. In a bipartite system AB, the logarithmic negativity is defined as

$$N(\rho) = \log_2 ||\rho^{T_B}||_1, \tag{4}$$

where ρ^{T_B} is the state that is obtained from ρ via a partial transposition with respect to system B and $||.||_1$ denotes the trace-norm.

On the level of covariance matrices, transposition amounts to time reversal, which is a transformation that leaves the positions invariant but reverses all momenta, $q \mapsto q$, $p \mapsto -p$. Partial transposition is correspondingly the application of time reversal to only one subsystem. Let us consider a system made up of $m+n$ degrees of freedom, where the first m oscillators form system A and the remaining n form part B. Applying time reversal to the latter, the covariance matrix is transformed as

$$\gamma \mapsto \gamma^{T_B} = \mathcal{P}\gamma\mathcal{P}, \tag{5}$$

where

$$\mathcal{P} = \mathbb{I}_{m+n} \oplus P, \quad P = \begin{bmatrix} \mathbb{I}_m & 0 \\ 0 & -\mathbb{I}_n \end{bmatrix}. \tag{6}$$

The *partition matrix* P is the diagonal matrix describing the subdivision of oscillators into parties. In the most general setting, $P_{j,j}$ is 1 if oscillator j belongs to system A, and -1 if it belongs to system B. The logarithmic negativity is then given by[9]

$$N = -\sum_{j=1}^{m+n} \log_2 \left(\min(1, |\gamma_j|)\right), \tag{7}$$

where the γ_j, $j = 1, ..., m+n$, are the symplectic eigenvalues of γ^{T_B}. This procedure of computing functionals of states via the computation of the symplectic spectrum is applicable for all unitarily invariant functionals. So similarly, the entropy of entanglement can be expressed in terms of the symplectic eigenvalues, now of the covariance matrix γ_A of the reduced state ρ_A with respect to part A. Given the symplectic eigenvalues γ_i of γ_A, $i = 1, ..., m$, the entropy of entanglement is

$$E = \sum_{i=1}^{m} \left(\frac{\gamma_i + 1}{2} \log_2 \frac{\gamma_i + 1}{2} - \frac{\gamma_i - 1}{2} \log_2 \frac{\gamma_i - 1}{2} \right). \tag{8}$$

In the following we will be particularly interested in states that arise naturally in physical systems, *e.g.*, ground states and thermal states for systems governed by a specific Hamiltonian. In general, such states are not Gaussian, but, fortunately, for Hamiltonians that are quadratic in the canonical position and momentum operators, they are. For a Hamiltonian of the form

$$H = \frac{1}{2} R^T \begin{bmatrix} V & 0 \\ 0 & T \end{bmatrix} R, \tag{9}$$

with V the potential matrix and T the kinetic matrix, we find that the covariance matrix of the ground state is given by

$$\gamma = (TV^{-1})^{1/2} \oplus (VT^{-1})^{1/2}. \tag{10}$$

In the important case where the interactions arise due to a coupling between the position operators only, $T = \mathbb{I}_n$, and γ reduces to[9]

$$\gamma = V^{-1/2} \oplus V^{1/2}. \tag{11}$$

Note that this expression captures nothing but the *classical two-point correlation functions*,

$$G_{i,j} = \langle 0|q_i q_j|0\rangle = (V^{-1/2})_{i,j}, \ H_{i,j} = \langle 0|p_i p_j|0\rangle = (V^{1/2})_{i,j}. \tag{12}$$

If $T = V$, corresponding to a Hamiltonian originating from the rotating wave approximation in quantum optical settings, then the ground state is given by $\gamma = \mathbb{I}_n \oplus \mathbb{I}_n$, which is the same as the ground-state of n non-interacting harmonic oscillators. Translationally invariant closed chains have a potential matrix V that is circulant, *i.e.*, $V_{jk} = v_{j-k \bmod n}$. We write $V = \mathrm{Circ}(v)$ for the circulant matrix whose first row is given by the n-vector v, and also for a block circulant matrix where the first block column is specified by a vector of matrices.

Another aim of this chapter is the investigation of the dynamical properties of the system of harmonic oscillators[16–18] and the evolution of entanglement properties under such dynamics. The dynamics of the covariance matrix under a Hamiltonian quadratic in position and momentum operators can be obtained straightforwardly from the Heisenberg equation

$$\partial_t X(t) = i[H, X(t)]. \tag{13}$$

For our time-independent Hamiltonian Eq. (9), this leads to the covariance matrix at time t as

$$\begin{bmatrix} \gamma_{XX}(t) & \gamma_{XP}(t) \\ \gamma_{PX}(t) & \gamma_{PP}(t) \end{bmatrix} = U_t \begin{bmatrix} \gamma_{XX} & \gamma_{XP} \\ \gamma_{PX} & \gamma_{PP} \end{bmatrix} U_t^\dagger \tag{14}$$

$$U_t = \exp\left(\begin{bmatrix} 0 & T \\ -V & 0 \end{bmatrix} t \right). \tag{15}$$

Equipped with these tools we can now proceed to the analysis of the static and dynamical properties of harmonic systems.

3. Static Properties of Harmonic Chains

In the absence of degeneracies, the ground state of interacting quantum systems is a pure state that, due to the interactions, can be expected to exhibit correlations between constituents. In recent years it has become a subject of intensive research to investigate the quantum correlations present in ground states of strongly-correlated many-body systems.[9–15] The motivations for such studies are manifold: from a fundamental point of view, it is interesting to identify genuine quantum correlations, as opposed to mere classical correlations. Then, the picture of the general relationship between systems being critical in the sense of a divergent classical correlation length or of a vanishing gap in the spectrum of the Hamiltonian and having divergent entanglement properties is just slowly emerging. The entanglement properties of ground states can in fact also be related to the success or failure of numerical methods such as DMRG to study ground state properties. Conformal field theory provides tools to assess the continuum limit of discrete theories, but is not applicable to higher dimensional systems. Such higher-dimensional systems are particularly relevant in the context the question of *geometric entropy* in quantum field theory. Here, the question is: how does the entropy of some distinguished region of a scalar free field in some dimension depend on the size of the region.

Initially, such studies of entanglement properties of ground states were largely undertaken as numerical studies, at hand of systems which were natural first starting points, such as systems with nearest-neighbor interactions. Exact results for infinite systems are much harder. This motivates to study systems that allow for a reduced description in terms of moments. For fermionic Gaussian systems, the exact solution in mono-partite systems are classic results. There, the ground state is obtained by investigating second moments of Majorana operators resembling the canonical coordinates in the context of the present book. In the light of entanglement questions, harmonic bosonic systems have recently been the subject of intense scrutiny and this has resulted in the first exact results for such questions. The simplest instance of an analytical result is the study of the entanglement of a simply connected subset of oscillators with the rest of the chain[9] in a one-dimensional system. Yet, similar albeit more involved methods can actually be applied for higher-dimensional systems. This applies, *e.g.*, to the scaling behaviour of the entanglement between a region and its surrounding with the size of the region,[15,19] similar to further investigations in spin systems. In the following we present some aspects of this work without presenting detailed analytical proofs.

To investigate the entanglement properties of harmonic chains, we focus on bipartite entanglement. Specifically, the harmonic oscillators are partitioned into two groups, and the bipartite entanglement between those two groups is calculated. As mentioned in the previous section, the partitioning into groups can be described using the partition matrix P.

We start by investigating a system of harmonic oscillators configured as a 1-dimensional closed chain.

It can be shown that the logarithmic negativity of an oscillator chain of length n and with potential matrix V with respect to the partitioning P is given by

$$N = -\sum_{j=1}^{n} \log_2 \left(\min(1, \lambda_j(Q))\right), \quad Q = V^{-1/2} P V^{1/2} P. \tag{16}$$

A case of special interest is the *symmetrically bisected* chain. Here n is even, oscillators 1 to $n/2$ form group 1 and oscillators $n/2+1$ to n form group 2. In this case, the partition matrix P is given by $P = \mathbb{I}_{n/2} \oplus (-\mathbb{I}_{n/2})$.

Expression (16) is still difficult to compute in general as it involves an eigenvalue decomposition. However, for a closed chain with translationally invariant nearest-neighbour Hamiltonian $V = \text{Circ}(1+2c, -c, 0, \ldots, 0, -c)$, where $c \geq 0$ is the coupling constant, the symmetry with respect to a flip F of the parts of the bisected chain plays a crucial role in the computation of the degree of entanglement. Then, the expression $\log_2(\text{Det}\,(\mathbb{I}+(Q-\mathbb{I})_+))$ can indeed be identified, where $(.)_+$ denotes the positive part of a matrix. So in fact, while the symplectic spectrum of the partially transposed covariance matrix itself cannot be computed analytically, the degree of entanglement can be formulated in a closed form. Surprisingly, one finds a very appealing simple expression for the log-negativity of the symmetrically bisected chain:[9]

For the nearest-neighbour Hamiltonian with coupling coefficient $c \geq 0$, the logarithmic negativity of the symmetrically bisected closed chain of length n is given by

$$N = \frac{1}{2} \log_2(1 + 4c). \tag{17}$$

It is remarkable that the negativity is independent of n, the chain length.

The entanglement properties of such a bipartite system in a pure Gaussian state can after all be captured entirely in terms of two-mode squeezed states. That is, the familiar *Schmidt decomposition* can be carried out, and the unitaries with respect to which one achieves the Schmidt decomposition can without loss of generality be chosen to be Gaussian unitaries. More specifically, on the level of second moments, we have the following: for any covariance matrix γ of a bipartite system AB consisting of n degrees of freedom each, there exist matrices S_A and S_B which are elements of the symplectic group such that

$$(S_A \oplus S_B)\gamma(S_A \oplus S_B)^T = \bigoplus_{i=1}^{n} \begin{bmatrix} \cosh(r_i) & \sinh(r_i) & 0 & 0 \\ \sinh(r_i) & \cosh(r_i) & 0 & 0 \\ 0 & 0 & \cosh(r_i) & -\sinh(r_i) \\ 0 & 0 & -\sinh(r_i) & \cosh(r_i) \end{bmatrix}, \tag{18}$$

where $r_i \geq 0$, $i = 1, ..., n$, are the *two-mode squeezing parameters.* That is to say, any pure Gaussian state is up to local Gaussian unitaries identical to a tensor product of two-mode squeezed states.[20–22] Bipartite pure state entanglement of Gaussian can always be quantified in terms of the numbers $(r_1, ..., r_n)$, although to find these numbers again amounts to a symplectic diagonalization.

It is worth noting that the above independence of the chain length is a specific property of the logarithmic negativity. Also, this result is strictly true for the bisected chain. For a chain where the first m degrees of freedom form system A, and the remaining n form system B, it is still the case that the log-negativity saturates to a finite value for large m.

The entropy of entanglement in turn does depend on the chain length. Yet, the saturation for large contiguous groups follows immediately from the result for the log-negativity. This might appear as a contradiction to the result from conformal field theory for the bosonic universality class, which states that

$$E = \frac{1}{3}\log(m/\varepsilon) + o(\log(m/\varepsilon)) \tag{19}$$

for the entropy of entanglement, with ε being a constant. This is the interesting limit in case of a continuum limit to the real Klein Gordon field in one dimension, as a continuum bosonic theory. Yet, this behaviour can be re-obtained – at least in a numerical study[23] and as an analytical bound[19] – in the strong coupling limit of the harmonic chain. The strong coupling limit is the one for which $c \to \infty$, corresponding to the limit where $\lambda_{\max}(V)/\lambda_{\min}(V) \to \infty$. For any value of c, the dependence of E as a function of m is then well-approximated by $E = \log(m/\varepsilon)/3 + O(\log\log(m/\varepsilon))$, until to the point where m is of the order of the classical correlation length.

The symmetrically bisected chain is an example where groups are contiguous, touching in their endpoints only, which is the minimal number possible. At the other end of the contiguity spectrum is the case of entanglement between the group of even oscillators and the group of odd ones, which touch each other in the maximal number of n points.

From numerical calculations shown in Fig. 1, this partitioning exhibits a log-negativity roughly proportional to n. This can also be shown analytically. The diagonal elements of the partition matrix for this configuration are $+1$ for odd index values, and -1 for even index values. For the nearest-neighbour Hamiltonian, Eq. (16) yields

$$N = \sum_{k=0}^{n/4} \log_2 \frac{1 + 2c(1 + \cos(2\pi k/n))}{1 + 2c(1 - \cos(2\pi k/n))} \tag{20}$$

for n being a multiple of 4. For large n, we can approximate the discrete sum by an integral,

$$N \approx \frac{n}{2\pi} \int_0^{\pi/2} dx \log_2 \frac{1 + 2c(1 + \cos(x))}{1 + 2c(1 - \cos(x))}, \tag{21}$$

which is indeed proportional to n.

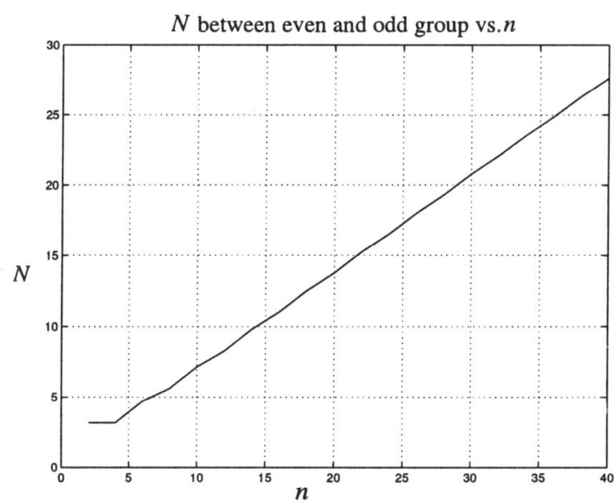

Fig. 1. Entanglement between the group of even oscillators and the group of odd oscillators, in function of the chain length n (even n only). Interaction is again nearest-neighbour with coupling $c = 20$. The log-negativity is seen to quickly converge to a constant times n.

These two extreme examples are suggestive of a general rule that entanglement may be proportional to the surface area of the boundary between the two groups of oscillators, which in the 1-dimensional case is just the number of points where the two groups touch. This intuition carries much further indeed, as a connection between entanglement and area of parts of systems that holds in a surprisingly general context for harmonic non-critical systems.[15,19] In the following we will briefly discuss the main result and refer the reader to the original literature for further details, generalizations and proofs.

The original question that initiated the study of geometric entropy in harmonic lattice systems is the following one: Consider a free quantum scalar field in the ground state, in any spatial dimension. Now one distinguishes a certain geometrical region, and look at the entropy of this region. This is obtained by tracing out over those degrees of freedom that elude the observation. How does this entropy – *i.e.*, the entropy of entanglement of the region with the remainder – relate to properties of the region? More specifically, does it depend on the volume of the region, or the boundary area? In more than one-dimensional systems, this question – albeit innocent-looking – constitutes a quite long-standing issue. Part of the motivation for the question of the area-dependence of quantum correlations originates from the suggestion that this entropy might be a source for the Bekenstein-Hawking black hole entropy.[24,25] Subsequently to this intriguing suggestion, this problem was numerically studied for lattice instances, notably by Bombelli *et al.*[26] and Srednicki.[27] The findings supported this view of an area-dependence of quantum correlations in arbitrary dimensions. In one-dimensional systems and for half-spaces, other powerful methods are accessible to solve this problem,[28] in particular for the continuum theory those of conformal field theory.

Recently, there has been renewed interest in questions of this kind[9,23,29–33] employing recently developed methods from entanglement theory in the context of quantum information science.[1] Indeed, such methods allowed to provide analytical answers to the scaling behaviour of the degree of entanglement between a region and its exterior for harmonic lattice Hamiltonians, such as discrete versions of the free scalar Klein-Gordon field, in arbitrary spatial dimensions. It is remarkable that this intuition, that in non-critical systems only a layer up to the correlation length contributes to the entanglement, carries over to such general situations. So for general finite-ranged harmonic lattice systems, one finds such a connection between entanglement and the area of distinguished regions.

The Hamiltonian under consideration is the d-dimensional analogue of the harmonic chain, that can be conceived as a discrete lattice version of a free real scalar quantum field. For any $d \geq 1$ we consider a d-dimensional simple cubic lattice $n^{\times d}$ comprising n^d degrees of freedom. The Hamiltonian can as before be written as

$$H = \frac{1}{2} R^T \begin{bmatrix} V & 0 \\ 0 & \mathbb{I} \end{bmatrix} R, \tag{22}$$

where now $R^T = (x_1, \ldots, x_{n^d}, p_1, \ldots, p_{n^d})$, in terms of the $n^d \times n^d$ potential matrix V. We choose V such that in the continuum limit one obtains

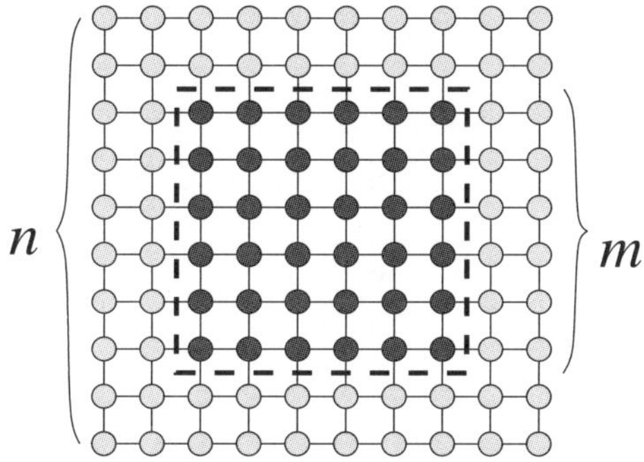

Fig. 2. The harmonic lattice in $d = 2$ with a distinguished $m \times m$ region in an $n \times n$ lattice.

the Hamiltonian of the real Klein-Gordon field. We will therefore consider the harmonic lattice Hamiltonian with nearest-neighbour interaction. More general types of interactions can also be treated,[19] with the very same conclusion. Here, the central result for this specific case,[15] together with a number of extensions.[19]

We denote the ground state of the system by ρ. For a distinguished cubic region $m^{\times d}$ in the $n^{\times d}$ lattice (see Fig. 2) its entropy of entanglement is $E_{n,m} = -\operatorname{Tr}[\rho_{n,m} \log_2 \rho_{n,m}]$. The reduced density matrix $\rho_{n,m}$ is formed by tracing out the variables outside the region $m^{\times d}$. One then obtains the following relationship:

The entropy of entanglement of the distinguished region $m^{\times d}$ in the lattice $n^{\times d}$ satisfies

$$\lim_{n \to \infty} E_{n,m} = \Theta(m^{d-1}), \tag{23}$$

where Θ is the Landau theta.

That is, asymptotically the entropy of entanglement is bounded by functions linear in the boundary area of the distinguised region, as result that holds true in any dimension d. This result can be extended in a variety of directions that can be obtained by similar techniques, including for example more general finite-range interactions. Essentially, there are three key ingredients to such a more general proof. Firstly– with more formal methods – one has to rigorously confirm the expectation that the classical two-point correlation functions are exponentially decaying. Then, a bound has to be formulated that depends only on joint properties between the region and the rest of the lattice. This can be expressed in terms of the log-negativity, by bounding after a number of steps $N \leq \|V^{-1/2}\Delta V^{1/2}\|_1$, where Δ is an appropriate indicator matrix. Finally, the result is obtained by bounding the relevant terms using the l_1-norm, always being larger than or equal to the trace-norm of a matrix, exploiting the exponential decay. Again, the strong coupling limit is interesting when considering the limit to a continuous theory.

Interestingly, the same ideas even carry over – only with minor complications – to the case of *Gibbs (thermal) states*

$$\rho = \frac{e^{-\beta H}}{\mathrm{Tr}[e^{-\beta H}]} \tag{24}$$

with respect to some temperature $1/\beta > 0$, with covariance matrix

$$\gamma = (V^{-1/2}M_\beta) \oplus (V^{1/2}M_\beta), \;\; M_\beta = \mathbb{I} + 2(\exp(\beta V^{1/2}) - \mathbb{I})^{-1}. \tag{25}$$

Note that in this case, the part corresponding to the position coordinates is no longer the inverse of the part associated with the momentum coordinates. Then, the distillable entanglement is bounded, in a situation where the entropy of a reduction is no longer a meaningful measure of entanglement. Asking for asymptotic rates for distillation is after all a genuine quantum information question, and can only be answered with the more detailed understanding of quantum entanglement that has been obtained over the last few years. The requirement that the distinguished region is cubic is also unnecessary, and a result can be obtained that is valid for regions of arbitrary shape. Details of these results can be found in the literature.[15,19]

As a very simple special case, yet one that already contains some of the intuition of the more formal proof, is provided by the class of interactions for which the potential matrix V is the square of a circulant band-matrix W. Then, the covariance matrix of the ground state is nothing but $\gamma = W^{-1} \oplus W$. In this case one immediately arrives at the connection between entanglement and area of the region since one can show that the number of terms contributing to the symplectic spectrum of the reduced covariance matrix is linear in the number of degrees of freedom at the boundary of the region, and the respective symplectic eigenvalues are bounded from above and from below, independently of n and m. Note in this simple case, there exists a 'disentangling' Gaussian unitary transformation, local to inside and outside of the region, such that only oscillators near to the boundary remain entangled, albeit with a modified coupling strength.[19]

It should finally be noted that, perhaps surprisingly, an 'area-dependence' can also be established analytically for classical correlations, quantified in terms of the mutual information, in *classical harmonic lattice systems.*[19] So one finds that

$$\lim_{n\to\infty} I^C_{n,m} = \Theta(m^{d-1}), \tag{26}$$

where $I^C_{n,m}$ is the classical mutual information with respect to a cubic region $m^{\times d}$ within a lattice $n^{\times n}$ in case of a thermal state. This result on classical systems can interestingly be established most economically using the techniques developed in the quantum case. Namely, by mapping the problem with interaction matrix V onto that of a quantum harmonic lattice with a squared interaction V^2 as has been described above.

4. Dynamical Properties of Harmonic Chains

In the preceding section we have studied properties of entanglement in ground and thermal states of a harmonic system. The findings suggest that entanglement between two blocks is not very far-reaching. Numerical studies indicate that the degree of entanglement between non-contiguous sets of degrees of freedom decreases exponentially with distance.[9,23] In particular, the degree of entanglement between two degrees of freedom is always zero, except for nearest neighbors.[9] This is a property not specific for harmonic chains. In all known spin models, there exists a 'small' k that such the two-point entanglement between two sites i and j vanishes for all $|i-j| > k$. In the light of this observation, the question arises naturally whether it is possible to create long range entanglement can be generated without very detailed local control of the system. As static systems appear to be unsuitable for this purpose research focused on dynamical situations.

To be brief, in the following, we will exemplify these ideas in an abstract setting.[16–18] It should be clear, yet, that similar settings apply to a number of practical situations, including nano-electromechanical systems,[16] chains of coupled atom-cavity systems, photonic crystals, or coupled vibrational modes of appropriately prepared molecules. To a very good approximation, these systems can be described harmonic oscillators, albeit ones that feel the presence of noise processes.

The main question is how it is possible to entangle two parts of an interacting chain of systems that are well separated in space, without the need for making them interact directly and with a minimum requirement for individual local control. In many contexts, this degree of local control is difficult to achieve. We will show that this is possible by triggering squeezing and entanglement by a global non-adiabatic change of the interaction strength in a linear array of oscillators, but without individually addressing any of the oscillators of the array. In this way, one can achieve *long-range entanglement* that will persist over length scales that are much larger than the typical entanglement length for the ground state of the system.[9] The created correlations can be understood in terms of propagating excitations that are created.

Consider a setup that is well-described by an array of coupled harmonic oscillators We assume that we have the ability to adjust the strength of a nearest neighbour interaction globally. The harmonic oscillators are assumed to be cooled to temperatures $1/\beta$ such that $1/(\beta\omega)\omega \ll 1$ with ω being the fundamental frequency of the oscillators, such that the array is operated deeply in the quantum regime.

Coupled oscillators with any mass M and eigenfrequency ω can straightforwardly be mapped onto the harmonic chain with unit frequency and mass, using $q_k = q'_k(m\omega)^{1/2}$, and $p_k = p'_k(m\omega)^{-1/2}$. Hence, the starting point is nothing but the one-dimensional harmonic chain,

$$H = \frac{\omega}{2}\sum_{k=1}^{N}\Big(p_k^2 + q_k^2(1+2c) - 2cq_kq_{k+1}\Big).$$

For the moment, we assume for simplicity periodic boundary conditions, *i.e.*, $q_{N+1} = q_1$, but this requirement can be relaxed without affecting the essential physics, and set $\omega = 1$, as in this ideal treatment this merely corresponds to a rescaling of the time scale. The normal coordinates are related to the previous ones by a discrete Fourier transform,

$$q_k = \frac{1}{\sqrt{N}}\sum_{l=1}^{N}\exp(2\pi ikl/N)Q_l, \quad p_k = \frac{1}{\sqrt{N}}\sum_{l=1}^{N}\exp(-2\pi ikl/N)P_l.$$

In these normal coordinates, satisfying $Q_k = Q_{N-k}^\dagger$ and $P_k = P_{N-k}^\dagger$, the Hamiltonian can be written in the form

$$H = \frac{1}{2}\sum_{k=1}^{N}\left(P_kP_k^\dagger + \omega_kQ_kQ_k^\dagger\right),$$

with $\omega_k := (1 + 4c\sin^2(\pi k/N))^{1/2}$. Introducing the annihilation operator $a_k = (\omega_k Q_k + iP_k^\dagger)/\sqrt{2\omega_k}$, and noting that in the Heisenberg picture the annihilation operator evolves according to $a_k(t) = \exp(-i\omega_k t)a_k(0)$, one arrives at the Heisenberg equations of motion for the original canonical coordinates

$$q_k(t) = \sum_{r=1}^{N} (q_r(0) f_{r-k}(t) + p_r(0) g_{r-k}(t)), \quad p_k(t) = \partial_t q_k(t), \tag{28}$$

where we have defined the functions

$$g_k(t) = \sum_{l=1}^{N} \exp(2\pi i k l/N) \frac{\sin(\omega_l t)}{N\omega_l}, \quad f_k(t) = \partial_t g_k(t). \tag{29}$$

We assume that for times $t < 0$, the oscillators are not interacting and are in the ground state (an assumption that will be relaxed later). This implies that $\Gamma_{q_n q_m} = \Gamma_{p_n p_m} = \delta_{n,m}$, and $\Gamma_{q_n p_m} = 0$, for $n, m = 1, \ldots, N$. At time $t = 0$ the interaction is suddenly, non-adiabatically switched on, which results in the following equations of motion for the second moments,[17]

$$\Gamma_{q_n q_m}(t) = (a_{n,m}(t) + d_{n,m}(t))/2, \quad \Gamma_{q_n p_m}(t) = (b_{n,m}(t) + e_{n,m}(t))/2,$$
$$\Gamma_{p_n p_m}(t) = (c_{n,m}(t) + a_{n,m}(t))/2, \tag{30}$$

where

$$a_{n,m} = \sum_{k=1}^{N} f_{k-n} f_{k-m}, \quad b_{n,m} = \partial_t a_{m,n}/2, \tag{31}$$
$$c_{n,m} = \sum_{k=1}^{N} \partial_t f_{k-n} \partial_t f_{k-m}, \quad d_{n,m} = \sum_{k=1}^{N} g_{k-n} g_{k-m}, \quad e_{n,m} = \partial_t d_{n,m}/2.$$

Using these equations of motion, the entanglement properties of two distant oscillators can now immediately be assessed, when we trace out all the others. The chain is translationally invariant, and hence, any single oscillator, say labeled 1, can be singled out, and we may look at the degree of entanglement as a function of time and discrete distance. We quantify the degree of entanglement in terms of the log-negativity.

Quite surprisingly, one finds that even very distant oscillators become significantly entangled over time. Fig. 3 depicts the behaviour of the degree of entanglement, additionally assuming a noise model the oscillators are being exposed to. Here, decoherence mechanisms are taken into account by linearly coupling each degree of freedom to a local heat bath consisting of

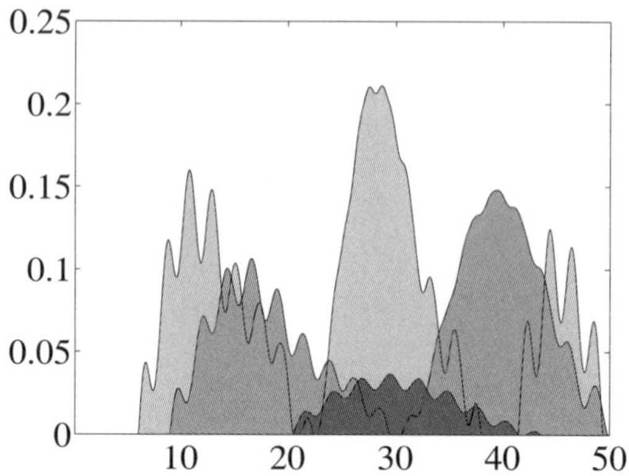

Fig. 3. The degree of entanglement including some decoherence and for non-zero temperature. Shown is the situation of a chain of length 8 for periodic boundary conditions. In this plot (up to rescaling of the time axis, and a quantification of the coupling strength $c = 0.3$, $c = 0.2$, and $c = 0.1$ in terms of the fundamental frequency), values are chosen that correspond to the Q-factor $Q = 10^3$, fundamental frequency 5 GHz, and temperature of 10 mK.

a finite number M of modes. For very weak coupling, the resulting decoherence process does not depend strongly on the chosen spectral density. Here an Ohmic spectral density is chosen, so the one in which case the Langevin equation for the Heisenberg picture position becomes the one of classical Brownian motion in the classical limit. The coupling strength with the local environments is chosen such that it results in an energy dissipation with a certain chosen rate $1/Q$. The initial state before switching on the interaction is then the Gibbs state of the canonical ensemble of the whole chain including the appended heat baths. The resulting map is then still a Gaussian operation, such that it is sufficient to know the second moments to specify entanglement properties.

For a time interval $[0, t_0)$, $t_0 > 0$, the state of the oscillators with labels 1 and n is separable. Then, for times $t > t_0$ it becomes entangled. This time t_0 is approximately given by $t_0 \approx n/(2\gamma\Omega)$. The amount of entanglement decays as $1/n$, but becomes strictly zero after a finite distance. For $c = 0.1$, for example, this happens for n larger than 500. This long-range nature of the entanglement is quite remarkable. We see that such ideas facilitate the generation of entanglement in the presence of noise processes and non-zero temperatures.

A more careful analysis reveals that comparably low Q-factors are not particularly harmful given the large speed of propagation. Too high temperatures, however, turn the correlations into merely classical correlations.

This effect is clearly more harmful for longer chains. For two oscillators, quite large values of the degree of entanglement can be achieved. For example, for a two-oscillator system, with system parameters as in Fig. 3, the degree of entanglement as quantified in terms of the log-negativity reaches values larger than 0.6 for $c = 0.4$.

The non-adiabatic character of the change of the interaction is crucial in such a setup. With slower speed of the rise of the interaction strength the amount of entanglement decreases rapidly.[16] This is not surprising in the light of the adiabatic theorem and the results obtained earlier in this chapter. If, starting with the system in the ground state, one changes a parameter in the Hamiltonian slowly, then the system will remain in its respective ground state. In the ground state, however, we have no long range entanglement in the sense required here.

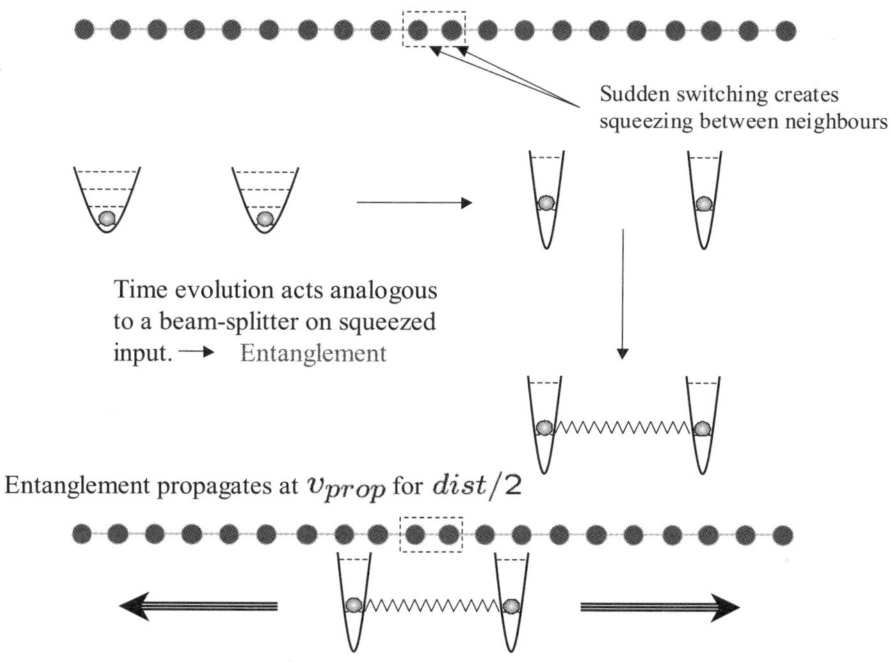

Fig. 4. Schematic picture outlining the origin of the entanglement in the system. The non-adiabatic of the coupling strength leads to a squeezing of the harmonic oscillator state. The interaction between neighbouring oscillators then leads to a time evolution analogous to that of a beam-splitter in quantum optics transforming squeezed states into entangled states. Subsequently this entanglement is propagated along the chain resulting in long distance entanglement.

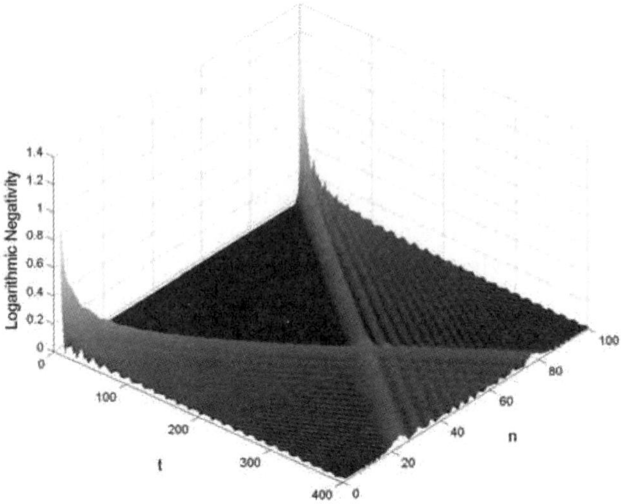

Fig. 5. For a ring of 40 oscillators and a coupling constant of $c = 0.1$, the time evolution of the logarithmic entanglement between an uncoupled oscillator and the n-th oscillator is shown when initially the uncoupled oscillator and the first oscillator are in a two-mode squeezed state with two-mode squeezing parameter $r = 0.8$. With increasing time, more and more distant oscillators are becoming entangled. Entanglement propagates both clockwise and anti-clockwise around the ring. After a sufficiently long time, the two counter-propagating "entanglement waves" meet at the opposite end of the ring and we observe some entanglement enhancement.

We hence arrive at the following picture explaining the origin of entanglement (see Fig. 4). The non-adiabatic character of the change of the interaction drives the system away from equilibrium, its state becoming squeezed. The interaction between neighboring oscillators then leads to a time evolution analogous to that of a beam-splitter in quantum optics, transforming squeezed states into entangled states.[34] Subsequently this entanglement is propagated along the chain resulting in long distance entanglement. This propagation is essential in the creation of the long distance entanglement and is also a significant source of loss of entanglement due to the spreading of the excitation packages over the lattice.

This loss of entanglement can be observed in Fig. 5, where the degree of entanglement is depicted, which is created between a distinguished oscillator and any other oscillator of a ring. As before, and not surprisingly, the degree of entanglement decreases with time and therefore distance. Note that because of the periodic boundary conditions, propagation occurs in a clockwise as well as in a counterclockwise manner. As a general conclusion, ideas along these lines allow for the generation of entanglement over large distances. They may well be provide an alternative to schemes that require detailed local control of individual quantum oscillators, asking for sophisticated technological tools concerning quantum state manipulation.

5. Summary and Conclusions

In this chapter we have considered entanglement properties in static and dynamic arrays of interacting harmonic oscillators. A theorem connecting entanglement between a region and the rest of the lattice with the surface area of the boundary between the two regions has been presented both for systems in one and many spatial dimensions. Several generalizations of this statement have been outlined. Methods for the generation of entanglement without the need of detailed local control employing dynamical properties of the system have been discussed and the physical origin of this entanglement have been explained. The dynamical properties of interacting quantum systems are largely unexplored and may provide the basis for novel ways to generate, distribute and manipulate quantum information and quantum entanglement.

Acknowledgments

This work is part of the QIP-IRC (www.qipirc.org) supported by EPSRC (GR/S82176/0) as well as the Leverhulme Trust F/07 058/U, the European Union Thematic Network QUPRODIS (IST-2001-38877), the DFG (Schwerpunktprogramm QIV), and the European Research Councils (EURYI).

References

1. J. Eisert and M.B. Plenio, Int. J. Quant. Inf. **1**, 479 (2003).
2. A. Ferraro, S. Olivares, and M.G.A. Paris, *Gaussian states in continuous variable quantum information*, Lecture Notes, Bibliopolis, Napoli, 2005, ISBN 88-7088-483-X.
3. C.H. Bennett, D.P. DiVincenzo, J.A. Smolin, and W.K. Wootters, Phys. Rev. A **54**, 3824 (1996).
4. G. Vidal and R.F. Werner, Phys. Rev. A **65**, 032314 (2002).
5. K. Audenaert, M.B. Plenio, and J. Eisert, Phys. Rev. Lett. **90**, 027901 (2003).
6. K. Zyczkowski, P. Horodecki, A. Sanpera, and M. Lewenstein, Phys. Rev. A **58**, 883 (1998).
7. J. Eisert and M.B. Plenio, J. Mod. Opt. **46**, 145 (1999).

8. J. Eisert, PhD thesis, Potsdam, February 2001.
9. K. Audenaert, J. Eisert, M.B. Plenio, and R.F. Werner, Phys. Rev. A **66**, 042327 (2002).
10. T.J. Osborne and M.A. Nielsen, Phys. Rev. A **66**, 032110 (2002).
11. P. Stelmachovic and V. Buzek, Phys. Rev. A **70**, 032313 (2004).
12. P. Calabrese and J. Cardy, J. Stat. Mech. **06**, 002 (2004).
13. F. Verstraete, M. Popp, and J.I. Cirac, Phys. Rev. Lett. **92**, 027901 (2004).
14. A.R. Its, B.-Q. Jin, and V.E. Korepin, J. Phys. A **38**, 2975 (2005).
15. M.B. Plenio, J. Eisert, J. Dreissig, and M. Cramer, Phys. Rev. Lett. **94**, 060503 (2005).
16. J. Eisert, M.B. Plenio, S. Bose, and J. Hartley, Phys. Rev. Lett. **93**, 190402 (2004).
17. M.B. Plenio, J. Hartley, and J. Eisert, New J. Phys. **6**, 36 (2004).
18. M.B. Plenio and F. Semião, New J. Phys. **7**, 73 (2005).
19. M. Cramer, J. Dreissig, J. Eisert, and M.B. Plenio, Phys. Rev. A **73**, 012309 (2006).
20. A.S. Holevo and R.F. Werner, Phys. Rev. A **63**, 032312 (2001).
21. G. Giedke, J. Eisert, J.I. Cirac, and M.B. Plenio, Quant. Inf. Comp. **3**, 211 (2003).
22. A. Botero and B. Reznik, Phys. Rev. A **67**, 052311 (2003).
23. A. Botero and B. Reznik, Phys. Rev. A **70**, 052329 (2004).
24. J.M. Bardeen, B. Carter, and S.W. Hawking, Commun. Math. Phys. **31**, 161 (1973).
25. J.D. Bekenstein, Lett. Nuovo Cimento **4**, 737 (1972).
26. L. Bombelli, R. K. Koul, J. Lee, and R.D. Sorkin, Phys. Rev. D **34**, 373 (1986).
27. M. Srednicki, Phys. Rev. Lett. **71**, 666 (1993).
28. C. Holzhey, F. Larsen, and F. Wilczek, Nucl. Phys. B **424**, 443 (1995).
29. S.J. Summers and R.F. Werner, Phys. Lett. A **110**, 257 (1985).
30. H. Halvorson and R. Clifton, J. Math. Phys. **41**, 1711 (2000).
31. R. Verch and R.F. Werner, Rev. Math. Phys. **17**, 545-576 (2005).
32. B. Reznik, A. Retzker, and J. Silman, J. Mod. Opt. **51**, 833 (2004).
33. M.M. Wolf, F. Verstraete, and J.I. Cirac, Phys. Rev. Lett. **92**, 087903 (2004).
34. M.M. Wolf, J. Eisert, and M.B. Plenio, Phys. Rev. Lett. **90**, 047904 (2003).

Chapter 4

Continuous-Variable Quantum Key Distribution

F. Grosshans

Laboratoire de Photonique Quantique et Moléculaire, UMR CNRS 8537, Ecole Normale Supérieure de Cachan, 94235 Cachan Cedex, France

A. Acín

ICFO - Institut de Ciencies Fotoniques, Mediterranean Technology Park, 08860 Castelldefels (Barcelona), Spain

N. J. Cerf

Centre for Quantum Information and Communication, Ecole Polytechnique, CP 165, Université Libre de Bruxelles, 1050 Brussels, Belgium

Quantum key distribution is a technique in which secret key bits are encoded into quantum states, which are transmitted over a quantum channel (*e.g.*, an optical link) so that the security is guaranteed by the laws of quantum physics. Most experimental realizations to date have relied on discrete protocols, involving ideally single-photons states (or, in practice, strongly attenuated light pulses) as well as single-photon detectors. In this chapter, we present an overview of the recent continuous-variable quantum cryptosystems, which rely on continuously-modulated Gaussian states (*e.g.*, coherent states) and homodyne (or heterodyne) detection. The series of security proofs of these protocols against increasingly powerful attacks will be reviewed. A particular emphasis will be put on the optimality of Gaussian attacks in this context, which holds provided that the second-order moments of the relevant variables are monitored.

1. Introduction

Quantum key distribution (QKD) is the most mature practical application of quantum information sciences today. Its provable security against arbitrarily powerful adversaries – even for parties exchanging a secret key using only present day's technology – allowed it to leave the laboratory and become already commercially available.[1] Although essentially all the currently deployed QKD systems are discrete, hence based on single-photon detectors following the original proposal by Bennett and Brassard (BB84),[2] continuous variables (CV) will probably also have a role to play because the detectors they rely on are technologically more developed. Indeed, while the photon counters used in BB84 seem to limit the detection rates to a few megahertz in ideal conditions,[3] the homodyne or heterodyne detectors used in continuous-variable QKD can easily operate in the gigahertz range. For instance, the use of homodyning allowed the very first proof-of-principle CV-QKD experiment to distribute keys at a rate of 1.7 Mbit/s.[4]

The field of CV-QKD is evolving very quickly, due to the relative simplicity of the experimental setups but also certainly thanks to the theoretical knowledge that was inherited from photon counting-based QKD. The security proofs have greatly improved over the last few years, going from the security against simple beamsplitting attacks as analyzed in the early paper by Hillery[5] to the security against very general (collective) attacks as exposed later in this chapter. This is certainly not the end of the story, and we are confident that a complete unconditional security proofs for CV-QKD including all experimental imperfections is not very far. A main simplification may come from the work of Renner,[6] indicating that the security against collective attacks actually ensures the security against general coherent attacks. The present chapter aims at providing a broad overview of the various security proofs that have been developed for CV-QKD, in particular for the so-called Gaussian protocols.[4,7–13]

2. Generic Description of Continuous-Variable Protocols

The objective of a QKD protocol is for two partners, traditionally named Alice and Bob, to agree on a secret random string (the key). This secret key has to be kept unknown to an eavesdropper (Eve) who is assumed to have access to a much more advanced technology than Alice and Bob. If Eve has unlimited resources and is able to do everything but violate the laws of quantum physics, one speaks about unconditional security.

QKD protocols can be divided into two main categories, the *prepare-and-measure* (P&M) and *entanglement-based* (E-B) schemes. A P&M protocol generally works as follows: Alice prepares quantum systems (usually light pulses) in some states and sends them to Bob through a quantum channel which is supposed to be controlled by Eve. After Bob has measured the received systems, Alice and Bob share correlated classical information, from which they extract the secret key by means of classical communication over a public authenticated channel. Of course, Eve is supposed to have interacted as much as she wanted with the quantum systems on their way from Alice to Bob. She also has listened to all communicated messages over the classical channel.

In an E-B protocol, Alice and Bob initially share an entangled state (which could even have been prepared by Eve) and perform both a measurement on their part of it. Everything else is identical to a P&M scheme. Since Alice's measurement can be viewed as a "preparation via measurement", these protocols are indeed equivalent to P&M schemes.[14] While E-B protocols are more difficult to realize experimentally, they are easier to study theoretically, not only because of the symmetry between Alice and Bob, but also because the "monogamy" of entanglement allows us to study Eve's attack more generally. In this chapter, we will use this point of view, and study the continuous-variable P&M protocols through their E-B counterparts.

The classical communication between Alice and Bob allows them to distill a secret key from their correlated data. It is usually divided into three steps: (i) *Channel evaluation*: Alice and Bob publish a random sample of their measurements and compare them to evaluate the characteristics of the quantum channel (and infer Eve's potential action from it); (ii) *Reconciliation*: they use error-correction techniques to correct the transmission errors and agree on a common bit string, partially known by Eve; (iii) *Privacy amplification*: they use a technique based on hash functions to extract, from this common string, a secret key unknown of Eve.

When turning to continuous variables, the above general description of QKD remains valid. But, in addition, CV-QKD can be understood in a restricted or a more general manner. In a restricted P&M version of CV-QKD, Bob is using homodyne detection, hence he measures continuous data, but Alice is sending states selected from a finite alphabet, typically made of just a few non-orthogonal states, see *e.g.* Refs. 5, 15. In a more general P&M version of CV-QKD as introduced in Ref. 7, Alice prepares randomly chosen Gaussian states drawn from an arbitrary continuous (*e.g.* Gaussian) distribution. In such fully-continuous schemes, the prepared states can be either squeezed[7] or coherent[8]. Bob then measures them with an homodyne or heterodyne detection. In addition, Bob can keep all his measurements or discard some part of it (postselection). In this chapter, we will limit ourselves to a Gaussian modulation and full measurement (no postselection), because the resulting family of protocols is better understood and easier to study. Restricted (or discretely-modulated) protocols with postselection such as in Ref. 15 seem to be easy to implement and robust to losses, but, to the best of our knowledge, no study has been carried out beyond Gaussian attacks, which are likely not the optimal attacks in this case.

In what follows, it will be more convenient to consider the E-B version of these fully-continuous Gaussian P&M protocols, as introduced in Ref. 16. In such a protocol, Alice prepares her state by measuring half of a two-mode vacuum squeezed state of parameter r_A, which was initially shared with Bob (see Fig. 1). For a coherent-state protocol, this means that Alice measures both quadratures, X_A and P_A, by using a beam-splitter of transmittance $T_A = 1/2$ (heterodyne detection). Denote by x_A and p_A the obtained outcomes. This effectively projects Bob's mode onto a coherent state centered on

$$x = \sqrt{2}\tanh\left(\frac{r_A}{2}\right)x_A \quad p = -\sqrt{2}\tanh\left(\frac{r_A}{2}\right)p_A, \tag{1}$$

and modulated according to a Gaussian distribution centered on the origin and of variance $\langle x^2\rangle = \langle p^2\rangle = [\cosh(r_A) - 1]/2$. In contrast, if $T_A = 1$ (homodyne detection) and Alice chooses randomly the measured quadrature, she is effectively preparing squeezed states of squeezing parameter $\cosh(r_A)$ which are modulated with a Gaussian distribution of variance $\langle x^2\rangle = \langle p^2\rangle = \sinh(r)^2/[2\cosh(r)]$.

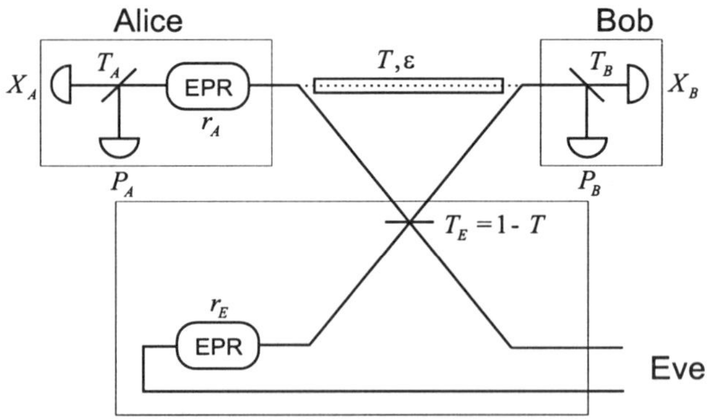

Fig. 1. Entanglement-based protocol under consideration. After Alice's effective preparation by measuring her part of an entangled state, the resulting Gaussian-modulated coherent or squeezed state of light is measured by Bob. Eve replaces the channel of transmittance T and excess noise ε by an entangling cloner of parameters r_E and T_E. The excess noise is defined as the noise that goes beyond the loss-induced noise.

Now, since Alice can arbitrarily delay her measurement, we can interpret this protocol as if Alice was sending half of a two-mode squeezed state to Bob through an insecure channel before performing her measurement. As a result, Alice and Bob would share a noisy entangled state that is mapped into correlated classical data by Gaussian measurements. Both protocols are of course equivalent from Eve's point of view, but the theoretical analysis turns out to be simpler for this E-B scheme. Nevertheless, the obtained security bounds automatically apply to the corresponding P&M scheme.

3. Structure of the Security Proofs

3.1. *Eve's physical attack*

In the E-B scheme, Alice and Bob share a mixed entangled state ρ_{AB}. The best situation for Eve is when she can "purify" this state: the global state between Alice, Bob, and Eve is then pure $|\Psi\rangle_{ABE}$, with $\mathrm{tr}_E|\Psi\rangle\langle\Psi| = \rho_{AB}$. If Alice and Bob could perform a full tomography of their state, they could know ρ_{AB} and deduce Eve's state up to an irrelevant transformation. However, this strategy, which is standard in discrete-variable QKD, is not realistic with continuous variables because of the infinite dimensionality of the Hilbert space. Therefore, tomography must be limited to a few parameters, usually the coefficients of the covariance matrix γ_{AB} of the state ρ_{AB}. Fortunately, if it is Gaussian, the state ρ_{AB} – and Eve's attack – becomes fully characterized by γ_{AB}. Of course, this is not true in the general case, so it will be crucial, in the security analysis, to show that the Gaussian attack is optimal for a given covariance matrix γ_{AB}.

It is convenient at this point to introduce Eve's Gaussian attack for a given symmetric Gaussian channel, characterized by its transmission T and excess noise ε, called the "entangling cloner" (see Fig. 1). As proven in Ref. 17, Eve can simulate the channel (T, ε) by combining at a beam-splitter of transmittance $T_E = 1 - T$ the intercepted state together with half of a two-mode squeezed vacuum state. The squeezing parameter r_E has to be chosen such that $(1 - T)\cosh r_E = 1 - T + \varepsilon T$. If the initial covariance matrix between Alice and Bob (*before* transmission) is γ_0, it becomes *after* transmission $\gamma_{AB} = M^{\mathrm{T}}\gamma_0 M + N$, with $M = \mathrm{diag}(1, 1, \sqrt{T}, \sqrt{T})$ and $N = \mathrm{diag}(0, 0, 1 - T + \varepsilon T, 1 - T + \varepsilon T)$.

3.2. *Eve's measurement*

As mentioned above, Eve's physical attack allows her to hold a (Gaussian or non-Gaussian) purification of ρ_{AB}. This is not enough *per se* to give her information about the secret key. She necessarily needs to perform some measurement in order to extract this information. This measurement can be divided into three categories of increasing power, namely (1) individual, (2) collective, and (3) coherent attacks.

In individual attacks, Eve makes one ancilla interact with each pulse, individually, and performs a measurement on it. This measurement cannot depend on Alice and Bob's classical communication (except for the basis choice, if any). Since this measurement outcome is classical, Eve's information is then measured by the Shannon (classical) mutual information. A variation of this attack is the "finite-size attack", where the interaction encompasses several pulses. The size of the block, however, has to be much smaller than the length of the codewords used in the key extraction stage, and, even more importantly, the joint measurement of the ancillas cannot depend on the exchanged messages during this key extraction stage.

In collective attacks, the interaction with the ancillas stays individual (or, at least, of a finite size), but the ancillas are stored in a quantum memory and measured only after Alice and Bob have communicated to perform the key extraction stage. At this point, a complex collective measurement is performed on the quantum memory. The information gained by Eve using this strategy is computed using the Von Neumann entropies instead of Shannon entropies, which leads to the Holevo information. This strategy potentially gives Eve more information than an individual attack.

Coherent attacks are, by definition, the most powerful attacks allowed by quantum mechanics: Eve interacts globally with all pulses and then performs a delayed global measurement. This global interaction renders any statistical assumption difficult, since Alice, Bob, and Eve now share a single high-dimensional quantum system. However, the collective attacks, which are currently known to be optimal within a restricted class of explicit attacks, are likely to be fully optimal within the set of all coherent attacks,[6] although there is no rigorous proof of it yet for continuous variables.

3.3. *Eve's knowledge*

To extract the secret key from their correlated data, Alice and Bob need to apply reconciliation (error correction) and privacy amplification. The reconciliation step of a continuous-variable protocol is somewhat distinct from its discrete-variable counterpart. Specific protocols, known as "slice reconciliation", have been developed for this purpose.[18] We will not discuss this point any further here; we only need to know that a common bit string can be extracted from Alice's and Bob's continuous data. Then, one applies privacy amplification, which, roughly speaking, allows them to filter out the bits known to Eve. All they need to know to apply privacy amplification is an upper bound I_E on Eve's information. Once this bound is known, they can extract a secret key whose length is at least $I(A{:}B) - I_E$, where $I(A{:}B)$ is the mutual information between Alice's and Bob's data.

The expression of I_E of course depends on Eve's strategy, but also on the direction of the classical information flow: if the classical communication is one-way and flows from Alice to Bob in order for him to error correct his data, it means that Alice's data form the secret key so that I_E is the amount of information Eve has gained on Alice's data. This is known as *Direct Reconciliation* (DR). For obvious symmetry reasons, such a strategy cannot succeed when the physical channel is a lossy channel with more than 50% losses. The symmetry between Bob and Eve has to be broken, which can only be done with a feedback, that is, with some classical communication flowing from Bob to Alice. This can be done using one-way backward classical communication (and no forward communication): this is the *Reverse Reconciliation* (RR) scenario, where the secret key is based on Bob's data. In this case I_E represents the amount of information gained by Eve on Bob's data. It is also possible to use two-way classical communication (*e.g.*, in postselection-based protocols[15,19]), but this strategy will not be discussed here.

4. Individual Attacks

4.1. *Preliminaries*

For individual attacks, Eve is assumed (i) to interact individually and in a same way with each quantum state sent over the channel, and (ii) to measure before the error correction and privacy amplification procedures have taken place. These two assumptions are realistic within the present-day technology, even though more general attacks may be imagined. The results shown in this subsection were published (with more details) in Refs. 4, 7, 8 for individual attacks and in Ref. 9 for finite-size attacks.

If Eve interacts individually and in a same way with all the states, this corresponds, in the E-B picture, to a situation where Alice, Bob, and Eve share many copies of the state $|\Psi\rangle_{ABE}$, resulting from Eve's interaction on half of a two-mode squeezed state, $|\psi(r_A)\rangle$, and a reference state $|R\rangle_E$, that is,

$$|\Psi\rangle_{ABE} = (\mathbb{1}_A \otimes U_{BE})|\psi(r_A)\rangle_{AB}|R\rangle_E. \tag{2}$$

After their measurements, Alice and Bob map their shared state into correlated random variables, A and B. Eve is also assumed to measure at this point, so she has a random variable E correlated with Alice's and Bob's outputs. Therefore, the three parties share correlated Classical-Classical-Classical information (CCC correlations). This results in the diagonal density operator

$$\rho_{ABE} = \sum_{A,B,E} p(A,B,E)|A\rangle\langle A| \otimes |B\rangle\langle B| \otimes |E\rangle\langle E|. \tag{3}$$

The process of distilling a secret key out of CCC correlations using one-way communication protocols was studied in Ref. 20. There, it was shown that given a CCC correlation with distribution $p(A,B,E)$, the achievable direct one-way secret-key rate satisfies

$$K^{\rightarrow} \geq I(A{:}B) - I(A{:}E) = \bar{K}_D. \tag{4}$$

In this formula, it is assumed that the flow of information in the error correction and privacy amplification stages goes from Alice to Bob (DR). Also, I stands for Shannon mutual information between the classical random variables,

$$I(X{:}Y) = H(X) - H(X|Y), \tag{5}$$

where $H(X)$ denotes a Shannon entropy and $H(X|Y)$ is a Shannon conditional entropy.[21]

The maximal information Bob can extract about Alice's variable A from his variable B is equal to the mutual information $I(A{:}B)$. The same holds for Eve, so her accessible information on Alice's data is given by $I(A{:}E)$. Therefore, the bound (4) compares the information on Alice's preparation accessible to Bob and Eve. The Csiszár-Körner bound (4), is thus quite intuitive as it reflects Bob's advantage over Eve, but its proof is actually rather involved! Very naturally, in the case of reverse reconciliation (RR), the previous bound becomes

$$K^{\leftarrow} \geq I(A{:}B) - I(B{:}E) = \bar{K}_R, \tag{6}$$

as it is the advantage of Alice over Eve which is relevant. We are now ready to analyze the rate of key extraction against individual attacks using these simple bounds. In particular, it will analyzed how it depends on the channel parameters, T and ε, for different protocols.

4.2. *Secure key rates against individual attacks*

Let us show how to compute the bounds of Eqs. (4) and (6) for the CV-QKD protocols using squeezed or coherent states, and homodyne or heterodyne measurements (a more detailed calculation can be found in Ref. 16). We restrict our considerations to Gaussian attacks, as in Fig. 1. It will be proven in Sec. 7 that these attacks minimize all the bounds, so they are maximally pessimistic (*i.e.*, optimal for Eve).

As explained above, Alice's state preparation can be done by means of a two-mode squeezed vacuum state of squeezing parameter r_A and a beam-splitter of transmittance T_A. After propagating through the insecure channel, Alice, Bob, and Eve share a tripartite state, $|\Psi\rangle_{ABE}$. This state depends on Alice's preparation, the channel properties (T, ε), and Bob's measurement, either homodyne ($T_B = 1$) or heterodyne ($T_B = 1/2$). Since Eve's attack is Gaussian, the state is completely specified by its covariance matrix γ_{ABE}, while the displacement vector is zero. It is relatively simple to calculate (4) and (6) from γ_{ABE}. The mutual information between Alice and Bob can be found through the Wigner function of their reduced state ρ_{AB}. The Wigner function indeed defines the Gaussian probability distribution of the quadrature measurements of Alice and Bob, from which $I(A{:}B)$ can be obtained. The same reasoning gives $I(A{:}E)$ (or $I(B{:}E)$).

Using this formalism, one can compute the key rates that are secure against any Gaussian individual attack for a given protocol (Alice's preparation and Bob's measurement) and channel parameters (T, ε). Not surprisingly, the obtained key rate turns out to be an increasing function of the modulation in the state preparation, that is, of r_A. Moreover, in some particular cases, one obtains relatively simple formulas. For instance, consider the situation where the excess noise ε in the channel is zero while r_A is large. Then, for the coherent-state protocols,[4,8]

$$\bar{K}_D \approx \frac{1}{2}\log\left(\frac{T}{1-T}\right) \qquad \bar{K}_R \approx \frac{1}{2}\log\left(\frac{1}{1-T}\right), \tag{7}$$

while for the squeezed-state protocol,[7]

$$\bar{K}_D \approx \log\left(\frac{T}{1-T}\right) \qquad \bar{K}_R \approx \log\left(\frac{1}{1-T}\right), \tag{8}$$

that is, they are twice as large as with coherent states. In the case of the protocol where both Alice and Bob perform heterodyne measurements,[22] *i.e.*, $T_A = T_B = 1/2$, one has

$$\bar{K}_D \approx \log\left(\frac{T}{1+T}\right) \qquad \bar{K}_R \approx \log\left(\frac{1}{1-T}\right). \tag{9}$$

All these bounds on the secret key rate define security conditions for lossy but noiseless channels, which guarantee provable security against individual attacks. For all direct protocols, we always have the constraint $T > 1/2$, which correspond to 3 dB of losses. In contrast, for all reverse protocols, arbitrarily high losses are tolerable, in principle, since the rate is positive for any non-zero value of T.

These bounds can also be computed for noisy channels, that is for a non-zero excess noise ε. Direct protocols (with coherent or squeezed states) are secure provided that the total equivalent input noise is smaller than the shot noise. This corresponds to a maximal excess noise $\varepsilon < 2 - 1/T$, which can only be positive for $T < 1/2$. The maximal tolerable excess noise for reverse protocols depends on the allowed squeezing: if Alice can send arbitrarily squeezed states, one has $\varepsilon < 2$, while if she can only send coherent states, $\varepsilon < \frac{1}{2} - \frac{1}{T} + \sqrt{\frac{1}{T^2} + \frac{1}{4}}$, which varies between $\frac{1}{2}$ (for $T \to 0$) and $(\sqrt{5}-1)/2 \simeq 0.61$ (for $T = 1$). Thus, for the practically important class of coherent-state protocols, reverse reconciliation is more appropriate for lossy channels with little noise, while direct reconciliation has an advantage for noisy channels with few losses.

5. Collective Attacks

5.1. *Preliminaries*

Even though the two restrictions that we have put on Eve in the analysis of individual attacks [(i) interaction with each pulse individually, and (ii) measurement before the classical key distillation procedure] are very realistic taking into account the present-day technology, they are unsatisfactory from a theoretical point of view. What we want to achieve in quantum cryptography is provable security without imposing any limitation on Eve's technological power. For instance, the second assumption seems to be particularly strong. After having interacted with the states, Eve holds a quantum system that is correlated with Alice's preparation and Bob's measurement results. During the reconciliation process, the honest parties exchange information through the classical public channel in order to increase their correlations. This information is also available to Eve, so it appears quite reasonable that her correlations with Alice's and Bob's data may also increase. Therefore, she can adapt and improve the measurement on her quantum state according to the exchanged messages. The aim of the next two sections is to extend the previous security analysis to such general attacks. We first get rid of assumption (ii) and allow Eve to delay her measurement until the end of the reconciliation process. This corresponds to the case of collective attacks, treated in this section (more details can be found in Refs. 10, 11, 12, 13). In the next section, we also get rid of assumption (i), which gives secure key rates against any attack consistent with quantum mechanics.

In collective attacks, Eve's interaction remains the same for each pulse as with individual attacks, so that Eq. (2) stays valid. After their measurements, Alice and Bob map again their shared state ρ_{AB} into correlated random variables, A and B. But, in contrast with individual attacks, Eve is not assumed to measure at this point, so she keeps a quantum state that is correlated with Alice and Bob's outputs, ρ_E^{AB}. Here, the three parties share correlated Classical-Classical-Quantum information (CCQ correlations). This can be summarized by means of the quantum state

$$\rho_{ABE} = \sum_{A,B} p\,(A,B)|A\rangle\langle A| \otimes |B\rangle\langle B| \otimes \rho_E^{AB}. \tag{10}$$

where the fact that Eve has not performed a measurement translates into the fact that the density operator is not diagonal in E.

The process of distilling a secret key out of CCQ correlations (and, even more generally, out of CQQ correlations) using one-way communication protocols has been studied in Ref. 23 (see also Ref. 24). There, it was shown that given a CQQ state

$$\rho_{ABE} = \sum_{A} p\,(A)|A\rangle\langle A| \otimes \rho_{BE}^{A}, \tag{11}$$

the one-way secret key rate in direct reconciliation satisfies

$$K^{\rightarrow} \geq \chi(A{:}B) - \chi(A{:}E). \tag{12}$$

Here, χ stands for the Holevo bound,[25] which gives the accessible classical information encoded into an ensemble of quantum states $\{p\,(x), \rho^x\}$,

$$\chi = S(\rho) - \sum_{x} p\,(x) S(\rho^x), \tag{13}$$

where $\rho = \sum_x p\,(x)\rho^x$ and $S(\rho) = -\mathrm{tr}(\rho \log \rho)$ denotes the von Neumann entropy. Equation (12) looks like a very intuitive extension of Eq. (4) since χ is the extension of Shannon information in a classical-quantum case. However, its proof is again rather involved!

Tracing out Eve, Bob is effectively receiving quantum states encoding Alice's classical data, $\rho_B^A = \mathrm{tr}_E \rho_{BE}^A$, with probability $p(A)$. The maximal information he can extract is equal to the Holevo bound $\chi(A{:}B)$, computed for the ensemble $\{p(A), \rho_B^A\}$. The Holevo bound can indeed be interpreted very naturally as the quantum mutual information between the internal state of the preparer (Alice) and the state arriving at the receiver (Bob).[26] The same holds for Eve, her accessible information on Alice's data being given by $\chi(A{:}E)$. Therefore, the bound (12) compares the information on Alice's preparation accessible to Bob and Eve, generalizing to the CQQ case the well-known Csiszár-Körner bound (4) for CCC correlations.

In any P&M protocol, Bob also holds classical data since he measures his quantum state upon receiving it, as described by Eq. (10). This simply represents a special subset of the more general scenario analyzed in Ref. 23, so that the same reasoning holds for CCQ correlations. The Holevo quantity $\chi(A{:}B)$ then simply coincides with the standard mutual information between Alice and Bob, $I(A{:}B)$.[26] Thus, the extractable secret key rate in direct reconciliation satisfies

$$K^{\rightarrow} \geq I(A{:}B) - \chi(A{:}E) = \hat{K}_D. \tag{14}$$

In the case of reverse reconciliation, this bound reads,

$$K^{\leftarrow} \geq I(A{:}B) - \chi(B{:}E) = \hat{K}_R. \tag{15}$$

5.2. *Secure key rates against collective attacks*

Let us analyze how the bounds (14) or (15) depend on the channel parameters, T and ε, for different QKD protocols (using squeezed or coherent states, and heterodyne or homodyne measurements). A more detailed calculation can be found in Ref. 10, 11. We again restrict our considerations to Gaussian attacks, as in Fig. 1, knowing that these attacks minimize all the bounds considered here (see Sec. 7).

The information $I(A{:}B)$ is calculated exactly as for individual attacks, while the calculation of $\chi(A{:}E)$ is slightly more involved. After tracing out Bob, one has the Gaussian state ρ_{AE} of covariance matrix γ_{AE} and zero displacement vector, completely specifying the correlations between Alice and Eve. This covariance matrix has the form

$$\gamma_{AE} = \begin{pmatrix} \gamma_A & C_{AE} \\ C_{AE}^T & \gamma_E \end{pmatrix}, \tag{16}$$

where γ_A (γ_E) is the covariance matrix of Alice's (Eve's) local state, and C_{AE} characterizes their correlations. Alice's measurement projects her state into a Gaussian state of covariance matrix $\gamma_A^{m_A}$ and displacement vector $\vec{d}_A^{m_A}$, depending on the obtained outcome m_A, and on the type of measurement. For instance, $\gamma_A^{m_A} = \mathbb{1}$ for a coherent state protocol. This measurement is also effectively preparing a Gaussian state on Eve's side, with covariance matrix $\gamma_E^{m_A}$ and displacement vector $\vec{d}_E^{m_A}$. These two quantities can be calculated using the Gaussian formalism developed in Refs. 27, 28, namely

$$\begin{aligned} \gamma_E^{m_A} &= \gamma_E - C_{AE}^T(\gamma_A + \gamma_A^{m_A})^{-1} C_{AE} \\ \vec{d}_E^{m_A} &= C_{AE}^T(\gamma_A + \gamma_A^{m_A})^{-1} d_A^{m_A}. \end{aligned} \tag{17}$$

In all the considered protocols, with squeezed or coherent states, $\gamma_A^{m_A}$ does not depend on the measurement outcome m_A, so $\gamma_E^{m_A}$ is also independent of m_A, $\gamma_E^{m_A} \equiv \gamma_E^A$. Therefore, $\chi(A{:}E)$ is simply equal to

$$\chi(A{:}E) = S(\gamma_E) - S(\gamma_E^A), \tag{18}$$

where we explicitly use the fact that the von Neumann entropy of a Gaussian state only depends on its covariance matrix. The same reasoning can be applied to the calculation of $\chi(B{:}E)$ for reverse reconciliation.

Using this formalism, one can compute secure key rates against any collective attack, for a given protocol (Alice's preparation and Bob's measurement) and channel parameters. The calculation of the bounds is lengthy but straightforward. In the case where the excess noise ε in the channel is zero and r_A is large, one obtains simple results. Then, for the coherent-state protocols of Refs. 4, 8, one has

$$\hat{K}_D \approx \frac{1}{2}\log\left(\frac{T}{1-T}\right) \qquad \hat{K}_R \approx \frac{1}{2}\log\left(\frac{1}{1-T}\right), \tag{19}$$

while, for the squeezed-state protocols of Ref. 7, one has

$$\hat{K}_D \approx \log\left(\frac{T}{1-T}\right) \qquad \hat{K}_R \approx \log\left(\frac{1}{1-T}\right). \tag{20}$$

In the case of the protocol of Ref. 22, where $T_A = T_B = 1/2$, one has

$$\hat{K}_D \approx \log\left(\frac{T}{1+T}\right) - \log e \qquad \hat{K}_R \approx \frac{1}{T}\log\left(\frac{1}{1-T}\right) - \log e. \tag{21}$$

All these bounds on the extractable secret key rate define conditions for provable security against collective attacks, as summarized in Fig. 2.

The previous formalism is also useful to establish the critical value of the excess noise in the line, above which no key distribution is possible, independently of Alice's modulation, see Ref. 11. These values have to be

Protocol	Direct	Reverse
Coherent states	3 dB	no limit
Squeezed states	3 dB	no limit
Heterodyne measurements	1.4 dB	no limit

Fig. 2. Critical values of the channel transmission for provable security against collective attacks in the case of zero excess noise and large modulation variance. The coherent-state protocol, squeezed-state protocol, and the protocol with heterodyne measurements are compared.

understood as simply testable sufficient conditions for secure key distribution. For squeezed-state protocols, it is always more convenient to employ reverse reconciliation. In contrast, for coherent-state protocols, direct reconciliation turns out to be more resistant against excess noise down to a channel transmission of ≈ 0.65. Note also that there exist limiting values of the excess noise, ε_c, for which the considered secret key rates are zero, independently of the modulation and the losses. These values can be computed analytically. For coherent states and direct reconciliation, one has that ε_c is the solution to the equation

$$\frac{1}{1+\varepsilon}\left(\frac{\sqrt{1+\varepsilon}+1}{\sqrt{1+\varepsilon}-1}\right)^{\sqrt{1+\varepsilon}} = e^2, \qquad (22)$$

that gives $\varepsilon_c \approx 0.8$, while for reverse reconciliation

$$\varepsilon_c = \frac{1}{2}\left(\sqrt{1+\frac{16}{e^2}}-1\right) \approx 0.39. \qquad (23)$$

In the case of squeezed states, the critical excess noise is equal to $2/e \approx 0.7$ for both reconciliation protocols. A similar picture can be obtained for the heterodyne measurement-based protocol of Ref. 22.

6. Coherent Attacks

A first approach to analyze the resistance of CV-QKD against the most general (coherent) attacks consists in exploiting the equivalence between quantum error correcting codes and one-way entanglement purification protocols, exactly as for discrete-variable QKD. This approach was followed in Ref. 29 in order to prove that Gaussian-modulated squeezed-state protocols can be made unconditionally secure provided that the squeezing exceeds some threshold $r \approx 0.3$. It was extended in Ref. 30 to the case of coherent-state protocols, although the tolerable loss is only of 0.4 dB in this case. In

Ref. 31, an explicit encoding scheme was proposed to distill more than one qubit (i.e., more than one secret bit) per entangled pair, extending the tolerable loss to 1.4 dB in some case. These results can be viewed as proofs of principle that unconditional security is achievable with continuous-variable protocols, but unfortunately they do not yield useful secret key rates.

Recently, however, powerful techniques for the analysis of general security proofs of QKD have been presented in Ref. 32, which can predict secret key rates. In any QKD scheme, there is a tomographic process that partly characterizes the insecure channel connecting Alice and Bob. It allows the honest parties to evaluate their mutual information, $I(A{:}B)$. Moreover, it puts a bound on Eve's knowledge: it was shown in Ref. 32 that, using the information collected during this process, one can construct a secure reconciliation protocol that allows one to extract

$$\tilde{K} = I(A{:}B) - \max_{\rho_{AB} \in \mathcal{R}} S(\rho_{AB}), \tag{24}$$

secret bits, where $\mathcal{R}$ is the set of quantum states consistent with the measured probabilities (see Ref. 32 for more details). Thus, this quantity represents a lower bound to the achievable key rate, $K \geq \tilde{K}$. For all the QKD schemes analyzed here, the attack minimizing $\tilde{K}$ for fixed first- and second-order moments of ρ_{AB} is Gaussian (see Sec. 7). Unfortunately, this bound does not make any distinction between direct and reverse reconciliation while it plays an important role in continuous-variable QKD protocols.

The calculation of $\tilde{K}$ proceeds along the same lines as above for $\hat{K}_D$ or $\hat{K}_R$. Consider first the coherent-state or squeezed-state protocol. For the case of a lossy but noiseless line, $\varepsilon = 0$, one can numerically see that there exists an optimal squeezing r_A^{opt} which is the same for coherent-state and squeezed-state protocols.[11] A reason for this counter-intuitive result may be that $\tilde{K}$ is known to be a non-tight bound to the optimal key rate.[32] This optimal squeezing, $r_A^{opt} \approx 1.5$, defines a critical value for the tolerable losses of approximately 1.7 and 0.83 dB for squeezed-state and coherent-state protocols, respectively.

As discussed in Ref. 32, it is possible to improve the bound (24) by conditioning the privacy amplification process on a classical random variable W (see Ref. 32 for more details), decreasing Eve's entropy. For the case of coherent states, Alice and Bob can make public the value of the second measured quadrature, instead of discarding it. This process does not modify Alice and Bob's mutual information but changes Eve's entropy. The obtained critical transmission, T_c, is now a decreasing function of the squeezing, as expected. One can see that in the limit of high modulation, $r_A \to \infty$,

$$T_c = \frac{e^2}{e^2 + 4}. \tag{25}$$

That is, the protocol using coherent states and homodyne measurements is secure up to 1.9 dB of losses.

Finally, let us consider the heterodyne measurement-based protocol of Ref. 22 in the case of a lossy line. Recall that the two quadratures measured by Alice contribute to the key. These two homodyne measurements effectively prepare a coherent state that propagates through the insecure channel, and Eve keeps a fraction $1 - T$ of it. Therefore, Eve receives pure coherent states, depending on x_A and p_A. This implies that $S(\rho_E)$ is actually equal to $\chi(A{:}E)$, which means that $\tilde{K} = \hat{K}_D$. Thus, a secure key distribution against general attacks is possible up to 1.4 dB of losses (see Fig 2).

7. Optimality of Gaussian Attacks

7.1. *Preliminaries*

The derivation of all the previous bounds on the extractable secret key rates has been done assuming that Eve's optimal attack was Gaussian. The goal of this section is to prove this optimality, that is, to show that for given first- and second-order moments of the measured quadratures by Alice and Bob, the attack minimizing $\bar{K}_D$, $\bar{K}_R$, $\hat{K}_D$, $\hat{K}_R$, and $\tilde{K}$ is Gaussian. A proof of this result has first been given in Ref. 9 for finite-size individual attacks (but assuming that Eve's measurement takes place before the key distillation procedure), and has recently been generalized in Refs. 12, 13 to collective and coherent attacks (when Eve's measurement is allowed to depend on the exchanged messages during the key distillation procedure).

The details of the proof are different for finite-size and collective attacks, but the generic idea is the same: Gaussian attacks are the ones which induce the less structured (*i.e.*, more entropic) noise on Bob's measurement outcomes for a given covariance matrix. Roughly speaking, since Eve is constrained by quantum mechanics, the more structure she induces on Bob's noise, the less freedom she has on her attack. More rigorously, the amount of information I_E she gains can be upper bounded by an entropic quantity that is calculated from the (experimentally accessible) covariance matrix of the state ρ_{AB} shared by Alice and Bob, and this maximum is attained for a Gaussian attack.

Note that, for all practical purposes, one only needs to bound I_E since $I(A{:}B)$ depends on the practical error-correcting codes used by Alice and Bob in the reconciliation stage. Even if these codes would yield a rate that is close to Shannon's limit for a Gaussian channel, the evolution of this rate for an arbitrary non-Gaussian attack would be difficult to predict. Nevertheless, this is not a problem in practical CV-QKD because Alice and Bob can always measure $I(A{:}B)$ by comparing a sample of their reconciliated keys, so there is no need to predict it from γ_{AB}.

7.2. *Entropy of Gaussian states $\tilde{\rho}$ — general attacks*

Let $\rho \in B(\mathcal{H}^2)$ denote an arbitrary density operator, and $\tilde{\rho}$ the density operator corresponding to a Gaussian state characterized by the same covariance matrix (or second-order moments) and displacement vector (or first-order moments) as ρ. Similarly, if $p\,(x)$ is a probability distribution for a random variable X, then $\tilde{p}\,(x)$ denotes the Gaussian probability distribution with the same first- and second-order moments as $p\,(x)$. Finally, if $F(x)$ represents any function of a random variable x, whose probability distribution is $p\,(x)$, then $F(\tilde{x})$ has to be understood as the same function F applied to the distribution $\tilde{p}\,(x)$. It can be shown that, for any state ρ, one has

$$\begin{aligned} S(\tilde{\rho}) - S(\rho) &= \mathrm{tr}(\rho \log \rho) - \mathrm{tr}(\rho \log \tilde{\rho}) + \mathrm{tr}(\rho \log \tilde{\rho}) - \mathrm{tr}(\tilde{\rho} \log \tilde{\rho}) \\ &= S(\rho || \tilde{\rho}), \end{aligned} \tag{26}$$

where the first two terms in the r.h.s. of Eq. (26) sum to the quantum relative entropy $S(\rho||\tilde{\rho})$. The sum of the last two terms in the r.h.s. of Eq. (26) vanishes because $\log \tilde{\rho}$ is a polynomial of second order in the field operators and because ρ and $\tilde{\rho}$ have, by definition, the same first- and second-order moments.

As a consequence, since the quantum relative entropy is positive semi-definite,[33] the state of maximal entropy for fixed first- and second-order moments is indeed Gaussian. In particular, if Alice and Bob share a state ρ_{AB}, they can bound its entropy from its covariance matrix, that is, $S(\rho_{AB}) \leq S(\tilde{\rho}_{AB})$. Using similar arguments, it can be seen that the same property holds for classical probability distributions

$$H(\tilde{x}) - H(x) = H(x||\tilde{x}) \geq 0, \tag{27}$$

where $H(x||\tilde{x}) = \sum_x p\,(x) \log[p\,(x)/\tilde{p}\,(x)]$ is the classical relative entropy.

The simple fact that the states with maximal entropy are Gaussian combined with the bound (24) gives us immediately the optimal general attack (for this bound): it is a Gaussian attack.

7.3. *Conditional entropy of $\tilde{\rho}$ — individual attacks*

The von Neumann conditional entropy[34] is

$$S(A|B) = S(\rho_{AB}) - S(\rho_B) = S(\tilde{A}|\tilde{B}) - S(\rho_{AB}||\tilde{\rho}_{AB}) + S(\rho_B||\tilde{\rho}_B). \quad (28)$$

The relative entropy is a discrimination measure between two states and can only decrease under a physical (*i.e.* trace preserving) map. That is, for any such map, denoted by $\mathcal{T}$, and any two states, ρ_1 and ρ_2,

$$S(\rho_1||\rho_2) \geq S(\mathcal{T}(\rho_1)||\mathcal{T}(\rho_2)). \quad (29)$$

Tracing out A is a particular instance of such a trace-preserving map, with $\mathcal{T}(\rho_{AB}) = \rho_B$ and $\mathcal{T}(\tilde{\rho}_{AB}) = \tilde{\rho}_B$. Therefore $S(\rho_B||\tilde{\rho}_B) \leq S(\rho_{AB}||\tilde{\rho}_{AB})$, which implies that quantum conditional entropy is also maximized for a Gaussian state

$$S(A|B) \leq S(\tilde{A}|\tilde{B}). \quad (30)$$

Naturally, the same reasoning applies to classical probability distributions, substituting von Neumann conditional entropies with Shannon conditional entropies, and replacing trace-preserving maps by stochastic maps:

$$H(x|y) \leq H(\tilde{x}|\tilde{y}). \quad (31)$$

In order to find the optimal individual attacks, one needs to combine this inequality with the entropic uncertainty principle,[35] which states that $H(p_A|p_B) + H(x_A|x_E) \geq 0$ where x_A and p_A are the two quadratures of Alice's state, inferred from Bob's (p_B) or Eve's (x_E) measurements. Note that that x_A and p_A are expressed here in the appropriate units so that the r.h.s. term is 0 (in other units, it would simply be a constant). Thus, Alice and Eve's mutual information can been rewritten as

$$I(A{:}E) = H(x_A) - H(x_A|x_E) \leq H(x_A) + H(p_A|p_B) \quad (32)$$

which is optimal (maximum) for a Gaussian attack as a consequence of Eqs. (27) and (31). Of course, the same reasoning applies to $I(B{:}E)$ in the case of reverse reconciliation. This confirms that the attack which minimizes the bounds $\bar{K}_D$ and $\bar{K}_R$ for individual (finite-size) attacks is Gaussian.

7.4. Effect of Alice's measurement — collective attacks

Let $\rho \in B(\mathcal{H}^2)$ be any physical state and ρ' the result of a measurement on a part of it by projection onto a given basis, say X,

$$\rho' = \sum_x p(x)|x\rangle\langle x| \otimes \rho^x. \tag{33}$$

where ρ^x is the resulting state knowing that the outcome of the measurement is x. More precisely, ρ' can be thought of as the joint state of the system under investigation and an ancilla, which, after measurement, contains the measurement outcome. It is straightforward to check that[26]

$$S(\rho') = H(x) + \sum_x p(x) S(\rho^x), \tag{34}$$

where H denotes the usual Shannon entropy. Now, Eve's accessible information on Alice's measurement outcome, which is needed to calculate the bound (14), can be written

$$\chi(A{:}E) = S(\rho_E) - \sum_{x_A} p(x_A) S(\rho_E^{x_A}), \tag{35}$$

see Eq. (13). Since the state of Alice, Bob, and Eve before Alice's measurement $|\Psi\rangle_{ABE}$ is pure, $S(\rho_E) = S(\rho_{AB})$. Similarly, since the state of Bob and Eve conditioned on Alice's measurement outcome x_A, *i.e.* $\langle x_A|\Psi\rangle_{ABE}$, is pure, $S(\rho_E^{x_A}) = S(\rho_B^{x_A})$. Thus

$$\chi(A{:}E) = S(\rho_{AB}) - \sum_{x_A} p(x_A) S(\rho_B^{x_A}) = S(\rho_{AB}) - S(\rho'_{AB}) + H(x_A) \tag{36}$$

where we have used Eq. (34) and ρ'_{AB} denotes the joint state of Alice and Bob after Alice's measurement. Now, using Eq. (29) with the map $\mathcal{T}$ being defined as Alice's measurement (Bob system staying unchanged), we have

$$S(\rho'_{AB}||\tilde{\rho}'_{AB}) \leq S(\rho_{AB}||\tilde{\rho}_{AB}). \tag{37}$$

Using Eq. (26), this can be expressed as

$$S(\rho_{AB}) - S(\rho'_{AB}) \leq S(\tilde{\rho}_{AB}) - S(\tilde{\rho}'_{AB}) \tag{38}$$

Therefore, we see that the r.h.s of Eq. (36) is again optimal (maximum) for a Gaussian attack as a result of Eqs. (27) and (38). The same reasoning also applies to $\chi(B{:}E)$ in reverse reconciliation. Note that $S(\rho'_{AB})$ and $S(\tilde{\rho}'_{AB})$ in the above expressions are defined up to an arbitrary constant, cfr. the caveat after Eq. (31). However, this arbitrary constant cancels with the one in $H(x_A)$ and $H(\tilde{x}_A)$, so that this Gaussian upper bound is well defined.

Note that if the states of Bob and Eve conditional on Alice's measurement are not pure (if Alice sends mixed states in the corresponding P&M protocol), a fourth party, Fred, can always be assumed to hold its purification. One then has $S(\rho_E) = S(\rho_{ABF})$, and the same reasoning as above applies, which maintains this optimality result. This confirms that the attack which minimizes the bounds $\hat{K}_D$ and $\hat{K}_R$ for collective attacks is Gaussian.

More generally, if Eve's attack is not identical from pulse to pulse, the above reasoning still holds with multimode Gaussian states instead of single-mode ones. However, Alice and Bob will not measure the full covariance matrix, but an averaged one, so they will overlook the pulse-to-pulse correlations. Fortunately, it is straightforward to show that this averaging will make them overestimate Eve's information, so they remain on the safe side. In other words, the optimal attack for a given estimated "single-pulse" covariance matrix remains the Gaussian attack described in Fig. 1.

8. Conclusion

We have outlined the main security proofs obtained today for assessing the security of continuous-variable quantum key distribution based on Gaussian-modulated Gaussian states and Gaussian measurements. We have discussed the increasingly difficult analyses of individual, collective, and coherent attacks. We have shown that, for a given estimated covariance matrix of Alice's and Bob's quadrature components, the Gaussian attacks are optimal; hence, they provide a tight bound on the attainable secure key rates. We hope that these theoretical progresses will further encourage bringing CV-QKD closer to practice. A promising recent step in this direction is the realization of an all-fibered coherent-state setup working at telecom wavelength (1550 nm) at a rate exceeding 1 Mbit/s, as reported in Ref. 36.

Acknowledgments

We would like to thank R. García-Patrón, P. Grangier, S. Iblisdir, M. Navascués, R. Tualle-Brouri, G. Van Assche, and J. Wenger for many stimulating discussions.

References

1. For a recent review, see N. Gisin, G. Ribordy, W. Tittel, and H. Zbinden, Rev. Mod. Phys. **74**, 145 (2002).

2. C. H. Bennett and G. Brassard, *Proc. IEEE Int. Conference on Computers, Systems and Signal Processing*, IEEE, New York, (1984).
3. K. J. Gordon, V. Fernandez, G. S. Bulleri, I. Rech, S. D. Cova, and P. D. Townsend, Optics Express **13**, 3015 (2005).
4. F. Grosshans, G. Van Assche, J. Wenger, R. Brouri, N. J. Cerf, and Ph. Grangier, Nature **421**, 238 (2003).
5. M. Hillery, Phys. Rev. **63**, 022309 (2000).
6. R. Renner, Ph.D. Thesis (ETH Zurich, 2005).
7. N. J. Cerf, M. Lévy, and G. Van Assche, Phys. Rev. A **63**, 052311 (2001).
8. F. Grosshans and Ph. Grangier, Phys. Rev. Lett. **88**, 057902 (2002).
9. F. Grosshans and N. J. Cerf, Phys. Rev. Lett. **92**, 047905 (2004).
10. F. Grosshans, Phys. Rev. Lett. **94**, 020504 (2005).
11. M. Navascués and A. Acín, Phys. Rev. Lett. **94**, 020505 (2005).
12. M. Navascués, F. Grosshans, and A. Acín, arXiv quant-ph/0608034.
13. R. García-Patrón and N. J. Cerf, arXiv quant-ph/0608032.
14. C. H. Bennett, G. Brassard, and N. D. Mermin, Phys. Rev. Lett. **68**, 557 (1992).
15. Ch. Silberhorn, T. C. Ralph, N. Lütkenhaus, and G. Leuchs, Phys. Rev. Lett. **89**, 167901 (2002).
16. F. Grosshans, N. J. Cerf, J. Wenger, R. Tualle-Brouri, and Ph. Grangier, Quant. Inf. Comp. **3**, 535 (2003).
17. F. Grosshans and Ph. Grangier, arXiv quant-ph/0204127.
18. G. Van Assche, J. Cardinal, and N. J. Cerf, IEEE Trans. Inf. Theory **50**, 394 (2004).
19. M. Navascués, J. Bae, J. I. Cirac, M. Lewenstein, A. Sanpera, and A. Acín, Phys. Rev. Lett. **94**, 010502 (2005).
20. Csiszár and Körner, IEEE Trans. Inf. Theory **24**, 339 (1978).
21. C. E. Shannon, Bell Syst. Tech. J. **27**, 479 and 623 (1948).
22. C. Weedbrook, A. M. Lance, W. P. Bowen, T. Symul, T. C. Ralph, and P. K. Lam, Phys. Rev. Lett. **93**, 170504 (2004).
23. I. Devetak and A. Winter, Phys. Rev. Lett. **93**, 080501 (2004); Proc. R. Soc. Lond. A, **461**, 207 (2005).
24. R. Renner and R. König, quant-ph/0403133.
25. A. S. Holevo, Probl. Inf. Trans. **9**, 177 (1973).
26. N. J. Cerf and C. Adami, arXiv quant-ph/9611032.
27. G. Giedke and J. I. Cirac, Phys. Rev. A **66**, 032316 (2002).
28. J. Fiurášek, Phys. Rev. Lett. **89**, 137904 (2002).
29. D. Gottesman and J. Preskill, Phys. Rev. **63**, 022309 (2001).
30. S. Iblisdir, G. Van Assche, and N. J. Cerf, Phys. Rev. Lett. **93**, 170502 (2004).
31. G. Van Assche, S. Iblisdir, and N. J. Cerf, Phys. Rev. A **71**, 052304 (2005).
32. M. Christandl, R. Renner, and A. Ekert, arXiv quant-ph/0402131.
33. A. Wehrl, Rev. Mod. Phys. **50**, 221 (1978).
34. N. J. Cerf and C. Adami, Phys. Rev. Lett. **79**, 5194 (1997).
35. I. Bialynicki-Birula and J. Mycielski , Commun. Math. Phys. **44**, 129 (1975).
36. J. Lodewyck, T. Debuisschert, R. Tualle-Brouri, and P. Grangier, Phys. Rev. A **72**, 050303(R) (2005).

Chapter 5

Gaussian Quantum Cellular Automata

Ole Krüger and Reinhard F. Werner
Institut für Mathematische Physik, TU Braunschweig
Mendelssohnstrasse 3, D-38106 Braunschweig, Germany
www.imaph.tu-bs.de/qi

1. Introduction

The idea of extending the concept of classical cellular automata to the quantum regime has been considered since long. It is already mentioned in Feynman's famous paper[1] from 1982 on the power of quantum computation. A cellular automaton is chiefly characterized by a discrete cell structure with a finite system for every cell and translational symmetry, a discrete, synchronous time evolution and finite propagation speed. The theoretical concept of its quantum version has connections to several fields: most obviously to the statistical mechanics of lattice systems, potentially in the ultraviolet regularization of quantum field theories but also as a model of computation.

Experimentally, it might prove useful for the realization of quantum computing in optical lattices[2] and arrays of microtraps.[3] The technology of theses systems is highly developed and they are promising candidates for quantum computers capable of useful applications. However, most computational concepts today rely on individual addressing of specific »registers«, which is difficult in these systems. It is much easier to change external parameters for all cells equally, which is exactly a characteristic of a cellular automaton. In fact, recently Vollbrecht *et al.*[4] have introduced a scheme for reversible, universal quantum computing in translationally invariant systems which proved to be a QCA.

Despite the popularity of quantum computational tasks involving the application of the quantum Fourier transform (like in Shor's algorithm), an even more useful employment of quantum computers from the point of view of general physics might be the simulation of quantum systems. Due to dimension and complexity explosion with growing size of these systems, classical computers face serious performance problems even for moderate system sizes. Since quantum computers convert this scaling into a feature, they could overcome the obstacle. The inherent translational symmetry would make quantum cellular automata especially suited for the simulation of models in solid state physics.

A first step towards more complex systems could be the toy application of simulating a one-dimensional quantum random walk[5] on a QCA. Each cell would correspond to the combination of a "slot" to host a particle and a coin to flip for the direction of the next step. If a particle is present in the respective cell, the dynamics of the QCA unitarily maps the state of the coin onto the direction of the particle and moves it to a neighboring cell accordingly. Running the QCA from an initial state with one particle and the coins on every site in a superposition of "left" and "right" then results in a quantum random walk on the line. An obvious extension of this model to *quantum diffusion* is to populate the lattice with additional particles. However, in this case it is necessary to specify a treatment of collisions between particles. One possible solution limits the number of particles per site, *e.g.*, to a maximum of one particle moving left and one moving right. This corresponds to a "hard core interaction". Another solution allows for an arbitrary number of particles per site by second quantization of the random walk. This attaches to every cell a Fock space equipped with an occupation number state basis. Equivalently, every cell can be described as a quantum harmonic oscillator in an excited state according to the number of particles occupying the cell. The movement of particles over the lattice corresponds to the exchange of excitations between the oscillators. This Boson system naturally gives rise to Gaussian QCAs, *i.e.*, continuous variable QCAs which map Gaussian states onto Gaussian states in the Schrödinger picture and which start from a Gaussian initial state. An experimental realization of a Gaussian QCA might use the vibrational degrees of freedom of atoms in an optical lattice.

After a short digression on classical cellular automata, we discuss the definition and properties of their quantum analog, including the problem of quantizing them in the first place. The remaining part of this chapter is devoted to Gaussian quantum cellular automata and the special instance of a one-dimensional chain of harmonic oscillators complete with Gaussian dynamics and Gaussian initial states. We present methods to deal with an infinite number of modes and investigate this system by decomposition into plane wave modes. As a result, we show that the system exhibits properties typically related to irreversibility: Although the system evolves from a pure product state under a reversible dynamics, the correlation function describing the state converges. The reflection symmetric limit states are thermal equilibrium states determined by the correlation function of a pure state and a mode-wise temperature parameter. Furthermore, numerical results suggest that the entanglement built up during time evolution is long-ranged while the correlation function decays exponentially with distance between lattice sites.

2. Classical Cellular Automata

Cellular automata are a model of computation from classical information theory. They belong to the class of discrete finite state automata, which are characterized by finite sets of internal states and inputs together with a transition function mapping these to the internal states in a discrete time step. A classical cellular automaton (a CA for short) combines an infinite lattice array of such automata into a dynamical, interacting system. The defining features are uniformity and locality of the dynamics: each cell of the lattice is described by the same type of automaton and the input is given by the states of the cells in a finite neighborhood. Consequently, the set of possible states and the transition rule is the same for each cell. The neighborhood of any cell is defined by a finite *neighborhood scheme* relative to an arbitrary cell (*cf.* Fig. 1). A CA thus comprises the following:

▷ a regular lattice of discrete cells,
▷ a finite set of internal states for all cells,
▷ a uniform finite neighborhood scheme,
▷ a uniform local transition rule acting
▷ at discrete time steps.

As an example, consider the toy CA of Fig. 1. It features two states per cell ("occupied" or "empty"), a neighborhood scheme consisting of the respective nearest neighbors and a transition rule that flips the state from empty to occupied if the left neighbor is occupied and *vice versa* if the right neighbor is empty; otherwise the state of a cell is not changed. After populating the lattice with an initial distribution of empty and occupied cells, the dynamics will straighten out the occupied cells and shift them to the right.

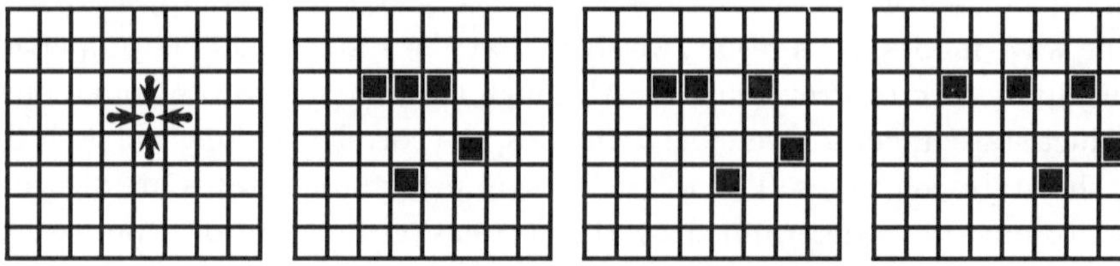

Fig. 1. Simple example of a classical cellular automaton which implements a conditional shift (showing only a finite pane of the infinite lattice). From left to right: neighborhood scheme, initial state, after first and after second time step. See text for details.

More complex CAs have a variety of uses. They became widely known when the "Game of Life" by John Conway[6] proved to be a popular diversion. In this CA a host of dynamic configurations emerges from simple rules. CAs can also be used as a model for universal computation. For example, a Turing machine can be simulated by a CA. Another prominent application is the simulation of appropriate real-world systems. Within physics CAs can be employed to study problems especially from statistical mechanics, *e.g.*, Ising spin dynamics, point particle gases, percolation or annealing.[7] Other problems include the dynamics of bacteria colony growth, forest fires, sand piles or road traffic.

The simple concept of a CA can easily be adopted to more complex models. By partitioning the set of states and the transition rule into two parts, it can distinguish between a "system" and an "environment" part. With a suitable initial state the implementation of boundaries is then possible. A (finite) memory can be accomplished by using compound internal states as a shift register. Another type of extension is the introduction of probabilities into the transition rule or the use of an infinite set of internal states. While this effectively leaves the domain of CAs, it is a useful feature for simulation.

3. Going Quantum

Any attempt to define a quantum cellular automaton has to deal with the infinite number of quantum systems at the lattice sites. For example, the basic operation of applying the same unitary transformation to each cell separately would require to multiply an infinite number of phases, which does not allow for a well-defined unitary operator describing the global state change. It is also problematic to assign different state vectors to the uncountably many classical configurations, which would be a way to introduce the configurations of QCA. (For a discussion of further definitions, see the paper by Schumacher and Werner.[8]) Both concepts suffer from the difficulties of defining states on infinite quantum system. Motivated by ideas from statistical mechanics of quantum spin systems, which deals exactly with infinite arrays of simple quantum systems, we choose a different approach. In contrast to the above, we focus on observables and thus work in the Heisenberg picture instead. The main advantage is that opposed to localized states there exist localized observables, *i.e.*, observables requiring a measurement only of a finite collection of cells. One time step in the global evolution of the QCA is a transformation T on the observable algebra of the infinite system. T is required to be a quantum channel, *i.e.*, it has to be completely positive and trace preserving. The localization property of classical CAs can be stated as a locality property of T in the following sense: If A is an observable localized on a finite region $\Lambda \in \mathcal{L}$ of the lattice $\mathcal{L}$ and $\mathcal{N}$ is the finite neighborhood scheme of the QCA, then $T(A)$ should be localized in $\Lambda + \mathcal{N} \equiv \{x + n \,|\, x \in \Lambda, n \in \mathcal{N}\} \in \mathcal{L}$.

A satisfactory theory of QCAs should connect the *global transition rule* T and the *local transition rule* such that either can be uniquely inferred from the other. The class of global rules should have an axiomatic specification, with the locality described above as the most important aspect, *i.e.*, the existence of a finite neighborhood scheme. On the other hand, for the local transition rules we prefer a constructive characterization. It turns out that this is simple for the class of reversible QCAs:[8] The global rule must fulfill the above locality property and have an inverse which is again an admissible quantum channel. This is equivalent to T being an automorphism of the observable algebra. The local rule is the restriction of T to the algebra of a single cell. Conversely, since every observable can be obtained as a linear combination of products of single-cell observables, the local rule determines the global transformation. In a nutshell, this notion of a QCA complies with the characteristics of a classical CA in the following sense:

- ▷ lattice of discrete cells: an infinite lattice labeled by $x \in \mathbb{Z}^s$ with local observable algebras $\mathcal{A}_x$, *e.g.*, the algebra of complex $d \times d$ matrices
- ▷ discrete global time evolution: an automorphism[a] $T\colon \mathcal{A}(\mathbb{Z}^s) \to \mathcal{A}(\mathbb{Z}^s)$ of the quasi-local algebra[b] $\mathcal{A}(\mathbb{Z}^s)$
- ▷ uniformity: translational invariance of the dynamics and the initial state

[a]For reversible QCAs, a homomorphism suffices.

[b]The quasi-local algebra is the closure of the algebra generated by all observables depending only on the state of cells on finitely many lattice sites.

- ▷ locality and finite propagation speed: for every finite set $\Lambda \in \mathbb{Z}^s$ and the algebra of observables $\mathcal{A}(\Lambda)$ localized on this region, $T(\mathcal{A}(\Lambda)) \subset \mathcal{A}(\Lambda + \mathcal{N})$ with the finite neighborhood scheme $\mathcal{N}$
- ▷ local transition rule: the restriction of T to a single site, $T_0 \colon \mathcal{A}_0 \to \mathcal{A}(\mathcal{N})$

For details and discussion of this definition, see the paper by Schumacher and Werner.[8] An important property of T (as well as T_0) follows from the properties of a homomorphism: if two observables are localized on disjoint regions of the lattice and thus commute, their pictures under the homomorphism T have to commute as well.

While the structure of the neighborhood scheme can in principle be arbitrarily complex, we will, however, restrict to a nearest neighbor scheme. In addition to being the most important, it also offers important theoretical results. Probably the strongest result is a theorem[8] which states that every nearest neighbor QCA is *structurally reversible*, *i.e.*, its inverse exists and is again a nearest neighbor QCA. This is accomplished by partitioning the cells into blocks and applying the same unitary operation to each block. In order to implement any global transition rule, this may be repeated with different blocks.

4. Gaussian Quantum Cellular Automata

By a Gaussian quantum cellular automaton we mean a continuous variable system which conforms to the above properties of a QCA and evolves under a quasi-free dynamics from a translationally invariant Gaussian initial state. For the sake of illustration, we will consider a simple example: a one-dimensional chain of one-mode harmonic oscillators with nearest neighbor coupling and translational invariance. However, the methods presented are valid for arbitrary lattices with translational symmetry and a suitable elementary cell. The phase space of this system is the set $\Xi = \{\xi \colon \mathbb{Z} \to \mathcal{R}^2 \,|\, \xi_x \equiv \xi(x) = 0 \text{ almost everywhere}\}$, where $\mathcal{R}^2$ is the phase space of a single oscillator and the functions ξ vanish everywhere except for a finite number of sites. The global phase space "vectors" ξ relate every site x with a proper local phase space vector $\xi_x \in \mathcal{R}^2$ for a single mode. This generalizes the concept of a direct sum of one-site phase spaces to an infinite set of such systems. The symplectic form on this phase space is defined in terms of the symplectic form on the one-mode phase space, σ_s, as $\sigma(\xi,\eta) = \sum_{x\in\mathbb{Z}} \sigma_s(\xi_x,\eta_x)$. Similarly, Weyl operators $W(\xi)$ on the whole system are defined as tensor products of single-site Weyl operators $w_x(\xi_x)$,

$W(\xi) = \bigotimes_{x\in\mathbb{Z}} w_x(\xi_x)$. Both definitions are well-formed even on the infinite lattice, since the ξ_x are zero except for finitely many sites.

Transition rule: The global transition rule T is supposed to map Gaussian systems into Gaussian systems. In the Heisenberg picture, this is accomplished by mapping the Weyl operators to Weyl operators subject to a symplectic transformation Γ:

$$T\big(W(\xi)\big) = W(\Gamma\,\xi). \tag{1}$$

The transformation Γ acts on phase space functions by site-wise applying suitable real 2×2 matrices $\Gamma_{x,z}$,

$$(\Gamma\,\xi)_x = \sum_{z\in\mathbb{Z}} \Gamma_{x,z}\cdot\xi_z.$$

In order to guarantee uniformity, we require translational invariance of the dynamics. For T to be invariant under lattice translations τ_Δ, where $(\tau_\Delta\,\xi)_x = \xi_{x-\Delta}$, Γ has to be invariant and thus has to commute with τ_Δ for all $\xi\in\Xi$ and all $x,\Delta\in\mathbb{Z}$:

$$(\Gamma\,\tau_\Delta\,\xi)_x = (\tau_\Delta\,\Gamma\,\xi)_x \iff \Gamma_{x,z} = \Gamma_{x-z}.$$

Nearest neighbor coupling forces $\Gamma_{x-z} = 0$ unless $|x-z|\le 1$. Consequently, Γ is completely determined by three real 2×2 matrices $\Gamma_- \equiv \Gamma_{-1}$, $\Gamma_+ \equiv \Gamma_{+1}$ and Γ_0, acting on phase space functions according to:

$$(\Gamma\,\xi)_x = \sum_{z=-1,0,+1} \Gamma_z\cdot\xi_{x-z}.$$

Comparing this with a usual matrix, Γ might be depicted as an "infinite matrix" of the form

$$\Gamma = \begin{pmatrix} \ddots & & & & & & \\ 0 & \Gamma_+ & \Gamma_0 & \Gamma_- & 0 & & \\ & 0 & \Gamma_+ & \Gamma_0 & \Gamma_- & 0 & \\ & & 0 & \Gamma_+ & \Gamma_0 & \Gamma_- & 0 \\ & & & & & & \ddots \end{pmatrix}.$$

A single time step of the system is implemented by applying T to the respective observable. For Weyl operators, this is by the definition in (1) the same as applying Γ to the phase space argument ξ. Further iteration of the dynamics for t time steps is equivalent to a transformation $\Gamma_{t+1} = \Gamma\,\Gamma_t$. Due to the translational invariance, this is a convolution-style operation,

$$(\Gamma_{t+1})_{x,z} = (\Gamma_{t+1})_{x-z} = \sum_{y=-1,0,+1} \Gamma_{(x-z)-y}\cdot(\Gamma_t)_y. \tag{2}$$

These definitions result in a QCA in the sense of Sec. 3: The local observable algebra $\mathcal{A}_x$ is spanned by the Weyl operators on single lattice sites, $w_x(\xi_x)$ where $\xi_x \in \mathcal{R}^2$. The global Weyl operators $W(\xi)$ with $\xi \in \Xi$ span the quasi-local algebra $\mathcal{A}(\mathbb{Z})$. Since Γ is supposed to be a symplectic transformation and is translationally invariant, T as defined above is a translationally invariant automorphism of $\mathcal{A}(\mathbb{Z})$. The requirement of locality and finite propagation speed is met by the nearest neighbor coupling inherent in Γ. Indeed, if $W(\delta)$ is localized on a single site x, then $T\big(W(\delta)\big) = W(\Gamma\,\delta)$ is localized on the finite region $\{x-1, x, x+1\} \subset \mathbb{Z}$. The local rule is the restriction of T to the algebra of single-site observables.

In order to make T an automorphism of the quasi-local algebra, Γ has to be a symplectic transformation, $\sigma(\Gamma\,\xi, \Gamma\,\eta) = \sigma(\xi, \eta)$. We introduce the *symplectic transpose* Γ^+ of a transformation such that $\sigma(\Gamma\,\xi, \eta) = \sigma(\xi, \Gamma^+\,\eta)$ and $(\Gamma^+)_{x,z} = (\Gamma_{x,z})^+ = -\sigma_{\mathrm{s}} \cdot (\Gamma_{x,z})^{\mathrm{T}} \cdot \sigma_{\mathrm{s}}$. The condition on Γ is then equivalent to

$$\Gamma^+\,\Gamma = \mathbb{1}. \tag{3}$$

Another restriction on Γ stems from the fact that commuting observables have to commute even after the transformation, as mentioned in Sec. 3. Evaluating this property on single-site Weyl operators one site apart from each other imposes the condition $\Gamma_+{}^+\,\Gamma_- = 0$ which together with Eq. (3) immediately yields that $\mathrm{Det}\,\Gamma_0 = 1$. All this can easily be generalized to multiple modes per lattice site.

In the case of one mode per site, the above is equivalent to Γ_0 being any symplectic transformation and Γ_+ and Γ_- projectors into the same one-dimensional subspace, say the configuration coordinate. They are essentially determined by Γ_0 since the above conditions require that

$$\Gamma_- = f_-\,\Gamma_\pm \text{ and } \Gamma_+ = f_+\,\Gamma_\pm, \text{ where } \Gamma_\pm = \begin{pmatrix} (\Gamma_0)_{2,1} & (\Gamma_0)_{2,2} \\ 0 & 0 \end{pmatrix} \tag{4}$$

and f_-, f_+ are arbitrary, real-valued coupling parameters.

Examples: of Gaussian QCA include the free evolution, the "left-" or "right-shifter", a continued squeezing (see below) and the second quantization of the quantum random walk described in the introduction.

States: In the Heisenberg picture, a state is a positive, normalized, linear functional ω on the observables, yielding a positive expectation value $\omega(A)$

for positive observables A. In sufficiently well-behaved cases,[c] *i.e.*, for *normal* states, the functional can be represented by a density matrix ρ and the expectation value with an observable A is $\mathrm{tr}\mathbb{1}(\rho A)$. However, any state in our setting can be described by its characteristic function χ, the expectation value of all Weyl operators, $\chi(\xi) = \omega\big(W(\xi)\big)$. A Gaussian state is a state with a Gaussian characteristic function,[d]

$$\chi(\xi) = \exp\Big(-\tfrac{1}{4}\gamma(\xi,\xi) + \mathbb{1}\sum_{x\in\mathbb{Z}} d_x\,\xi_x\Big),$$

which is determined by its displacements $d_x \in \mathcal{R}^2$ and the correlation function $\gamma(\xi,\xi) = \sum_{x,z\in\mathbb{Z}} \xi_x^{\mathrm{T}} \cdot \gamma_{x,z} \cdot \xi_z$, where again the sum is finite. For translationally invariant states, the displacement must be independent of the position, $d_x \equiv d$, and can be interpreted as a global "amplitude". Also, the real 2×2 matrices $\gamma_{x,z}$ depend only on the distance between the two sites x and z, $\gamma_{x,z} = \gamma(x-z)$. Since as a correlation function γ has to be symmetric, require $\gamma(-x) = \big(\gamma(x)\big)^{\mathrm{T}}$.

In order to describe an admissible Gaussian quantum state, the correlation function has to fulfill the necessary and sufficient condition $\gamma + \mathbb{1}\,\sigma \geq 0$, meaning

$$\sum_{x,y\in\mathbb{Z}} \overline{\mu}_x^{\mathrm{T}} \cdot \big(\gamma(x-y) + \mathbb{1}\,\delta(x-y)\,\sigma_{\mathrm{s}}\big) \cdot \mu_y \geq 0 \tag{5}$$

for all $\mu\colon \mathbb{Z} \mapsto \mathcal{C}^2$ with $\mu \neq 0$ for only finitely many sites. Here $\delta(x) = 1$ for $x = 0$ while $\delta(x) = 0$ otherwise and the bar denotes complex conjugation.

Fourier Transform: Since the system obeys a translational invariance, we can simplify the expressions by turning to the Fourier transform of the phase space, *i.e.*, we decompose the phase space elements ξ into plane wave modes and consider the resulting weight functions $\hat{\xi}$ with values $\hat{\xi}(k) \in \mathcal{R}^2$:

$$\xi(x) = \frac{1}{2\pi}\int_{-\pi}^{\pi} \mathrm{d}k\,\hat{\xi}(k)\,\mathrm{e}^{+\mathbb{1}kx} \quad \text{and} \quad \hat{\xi}(k) = \sum_{x\in\mathbb{Z}} \xi(x)\,\mathrm{e}^{-\mathbb{1}kx}. \tag{6}$$

This casts the iteration relation (2) into an ordinary multiplication of matrices, $\hat{\Gamma}_t(k) = \big(\hat{\Gamma}(k)\big)^t$, where $\hat{\Gamma}(k) = \Gamma_0 + (f_+\mathrm{e}^{-\mathbb{1}k} + f_-\mathrm{e}^{+\mathbb{1}k})\,\Gamma_\pm$ is the Fourier transform of $\Gamma(x)$ according to (6).

---

[c] For a short discussion see, *e.g.*, Keyl *et al.*[9] and references therein.

[d] We use the Wigner-Weyl characteristic function which is related to the Wigner function by a Fourier transform. A Gaussian characteristic function is thus equivalent to a Gaussian Wigner function. For the theory of Gaussian states, see, *e.g.*, the book by Holevo.[10]

Fourier transform also simplifies the state condition (5) on γ. To properly define the transformed $\widehat{\gamma}(k)$, we restrict $\gamma(x)$ to be absolute-summable, $\sum_{x\in\mathbb{Z}}\|\gamma\| \leq \infty$. This condition excludes problematic correlation functions, *e.g.*, with singular portions but retains the important cases of product and clustering initial states. From a mathematical point of view, it requires $\gamma(x)$ to decrease faster than $1/|x|$ and makes $\widehat{\gamma}(k)$ continuous and square-summable. With this, the state condition on the correlation function reads in terms of Fourier transforms

$$\frac{1}{2\pi}\int_{-\pi}^{\pi}\mathrm{d}k\,\overline{\widehat{\mu}^{\mathrm{T}}(k)}\cdot\big(\widehat{\gamma}(k)+\mathbb{1}\,\sigma_{\mathrm{s}}\big)\cdot\widehat{\mu}(k)\geq 0. \tag{7}$$

This is equivalent to the condition on $2{\times}2$ matrices that $\widehat{\gamma}(k)+\mathbb{1}\,\sigma_{\mathrm{s}}\geq 0$ for all $k\in[-\pi,\pi]$: if this condition holds for all k, then the l. h. s. of (7) is indeed positive semi-definite; if on the other hand $\widehat{\gamma}(k_0)+\mathbb{1}\,\sigma_{\mathrm{s}}$ is not positive semi-definite for some k_0, then the l. h. s. of (7) can be made negative by choosing an appropriate $\widehat{\mu}(k)$, *e.g.*, the Fourier transform of a flat Gaussian centered around k_0 and restricted to a finite support. By an argument along the same lines it is clear that if $\widehat{\gamma}(k_0)+\mathbb{1}\,\sigma_{\mathrm{s}}$ is strictly positive for some k_0, then γ determines the characteristic function of a pure Gaussian state and additional Gaussian noise and therefore corresponds to a mixed state. The state condition on the bilinear form γ over the infinite lattice is thus transformed into a condition of the same form on finite matrices under Fourier transform. The above is summarized in the following

Lemma: A function γ which maps $x\in\mathbb{Z}$ to real $2{\times}2$ matrices and obeys $\sum_{x\in\mathbb{Z}}\|\gamma\|\leq\infty$ and $\gamma(-x)=\big(\gamma(x)\big)^{\mathrm{T}}$

(1) defines a translationally invariant Gaussian state on the linear chain labeled by $\mathbb{Z}$ if and only if the Fourier transform $\widehat{\gamma}(k)$ fulfills $\widehat{\gamma}(k)+\mathbb{1}\,\sigma_{\mathrm{s}}\geq 0$ for all $k\in[-\pi,\pi]$ and

(2) corresponds to a pure Gaussian state if and only if this inequality is not strict for any k.

During time evolution of the system, the correlation function γ changes according to the symplectic transformation Γ of the phase space argument in (1) as

$$\begin{aligned}\gamma_t(x)&=\sum_{y,z\in\mathbb{Z}}\big(\Gamma_t(y)\big)^{\mathrm{T}}\cdot\gamma_0(x+y-z)\cdot\Gamma_t(z)\quad\text{or}\\ \widehat{\gamma}_t(k)&=\overline{\hat{\Gamma}_t^{\mathrm{T}}(k)}\cdot\widehat{\gamma}_0(k)\cdot\hat{\Gamma}_t(k),\end{aligned}\tag{8}$$

where γ_0 denotes the correlation function of the initial state.

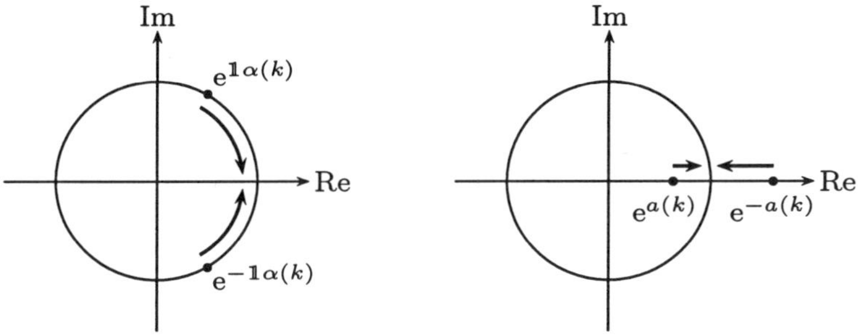

Fig. 2. Depicting the eigenvalues of $\hat{\Gamma}(k)$: for small coupling parameter $|f| < f_{\text{crit}} = |\tan(\phi/2)/\cos(k)|$ the two eigenvalues are complex phases and conjugated to each other (left figure); for large coupling $|f| > f_{\text{crit}}$, the eigenvalues are real and inverse to each other (right figure). All eigenvalues meet at a value of 1 for $f \to f_{\text{crit}}$.

Example: As an instance, consider the above system in a coherent product state described by the correlation function $\gamma_0(0) = \mathbb{1}$ and $\gamma_0(x) = 0$ otherwise. The resulting Fourier transform is $\widehat{\gamma}_0(k) = \mathbb{1}$. Let the interaction be determined by the on-site rotation $\Gamma_0 = \left(\begin{smallmatrix} \cos\phi & \sin\phi \\ -\sin\phi & \cos\phi \end{smallmatrix}\right)$ and equal coupling parameters $f_+ = f_- = f$. This yields for the Fourier transform $\widehat{\Gamma}$:

$$\hat{\Gamma}(k) = \Gamma_0 + 2\,f\cos(k)\,\Gamma_\pm.$$

Since $\Gamma_\pm$ contains a row of Γ_0, we get $\operatorname{Det}\hat{\Gamma}(k) = \operatorname{Det}\Gamma_0 = 1$. $\hat{\Gamma}(k)$ thus induces a symplectic transformation on every single mode k. The value of the coupling parameter f determines whether the eigenvalues of $\hat{\Gamma}(k)$ are real and inverse or complex and conjugate to each other. This can be seen from the trace as the sum of both eigenvalues,

$$\begin{aligned} \mathrm{tr}\mathbf{1}\hat{\Gamma}(k) &= e^{\mathbf{1}\alpha(k)} + e^{-\mathbf{1}\alpha(k)} = \\ &\qquad 2\cos\alpha(k) = 2\cos\phi + 2\,f\cos(k)\,\sin\phi. \end{aligned} \tag{9}$$

If $|\mathrm{tr}\mathbf{1}\hat{\Gamma}(k)| \leq 2$, $\alpha(k)$ is real-valued, $|e^{\mathbf{1}\alpha(k)}| = 1$ and $\hat{\Gamma}(k)$ is rotation-like; otherwise $\alpha(k)$ is purely imaginary-valued, the eigenvalues are real and $\hat{\Gamma}(k)$ is of squeezing type. The critical value of f between the two regimes, *i.e.*, for eigenvalues of 1, is given by $|f| = |\tan(\phi/2)/\cos(k)|$ with eigenvalues complex for $|f|$ below this value and real eigenvalues for larger $|f|$. These relations are illustrated in Fig. 2, for $\alpha(k)$ confer Fig. 3.

If some $\hat{\Gamma}(k_0)$ had real eigenvalues larger than 1, the resulting dynamics would constantly squeeze the respective mode, transforming any input state over time into an "infinitely squeezed state".[9] The limit state of such dynamics is highly singular. For example, the probability for any oscillator in the chain to be finitely excited is zero. We thus concentrate on the case of small couplings $|f| < |\tan(\phi/2)|$.

The following argument builds upon the spectral decomposition of $\hat{\Gamma}$ written as $\hat{\Gamma}(k) = \mathrm{e}^{\mathbb{1}\alpha(k)}\, P_k + \mathrm{e}^{-\mathbb{1}\alpha(k)}\, \overline{P_k}$ where $\alpha(k) \in \mathcal{R}$ and P_k, $\overline{P_k}$ are (non-orthogonal) projections on disjoint subspaces, hence $P_k \cdot \overline{P_k} = 0$. Using that $P_k + \overline{P_k} = \mathbb{1}$ and $\widehat{\Gamma}(k)$ is real, it follows that $\overline{P_k}$ is complex conjugate to P_k, where

$$P_k = \frac{1}{2}\,\mathbb{1} + \frac{\mathbb{1}}{2\sin\alpha(k)}\big(\cos\alpha(k)\,\mathbb{1} - \Gamma(k)\big). \tag{10}$$

The time-dependent correlation function $\gamma_t(x)$ is obtained by inverse Fourier transform from (8):

$$\begin{aligned}
\gamma_t(x) &= \frac{1}{2\pi}\int_{-\pi}^{\pi} \mathrm{d}k\, \mathrm{e}^{\mathbb{1}kx}\, \overline{\hat{\Gamma}_t^{\mathrm{T}}(k)} \cdot \widehat{\gamma}_0(k) \cdot \hat{\Gamma}_t(k) \\
&= \frac{1}{2\pi}\int_{-\pi}^{\pi} \mathrm{d}k\, \mathrm{e}^{\mathbb{1}kx} \left(\mathrm{e}^{2\mathbb{1}t\alpha(k)}\, P_k^{\mathrm{T}} \cdot \widehat{\gamma}_0(k) \cdot P_k + \mathrm{e}^{-2\mathbb{1}t\alpha(k)}\, \overline{P_k}^{\mathrm{T}} \cdot \widehat{\gamma}_0(k) \cdot \overline{P_k}\right) \\
&\quad + \frac{1}{2\pi}\int_{-\pi}^{\pi} \mathrm{d}k\, \mathrm{e}^{\mathbb{1}kx} \left(P_k^{\mathrm{T}} \cdot \widehat{\gamma}_0(k) \cdot \overline{P_k} + \overline{P_k}^{\mathrm{T}} \cdot \widehat{\gamma}_0(k) \cdot P_k\right).
\end{aligned} \tag{11}$$

In the limit of large time t, the oscillating terms vanish and the correlation function converges: Starting from a product state (or any clustering state), $\widehat{\gamma}_0(k)$ is continuous; as $\hat{\Gamma}(k)$ is well-behaved, so is the whole integrand. Hence, the oscillatory part of the integral vanishes as $t \to \infty$, where the leading asymptotics is governed by the stationary phase in t and $\alpha(k)$. This argument would be spoiled by any $\hat{\Gamma}(k_0)$ with real eigenvalues, resulting in continued squeezing of the respective mode. This explains the restriction to small coupling parameter f.

It is remarkable that while the initial state is a pure product state and the dynamics is reversible both for the whole system as well as for every mode, the system exhibits converges. This is a sign of emerging *irreversibility* as the whole range of intermediate states is mapped to the same limit state.

From the last part of the integral in (11) it follows that the limit state is determined by a single parameter for each mode:

$$\begin{aligned}
\gamma_\infty(x) &= \frac{1}{2\pi}\int_{-\pi}^{\pi} \mathrm{d}k\, \mathrm{e}^{\mathbb{1}kx} \left(P_k^{\mathrm{T}} \cdot \widehat{\gamma}_0(k) \cdot \overline{P_k} + \left(P_k^{\mathrm{T}} \cdot \widehat{\gamma}_0(-k) \cdot \overline{P_k}\right)^{\mathrm{T}}\right) \\
&= \frac{1}{2\pi}\int_{-\pi}^{\pi} \mathrm{d}k\, \mathrm{e}^{\mathbb{1}kx}\, c(k) \left(P_k^{\mathrm{T}} \cdot \overline{P_k} + \overline{P_k}^{\mathrm{T}} \cdot P_k\right)
\end{aligned}$$

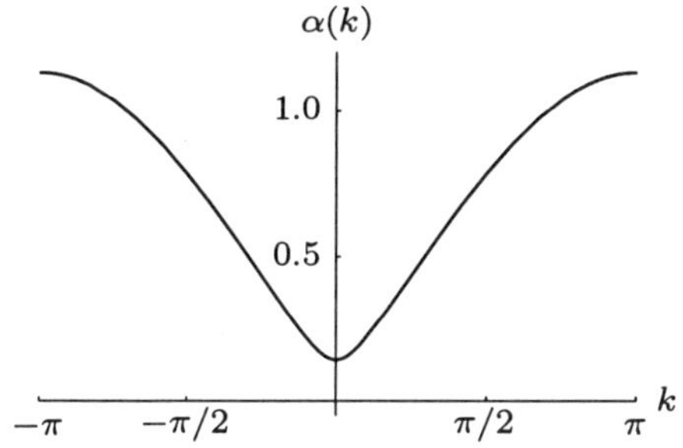

Fig. 3. Plot of $\alpha(k) = \arccos(\cos\phi + f\cos(k)\sin\phi)$ according to Eq. (9), for $f = 0.4$ and $\phi = \frac{\pi}{4}$.

since in our example $\widehat{\gamma}_0(-k) = \widehat{\gamma}_0(k)$ and the projection character of P_k, $\overline{P_k}$ effectively reduces $\widehat{\gamma}_0(k)$ to a single matrix element $c(k)$. This is true for all reflection symmetric states, *i.e.*, states with $\gamma(x) = \gamma(-x)$ and thus $\widehat{\gamma}_0(k) = \widehat{\gamma}_0(-k)$. Reversing the argument, we can describe any stationary, reflection symmetric state by a unique pure such state and a mode-wise "temperature" parameter. Casting the expression into a different form gives rise to the following

Theorem: All stationary, translationally invariant and reflection symmetric Gaussian states of the above system are *thermal equilibrium states*, described by their Fourier transformed correlation function $\widehat{\gamma}_{\text{stat}}(k) = g(k)\,\widehat{\varepsilon}(k)$ comprising

- ▷ the correlation function of a pure state with Fourier transform $\widehat{\varepsilon}(k) = \mathbb{1}\,\sigma_{\text{s}}(P_k - \overline{P_k})$ and
- ▷ a function $g(k)$ of mode-wise "temperature" parameters with $g(k) = g(-k) \geq 1$.

The Fourier transformed correlation function $\widehat{\varepsilon}(k)$ is indeed symmetric under interchange of k and $-k$ since $\widehat{\Gamma}(-k) = \widehat{\Gamma}(k)$ as well as $\alpha(-k) = \alpha(k)$ and thus $P_{-k} = P_k$, $\overline{P_{-k}} = \overline{P_k}$ from (10). Noting that $\widehat{\Gamma}(k)$ commutes with P_k, $\overline{P_k}$ and as a symplectic transformation preserves σ_{s}, $\widehat{\varepsilon}(k)$ is immediately shown to be invariant under the dynamics, *i.e.*, it does not change under the action of Γ:

$$\hat{\Gamma}^{\mathrm{T}}(k)\cdot\widehat{\varepsilon}(k)\cdot\hat{\Gamma}(k) = \mathbb{1}\,\hat{\Gamma}^{\mathrm{T}}(k)\cdot\sigma_{\text{s}}\cdot\hat{\Gamma}(k)\cdot(P_k - \overline{P_k}) = \mathbb{1}\,\sigma_{\text{s}}\cdot(P_k - \overline{P_k}) = \widehat{\varepsilon}(k).$$

$\widehat{\varepsilon}(k)$ also fulfills the state condition $\widehat{\varepsilon}(k) + \mathbb{1}\,\sigma_{\text{s}} \geq 0$ by the following argument: We have $\widehat{\varepsilon}(k) + \mathbb{1}\,\sigma_{\text{s}} = 2\mathbb{1}\,\sigma_{\text{s}}\cdot P_k$. Since $\hat{\Gamma}(k)$ is a symplectic transformation, expanding the identity $\overline{P_k}^{*}\cdot(\sigma_{\text{s}}\cdot P_k) = \overline{P_k}^{*}\cdot(\widehat{\Gamma}^{\mathrm{T}}(k)\,\sigma_{\text{s}}\,\widehat{\Gamma}(k)\cdot P_k)$

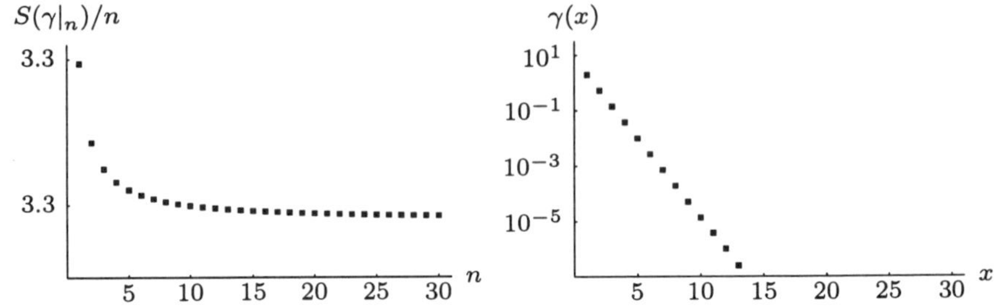

Fig. 4. Numerical results for the limit state of the example system: (left) entropy density $S(\gamma|_n)/n$ of a sub-chain of length n and (right) correlation function $\gamma_\infty(x)$ for $\phi = \frac{\pi}{2}$ and $f = 0.5$.

implies $\overline{P_k}^* \cdot \sigma_s \cdot P_k = 0$ and $\mathbb{1}\,\sigma_s \cdot P_k = P_k^* \cdot \mathbb{1}\,\sigma_s \cdot P_k$. The r. h. s. of this equation is a real multiple of a projector and hence renders $\mathbb{1}\,\sigma_s \cdot P_k$ either positive- or negative-semidefinite. In the latter case, replace $\widehat{\varepsilon}(k)$ by $-\widehat{\varepsilon}(k)$. Finally, $\widehat{\varepsilon}(k)$ corresponds to a pure state since $\left(\sigma_s \cdot \widehat{\varepsilon}(k)\right)^2 = -\mathbb{1}$. In addition, $\widehat{\varepsilon}(k)$ can be modified mode-wise by a factor $g(k) = g(-k) \geq 1$ without affecting the above relations, expect for the pure state condition. Hence, $g(k)$ plays the role of a temperature. This completes the theorem.

As another interesting aspect, numerical results suggest that the entanglement between a sub-chain and the rest of the system is long-ranged. However, the correlation functions decreases rapidly with distance between lattice sites, corresponding to short-range correlations. This is illustrated in Fig. 4: The entropy density $S(\gamma|_n)/n$ of a sub-chain of length n in the limit state obtained from the initial pure product state above does not decreases for large n; this suggests that part of the entanglement is "radiated to infinity". In contrast, the corresponding correlation function decays rapidly with distance x. In fact, since $\widehat{\gamma}_0(k)$ and $\hat{\Gamma}(k)$ are analytical, all $\gamma_t(x)$ and $\gamma_\infty(x)$ obtained by inverse Fourier transform decrease faster than any polynomial with distance x between any two lattice sites.

5. Summary

In this chapter, we have introduced the concept of Gaussian quantum cellular automata. Quantum cellular automata (QCA) provide a new computational model on lattice systems. Their Gaussian instances arise for example in the second quantization of such systems. After a short digest of the issues connected to general QCAs, we have presented methods suitable for dealing with Gaussian QCAs. This concerns an infinite number of modes as well as translational invariance. For illustration, we choose the example of a one-dimensional chain of harmonic oscillators with quasi-free dynamics and an initial coherent product state.

This systems exhibits remarkable aspects of irreversibility: Although it evolves from a pure product state under a reversible dynamics, the correlation function converges. The invariant states of the system with reflection symmetry are thermal equilibrium states. Moreover, numerical results suggest that the generated entanglement is long-ranged while the correlations decrease on a short scale.

References

1. R. Feynman, *Simulating physics with computers*, Int. J. Theor. Phys. **21**, 467 (1982); reprinted in: *Feynman and Computation – Exploring the Limits of Computers*, A. J. G. Hey (Ed.), Perseus Books, Reading/Mass., 1999.
2. O. Mandel, M. Greiner, A. Widera, T. Rom, T. W. Hänsch, and I. Bloch, *Coherent transport of neutral atoms in spin-dependent optical lattice potentials*, Phys. Rev. Lett. **91**, 010407 (2003).
3. R. Dumke, M. Volk, T. Muehter, F. B. J. Buchkremer, G. Birkel, and W. Ertmer, *Microoptical Realization of Arrays of Selectively Addressable Dipole Traps: A Scalable Configuration for Quantum Computation with Atomic Qubits*, Phys. Rev. Lett. **89**, 097903 (2002).
4. K. G. H. Vollbrecht, E. Solano, and J. I. Cirac, *Ensemble quantum computation with atoms in periodic potentials*, Phys. Rev. Lett. **93**, 220502 (2004); K. G. H. Vollbrecht and J. I. Cirac, *Reversible universal quantum computation within translation invariant systems*, quant-ph/0502143 (2005).
5. J. Kempe, *Quantum random walks: an introductory overview*, Contemp. Phys. **44**, 307 (2003).
6. E. R. Berlekamp, J. H. Conway, and R. K. Guy, *Winning Ways for your mathematical plays*, Academic Press, London and New York 1982.
7. B. Chopard and M. Droz, *Cellular Automata Modeling of Physical Systems*, Cambridge University Press, Cambridge 1998.
8. B. Schumacher and R. F. Werner, *Reversible quantum cellular automata*, quant-ph/0405174 (2004).
9. M. Keyl, D. Schlingemann, and R. F. Werner, *Infinitely entangled states*, Quant. Inf. Comp. **3**, 281 (2003).
10. A. S. Holevo, *Probabilistic and statistical aspects of quantum theory*, North-Holland, Amsterdam 1982.

Chapter 6

Distillation of Continuous-Variable Entanglement

Jaromír Fiurášek, Ladislav Mišta and Radim Filip
Department of Optics, Palacký University
17. listopadu 50, 77200 Olomouc, Czech Republic

email: fiurasek@optics.upol.cz

We review the protocols for distillation and concentration of continuous-variable entanglement. After proving that it is impossible to distill the entanglement of Gaussian states with Gaussian operations, we describe entanglement concentration protocols that involve either cross-Kerr interaction or photon number measurements. If combined with an iterative gaussification procedure, these techniques would allow to implement an entanglement distillation protocol for continuous variables.

1. Introduction

The entanglement lies at the heart of quantum mechanics and it is an essential resource for many quantum information processing tasks such as quantum cryptography, quantum teleportation or quantum dense coding. The advantage of quantum continuous variables is that continuous-variable (CV) entangled states of light can be relatively easily generated in a deterministic way in optical parametric oscillators or amplifiers that can produce entangled two-mode squeezed states. Moreover, the Bell measurement on two modes, required for tasks such as teleportation or dense coding, can be performed deterministically with the help of a balanced beam splitter and two homodyne detectors, in contrast to single-photon approaches where only a partial Bell measurement can be performed without using extra ancillary photons and complicated multiphoton interferometers.

All these features combined with very high detection efficiency and speed of balanced homodyne detectors render approaches based on continuous variables a very promising candidate for quantum communications. However, secure long-distance quantum communication has to overcome the problems with losses and decoherence which grow exponentially with distance. The way to efficiently combat these effects is to employ the quantum repeaters[1] that combine quantum memory,[2] entanglement distillation[3] and quantum teleportation[4,5] to distribute almost pure entangled states over long distances in such a way that the total number of operations grows only polynomially with the distance.

However, it has been proved that the crucial element of this scheme, namely the entanglement distillation by means of local operations and classical communication cannot be performed if only Gaussian entangled states are available and only Gaussian operations can be performed.[6–8] The Gaussian operations on light beams can be implemented with the help of passive linear interferometers, squeezers, homodyne detectors and auxiliary modes prepared in vacuum states.

In this chapter, we will first present the proof of the no-go theorem for entanglement distillation of Gaussian states with Gaussian operations and then discuss the possible ways to circumvent this limitation. One way to go beyond the class of Gaussian operations is to explore the Kerr interaction. We will describe two techniques for entanglement concentration which rely on cross-Kerr interaction, one proposed by Duan *et al.*[9,10] and the other suggested by Fiurášek *et al.*[11] While being of fundamental interest, these methods would be very difficult to implement experimentally as the required Kerr nonlinearity is very high. We will therefore turn our attention to a much more practical technique which employs a projection on a single photon state. This has been first suggested by Opatrný *et al.*[12] and very recently a de-Gaussification of single-mode squeezed vacuum by conditional subtraction of a single photon has been demonstrated experimentally,[13] which clearly illustrates the great potential of this approach. The entanglement distillation is a recursive procedure which can extract a few highly entangled states from many copies of weakly entangled states. We will sketch a simple method proposed by Browne *et al.* to distill entanglement of continuous vatriables.[14,15] Their scheme converts the initial (non-Gaussian) entangled states into a final Gaussian entangled state which can be highly entangled. Finally, we conclude with brief Conclusions.

2. Entanglement Distillation of Gaussian States with Gaussian Operations is Impossible

In this section we show that it is impossible to distill the entanglement of Gaussian states using only local Gaussian operations and classical communication.[6–8] This no-go theorem[6] reveals that in order to distill the continuous-variable entanglement one has to go beyond the class of Gaussian states and Gaussian operations. These options will be discussed in the subsequent sections.

We begin by briefly reviewing the formalism of Gaussian states and Gaussian operations. The Gaussian states are fully characterized by the first and second moments of the quadrature operators x_j, p_j which satisfy the canonical commutation relations $[x_j, p_k] = i\delta_{jk}$. We define the (column) vector of quadrature operators $R = (x_1, p_1, \ldots, x_N, p_N)^T$, the vector of first moments $d = \langle R \rangle$ and the covariance matrix $\gamma_{jk} = \langle \Delta R_j \Delta R_k + \Delta R_k \Delta R_j \rangle$, where $\Delta R_j = R_j - d_j$. The Wigner function of the N-mode Gaussian state is given by

$$W(r) = \frac{1}{\pi^N \sqrt{\operatorname{Det} \gamma}} \exp[-(r-d)^T \gamma^{-1} (r-d)], \tag{1}$$

where r is the coordinate vector in phase space, and its Fourier transform is the characteristic function $C(\xi) = \exp(-i\xi^T d - \frac{1}{4}\xi^T \gamma \xi)$. The covariance matrix γ must satisfy the generalized Heisenberg inequality $\gamma + iJ^{\oplus N} \geq 0$, where

$$J = \begin{pmatrix} 0 & 1 \\ -1 & 0 \end{pmatrix} \tag{2}$$

is the symplectic form, and the matrix $iJ^{\oplus N}$ comprises the commutators $[R_j, R_k]$.

Gaussian operations are defined as the operations which preserve the Gaussian shape of Wigner functions, *i.e.* they map Gaussian states onto Gaussian states. Deterministic Gaussian operations are fully characterized by two matrices S and G and the mapping is linear on the level of mean values and covariance matrices,

$$\gamma_{\text{out}} = S\gamma_{\text{in}} S^T + G, \qquad d_{\text{out}} = S d_{\text{in}}. \tag{3}$$

Physical transformations must be completely positive (CP) maps, which imposes the constraint[16] $G + iJ^{\oplus M} - iSJ^{\oplus N} S^T \geq 0$, where N and M is the number of input and output modes, respectively.

The entanglement distillation is a probabilistic operation so we have to employ the more general formalism of trace-decreasing (*i.e.* probabilistic) Gaussian completely positive maps. A full description of this class of operations is possible with the help of Jamiolkowski-Choi isomorphism,[17,18] according to which every linear CP map $\mathcal{E}$ that maps operators on $\mathcal{H}_{\text{in}}$

onto operators on $\mathcal{H}_{\text{out}}$ is isomorphic to an operator E on the tensor-product Hilbert space $\mathcal{H}_{\text{in}} \otimes \mathcal{H}_{\text{out}}$. The operator E can be obtained by preparing a maximally entangled state on $\mathcal{H}_{\text{in}}^{\otimes 2}$ and sending one part of this state through the quantum channel $\mathcal{E}$. The input-output transformation $\rho_{\text{out}} = \mathcal{E}(\rho_{\text{in}})$ can be then rewritten as $\rho_{\text{out}} = \text{Tr}_{\text{in}}[E\rho_{\text{in}}^T \otimes I_{\text{out}}]$, where T denotes transposition, Tr_{in} indicates partial trace over the input Hilbert space and I is an identity operator. When dealing with continuous-variable states, it is convenient to use the Wigner representation of the operator E, $W_E(r_{\text{in}}, r_{\text{out}})$ and express the CP map in terms of Wigner functions,

$$W_{\text{out}}(r_{\text{out}}) = (2\pi)^N \int_{-\infty}^{\infty} \cdots \int_{-\infty}^{\infty} W_E(r_{\text{in}}, r_{\text{out}}) W_{\text{in}}(\Lambda r_{\text{in}}) d^{2N} r_{\text{in}}, \tag{4}$$

where the integration is carried over the whole input phase space. The diagonal matrix $\Lambda = \text{diag}(1, -1, \ldots, 1, -1)$ represents the transposition in the phase space because the Wigner function of the transposed density matrix ρ^T is obtained by changing the sign of all p quadratures.

It can be easily seen that the transformation (4) preserves the Gaussian shape of $W_{\text{in}}(r_{\text{in}})$ for any admissible $W_{\text{in}}(r_{\text{in}})$ if and only if $W_E(r_{\text{in}}, r_{\text{out}})$ is itself a Gaussian function. Thus the Gaussian CP maps are isomorphic to (generally mixed) $N + M$-mode Gaussian states. The coherent displacements (first moments), can be always set to zero by local displacement operations and are irrelevant as far as the entanglement properties are concerned. Without loss of generality, we can therefore consider only Gaussian maps $\mathcal{E}_G$ isomorphic to Gaussian states with vanishing first moments. All these maps are then fully specified by the covariance matrix Γ. It is convenient to divide this matrix into four sub-matrices with respect to in-out splitting,

$$\Gamma = \begin{pmatrix} A & C \\ C^T & B \end{pmatrix}, \tag{5}$$

where A is the covariance matrix of the "input" modes, B is the covariance matrix of the "output" modes and C comprises the in-out correlations. We can obtain the generalization of the covariance-matrix transformation formula (3) to the probabilistic Gaussian operations if we evaluate the integral (4) for Gaussian input state. It is convenient to Fourier transform the formula (4) and work with the characteristic functions instead of Wigner functions, where

$$C_E(\xi_{\text{in}}, \xi_{\text{out}}) = \exp\left[-\frac{1}{4}(\xi_{\text{in}}^T A \xi_{\text{in}} + \xi_{\text{out}}^T B \xi_{\text{out}} + \xi_{\text{in}}^T C \xi_{\text{out}} + \xi_{\text{out}}^T C^T \xi_{\text{in}})\right],$$

and the map (4) expressed in terms of characteristic functions reads

$$C_{\text{out}}(\xi_{\text{out}}) = \frac{1}{(2\pi)^N} \int_{-\infty}^{\infty} \cdots \int_{-\infty}^{\infty} C_E(\xi_{\text{in}}, \xi_{\text{out}}) C_{\text{in}}(-\Lambda \xi_{\text{in}}) d^{2N} \xi_{\text{in}}. \tag{6}$$

After some algebra we arrive at,

$$\gamma_{\text{out}} = B - C^T \frac{1}{A + \Lambda^T \gamma_{\text{in}} \Lambda} C, \tag{7}$$

and

$$d_{\text{out}} = C^T \frac{1}{A + \Lambda^T \gamma_{\text{in}} \Lambda} \Lambda d_{\text{in}}. \tag{8}$$

Note that in contrast to deterministic Gaussian operations, the transformation law (7) for the covariance matrix is non-linear.

Entanglement distillation has to be performed with local operations and classical communication (LOCC). A necessary condition that the map $\mathcal{E}$ is LOCC is that the operator E is separable with respect to Alice|Bob splitting. Indeed, if this were not the case, then Alice and Bob could prepare an entangled state by preparing locally maximally entangled states and applying the map $\mathcal{E}$ to parts of their entangled states. The separability criterion for Gaussian operators that we shall use in what follows was derived by Werner and Wolf[19], who showed that a Gaussian state with covariance matrix γ_{AB} is separable if and only if there exist covariance matrices $\tilde{\gamma}_A$ and $\tilde{\gamma}_B$ of the subsystems A and B, respectively, such that

$$\gamma_{AB} \geq \tilde{\gamma}_A \oplus \tilde{\gamma}_B \tag{9}$$

holds. Note that the matrices satisfy the generalized Heisenberg inequality $\tilde{\gamma}_A + iJ^{\oplus N_A} \geq 0$, $\tilde{\gamma}_B + iJ^{\oplus N_B} \geq 0$, where N_A and N_B denote the number of modes on Alice's and Bob's sides, respectively. It follows that the necessary condition that a Gaussian CP map acting on two subsystems A and B and characterized by covariance matrix Γ_{AB} is LOCC is given by $\Gamma_{AB} \geq \tilde{\Gamma}_A \oplus \tilde{\Gamma}_B$.

In order to prove the impossibility of distillation of Gaussian entangled states with Gaussian operations it is necessary to introduce a suitable measure of entanglement. Such a measure can be constructed starting from the separability criterion (9). This inequality is not satisfied for entangled states, but one can consider a generalized inequality

$$\gamma_{AB} \geq s\tilde{\gamma}_A \oplus \tilde{\gamma}_B \tag{10}$$

and ask what is the maximum s such that there exist covariance matrices $\tilde{\gamma}_A$ and $\tilde{\gamma}_B$ satisfying (10). This maximum s_m exhibits some properties of measure of the entanglement of Gaussian states and it is intuitively clear that the smaller the s_m the stronger is the entanglement. In particular, let us consider pure two-mode squeezed vacuum state with covariance matrix

$$\gamma_{AB} = \begin{pmatrix} \cosh(2r) & 0 & \sinh(2r) & 0 \\ 0 & \cosh(2r) & 0 & -\sinh(2r) \\ \sinh(2r) & 0 & \cosh(2r) & 0 \\ 0 & -\sinh(2r) & 0 & \cosh(2r) \end{pmatrix}, \tag{11}$$

where r is the squeezing parameter. It follows from the symmetry of the state that the optimal $\tilde{\gamma}_{A,B}$ leading to maximum s for which inequality (10) is still satisfied are covariance matrices of vacuum, $\tilde{\gamma}_A = \tilde{\gamma}_B = \mathrm{diag}(1,1)$. A straightforward calculation yields $s_m = \exp(-2r)$ and s_m exponentially decreases with increasing squeezing. Since the entropy of entanglement is a monotonically increasing function of r, the s_m monotonically decreases as the entanglement grows and $s_m \to 0$ in the limit of infinite squeezing.

We will now prove that s_m cannot be decreased by means of LOCC Gaussian operations which shows that it is impossible to distill the entanglement of Gaussian states using only Gaussian operations. First observe that it suffices to consider only LOCC Gaussian maps which satisfy $\Gamma_{AB,0} = \tilde{\Gamma}_A \oplus \tilde{\Gamma}_B$. This corresponds to the situation when Alice and Bob independently apply some local Gaussian filtering operations and then only exchange the information whether their operations succeeded. Since for any LOCC map $\Gamma_{AB} \geq \tilde{\Gamma}_A \oplus \tilde{\Gamma}_B$, we can obtain the covariance matrix Γ_{AB} from the covariance matrix $\tilde{\Gamma}_A \oplus \tilde{\Gamma}_B$ by applying correlated Gaussian distributed displacements to the subsystems A and B. Alice and Bob can keep the record of these displacements and the resulting Gaussian CP map is a Gaussian mixture of maps with covariance matrix $\Gamma_{AB,0}$ and varying displacements. Note that these displacements do not change the transformation law for the covariance matrix, they will only modify the displacement of the output state. Consequently, the resulting state will be a Gaussian convex mixture of states with covariance matrix $\gamma_{AB,0,\mathrm{out}}$ corresponding to the map with matrix $\Gamma_{AB,0}$, and varying displacements. This mixture of states cannot exhibit better entanglement properties than each single state in the mixture, hence it suffices to restrict ourselves to the maps of the form $\Gamma_{AB,0}$.

Let us further decompose each matrix $\tilde{\Gamma}_{A,B}$ with respect to the input and output modes on Alice's, and Bob's sides,

$$\tilde{\Gamma}_A = \begin{pmatrix} A_A & C_A \\ C_A^T & B_A \end{pmatrix}, \qquad \tilde{\Gamma}_B = \begin{pmatrix} A_B & C_B \\ C_B^T & B_B \end{pmatrix}. \tag{12}$$

It can be shown that the matrix transformation (7) preserves the inequality $\gamma_{\text{in},1} \geq \gamma_{\text{in},2}$. On applying the transformation to the inequality $\gamma_{AB,\text{in}} \geq s_m\tilde{\gamma}_A \oplus \tilde{\gamma}_B$, with the map $\Gamma_{AB,0}$, we find that

$$\gamma_{AB,\text{out}} \geq \tilde{\gamma}_{A,\text{out}} \oplus \tilde{\gamma}_{B,\text{out}}, \tag{13}$$

where

$$\tilde{\gamma}_{A,\text{out}} = B_A - C_A^T \frac{1}{A_A + s_m\Lambda\tilde{\gamma}_A\Lambda} C_A,$$
$$\tilde{\gamma}_{B,\text{out}} = B_B - C_B^T \frac{1}{A_B + s_m\Lambda\tilde{\gamma}_B\Lambda} C_B. \tag{14}$$

Let us consider the first term $\tilde{\gamma}_{A,\text{out}}$, the second term can be dealt with similarly. First note that from $\tilde{\gamma}_A + iJ^{\oplus N_A} \geq 0$ we get

$$\tilde{\gamma}_{A,\text{out}} \geq B_A - C_A^T \frac{1}{A_A + s_m iJ^{\oplus N_A}} C_A, \tag{15}$$

where we took into account that $\Lambda J^{\oplus N_A}\Lambda = -J^{\oplus N_A}$. Next we exploit the complete positivity of the Gaussian map, namely, that $\tilde{\Gamma}_A$ must satisfy $\tilde{\Gamma}_A + iJ^{\oplus(N_A+M_A)} \geq 0$. If this inequality holds then also

$$\tilde{\Gamma}_A + s_m iJ^{\oplus(N_A+M_A)} \geq 0 \tag{16}$$

is satisfied for any $0 \leq s_m \leq 1$. The inequality (16) implies that

$$B_A + s_m iJ^{\oplus M_A} - C_A^T \frac{1}{A_A + s_m iJ^{\oplus N_A}} C_A \geq 0. \tag{17}$$

If we combine together the inequalities (15) and (17) we finally arrive at

$$\tilde{\gamma}_{A,\text{out}} \geq -s_m iJ^{\oplus M_A}, \tag{18}$$

and, consequently, $\bar{\gamma}_A \equiv \tilde{\gamma}_{A,\text{out}}/s_m$ is a covariance matrix which satisfies the generalized Heisenberg inequality. Going back to the inequality (13), we find that

$$\gamma_{AB,\text{out}} \geq s_m\bar{\gamma}_A \oplus \bar{\gamma}_B, \tag{19}$$

which concludes the proof that s_m cannot be decreased under LOCC Gaussian operations.

3. Entanglement Concentration Based on Cross-Kerr Effect

Experimental implementations of quantum information protocols with CVs most widely use as an entanglement resource the two-mode squeezed vacuum state

$$|\psi_{\rm in}(\lambda)\rangle_{AB} = \sum_{n=0}^{\infty} c_n |n,n\rangle_{AB}, \qquad c_n = \sqrt{1-\lambda^2}\lambda^n, \tag{20}$$

where $\lambda = \tanh(r)$ and r is the squeezing parameter. In practice, the state is generated in the process of the spontaneous parametric down-conversion in the nondegenerate optical parametric amplifier[20] or by superimposing on a balanced beam splitter two modes equally squeezed in complementary quadratures.[5] The state (20) is a two-mode Gaussian state with the covariance matrix (11). Since it is a pure state its entanglement can be uniquely quantified by the entropy of entanglement that is for any state of the form $|\phi\rangle = \sum_{n=0}^{\infty} a_n|n,n\rangle_{AB}$ defined as the Shannon entropy of the squares of the Schmidt coefficients a_n

$$S_E(|\phi\rangle) = -\sum_{n=0}^{\infty} a_n^2 \log_2 a_n^2. \tag{21}$$

The entropy of entanglement for the state (20) then can be calculated in the form $S_E(|\psi_{\rm in}(\lambda)\rangle_{AB}) = \cosh^2(r)\log_2\left[\cosh^2(r)\right] - \sinh^2(r)\log_2\left[\sinh^2(r)\right]$. Since for $r > 0$ the entropy $S_E(|\psi_{\rm in}(\lambda)\rangle_{AB})$ is a positive monotonically increasing function of r the state (20) is an entangled Gaussian state whose entanglement increases with increasing squeezing. Hence, it follows directly from the above proved no-go theorem that the two-mode squeezed vacuum state (20) can be distilled only if we resort to the non-Gaussian operations.

The first non-Gaussian distillation protocol to be considered here was proposed by Fiurášek *et al.*[11] The non-Gaussian part of the protocol is the cross-Kerr interaction. The scheme allows to prepare with a certain probability from only a single copy of $|\psi_{\rm in}(\lambda)\rangle_{AB}$ a state with a higher entanglement. The basic principle of the protocol closely follows the Procrustean method[21] developed for discrete variables and can be sought as its CV extension. Its essence lies in the probabilistic application of a suitable filtering operation on one side of the shared entangled state that modulates the Schmidt coefficients c_n in such a way that the resulting state is a better entanglement resource than the original one. The setup of the protocol is schematically depicted in Fig. 1. In that setup, an auxiliary mode C is prepared in a (strong) coherent state $|\alpha\rangle$ (α is real positive) and then interacts

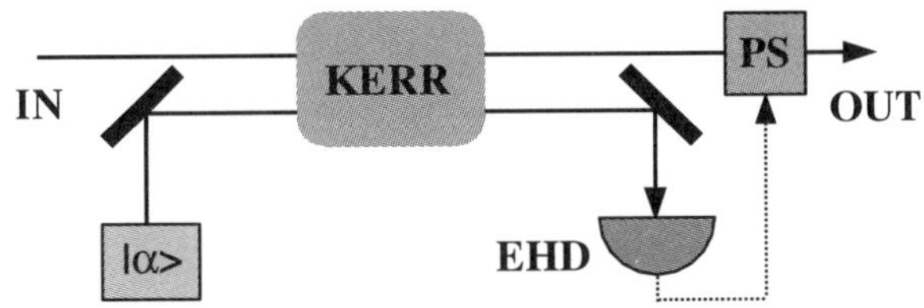

Fig. 1. Schematic of the entanglement concentration setup for traveling light fields that is based on auxiliary coherent states, cross-Kerr interaction, eight-port homodyne detection (EHD), and a linear phase shift depending on the outcome of the measurement (PS).

with Alice's mode A of the state (20) in the medium exhibiting cross-Kerr effect described by the interaction Hamiltonian

$$H_{\text{Kerr}} = \hbar\kappa a^\dagger a c^\dagger c, \tag{22}$$

where a and c denote annihilation operators of Alice's and the auxiliary modes, respectively.

After the interaction, we project the output state of auxiliary mode onto coherent state $|\beta\rangle$ in the eight-port homodyne detector (EHD). The protocol is completed by the phase shift $\exp(-in\varphi|\alpha\beta|\cos\varphi_0)$ of mode A, where $n = a^\dagger a$ is the photon number operator, $\varphi_0 = \arg\beta$ and $\varphi = -\kappa t$, where t is the effective interaction time. If the EHD gives the outcome β the proposed distillation scheme transforms the state (20) to

$$|\psi_{\text{out}}(\beta)\rangle = \sum_{n=0}^{\infty} d_n(\beta)|n, n\rangle_{AB}, \tag{23}$$

where

$$d_n(\beta) = \frac{\sqrt{1-\lambda^2}\lambda^n \exp\left[\alpha\beta^* e^{in\varphi} - in\varphi|\alpha\beta|\cos\varphi_0\right]}{\sqrt{\pi Q(\beta)}\exp\left(|\alpha|^2/2 + |\beta|^2/2\right)} \tag{24}$$

is the normalized Schmidt coefficient corresponding to the measurement outcome β and

$$Q(\beta) = \frac{1-\lambda^2}{\pi}\sum_{n=0}^{\infty}\lambda^{2n}\exp(-|\alpha e^{in\varphi} - \beta|^2) \tag{25}$$

is the probability density of detecting the outcome β. We need to establish a criterion according to which we will accept or reject the state (23) depending on the measurement outcome β. The most natural approach is to choose some reasonable figure of merit $F(\beta)$ that has to be evaluated for each β and then specify the region Ω in the phase space where this figure of merit is sufficiently large. The entanglement concentration succeeds only if $\beta \in \Omega$ and fails otherwise.

The usefulness of the proposed concentration scheme can be illustrated in the Braunstein-Kimble (BK) teleportation scheme.[22] In this case a natural figure of merit is the fidelity of teleportation of coherent states. If the quantum channel of teleportation is in the two-mode squeezed vacuum state (20) the fidelity reads as

$$F = \frac{1+\lambda}{2}. \tag{26}$$

Assuming on the other hand the state (23) as a quantum channel the fidelity can be expressed as

$$F(\beta) = \frac{1}{2}\sum_{m=0}^{\infty}\sum_{n=0}^{\infty}\binom{m+n}{n}\frac{d_m(\beta)d_n^*(\beta)}{2^{m+n}}, \tag{27}$$

where the coefficients $d_n(\beta)$ are given in Eq. (24). We define the region Ω as the set of measurement outcomes β for which $F(\beta) \geq F_{\mathrm{th}}$, where F_{th} is some chosen threshold fidelity. Hence, on the sub-ensemble of states (23) for which $\beta \in \Omega$ the performance of the BK teleportation scheme can be characterized by the average fidelity

$$F = \frac{1}{P}\int_{\Omega} d^2\beta Q(\beta)F(\beta), \tag{28}$$

where $P = \int_{\Omega} d^2\beta Q(\beta)$ is the probability of success of concentration, *i.e.*, the probability that the measurement outcome will fall into Ω. The dependence of the probability of success P on the average fidelity F is depicted in Fig. 2. Note that for the parameters in Fig. 2 the input two-mode squeezed vacuum state gives the fidelity $F_{\mathrm{in}} = 0.75$ as can be seen from the formula (26). The Fig. 2 reveals that the proposed concentration scheme allows to attain higher fidelities with the reasonably high success probability. For example, the fidelity $F = 0.8$ can be achieved with the success probability $P \approx 0.2$. This example clearly demonstrates the practical utility of the proposed concentration scheme.

The second distillation protocol that will be mentioned here was proposed by Duan *et al.*[9] The distillation method used in this protocol is reminiscent of the Schmidt projection method[21] and can be considered as its extension to infinite-dimensional Hilbert space. It allows to distill with a certain probability from several copies of the two-mode squeezed vacuum state (20) a state with a higher entanglement. The central part of this protocol is the local nondemolition measurement of the total photon number in

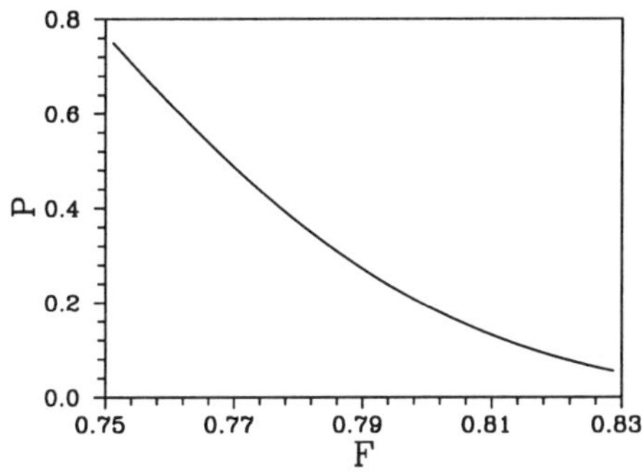

Fig. 2. The probability P of success of the entanglement concentration is plotted in dependence on the fidelity F of teleportation of coherent states that can be achieved with the entangled state after the concentration. The parameters are $\lambda = 1/2$, $\alpha = 10$, $\varphi = \pi/100$.

two (or more) modes which projects the joint shared state onto a maximally entangled state in some finite-dimensional Hilbert space. If the measured total number of photons is sufficiently large the distilled state has a larger entropy of entanglement (21) than the input one.

The non-demolition measurement of the total photon number $n_A = \sum_{i=1}^{K} n_{A_i}$ can be performed using the cross-Kerr interaction by letting all modes A_i sequentially interact with an auxiliary mode C prepared in strong coherent state. The phase shift experienced by the mode C is proportional to the total number of photons n_A and can be determined by measuring the phase-quadrature of the mode C after it interacted with the modes A_i. The total photon number n_A can then be inferred from the value of the measured quadrature.

To describe the performance of Duan's protocol quantitatively, assume that Alice and Bob share K pairs of modes A_i, B_i $(i = 1, 2, \ldots, K)$ all prepared in the two-mode squeezed state (20). The joint state shared by them is then of the form

$$|\Psi_{\text{in}}(\lambda)\rangle_{\{A_i,B_i\}} = \bigotimes_{i=1}^{K} |\psi_{\text{in}}(\lambda)\rangle_{A_i B_i}, \tag{29}$$

where the symbol $\{A_i, B_i\}$ stands for $A_1, B_1, \ldots, A_K, B_K$. Owing to the specific structure of the state (20) the state (29) can be rewritten as

$$|\Psi_{\text{in}}(\lambda)\rangle_{\{A_i,B_i\}} = (1-\lambda^2)^{\frac{K}{2}} \sum_{L=0}^{\infty} \lambda^L \sqrt{d_L^{(K)}} |L\rangle_{\{A_i B_i\}}, \tag{30}$$

where the normalized state $|L\rangle_{\{A_iB_i\}}$ is defined by the formula

$$|L\rangle_{\{A_i,B_i\}} = \frac{1}{\sqrt{d_L^{(K)}}} \sum_{i_1,i_2,\ldots,i_K}^{i_1+i_2+\ldots+i_K=L} |i_1,i_2,\ldots,i_K\rangle_{\{A_i\}}|i_1,i_2,\ldots,i_K\rangle_{\{B_i\}} \tag{31}$$

and the normalization factor $d_L^{(K)}$ reads as $d_L^{(K)} = \frac{(L+K-1)!}{L!(K-1)!}$. The distillation is accomplished by the nondemolition measurement of the total photon number in Alice's modes $A_1, A_2, \ldots, A_K$. With probability

$$p_L^{(K)} = (1-\lambda^2)^K \lambda^{2L} d_L^{(K)}, \tag{32}$$

this non-Gaussian measurement gives the result L and the joint state (29) collapses into the state (31). The distilled state (31) is a pure bipartite maximally entangled state in the finite-dimensional Hilbert space with dimension $d_L^{(K)} \times d_L^{(K)}$ and therefore its entropy of entanglement (21) is equal to $S_E(|L\rangle_{\{A_i,B_i\}}) = \log_2(d_L^{(K)})$. It is easy to see that the obtained entropy exceeds the entropy of entanglement of the original state $S_E(|\psi_{\text{in}}(\lambda)\rangle_{AB})$ only if the measured total photon number L is sufficiently large. Therefore, in order to distill a state with a higher entanglement we have to perform a postselection of the distilled states according to the measurement outcome L. Nevertheless, even in the case of small number of shared pairs K the probability of distilling a state with a higher entanglement is quite high. To illustrate this, consider two copies ($K = 2$) of the state (20) with $\lambda = 1/3$ for which the entropy of entanglement is equal to $S_E(|\psi_{\text{in}}(\lambda = 1/3)\rangle_{AB}) \approx 0.566$ ebits. The probability of distilling a state $|L = 1\rangle_{\{A_i,B_i\}}$ having the entropy of entanglement $S_E(|L = 1\rangle_{\{A_i,B_i\}}) = 1$ ebit is $p_1^{(2)} \approx 0.18$. This result confirms that Duan's scheme is highly efficient in distillation of higher entanglement from several copies of the state (20).

4. Entanglement Concentration by Subtraction of Photons

The nonlinear interactions such as Kerr effect required for the distillation of CV entanglement are usually very weak in practice and it is notoriously difficult to achieve sufficiently strong coupling of quantum beams which contain only a few photons. An elegant and very promising way to circumvent this obstacle consists in exploiting the measurement-induced nonlinearities. In the context of quantum information processing with discrete variables, this approach has been suggested by Knill, Laflamme and Milburn[23] who showed how to construct a scalable all-optical quantum computer using

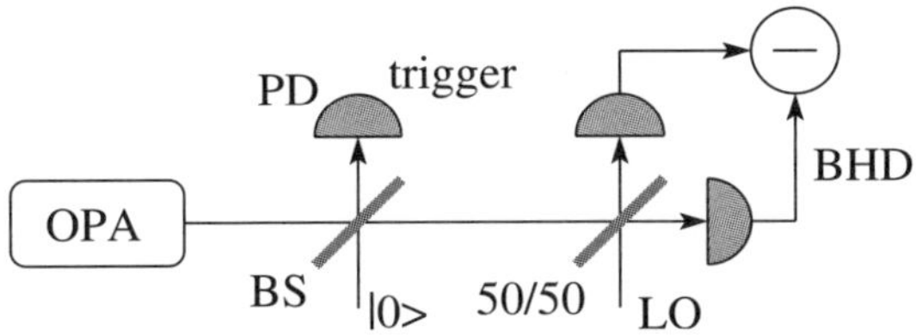

Fig. 3. Scheme for conditional subtraction of a single photon from pulsed single-mode squeezed vacuum generated in optical parametric amplifier (OPA). The beam passes through a strongly unbalanced beam splitter (BS) with transmittance T. The reflected beam impinges on a single photon detector (PD) whose click heralds the successful removal of a photon from the signal beam. The generated state can be probed, *e.g.* in a balanced homodyne detector (BHD).

only passive linear optics, single photon sources, and photodetectors with single photon resolution.

This technique can be extended also to CV regime where it allows to go beyond the class of Gaussian operations. The basic conditional non-Gaussian operation is a subtraction of a single photon from the mode, see Fig. 3. The photon can be removed by sending the beam on a strongly unbalanced beam splitter BS with transmittance T where a tiny portion of the signal is reflected and then impinges upon a single-photon detector PD. The removal of a photon is heralded by the click of the detector. In practice, the commonly employed avalanche photodiodes exhibit only a single photon sensitivity but not a single photon resolution. Nevertheless, the scheme still works correctly and results in conditional removal of a single photon provided that the beam splitter transmittance is high enough. In this case, the most probable event leading to a click of the detector is that a single photon has been reflected on a beam splitter.

Assuming for simplicity ideal detectors and limit $T \to 1$, the conditional photon subtraction can be mathematically described by a non-unitary transformation $|\psi\rangle_{\text{out}} = a|\psi\rangle_{\text{in}}$, where a is the annihilation operator. For instance, for single-mode squeezed vacuum

$$|\psi\rangle_{SMS} = (1-\lambda^2)^{1/4} \sum_{n=0}^{\infty} \frac{\sqrt{(2n)!}}{n!} \left(\frac{\lambda}{2}\right)^n |2n\rangle \tag{33}$$

we get

$$|\psi\rangle_{SMS,-} \propto \sum_{n=1}^{\infty} \frac{\sqrt{(2n)!}}{n!} \sqrt{2n} \left(\frac{\lambda}{2}\right)^n |2n-1\rangle. \tag{34}$$

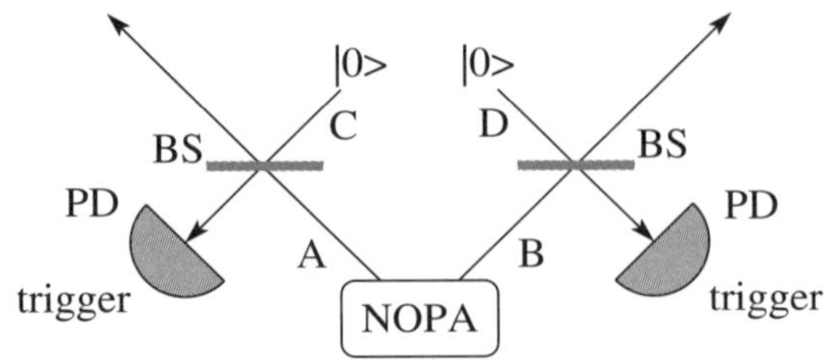

Fig. 4. Scheme for entanglement concentration via subtraction of photons. A two-mode squeezed vacuum is generated in non-degenerate optical parametric amplifier (NOPA) and a photon is removed from each mode using the method shown in Fig. 3. A successful joint photon subtraction is indicated by clicks of both single-photon detectors PD.

This state contains only odd Fock states hence its Wigner function at the origin of the phase space $W(0)$ is negative, because $W(\alpha)$ can be defined as the average of the displaced parity operator, and in particular $W(0) = \frac{1}{\pi}\sum_{n=0}^{\infty}(-1)^n p_n$, where p_n is the photon number distribution. The Wigner function of the squeezed vacuum (33) is a Gaussian peaked at the origin, $W_{SMS}(0) = 1/\pi$, while the Wigner function of the photon subtracted state (34) exhibits a deep dip and is negative at the origin, $W_{SMS,-}(0) = -1/\pi$. The states with non-positive Wigner function are clearly non-Gaussian and very strongly non-classical.

The de-Gaussification of a single-mode squeezed vacuum via subtraction of a single photon has been recently demonstrated experimentally[13] using the scheme schematically depicted in Fig. 3. In the experiment, the dip of the Wigner function at the origin was clearly observed, however, the Wigner function did not reach negative value. The main reason is that the OPA emits squeezed states into many modes and the photodetector can be sometimes triggered by photons coming from other modes than the mode which is observed in the balanced homodyne detector BHD. The suppression of this undesired effect will require better spatial and spectral filtering of the mode fed to the avalanche photodiode PD, to suppress the contribution from the other modes.

The entanglement of Gaussian two-mode squeezed vacuum state (20) can be concentrated by performing local single photon subtractions on each mode. This procedure has been first proposed by Opatrný *et al.*[12] and has been subsequently further investigated and generalized by several other authors.[24,25] The setup is schematically depicted in Fig. 4. An unbalanced beam splitter is used on each side to reflect a part of the beam which is sent onto a single-photon detector. The entanglement concentration succeeds if both photodetectors click. Let us first give a simple idealized description assuming perfect detectors. Similarly as in the single-mode case, we can write $|\psi_{\text{out}}\rangle_{AB} \propto ab|\psi_{\text{in}}(T\lambda)\rangle_{AB}$, where a and b are annihilation operators of modes A and B. The replacement of the squeezing factor $\lambda = \tanh(r)$ by $T\lambda$, where T is the transmittance of the beam splitter, accounts for the reduction of squeezing due to the passage through BS. The normalized conditionally prepared state is given by

$$|\psi_{\text{out}}\rangle_{AB} = \sqrt{\frac{(1-T^2\lambda^2)^3}{1+T^2\lambda^2}} \sum_{n=0}^{\infty}(n+1)(T\lambda)^n|n,n\rangle_{AB}. \tag{35}$$

Notice that this procedure preserves the perfect photon-number correlation between the modes A and B. The probability of successful entanglement concentration, *i.e.* the probability that both photodetectors click, can be expressed as

$$P = (1-T)^2\lambda^2(1-\lambda^2)\frac{1+T^2\lambda^2}{(1-T^2\lambda^2)^3}, \tag{36}$$

where perfect detectors with unit efficiency are assumed. The state (35) is pure and therefore its entanglement can be again quantified by the entropy of entanglement (21). The entropy of entanglement as a function of squeezing λ is plotted in Fig. 5(a) for states (20) and (35). In the latter case, $T = 1$ was set to allow for a direct comparison. We can see that the de-Gaussification indeed increases the amount of entanglement because the pre-factor $n+1$ appearing in the formula (35) partially compensates for the exponential decreasing of the Schmidt coefficients and makes the coefficients c_n more balanced. It can be shown that the Wigner function of the photon-subtracted two-mode squeezed vacuum is negative in some regions of phase space.

Another useful entanglement characterization, that is often considered in the literature for its practical relevance and computational simplicity, is the fidelity of teleportation of coherent states in the Braunstein-Kimble scheme with unity gain, where the entangled state is used as a quantum channel. The fidelity of teleportation with two-mode squeezed vacuum is given by the formula (26) while the fidelity achieved with the non-Gaussian photon subtracted two-mode squeezed state reads

$$F_- = \frac{(1+T\lambda)^3}{4(1+T^2\lambda^2)}(2-2T\lambda+T^2\lambda^2). \tag{37}$$

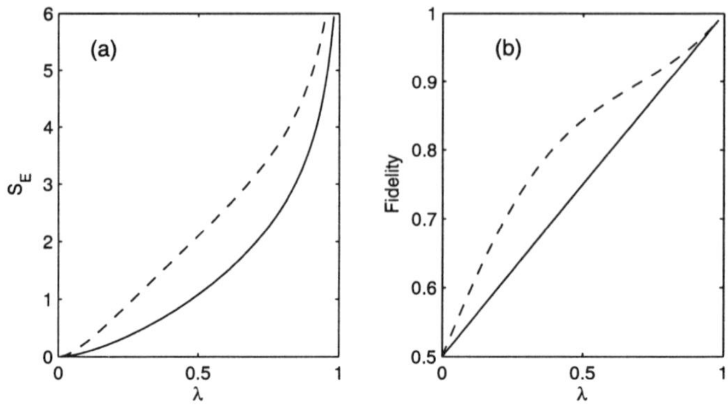

Fig. 5. The dependence of the entropy of entanglement (a) and of the fidelity of teleportation of coherent states (b) on the squeezing parameter λ is plotted for the two-mode squeezed vacuum state (solid line) and for the non-Gaussian state obtained by removing a single photon from each mode of the two-mode squeezed vacuum (dashed line).

The two fidelities are plotted in Fig. 5(b) and we can see that, again, the fidelity of teleportation with non-Gaussian state is higher than the fidelity achievable with the two-mode squeezed vacuum.

As noted above, in practice the detectors can only resolve the presence or absence of photons in the beam but cannot count the number of photons, moreover the detection efficiency η is typically much less then unity. It is therefore important to develop a more realistic description of the experimental setup that can take into account all these imperfections. The realistic detectors with efficiency η which respond with only two outcomes, either a click or a no-click, can be modeled as a sequence of lossy channel with transmittance η followed by idealized detectors which perform projection onto the vacuum, $\Pi_0 = |0\rangle\langle 0|$ (no click) and on the rest of the Hilbert space, $\Pi_1 = I - |0\rangle\langle 0|$ (click). The mixing of the modes A and B with vacua on unbalanced beam splitters BS, as well as the subsequent passage of the modes C and D through the lossy channels with transmittance η can be described by a Gaussian CP map $\mathcal{G}$ and the resulting conditionally generated state reads,

$$\rho_{AB,\mathrm{out}} = \mathrm{Tr}_{CD}[\mathcal{G}(\rho_{AB,\mathrm{in}})(I_{AB} \otimes \Pi_{1,C} \otimes \Pi_{1,D})]. \tag{38}$$

For input Gaussian states, the state $\mathcal{G}(\rho_{AB,\mathrm{in}})$ is still Gaussian. Furthermore, the projector Π_1 is a difference of the Gaussian operators - the identity and the projector onto vacuum. The Wigner representation of Π_1 is therefore given by

$$W_{\Pi_1}(x,p) = \frac{1}{2\pi} - \frac{1}{\pi}e^{-x^2-p^2}. \tag{39}$$

Similarly as in Sec. 2 we explore that the trace of a product of two operators can be evaluated as the overlap integral of the Wigner representations of the operators. It follows from (39) that the Wigner representation of $I_{AB} \otimes \Pi_{1,C} \otimes \Pi_{1,D}$ consists of a linear combination of four Gaussian functions. For Gaussian input states, the partial trace (38) preserves this property and the conditionally generated state can be expressed as a linear non-convex combination of four Gaussians,

$$W_{\mathrm{out}}(r_{AB}) = \frac{1}{\pi^2}\sum_{j=1}^{4} C_j \exp\left[-r_{AB}^T \gamma_j^{-1} r_{AB}\right]. \tag{40}$$

We note that two of the coefficients C_j are positive and two are negative, so the state ρ_{out} cannot be obtained as a mixture of Gaussian states. The explicit dependence of the coefficients C_j and covariance matrices γ_j on the various parameters of the setup, such as the squeezing λ, the beam splitter transmittance T and the detector efficiency η can be expressed analytically, however, the resulting formulas are rather complicated. Numerical calculations confirm that if the transmittance T is high enough then the conditionally prepared states are very close to those that would be obtained with the use of the ideal setup with perfect photon number discriminating detectors. The low efficiency of the detectors does not significantly decrease the fidelity of the generated states and it only reduces the success probability which scales approximately as η^2. This robustness of the setup with respect to detector imperfections is very important as it renders the scheme experimentally feasible with current technology.

5. Gaussification by Means of LOCC Operations

In the preceding sections we have seen that the continuous-variable entanglement can be increased and modified by conditional photon subtraction or by a filtering operation which relies on cross-Kerr interaction. These procedures allow to increase the entanglement by means of LOCC operations so they can be applied after the entanglement has been distributed to spatially distant parties. In this section we show that it is possible to extend these methods to a full continuous-variable entanglement distillation protocol[15] if we combine them with the gaussification procedure that can be realized by means of LOCC Gaussian operations and was proposed by Browne *et al.*[14]

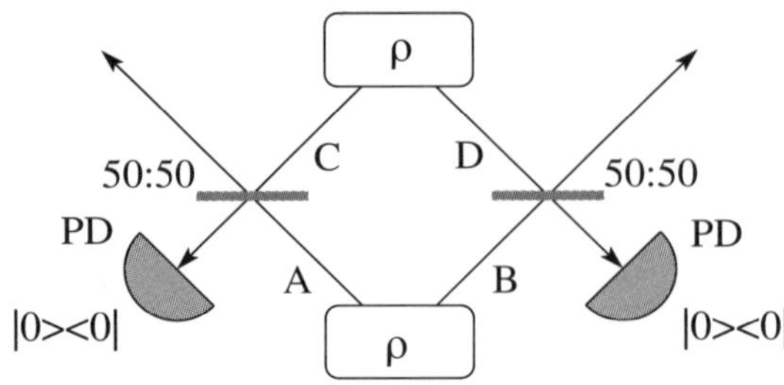

Fig. 6. A single step of the iterative gaussification procedure. The corresponding modes of two copies of a two-mode state ρ are locally combined on a balanced beam splitter and one of the output modes is measured on each side. The gaussification step succeeds if both modes are projected onto vacuum.

The resulting entanglement distillation protocol involves only a single step where the procedure goes beyond the class of Gaussian operations – the de-Gaussification of the entangled states using, *e.g.*, single-photon subtractions. The gaussification is an iterative procedure which from many copies of non-Gaussian states extracts a single Gaussian state which may exhibit much higher entanglement (and purity) than the initial states.

The scheme of one step of the gaussification protocol on two copies of a two-mode state $\rho^{(j)}$ distributed to two parties Alice and Bob is depicted in Fig. 6. Both Alice and Bob combine their parts of the two two-mode states on a balanced beam splitter and subsequently they perform a measurement on one output port with a photodetector that can distinguish the presence or absence of photons in the mode. The gaussification step succeeds if both Alice and Bob project the mode onto vacuum. Let us denote by $\mathcal{G}$ the resulting trace-decreasing Gaussian map that is applied to the two copies of the input two-mode state $\rho^{(j)}$. The iteration produces with certain probability a single output two-mode state

$$\rho^{(j+1)} = \frac{\mathcal{G}(\rho^{(j)} \otimes \rho^{(j)})}{\mathrm{Tr}[\mathcal{G}(\rho^{(j)} \otimes \rho^{(j)})]}, \tag{41}$$

which is then stored in a quantum memory[2] and later on used as an input at the next iteration stage.

It can be proved that the only fixed points of the nonlinear iteration map (41) are Gaussian states with zero first moments.[15] Indeed, it can be easily shown that if two such identical Gaussian states ρ_G are combined on balanced beam splitters, then the output modes are in exactly the same state $\rho_G \otimes \rho_G$ as the inputs so the map $\mathcal{G}$ simply preserves the state ρ_G with probability $\langle 00|\rho_G|00\rangle$. The covariance matrix of the asymptotic Gaussian state obtained after many iterations of the map (41) can be expressed in terms of the density matrix $\rho^{(1)}$ of the state after one iteration of the map (41). Remarkably, the covariance matrix only depends on the density matrix elements $\rho^{(1)}_{00,00}$, $\rho^{(1)}_{10,10}$, $\rho^{(1)}_{01,01}$, $\rho^{(1)}_{10,01}$, $\rho^{(1)}_{11,00}$, $\rho^{(1)}_{20,00}$ and $\rho^{(1)}_{02,00}$. So it is the structure of the density matrix at low Fock numbers that determines the final state. The general formula for the covariance matrix of the asymptotic state is rather complicated.[15] Here we will present a simplified formula[14] which is valid for initial pure state $|\psi\rangle = \sum_{m,n=0}^{\infty} c_{mn}|m,n\rangle$. Since the conditional Gaussian operation $\mathcal{G}$ preserves the purity, the asymptotic output Gaussian state is also pure and its coefficients c_{mn}^{∞} can be expressed as follows,

$$c_{mn}^{\infty} = \langle m,n| \exp\left[\frac{1}{2}g_{aa}(a^\dagger)^2 + \frac{1}{2}g_{bb}(b^\dagger)^2 + g_{ab}a^\dagger b^\dagger\right]|0,0\rangle,$$

where $g_{aa} = \sqrt{2}\tilde{c}_{20} - \tilde{c}_{10}^2$, $g_{bb} = \sqrt{2}\tilde{c}_{02} - \tilde{c}_{01}^2$, $g_{ab} = \tilde{c}_{11} - \tilde{c}_{10}\tilde{c}_{01}$ and $\tilde{c}_{jk} = c_{jk}/c_{00}$. Note that $c_{00} > 0$ is a necessary condition for the iterative map $\mathcal{G}$ to converge to a fixed point. The iterative gaussification combined with de-Gaussification of the distributed entangled states forms an entanglement distillation and purification protocol which can produce a single copy of highly entangled state from several copies of weakly entangled mixed states.[15]

6. Conclusions

In this paper we have presented the entanglement concentration and distillation protocols for continuous variables. This field of research is still very active and many new exciting developments can be expected. In particular, the recent spectacular experimental advances such as the de-Gaussification of a single-mode squeezed vacuum by means of single-photon subtraction[13] or the first demonstration of the atomic quantum memory for light[2] provide a strong motivation for further theoretical and experimental research in this direction with the ultimate goal to develop quantum repeaters that would allow an efficient long-distance distribution of entangled states of light over lossy and noisy quantum channels.

Acknowledgments

We acknowledge financial support from the EU under the project COVAQIAL (FP6-511004) and from the Czech Ministry of Education under the project Information and Measurement in Optics (MSM 6198959213). JF and RF acknowledge support from the grants 202/05/0498 and 202/03/D239 of the Czech Grant Agency.

References

1. H. J. Briegel, W. Dür, J. I. Cirac, and P. Zoller, Phys. Rev. Lett. **81**, 5932 (1998).
2. B. Julsgaard, J. Sherson, J. I. Cirac, J. Fiurášek, and E. S. Polzik, Nature (London) **432**, 482 (2004).
3. C. H. Bennett, G. Brassard, S. Popescu, B. Schumacher, J. A. Smolin, and W. K. Wootters, Phys. Rev. Lett. **76**, 722 (1996).
4. C. H. Bennett, G. Brassard, C. Crepeau, R. Jozsa, A. Peres, and W. K. Wootters, Phys. Rev. Lett. **70**, 1895 (1993).
5. A. Furusawa, J. L. Sørensen, S. L. Braunstein, C. A. Fuchs, H. J. Kimble, and E. S. Polzik, Science **282**, 706 (1998).
6. G. Giedke and J. I. Cirac, Phys. Rev. A **66**, 032316 (2002).
7. J. Eisert, S. Scheel, and M. B. Plenio, Phys. Rev. Lett. **89**, 137903 (2002).
8. J. Fiurášek, Phys. Rev. Lett. **89**, 137904 (2002).
9. L.-M. Duan, G. Giedke, J. I. Cirac, and P. Zoller, Phys. Rev. Lett. **84**, 4002 (2000).
10. L.-M. Duan, G. Giedke, J. I. Cirac, and P. Zoller, Phys. Rev. A **62**, 032304 (2000).
11. J. Fiurášek, L. Mišta, Jr., and R. Filip, Phys. Rev. A **67**, 022304 (2003).
12. T. Opatrný, G. Kurizki, and D.-G. Welsch, Phys. Rev. A **61**, 032302 (2000).
13. J. Wenger, R. Tualle-Brouri, and P. Grangier, Phys. Rev. Lett. **92**, 153601 (2004).
14. D. E. Browne, J. Eisert, S. Scheel, and M. B. Plenio, Phys. Rev. A **67**, 062320 (2003).
15. J. Eisert, D. E. Browne, S. Scheel, and M. B. Plenio, Ann. Phys. **311**, 431 (2004).
16. G. Lindblad, J. Phys. A **33**, 5059 (2000).
17. A. Jamiolkowski, Rep. Math. Phys. **3**, 275 (1972).
18. M.-D. Choi, Linear Algebr. Appl. **10**, 285 (1975).
19. R. F. Werner and M. M. Wolf, Phys. Rev. Lett. **86**, 3658 (2001).
20. Z. Y. Ou, S. F. Pereira, H. J. Kimble, and K. C. Peng, Phys. Rev. Lett. **68**, 3663 (1992).
21. C. H. Bennett, H. J. Bernstein, S. Popescu, and B. Schumacher, Phys. Rev. A **53**, 2046 (1996).
22. S. L. Braunstein and H. J. Kimble, Phys. Rev. Lett. **80**, 869 (1998).
23. E. Knill, R. Laflamme, and G. J. Milburn, Nature (London) **409**, 46 (2001).
24. P. T. Cochrane, T. C. Ralph, and G. J. Milburn, Phys. Rev. A **65**, 062306 (2002).
25. S. Olivares, M. G. A. Paris, and R. Bonifacio, Phys. Rev. A **67**, 032314 (2003).

Chapter 7

Loophole-Free Test of Quantum Nonlocality with Continuous Variables of Light

R. García-Patrón and N. J. Cerf

QUIC, Ecole Polytechnique, CP 165, Université Libre de Bruxelles
1050 Brussels, Belgium

J. Fiurášek

Department of Optics, Palacký University
17. listopadu 50, 77200 Olomouc, Czech Republic

It is shown that a loophole-free Bell test can be achieved using continuous variables of light. A feasible optical setup is proposed for this purpose, based on a non-Gaussian state of light and high-efficiency homodyne detectors. The non-Gaussian entangled state can be generated from a two-mode squeezed vacuum state by subtracting a single photon from each mode using beam splitters and standard low-efficiency single-photon detectors. A Bell violation exceeding 1% can be attained with 6 dB squeezed light and an homodyne efficiency around 95%. A detailed feasibility analysis, including the effect of the detector efficiency, the electronic noise, the impurity of the non-Gaussian state, and the probability of false triggers, suggests that this method opens a promising avenue towards a complete experimental Bell test.

1. Introduction

Over the last century, quantum physics has developed into a powerful tool, allowing the description of a wide range of phenomena at the microscopic scale. Technologies such as lasers, NMR, or semi-conductor based systems would be impossible without quantum physics. Even if quantum theory has reached a high level of maturity, some of its basic concepts still are very counterintuitive and have puzzled physicists since the early days of the theory. Feynman's famous expression "I think I can safely say that nobody understands quantum mechanics" is a good illustration of this opinion.

Since the inception of quantum mechanics, several physicists have considered this counterintuitive aspect as an evidence of the incompleteness of the theory. There have been repeated suggestions that its probabilistic features may possibly be described by an underlying deterministic substructure. The first attempt in this direction originates from the famous paper by Einstein, Podolsky, and Rosen (EPR)[1] in 1935. There, it was advocated that if "local realism" (causality + deterministic substructure) is taken for granted, then quantum theory is an incomplete description of the physical world.

The EPR argument gained a renewed attention in 1964, when John Bell derived his famous inequalities, which must be satisfied within the framework of any local realistic theory.[2] Bell showed that any such deterministic substructure model (also called "hidden-variable model"), if local, yields predictions that significantly differ from those of quantum mechanics. The merit of Bell inequalities lies in the possibility to test them experimentally, allowing physicists to verify whether either quantum mechanics or local realism is the correct description of Nature.

2. Bell Inequalities

In this chapter, we will use the Clauser-Horne-Shimony-Holt inequality (called Bell-CHSH inequality in the following), originally devised for a two-qubit system.[3] Let us consider the following thought experiment, which we will analyze from the point of view of local realism. The experiment involves three distant parties, Sophie, Alice, and Bob. Sophie (the source) prepares a bipartite state and distributes it to Alice and Bob (the two usual partners), see Fig. 1. Then, Alice and Bob randomly and independently decide between one of two possible quantum measurements A_1 or A_2 (B_1 or B_2), which should have only two possible outcomes $+1$ or -1. The timing of the experiment should be arranged in such a way that Alice and Bob do their measurements in a causally disconnected manner. Thereby, Alice's measurement cannot influence Bob's, and vice-versa. Local realism implies two assumptions:

(1) Realism: the physical properties A_1, A_2, B_1, B_2 have definite values a_1, a_2, b_1, b_2, which exist independently of their observation. This implies the existence of a probability distribution $P(a_1, a_2, b_1, b_2)$, dependent on how Sophie generates the bipartite state.
(2) Locality: Alice's measurement choice and outcome do not influence the result of Bob's measurement, and vice-versa. The measurement events are separated by a spacelike interval.

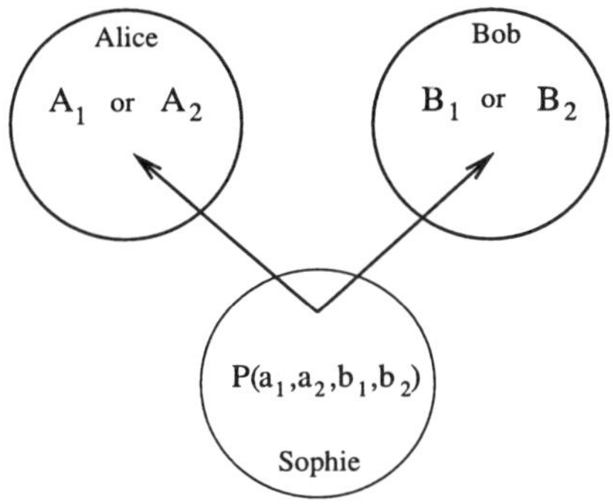

Fig. 1. Sophie prepares a bipartite state and distributes it to Alice and Bob, who perform each a measurement. Alice measures either A_1 or A_2, while Bob measures B_1 or B_2. In a local realistic theory, there must exist an underlying probability distribution $p(a_1, a_2, b_1, b_2)$, generated by Sophie.

If we consider local realism as the correct description of the physical world, then we obtain the Bell-CHSH inequality

$$S = |\langle a_1 b_1 \rangle + \langle a_1 b_2 \rangle + \langle a_2 b_1 \rangle - \langle a_2 b_2 \rangle| \leq 2, \tag{1}$$

where $\langle a_j b_k \rangle$ denotes the average over the subset of experimental data where Alice measured a_j and, simultaneously, Bob measured b_k. Indeed, if there is an underlying probability distribution $p(a_1, a_2, b_1, b_2)$, then each realization of it contributes by $a_1(b_1 + b_2) + a_2(b_1 - b_2) = \pm 2$ to the average, implying Eq. (1).

Now, if we consider that Sophie generates and distributes an entangled pair of qubits, quantum mechanics predicts $S \leq 2\sqrt{2}$, which is in contradiction with local realism. Thus, an experimental test of Bell-CHSH inequalities where a violation of $S \leq 2$ is observed disproves any classical (local realistic) description of Nature.

3. Experimental Bell Test and Related Loopholes

From the beginning of the 80's, many experimental Bell tests[4–10] have been performed, observing the violation of Bell inequalities predicted by quantum mechanics. All these schemes used optical setups because, at that time, it was the only known way of generating and distributing entangled particles (photons) at a distance in order to make Alice's and Bob's measurements causally disconnected. Unfortunately, the available single-photon detectors suffer from a low efficiency η_{PD}, which can be exploited by a local realistic model to yield a violation. Thus, to reject local realism, it is necessary to make the extra assumption that the registered pairs form a fair sample of the emitted pairs. So, from a logical point of view, these experiments do not succeed in ruling out a local realistic model; this is the so-called *detector-efficiency loophole.*[11–13] This loophole has been closed in a recent experiment with trapped ions,[14] thanks to the high efficiency of the measurement of the ion states. However, the ions were held in a single trap, only several micrometers apart, so that the measurement events were not spacelike separated, opening in turn the so-called *locality loophole.*[15,16]

So far, no experimental test has succeeded to close both loopholes at the same time, that is, the measured correlations may be explained in terms of local realistic theories exploiting the low detector efficiency or the timelike interval between the two detection events. It was suggested that two distant trapped ions can be entangled via entanglement swapping by first preparing an entangled state of an ion and a photon on each side and then projecting the two photons on a maximally entangled singlet state.[17–20] Very recently, the first step toward this goal, namely the entanglement between a trapped ion and a photon emitted by the ion, has been observed experimentally.[21] However, the entanglement swapping would require the interference of two photons emitted by two different ions, which is experimentally very challenging. An interesting alternative to the atom-based approaches[17,22,23] consists of all-optical schemes based on continuous variables of light. Indeed, the balanced homodyne detection used in these schemes can exhibit a high detection efficiency,[24] sufficient to close the detection loophole.

4. Bell Test with Continuous Variables of Light

Quantum continuous variables of light have been successfully used to realize some of the standard informational tasks traditionally based on qubits. Unfortunately, the entangled two-mode squeezed state that can easily be generated experimentally[25–27] cannot be directly employed to test Bell inequalities with homodyning. Indeed, as noted by Bell himself, this state is described by a positive-definite Gaussian Wigner function, which thus provides a local realistic model that can explain all correlations between quadrature measurements (carried out by balanced homodyne detectors). Thus, similarly as in the case of the purification of continuous variable entanglement,[28–32] one has to go beyond the class of Gaussian states or Gaussian operations.

In particular, it is possible to obtain a Bell violation with a Gaussian two-mode squeezed vacuum state by performing a non-Gaussian measurement, for example a photon-counting measurement.[33] As shown in Fig. 2, Sophie prepares an entangled state and distributes it to Alice and Bob. The two possible measurements on Alice's and Bob's sides consist in randomly choosing between applying the displacement $D(\alpha)$ or no displacement, followed by a measurement of the parity of the number of photons n impinging on the single-photon detector. The resulting parity $a_i = (-1)^n$ gives the binary result used in the Bell-CHSH inequality. It can be shown[33] that

$$S = |W(0,0) + W(\alpha,0) + W(0,\alpha) - W(\alpha,\alpha)|, \tag{2}$$

where $W(x,p)$ is the Wigner function of the entangled state, violates the Bell-CHSH inequality $S \leq 2$ by about 10% for an appropriate choice of α. Recent proposals using more abstract measurements described in Refs. 34-36 gave similar results. Note, however, that these measurements are either experimentally infeasible or suffer from a very low detection efficiency, thereby re-opening the detection loophole.

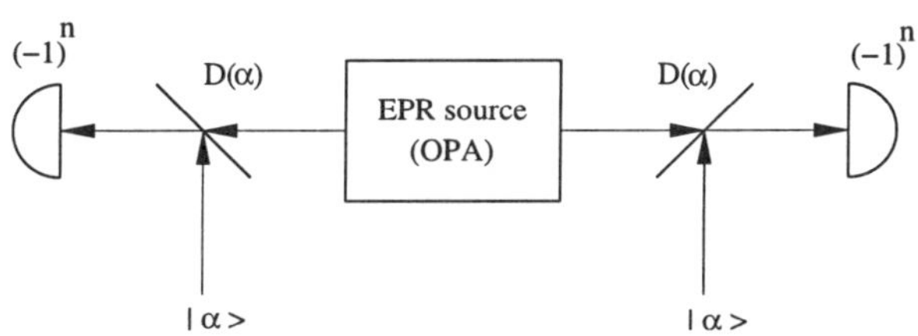

Fig. 2. Bell test using the parity of the number of photons impinging on each photodetector. Sophie prepares an entangled state (EPR) and distributes it to Alice and Bob. Each of them either applies a displacement $D(\alpha)$ or not, and uses the parity of the number of photons measured using a photodetector with single-photon resolution.[33]

Considering the current state of the art in quantum optics technologies, the scheme based on high-efficiency homodyne detection seems to be the most promising way of closing the detection loophole. However, since homodyning is a Gaussian measurement, it is then necessary to generate highly non-classical non-Gaussian entangled states, whose Wigner function is not positive definite. In addition, one has to develop a method for converting the continuous result obtained by homodyne measurement into a binary result (the so-called "binning" method).

Several recent theoretical works have demonstrated that a violation of Bell inequalities can be observed using balanced homodyning provided that specific entangled light states such as pair-coherent states, squeezed

Schrödinger cat-like states, or specifically tailored finite superpositions of Fock states, are available.[37–40] More specifically, the violation of the Bell-CHSH inequality was derived in Ref. 39 for a state of the form

$$|\psi_{\text{in}}\rangle_{AB} = \sum_{n=0}^{\infty} c_n |n, n\rangle_{AB}, \tag{3}$$

with $|n\rangle$ denoting Fock states, and a binning based on the sign of the measured quadrature. Optimizing over the quadrature angles and probability amplitudes c_n (see Fig. 3), one obtains a maximal Bell-CHSH inequality violation of $S = 2.076$. Interestingly, it was shown in Ref. 40 that the highest possible violation of $S = 2\sqrt{2}$ can be obtained with the bipartite state

$$|\psi_{\text{in}}\rangle_{AB} = |f, f\rangle + e^{i\theta}|g, g\rangle, \tag{4}$$

where $f(q)$ and $g(q)$ are the wave functions of some specific states, and a more complicated binning based on the roots of $f(q)$ and $g(q)$ is used. Unfortunately, no feasible experimental scheme is known today that could generate the states required in Refs. 37-40.

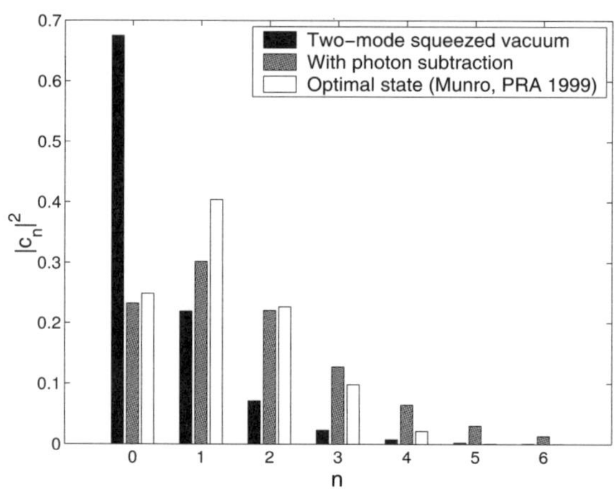

Fig. 3. Probabilities $|c_n|^2$ in the Fock basis of the two-mode squeezed vacuum state with $\lambda = 0.57$ (black), the non-Gaussian state obtained from the previous state by subtracting one photon from each mode (grey), and the optimal state of Ref. 39 (white).

Recently, it was shown by us together with J. Wenger, R. Tualle-Brouri and P. Grangier,[41] and independently by Nha and Carmichael,[42] that a very simple non-Gaussian state obtained by subtracting a single photon from each mode of a two-mode squeezed vacuum state can exhibit a Bell violation with homodyning. Note that this non-Gaussian state is close to the optimal state obtained in Ref. 39, as is visible in Fig. 3, and gives a violation of $S = 2.046$.

An essential feature of this proposal is that the photon subtraction can be successfully performed with low-efficiency single-photon detectors,[43–45] which renders the setup experimentally feasible. In fact, the basic building block of the scheme, namely the de-Gaussification of a single-mode squeezed vacuum via single-photon subtraction, has recently been demonstrated experimentally.[46]

5. Loophole-Free Bell Test Using Homodyne Detectors

The conceptual scheme of the proposed experimental setup is depicted in Fig. 4. A source generates a two-mode squeezed vacuum state in modes A and B. This can be accomplished, *e.g.*, by means of non-degenerate optical parametric amplification in a $\chi^{(2)}$ nonlinear medium or by generating two single-mode squeezed vacuum states and combining them on a balanced beam splitter. Subsequently, the state is de-gaussified by conditionally subtracting a single photon from each beam. A tiny part of each beam is reflected from a beam splitter BS_A (BS_B) with a high transmittance T. The reflected portions of the beams impinge on single-photon detectors such as avalanche photodiodes. A successful photon subtraction is heralded by a click of each photodetector PD_A and PD_B.[45]

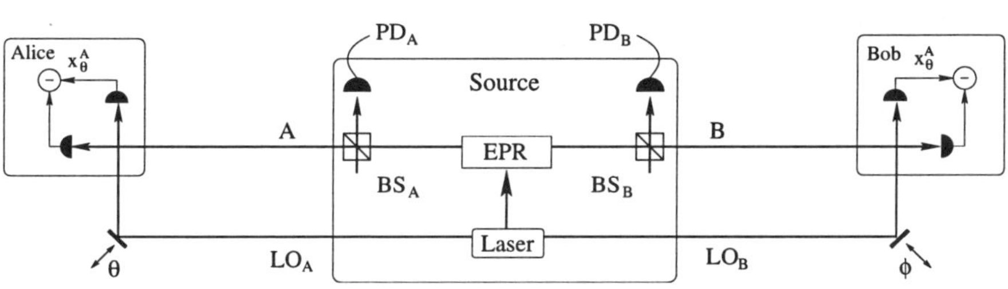

Fig. 4. Proposed experimental setup for performing a Bell test with balanced homodyning. The source emits a two-mode squeezed vacuum state in modes A and B. A small part of the beams is diverted by two highly unbalanced beam splitters BS_A and BS_B, and sent to the single-photon detectors PD_A and PD_B. The two remaining beams A and B, which are conditionally prepared in a non-Gaussian entangled state, are sent to Alice and Bob, who perform each a balanced homodyne detection using their local oscillator LO_A and LO_B.

In practice, the available photodetectors exhibit a single-photon sensitivity but not a single-photon resolution, that is, they can distinguish the absence and presence of photons but cannot measure the number of photons in the mode. Nevertheless, this is not a problem here because in the limit of high T, the most probable event leading to the click of a photodetector is precisely that a single photon has been reflected by the beam splitter. The probability of an event where two or more photons are subtracted from a single mode is smaller by a factor of $\approx 1-T$ and becomes totally negligible in the limit of $T \to 1$. Another important feature of the scheme is that the detector efficiency η_{PD} can be quite low because a small η_{PD} only reduces the success rate of the conditional single-photon subtraction, but does not significantly decrease the fidelity of this operation.

After generation of the non-Gaussian state, the beams A and B together with the appropriate local oscillators LO_A and LO_B are sent to Alice and Bob, who then randomly and independently measure one of two quadratures $x^A_{\theta_j}$, $x^B_{\phi_k}$ characterized by the relative phases θ_1, θ_2 and ϕ_1, ϕ_2 between the measured beam and the corresponding local oscillator.

To avoid the locality loophole, the whole experiment has to be carried out in the pulsed regime and a proper timing is necessary. In particular, the measurement events on Alice's and Bob's sides (including the choice of phases) have to be spacelike separated. A specific feature of the proposed setup is that the non-Gaussian entangled state is conditionally generated when both "event-ready" (see Ref. 15 p. 29 and 105) detectors PD_A and PD_B click. This can be viewed as some preselection of the non-Gaussian state at the source. However, we would like to stress that this does not open any causality loophole if proper timing is satisfied. Namely, in each experimental run, the detection of the clicks (or no-clicks) of photodetectors PD_A and PD_B at the source should be spacelike separated from Alice's and Bob's measurements. This guarantees that the choice of the measurement on Alice's and Bob's sides cannot in any way influence the conditioning "event-ready" measurement.[15,17,41]

In the proposed experiment, Alice and Bob measure quadratures which have a continuous spectrum. These quadratures can be discretized by postulating that the outcome is $+1$ when $x \geq 0$ and -1 otherwise. The two different measurements on each side correspond to the choices of two relative phases θ_1, θ_2 and ϕ_1, ϕ_2. Thus, the quantum correlation $E(\theta_j, \phi_k) \equiv \langle a_j b_k \rangle$ can be expressed as

$$E(\theta_j, \phi_k) = \int_{-\infty}^{\infty} \mathrm{sign}(x^A_{\theta_j} x^B_{\phi_k}) P(x^A_{\theta_j}, x^B_{\phi_k}) dx^A_{\theta_j} dx^B_{\phi_k}, \tag{5}$$

where $P(x_{\theta_j}^A, x_{\phi_k}^B) \equiv \langle x_{\theta_j}^A, x_{\phi_k}^B|\rho_{\mathrm{out},AB}|x_{\theta_j}^A, x_{\phi_k}^B\rangle$ is the joint probability distribution of the two commuting quadratures $x_{\theta_j}^A$ and $x_{\phi_k}^B$, and $\rho_{\mathrm{out},AB}$ denotes the (normalized) conditionally generated non-Gaussian state of modes A and B. The entire data analysis must be performed on a pulsed basis, with Sophie sending time-tagged light pulses (local oscillator and squeezed light) to Alice and Bob. In each experimental run, Sophie records whether her two photodetectors PD_A and PD_B clicked, while Alice and Bob carry out space-like separated measurements of one of two randomly chosen quadratures. After registering a large number of events, the three partners discard all events obtained in measurement runs where either PD_A or PD_B did not click. The correlation coefficients $\langle a_j b_k\rangle$ are then evaluated from all remaining events, and plugged into the S parameter (1).

6. Simplified Model with Ideal Photodetectors

First, we consider a simplified description of the setup, assuming ideal photodetectors ($\eta_{\mathrm{PD}} = 1$) with single-photon resolution and conditioning on detecting exactly one single photon at each detector.[43,44] This idealized treatment is valuable since it provides an upper bound on the practically achievable Bell factor S. Moreover, as noted above, in the limit of high transmittance $T \to 1$, a realistic (inefficient) detector with single-photon sensitivity is practically equivalent to this idealized detector.

The two-mode squeezed vacuum state is expressed in the Fock basis as

$$|\psi_{\mathrm{in}}(\lambda)\rangle_{AB} = \sqrt{1-\lambda^2}\sum_{n=0}^{\infty}\lambda^n|n,n\rangle_{AB}, \tag{6}$$

where $\lambda = \tanh(s)$ and s is the squeezing constant. In the case of ideal photodetectors, the single-photon subtraction results in the state

$$|\psi_{\mathrm{out}}\rangle_{AB} \propto \hat{a}_A\hat{a}_B|\psi_{\mathrm{in}}(T\lambda)\rangle_{AB}, \tag{7}$$

where $\hat{a}_{A,B}$ are annihilation operators and the parameter λ is replaced by $T\lambda$ in order to take into account the transmittance of BS_A and BS_B. A detailed calculation shows that this non-Gaussian state has the form

$$|\psi_{\mathrm{out}}\rangle_{AB} = \sqrt{\frac{(1-T^2\lambda^2)^3}{1+T^2\lambda^2}}\sum_{n=0}^{\infty}(n+1)(T\lambda)^n|n,n\rangle_{AB}. \tag{8}$$

For pure states exhibiting perfect photon-number correlations such as Eq. (8), the correlation coefficient (5) depends only on the sum of the angles, $E(\theta_j, \phi_k) = \mathcal{E}(\theta_j + \phi_k)$. With the help of the general formula derived in Ref. 39, we obtain for the state (8)

$$\mathcal{E}(\varphi) = \frac{(1-T^2\lambda^2)^3}{1+T^2\lambda^2} \sum_{n>m} \frac{8\pi(2T\lambda)^{n+m}}{n!m!(n-m)^2}(n+1)(m+1) \times [\mathcal{F}(n,m) - \mathcal{F}(m,n)]^2 \cos[(n-m)\varphi], \tag{9}$$

where $\mathcal{F}(n,m) = \Gamma^{-1}((1-n)/2)\Gamma^{-1}(-m/2)$ and $\Gamma(x)$ stands for the Euler gamma function.

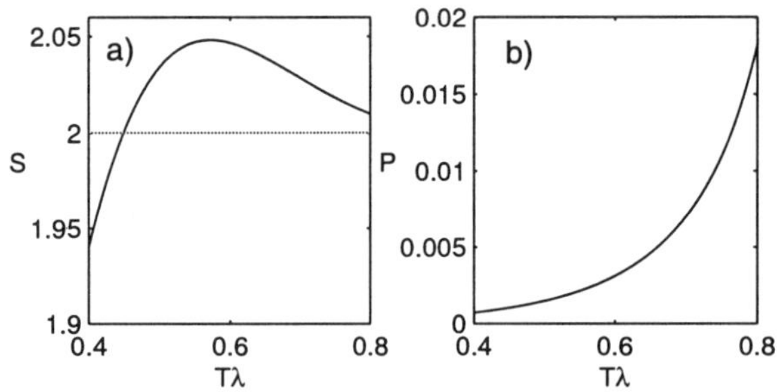

Fig. 5. (a) Bell factor S as a function of the effective squeezing parameter $T\lambda$ for $\theta_1 = 0$, $\theta_2 = \pi/2$, $\phi_1 = -\pi/4$ and $\phi_2 = \pi/4$. (b) Probability P of successful conditional generation of the state $|\psi_{\text{out}}\rangle$ as a function of the effective squeezing parameter $T\lambda$, assuming $T = 0.95$.

We have numerically optimized the angles $\theta_{1,2}$ and $\phi_{1,2}$ to maximize the Bell factor S. It turns out that, for any λ, it is optimal to choose $\theta_1 = 0$, $\theta_2 = \pi/2$, $\phi_1 = -\pi/4$ and $\phi_2 = \pi/4$. The Bell factor S for this optimal choice of angles is plotted as a function of the effective squeezing parameter $T\lambda$ in Fig. 5(a), and the corresponding probability of success of the conditional preparation of the state $|\psi_{\text{out}}\rangle$ is plotted in Fig. 5(b). We can see that S is higher than 2 so the Bell-CHSH inequality is violated when $T\lambda > 0.45$. The maximal violation is achieved for $T\lambda \approx 0.57$, giving $S \approx 2.048$. This violation is quite close to the maximum Bell factor $S = 2.076$ that can be reached with homodyne detection, sign binning, and arbitrary states exhibiting perfect photon-number correlations $|\psi\rangle = \sum_n c_n|n,n\rangle$.[39]

7. Realistic Model

Here, we consider a realistic scheme with inefficient ($\eta_{\text{PD}} < 1$) photodetectors exhibiting single-photon sensitivity but no single-photon resolution, and realistic balanced homodyning with efficiency $\eta_{\rm BHD} < 1$. The mathematical description of this realistic model is simplified by working in the phase-space representation and using the Wigner function formalism. Even though the state used for the Bell test is intrinsically non-Gaussian, it can be expressed as a linear combination of Gaussian states, so all the powerful Gaussian tools may still be used.[47]

7.1. *Calculation of the Wigner function*

As shown in Fig. 6, the modes A and B are initially prepared in a two-mode squeezed vacuum state associated with the Wigner function

$$W_{AB}(r) = W_G(r_{AB}; \Gamma_{\rm in}) = \frac{\sqrt{\operatorname{Det}\Gamma_{\rm in}}}{\pi^2} e^{-r_{AB}^T \Gamma_{\rm in} r_{AB}}, \tag{10}$$

where W_G means Gaussian Wigner function, $r_{AB} = (x_A, p_A, x_B, p_B)^T$, and $\Gamma_{\rm in}$ is the inverse of the covariance matrix of a two-mode squeezed vacuum state.

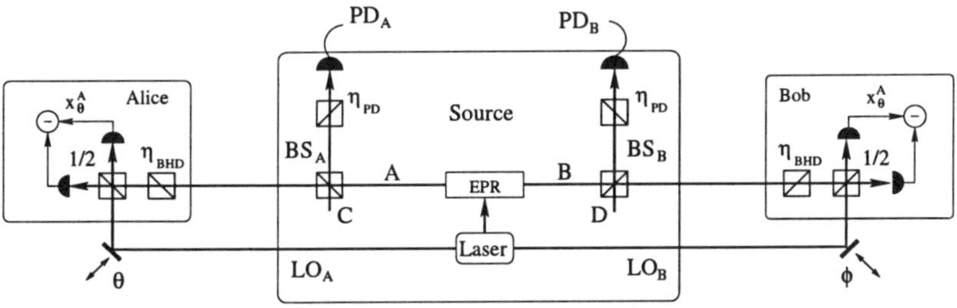

Fig. 6. Detailed optical setup of the proposed Bell test taking into account the realistic single-photon ($\eta_{\rm PD} < 1$) and homodyne ($\eta_{\rm BHD} < 1$) detectors.

The output state $\rho_{{\rm out},AB}$ is prepared by conditioning on observing clicks at both photodetectors PD$_A$ and PD$_B$. These detectors respond with two different outcomes, either a click, or no click. Mathematically, an ideal detector with single-photon sensitivity is described by a two-component positive operator valued measure (POVM) consisting of projectors onto the vacuum state and the rest of the Hilbert space, $\Pi_0 = |0\rangle\langle 0|$ and $\Pi_1 = I - |0\rangle\langle 0|$. The resulting conditionally prepared state $\rho_{{\rm out},AB}$ is thus

$$\rho_{{\rm out},AB} = \operatorname{Tr}_{CD}[\mathcal{M}(\rho_{AB} \otimes |0\rangle_{C,D}\langle 0|)(I_{AB} \otimes \Pi_{1,C} \otimes \Pi_{1,D})], \tag{11}$$

where $\mathcal{M}$ denotes the Gaussian CP map that describes the mixing of modes A with C (and B with D) on the unbalanced beam splitters BS$_A$ (and BS$_B$), followed by the "virtual" lossy channels of transmittance η_{PD} (and η_{BHD}) modeling the inefficiency of the single-photon detectors (and homodyne detectors). As a result, the Wigner function of the state $\rho_{\mathrm{out},AB}$ can be written as a linear combination of 4 Gaussian functions,

$$W_{\mathrm{out},AB}(r) = \frac{1}{P_G} \sum_{j=1}^{4} C_j W_G(r;\Gamma_j). \tag{12}$$

where P_G is the probability of successful photon subtractions. The correlation matrices Γ_j and the coefficients C_j can be expressed in terms of Γ_{in}, see Ref. 47 for a detailed derivation.

7.2. *Resulting Bell violation*

The joint probability distribution $P(x^A_{\theta_j}, x^B_{\phi_k})$ of the quadratures $x^A_{\theta_j}$ and $x^B_{\phi_k}$ appearing in Eq. (5) for the correlation coefficient $E(\theta_j, \phi_k)$ can be obtained from the Wigner function (12) as a marginal distribution:

$$P(x^A_{\theta_j}, x^B_{\phi_k}) = \int_{-\infty}^{\infty}\int_{-\infty}^{\infty} W_{\mathrm{out},AB}(S^T_{\mathrm{sh}} r_{\theta_j,\phi_k})\, dp^A_{\theta_j}\, dp^B_{\phi_k}, \tag{13}$$

where $r_{\theta_j,\phi_k} = [x^A_{\theta_j}, p^A_{\theta_j}, x^B_{\phi_k}, p^B_{\phi_k}]$ and the symplectic matrix S_{sh} describes local phase shifts that must be applied to modes A and B in order to map the measured quadratures $x^A_{\theta_j}$ and $x^B_{\phi_k}$ onto the quadratures x^A and x^B, respectively. As can be seen in Fig. 7(a,b), the joint probability P exhibits two peaks, both located in a quadrant where Alice's and Bob's measured quadratures have the same sign. This double-peak structure is a clear signature of the non-Gaussian character of the state. The plots for the corresponding Gaussian state (before photon subtraction) are also shown in Fig. 7(c,d) for comparison.

The Bell factor can be expressed as

$$S = E(\theta_1, \phi_1) + E(\theta_1, \phi_2) + E(\theta_2, \phi_1) - E(\theta_2, \phi_2) \tag{14}$$

where, taking into account the sign binning, the normalization of the joint probability distribution $P(x^A_{\theta_j}, x^B_{\phi_k})$, and its symmetry $P(x^A_{\theta_j}, x^B_{\phi_k}) = P(-x^A_{\theta_j}, -x^B_{\phi_k})$, we can express the correlation coefficient as

$$E(\theta_j, \phi_k) = 4\int_0^{\infty}\int_0^{\infty} P(x^A_{\theta_j}, x^B_{\phi_k})\, dx^A_{\theta_j}\, dx^B_{\phi_k} - 1. \tag{15}$$

This last integral can be easily evaluated analytically.[47]

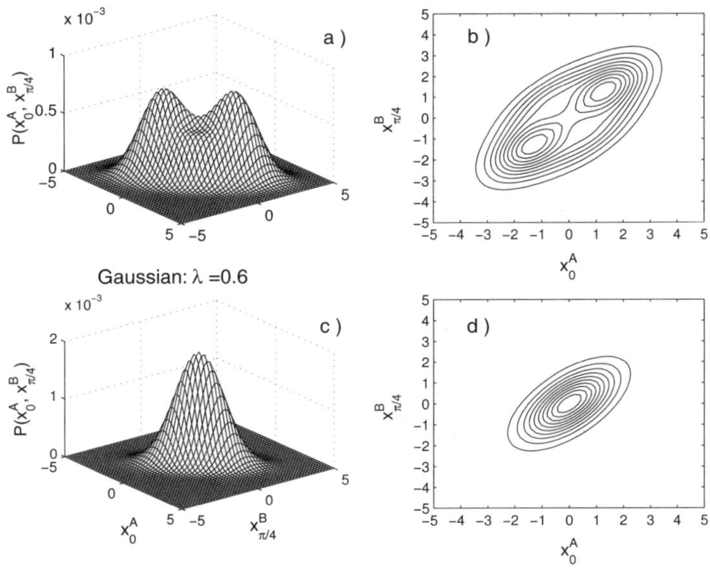

Fig. 7. Joint probability distribution $P(x^A_{\theta_j}, x^B_{\phi_k})$. Panels (a) and (b) show the distribution for the conditionally-prepared non-Gaussian state with $T = 0.99$. Panels (c) and (d) correspond to the initial Gaussian two-mode squeezed vacuum state. The curves are plotted for perfect detectors $\eta_{\rm PD} = \eta_{\rm BHD} = 100\%$, squeezing $\lambda = 0.6$ and $\theta_{\rm Alice} = 0$ and $\phi_{\rm Bob} = \pi/4$.

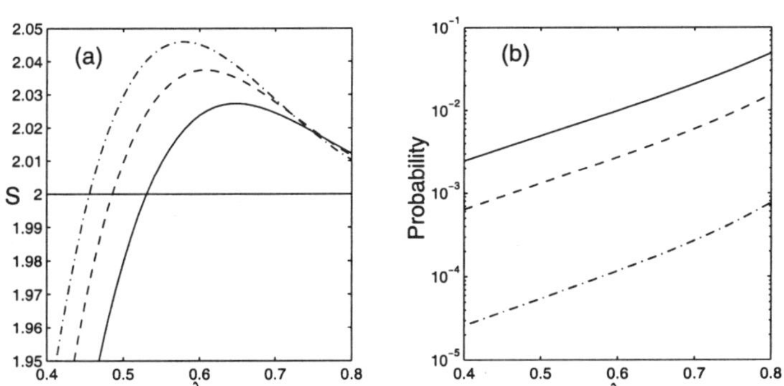

Fig. 8. Violation of the Bell-CHSH inequality with the conditionally-prepared non-Gaussian state.[47] (a) Bell factor S as a function of the squeezing λ. (b) Probability of success P_G of the generation of the non-Gaussian state as a function of the squeezing λ. The curves are plotted for perfect detectors ($\eta_{\rm PD} = \eta_{\rm BHD} = 100\%$) with $T = 0.9$ (solid line), $T = 0.95$ (dashed line), and $T = 0.99$ (dot-dashed line).

Figure 8(a) confirms that the Bell-CHSH inequality $|S| \leq 2$ can indeed be violated with the proposed set-up, and shows that there is an optimal squeezing λ_{opt} which maximizes S. This optimal squeezing is well predicted by the simplified model assuming ideal detectors with single-photon resolution, that is, $\lambda_{\text{opt}} T \approx 0.57$. The maximum achievable Bell factor is $S_{\max} \approx 2.045$, which represents a violation of about 2.2%. To get close to $S_{\max}$, one needs sufficiently high (but not too strong) squeezing. In particular, the value $\lambda \approx 0.57$ corresponds to approximately 5.6 dB of squeezing. Figure 8(b) illustrates that there is a clear trade-off between S and the probability of success P_G. To maximize S, one should use highly transmitting beam splitters ($T \approx 1$), but this would drastically reduce P_G. The optimal T must be chosen depending on the details of the experimental implementation.

7.3. *Sensitivity to experimental imperfections*

Let us now study the sensitivity of this Bell test to the different imperfections that would necessarily occur in a realistic optical experiment, namely the non-unity efficiency and non-zero probability of false triggers of the photodetectors, the non-unity efficiency and noise of the homodyne detection, and the thermal noise in the two-mode squeezed vacuum state.

First, the Bell factor S depends only very weakly on the efficiency η_{PD} of the single-photon detectors, so the Bell-CHSH inequality can be violated even if $\eta_{\text{PD}} \approx 1\%$. This is very important from the experimental point of view because, although the quantum detection efficiencies of the avalanche photodiodes may be of the order of 50%, the necessary spectral and spatial filtering which selects the mode that is detected by the photodetector may reduce the overall detection efficiency to a few percent. In practice, the minimum necessary η_{PD} will be determined mainly by the constraints on the total time of the experiment and by the dark counts of the detectors.

In contrast, the Bell factor S very strongly depends on the efficiency of the homodyne detectors, and η_{BHD} must be above $\approx 90\%$ in order to observe a Bell violation. However, this is not an obstacle because such (and even higher) homodyne efficiencies have already been achieved experimentally (see *e.g.* Ref. 48). Interestingly, it was found in Ref. 47 that it is possible to partially compensate for the low homodyning efficiency by increasing the squeezing of the initial state.

The electronic noise of the homodyne detector is another factor that may reduce the observed Bell violation. As shown in Ref. 47, the electronic noise should be $15 - 20$ dB below the shot noise, which is currently attainable with low-noise charge amplifiers. Again, higher squeezing can partially compensate for an increasing noise. Another source of noise originates from the state generation. In the analysis, it was assumed that the source emits a pure two-mode squeezed vacuum state. However, experimentally, it is very difficult to generate pure squeezed vacuum saturating the Heisenberg inequality. It is more realistic to consider a mixed Gaussian state. Here again, the added noise in the initial Gaussian state should be $15 - 20$ dB below the shot noise for a successful Bell test.[47]

Finally, a main source of imperfection that was observed in the experimental demonstration of single-photon subtraction in Ref. 46 comes from the false triggering of single-photon detectors. Indeed, a single-photon detector may be triggered by a photon coming from another mode than the one detected in the balanced homodyne detector. The single-mode description of a parametric amplifier is only an approximation, and the amplifier produces squeezed vacuum in several modes. A balanced homodyne detector very efficiently selects a single mode defined by the spatiotemporal profile of the local oscillator pulse. However, such a reference is missing in case of a single-photon detector, where the effective single mode has to be selected by spatial and spectral filtering, which reduces the overall detection efficiency η_{PD}. In practice, this filtering is never perfect, hence the photodetector PD_A (PD_B) can sometimes click although no photon was removed from mode A (B). This false triggering can be modeled by re-defining the POVM element $\Pi_{1,C}$ (and $\Pi_{1,D}$) as a convex mixture of the original POVM element $I - |0\rangle\langle 0|$, which corresponds to a triggering by a photon coming from the mode A (B), and the identity operator I, which corresponds to a false triggering. As expected, the achievable Bell factor decreases with increasing probability of false triggers P_f. For a transmittance $T = 0.95$, up to 6% of false triggers can be tolerated.[47] In the experiment reported in Ref. 46, the estimated fraction of false triggers was $P_f \approx 30\%$, which should thus be significantly reduced in order to realize a Bell test experiment. Possible ways of suppressing false triggers include better filtering and/or using sources that produce squeezed light in well defined spatial modes, such as nonlinear periodically poled waveguides.

8. Alternative Schemes

It is interesting to analyze whether alternative schemes to the one studied so far may possibly lead to a larger violation of the Bell-CHSH inequalities, therefore being more promising. Let us consider alternative schemes which involve from one to four photon subtractions. Since the probability of successful generation of a non-Gaussian state significantly decreases with the number of photon subtractions (while the complexity of the optical implementation increases with the number of photon subtractions), it is natural that the most interesting schemes for a Bell test are those involving only one photon subtraction. Unfortunately, none of the schemes with a single photon subtraction that were considered in Ref. 47 leads to a Bell violation, and it is unknown whether such a scheme can be found. The simplest schemes are then those involving two photon subtractions. In the preceding sections, it was shown that it is indeed possible to violate the Bell-CHSH inequality with the scheme of Fig. 4 involving two photon subtractions, giving $S_{\mathrm{max,2ph}} = 2.048$. Several other schemes with two photon subtractions may also be devised which violate the Bell-CHSH inequality, but the achievable Bell factor S is smaller, see Ref. 47.

By adding one more photon subtraction, one can construct an ensemble of schemes with three photon subtractions. After numerical optimization, it was found in Ref. 47 that none of these schemes succeeds in violating the Bell-CHSH inequality. This striking result together with the fact that no interesting scheme based on a single photon subtraction has been found suggests that it may be necessary to subtract an even number of photons in order to observe $S > 2$.

Among the various schemes with four photon subtractions that were studied in Ref. 47, the most interesting one is obtained by applying two photon subtractions on each mode. Numerical calculations show that the maximum Bell violation is achieved for $T^2\lambda = 0.40$ and yields $S_{\mathrm{max,4ph}} = 2.064$, which is indeed higher than the maximum achievable with two-photon subtraction, $S_{\mathrm{max,2ph}} = 2.048$, and very close to the maximum value $S = 2.076$ obtained in Ref. 39. Unfortunately, a more realistic description of the four-photon subtraction scheme that takes into account realistic imperfect detectors shows that, for $T < 0.95$, the fact that the photodetectors do not distinguish the number of photons reduces the Bell factor and dramatically decreases the probability of generating the non-Gaussian state ($P_G \approx 10^{-6}$). Therefore, it seems that, from a practical point of view, there is no advantage in using a scheme with four photon subtractions instead of the much simpler scheme with two photon subtractions shown in Fig. 4.

In a recent paper,[49] another scheme has been proposed for generating a state of the form (3) reaching $S = 2.071$, which is very close to the maximum $S = 2.076$ of Ref. 39. The state generation procedure needs several successive pairs of photon subtractions interspersed by some Gaussification operation.[31] Unfortunately, a realistic description of the scheme that takes into account realistic imperfect detectors should necessarily bring the same conclusions as for the schemes with four photon subtractions.

9. Conclusions

In this Chapter, we have described an experimentally feasible setup allowing for a loophole-free Bell test with efficient homodyne detection. This scheme is based on a non-Gaussian entangled state which is conditionally generated from a two-mode squeezed vacuum state by subtracting a single photon from each mode. We have discussed the influence on the achievable Bell violation of the detector inefficiencies, the electronic noise of the homodyne detector, the impurity of the input state, and the effect of false triggers in the single-photon detectors. The main advantage of this scheme is that it is largely insensitive to the detection efficiency of the avalanche photodiodes that are used for the conditional preparation of the non-Gaussian state, so that efficiencies of the order of a few per cent are sufficient. We also have discussed several alternative schemes that involve the subtraction of one, two, three or four photons. The current conclusion is that there seems to be no advantage in using these other schemes instead of the above two-photon subtraction scheme.

This analysis makes it possible to define a set of realistic parameter values, which should be reached in a loophole-free Bell test: with $\eta_{PD} = 30\%$, $T = 95\%$, and 6 dB of squeezing, a violation of the Bell-CHSH inequality by about 1% should be observable if the homodyne efficiency η_{BHD} is larger than 95% and less than 6% of false triggers impinge on the single-photon detectors. With a repetition rate of 1 MHz and $P \approx 2.6 \times 10^{-4}$, the number of data samples would be several hundreds per second, so that the required statistics to see a violation in the percent range could be obtained in a reasonable time (a few hours). In addition, the electronic noise of the homodyne detectors should be 15-20 dB below shot noise, which is attainable with low-noise charge amplifiers. All these numbers have already been reached separately in various experiments, but attaining them simultaneously certainly represents a serious challenge.

The very recent experimental demonstration of a single-photon subtraction from a single-mode squeezed vacuum state provides a strong incentive for further theoretical and experimental developments along these lines. In particular, the issues of more complex binnings, generalized Bell inequalities in higher dimension, or multipartite Bell inequalities deserve further investigations. Any improvement of the amount of violation would certainly help making such a loophole-free Bell test possible with the present technology.

Acknowledgments

We would like to thank J. Eisert, Ph. Grangier, R. Tualle-Brouri, and J. Wenger for many stimulating discussions. We acknowledge financial support from the Communauté Française de Belgique under grant ARC 00/05-251, from the IUAP programme of the Belgian government under grant V-18, and from the EU under projects COVAQIAL (FP6-511004). R.G-P. acknowledges support from the Belgian foundation FRIA. JF also acknowledges support from the Project Center of Modern Optics (LC06007) of the Czech Ministry of Education.

References

1. A. Einstein, B. Podolsky, and N. Rosen, Phys. Rev. **47**, 777 (1935).
2. J. S. Bell, Physics (Long Island City, N.Y.) **1**, 195 (1964).
3. J. F. Clauser, M.A. Horne, A. Shimony and R.A. Holt, Phys. Rev. Lett. **23**, 880 (1969).
4. S. J. Freedman and J. F. Clauser, Phys. Rev. Lett. **28**, 938 (1972).
5. A. Aspect, P. Grangier, and G. Roger, Phys. Rev. Lett. **47**, 460 (1981).
6. A. Aspect, P. Grangier, and G. Roger, Phys. Rev. Lett. **49**, 91 (1982).
7. A. Aspect, J. Dalibard, and G. Roger, Phys. Rev. Lett. **49**, 1804 (1982).
8. P. G. Kwiat, K. Mattle, H. Weinfurter, A. Zeilinger, A. V. Sergienko, and Y. Shih, Phys. Rev. Lett. **75**, 4337 (1995).
9. G. Weihs, T. Jennewein, C. Simon, H. Weinfurter, and A. Zeilinger, Phys. Rev. Lett. **81**, 5039 (1998).
10. W. Tittel, J. Brendel, B. Gisin, T. Herzog, H. Zbinden, and N. Gisin, Phys. Rev. A **57**, 3229 (1998).
11. Philip M. Pearle, Phys. Rev. D **2**, 1418 (1970).
12. E. Santos, Phys. Rev. A **46**, 3646 (1992).
13. P. G. Kwiat, P. H. Eberhard, A. M. Steinberg, and R. Y. Chiao, Phys. Rev. A **49**, 3209 (1994).
14. M.A. Rowe, D. Kielpinski, V. Meyer, C.A. Sackett, W. M. Itano, C. Monroe, and D.J. Wineland, Nature (London) **409**, 791 (2001).
15. J.S. Bell, Speakable and Unspeakable in Quantum Mechanics (Cambridge University Press, Cambridge, 1988).
16. E. Santos, Phys. Lett. A **200**, 1 (1995).
17. C. Simon and W.T.M. Irvine, Phys. Rev. Lett. **91**, 110405 (2003).
18. X.-L. Feng, Z.-M. Zhang, X.-D. Li, S.-Q. Gong, and Z.-Z. Xu, Phys. Rev. Lett. **90**, 217902 (2003).
19. L. M. Duan and H. J. Kimble, Phys. Rev. Lett. **90**, 253601 (2003).

20. D. E. Browne, M. B. Plenio, and S. F. Huelga, Phys. Rev. Lett. **91**, 067901 (2003)
21. B. B. Blinov, D. L. Moehring, L. M. Duan, and C. Monroe, Nature (London) **428**, 153 (2004).
22. E. S. Fry, T. Walther, and S. Li, Phys. Rev. A **52**, 4381 (1995).
23. M. Freyberger, P. K. Aravind, M. A. Horne, and A. Shimony, Phys. Rev. A **53**, 1232 (1996).
24. E. S. Polzik, J. Carri, and H. J. Kimble, Phys. Rev. Lett. **68**, 3020 (1992).
25. Z. Y. Ou, S. F. Pereira, H. J. Kimble, and K. C. Peng, Phys. Rev. Lett. **68**, 3663-3666 (1992).
26. C. Schori, J. L. Sørensen, and E. S. Polzik, Phys. Rev. A **66**, 033802 (2002).
27. W. P. Bowen, R. Schnabel, P. K. Lam, and T. C. Ralph, Phys. Rev. A **69**, 012304 (2004).
28. J. Eisert, S. Scheel, and M.B. Plenio, Phys. Rev. Lett. **89**, 137903 (2002).
29. J. Fiurášek, Phys. Rev. Lett. **89**, 137904 (2002).
30. G. Giedke and J.I. Cirac, Phys. Rev. A **66**, 032316 (2002).
31. D. E. Browne, J. Eisert, S. Scheel, and M. B. Plenio, Phys. Rev. A **67**, 062320 (2003).
32. J. Eisert, D. Browne, S. Scheel, and M. B. Plenio, Annals of Physics (NY) **311**, 431 (2004).
33. K. Banaszek and K. Wódkiewicz, Phys. Rev. A **58**, 4345 (1998).
34. Z.-B. Chen, J.-W. Pan, G. Hou, and Y.-D. Zhang, Phys. Rev. Lett. **88**, 040406 (2002).
35. L. Mišta, Jr., R. Filip, and J. Fiurášek, Phys. Rev. A **65**, 062315 (2002).
36. R. Filip and L. Mišta, Jr., Phys. Rev. A **66**, 044309 (2002).
37. A. Gilchrist, P. Deuar, and M. D. Reid, Phys. Rev. Lett. **80**, 3169 (1998).
38. A. Gilchrist, P. Deuar, and M. D. Reid, Phys. Rev. A **60**, 4259 (1999).
39. W. J. Munro, Phys. Rev. A **59**, 4197 (1999).
40. J. Wenger, M. Hafezi, F. Grosshans, R. Tualle-Brouri, and P. Grangier, Phys. Rev. A **67**, 012105 (2003).
41. R. García-Patrón, J. Fiurášek, N. J. Cerf, J. Wenger, R. Tualle-Brouri, and Ph. Grangier, Phys. Rev. Lett. **93**, 130409 (2004).
42. H. Nha and H. J. Carmichael, Phys. Rev. Lett. **93**, 020401 (2004).
43. T. Opatrný, G. Kurizki, and D.-G. Welsch, Phys. Rev. A **61**, 032302 (2000).
44. P. T. Cochrane, T. C. Ralph, and G. J. Milburn, Phys. Rev. A **65**, 062306 (2002).
45. S. Olivares, M. G. A. Paris, and R. Bonifacio, Phys. Rev. A **67**, 032314 (2003).
46. J. Wenger, R. Tualle-Brouri, and Ph. Grangier, Phys. Rev. Lett. **92**, 153601 (2004).
47. R. García-Patrón, J. Fiurášek, N. J. Cerf, Phys. Rev. A **71**, 022105 (2005).
48. T. C. Zhang, K. W. Goh, C. W. Chou, P. Lodahl, and H. J. Kimble, Phys. Rev. A **67**, 033802 (2003).
49. S. Daffer and P. L. Knight, Phys. Rev. A **72**, 034101 (2005).

Chapter 8

Homodyne Tomography and the Reconstruction of Quantum States of Light

Giacomo Mauro D'Ariano, Lorenzo Maccone and Massimiliano Federico Sacchi
QUIT - Quantum Information Theory Group Dip. di Fisica "A. Volta", Università di Pavia, via A. Bassi 6, I-27100 Pavia, Italy

Quantum tomography is a procedure to determine the quantum state of a physical system, or equivalently, to estimate the expectation value of any operator. It consists in appropriately averaging the outcomes of the measurement results of different observables, obtained on identical copies of the same system. Alternatively, it consists in maximizing an appropriate likelihood function defined on the same data. The procedure can be also used to completely characterize an unknown apparatus. Here we focus on the electromagnetic field, where the tomographic observables are obtained from homodyne detection.

1. Introduction

The properties of each physical system are, by definition, completely determined by its quantum state. Its mathematical description is given in form of a density operator ϱ. Bohr's principle of complementarity,[1] which is in many ways connected with the uncertainty relations,[2] forbids one to recover the quantum state from a single physical system. In fact, the precise knowledge of one property of the system implies that the measurement outcomes of the complementary observables are all equiprobable: the properties of a single system related to complementary observables are simultaneously unknowable. Moreover, the no-cloning principle[3] precludes to obtain many copies of a state starting from a single one, unless it is already known. Hence, complementarity and no-cloning prevent one to recover a complete information starting from a single quantum system, *i.e.* to recover its state. The only possibility is to recover it from multiple copies of the system. (Notice that, if the multiple copies are not all in the same quantum state, we will recover the mixed state of the ensemble). Given N copies of a system, we can either perform a collective measurement on all (or on subsets), or perform measurements separately on each system and combine the measurement results at the data analysis stage. Even though the former strategy would probably increase the speed of the statistical convergence of the measured state to the true one, it is quite impractical. Tomography thus adopts the latter strategy, which is the simplest to perform experimentally.

What is quantum tomography? It is the name under which all state reconstruction techniques are denoted. It derives from the fact that the first tomographic method (see Sec. 7) employed the same concepts of Radon-transform inversion we find in conventional medical tomographic imaging. Since then, better methods have evolved which eliminate the bias that the Radon-transform necessarily entails. These fall into two main categories: the plain averaging method and the maximum likelihood method. As will be seen in detail, the first method requires a simple averaging of a function calculated on the N measurement outcomes x_n of the homodyne quadratures X_{ϕ_n}. Thus, the statistical error which affects the estimated quantity can be easily evaluated through the variance of the data. The second method, *i.e.* the maximum likelihood method, is based on the assumption that the data we obtained is the most probable. Hence, we need to search for the state that maximizes the probability of such data, *i.e.* the state ϱ for which $\prod_{i=1}^{N} {}_{\phi_n}\langle x_n|\varrho|x_n\rangle_{\phi_n}$ is maximum, where ${}_{\phi_n}\langle x_n|\varrho|x_n\rangle_{\phi_n}$ is the probability of obtaining the result x_n when measuring the quadrature X_{ϕ_n} (which has eigenstates $|x\rangle_{\phi_n}$).

Their involved mathematical derivation has given these tomographic techniques a false aura of being complicated procedures. This is totally unjustified: the reader only interested in applying the method can simply skip all the mathematical details and proceed to Sec. 5, where we present only the end result, *i.e.* the procedure needed in practice for a tomography experiment (the experimental setup is, instead, given in Sec. 2.1).

The chapter starts by introducing the method of homodyne tomography in Sec. 2, along with the description of homodyne detectors, noise deconvolution and adaptive techniques to reduce statistical errors. Then, in Sec. 3 we present the Monte Carlo integration methods and the statistical error calculations that are necessary for the plain averaging technique. In Sec. 4, the maximum likelihood methods are presented and analyzed. In Sec. 5, the step-by-step procedure to perform in practice a tomography experiment is presented. In Sec. 6, a tomographic method to calibrate (*i.e.* completely characterize) an unknown measurement device is presented. Finally, in Sec. 7, a historical excursus on the development of quantum tomography is briefly given.

2. Homodyne Tomography

The method of homodyne tomography is a direct application of the fact that the displacements operators $\mathcal{D}(\alpha) = e^{\alpha a^\dagger - \alpha^* a}$ are a complete orthonormal set for the linear space of operators. Recalling that the scalar product in a space of operators takes the Hilbert-Schmidt form $\langle A|B\rangle = \mathrm{Tr}[A^\dagger B]$, this means that

$$A = \int_{\mathbb{C}} \frac{d^2\alpha}{\pi} \mathrm{Tr}[A\,\mathcal{D}^\dagger(\alpha)]\mathcal{D}(\alpha) = \int_0^\pi \frac{d\phi}{\pi} \int_{-\infty}^{+\infty} dr \frac{|r|}{4} \mathrm{Tr}[A\, e^{irX_\phi}] e^{-irX_\phi}, \quad (1)$$

where the polar variables $\alpha \equiv -ir\, e^{i\phi}/2$ were used in the second equality. Upon introducing the probability $p(x,\phi) = {}_\phi\langle x|\varrho|x\rangle_\phi$ of obtaining x when measuring the quadrature $X_\phi = (a^\dagger e^{i\phi} + a e^{-i\phi})/2$, one obtains the tomographic formula

$$\langle A\rangle = \mathrm{Tr}[A\varrho] = \int_0^\pi \frac{d\phi}{\pi} \int_{-\infty}^{+\infty} dx\; p(x,\phi)\; K_A(x,\phi)\;, \quad (2)$$

where

$$K_A(x,\phi) \equiv \int_{-\infty}^{+\infty} dr \frac{|r|}{4} \mathrm{Tr}[A\; e^{ir(X_\phi - x)}], \quad (3)$$

defines the *kernel* of homodyne tomography. In the case of the density matrix reconstruction in the Fock basis $|n\rangle$ (*i.e.* when $A = |n\rangle\langle m|$), the kernel function is[4]

$$K_A(x,\phi) = 2e^{i(m-n)\phi} \sqrt{\frac{m!}{n!}} e^{-x^2} \sum_{j=m-n}^{n} \frac{(-1)^j}{j!} \binom{n}{m-j} \times (2j+n-m+1)!\; \mathsf{Re}\Big[(-1)^{n-m} \mathcal{D}_{-2(2j+n-m+2)}(-2ix)\Big], \quad (4)$$

where Re denotes the real part and $\mathcal{D}_l(x)$ denotes the parabolic cylinder function (which can be easily calculated through its recursion formulas).

The multimode case is immediately obtained by observing that the quadrature operators for different modes commute, so that for an operator A_M (acting on the Hilbert space of M modes) we find

$$\langle A_M\rangle = \int_0^\pi \frac{d\phi_1 \cdots d\phi_M}{\pi^M} \int_{-\infty}^{+\infty} dx_1 \cdots dx_M\; p(x_1,\phi_1,\cdots,x_M,\phi_M) \times K_{A_M}(x_1,\phi_1,\cdots,x_M,\phi_M), \quad (5)$$

where $p(x_1, \phi_1, \cdots, x_M, \phi_M)$ is the joint probability of obtaining the results $\{x_m\}$ when measuring the quadratures $\{X_{\phi_m}\}$, and where

$$K_{A_M}(x_1, \phi_1, \cdots) \equiv \int_{-\infty}^{+\infty} dr_1 \cdots dr_M \prod_{m=1}^{M} \frac{|r_m|}{4} \mathrm{Tr}[A_M \, e^{ir_m(X_{\phi_m} - x_m)}]. \quad (6)$$

However, such a simple generalization to multimode fields requires a separate homodyne detector for each mode, which is unfeasible when the modes of the field are not spatio-temporally separated. This is the case, for example of pulsed fields, for which a general multimode tomographic method is especially needed, because of the problem of mode matching between the local oscillator and the detected fields (determined by their relative spatio-temporal overlap), which produces a dramatic reduction of the overall quantum efficiency. A general method for multimode homodyne tomography can be found[5] that uses a *single* local oscillator that randomly scans all possible linear combinations of incident modes.

2.1. *Homodyne detection*

The balanced homodyne detector[6] measures the quadratures $X_\phi \equiv (a^\dagger e^{i\phi} + ae^{-i\phi})/2$. The experimental setup is described in Fig. 1. The input-output transformations of the modes a and b that impinge into a 50-50 beam-splitter are $c = (a+b)/\sqrt{2}$, $d = (a-b)/\sqrt{2}$ where c and d are the two beam-splitter output modes, each of which impinge into a different photodetector. The difference of the two photocurrents is the homodyne detector's output, and thus is proportional to $c^\dagger c - d^\dagger d = a^\dagger b + b^\dagger a$. In the strong local oscillator limit, with mode b in an excited coherent state $|\beta\rangle$ ($|\beta| \gg 1$), the expectation value of the output is $I_H \propto \langle a^\dagger \rangle \beta + \langle a \rangle \beta^*$ which is proportional to the expectation value of the quadrature X_ϕ, with ϕ the relative phase of the local oscillator.

A detector with non-unit quantum efficiency η is equivalent[7] to a perfect $\eta = 100\%$ detector, preceded by a beam-splitter with transmissivity η. Inserting two beam-splitters in front of the two photodiodes of the homodyne scheme, the modes c and d evolve as $c' = \sqrt{\eta}\, c + \sqrt{1-\eta}\, u$ and $d' = \sqrt{\eta}\, d + \sqrt{1-\eta}\, v$, where u and v are vacuum noise modes. The homodyne output, is now proportional to $c'^\dagger c' - d'^\dagger d'$, *i.e.* to $L \equiv \eta\, (a^\dagger b + b^\dagger a) + (1-\eta)(u^\dagger u - v^\dagger v) + \sqrt{(1-\eta)\eta/2}[a(u^\dagger - v^\dagger) + b(u^\dagger + v^\dagger) + a^\dagger(u-v) + b^\dagger(u+v)]$. As before, we take the limit $|\beta| \gg 1$ of strong pump in b, and rescale the

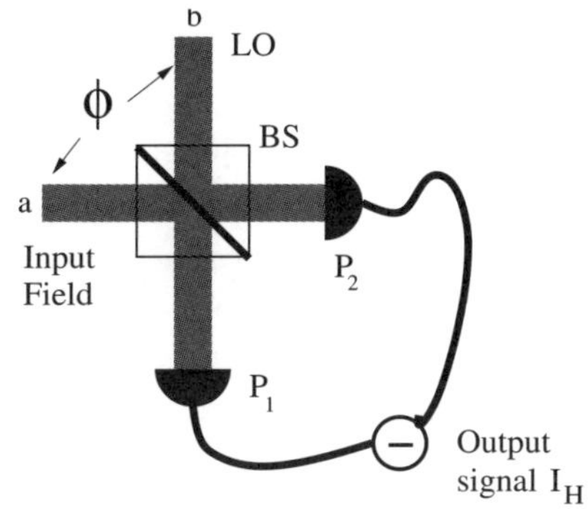

Fig. 1. Homodyne detector. The input signal (in mode a) is mixed by a 50-50 beam-splitter (BS) with a strong local oscillator (LO), which is coherent with the input field and is in a strong coherent state. The relative phase ϕ between the signal and the LO must be known and should be varied in $[0, \pi]$ with uniform probability. Two identical high efficiency linear photodetectors P_1 and P_2 measure the field. The photocurrents are then accurately subtracted electronically yielding the output I_H. Since the LO amplifies the weak quantum signals of the input, one can use high efficiency detectors that work only with strong signals.

output difference photocurrent by $2|\beta|\eta$, obtaining

$$I_H(\eta) = \lim_{|\beta|\to\infty} \frac{\langle L\rangle}{2|\beta|\eta} = \langle X_\phi\rangle + \sqrt{\frac{1-\eta}{2\eta}}\langle u_\phi + v_\phi\rangle, \tag{7}$$

where the modes u and v are in the vacuum state. Since the quadrature outcome for each vacuum state is Gaussian-distributed with variance $1/4$, this means that the distribution of the noisy data are a convolution of the clean data with a Gaussian of variance $\Delta_\eta^2 = (1-\eta)/(4\eta)$, namely

$$p_\eta(x,\phi) = \frac{1}{\sqrt{2\pi\Delta_\eta^2}} \int_{-\infty}^{+\infty} dx' e^{-(x-x')^2/(2\Delta_\eta^2)}\, p(x',\phi). \tag{8}$$

2.2. *Noise deconvolution*

The data-analysis procedure can be modified to yield the result we would obtain from perfect detectors, even though the data was collected with noisy ones.[8] In fact, depending on which operator A we consider and on the value of the quantum efficiency η, the noise may be numerically deconvolved. The output of the noisy homodyne is distributed according to Eq. (8), and one can rewrite Eq. (2) as follows

$$\langle A\rangle = \int_0^\pi \frac{d\phi}{\pi} \int_{-\infty}^{\infty} dx\, p_\eta(x,\phi) \int_{-\infty}^{\infty} dr\, \frac{|r|}{4}\, e^{r^2\Delta_\eta^2/2}\, \mathrm{Tr}[A\, e^{ir(X_\phi - x)}], \quad (9)$$

where $p_\eta(x,\phi)$ is the probability of the noisy data. In the case when all the integrals are convergent, the noise inversion can be performed successfully.

It is clear the possibility of noise deconvolution depends on the quantum efficiency of the detectors and the operator to be estimated. For example, there is a bound $\eta > 50\%$ for the reconstruction of the density matrix in the Fock basis (*i.e.* for $A = |n\rangle\langle m|$). In fact, one can see that for $\eta < 50\%$ Eq. (9) has an unbounded kernel. Notice that actual homodyne detectors have efficiencies ranging between 70% and 90%.

2.3. *Adaptive tomography*

Adaptive tomography[9] exploits the existence of *null estimators* to reduce statistical errors. In fact, the addition of a null estimator in the ideal case of infinite statistics does not change the average of the data since, by definition, the mean value of a null estimator is zero. However, it can change the variance of the data. Thus, one can look for a procedure to reduce the variance by adding suitable null functions.

In homodyne tomography null estimators are obtained as linear combinations of the following operators

$$\mathcal{N}_{k,n}(X_\phi) = X_\phi^k\, e^{\pm i(k+2+2n)\phi}\,, \qquad k, n \geq 0. \quad (10)$$

One can easily check that such functions have zero average over ϕ, independently on ϱ. Hence, for every operator A one actually has an equivalence class of infinitely many unbiased estimators, which differ by a linear combination of functions $\mathcal{N}_{k,n}(X_\phi)$. It is then possible to minimize the rms error in the equivalence class by the least-squares method. This yields an optimal estimator that is adapted to the particular set of experimental data. Examples of simulations of the adaptive technique that efficiently reduce statistical noise of homodyne tomographic reconstructions can be found in Ref. 9.

3. Monte Carlo Methods for Tomography

In this section we will very briefly review the basics of the Monte Carlo integration techniques that are needed and we show how to evaluate the statistical error bars of the tomographically estimated quantities.

A tomographic technique is based on an integral of the form

$$F = \int_{-\infty}^{+\infty} dx\, p(x)\, f(x), \tag{11}$$

where $p(x)$ is a probability. Since we have experimental outcomes $\{x_n,\ n = 1, \cdots N\}$ distributed according to the probability $p(x)$, we sample the integral (11) using

$$\int_{-\infty}^{+\infty} dx\, p(x)\, f(x) = \lim_{N\to\infty} \frac{1}{N} \sum_{n=1}^{N} f(x_n). \tag{12}$$

For finite N, the sum will be an unbiased estimator for the integral, affected by statistical errors only (which can be made arbitrarily small by increasing N). The central limit theorem guarantees that the finite sum $F_N = \sum_{n=1}^{N} f(x_n)/N$ is a statistical variable distributed as a Gaussian (for sufficiently high N) with mean value F and variance

$$\sigma^2 = \frac{1}{N^2} \sum_{n=1}^{N} \lim_{M\to\infty} \frac{1}{M} \sum_{j=1}^{M} [f(x_j)]^2 - F^2 = \frac{\sigma^2(F)}{N}. \tag{13}$$

Hence, the tomographic estimated quantity converges with a statistical error that decreases as $1/\sqrt{N}$. It can be estimated from the data as

$$s^2(F_n) = \frac{1}{N-1} \sum_{n=1}^{N} (F_n - m)^2\ . \tag{14}$$

(Remember that the factor $N-1$ in the variance denominator arises from the fact that we are using the experimental estimated mean value m in place of the real one F.) The variance of the statistical variable 'mean m' is then given by $\sigma^2(m) = \sigma^2(F_N)/N$, and thus the error bar on the mean m estimated from the data is given by

$$\epsilon = \frac{1}{\sqrt{N}}\, s(F_N) = \Big[\sum_{n=1}^{N} \frac{(F_n - m)^2}{N(N-1)}\Big]^{1/2}. \tag{15}$$

From the Gaussian integral one recovers the usual statistical interpretation to the obtained results: the "real" value F is to be found in the interval $[m-\epsilon, m+\epsilon]$ with $\sim 68\%$ probability, in the interval $[m-2\epsilon, m+2\epsilon]$ with $\sim 95\%$ probability and in $[m-3\epsilon, m+3\epsilon]$ with $\sim$ unit probability.

In order to test that the confidence intervals are estimated correctly and that errors in the data analysis or systematic errors in the experimental data do not undermine the final result, one may check the F_n distribution, to see if it actually is a Gaussian distribution. This can be done by comparing a histogram of the data to a Gaussian, or by using the χ^2 test. Notice that when we have very low statistics it may be useful to use also bootstrapping techniques to calculate the variance of the data.

For a more rigorous treatment of the statistical properties of quantum tomography, and also some open statistical questions, see Ref. 10.

4. Maximum Likelihood Tomography

The maximum likelihood tomography is based on the assumption that the data obtained from the measurements is the most likely.[11] In contrast to the plain averaging method presented above, the outcome is not a simple average of functions of the data, but a Lagrange-multiplier maximization is usually involved. The additional complexity introduced is compensated by the fact that the results are statistically less noisy. Estimation of operator expectation values is, however, indirect: one must first estimate the state ϱ and then calculate the expectation value as $\mathrm{Tr}[\varrho A]$.

Consider a known probability distribution $p_\gamma(x)$ parametrized by a parameter γ (which may also be a multidimensional parameter). We want to estimate the value of γ from the data set $\{x_1, \cdots, x_N\}$. The joint probability of obtaining such data is given by the likelihood function

$$\mathcal{L}(x_1, \cdots, x_N; \gamma) = \prod_{i=1}^{N} p_\gamma(x_k). \tag{16}$$

The maximum likelihood procedure consists essentially in finding the $\gamma_0(x_1, \cdots, x_N)$ which maximizes the likelihood function $\mathcal{L}(x_1, \cdots, x_N; \gamma)$. Equivalently, it may be convenient to maximize its logarithm $\log \mathcal{L}(x_1, \cdots, x_N; \gamma)$, in order to convert into a sum the product in Eq. (16). Usually, various constraints are known on the parameters γ, which can be taken into account by performing a constrained maximization. The confidence interval for the estimated γ_0 can be evaluated from the data using a bootstrapping technique: we can extract a rough estimate of the probability distribution of the $\{x_i\}$ from the data set, generate M simulated sets of N data points, and repeat the procedure to obtain a set of M parameters $\gamma_0^{(m)}$. Their variance estimates the variance of the reconstruction. Moreover, if a sufficiently large data set is present, we can attain the Cramer-Rao bound $\sigma_\gamma^2 \geqslant 1/NF_\gamma$, where F_γ is the Fisher information relative to $p_\gamma(x)$, *i.e.*

$$F_\gamma \equiv \int dx \frac{1}{p_\gamma(x)} \left(\frac{\partial}{\partial \gamma} \, p_\gamma(x) \right)^2 . \tag{17}$$

Since the Cramer-Rao bound is achieved only for the optimal estimator,[12] the maximum likelihood is among the best (*i.e.* least statistically noisy) estimation procedures.

The maximum likelihood method can be extended to the quantum domain.[11] The probability distribution of a measurement is given by the Born rule as $p_i = \text{Tr}[\Pi_i \varrho]$ where $\{\Pi_i\}$ is the positive operator-valued measurement (POVM) that describes the measurement. Thus we need to maximize the log-likelihood function $L(\varrho) \equiv \sum_i \log \text{Tr}[\Pi_i \varrho]$ over the set of density operators ϱ. In the case of finite Hilbert space, $L(\varrho)$ is a concave function defined on a convex set of density operators: its maximum is achieved on a single point or on a convex subset. The main difficulty of this procedure consists in finding a simple parameterization for the density matrix, that enforces both the positivity and the normalization $\text{Tr}[\varrho] = 1$. The former is guaranteed by requiring that $\varrho = T^\dagger T$, the latter must be taken into account through an appropriate Lagrange multiplier. In order to employ the minimum number of parameters, it is sufficient to consider T as an upper complex triangular matrix with nonnegative diagonal elements—so called Cholesky decomposition. This decomposition achieves minimal parameterization (up to the normalization condition), as it requires d^2 real parameters for a $d \times d$ Hermitian matrix. Thus, in practice we need to maximize the operator $L_\lambda[\varrho] \equiv \sum_i \log\text{Tr}[\Pi_i T^\dagger T] - \lambda[T^\dagger T]$, where λ is a Lagrange multiplier that accounts for the normalization. By expressing ϱ in terms of its eigenstates as $\varrho = \sum_m y_m^2 |\psi_m\rangle\langle\psi_m|$, the condition for the maximum, $\partial L_\lambda / \partial y_m = 0$, becomes

$$\sum_i \{y_m \langle\psi_m|\Pi_i|\psi_m\rangle / \text{Tr}[\varrho \Pi_i]\} - \lambda\, y_m = 0 \qquad \forall m. \tag{18}$$

Multiplying both members by y_m and summing over m, through the Born rule and the normalization of ϱ, we find that λ is equal to the number of measurements employed. Thus, we are left with the problem of finding the maximum of the d^2-parameter function $L_{\lambda=N}[\varrho = T^\dagger T]$, which can be tackled with conventional numerical techniques such as expectation-maximization or downhill simplex.[11] By using the ML method only small samples of data are required for a precise determination, even in the presence of low quantum efficiency at the detectors. However, we want to emphasize that such method is not always the optimal solution of the tomographic problem, since it suffers from some major limitations. Besides being biased due to the Hilbert space truncation — even though the bias can be very small if, from other methods, we know where to truncate — it cannot be generalized to the estimation of any ensemble average, but just of a set of parameters from which the density matrix depends. In addition, for the multi-mode case, the method has exponential complexity versus the number of modes.

5. Tomography for Dummies

In this section we just give the step-by-step procedure to implement a tomography experiment, employing all the results obtained in the previous sections.

- Plain averaging method:

(1) Calculate the Kernel function K_A for the operator A whose expectation value we want to estimate through Eq. (3). For example, to estimate the density matrix in the Fock basis, we need the K_A defined in Eq. (4).
(2) The experimental apparatus, described in Sec. 2.1, yields a set of N data points $\{\phi_n, x_n\}$: each datum is composed by the quadrature phase ϕ_n that was measured and by the corresponding measurement result x_n .
(3) Evaluate $\frac{1}{N}\sum_n K_A(x_n, \phi_n)$. In the limit $N \to \infty$ this average yields the expectation value $\langle A \rangle$ we are looking for.
(4) For finite N, we can estimate the purely statistical error on the result through Eq. (15), replacing m with the average obtained at the previous point and F_n with the nth Kernel function evaluation, $K_A(x_n, X_{\phi_n})$.

Further data massaging is also possible: we can employ adaptive tomography to reduce the statistical noise (see Sec. 2.3). Moreover, we can remove the detector noise due to homodyne measurements with non unit quantum efficiency η, as long as $\eta > 1/2$ (see Sec. 2.2).

- Maximum likelihood method:

(1) Parametrize the unknown quantum state through the upper triangular $d \times d$ matrix T as $\varrho = T^\dagger T$.
(2) Use the same experimental apparatus (homodyne detection) to obtain N data points $\{\phi_n, x_n\}$. Calculate the log likelihood function on the experimental data as $\log \sum_{n=1}^{N} {}_{\phi_n}\langle x_n | T^\dagger T | x_n \rangle_{\phi_n}$.
(3) Numerically maximize this quantity over the d^2 parameters of T with the additional constrain $\mathrm{Tr}[T^\dagger T] = 1$. This maximum is achieved on our best estimate for the state $\varrho = T^\dagger T$.
(4) The confidence intervals for our estimation can be obtained using bootstrapping techniques, or employing the Cramer-Rao bound of Eq. (17).

6. Quantum Calibration of Measurement Devices

In this section we review the method to measure the POVM of an unknown measurement apparatus presented in Ref. 13. The method is based on analyzing the correlations in measurements on a bipartite system: one of the two parts is fed into the unknown apparatus A, while the other is measured with a known set B of detectors that measures a quorum of observables (see Fig. 2). As will be shown in the following, there is ample freedom in the choice of both the input bipartite states and the set of observables. The procedure is repeated many times and the joint measurement outcomes are analyzed using the tomographic algorithms described above, which (in the limit of infinite input data) yield the POVM of the unknown apparatus. For finite data, the reconstructed POVM will be affected only by statistical errors which can be easily estimated. For the sake of illustration, a Monte-Carlo simulation of the procedure is given at the end of this section. It aptly illustrates the advantage of using maximum likelihood techniques over plain averaging: the maximum likelihood reconstruction is significantly less noisy.

The following simple example illustrates how the procedure works. Suppose we want to evaluate the POVM of a von Neumann measurement of the observable O which acts on a d-dimensional Hilbert space $\mathcal{H}_A$ and has

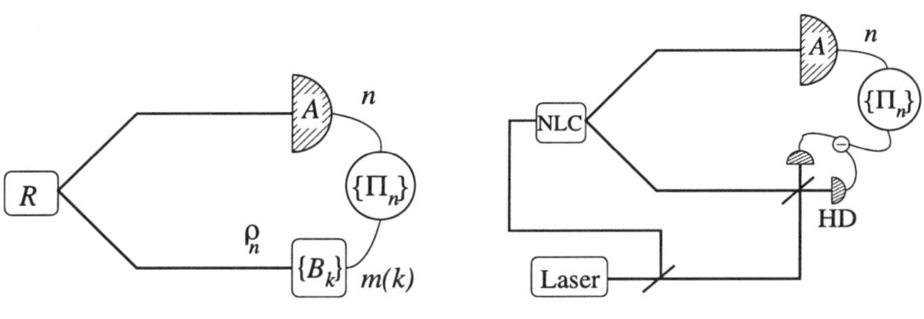

Fig. 2. (Left) Experimental setup to determine the POVM of the unknown measurement apparatus A: one part of the bipartite input state R is sent to the apparatus A which yields the measurement result n; the other part (with quantum state ϱ_n) is sent to the known detector B which performs a projective measurement of an observable B_k from the complete set $\{B_k\}$ yielding the result $m(k)$. The joint measurement results are processed using a tomographic algorithm to obtain the POVM $\{\Pi_n\}$ of A. (Right) Example of application of the scheme to the radiation field. The bipartite state R is generated via a non-linear crystal through spontaneous parametric down-conversion. The tomographer B is, in this case, a homodyne detector (HD) which measures the quadratures, a complete set of observables.

spectral decomposition $\sum_n o_n|o_n\rangle\langle o_n|$. We can use the maximally entangled input state $|\Psi\rangle = \sum_{i=1}^d |i\rangle|i\rangle/\sqrt{d}$, which lives in the space $\mathcal{H}_A \otimes \mathcal{H}_B$. In fact, this state can be also written as $\frac{1}{\sqrt{d}}\sum_{j=1}^d |o_j\rangle|o_j^*\rangle$, where $*$ denotes complex conjugation with respect to the basis $|i\rangle$. Clearly the outcome o_n at detector A (corresponding to the state $|o_n\rangle$ in $\mathcal{H}_A$) means that the state $\varrho_n = |o_n^*\rangle\langle o_n^*|$ in $\mathcal{H}_B$ impinges in detector B. The POVM can be recovered using tomographical state reconstruction at B, since in this simple case $\Pi_n = \varrho_n^*$.

It is not difficult to generalize the above example to arbitrary POVMs and measurement procedures. Let the unknown apparatus A be described by the POVM $\{\Pi_n\}$ we want to estimate, and let the apparatus B measure the quorum observables O_k described by the von Neumann projections $\{|k_m\rangle\langle k_m|\}$ (with $\{|k_m\rangle\}$ basis for all k). From the Born statistical formula we can derive the state that impinges into the known detector B if the unknown detector A gave result n for the measurement on the initial bipartite state R, as

$$\varrho_n = \frac{\mathrm{Tr}_1[(\Pi_n \otimes \mathbb{1})R]}{\mathrm{Tr}[(\Pi_n \otimes \mathbb{1})R]}. \tag{19}$$

It describes the state reduction at B stemming from a measurement at A with outcome n. The denominator is the probability $p(n)$ of obtaining the result n at B. The state ϱ_n contains some information on the POVM element Π_n. It can be recovered by introducing the map $\mathcal{R}(X) \equiv \mathrm{Tr}_1[(X \otimes \mathbb{1})R]$, so that Eq. (19) rewrites as $\varrho_n = \mathcal{R}[\Pi_n/p(n)]$. This implies that the POVM can be recovered as $\Pi_n = p(n)\mathcal{R}^{-1}(\varrho_n)$, where the map $\mathcal{R}$ depends only on the input state R: the input state R allows the POVM reconstruction if the inverse map $\mathcal{R}^{-1}$ exists. This condition can be cast in a more transparent form by rewriting the map $\mathcal{R}$ in a multiplicative form via isomorphism between operators on $\mathcal{H} \otimes \mathcal{H}$ and maps.[14] We can obtain an operator of this form by considering $S = R^{T_1}$, *i.e.* the partial transposition on the first space of the input state R. In fact, taking two operators X and Y such that $Y = \mathcal{R}(X)$, we see that

$$Y_{il} = \sum_{jk} X_{jk}\langle i|\mathcal{R}\Big(|j\rangle\langle k|\Big)|l\rangle = \sum_{jk} X_{jk}(R^{T_1})_{jk,il}, \tag{20}$$

where $Y_{il} = \langle i|Y|l\rangle$, $X_{jk} = \langle j|X|k\rangle$, and $(R^{T_1})_{jk,il} = \langle j|\langle i|R^{T_1}|k\rangle|l\rangle$, the set $\{|n\rangle\}$ being a basis in $\mathcal{H}$. In matrix notation (considering jk and il as collective indexes), Eq. (20) rewrites as $Y = SX$. It follows immediately that the map $\mathcal{R}$ is invertible if S^{-1} exists so that $X = S^{-1}Y$. In this case we say that the input state R is faithful.[14] Since invertibility is a condition satisfied by a dense set of operators, the set of input states R that allow the POVM reconstruction is also dense, *i.e.* almost any bipartite state will do. In particular, all Gaussian bipartite states—with the trivial exception of product states—are faithful.[15] To recapitulate: in order to check whether the state ϱ_n allows to obtain the POVM (*i.e.* whether the input state R is faithful) we must verify that the operator $(R^{T_1})_{jk,il}$ is invertible when jk and il are considered as collective indexes. As an illustration of this check, take the simple example given above: the state $|\Psi\rangle = \sum_i |ii\rangle/\sqrt{d}$ is faithful since $|\Psi\rangle\langle\Psi|^{T_1} = \sum_{ij} |ji\rangle\langle ij|/d$ is invertible: it is a multiple of the swap operator $E \equiv \sum_{ij} |ji\rangle\langle ij|$.

To recover ϱ_n from the measurements at B (and hence the POVM if the input R is faithful), we can use the quantum tomographic techniques described in the previous sections. If we employ the plain averaging technique, we may recover the density matrix elements ϱ_{ij} in some basis and then calculate the POVM using the inverse map $\mathcal{R}^{-1}$, as

$$\langle j|\Pi_n|k\rangle = p(n)\sum_{il} \varrho^{(n)}_{il}\;(R^{T_1})^{-1}_{jk,il},$$

where the inverse of R^{T_1} must be calculated considering jk and il as collective indexes. On the other hand, if we employ maximum likelihood we may directly maximize the probability of acquiring the data we obtained from the measurements,[11] *i.e.* the joint probability $p_k(n,m) = \mathrm{Tr}[(\Pi_n \otimes |k_m\rangle\langle k_m|)R]$. Equivalently, one can maximize the logarithm of this quantity and consider simultaneously all the N joint measurement outcomes $\{n_1, m_1\}, \cdots, \{n_N, m_N\}$ of the quorum operators $O_{k^{(i)}}$ at detector A and of the unknown detector B. Thus, the POVM $\{\Pi_n\}$ is the one that maximizes the quantity

$$\mathcal{L}(\{\Pi_n\}) \equiv \sum_{i=1}^{N} \log \mathrm{Tr}\Big[(\Pi_{n_i} \otimes |k^{(i)}_{m_i}\rangle\langle k^{(i)}_{m_i}|)R\Big], \tag{21}$$

with the additional constraints $\Pi_n \geqslant 0$ and $\sum_n \Pi_n = \mathbb{1}$. Other prior knowledge on the quantities to be estimated can be easily introduced adding further constraints to the maximization. Also in this case it is possible to take into account a known source of noise at the detector B: if we replace the term $|k^{(i)}_{m_i}\rangle\langle k^{(i)}_{m_i}|$ in Eq. (21) with the noise-evolved $\mathcal{N}(|k^{(i)}_{m_i}\rangle\langle k^{(i)}_{m_i}|)$, then the maximization yields the POVM that maximizes the *noisy* measurement results.

For the sake of illustration, we give a Monte-Carlo simulation of the calibration procedure in which we recover the POVM of a simple inefficient photodetector.[13] An inefficient photodetector is aptly modeled by a perfect photodetector (which is a device which measures the observable "number of photons" $a^\dagger a = \sum_n n|n\rangle\langle n|$), preceded by a beam-splitter with a transmissivity equal to the quantum efficiency η of the detector. Possible dark counts can be considered by feeding the other beam-splitter port with a thermal state with $\bar{n}$ average photons. In this case, the theoretical POVM is given by

$$\Pi_n = \sum_{p=0}^{\infty} |p\rangle\langle p| \tag{22}$$

$$\times \sum_{k=0}^{\infty} \sum_{j=0}^{\min(p,k+n)} \binom{p}{j} \binom{-n-1}{k} \binom{k+n}{j} \eta^j (1-\eta)^{k+n-j}\, \bar{n}^{k+n-j}.$$

Since this POVM is diagonal in the Fock basis, we can limit the reconstruction to the diagonal elements. As input state R we employ a twin beam state $|TB\rangle$, *i.e.* the result of spontaneous parametric down-conversion:

$$|TB\rangle \equiv \sqrt{1-|\xi|^2} \sum_m \xi^m |m\rangle_a |m\rangle_b, \tag{23}$$

where ξ is the parametric amplifier gain and $|m\rangle_a$ and $|m\rangle_b$ are Fock states of the modes a and b that impinge in the detectors A and B respectively. This is a faithful state since $|TB\rangle\langle TB|^{T_1} = (1-|\xi|^2) E \xi^{a^\dagger a} \otimes \xi^{* b^\dagger b}$ (where E is the swap operator) is invertible. The photon counter measures the mode a at position A, while homodyne detection with quantum efficiency η_h measures the mode b at position B acting as tomographer (see Fig. 2). Since only the diagonal part of the POVM is needed, we can use a homodyne detector with uniformly distributed local oscillator phase. (A phase-controlled homodyne detector would allow to recover also the off-diagonal elements of the POVM, ensuring a complete characterization of the device.)

In Figs. 3 and 4 we present the results of the POVM reconstruction deriving from the two tomographic methods described above (simple averaging and maximum likelihood, respectively). The convergence of the maximum likelihood procedure is assured since the likelihood functional $\mathcal{L}$ is convex over the space of diagonal POVMs. However, the convergence speed can become very slow: in the simulation of Fig. 4 a mixture of sequential quadratic programming (to perform the constrained maximization) and expectation-maximization techniques were employed. From the graphs it is evident that the maximum likelihood estimation is statistically more efficient since it needs much less experimental data than tomography. This is a general characteristic of this method, since if the optimal estimator (*i.e.*

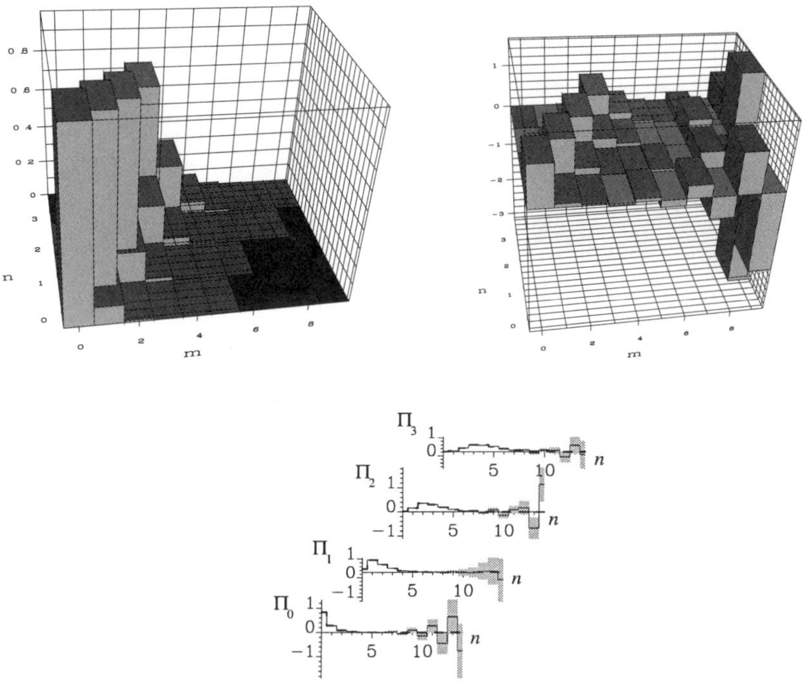

Fig. 3. (Above left) Theoretical value of the diagonals of the POVM elements $\langle m|\Pi_n|m\rangle$ of the inefficient photodetector described by Eq. (22), with parameters $\bar{n} = 1$, $\eta = 80\%$. (Above right) Simulated reconstruction of the same quantity. The data are simulated as coming from an input twin-beam state $|TB\rangle$ with $\xi = 0.88$, and as being detected from a phase insensitive homodyne detector with quantum efficiency $\eta_h = 90\%$. Here 5×10^6 simulated homodyne measurements are employed. (Below) The same data is plotted separately for each POVM element to emphasize the error bars. They are obtained from the root-mean-square of the recovered POVM matrix elements. (The theoretical value is plotted as the thick dashed line.) Plain tomographic averaging with noise deconvolution has been employed here, since the noise map of inefficient homodyne detection can be inverted for $\eta_h > 50\%$.

the one achieving the Cramer-Rao bound) exists, then it is equal to the maximum likelihood estimator.[11] An added bonus, evident from Eq. (21), is that the maximum likelihood recovers all the POVM elements at the same time additionally increasing the statistical efficiency. On the other hand, the tomographic reconstruction is completely unbiased: no previous information on the quantity to be recovered is introduced.

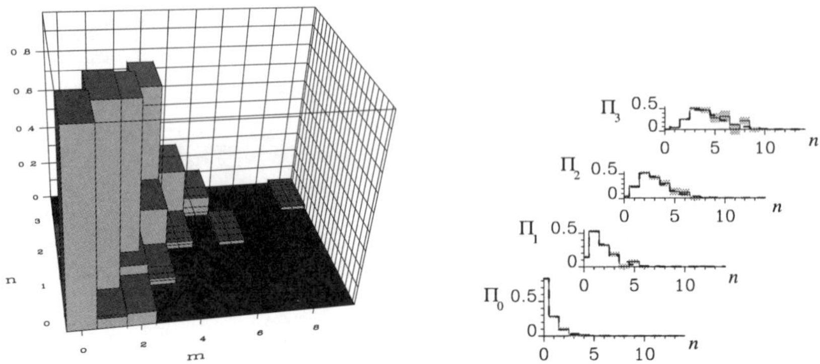

Fig. 4. Maximum likelihood reconstruction of the same POVM of Fig. 3 with the same parameters, but here only 5×10^4 simulated homodyne measurements are employed. The statistical error bars are obtained by bootstrapping, *i.e.* by calculating the variance using the data of 50 numerical experiments. Notice that the result is statistically less noisy than the results presented in Fig. 3 even if here less measurements are employed: maximum likelihood is usually a better estimator.

This simulated experiment uses realistic parameters and is feasible in the lab with currently available technology.[16] The major experimental challenge lies in the phase matching of the detectors, *i.e.* in ensuring that the modes detected at A and B actually correspond to the modes a and b of the state $|TB\rangle$.

7. History of Quantum Tomography

In this section a brief historical perspective (see also Refs. 17, 18) on quantum tomography is presented. Already in 1957 Fano[19] stated the problem of quantum state measurement, followed by rather extensive theoretical work. It was only with the proposal by Vogel and Risken,[20] however, that homodyne tomography was born. The first experiments followed[21] by showing reconstructions of coherent and squeezed states. The main idea at the basis of these works, is that it is possible to extend to the quantum domain the algorithms that are conventionally used in medical tomographic imaging to recover two-dimensional distributions (say of mass) from unidimensional projections in different directions. However, these first tomographic methods are unreliable for determining an unknown quantum state, since arbitrary smoothing parameters need to be used, which introduce an uncontrollable bias that depends on the state.

A first exact approach to quantum homodyne tomography was then proposed[22,23] which bypasses the Radon transform inversion, allowing to recover the quantum state of the field ϱ directly from the data, without systematic errors. Only statistical errors (that can be reduced arbitrarily by collecting more experimental data) are left. Quantum tomography has been then generalized to the estimation of arbitrary observable of the field[24] (in infinite dimensions the estimation of unbounded observables from the measured density matrix is limited by statistical errors on matrix elements), to any number of modes,[5] and to arbitrary quantum systems via group theory,[25] with further improvements such as noise deconvolution,[8] adaptive tomographic methods,[9] and the use of max-likelihood strategies,[11] which has made possible to reduce dramatically the number of experimental data, with negligible bias for most practical cases of interest. The latest developments are based on a general method,[26] where the tomographic reconstruction is based on the existence of spanning sets of operators, of which group tomography[25] is just a special case.

Acknowledgments

We acknowledge financial support by INFM PRA-2002-CLON and MIUR for Cofinanziamento 2003 and ATESIT project IST-2000-29681.

References

1. N. Bohr, Naturwissenschaften **16**, 245 (1928).
2. See, for example, S. Dürr and G. Rempe, Am. J. Phys. **68**, 1021 (2000); O. Steuernagel, Eprint quant-ph/9908011 (1999).
3. W. K. Wootters and W. H. Zurek, Nature **299**, 802 (1982); H. P. Yuen, Phys. Lett. **A** 113, 405 (1986).
4. G. M. D'Ariano, *Tomographic methods for universal estimation in quantum optics*, Scuola 'E. Fermi' on Experimental Quantum Computation and Information, F. De Martini and C. Monroe (Eds.) IOS Press, Amsterdam 2002, p. 385.
5. G. M. D'Ariano, P. Kumar, and M. F. Sacchi, Phys. Rev. A **61**, 013806 (2000).
6. G. M. D'Ariano, *Quantum estimation theory and optical detection*, in: *Quantum Optics and the Spectroscopy of Solids*, T. Hakioğlu and A.S. Shumovsky (Eds.), Kluwer Academic Publishers 1997, p. 139.
7. L. Mandel, Proc. Phys. Soc. **72**, 1037 (1958); *ibid.* **74**, 233 (1959); P. L. Kelley and W. H. Kleiner, Phys. Rev. A **30**, 844 (1964).
8. G. M. D'Ariano, Phys. Lett. A **268**, 151 (2000).
9. G. M. D'Ariano and M. G. A. Paris, Phys. Rev. A **60**, 518 (1999); G. M. D'Ariano and M. G. A. Paris, Acta Phys. Slov. **48**, 191 (1998).

10. L. M. Artiles, R. D. Gill, and M. I. Guta, J. Royal Stat. Soc. B **67**, 109 (2005).
11. K. Banaszek, G. M. D'Ariano, M. G. A. Paris, and M. F. Sacchi, Phys. Rev. A **61**, R10304 (2000).
12. H. Cramer, *Mathematical Methods of Statistics*, Princeton University Press, Princeton 1946.
13. G. M. D'Ariano, P. Lo Presti, and L. Maccone, Phys. Rev. Lett. **93**, 250407 (2004).
14. G. M. D'Ariano and P. Lo Presti, Phys. Rev. Lett. **86**, 4195 (2001); **91**, 47902 (2003).
15. G. M. D'Ariano and M. F. Sacchi, J. Opt. B **7**, S408 (2005).
16. See, for example: G. M. D'Ariano, M. Vasilyev, and P. Kumar Phys. Rev. A **58**, 636 (1998); A. I. Lvovsky and S. A. Babichev, Phys. Rev. A **66**, 011801R (2002); J. Wenger, R. Tualle-Brouri, and P. Grangier, Opt. Lett. **29**, 1267 (2004); A. Zavatta, S. Viciani, and M. Bellini, Phys. Rev. A **70**, 053821 (2004).
17. G. M. D'Ariano, *Measuring quantum states*, in: *Quantum Optics and the Spectroscopy of Solids*, T. Hakioğlu and A. S. Shumovsky (Eds.), Kluwer Academic Publishers 1997, p. 175.
18. G. M. D'Ariano, M. G. A. Paris, and M. F. Sacchi, *Quantum Tomography*, in: *Advances in Imaging and Electron Physics* **128**, 205 (2003); G. M. D'Ariano, M. G. A. Paris, and M. F. Sacchi, *Quantum tomographic methods*, in: *Quantum State Estimation*, Lecture Notes in Physics **649**, 7 (2004), Springer-Verlag, Berlin.
19. U. Fano, Rev. Mod. Phys. **29**, 74 (1957), Sec. 6.
20. K. Vogel and H. Risken, Phys. Rev. A **40**, 2847 (1989).
21. D. T. Smithey, M. Beck, M. G. Raymer, and A. Faridani, Phys. Rev. Lett. **70**, 1244 (1993); M. G. Raymer, M. Beck, and D. F. McAllister, Phys. Rev. Lett. **72**,1137 (1994); D. T. Smithey, M. Beck, J. Cooper, and M. G. Raymer, Phys. Rev. A, **48**, 3159 (1993).
22. G. M. D'Ariano, C. Macchiavello, and M. G. A. Paris, Phys. Rev. A **50**, 4298 (1994).
23. G. M. D'Ariano, U. Leonhardt, and H. Paul, Phys. Rev. A **52**, R1801 (1995).
24. G. M. D'Ariano, in *Quantum Communication, Computing, and Measurement*, O. Hirota, A. S. Holevo, and C. M. Caves (Eds.), Plenum Publishing, New York and London 1997, p. 253.
25. G. M. D'Ariano, in *Quantum Communication, Computing, and Measurement*, P. Kumar, G. M. D'Ariano, and O. Hirota(Eds.), Kluwer Academic/Plenum Publishers, New York and London 2000, p. 137; G. Cassinelli, G. M. D'Ariano, E. De Vito, and A. Levrero, J. Math. Phys. **41**, 7940 (2000); G. M. D'Ariano, L. Maccone, and M. Paini, J. Opt. B **5**, 77 (2003).
26. G. M. D'Ariano, L. Maccone, and M. G. A. Paris, Phys. Lett. A **276**, 25 (2000); J. Phys. A **34**, 93 (2001).

Chapter 9

Schrödinger Cat States for Quantum Information Processing

Hyunseok Jeong and Timothy C. Ralph
Centre for Quantum Computer Technology
Department of Physics, University of Queensland
St Lucia, Qld 4072, Australia

In this chapter, we discuss how Schrödinger cat states (superpositions of well-separated coherent states) can be used for quantum information processing.

1. Introduction

In the early days of quantum mechanics many of its founders became very worried by some of the paradoxical predictions that emerged from thought experiments based on the new theory. Now, eighty years on, some of these early thought experiments are being experimentally realized, and more than just confirming the fundamentals of the theory they are also being recognized as the basis of 21st century technologies.[1] An example is the EPR paradox, proposed by Einstein, Podolsky and Rosen in 1935,[2] which discussed the strange properties of quantum entanglement. Today, entanglement has been observed in optical[3,4] and ion[5] systems and is recognized as a resource for many quantum information processing tasks.[6]

About the same time as the EPR discussion, Schrödinger proposed his famous cat paradox[7] that highlighted the unusual consequences of extending the concept of superposition to macroscopically distinguishable objects. From a quantum optics view point, the usual paradigm is to consider superpositions of coherent states with amplitudes sufficiently different that they can be resolved using homodyne detection.[8,9] In this chapter we discuss how, beyond their fundamental interest, these types of states can be used in quantum information processing. We then look at the problem of producing such states with the required properties.

2. Quantum Information Processing with Schrödinger Cat States

2.1. *Coherent-state qubits*

We now introduce qubit systems using coherent states. A coherent state can be defined as[10,11]

$$|\alpha\rangle = e^{-|\alpha|^2/2} \sum_{n=0}^{\infty} \frac{\alpha^n}{\sqrt{n!}} |n\rangle, \tag{1}$$

where $|n\rangle$ is a number state and α is the complex amplitude of the coherent state. The coherent state is a very useful tool in quantum optics and a laser field is considered a good approximation of it. Let us consider two coherent states $|\alpha\rangle$ and $|-\alpha\rangle$. The two coherent states are not orthogonal to each other but their overlap $|\langle\alpha|-\alpha\rangle|^2 = e^{-4|\alpha|^2}$ decreases exponentially with $|\alpha|$. For example, when $|\alpha|$ is as small as 2, the overlap is $\approx 10^{-7}$, *i.e.*, $|\langle\alpha|-\alpha\rangle|^2 \approx 0$. We identify the two coherent states of $\pm\alpha$ as basis states for a logical qubit as $|\alpha\rangle \to |0_L\rangle$ and $|-\alpha\rangle \to |1_L\rangle$, so that a qubit state is represented by

$$|\phi\rangle = \mathcal{A}|0_L\rangle + \mathcal{B}|1_L\rangle = \mathcal{A}|\alpha\rangle + \mathcal{B}|-\alpha\rangle. \tag{2}$$

The basis states, $|\alpha\rangle$ and $|-\alpha\rangle$, can be unambiguously discriminated by a simple measurement scheme with a 50-50 beam splitter, an auxiliary coherent field of amplitude α and two photodetectors.[12] At the beam splitter, the qubit state $|\phi\rangle_1$ is mixed with the auxiliary state $|\alpha\rangle_2$ and results in the output

$$|\phi_R\rangle_{ab} = \mathcal{A}|\sqrt{2}\alpha\rangle_a|0\rangle_b + \mathcal{B}|0\rangle_a|-\sqrt{2}\alpha\rangle_b. \tag{3}$$

The two photodetector are set for modes a and b respectively. If detector A registers any photon(s) while detector B does not, we know that $|\alpha\rangle$ was measured. On the contrary, if A does not click while B does, the measurement outcome was $|-\alpha\rangle$. Even though there is non-zero probability of failure $P_f(\phi_R) = |\langle 00|\phi_R\rangle|^2 = |\mathcal{A}+\mathcal{B}|^2 \mathrm{e}^{-2|\alpha|^2}$ in which both of the detectors do not register a photon, the failure is known from the result whenever it occurs, and P_f approaches to zero exponentially as α increases. Note that the detectors do *not* have to be highly efficient for unambiguous discrimination. Alternatively, homodyne detection can also be very efficient for the qubit readout because the overlap between the coherent states $|\alpha\rangle$ and $|-\alpha\rangle$ would be extremely small for an appropriate value of α.

It is possible to construct an exactly orthogonal qubit basis with the equal superposition of two linear independent coherent states $|\alpha\rangle$ and $|-\alpha\rangle$.

Consider the basis states

$$|e\rangle = \mathcal{N}_+(|\alpha\rangle + |-\alpha\rangle) \rightarrow |0_L\rangle, \tag{4}$$

$$|d\rangle = \mathcal{N}_-(|\alpha\rangle - |-\alpha\rangle) \rightarrow |1_L\rangle, \tag{5}$$

where $\mathcal{N}_\pm = 1/\sqrt{2(1 \pm \exp[-2|\alpha|^2])}$. It can be simply shown that they form an orthonormal basis as $\langle e|d\rangle = \langle d|e\rangle = 0$ and $\langle e|e\rangle = \langle d|d\rangle = 1$. The basis state $|e\rangle$ ($|d\rangle$) is called "even cat state" ("odd cat state") because it contains only even (odd) number of photons as

$$|e\rangle = 2\mathcal{N}_+ e^{-\frac{|\alpha|^2}{2}} \sum_{n=0}^{\infty} \frac{\alpha^{2n}}{\sqrt{(2n)!}}|2n\rangle, \tag{6}$$

$$|d\rangle = 2\mathcal{N}_- e^{-\frac{|\alpha|^2}{2}} \sum_{n=0}^{\infty} \frac{\alpha^{(2n+1)}}{\sqrt{(2n+1)!}}|2n+1\rangle. \tag{7}$$

The even and odd cat states can thus be discriminated by a photon parity measurement which can be represented by $O_\Pi = \sum_{n=0}^{\infty}(|2n\rangle\langle 2n| - |2n+1\rangle\langle 2n+1|)$. As α goes to zero, the odd cat state $|d\rangle$ approaches a single photon state $|1\rangle$ while the even cat state $|e\rangle$ approaches $|0\rangle$. No matter how small α is, there is no possibility that no photon will be detected from the state $|d\rangle$ at an ideal photodetector.

2.2. *Quantum teleportation*

Quantum teleportation is an interesting phenomenon for demonstrating quantum theory and a useful tool in quantum information processing.[13] By quantum teleportation, an unknown quantum state is disentangled in a sending place and its perfect replica appears at a distant place via dual quantum and classical channels. The key ingredients of quantum teleportation are an entangled channel, a Bell-state measurement and appropriate unitary transformations. In what follows we shall explain how teleporation can be performed for a coherent-state qubit.[14,15]

Let us assume that Alice wants to teleport an unknown coherent-state qubit $|\phi\rangle_a$ via a pure entangled coherent channel

$$|\Psi_-\rangle_{bc} = N_-(|\alpha\rangle_b|-\alpha\rangle_c - |-\alpha\rangle_b|\alpha\rangle_c), \tag{8}$$

where N_- is the normalization factor. After sharing the quantum channel $|\Psi_-\rangle$, Alice should perform a Bell-state measurement on her part of the quantum channel and the unknown qubit $|\phi\rangle$ and send the outcome to

Bob. The Bell-state measurement is to discriminate between the four Bell-cat states which can be defined with coherent states as[16–19]

$$|\Phi_\pm\rangle = N_\pm(|\alpha\rangle|\alpha\rangle \pm |-\alpha\rangle|-\alpha\rangle), \tag{9}$$
$$|\Psi_\pm\rangle = N_\pm(|\alpha\rangle|-\alpha\rangle \pm |-\alpha\rangle|\alpha\rangle), \tag{10}$$

where $N_\pm$ are normalization factors. The four Bell-cat states defined in our framework are a very good approximation of the Bell basis. These states are orthogonal to each other except $\langle\Psi_+|\Phi_+\rangle = 1/\cosh 2|\alpha|^2$, and $|\Psi_+\rangle$ and $|\Phi_+\rangle$ rapidly become orthogonal as $|\alpha|$ grows.

A Bell-state measurement, or simply Bell measurement, is very useful in quantum information processing. It was shown that a complete Bell-state measurement on a product Hilbert space of two two-level systems is not possible using linear elements.[20] A Bell measurement scheme using linear optical elements[21] has been used to distinguish only up to two of the Bell states for teleportation[22] and dense coding.[23] However, a remarkable feature of the Bell-cat states is that each one of them can be unambiguously discriminated using only a beam splitter and photon-parity measurements.[15,24] Suppose that the modes, a and b, of the entangled state are incident on a 50-50 beam splitter. After passing the beam splitter, the Bell-cat states become

$$\begin{aligned}
|\Phi_+\rangle_{ab} &\longrightarrow |E\rangle_f|0\rangle_g,\\
|\Phi_-\rangle_{ab} &\longrightarrow |D\rangle_f|0\rangle_g,\\
|\Psi_+\rangle_{ab} &\longrightarrow |0\rangle_f|E\rangle_g,\\
|\Psi_-\rangle_{ab} &\longrightarrow |0\rangle_f|D\rangle_g,
\end{aligned} \tag{11}$$

where the even cat state $|E\rangle \propto |\sqrt{2}\alpha\rangle + |-\sqrt{2}\alpha\rangle$ definitely contains an even number of photons, while the odd cat state $|D\rangle \propto |\sqrt{2}\alpha\rangle - |-\sqrt{2}\alpha\rangle$ definitely contains an odd number of photons. By setting two photodetectors for the output modes f and g respectively to perform number parity measurement, the Bell-cat measurement can be simply achieved. For example, if an odd number of photons is detected for mode f, the state $|\Phi_-\rangle$ is measured, and if an odd number of photons is detected for mode g, then $|\Psi_-\rangle$ is measured. Even though there is non-zero probability of failure in which both of the detectors do not register a photon due to the non-zero overlap of $|\langle 0|E\rangle|^2 = 2e^{-2|\alpha|^2}/(1+e^{-4|\alpha|^2})$, it is small for an appropriate choice of α and the failure is known from the result whenever it occurs.

To complete the teleportation process, Bob performs a unitary transformation on his part of the quantum channel according to the measurement result sent from Alice via a classical channel. The required transformations are σ_x and σ_z on the coherent-state qubit basis, where σ's are Pauli operators. When the measurement outcome is $|B_4\rangle$, Bob obtains a perfect replica of the original unknown qubit without any operation. When the measurement outcome is $|B_2\rangle$, Bob should perform $|\alpha\rangle \leftrightarrow |-\alpha\rangle$ on his qubit. Such a phase shift by π can be done using a phase shifter whose action is described by $P(\varphi) = e^{i\varphi a^\dagger a}$, where a and $a^\dagger$ are the annihilation and creation operators. When the outcome is $|B_3\rangle$, the transformation should be performed as $|\alpha\rangle \rightarrow |\alpha\rangle$ and $|-\alpha\rangle \rightarrow -|-\alpha\rangle$. This transformation is more difficult but can be achieved most straightforwardly by simply teleporting the state again (locally) and repeating until the required phase shift is obtained. Therefore, both of the required unitary transformation, σ_x and σ_z, can be performed by linear optics elements. When the outcome is $|B_1\rangle$, σ_x and σ_z should be successively applied.

2.3. *Quantum computation*

We now describe how a universal set of quantum gates can be implemented on coherent state qubits using only linear optics and photon detection, provided a supply of cat states is available as a resource. The idea was originally due to Ralph, Munro and Milburn[25] and was later expanded on by Ralph *et al.*[26]

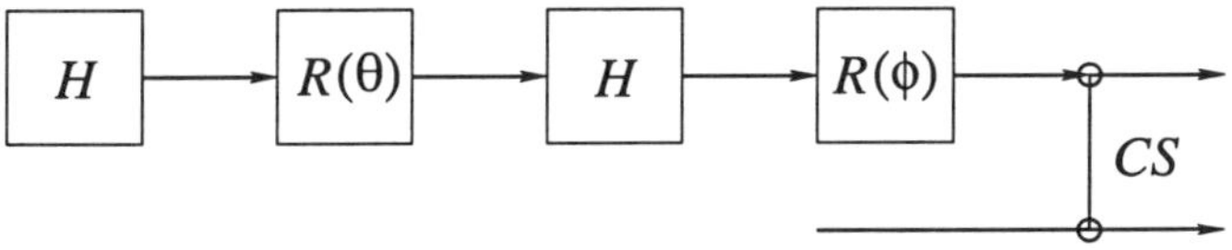

Fig. 1. A set of Hadamard (H) gates, rotations (R) about the Z-axis and control sign (CS) gates can provide universal gate operations.

A universal single qubit quantum gate element can be constructed from the following sequence of gates: Hadamard (H); rotation about the Z-axis by angle θ ($R(\theta)$); Hadamard (H) and; rotation about the Z-axis by angle ϕ ($R(\phi)$). If the two qubit gate, control sign (CS), is also available then universal processing is possible (See Fig. 1). We now describe how these gates can be implemented. We will assume that deterministic single qubit measurements can be made in the computational basis, $|\alpha\rangle, |-\alpha\rangle$ and the phase superposition basis $|\alpha\rangle \pm \exp[i\epsilon]|-\alpha\rangle$. As described in the previous section, computational basis measurements can be achieved using either homodyne or photon counting techniques. The phase superposition basis can be measured using photon counting in a Dolinar receiver type arrangement.[27,28] The simplest case is for $\epsilon = 0$ where we need to differentiate only between odd or even photon numbers in direct detection. We also assume we can make two qubit Bell-measurements and, more generally, perform teleportation, as described in the previous section.

Hadamard Gate: The Hadamard gate (H) can be defined by its effect on the computational states: $H|\alpha\rangle = |\alpha\rangle + |-\alpha\rangle$ and $H|-\alpha\rangle = |\alpha\rangle - |-\alpha\rangle$ where for convenience we have dropped normalization factors. One way to achieve this gate is to use the resource state $|HR\rangle = |\alpha,\alpha\rangle + |\alpha,-\alpha\rangle + |-\alpha,\alpha\rangle - |-\alpha,-\alpha\rangle$. This state can be produced non-deterministically from cat state resources, as will be described shortly. It is straight forward to show that if a Bell-state measurement is made between an arbitrary qubit state $|\sigma\rangle$ and one of the modes of $|HR\rangle$ then the remaining mode is projected into the state $H|\sigma\rangle$, where dependent on the outcome of the Bell-measurement a bit-flip correction, a phase-flip correction, or both may be necessary.

Phase Rotation Gate: The phase rotation gate ($R(\theta)$) can be defined by its effect on the computational states: $R(\theta)|\alpha\rangle = \exp[i\theta]|\alpha\rangle$ and $R(\theta)|-\alpha\rangle = \exp[-i\theta]|-\alpha\rangle$. One way to achieve this gate is the following: The arbitrary qubit, $\mu|\alpha\rangle + \nu|-\alpha\rangle$ is split on a 50:50 beam splitter giving the two mode state: $\mu|\alpha/\sqrt{2}\rangle|\alpha/\sqrt{2}\rangle + \nu|-\alpha/\sqrt{2}\rangle|-\alpha/\sqrt{2}\rangle$. One of the modes is then measured in the phase superposition basis $|\alpha/\sqrt{2}\rangle \pm \exp[-2i\theta]|-\alpha/\sqrt{2}\rangle$, thus projecting the other mode into the state $\mu\exp[i\theta]|\alpha/\sqrt{2}\rangle \pm \nu\exp[-i\theta]|-\alpha/\sqrt{2}\rangle$. The amplitude decrease can be corrected by teleportation in the following way.[26] The asymmetric Bell state entanglement, $|\alpha/\sqrt{2}\rangle|\alpha\rangle + |-\alpha/\sqrt{2}\rangle|-\alpha\rangle$ is produced by splitting the cat state $|\sqrt{3/2}\alpha\rangle + |-\sqrt{3/2}\alpha\rangle$ on a $1/3 : 2/3$ beam splitter. Teleportation is then carried out with the Bell state measurement being performed between the matching "$\alpha/\sqrt{2}$" modes and the teleported state ending up on the "α" mode. Dependent on the outcome of the phase basis measurement and the Bell-measurement a bit-flip correction, a phase-flip correction, or both may be necessary.

Control Sign Gate: The control-sign gate (CS) can be defined by its effect on the two qubit computational states: $CS|\alpha\rangle|\alpha\rangle = |\alpha\rangle|\alpha\rangle$; $CS|\alpha\rangle|-\alpha\rangle = |\alpha\rangle|-\alpha\rangle$; $CS|-\alpha\rangle|\alpha\rangle = |-\alpha\rangle|\alpha\rangle$ and; $CS|-\alpha\rangle|-\alpha\rangle = -|-\alpha\rangle|-\alpha\rangle$. One way to achieve this gate is the following: The two arbitrary qubits, $\mu|\alpha\rangle + \nu|-\alpha\rangle$ and $\gamma|\alpha\rangle + \delta|-\alpha\rangle$ are both split on 50:50 beam splitters giving the two mode states: $\mu|\alpha/\sqrt{2}\rangle|\alpha/\sqrt{2}\rangle + \nu|-\alpha/\sqrt{2}\rangle|-\alpha/\sqrt{2}\rangle$ and $\gamma|\alpha/\sqrt{2}\rangle|\alpha/\sqrt{2}\rangle + \delta|-\alpha/\sqrt{2}\rangle|-\alpha/\sqrt{2}\rangle$. A Hadamard gate is then performed on the second mode of the first qubit giving the state $\mu|\alpha/\sqrt{2}\rangle(|\alpha/\sqrt{2}\rangle +$

$|-\alpha/\sqrt{2}\rangle) + \nu|-\alpha/\sqrt{2}\rangle(|\alpha/\sqrt{2}\rangle - |-\alpha/\sqrt{2}\rangle)$. If a Bell-measurement is then carried out between the second mode of the first qubit and one of the modes of the second qubit a CS gate will be achieved. The amplitude reduction can be corrected as before using teleportation. Dependent on the outcome of the various Bell-measurements, bit-flip corrections, phase-flip corrections, or both may be necessary.

Resource State: The resource state $|HR\rangle$ can be produced in the following way. Consider the beam splitter interaction given by the unitary transformation

$$U_{ab} = \exp[i\frac{\theta}{2}(ab^\dagger + a^\dagger b)], \tag{12}$$

where a and b are the annihilation operators corresponding to two coherent state qubits $|\gamma\rangle_a$ and $|\beta\rangle_b$, with γ and β taking values of $-\alpha$ or α. It is well known that the output state produced by such an interaction is

$$U_{ab}|\gamma\rangle_a|\beta\rangle_b = |\cos\frac{\theta}{2}\gamma + i\sin\frac{\theta}{2}\beta\rangle_a|\cos\frac{\theta}{2}\beta + i\sin\frac{\theta}{2}\gamma\rangle_b, \tag{13}$$

where $\cos^2\frac{\theta}{2}$ ($\sin^2\frac{\theta}{2}$) is the reflectivity (transmissivity) of the beam splitter. Suppose two cat states are fed into the beam splitter and both output beams are then teleported, the output state will be:

$$e^{-\theta^2\alpha^2/4}(e^{i\theta\alpha^2}|-\alpha\rangle_a|-\alpha\rangle_b \pm e^{-i\theta\alpha^2}|\alpha\rangle_a|-\alpha\rangle_b \pm \\ e^{-i\theta\alpha^2}|-\alpha\rangle_a|\alpha\rangle_b + e^{i\theta\alpha^2}|\alpha\rangle_a|\alpha\rangle_b), \tag{14}$$

where the $\pm$ signs depend on the outcome of the Bell measurements. If we choose $\phi = 2\theta\alpha^2 = \pi/2$ then the resulting state is easily shown to be locally equivalent to $|HR\rangle$ (related by phase rotations). Preparation of this state is non-deterministic because of non-unit overlap between the state of Eq. (13) and the Bell states used in the teleporter. As a result the teleporter can fail by recording photons at both outputs in the Bell-measurement. The probability of success is $e^{-\theta^2\alpha^2/2}$. For $\alpha = 2$ this is about 92% probability of success.

Correction of Phase-flips: After each gate we have noted that bit flip and/or phase flip corrections may be necessary since our gate operations are based on the teleportation protocol. As discussed in the previous section, bit flips can be easily implemented using a phase shifter, $P(\pi)$, while phase-flips are more expensive. We now argue that in fact only active correction of bit-flips is necessary. This is because phase-flips commute with the phase rotation gate and the control sign gate but are converted into bit flips by the Hadamard gate. This suggests the following strategy: After each gate operation any bit-flips are corrected whilst phase-flips are noted. After the next Hadamard gate the phase flips are converted to bit-flips which are then corrected and any new phase-flips are noted. By following this strategy only bit-flips need to be corrected actively, with, at worst, some final phase-flips needing to be corrected in the final step of the circuit.

2.4. *Entanglement purification for Bell-cat states*

It is not possible to perfectly isolate a quantum state from its environment. A quantum state inevitably loses its quantum coherence in a dissipative environment. This process is called decoherence and has been known as the main obstacle to the physical implementation of quantum information processing. Quantum error correction[26,29,30] and entanglement purification[24,31] have been studied for quantum information processing using cat states to overcome this problem. Here we discuss an entanglement purification technique.

An entanglement purification for entangled coherent states (Bell-cat states) have been studied by several authors.[24,31] It has been found that certain types of mixed states including the Werner-type mixed states composed of the Bell-cat states can be purified by simple linear optics elements and inefficient detectors.[24] The other types of mixed states need to be transformed to the Werner type states by local operations. This scheme performs amplification of the Bell-cat states simultaneously with entanglement purification. This is an important observation because Bell-cat states of large amplitudes are preferred for quantum information processing while their generation is hard. A similar technique is employed to generate single-mode large cat states.[32]

We first explain the purification-amplification protocol for entangled coherent states by a simple example and then apply it to a realistic situation.[24] Let us suppose that Alice and Bob want to distill entangled coherent states $|\Phi_+\rangle$ from a type of ensemble

$$\rho_{ab} = F|\Phi_+\rangle\langle\Phi_+| + G|\Psi_+\rangle\langle\Psi_+|, \tag{15}$$

where $F + G \approx 1$ for $|\alpha| \gg 1$. We shall assume this condition, $|\alpha| \gg 1$, for simplicity. The purification-amplification process can be simply accomplished by performing the process shown in Fig. 2. Alice and Bob choose two pairs from the ensemble which are represented by the following density operator

$$\rho_{ab}\rho_{a'b'} = F^2|\Phi_+\rangle\langle\Phi_+| \otimes |\Phi_+\rangle\langle\Phi_+| + F(1-F)|\Phi_+\rangle\langle\Phi_+| \otimes |\Psi_+\rangle\langle\Psi_+|$$
$$+ F(1-F)|\Psi_+\rangle\langle\Psi_+| \otimes |\Phi_+\rangle\langle\Phi_+| + (1-F)^2|\Psi_+\rangle\langle\Psi_+| \otimes |\Psi_+\rangle\langle\Psi_+|. \tag{16}$$

The fields of modes a and a' are in Alice's possession while b and b' in Bob's. In Fig. 2(a), we show that Alice's action to purify the mixed entangled state. The same is conducted by Bob on his fields of b and b'.

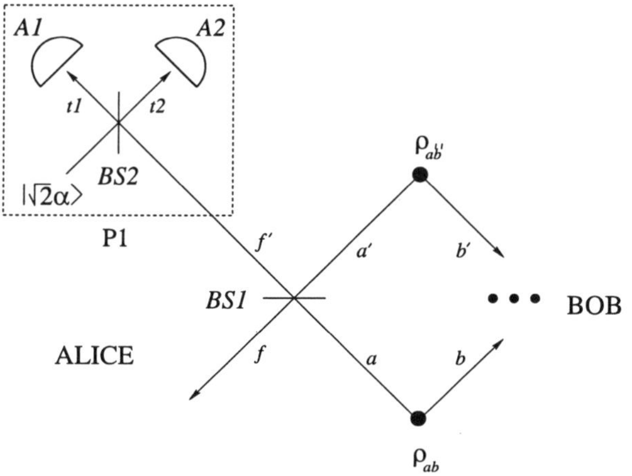

Fig. 2. A schematic of the entanglement purification scheme for mixed entangled coherent states. P1 tests if the incident fields a and a' were in the same state by simultaneous clicks at $A1$ and $A2$.

There are four possibilities for the fields of a and a' incident onto the beam splitter ($BS1$), which gives the output (In the following, only the cat part for a component of the mixed state is shown to describe the action of the apparatuses)

$$|\alpha\rangle_a|\alpha\rangle_{a'} \longrightarrow |\sqrt{2}\alpha\rangle_f|0\rangle_{f'}, \tag{17}$$
$$|\alpha\rangle_a|-\alpha\rangle_{a'} \longrightarrow |0\rangle_f|\sqrt{2}\alpha\rangle_{f'}, \tag{18}$$
$$|-\alpha\rangle_a|\alpha\rangle_{a'} \longrightarrow |0\rangle_f|-\sqrt{2}\alpha\rangle_{f'}, \tag{19}$$
$$|-\alpha\rangle_a|-\alpha\rangle_{a'} \longrightarrow |-\sqrt{2}\alpha\rangle_f|0\rangle_{f'}. \tag{20}$$

In the boxed apparatus P1, Alice checks if modes a and a' were in the same state by counting photons at the photodetectors $A1$ and $A2$. If both modes a and a' are in $|\alpha\rangle$ or $|-\alpha\rangle$, f' is in the vacuum, in which case the output field of the beam splitter $BS2$ is $|\alpha, -\alpha\rangle_{t1,t2}$. Otherwise, the output field is either $|2\alpha, 0\rangle_{t1,t2}$ or $|0, 2\alpha\rangle_{t1,t2}$. When both the photodetectors $A1$ and $A2$ register any photon(s), Alice and Bob are sure that the two modes a and a' were in the same state but when either $A1$ or $A2$ does not resister a photon, a and a' were likely in different states. The remaining pair is selected only when Alice and Bob's all four detectors click together. Of course, there is a probability not to resister a photon even though the two modes were in the same state, which is due to the nonzero overlap of $|\langle 0|\sqrt{2}\alpha\rangle|^2$. Note that inefficiency of the detectors does not degrade the the quality of the distilled entangled coherent states but decreases the success probability.

It can be simply shown that the second and third terms of Eq. (16) are always discarded by the action of P1 and Bob's similar apparatus. For example, at the output ports of $BS1$ and Bob's beam splitter corresponding to $BS1$, $|\Phi_+\rangle_{ab}|\Psi_+\rangle_{a'b'}$ becomes

$$\begin{aligned}|\Phi_+\rangle_{ab}|\Psi_+\rangle_{a'b'} \longrightarrow N_+^2\big(&|\sqrt{2}\alpha, 0, 0, \sqrt{2}\alpha\rangle + |0, \sqrt{2}\alpha, \sqrt{2}\alpha, 0\rangle \\ &+ |0, -\sqrt{2}\alpha, -\sqrt{2}\alpha, 0\rangle + |-\sqrt{2}\alpha, 0, 0, -\sqrt{2}\alpha\rangle\big)_{fgf'g'},\end{aligned} \tag{21}$$

where g and g' are the output field modes from Bob's beam splitter corresponding to $BS1$. The fields of modes f' and g' can never be in $|0\rangle$ at the same time; at least, one of the four detectors of Alice and Bob must not click. The third term of Eq. (16) can be shown to lead to the same result by the same analysis.

For the cases of the first and fourth terms in Eq. (16), all four detectors may register photon(s). After the beam splitter $BS1$, the ket of $(|\Phi_-\rangle\langle\Phi_-|)_{ab} \otimes (|\Phi_-\rangle\langle\Phi_-|)_{a'b'}$ of Eq. (16) becomes

$$|\Phi_-\rangle_{ab}|\Phi_-\rangle_{a'b'} \longrightarrow |\Phi'_+\rangle_{fg}|0,0\rangle_{f'g'} - |0,0\rangle_{fg}|\Phi'_+\rangle_{f'g'}, \tag{22}$$

where $|\Phi'_+\rangle = N'_+(|\sqrt{2}\alpha, \sqrt{2}\alpha\rangle + |-\sqrt{2}\alpha, -\sqrt{2}\alpha\rangle)$ with the normalization factor N'_+. The normalization factor in the right hand side of Eq. (22) is omitted. The first term is reduced to $(|\Phi'_+\rangle\langle\Phi'_+|)_{fg}$ after $(|0,0\rangle\langle 0,0|)_{f'g'}$ is measured out by Alice and Bob's P1's. Similarly, the fourth term of Eq. (16) yields $(|\Psi'_+\rangle\langle\Psi'_+|)_{fg}$, where $|\Psi'_+\rangle$ is defined in the same way as $|\Phi'_+\rangle$, after $(|0,0\rangle\langle 0,0|)_{f'g'}$ is measured. Thus the density matrix for the field of modes f and g conditioned on simultaneous measurement of photons at all four photodetectors is

$$\rho_{fg} = F'|\Phi'_+\rangle\langle\Phi'_+| + (1-F')|\Psi'_+\rangle\langle\Psi'_+|, \tag{23}$$

where $F' = F^2/\{F^2 + (1-F)^2\}$, and F' is always larger than F for any $F > 1/2$. By reiterating this process, Alice and Bob can distill some maximally entangled states $|\Phi_+\rangle$ of a large amplitude asymptotically. Of course, a sufficiently large ensemble and initial fidelity $F > 1/2$ are required for successful purification.[33]

We now apply our scheme to a realistic example in a dissipative environment. When the entangled coherent channel $|\Phi_-\rangle$ is embedded in a vacuum, the channel decoheres and becomes a mixed state of its density operator $\rho_{ab}(\tau)$, where τ stands for the decoherence time. By solving the master equation[34]

$$\frac{\partial \rho}{\partial \tau} = \hat{J}\rho + \hat{L}\rho \; ; \quad \hat{J}\rho = \gamma \sum_i a_i \rho a_i^\dagger, \quad \hat{L}\rho = -\frac{\gamma}{2}\sum_i (a_i^\dagger a_i \rho + \rho a_i^\dagger a_i), \tag{24}$$

where γ is the energy decay rate, the mixed state $\rho_{ab}(\tau)$ can be straightforwardly obtained as

$$\begin{aligned}\rho_{ab}(\tau) = \widetilde{N}(\tau)\Big\{&|t\alpha, t\alpha\rangle\langle t\alpha, t\alpha| + |-t\alpha, -t\alpha\rangle\langle -t\alpha, -t\alpha| \\ &- \Gamma(|t\alpha, t\alpha\rangle\langle -t\alpha, -t\alpha| + |-t\alpha, -t\alpha\rangle\langle t\alpha, t\alpha|)\Big\},\end{aligned} \tag{25}$$

where $|\pm t\alpha, \pm t\alpha\rangle = |\pm t\alpha\rangle_a |\pm t\alpha\rangle_b$, $t = e^{-\gamma\tau/2}$, $\Gamma = \exp[-4(1-t^2)|\alpha|^2]$, and $\widetilde{N}(\tau)$ is the normalization factor. The decohered state $\rho_{ab}(\tau)$ may be represented by the dynamic Bell-cat states defined as follows:

$$|\widetilde{\Phi}_\pm\rangle_{ab} = \widetilde{N}_\pm(|t\alpha\rangle_a|t\alpha\rangle_b \pm |-t\alpha\rangle_a|-t\alpha\rangle_b), \tag{26}$$

$$|\widetilde{\Psi}_\pm\rangle_{ab} = \widetilde{N}_\pm(|t\alpha\rangle_a|-t\alpha\rangle_b \pm |-t\alpha\rangle_a|t\alpha\rangle_b), \tag{27}$$

where $\widetilde{N}_\pm = \{2(1 \pm e^{-4t^2|\alpha|^2})\}^{-1/2}$. The decohered state is then

$$\begin{aligned}\rho_{ab}(\tau) &= \widetilde{N}(\tau)\Big\{\frac{(1+\Gamma)}{\widetilde{N}_-^2}|\widetilde{\Phi}_-\rangle\langle\widetilde{\Phi}_-| + \frac{(1-\Gamma)}{\widetilde{N}_-^2}|\widetilde{\Phi}_+\rangle\langle\widetilde{\Phi}_+|\Big\} \\ &\equiv F(\tau)|\widetilde{\Phi}_-\rangle\langle\widetilde{\Phi}_-| + (1-F(\tau))|\widetilde{\Phi}_+\rangle\langle\widetilde{\Phi}_+|,\end{aligned} \tag{28}$$

where, regardless of the decay time τ, $|\widetilde{\Phi}_-\rangle$ is maximally entangled and $|\widetilde{\Phi}_-\rangle$ and $|\widetilde{\Phi}_+\rangle$ are orthogonal to each other. The decohered state (28) is not in the same form as Eq. (15) so that we need some bilateral unitary transformations before the purification scheme is applied. A Hadamard gate H for coherent-state qubits can be used to transform the state (28) into a distillable form

$$H_a H_b \rho_{ab}(\tau) H_b^\dagger H_a^\dagger = F(\tau)|\widetilde{\Psi}_+\rangle\langle\widetilde{\Psi}_+| + (1-F(\tau))|\widetilde{\Phi}_+\rangle\langle\widetilde{\Phi}_+|,$$

which is now in the same form as Eq. (15).

The ensemble of state (28) can be purified successfully only when $F(\tau)$ is larger than 1/2. Because $F(\tau)$ is obtained as

$$F(\tau) = \frac{N_+^2(1+\Gamma)}{N_+^2(1+\Gamma) - N_-^2(1-\Gamma)}, \tag{30}$$

it is found that purification is successful when the decoherence time $\gamma\tau < \ln 2$ regardless of α. This result is in agreement with the decay time until which teleportation can be successfully performed via an entangled coherent state shown in Ref. 15.

3. Production of Schrödinger Cat States

A key requirement of quantum information processing with cat states is the generation of cat states in free-propagating optical fields. This has been known to be extremely demanding using current technology because strong nonlinearity[35] or precise photon counting measurements[36,37] are necessarily required. However, very recently, there has been remarkable progress which may enable one to generate free propagating cat states without strong nonlinearity or photon counting measurements. For example, it was shown that cat states of reasonably large amplitudes can be produced with simple linear optics elements and single photons.[32] Relatively small nonlinearity was shown to be still useful with conditioning homodyne detection[38] or with a single photon interacting with a coherent state[39] to generate cat states. It was shown that a deterministic cat-state source can be obtained using a single-atom cavity.[40] Recent experiments[41] could be directly improved by the cat-amplification scheme in Ref. 32 to generate a cat state of a larger amplitude and higher fidelity. The above proposals have now brought the generation of free-propagating cat states of $\alpha \approx 2$ within reach of current technology. Electromagnetically induced transparency (EIT) has also been studied as a method to obtain a giant Kerr nonlinearity,[42–44] and there has been an improved suggestion to generate cat states with it.[45] In what follows, some of these suggestions will be briefly covered.

3.1. *Schemes using linear optics elements*

Since it is extremely hard to generate cat states using $\chi^{(3)}$ nonlinearity, some alternative methods have been studied based upon conditional measurements.[36,37] A crucial drawback of these schemes is that a highly efficient photon counting measurement, which is extremely demanding in current technology, is necessary. However, it was shown recently that a free-propagating optical cat state can be generated with a single photon source and simple optical operations without efficient photon detection.[32] This suggestion contains two main points:

- ▷ An arbitrarily large cat state can be produced out of arbitrarily small cat states using the simple experimental set-up depicted in Fig. 3.
- ▷ A small odd cat state with $\alpha \leq 1.2$ is very well approximated by a squeezed single photon, $S(s)|1\rangle$, where $S(s)$ is the squeezing operator with the squeezing parameter s and $|1\rangle$ is the single-photon state.

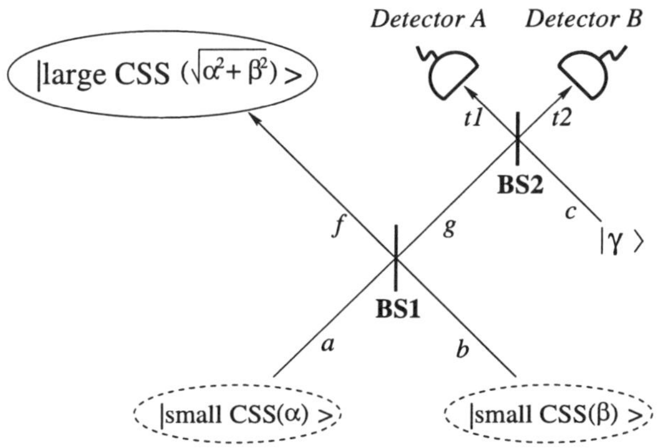

Fig. 3. A schematic of the non-deterministic cat-amplification process. See text for details.

Firstly, the cat-amplification process is summarized as follows. Suppose two small cat states, $|\mathrm{cat}_{\varphi}(\alpha)\rangle$ and $|\mathrm{cat}_{\phi}(\beta)\rangle$, with amplitudes α and β. The reflectivity r and transmitivity t of BS1 are set to $r = \beta/\sqrt{\alpha^2+\beta^2}$ and $t = \alpha/\sqrt{\alpha^2+\beta^2}$, where the action of the beam splitter is represented by $\hat{B}_{a,b}(r,t)|\alpha\rangle_a|\beta\rangle_b = |t\alpha + r\beta\rangle_f| - r\alpha + t\beta\rangle_g$. The other beam splitter BS2 is a 50:50 beam splitter ($r = t = 1/\sqrt{2}$) regardless of the conditions and the amplitude γ of the auxiliary coherent field is determined as $\gamma = 2\alpha\beta/\sqrt{\alpha^2+\beta^2}$. The resulting state for mode f then becomes $|\mathrm{cat}_{\varphi+\phi}(\mathcal{A})\rangle \propto |\mathcal{A}\rangle + e^{i(\varphi+\phi)}|-\mathcal{A}\rangle$, whose coherent amplitude $\mathcal{A} = \sqrt{\alpha^2+\beta^2}$ is larger than both α and β. The relative phase of the resulting cat state is the sum of the relative phases of the input cat states. The success probability $P_{\varphi,\phi}(\alpha,\beta)$ for a single iteration is

$$P_{\varphi,\phi}(\alpha,\beta) = \frac{(1-e^{-\frac{2\alpha^2\beta^2}{\alpha^2+\beta^2}})^2[1+\cos(\varphi+\phi)e^{-2(\alpha^2+\beta^2)}]}{2(1+\cos\varphi e^{-2\alpha^2})(1+\cos\phi e^{-2\beta^2})},$$

which approaches $1/2$ as the amplitudes of initial cat states becomes large. Note that the probabilities depend on the type of cat states (odd or even) used. The effect of detector inefficiency is just to decrease this success probability.

Secondly, the fidelity between the squeezed single photon and the cat state is

$$F(s,\alpha) = \frac{2\alpha^2 \exp[\alpha^2(\tanh s - 1)]}{(\cosh s)^3(1-\exp[-2\alpha^2])},$$

where α is the amplitude of the cat state. Fig. 4 shows the maximized fidelity on the y-axis plotted against a range of possible values for α for the desired odd cat state. Some example values are: $F = 0.99999$ for amplitude $\alpha = 1/2$, $F = 0.9998$ for $\alpha = 1/\sqrt{2}$, and $F = 0.997$ for $\alpha = 1$, where the maximizing squeezing parameters are $s = 0.083$, $s = 0.164$, and $s = 0.313$ respectively. The fidelity approaches unity for α very close to zero while the fidelity tends to zero as α increases.

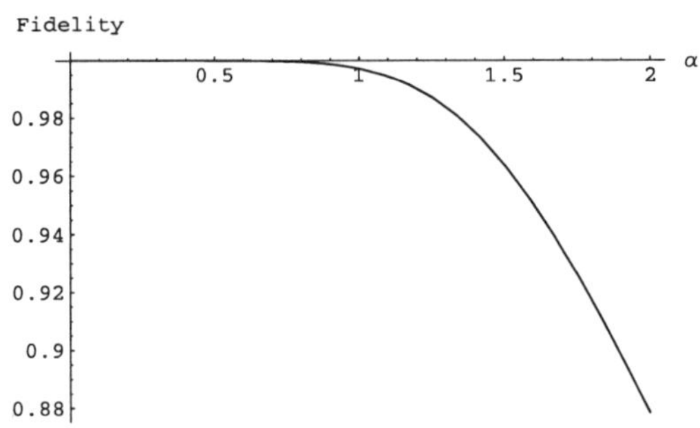

Fig. 4. The fidelity between an odd cat state and squeezed single photon. The odd cat state is extremely well approximated by the squeezed single photon for a small coherent amplitude, $\alpha \leq 1.2$.

The two points explained above can be efficiently combined. If one uses squeezed single photons as small cat states, a cat state of high fidelity ($F > 0.99$) can be obtained up to $\alpha = 2.5$. Another interesting aspect of this process is that it is somewhat resilient to the photon production inefficiency because its first iteration purifies the mixed cat states while amplifying them. For example, if the inefficiency of the single photon source is about 40%, the fidelity of the initial cat, which is a mixture with a squeezed vacuum, is $F \approx 0.60$ but it will become $F \approx 0.89$ by the first iteration.

It is also important to note that there is an alternative method to obtain a squeezed single photon even without a single photon source. An interesting observation is that a squeezed single photon can be obtained by subtracting a photon from a squeezed vacuum. This can be shown by applying the annihilation operator to a squeezed single photon as $\hat{a}S(s)|0\rangle = \cosh s S(s)|1\rangle$. In recent experiments,[41] the single photon subtraction was approximated by a beam splitter of low reflectivity and a single photon detector. Such experiments could be immediately linked to our suggestion to experimentally generate a larger cat state. One can then generate a cat state of $\alpha > 2$ using our scheme without a single photon source.

Since this scheme uses at least two beam splitters to mix propagating fields, good mode matching is required to obtain a cat state of high quality. Highly efficient mode matching of a single photon from parametric down conversion and a weak coherent state from an attenuated laser beam at a beam splitter has been experimentally demonstrated using optical fibers.[46] Such techniques could be employed for the implementation of our scheme. The success probability will rapidly drop down and the required resources will exponentially increase as the number of steps increases. However, if quantum optical memory is available, one can considerably boost up the success probability by holding the resulting states for every step.[47]

3.2. *Schemes using cavity quantum electrodynamics*

Cavity quantum electrodynamics (QED) has been studied to enhance nonlinear effects to generate macroscopic superpositions.[48] Some success has been reported in creating such superposition states within high Q cavities in the microwave[49] and optical[50] domains. Simplified versions of cavity QED schemes have been developed for deterministic generation of cat states in a cavity.[51] While this method is relatively effective to generate cat states in cavity, most of the schemes suggested for quantum information processing with coherent states require *free propagating* cat states.

Recently, a method was proposed to generated free propagating cat states by using cavity-assisted interactions of coherent optical pulses.[40] This suggestion employs an atom of three relevant levels trapped in an optical cavity with a coherent-state pulse $|\alpha\rangle$ incident onto the cavity. One of the atomic level, $|e\rangle$, is its excited state, and the other two levels, $|g_0\rangle$ and $|g_1\rangle$, are levels in the ground state with different hyperfine spins. The transition from $|g_1\rangle$ to $|e\rangle$ is resonantly coupled to a cavity mode while $|g_0\rangle$ is decoupled from the cavity mode. In such a preparation, if the trapped atom was prepared in state $|g_0\rangle$, the input field becomes $|-\alpha\rangle$ after a resonant reflection as the input pulse is resonant with the bare cavity mode. On the other hand, if the atom was prepared in state $|g_1\rangle$, it remains $|\alpha\rangle$ due to a strong atom-cavity coupling. Therefore, if the trapped atom was prepared in a superposition state such as $(|g_0\rangle+|g_1\rangle)/\sqrt{2}$, the reflected field becomes an entangled state $(|g_0\rangle|\alpha\rangle+|g_1\rangle|-\alpha\rangle)/\sqrt{2}$, which can be projected to a single mode cat state by a measurement on a superposed basis $|g_0\rangle \pm |g_1\rangle$. An advantage of this scheme is weak dependence on dipole coupling, but wave front distortion due to difference between resonant and non-resonant interactions could be a problem in a real experiment. This work[40] concludes that a cat state with a quite large amplitude ($\alpha \approx 3.4$) could be generated in this way with a 90% fidelity using current technology.

3.3. *Schemes using weak nonlinearity*

There has been a suggestion to use relatively weak nonlinearity with beam splitting with a vacuum and conditioning by homodyne detection to generate cat states.[38] As beam splitting with a vacuum and homodyne measurement can be highly efficient in quantum optics laboratories, this shows that relatively weak nonlinearity can still be useful to generate cat states.

The Hamiltonian of a single-mode Kerr nonlinear medium is $\mathcal{H}_{NL} = \omega a^\dagger a+\lambda(a^\dagger a)^2$, where a and $a^\dagger$ are annihilation and creation operators, ω is the energy level splitting for the harmonic-oscillator part of the Hamiltonian and λ is the strength of the Kerr nonlinearity.[35] Under the influence of the nonlinear interaction the initial coherent state $|\alpha\rangle$ evolves to the following state at time $\tau=\pi/\lambda N$:[52]

$$|\psi_N\rangle = \sum_{n=1}^{N} C_{n,N}|-\alpha e^{2in\pi/N}\rangle, \tag{33}$$

where

$$C_{n,N} = \frac{1}{N}\sum_{k=0}^{N-1}(-1)^k \exp[-\frac{i\pi k}{N}(2n-k)]. \tag{34}$$

The length L of the nonlinear cell corresponding to τ is $L=v\pi/2\lambda N$, where v is the velocity of light. For $N=2$, we obtain a desired cat state of relative phase $\varphi=\pi/2$. We again emphasize the nonlinear coupling λ is typically very small such that $N = 2$ cannot be obtained in a length limit where the decoherence effect can be neglected.

If λ is not as large as required to generate the cat state, the state (33) with $N > 2$ may be obtained by choosing an appropriate interaction time. From the state (33), it is required to remove all the other coherent component states except two coherent states of a π phase difference. First, it is assumed that the state (33) is incident on a 50-50 beam splitter with the vacuum onto the other input of the beam splitter. The initial coherent amplitude α_i is taken to be real for simplicity. The state (33) with initial amplitude α_i after passing through the beam splitter becomes

$$|\psi_N\rangle = \sum_{n=1}^{N} C_{n,N}|-\alpha_i e^{2in\pi/N}/\sqrt{2}\rangle|-\alpha_i e^{2in\pi/N}/\sqrt{2}\rangle, \tag{35}$$

where all $|C_{n,N}|$'s have the same value. The real part of the coherent amplitude in the state (35) is then measured by homodyne detection in order to produce the cat state in the other path. By the measurement result, the state is reduced to

$$|\psi_N^{(1)}\rangle = \sum_{n=1}^{N} C_{n,N}^{(1)}(\alpha_i)|-\alpha_i e^{2in\pi/N}/\sqrt{2}\rangle, \tag{36}$$

where $C_{n,N}^{(1)}(\alpha_i) = \mathcal{N}_\psi \sum_{n=1}^{N} C_{n,N}\langle X|-\alpha_i e^{2in\pi/N}/\sqrt{2}\rangle$ with $\mathcal{N}_\psi$ the normalization factor and $|X\rangle$ the eigenstate of $\hat{X} = (a + a^\dagger)/\sqrt{2}$. After the homodyne measurement, the state is selected when the measurement result is in certain values. If coefficients $|C_{N/2,N}^{(1)}(\alpha_i)|$ and $|C_{N,N}^{(1)}(\alpha_i)|$ in Eq. (36) have the same nonzero value and all the other $|C_{n,N}^{(1)}(\alpha_i)|$'s are zero, then the state becomes a desired cat state. Suppose $N = 4k$ where k is a positive integer number. If $X = 0$ is measured in this case, the coefficients $|C_{n,N}^{(1)}(\alpha_i)|$'s will be the largest when $n = N/4$ and $n = 3N/4$, and become smaller as n is far from these two points. The coefficients can be close to zero for all the other n's for an appropriately large α_i so that the resulting state may become a cat state of high fidelity. Using this technique, one may observe a conspicuous signature of a cat state even with a 1/100 times weaker nonliearity compared with the currently required level.[38] In particular, this approach can be useful to produce a cat state with a significantly large amplitude such as $\alpha \geq 10$.

Another scheme[39] proposed for linear optics quantum computation[53] uses weak cross-Kerr nonlinearity of the interaction Hamiltonian $H = \hbar\chi a_1^\dagger a_1 a_2^\dagger a_2$ to generate a cat state. The interaction between a coherent state, $|\alpha\rangle_2$, and a single-photon qubit, e.g., $|\psi\rangle_1 = (|0\rangle_1 + |1\rangle_1)/\sqrt{2}$, is described as

$$U_K|\psi\rangle_1|\alpha\rangle_2 = e^{iH_K t/\hbar}\frac{1}{\sqrt{2}}(|0\rangle_1 + |1\rangle_1)|\alpha\rangle_2 \tag{37}$$

$$= \frac{1}{\sqrt{2}}(|0\rangle_1|\alpha\rangle_2 + |1\rangle_1|\alpha e^{i\theta}\rangle_2), \tag{38}$$

where $|0\rangle$ ($|1\rangle$) is the vacuum (single-photon) state, α is the amplitude of the coherent state, and $\theta = \chi t$ with the interaction time t. If θ is π and one measures out mode 1 on a superposed basis $(|0\rangle \pm |1\rangle)/\sqrt{2}$, a macroscopic superposition state (so-called Schrödinger cat state), $(|\alpha\rangle \pm |-\alpha\rangle)/\sqrt{2}$. Using dual rail logic instead of the superposition between the single photon and the vacuum, the measurement on the superposed basis can be simply realized with a beam splitter and two photodetectors.[54] It is again extremely difficult to obtain $\theta = \pi$ using currently available nonlinear media. However, simply by increasing the amplitude α, one can gain an arbitrarily large separation between α and $\alpha e^{i\theta}$ within an arbitrarily short interaction time. It is possible to transform the state of the form of $|\alpha\rangle \pm |\alpha^{i\theta}\rangle$ to the symmetric form of $|\alpha'\rangle \pm |-\alpha'\rangle$ by the displacement operation which can be simply performed using a beam splitter with the transmission coefficient close to one and a strong coherent state being injected into the other input port. Therefore, weak cross-Kerr nonlinearity can also be useful to generate a cat state with a single photon, strong coherent states, beam splitters and two photodetectors. Remarkably, it can be shown that this approach can also reduce decoherence effects by increasing the initial amplitude α, which is also true for Ref. 38. It should also be noted that the detectors and the single photon source in Ref. 54 which can be directly combined with Ref. 39 do not have to be efficient for a conditional generation of a cat state because these factors only degrade the success probability to be less than unity.

4. Conclusion

We have discussed quantum processing tasks using qubits based on coherent states and Schrödinger cat states as resources. We have shown that a universal set of processing tasks can be achieved using only linear optics, feedforward and photon counting. This is a similar result to that of Knill Laflamme and Milburn for single photon qubits.[53] However, far fewer operations per gate are needed in the coherent state scheme and shortcuts are available for certain tasks. On the other hand these advantages are not useful unless a good way of producing cat states can be found. Thus we have spent some time discussing various proposals, both linear and non-linear, for producing cat states. We believe the near term prospects for demonstrating small travelling-wave cat states and basic processing tasks based on them are good. Whether coherent state qubits or single photon qubits will prove better for larger scale quantum optical processing in the long run remains an open question.

References

1. J. P. Dowling and G. J. Milburn, Phil. Trans. Roy. Soc. Lond. **A 361**, 1655-1674 (2003).
2. A. Einstein, B. Podolsky and N. Rosen, Phys. Rev. **47**, 777-780 (1935).
3. A. Aspect, P. Grangier and G. Roger, Phys. Rev. Lett. **47**, 460 (1981).
4. P.G.Kwiat, K.Mattle, H.Weinfurter, A.Zeilinger, A.V.Sergienko, and Y.Shih, Phys. Rev. Lett. **75**, 4337-4341 (1995).
5. C. A.Sackett, D. Kielpinski, B. E. King, C. Langer, V. Meyer, C. J. Myatt, M. Rowe, Q. A. Turchette, W. M. Itano, D. J. Wineland, and C. Monroe, Nature **404**, 256-259 (2000).
6. M.Nielsen and I.Chuang, *Quantum computation and quantum information* Cambridge University Press, Cambridge, UK, 2000.
7. E. Schrödinger, Naturwissenschaften **23**, 807-812, 823-828, 844-849 (1935).
8. A.J. Leggett and A. Garg, Phys. Rev. Lett. **54**, 857-860 (1985).
9. M. D. Reid, (1997). Quantum Semiclass. Opt. **9**, 489-499 (1997).
10. E. Schrödinger, Naturwissenschaften **14**, 664-666 (1935).
11. K. E. Cahill and R. J. Glauber, Phys. Rev. **177**, 1857-1881 (1969).
12. H. Jeong and M.S. Kim (2002), Phys. Rev. A **65**, 042305-042310 (2002).
13. C. H. Bennett, G. Brassard, C. Crépeau, R. Jozsa, A. Peres, and W. K. Wootters, Phys. Rev. Lett. **70**, 1895-1899 (1993).
14. S. J. van Enk and O. Hirota, Phys. Rev. A **64**, 022313-1-022313-6 (2001).
15. H. Jeong, M. S. Kim and J. Lee, Phys. Rev. A **64**, 052308-052314 (2001).
16. B. C. Sanders, Phys. Rev. A **45**, 6811-6815 (1992).
17. B. C. Sanders, K. S. Lee and M. S. Kim, Phys. Rev. A **52**, 735-741 (1995).
18. O. Hirota and M. Sasaki, quant-ph/0101018 (2001).
19. O. Hirota, S. J. van Enk, K. Nakamura, M. Sohma and K. Kato, quant-ph/0101096 (2001).
20. N. Lütkenhaus, J. Calsamiglia, and K.-A. Suominen, Phys. Rev. A **59**, 3295-3300 (1999).
21. S. L. Braunstein and A. Mann, Phys. Rev. A **51**, R1727-R1730 [1995).
22. D. Bouwmeester, J. W. Pan, K. Mattle, M. Eibl, H. Weinfurter, and A. Zeilinger, Nature **390**, 575-579 (1997).
23. K. Mattle, H. Weinfurter, P. G. Kwiat, and A, Zeilinger, Phys. Rev. Lett. **76**, 4656-4659 (1996).
24. H. Jeong and M. S. Kim, Quantum Information and Computation **2**, 208-221 (2002).

25. T. C. Ralph, W. J. Munro, and G. J. Milburn, quant-ph/0110115 Proceedings of SPIE **4917**, 1-12 (2002).
26. T. C. Ralph, A. Gilchrist, G. J. Milburn, W. J. Munro, and S. Glancy, Phys. Rev. A **68**, 042319-1-042319-11 (2003).
27. S. J. Dolinar, *Research Laboratory of Electronics, MIT, Quarterly* Progress Report No. 111, unpublished, pp. 115-120 (1973).
28. M.Takeoka and M.Sasaki, Phys. Rev. A **71**, 022318 (2005).
29. P. T. Cochrane, G. J. Milburn, and W. J. Munro, Phys. Rev. A **59**, 2631-2634 (1999).
30. S. Glancy, H. M. Vasconcelos, and T. C. Ralph, Phys. Rev. A **70**, 022317-1-022317-7 (2004).
31. J. Clausen, M. Dakna, L. Knöll and D.-G. Welsch, Optics Communications **179**, 189-196 (2000).
32. A. P. Lund, H. Jeong, T. C. Ralph, and M. S. Kim, Phys. Rev. A **70**, 020101(R)-1–020101(R)-4 (2004).
33. C. H. Bennett, D.P. DiVincenzo, J.A. Smolin, and W.K. Wootters, Phys. Rev. A **54**, 3824-3851 (1996).
34. S. J. D. Phoenix (1990), Phys. Rev. A. **41**, 5132–5138 (1990).
35. B. Yurke and D. Stoler, Phys. Rev. Lett. **57**, 13-16 (1986).
36. S. Song, C. M. Caves and B. Yurke, Phys. Rev. A **41**, R5261-R5264 (1990).
37. M. Dakna, T. Opatrný, L. Knöll and D. -G. Welsh, Phys. Rev. A **55**, 3184-3194 (1997).
38. H. Jeong, M. S. Kim, T. C. Ralph, and B. S. Ham, Phys. Rev. A **70**, 020101(R)-1-020101(R)-4 (2004).
39. K. Nemoto and W. J. Munro, Phys. Rev. Lett. **93**, 250502-1-250502-4 (2004).
40. B. Wang and L.-M. Duan, Phys. Rev. A **72**, 022320-1-022320-5 (2005).
41. J. Wenger, R. Tualle-Brouri, and P. Grangier, Phys. Rev. Lett. **92**, 153601-1-153601-4 (2004); J. S. Neergaard-Nielsen, B. M. Nielsen, C. Hettich, K. Mølmer, and E. S. Polzik, quant-ph/0602198; A. Ourjoumtsev, R. Tualle-Brouri, J. Laurat, and P. Grangier, Science **312**, 83-86 (2006).
42. H. Schmidt and A. Imamoğlu, Opt. Lett. **21**, 1936-1938 (2004).
43. L.V. Hau, S.E. Harris, Z. Dutton, and C.H. Behroozi, Nature **397**, 594-598 (1999).
44. M.D. Lukin and A. Imamoğlu, Phys. Rev. Lett. **84**, 1419-1422 (2000).
45. M. Paternostro, M. S. Kim, and G. M. Palma, Phys. Rev. A **67**, 023811-1-023811-15 (2003).
46. T. B. Pittman and J. D. Franson, Phys. Rev. Lett. **90**, 240401-1-240401 (2003).
47. H. Jeong, A. P. Lund, and T. C. Ralph, Phys. Rev. A **72**, 013801-1-013801-12 (2005).
48. Q. A. Turchette, C. J. Hood, W. Lange, H. Mabuchi, and H. J. Kimble, Phys. Rev. Lett. **75**, 4710-4713 (1995).
49. M. Brune, E. Hagley, J. Dreyer, X. Maître, A. Maali, C. Wunderlich, J.M. Raimond, and S. Haroche, Phys. Rev. Lett. **77**, 4887-4890 (1996).

50. C. Monroe, D. M. Meekhof, B. E. King, and D. J. Wineland, Science **272**, 1131-1135 (1996).
51. H. Ritsch, G. J. Milburn and T. C. Ralph, Phys. Rev. A **70**, 033804-1-033804-4 (2004).
52. K. S. Lee, M. S. Kim, S. -D. Lee and V. Buzěk, J. Kor. Phys. Soc. **26**, 197–204 (1993).
53. E. Knill, R. Laflamme, and G. J. Milburn, Nature **409**, 46–52 (2001).
54. C. C. Gerry, Phys. Rev. A **59**, 4095–4098 (1999).

Part II: Optical Continuous Variables

Chapter 10

Polarization Squeezing and Entanglement

Natalia Korolkova

School of Physics and Astronomy, University of St. Andrews
North Haugh, St. Andrews KY16 9SS, Scotland, UK

1. Introduction

Within the framework of the quantum continuous variables (CV), nonclassical polarization states have recently attracted a particular interest due to their compatibility with the spin variables of atomic systems and due to their simple detection schemes.[1–4] The relevant continuous polarization variables are hermitian Stokes operators (see Ref. 2 and references therein):

$$\hat{S}_0 = \hat{a}_x^\dagger \hat{a}_x + \hat{a}_y^\dagger \hat{a}_y, \qquad \hat{S}_1 = \hat{a}_x^\dagger \hat{a}_x - \hat{a}_y^\dagger \hat{a}_y,$$
$$\hat{S}_2 = \hat{a}_x^\dagger \hat{a}_y + \hat{a}_y^\dagger \hat{a}_x, \qquad \hat{S}_3 = i\left(\hat{a}_y^\dagger \hat{a}_x - \hat{a}_x^\dagger \hat{a}_y\right). \tag{1}$$

where the $\hat{a}_x$ and $\hat{a}_y$ denote the bosonic photon destruction operators associated with the x and y orthogonal polarization modes. The Stokes operator $\hat{S}_0$ commutes with all the others. The operators $\hat{S}_j, j \neq 0$ obey the commutation relations of the SU(2) Lie algebra:

$$[\hat{S}_k, \hat{S}_l] = \epsilon_{klm}\, 2i\hat{S}_m, \qquad k, l, m = 1, 2, 3. \tag{2}$$

Simultaneous exact measurements of these Stokes operators are thus impossible in general and their variances are restricted by the uncertainty relations:

$$V_2 V_3 \geq |\langle \hat{S}_1 \rangle|^2, \; V_3 V_1 \geq |\langle \hat{S}_2 \rangle|^2, \; V_1 V_2 \geq |\langle \hat{S}_3 \rangle|^2, \tag{3}$$

where $V_j = \langle \hat{S}_j^2 \rangle - \langle \hat{S}_j \rangle^2$ is a shorthand notation for the variance of the quantum Stokes parameter $\hat{S}_j$. The angle brackets denote expectation values with respect to the state of interest.

Within the last few years, successful generation of polarization squeezed[1,3,5,6] and polarization entangled[4,7] states has been reported (see also chapter 13 in this book). The respective definitions of polarization squeezing[1,8,9] and entanglement[2,4,9] were formulated. While dealing with continuous variable polarization states, one should pay particular attention to the subtleties arising due to the q-number, *i.e.* operator-valued, commutator (*cf.* Eq. (2)). This chapter reviews the definitions of polarization squeezing and polarization entanglement with the emphasis on the specific properties of nonclassical polarization states.

2. Polarization Squeezing

The analogy between quadrature squeezing and polarization squeezing is both elucidating and misleading. Squeezing in general refers to the suppression - squeezing - of quantum uncertainty in a particular variable below the respective reference level at the cost of increasing uncertainty in the conjugate variable. For quadratures, the level of quantum fluctuations of the coherent state conveniently serves as such a reference, which corresponds to the minimal possible quantum uncertainty equally distributed between the relevant conjugate variables. The important difference between quadrature squeezing and polarization squeezing is the discrepancy between coherent and minimum uncertainty states for the latter due to the specific form of the commutation relation.

A coherent polarization state is defined as a quantum state with both polarization modes having a coherent excitation α_x, α_y: $\psi_{coh} = |\alpha_x\rangle_x |\alpha_y\rangle_y$. The quantum uncertainty of such a state is equally distributed between the Stokes operators and their variances are all equal to $V_j = V^{coh} = |\alpha_x|^2 + |\alpha_y|^2 = \langle \hat{n} \rangle$. In analogy to quadrature squeezing, $V_j < V^{coh}$ seems at first glance to be a natural definition for polarization squeezing. However, due to the SU(2) commutation algebra, a coherent polarization state is not a minimum uncertainty state for all three Stokes operators simultaneously. This was known for atomic states, *i.e.* for spin coherent states[11] and angular momentum coherent states.[12] The construction of the minimum uncertainty product for the SU(2) algebra and the properties of atomic coherent states were broadly studied around early seventies.[11–14] Although a polarization state with a sub-shot-noise variance $V_j < V^{coh}$ is always a non-classical state, it implies nothing more than conventional quadrature or single-mode squeezing observed through the measurement of the Stokes parameters.

Hence a state is called *polarization squeezed* if:

$$V_k < |\langle \hat{S}_l \rangle| < V_m, \qquad k \neq l \neq m = 1, 2, 3. \tag{4}$$

Thus the reference state to quantify squeezing is chosen to be the corresponding minimum uncertainty state, not a coherent state.

This is, however, not the only possible definition of polarization squeezing. The intuitive idea of squeezing implies that polarization fluctuations are reduced below some prescribed level. In this regard, polarization is a more complex variable than quadrature, so it turns out that for polarization it is necessary to explicitly specify what is understood by polarization fluctuations. Unlike quadrature squeezing it appears that the precise quantitative formulation of polarization squeezing depends on the purpose, i.e., on the envisaged application of polarization squeezed light. However, there are some general requirements to a physically sensible squeezing criterion. For example, quadrature squeezing is knowingly invariant under linear unitary local transformations. It was suggested[15] that the SU(2) invariance should be an important property of a measure for polarization squeezing. It means, polarization squeezing criteria should be invariant under the action of linear energy conserving transformations of the complex amplitudes a_x, a_y, such as the ones produced by beam splitters and phase plates. This is because they are rotations of the Poincaré sphere which purely displace the polarization distribution, i.e., change the Stokes parameters but preserve the form of the polarization distribution. Since fluctuations depend exclusively on the form of the corresponding probability distribution irrespectively of its location, one can conclude that any criterion of polarization squeezing should be invariant under SU(2) transformations.[15]

The idea of invariance suggests that in general one should avoid formulating squeezing criteria in terms of the components of the Stokes vector along arbitrary unspecified Cartesian axes without further cautions. A simple procedure to guarantee SU(2) invariance is obtained by using specific components of the Stokes operators referred to the mean value $\langle \hat{S} \rangle$. These are the parallel component $S_{\parallel}$ along $\langle \hat{S} \rangle$, with $|\langle S_{\parallel} \rangle| = |\langle \hat{S} \rangle|$, and two components orthogonal to $\langle \hat{S} \rangle$, denoted by $S_{\perp}$, with $\langle S_{\perp} \rangle = 0$. Two orthogonal components $S_{\perp 1}$, $S_{\perp 2}$ define the so-called dark plane (see also Chapter 13 in this book). Actually, in many works on polarization squeezing the invariance requirement is implicitly taken into account. This occurs when the Cartesian axes are chosen so that $\langle \hat{S} \rangle$ is aligned with one of the $\hat{S}_j$ components.

Thus, the use of the operator basis $S_{\parallel}$, $S_{\perp 1,2}$ guarantees the SU(2) invariance of the polarization squeezing criterion. A further advantage of using this operator basis is that it allows one to reduce the three uncertainty relations for the Stokes operators to a single nontrivial one:

$$\Delta S_{\perp 1} \Delta S_{\perp 2} \geq |\langle S_{\parallel} \rangle|. \tag{5}$$

The other two become trivial

$$\Delta S_{\perp 1} \Delta S_{\parallel} \geq 0, \qquad \Delta S_{\perp 2} \Delta S_{\parallel} \geq 0. \tag{6}$$

Importantly, when uncertainty relations are expressed in the form of Eq. (5, 6), both quadrature coherent states and SU(2) coherent states are minimum uncertainty states in the sense of leading to an equality in Eq. (5), since for these states we have

$$(\Delta S_{\perp 1})^2 = (\Delta S_{\perp 2})^2 = \langle S_{\parallel} \rangle = \langle S_0 \rangle. \tag{7}$$

This resolves the discrepancy between quadrature coherent states and minimum uncertainty states for the Stokes operators mentioned in the beginning of this section. It is worth noting that the minimum uncertainty character of SU(2) coherent states no longer holds for arbitrary Cartesian projections of $\hat{S}$.

Several definitions of polarization squeezing can be found in the literature, which are listed below in the SU(2) invariant form (for more details on different squeezing criteria see Ref. 15 and references therein). The simpler one focus directly on the comparison of the fluctuations of the orthogonal components $\Delta S_{\perp}$ with the fluctuations of the same components for quadrature or SU(2) coherent states, for which $\Delta S_{\perp} = \sqrt{\langle S_0 \rangle}$ for every $S_{\perp}$. Therefore, polarization squeezing occurs if there are orthogonal components with

$$\Delta S_{\perp} < \sqrt{\langle S_0 \rangle}. \tag{8}$$

Another criterion based on precision measurements defines polarization squeezing whenever for a suitable orthogonal component $S_{\perp}$ it occurs that

$$\frac{\Delta S_{\perp}}{|\langle S_{\parallel} \rangle|} < \frac{1}{\sqrt{\langle S_0 \rangle}}. \tag{9}$$

The states fulfilling this criterion allow to perform measurements with a precision beyond the level achievable when using quadrature coherent states or SU(2) coherent states (that lead to the equality in the preceding relation). This can be also regarded as the SU(2) invariant form of another measure of spin squeezing or entanglement of spin 1/2 particles expressed in Cartesian components as[16]

$$\frac{(\Delta S_1)^2}{\langle S_2\rangle^2 + \langle S_3\rangle^2} < \frac{1}{\langle S_0\rangle}. \tag{10}$$

Finally, we consider a suitable analogy with quadrature squeezing derived from the uncertainty relation (5) by stating that there is polarization squeezing provided that[1,9]

$$(\Delta S_{\perp 1})^2 < |\langle S_{\parallel}\rangle| < (\Delta S_{\perp 2})^2, \tag{11}$$

for a suitable pair of components $S_{\perp j}$ in the dark plane. Equivalently,

$$\frac{(\Delta S_{\perp})^2}{|\langle S_{\parallel}\rangle|} < 1, \tag{12}$$

for a suitable orthogonal component $S_{\perp}$. This criterion might be useful for quantum information applications of polarization squeezing, e.g., for polarization entanglement generation, since the relevant entanglement criteria deal with squeezing properties derived from uncertainty relations for sums and differences of single party observables. This is also the SU(2) invariant form of the criterion (4), with which we have started the considereation of polarization squeezing in the beginning of this section.

Although these three criteria are not equivalent, there is a definite relationship between them since they can be definitely arranged in a series of increasing stringency. Since it always holds that $\langle S_0\rangle \geq |\langle S_{\parallel}\rangle|$, the fulfillment of criterion (9) implies that criterion (12) is satisfied, which in turn implies that criterion (8) holds,

$$(\Delta S_{\perp})^2 \leq \frac{\langle S_{\parallel}\rangle^2}{\langle S_0\rangle} \leq |\langle S_{\parallel}\rangle| \leq \langle S_0\rangle, \tag{13}$$

while the converse are not necessarily true. In other words, criterion (8) can be regarded as a necessary condition for polarization squeezing.

It can be appreciated that in these criteria the fluctuations of $S_{\parallel}$ are absent. It seems that the only way to include $\Delta S_{\parallel}$ should be by adding a condition similar to Eq. (8) in the form $\Delta S_{\parallel} < \sqrt{\langle S_0\rangle}$. In such a case it should be understood that for this last expression the reference level for polarization fluctuations would be quadrature coherent states exclusively, since for SU(2) coherent states $\Delta S_{\parallel} = 0$. The inclusion of $\Delta S_{\parallel}$ in other criteria of the form (9), (12) is excluded from the very beginning. On the one hand, in precision measurements the fluctuations of the component $S_{\parallel}$ play no role since only orthogonal components are measured. Moreover, the only uncertainty relations involving $\Delta S_{\parallel}$ are trivial as shown in Eqs. (5), (6).

Regarding the relation of polarization squeezing to other nonclassical properties, such as entanglement or quadrature squeezing, it was shown that polarization squeezing is always a nonclassical property with entanglement in a certain mode basis being a necessary condition for polarization squeezing.[15] In many situations (including those of most practical relevance) polarization squeezing is equivalent to quadrature squeezing. However, this relation is no universally valid: there are polarization squeezed states without quadrature squeezing[15] and quadrature squeezed states without polarization squeezing.[5]

3. Continuous Variable Polarization Entanglement

Along with polarization squeezing, CV polarization entanglement[2] has proven to be a useful tool in quantum communication. The nonseparability is a property of the state irrespective to the observables under consideration. All continuous variable entangled states have the same nature in this sense. In practice, some particular variables might be more advantageous to use. *Polarization entanglement* is a nonseparable state which implies correlations of the quantum uncertainties between one or more pairs of Stokes operators of two spatially-separated optical beams. To quantify the degree of these quantum correlation and to verify the nonseparability of the state, different criteria can be used: There is no unique criterion to quantify CV entanglement in general, in particular for mixed states. A useful reference for the generalization and comparison of different sum and product entanglement criteria for CVs is the paper by Giovannetti *et al.*[17]

EPR criterion for the Stokes operators: The demonstration of the Einstein-Podolsky-Rosen (EPR) paradox for continuous variables takes place when measurements carried out on one subsystem can be used to infer the values of non-commuting observables of another, spatially-separated subsystem to sufficient precision that an "apparent" violation of the uncertainty principle occurs.[19] The precision with which we can infer the value of an observable $\hat{Z}_D$ of subsystem D from the measurement of $\hat{Z}_C$ on subsystem C is given by the conditional variance

$$V_{\rm cond}(\delta\hat{Z}_D|\delta\hat{Z}_C) = V(\delta\hat{Z}_D)\left(1 - \frac{\left|\langle\delta\hat{Z}_D\,\delta\hat{Z}_C\rangle\right|^2}{V(\delta\hat{Z}_D)V(\delta\hat{Z}_C)}\right). \qquad (14)$$

Here the linearized approach is used, $\hat{Z}_j = \langle Z_j \rangle + \delta\hat{Z}_j$ ($j = C, D$), $\langle \delta\hat{Z}_j \rangle = 0$, and the variances $V(\delta\hat{Z}_j)$ are assumed to be inherently normalized to the shot noise level. For the quadrature components this normalization is, for instance, just equal to unity, but note that in an experiment the shot noise level should be always explicitly determined using respective measurement techniques. The EPR-entanglement of the Stokes parameters in the sense of the EPR-like correlations of their uncertainties is realized if for any of the conjugate pair of the Stokes operators, *e.g.* S_1, S_3, the following inequality holds:[2]

$$V_{\text{cond}}(\delta\hat{S}_{3D}|\delta\hat{S}_{3C})V_{\text{cond}}(\delta\hat{S}_{1D}|\delta\hat{S}_{1C}) < \left|\langle\hat{S}_{2C}\rangle\right|^2 . \tag{15}$$

The two other inequalities for S_1, S_2 and S_2, S_3 are obtained by the cyclic permutation of indices.

Nonseparability criterion: general form: The nonseparability criterion of Duan *et al.*[20] is derived for the canonical conjugate variables having a *c*-number commutator, like position and momentum or quadrature components of the light field. For a certain type of Gaussian states the criterion is necessary and sufficient. Unlike position and momentum, the Stokes operators are the variables with an operator-valued commutator. This provides a major obstacle in the derivation of the continuous variable nonseparability criterion in terms of polarization or spin. So far, it was only possible to derive the sufficient criterion of nonseparability for such variables with a *q*-number operator .[2,4,9,21] The formulation of the sufficient nonseparability criterion[20] for the Stokes operators was first presented in Ref. 2 in terms of the quantities measured in an experiment and further elaborated in Refs. 4, 18.

In Refs. 2, 4, 18 the underlying fundamental principle is the Heisenberg uncertainty relation. However, due to specific properties emerging from the *q*-number commutator, the generaized uncertainty relation[22] is more suitable as a starting point for the derivation of the general nonseparability criterion.[9] The generalized Heisenberg uncertainty relation has the following form:

$$V_A V_B \geq \frac{1}{4}\left|\left\langle\left\{\Delta\hat{A}, \Delta\hat{B}\right\}\right\rangle\right|^2 + \frac{1}{4}\left|\left\langle\left[\Delta\hat{A}, \Delta\hat{B}\right]\right\rangle\right|^2, \tag{16}$$

where the anticommutator and commutator of the two operators are defined by

$$\left\{\Delta\hat{A},\Delta\hat{B}\right\} = \hat{A}\hat{B}+\hat{B}\hat{A}-2\langle\hat{A}\rangle\langle\hat{B}\rangle,$$
$$\left[\Delta\hat{A},\Delta\hat{B}\right] = \hat{A}\hat{B}-\hat{B}\hat{A}=[\hat{A},\hat{B}]. \tag{17}$$

This differes from the frequently-used form

$$V_AV_B \geq \frac{1}{4}\left|\left\langle\left[\Delta\hat{A},\Delta\hat{B}\right]\right\rangle\right|^2. \tag{18}$$

by the presence of the anti-commutator term. For the canonical observables like position and momentum, the commutator is a c-number whereas the anti-commutator $\{\Delta\hat{A},\Delta\hat{B}\}$ is usually a q-number. Thus the *universal* general form of the Heisenberg inequality (18) can be obtained by retaining state-independent part, the commutator term. If the commutator is an operator quantity, like in the case of the Stokes operators, then both contributions on the right of (16) depend on the state of the system and there is no reason to remove any of them. This jusitifies the use of the generalized Heisenberg uncertainty relation (17) as the starting point in the derivation of the nonseparability criterion[9] along the lines suggested by Duan *et al.*. The inequality in (18) remains valid but the full form in (16) provides a stronger inequality with a higher minimum value of the variance product. There is no universal uncertainty relation in such cases, as in the examples of the angular momentum operators and of the Stokes-parameter operators considered here.

The derivation of the nonseparability criterion for CV position x and momentum p having a c-number commutator[20] considers an overall system composed of two subsystems, c and d, described by operators

$$\hat{A} = |a|\hat{x}_c + \frac{1}{a}\,\hat{x}_d, \qquad \hat{B} = |a|\hat{p}_c - \frac{1}{a}\,\hat{p}_d, \tag{19}$$

$$[\hat{x}_i,\hat{p}_j] = i\delta_{ij}\ (i,j=c,d), \quad [\hat{A},\hat{B}] = i\left(a^2-\frac{1}{a^2}\right). \tag{20}$$

The restrictions on the sum of the two variances are direct consequences of the uncertainty relation. With the use of the Schwarz inequality

$$V_AV_B \geq |\langle\Delta\hat{A}\Delta\hat{B}\rangle|^2, \qquad \Delta\hat{A}=\hat{A}-\langle\hat{A}\rangle, \Delta\hat{B}=\hat{B}-\langle\hat{B}\rangle \tag{21}$$

and the Cauchy inequality $V_A^2+V_B^2 \geq 2V_AV_B$ it follows that

$$V_A+V_B \geq 2|\langle\Delta\hat{A}\Delta\hat{B}\rangle|. \tag{22}$$

Thus, with the Heisenberg uncertainty relation taken in the form (18), *all* states must satisfy

$$V_A V_B \geq \frac{1}{4}\left(a^2 - \frac{1}{a^2}\right)^2 \text{ and } V_A + V_B \geq \left|a^2 - \frac{1}{a^2}\right|. \tag{23}$$

It is shown in[20] that *separable* states of the two subsystems must satisfy the stronger inequality

$$V_A + V_B \geq a^2 + \frac{1}{a^2}. \tag{24}$$

Nonseparable or *entangled* states thus exist in the region defined by

$$\left|a^2 - \frac{1}{a^2}\right| \leq V_A + V_B < a^2 + \frac{1}{a^2}, \tag{25}$$

where the lower limit on the left comes from the development of the Heisenberg uncertainty relation in (23) and the upper limit on the right comes from the nonseparability criterion in[20] in its sufficient form.

The derivation of[20] can be reworked for the basic operator commutation relations more general than those given in (19), (20):[9]

$$\hat{A} = \hat{A}_c + \hat{A}_d, \qquad \hat{B} = \hat{B}_c - \hat{B}_d, \tag{26}$$

$$[\hat{A}_c, \hat{B}_d] = [\hat{B}_c, \hat{A}_d] = 0, \qquad [\hat{A}, \hat{B}] = [\hat{A}_c, \hat{B}_c] - [\hat{A}_d, \hat{B}_d]. \tag{27}$$

Here the nonzero commutators may themselves be operators. The uncertainty relations (23) are generalized to

$$V_A V_B \geq \left|\left\langle \Delta\hat{A}_c \Delta\hat{B}_c \right\rangle - \left\langle \Delta\hat{A}_d \Delta\hat{B}_d \right\rangle\right|^2, \tag{28}$$

$$V_A + V_B \geq 2\left|\left\langle \Delta\hat{A}_c \Delta\hat{B}_c \right\rangle - \left\langle \Delta\hat{A}_d \Delta\hat{B}_d \right\rangle\right|. \tag{29}$$

Note that these relations reduce to those in (21) and (22) when there is only a single system, c or d. By substitution of (26) into (19,22), the Eqs. (3,4) in[20] can be reworked for the pair of variables with the q-number commutator giving the *sufficient* nonseparability criterion. The main difference to the derivation of[20] is the replacement of the universal limit in (25) by the state-dependent contribution containing the mean value of the operator-valued commutator (27) and the retainment of the state-dependent anticommutator contribution.

Theorem: For any separable state ρ_{sep} the following inequality holds:

$$V_A + V_B \geq 2\left|\left\langle \Delta\hat{A}_c \Delta\hat{B}_c \right\rangle\right| + 2\left|\left\langle \Delta\hat{A}_d \Delta\hat{B}_d \right\rangle\right|. \tag{30}$$

Proof: A separable quantum state ρ_{sep} can be written as a convex decomposition

$$\rho_{sep} = \sum_j p_j \rho_{jc} \otimes \rho_{jd}. \tag{31}$$

Using this decomposion we can directly compute the sum of the variances $V_A + V_B$. The averaging in the expressions below is performed over the product density matrix $\rho_{sep} = \sum_j p_j \rho_{jc} \otimes \rho_{jd}$. We obtain:

$$V_A + V_B = \sum_j p_j \left(\langle \hat{A}^2 \rangle_j + \langle \hat{B}^2 \rangle_j \right) - \left(\sum_j p_j \langle A \rangle_j \right)^2 - \left(\sum_j p_j \langle B \rangle_j \right)^2$$
$$= \sum_j p_j \left(\langle \hat{A}_c^2 \rangle_j + \langle \hat{A}_d^2 \rangle_j + \langle \hat{B}_c^2 \rangle_j + \langle \hat{B}_d^2 \rangle_j \right) +$$
$$+2 \left(\sum_j p_j \langle A_c \rangle_j \langle A_d \rangle_j - \sum_j p_j \langle B_c \rangle_j \langle B_d \rangle_j \right) - \left(\sum_j p_j \langle A \rangle_j \right)^2 - \left(\sum_j p_j \langle B \rangle_j \right)^2$$

$$= \sum_j p_j \left(\langle (\Delta \hat{A}_c)^2 \rangle_j + \langle (\Delta \hat{A}_d)^2 \rangle_j + \langle (\Delta \hat{B}_c)^2 \rangle_j + \langle (\Delta \hat{B}_d)^2 \rangle_j \right)$$
$$+ \sum_j p_j \left(\langle \hat{A} \rangle_j^2 + \langle \hat{B} \rangle_j^2 \right) - \left(\sum_j p_j \langle A \rangle_j \right)^2 - \left(\sum_j p_j \langle B \rangle_j \right)^2. \tag{32}$$

Let us estimate the limits for the last two lines in (32). We use the Schwarz inequality in the form (21) and $V_A^2 + V_B^2 \geq 2 V_A V_B$ and get:

$$\sum_j p_j \left(\langle (\Delta \hat{A}_c)^2 \rangle_j + \langle (\Delta \hat{B}_c)^2 \rangle_j + \langle (\Delta \hat{A}_d)^2 \rangle_j + \langle (\Delta \hat{B}_d)^2 \rangle_j \right)$$
$$\geq 2 \left| \left\langle \Delta \hat{A}_c \Delta \hat{B}_c \right\rangle \right| + 2 \left| \left\langle \Delta \hat{A}_d \Delta \hat{B}_d \right\rangle \right|.$$

Note that the application of the Schwarz inequality (21) corresponds to the use of the generalized uncertainty relation: Eq. (21) is readily re-expressed in the form Eq. (16) and the anti-commutator term is retained. Furthermore, it can be easily shown[20] using the Cauchy-Schwarz inequality $(\sum_j p_j)(\sum_j p_j \langle \hat{A} \rangle_j^2) \geq (\sum_j p_j |\langle \hat{A} \rangle_j|)^2$ that the lower bound for the last line in (32) is zero,

$$\sum_j p_j \left(\langle \hat{A} \rangle_j^2 + \langle \hat{B} \rangle_j^2 \right) - \left(\sum_j p_j \langle A \rangle_j \right)^2 - \left(\sum_j p_j \langle B \rangle_j \right)^2 \geq 0.$$

Hence, for any separable state (31) the inequality (30) holds, which proves our statement.

It follows from Eq. (30) and uncertainty relations (21), (16) that the nonseparable or entangled states have to satisfy Eq. (34):

$$2\left|\left\langle \Delta\hat{A}_c\Delta\hat{B}_c\right\rangle - \left\langle \Delta\hat{A}_d\Delta\hat{B}_d\right\rangle\right| \le V_A + V_B < \\ 2\left|\left\langle \Delta\hat{A}_c\Delta\hat{B}_c\right\rangle\right| + 2\left|\left\langle \Delta\hat{A}_d\Delta\hat{B}_d\right\rangle\right|. \tag{33}$$

In contrast to Eqs. (5-7) in Ref. 17 the lower limit in (30) and hence the upper limit in Eqs. (34), (33) does not depend on the particular form of the convex decomposition in (31). However, the lower bound for $V_A + V_B$ (30) and the limits in the nonseparability criterion Eqs. (34), (33) do depend on the quantum state under consideration. There is no universal separability limit for the sum or product of the two variances V_A, V_B in the case of the operator-valued commutators $[\hat{A}_c, \hat{B}_c]$ and $[\hat{A}_d, \hat{B}_d]$. Nevertheless, the inequalities of Eqs. (30), (34), (33) provide a sensible operational sufficient criterion for nonseparability which can be readily verified in an experiment.

Nonseparable or *entangled* states must then satisfy the condition

$$2\left|\left\langle \Delta\hat{A}_c\Delta\hat{B}_c\right\rangle - \left\langle \Delta\hat{A}_d\Delta\hat{B}_d\right\rangle\right| \le V_A + V_B < 2\left|\left\langle \Delta\hat{A}_c\Delta\hat{B}_c\right\rangle\right| \\ + 2\left|\left\langle \Delta\hat{A}_d\Delta\hat{B}_d\right\rangle\right|, \tag{34}$$

where the lower limit on the left comes from the development of the Heisenberg uncertainty relation in (29) and the upper limit on the right comes from the generalization of the *sufficient* nonseparability criterion. A derivation of the nonseparability criterion in its *necessary and sufficient form* in the case of the q-number commutator still remains a challenge. The sufficient general *product* criterion was obtained in Ref. 17, where the standard form of the Heisenberg uncertainty relation was used to derive an upper limit for the product of two variances.

Different formulations and application to polarization and spin variables: The inseparability and EPR crieria can also be formulated to provide a single number as a figure of merit to quantify entanglement. As a starting point, Bowen *et al.*[4,18] use the Heisenberg uncertainty relation in the form:

$$\Delta^2\hat{A}\Delta^2\hat{B} \ge \frac{|[\delta\hat{A}, \delta\hat{B}]|^2}{4}, \tag{35}$$

where $\Delta^2\hat{O} = \langle\hat{O}^2\rangle - \langle\hat{O}\rangle^2$ is the variance and $\delta\hat{O}$ is the noise operator in the linearized approch, as before. The minimum of the variance of the sum or difference of the operator $\hat{O}$ between two beams c and d (x and y in the notation of[4,18]) is given by:

$$\Delta^2_{c\pm d}\hat{O} = \min\langle(\delta\hat{O}_c \pm \delta\hat{O}_d)^2\rangle. \tag{36}$$

On the basis of Eqs. (35, 36), they define the *degree of inseparability* $\mathcal{I}(\hat{A},\hat{B})$:

$$\mathcal{I}(\hat{A},\hat{B}) = \frac{\Delta^2_{c\pm d}\hat{A} + \Delta^2_{c\pm d}\hat{B}}{2|[\delta\hat{A},\delta\hat{B}]|^2} \tag{37}$$

so that if $\mathcal{I}(\hat{A},\hat{B}) < 1$, the state is inseparable. Note the following difference to the criterion of Eq. (34): Although the formulation of the inseparability criterion in Refs. 4, 18 starts with the generalized Heisenberg uncertainty relation, the latter is only used to generalize the criterion to any pairs of variables $\hat{A},\hat{B}$ not restricted to the Stokes operators. The state-dependend correlation term is then omitted and the conventional form of the Heisenberg relation (18, 35) is used.

The criterion may be re-formulated in a product form, extending the product criterion for the quadrature operators first formulated in Ref. 23:

$$\mathcal{I}_{\text{product}}(\hat{A},\hat{B}) = \frac{\Delta^2_{c\pm d}\hat{A}\ \Delta^2_{c\pm d}\hat{B}}{|[\delta\hat{A},\delta\hat{B}]|^2}. \tag{38}$$

The product measures are independent from local squeezing operations performed on entangled beams (see also Ref. 17). Analogously, the (product) EPR criterion (15) can be re-written in the form of the *degree of EPR paradox* $\mathcal{E}(\hat{A},\hat{B})1.$[18]

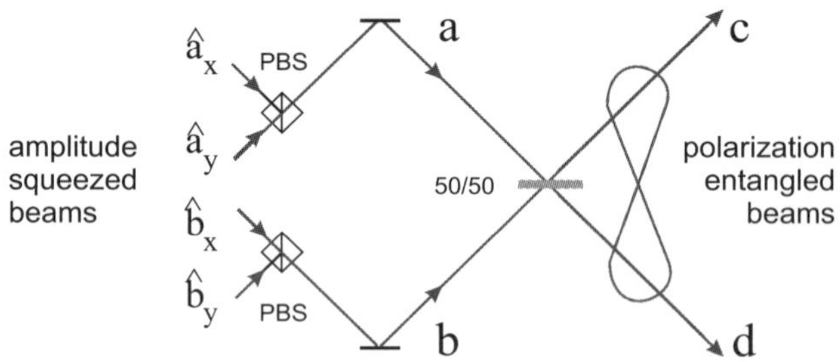

Fig. 1. Scheme for the generation of polarization entanglement (see Korolkova *et al.*[2]). PBS: polarization beam splitter; 50/50: 50/50 beam splitter.

To illustrate the role of the anti-commutator (correlation) term in the general form of the nonseparability criterion (34) as compared to the reduced forms (37, 38), let us consider generation of polarization entanglement by mixing two polarization squeezed beams a and b on a beam splitter as described in[2] and Fig. 1. Polarization squeezing is obtained by an approriate combination of two amplitude-squeezed beams of orthogonal polarizations denoted x, y as in Eq. (1). For this particular case the nonseparability condition (34) takes the form:[9]

$$0 < V_A + V_B = V(\hat{S}_{1c} + \hat{S}_{1d}) + V(\hat{S}_{3c} - \hat{S}_{3d}) <$$
$$< \{\alpha^4(V^+_{ax} - V^-_{ay} + V^+_{ax} - V^-_{ay} - V^+_{bx} + V^-_{bx} - V^+_{by} + V^-_{by})^2 + 64\alpha^4\}^{1/2}, \quad (39)$$

where α is the coherent amplitude of the four incoming beams taken to be equal. V^+ denotes the squeezed amplitude-quadrature variance and V^- the anti-squeezed phase-quadrature variance of the interfering amplitude-squeezed beams. The first term in the square root comes from the anti-commutator term and the second from the commutator. In the simple case where all four modes making up the polarization squeezed beams have same amplitude squeezing (all V^+ are equal and all V^- are equal), the correlation term vanishes and only the contribution of the commutator survives: $|\langle\Delta\hat{S}_{1c}\Delta\hat{S}_{3c}\rangle| = |\langle\Delta\hat{S}_{1d}\Delta\hat{S}_{3d}\rangle| = 2\alpha^2$. Then the criterion of Eq. (34) in terms of the experimentally measurable quantities reads:

$$V(\hat{S}_{1c} + \hat{S}_{1d}) + V(\hat{S}_{3c} - \hat{S}_{3d}) < 8\alpha^2 \quad (40)$$

or, in more general form:

$$V(\hat{S}^c_k \pm \hat{S}^d_k) + V(\hat{S}^c_l \mp \hat{S}^d_l) < 2|\langle\hat{S}^c_m\rangle| + 2|\langle\hat{S}^d_m\rangle|, \qquad \epsilon_{klm}, \quad (41)$$

where indices $k, l, m = 1, 2, 3$ denote the corresponding Stokes operator $\hat{S}_j$ and c, d refer to the respective subsystem. The nonseparability criterion for the Stokes operators (41) is being used in the Chapter 13 of J. Heersink and V. Josse *et al.* of this book to characterize the polarization entanglement generated using Kerr nonlinear interactions. The sum of the combined variances $V(\hat{S}^A_k \pm \hat{S}^B_k) + V(\hat{S}^A_l \mp \hat{S}^B_l)$ is denoted by $\mathcal{I}^S_{A,B}$ for the entanglement of the Stokes operator S between two subsystems A, B.

An exciting application of nonclassical polarization states is the light-matter interface. The Stokes operators obey the same commutation rules as the continuous spin variables, which are the components $\hat{J}_{x,y,z}$ of the collective spin $\hat{J}$ of an atomic ensemble. The possibility to produce spin squeezing of an atomic ensemble was first shown in Ref. 1, where the ensemble was illuminated with polarization squeezed light and in this way the quantum correlations were mapped from light onto atoms. Further work involved entangling two macroscopic atomic ensembles[24] and first steps towards implementation of a quantum memory.[25] The nonseparability criteria mentioned in this chapter are all directly transferrable to the continuous spin variables having the commutation relation of the same type, *e.g.*, $[\hat{J}_y, \hat{J}_z] = i\hat{J}_x$ (others are obtained by cyclic permutation of indices). In addition, Ref. 21 presents an alternative derivation of a separability criterion for spin variables. However, to quantify experiments with atomic ensembles mentioned above, a simpler criterion equivalent to that for position and momentum or for light quadratures was used. This possibility is linked to particular experimental settings which are most commonly used in real life, namely the light used is strongly linearly polarized and the collective spin $\hat{J}$ is polarized as well along one of the directions $\hat{J}_j$, say along $\hat{J}_x$. In this case the x-component of the spin is dominant and can be treated classically, $\hat{J}_x \to J_x$. This allows one to renormalize the relevant quantities. One can introduce $\hat{J}'_j = \hat{J}_j/\sqrt{J_x}$ and take advantage of having the conventional position and momentum commutator: $[\hat{J}'_y, \hat{J}'_z] = i$. One can then apply to the new primed variables the nonseparability criteria for a c-number commutator derived by Duan *et al.*,[20] Simon[26] and further developed by others. For the particular experimental setting,[24] the following criterion for CV spin entanglement is obtained:

$$V(\hat{J}_z^c + \hat{J}_z^d) + V(\hat{J}_y^c + \hat{J}_y^d) < 2J_x, \tag{42}$$

where the classical component J_x is typically equal in both subsystems. In more general terms it reads (*cf.* Eq. (41)):

$$V(\hat{J}_k^c + \hat{J}_k^d) + V(\hat{J}_l^c + \hat{J}_l^d) < J_m^c + J_m^d. \tag{43}$$

The same holds for the Stokes operators, when strongly polarized light is used.

As we have seen, for certain experimental conditions the use of the reduced crtieria (37, 41, 43) is more reasonable than the full form (34). However, some new interesting features can be lost in cases when the reduced forms or the c-number form are sufficient for characterization. For example, the simultaneous entanglement between different pairs of Stokes operators is possible only if the q-number commutation relation holds. For different possible scenaria in this context see *e.g.* the discussion of Eqs. (39, 40), examples in [9,18] and the chapters on experiments with atomic ensembles and on experimental polarization squeezing and entanglement in this book.

References

1. J. Hald, J. L. Sorensen, C. Schori, and E. S. Polzik, J. Mod. Opt. **47**, 2599 (2001).
2. N. Korolkova, G. Leuchs, R. Loudon, T. C. Ralph, and Ch. Silberhorn, Phys. Rev. A **65**, 052306 (2002).
3. W.P. Bowen, R. Schnabel, and H.A. Bachor, P. K. Lam, Phys. Rev. Lett. **88**, 093601 (2002).
4. W. P. Bowen, N. Treps, and R. Schnabel, P. K. Lam, Phys. Rev. Lett. **89**, 253601 (2002).
5. J. Heersink, T. Gaber, S. Lorenz, O. Glöckl, N. Korolkova, and G. Leuchs, Phys. Rev. A **68**, 013815 (2003).
6. V. Josse, A. Dantan, A. Bramati, M. Pinard, and E. Giacobino, Phys. Rev. Lett. **91**, 10360 (2003); Quant. Semiclass. Opt. **6**, S532 (2004).
7. O. Glöckl, J. Heersink, N. Korolkova, G. Leuchs, and S. Lorenz, Quantum Semiclass. Opt. **5**, 492 (2003).
8. A. S. Chirkin, A. A. Orlov, and D. Yu. Paraschuk, Quantum Electron. **23**, 870 (1993).
9. N. Korolkova and R. Loudon, Phys. Rev. A **71**, 032343 (2005).
10. S. Lorenz, N. Korolkova, and G. Leuchs, Appl. Phys. B **79**, 273 (2004).
11. J. M. Radcliffe, J. Phys. A **4**, 313 (1971).
12. P. W. Atkins and J. C. Dobson, Proc. Roy. Soc. A (London) **321**, 321 (1971).
13. R. Jackiw, J. Math. Phys. **9**, 339 (1968).
14. F. T. Arecchi, E. Courtens, R. Gilmore, and H. Thomas, Phys. Rev. A **6**, 2211 (1972).
15. A. Luis and N. Korolkova, Phys. Rev. A (2006), submitted.
16. A. Sørensen, L.-M. Duan, J. I. Cirac, and P. Zoller, Nature **409**, 63 (2001); S. Raghavan, H. Pu, P. Meystre, and N. P. Bigelow, Opt. Commun. **188**, 149 (2001).
17. V. Giovannetti, S. Mancini, D. Vitali, and P. Tombesi, Phys. Rev. A **67**, 022320 (2003).
18. W. P. Bowen, N. Treps, R. Schnabel, T. C. Ralph, and P. K. Lam, J. Opt B: Quantum Semiclass. Opt. **5**, S467 (2003).
19. M. D. Reid and P. D. Drummond, Phys. Rev. Lett. **60**, 2731 (1988); M. D. Reid, Phys. Rev. A **40**, 913 (1989).
20. L.-M. Duan, G. Giedke, J. I. Cirac, P. Zoller, Phys. Rev. Lett. **84**, 2722 (2000).
21. M. G. Raymer, A. C. Funk, B. C. Sanders, and H. de Guise, Phys. Rev. A **67**, 052104 (2003).
22. E. Merzbacher, *Quantum Mechanics*, 3rd ed., Wiley, New York 1998, pp. 217-220.
23. S. Mancini, V. Giovannetti, D. Vitali, and P. Tombesi, Phys. Rev. Lett. **88**, 120401 (2002).
24. B. Julsgaard, A. Kozhekin, and E. S. Polzik, Nature **413**, 400 (2001); E. S. Polzik, Physics World, September 2002.

25. B. Julsgaard, J. Sherson, I. J. Cirac, J. Fiurasek, and E. S. Polzik, Nature **432**, 482 (2004).
26. R. Simon, Phys. Rev. Lett. **84**, 2726 (2000).

Chapter 11

Type-II Optical Parametric Oscillator: A Versatile Source of Quantum Correlations and Entanglement

Julien Laurat, Thomas Coudreau and Claude Fabre
Laboratoire Kastler Brossel
Université Pierre et Marie Curie, Case 74, 75005 Paris

email: fabre@spectro.jussieu.fr

Type-II Optical Parametric Oscillators are efficient sources of quadrature squeezed or polarization-squeezed light, intensity correlated beams, and entangled light. We review here the different levels of quantum correlations and entanglement that are reached in this device, and present some applications.

1. Introduction

Quantum correlations play a key role in quantum mechanics, in basic issues such as non-locality or decoherence and also in potential applications such as quantum information processing and computation. The existence of correlations between different physical systems is obviously not a specific property of quantum physics: it is simply the consequence of a former interaction, whatever its character, between the systems submitted to the measurement. Consequently, the observation or prediction of a correlation, even perfect, between the measurements of two variables is not at all a proof of the quantum character of the phenomenon under study. One can find in the literature a great deal of criteria setting a border between the classical and the quantum effects, differing by the definitions of the quantum character of a given physical situation. The purpose of this chapter is to review some criteria for quantum correlations (Sec. 2) and to describe how a single device, namely a type-II Optical Parametric Oscillator (Sec. 3.1) produces various kinds of correlations fulfilling these criteria (Sec. 3.2 and 3.3). We will also show that the same device can generate entangled states in a non-standard form (Sec. 4). This last section will provide a good insight into general properties of two-mode gaussian states, illustrated in terms of covariance matrices.

The results presented in this chapter are detailed in Refs. 1–6.

2. Correlation Criteria

Let us consider two light beams denoted by indices 1 and 2. We denote by $\delta X_{1,2}$ one quadrature component of these beams, which can be measured either by direct photodetection (amplitude quadrature) or by an homodyne detection, normalized in such a way that vacuum fluctuations have a variance equal to 1. We restrict ourselves in this chapter to the "balanced" case when the two beams have equal variances F on these quadratures, and also equal frequencies. More general criteria in the unbalanced case can be found in Ref. 1. Let us stress also that we are in the case where the quantum properties of the system are well described by a linearized approach of quantum fluctuations.

2.1. *"Gemellity"*

A first criterion of quantum correlations can be defined as follows: *the correlation measured in the system cannot be described by a semi-classical model involving classical electromagnetic fields having classical fluctuations.*

It is easy to show that the classical character of light fields is preserved by linear passive optical devices, which involve only linear, energy-preserving, optical elements like beamsplitters and free propagation. In order to ascertain the quantum character of correlations existing between δX_1 and δX_2, the simplest way is therefore to process the two beams by all possible linear passive optical devices: if one is able to produce in such a way a beam having fluctuations below the quantum noise limit, that is well-known to be "non-classical", the initial correlation will also be termed as non-classical.

For balanced beams, the best linear processing is simply to send them on a 50/50 beam-splitter: one obtains at one of its output ports a beam with quadrature fluctuations δX_{out} given by

$$\delta X_{out} = \frac{\delta X_1 - \delta X_2}{\sqrt{2}} \tag{1}$$

having a variance given by:

$$G = \frac{1}{2}\left\langle (\delta X_1 - \delta X_2)^2 \right\rangle . \tag{2}$$

The correlation will be said to be non-classical when this quantity, that can be called the "gemellity", is smaller than 1. G can also be written in terms of the noise variance of each beam F and of the normalized correlation coefficient C_{12}:

$$G = F(1 - |C_{12}|). \tag{3}$$

Therefore a correlation is non-classical when the normalized correlation function fulfills the following condition:

$$|C_{12}| > 1 - \frac{1}{F}. \tag{4}$$

Thus the larger the classical noise is on each beam, the more stringent the condition becomes.

Finally, let us stress that G can be easily measured experimentally: this is done in all homodyne detection schemes of squeezing, which actually measures the quantum character of the correlation existing between the two beams produced by mixing the field to measure with the local oscillator, and in all twin beams experiments.[2,7]

2.2. *Quantum Non Demolition correlation*

When two observables M_1 and M_2 are correlated, the measurement of M_2 gives some information about the value of M_1 without any interaction with system 1. Correlations provide therefore opportunities for Non Demolition measurements. One is led to a second criterion of quantum correlation: *the correlation is such that the information extracted from the measurement on one field provides a Quantum Non Demolition measurement of the other.*[8]

This criterion is related to the conditional variance given by:

$$V_{1|2} = F_1(1 - C_{12}^2), \tag{5}$$

where F_1 is the noise of beam 1 normalized to shot noise. QND correlations correspond to values of $V_{1|2}$ below 1, and therefore to a correlation satisfying the inequality:

$$|C_{12}| > \sqrt{1 - \frac{1}{F}}. \tag{6}$$

Equation 5 can also be expressed in terms of the gemellity:

$$V_{1|2} = V_{2|1} = G(1 + |C_{12}|) = 2G - \frac{G^2}{F}. \tag{7}$$

It is easy to show from these relations that all QND-correlated beams have a gemellity smaller than 1, whereas a gemellity smaller than 0.5 is required to have QND-correlated beams (in the limit of large individual noise).

2.3. *Inseparability*

Let us now define a new criterion related to entanglement: *the correlation cannot be described by separable quantum states.* Can the state be written as (a sum of) tensor products or not?

If one is sure that the system is in a pure state, separable or factorizable state vectors give rise to no correlations at all, whatever the observables: the existence of a non-zero correlation, even "classical", on a single quadrature is sufficient to prove the inseparability of the state.

When the state is mixed, which is the general case, this is no longer the case. Let us consider for example the mixed state described by the density matrix

$$\rho = \sum_n p_n \left(|1:n\rangle \otimes |2:n\rangle\right)\left(\langle 1:n| \otimes \langle 2:n|\right), \tag{8}$$

where $|1,2:n\rangle$ is a Fock state with n photons in mode 1, 2. This highly non-classical state has perfect intensity correlations ($C_{12} = 1$), so that $G = V_{1|2} = 0$). However, it is a separable state, being a statistical mixture of factorized state vectors.

In order to ascertain the separable character of the physical state of a system, one needs to make two joint correlation measurements on non-commuting observables on the system, and not only one, as was the case in the two previous criteria. More precisely, Duan *et al.*[9] have shown that, in the case of Gaussian states for which the covariance matrix is expressed in the so-called standard form, there exists a necessary and sufficient criterion of separability in terms of the quantity $\mathcal{I}$, that we will call separability, and is given by:

$$\mathcal{I} = \frac{1}{4}\left(\left\langle(\delta X_1 - \delta X_2)^2\right\rangle + \left\langle(\delta P_1 + \delta P_2)^2\right\rangle\right) = \frac{1}{2}\left(G_X + G_P\right). \tag{9}$$

The separability appears as the half-sum of the gemellity measuring the correlations between quadratures δX and the (anti)gemellity measuring the anticorrelations between δP. A state for which $\mathcal{I}$ is smaller than one will be a non-separable or entangled state. As a consequence, classically correlated beams, for which these two gemellities are larger than 1, are separable.

Let us note that in the case of symmetric gaussian states the entanglement can be quantified by a quantity called entropy of formation – or entanglement of formation EOF –, that was introduced in Ref. 10. It represents the amount of pure state entanglement needed to prepare the entangled state. This quantity is related to the value of the inseparability $\mathcal{I}$ by:

$$EOF = c_+ \log_2(c_+) - c_- \log_2(c_-) \tag{10}$$

with

$$c_{\pm} = (\mathcal{I}^{-1/2} \pm \mathcal{I}^{1/2})^2/4. \tag{11}$$

EOF takes a positive value only for entangled beams. Its interest is that it constitutes a real measure of the amount of entanglement. In addition, it is also used in the discrete variable regime.

A more general quantity has been introduced to characterize the entanglement: the logarithmic negativity.[11] This quantity can be calculated for any arbitrary bipartite system. We will consider it in more detail in the last section of this chapter where the generated two-mode state is not in a standard form.

2.4. *Einstein-Podolsky-Rosen correlations*

Two correlations give the opportunity of Non Demolition measurements on two non-commuting variables. As for a single quadrature, one can be interested in the quality of the information that one gets on one beam by measuring the other. This question is related to the question asked by Einstein, Podolsky and Rosen in their famous paper.[12] In particular, we will say that we have *"EPR beams" when the information extracted from the the measurement of the two quadratures of one field provide values for the quadratures of the other which "violate" the Heisenberg inequality.* This criterion was considered and discussed extensively by Reid and co-workers.[13] They showed that to characterize this property, one can use the product of the conditional variances,

$$\mathcal{V} = V_{X_1|X_2}.V_{P_1|P_2}. \tag{12}$$

When this quantity is smaller than one we will say that we have "EPR-correlated beams". Let us note that when this condition is fulfilled, one can perform double QND-measurements, that is two QND-measurements on non-commuting quadratures. One can show that all EPR-correlated beams are not separable, whereas the reverse is not true. EPR-correlation is therefore the strongest of the correlation criteria that we have listed here. One can envision other criteria which are even stronger, but not relevant for the problem of measuring correlated quadratures with Gaussian statistics that we are considering here.

3. Experimental Investigation of Quantum Correlations

In this section, we will show how these various criteria can be tested using the states produced by a triply resonant type-II Optical Parametric

Oscillator. Such a system consists of a triply-resonant optical cavity containing a type-II phase matched $\chi^{(2)}$ crystal. Spontaneous parametric down-conversion which occurs in such crystals is well known to produce twin photons, that is photons created in pairs. When such a crystal is placed inside a cavity and the system pumped above a critical value (threshold), one generates intense beams which are correlated. The system transfers the correlations existing in the discrete regime to the continuous variable one. The OPO is thus an ideal system to test the various criteria that we have mentioned above.

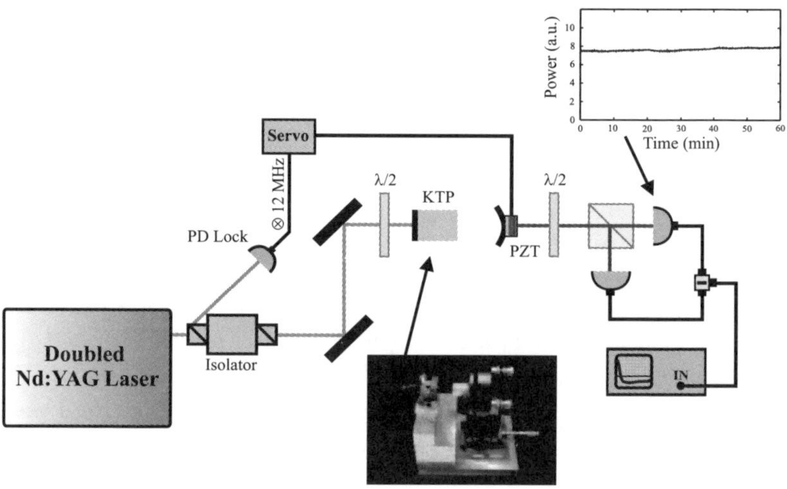

Fig. 1. A cw doubled Nd:YAG laser pumps above threshold a type-II OPO. Intensity correlations are directly measured by a balanced detection scheme. PD Lock: FND-100 photodiode for locking of the OPO.

3.1. *Experimental set-up*

The experimental setup is shown in Fig. 1. A continuous-wave frequency-doubled Nd:YAG laser pumps a triply resonant OPO above threshold, made of a semi-monolithic linear cavity. The intensity reflection coefficients for the input coupler are 95% for the pump at 532 nm and almost 100% for the signal and idler beams at 1064 nm. The output mirror is highly reflective for the pump and its transmission coefficient T can be chosen to be 5 or 10%. With $T = 5\%$, at exact triple resonance, the oscillation threshold is less than 15 mW. In spite of the triple resonance which generally makes OPOs much more sensitive to disturbances, length and temperature controls enable stable operation over more than one hour without mode-hopping.

3.2. "2×1 *quadrature*" *case*

3.2.1. *Twin beams*

Type II optical parametric oscillators are well-known to generate above threshold highly quantum correlated bright twin beams. Intensity correlations were experimentally observed several years ago and applied to measurements of weak physical effects.[7] We describe here a recent improvement of the observed correlation.

Intensity correlations are directly measured by a balanced detection scheme (Fig. 1). The signal and idler orthogonally polarized beams are separated on a polarizing beam splitter and detected on a pair of high quantum efficiency photodiodes. With a transmission $T = 10\%$ for the output mirror, we have obtained a noise reduction of 9.7 ± 0.5 dB (89%) around 5 MHz (Fig. 2), which corresponds to a gemellity of $G = 0.11$. To the best of our knowledge, this noise reduction is the strongest reported to date in the experimental quantum optics field.

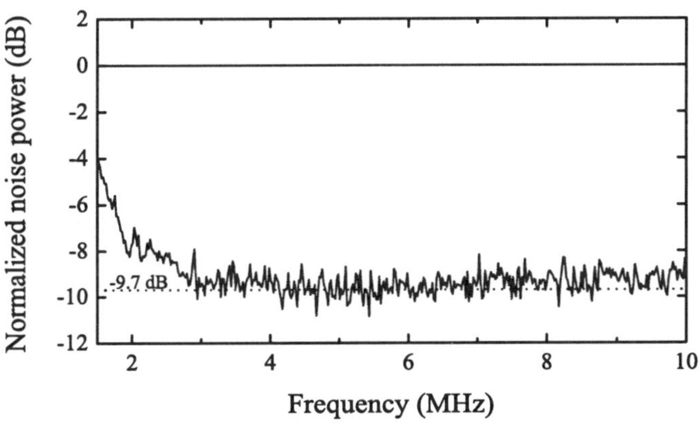

Fig. 2. Normalized noise power of the intensity difference of the signal and idler as a function of the frequency, after correction of the electronic noise.

3.2.2. *QND correlations and conditional preparation of a non-classical state*

The observed correlation is strong enough to yield a conditional variance well below 1. We will show now that such a QND-correlation can be used to produce a squeezed state *via* conditional preparation performed on continuous variables.

A well-known technique to generate a single photon state from twin photons is to use the method of conditional measurement: if one labels (1) and (2) the two modes in which the twin photons are emitted, it consists in retaining in the information collected in mode (1) only the counts occurring when a photon is detected in mode (2) within a given time window ΔT. State preparation by conditional measurement can be readily extended to the continuous variable regime, where the instantaneous values of the signal and idler photocurrents play the role of the occurrence of counts in the photon counting regime. The technique consists in selecting the signal photocurrent I_s only during the time intervals when the idler intensity I_i has a given value I_0 (within a band ΔI smaller than the photocurrent standard deviation). The measurements outside these time intervals are discarded. If the correlation is perfect and the interval ΔI close to zero, the recorded signal intensity is perfectly constant, and an intense number state is generated; in a real experiment, the correlation between the signal and idler photocurrents is not perfect, and the selection band ΔI is finite, so that the method will not prepare a perfect number state, but a sub-Poissonian state instead.

A theoretical analysis of this protocol[4] shows that in the limit where ΔI is very small the conditional measurement produces a beam characterized by a Fano factor equal to the conditional variance of the signal and idler beams. This means that the present protocol produces a sub-Poissonian beam when the signal and idler beams are QND-correlated. As shown in Eq. 7 in the limit of large correlations, the residual intensity noise F on the conditionally prepared state will be equal to twice the gemellity ($F = V \simeq 2G$).

Figure 3 sums up the experimental results. The Fano factor F of both the signal and idler beams exceeds 100 (20 dB above the shot noise level), and the measured gemellity G is equal to 0.18 (0.14 after correction of dark noise). The ensemble of values of the signal intensity for which the idler intensity falls within the selection band is given in Fig. 3 (c): one indeed observes a significant narrowing of the probability distribution below the

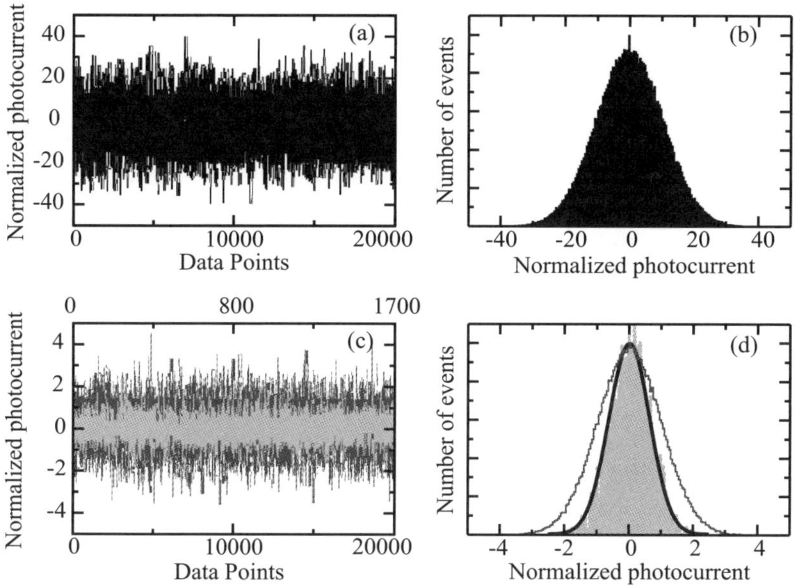

Fig. 3. Experimental results: (a) Idler intensity fluctuations: 200 000 acquired points at 3.5 MHz demodulation frequency (only 20 000 shown). (b) Corresponding probability distribution. The unit is the width σ_0 of the Poisson distribution of same mean intensity (shot noise). (c) Values of the signal intensity conditionally selected by the value of the idler intensity recorded at the same time (selection bandwidth ΔI equal to 0.1 σ_0 around the mean), superimposed to the corresponding experimentally measured shot noise. (d) Corresponding probability distribution, compared to the Poisson distribution (grey line), displaying the sub-Poissonian character of the conditionally generated state. The black line is a gaussian fit of the intensity distribution.

shot noise level. With a selection bandwidth ΔI equal to 0.1 times the standard deviation σ_0 of a coherent state having the same power (shot noise level), the conditionally prepared light state has a measured Fano factor $F = 0.36$, which turns out to be equal, as expected, to the conditional variance of the twin beams. The success rate of the conditional preparation is around 0.85% (1700 points out of 200 000 are accepted). An advantage of the conditional preparation using continuous variables is that one can use at the same time different selection non-overlapping bands on the idler beam. Each allows one to conditionally prepare a different sub-Poissonian state, each having a Fano factor $F = 0.36$. With 200 different selection bands, the overall success rate is close to 100%.

3.3. "2 × 2" *quadratures case*

3.3.1. *Entanglement below threshold*

Type-II OPO below threshold are well-known to generate entangled beams. The first experimental demonstration of EPR correlations in the continuous variable regime in 1992 was performed with such a device.[14] Our experimental setup is similar to the previous one (Fig. 1) but the detection system is now based on two simultaneous homodyne detections (Fig. 4). In order to measure the separability $\mathcal{I}$, one must characterize the noise of the superposition modes oriented $\pm 45°$ from the axes of the crystal:

$$A_+ = \frac{A_1 + A_2}{\sqrt{2}} \qquad \text{and} \qquad A_- = \frac{A_1 - A_2}{\sqrt{2}}.$$

Equation 9 shows that the signal and idler fields are entangled as soon as these two modes have squeezed fluctuations on orthogonal quadratures. The orthogonally polarized modes are separated on a first polarizing beam splitter at the output of the OPO. A half-wave plate inserted before this polarizing beam splitter enables us to choose the fields to characterize: the signal and idler modes which are entangled, or the $\pm 45°$ rotated modes which are squeezed. The detection setup is able to characterize simultaneously the two chosen modes with the same phase reference, and to measure the noise reductions either in phase ("in phase homodyne detection") or in quadrature ("in quadrature homodyne detection"), by inserting or not a $\lambda/4$ plate in the beam exiting the OPO. This configuration permits a di-

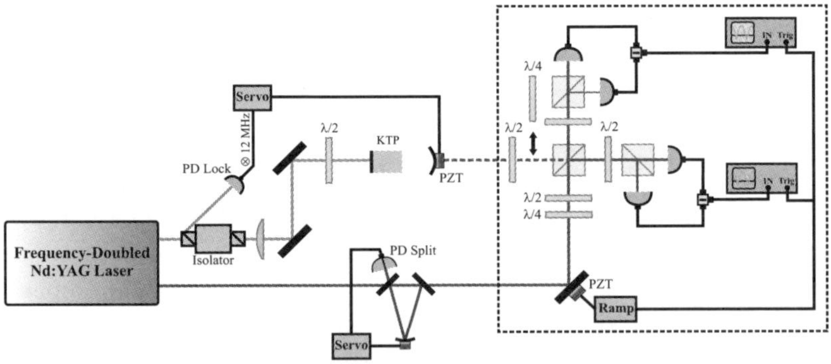

Fig. 4. A doubled Nd:YAG laser pumps a type II OPO, below or above threshold. The generated two-mode state is characterized by two simultaneous homodyne detections. PD Split: split two-element InGaAs photodiode for tilt-locking of the filtering cavity.

rect and instantaneous verification of the inseparability criterion by simply adding the two squeezed variances.

Typical spectrum analyzer traces while scanning the local oscillator phase are shown on Fig. 5. Normalized noise variances of the $\pm 45°$ vacuum modes at a given noise frequency of 3.5 MHz are superimposed for in-phase and in-quadrature homodyne detections. One indeed observes, as expected, correlations and anti-correlations of the emitted modes on orthogonal quadratures. The homodyne detection can be locked on the squeezed quadrature (Fig. 5). The observed amount of simultaneous squeezing for the

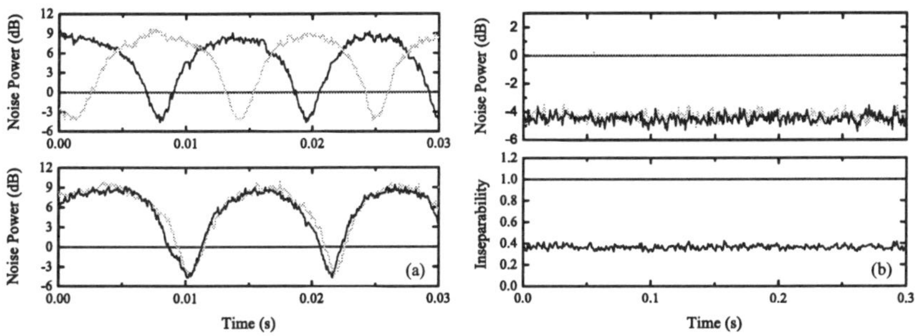

Fig. 5. (a) Normalized noise variances at 3.5 MHz of the $\pm 45°$ modes while scanning the local oscillator phase. The first plot corresponds to in-phase homodyne detections and the second one in-quadrature. Squeezing is well observed on orthogonal quadratures. (RBW 100 kHz, VBW 1 kHz) (b) Normalized noise variances at 3.5 MHz of the $\pm 45°$ modes and inseparability $\mathcal{I}$ for signal and idler modes. The homodyne detections are in-quadrature and locked on the squeezed quadratures. After correction of the electronic noise, the inseparability criterion reaches 0.33 ± 0.02. (RBW 100 kHz, VBW 300 Hz).

two rotated modes is -4.3 ± 0.3 dB and -4.5 ± 0.3 dB below the standard quantum limit (-4.7 ± 0.3 dB and -4.9 ± 0.3 dB after correction of the electronic noise). This gives a value of the separability of $\mathcal{I} = 0.33 \pm 0.02$, well below the unit limit for inseparability. With a measured value of the parameter F of 6.6, one obtains a product of conditional variances of 0.42 ± 0.05, well below 1, which establishes the EPR character of the measured correlations. The entanglement of formation EOF of the two beams is, according to formulae (10) and (11), equal to 1.1 ± 0.1 *ebits*. To the best of our knowledge, this setup generates the best EPR/entangled beams to date produced in the continuous variable regime.

Non-classical properties are generally measured in the MHz range of Fourier frequencies, because of the presence of large classical noise at lower frequencies. In the present device significant quantum correlations and EPR entanglement have been observed in from 50 kHz to 10 MHz. Fig. 6 gives the squeezed variances for low noise frequencies, between 40 kHz and 150 kHz. Let us mention that squeezing from a single type-I OPA was recently reported at a record Fourier frequency of 200 Hz.[15]

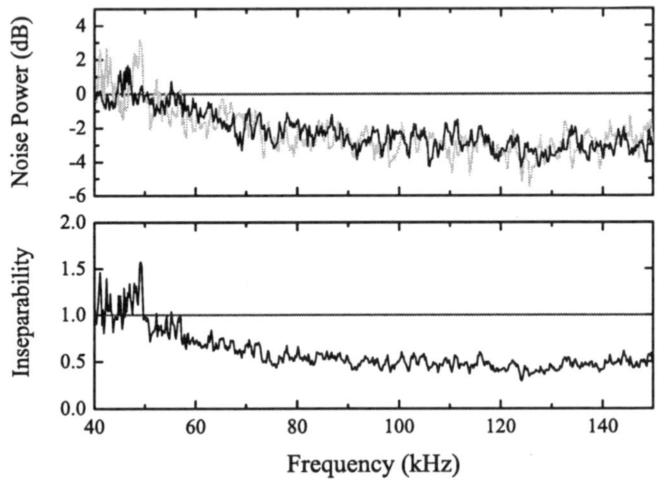

Fig. 6. Normalized noise variances from 40 kHz to 150 kHz of the ±45° modes after correction of the electronic noise and inseparability criterion for signal and idler modes. Squeezing and entanglement are observed down to 50 kHz. (RBW 3 kHz, VBW 10 Hz.)

3.3.2. *Bright EPR beams above threshold and polarization squeezing*

A type-II OPO pumped above threshold has been theoretically predicted to be a very efficient source of bright entangled and EPR beams. This means that, in addition to the already demonstrated intensity correlations, phase anticorrelations exist in the system. However, they can be easily measured by usual homodyne detection techniques only in the frequency-degenerate regime. Frequency degeneracy occurs only accidentally above threshold because it corresponds to a single point in the experimental parameter space. Actually, up to now, no direct evidence of such phase anti-correlations has been observed. In 1998, Mason and Wong proposed an elegant way to achieve frequency degenerate operation above threshold:[16,17] they inserted inside the OPO cavity a birefringent plate making an angle with the axis of the non-linear crystal. The induced linear coupling between the signal and idler results in a locking phenomenon.[18] It has been shown theoretically that in such a "self-phase-locked" OPO the quantum correlations are preserved for small angles of the plate and that the system produces entangled states in a wide range of parameters.[19,20]

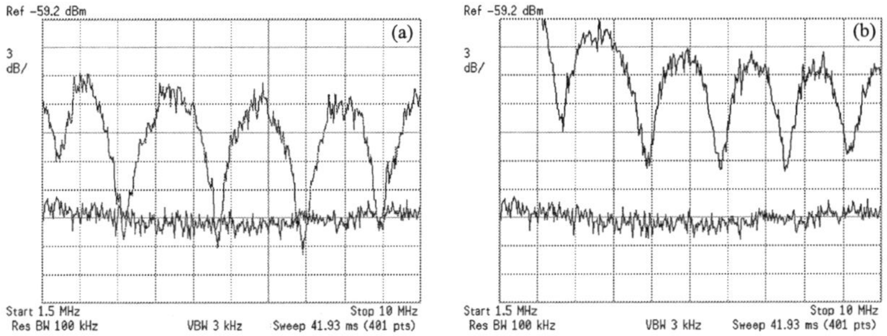

Fig. 7. (a) Noise power of the mode A_- while scanning simultaneously the phase of the local oscillator and the noise frequency between 1.5 and 10 MHz. The lower trace gives the shot noise level. (b) Noise power of the mode A_+ while scanning the phase of the local oscillator, for a noise frequency between 1.5 and 10 MHz. The shot noise level is given by the lower trace plus 3 dB.

In the experiment, the frequency locking phenomenon can be maintained during more than hour. Degenerate operation is confirmed by the fact that the generated mode has now a fixed polarization: at the minimum threshold point, the generated state is linearly polarized at $+45°$. Due to the defined phase relation existing now between the signal and idler fields, A_+ is a bright mode, and A_- has a zero mean value. Fig. 7 (a) gives the noise power of the mode A_- while scanning the local oscillator phase, for a transmission $T = 5\%$ and a plate angle of $0.1°$. A noise reduction of 4.5dB is observed. This strong noise reduction on the mode A_- confirms the quantum intensity correlation between the signal and idler modes. Fig. 7 (b) shows the noise power of the mode A_+ in the same condition. As the plate angle is very small, a similar amount of noise reduction is expected. However, a slight excess noise of 3 dB is measured for the minimal noise quadrature: the phase anticorrelations appear to be slightly degraded, probably by external noise sources.

Despite this slight excess noise which prevents from reaching the proof of entanglement in the OPO above threshold, the generated state turns out to be squeezed in the polarization orthogonal to the mean field: A_+ is the main mode and A_- the squeezed vacuum one. This condition is required to obtain a so-called "polarization squeezed" state.[21–23] 4.5 dB of polarization squeezing has been thus generated in the self-phase-locked OPO. Such states have recently raised great interest, in particular because of the possibility to map quantum polarization state of light onto an atomic ensemble.[24]

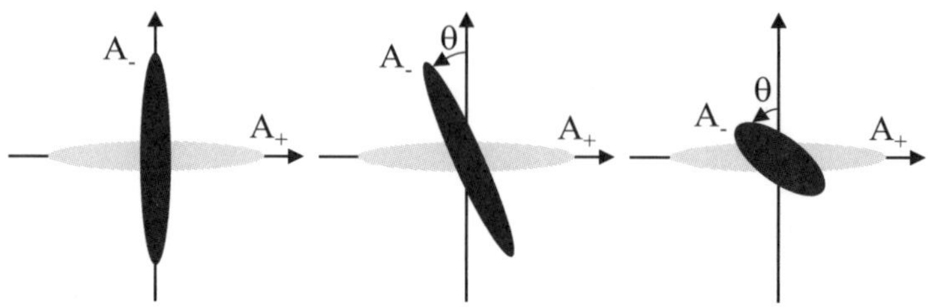

Fig. 8. Fresnel representation of the noise ellipse of the ±45° rotated modes when the plate angle is increased. Without coupling, squeezing is predicted on orthogonal quadratures. The noise ellipse of the −45° mode rotates and the noise reduction is degraded when the coupling increases while the +45° rotated mode is not affected.

4. Manipulating Entanglement with Polarization Elements

The self-phase-locked OPO can also be operated below threshold. It produces a two-mode state with strong quantum features which manifest themselves in terms of noise reduction properties in a given polarization basis, and in terms of entanglement and EPR correlations in another. This last section is devoted to the general study of this two-mode Gaussian quantum state.

4.1. *Manipulation of entanglement in the two-mode state produced by the type-II OPO with mode coupling*

In a standard OPO the correlated quadratures are orthogonal to the anti-correlated ones which results in squeezing of the rotated modes on orthogonal quadratures. It is no more the case when a linear coupling is introduced. When the plate angle increases, the correlated quadratures rotate and the correlations are degraded. The evolution is depicted in Fig. 8 through the noise ellipses of the rotated (squeezed) modes. In order to maximize the entanglement between the signal and idler modes, the optimal quadratures have to be made orthogonal.[25] Such an operation consists of a phase-shift of A_- relative to A_+. This transformation is passive and "non-local" in the sense of the EPR argument: it acts simultaneously on the two considered sub-systems. In the type II OPO, such "non-local" transformations are easy to perform by inserting polarizing birefringent elements in the total beam, because the two polarization modes are produced by the OPO in the same transverse spatial mode.

As the generated two-mode state is not in the standard form, we need to use a general measure of entanglement. Let us introduce the covariance matrix formalism and the logarithmic negativity. In a given mode basis, the quantum properties of the generated state, of zero mean value, are completely contained in the covariance matrix Γ_{AB} defined as:

$$\Gamma_{AB} = \begin{pmatrix} \gamma_A & \sigma_{AB} \\ \sigma_{AB}^T & \gamma_B \end{pmatrix}$$

γ_A and γ_B are the covariance matrix of the individual modes while σ_{AB} describes the intermodal correlations. The elements of the covariance matrix are written $\Gamma_{ij} = \langle \delta R_i \delta R_j + \delta R_j \delta R_i \rangle / 2$ where $R_{\{i,i=1,..,4\}} = \{X_A, Y_A, X_B, Y_B\}$. X and Y corresponds to an arbitrary orthogonal basis of quadratures. In order to measure the degree of entanglement of Gaussian states, a simple computable formula of the logarithmic negativity $E_{\mathcal{N}}$ has been obtained[11] (see also Ref. 26 for a general overview). $E_{\mathcal{N}}$ can be easily evaluated from the largest positive symplectic eigenvalue ξ of the covariance matrix which can be obtained from

$$\xi^2 = \frac{1}{2}(D - \sqrt{D^2 - 4\mathrm{Det}\,\Gamma_{AB}}) \tag{13}$$

with

$$D = \mathrm{Det}\,\gamma_A + \mathrm{Det}\,\gamma_B - 2\mathrm{Det}\,\sigma_{AB}. \tag{14}$$

The two-mode state is entangled if and only if $\xi < 1$. The logarithmic negativity can thus be expressed by $E_{\mathcal{N}} = -\log_2(\xi)$. The maximal entanglement which can be extracted from a given two-mode state by passive operations is related to the two smallest eigenvalues of Γ, λ_1 and λ_2, by $E_{\mathcal{N}}^{max} = -\log_2(\lambda_1\lambda_2)/2$.[25]

We give here a numerical example for realistic experimental values $\rho = 1.3°, \sigma = 0.9$ and $\Omega = 0$, where ρ stands for the plate angle, σ the pump power normalized to the threshold and Ω the noise frequency. The covariance matrices for the A_1/A_2 modes and for the A_+/A_- modes are given in Fig. 9 with and without the phase-shift. The matrix of the A_+/A_- modes are well-suited to understand the behavior of the device. At first, the intermodal blocks are zero, showing that these two modes are not at all correlated and consequently are the most squeezed modes of the system: there is no way to extract more squeezing. But one can also note that the diagonal blocks are not diagonalized simultaneously. This corresponds to the tilt angle θ of the squeezed quadrature of A_-. A phase-shift of the angle θ permits to diagonalize simultaneously the two blocks and to obtain squeezing on orthogonal quadratures. The logarithmic negativity $E_{\mathcal{N}}$ has increased in the transformation from 4.06 to 4.53. The maximal entanglement available has been extracted in this way as $E_{\mathcal{N}}^{max} = -\log_2(\lambda_1\lambda_2)/2 = 4.53$.

$$\Gamma_{A_1A_2} = \left(\begin{array}{cc|cc} 181.192 & 0 & 179.808 & -0.255 \\ 0 & 0.386 & -0.255 & -0.383 \\ \hline 179.808 & -0.255 & 181.192 & 0 \\ -0.255 & -0.383 & 0 & 0.386 \end{array}\right) \Longrightarrow \Gamma'_{A_1A_2} = \left(\begin{array}{cc|cc} 180.839 & 0 & 180.161 & 0 \\ 0 & 0.739 & 0 & -0.736 \\ \hline 180.161 & 0 & 180.839 & 0 \\ 0 & -0.736 & 0 & 0.739 \end{array}\right)$$

$$\Gamma_{A_+A_-} = \left(\begin{array}{cc|cc} 361 & 0 & 0 & 0 \\ 0 & 0.00277 & 0 & 0 \\ \hline 0 & 0 & 1.383 & -0.256 \\ 0 & 0 & -0.256 & 0.770 \end{array}\right) \Longrightarrow \Gamma'_{A_+A_-} = \left(\begin{array}{cc|cc} 361 & 0 & 0 & 0 \\ 0 & 0.00277 & 0 & 0 \\ \hline 0 & 0 & 0.677 & 0 \\ 0 & 0 & 0 & 1.476 \end{array}\right)$$

Fig. 9. Numerical example of covariance matrix of the A_1/A_2 modes and the A_+/A_- modes before and after the non-local operation for a plate angle of $\rho = 1.3°$. ($\sigma = 0.9$ and $\Omega = 0$).

4.2. *Experimental optimization of entanglement*

Let us now describe how to experimentally optimize the EPR entanglement generated by the self-phase-locked OPO below threshold.

In order to extract the maximal entanglement, one must perform an appropriate phase-shift on the rotated modes. This is achieved by using an association of one $\lambda/2$ and one $\lambda/4$ plates added at the output of the OPO. The double homodyne detection we have developed is necessary in order to be able to characterize simultaneously the two modes with the same phase reference.

Fig. 10 displays the normalized noise variances of the rotated modes for a plate angle of $\rho = 0.3°$, before and after the phase-shift. The homodyne detections are operated in quadrature so that squeezing on orthogonal quadratures is observed simultaneously on the spectrum analyzers. After the operation is performed, squeezing is obtained on orthogonal quadratures as in a standard type-II OPO without mode coupling. Experimentally, the logarithmic negativity goes from 1.13 to 1.32, showing that we are able to extract more quantum resource from the state after the operation.

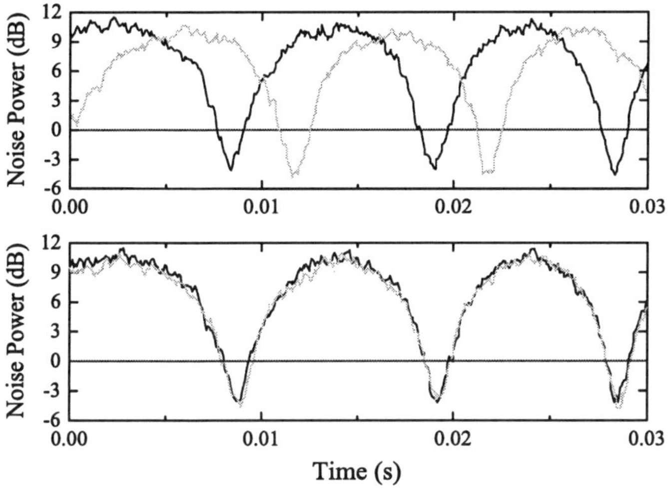

Fig. 10. Normalized noise variances at 3.5 MHz of the rotated modes while scanning the local oscillator phase for a plate angle $\rho = 0.3°$, before and after the non-local operation. The homodyne detections are in-quadrature. After this operation, squeezing is observed on orthogonal quadratures.

5. Conclusion

We have seen that type II triply resonant OPO produce in a very stable way the strongest intensity correlation and EPR entanglement to date. Phase-locked, frequency degenerate operation can be obtained using an intracavity birefringent plate. This locking permits the experimental realization of homodyne detection of the quadratures even when the system is operated above threshold. This result opens a very promising way to the direct generation of intense entangled beams and offers a new and simple method to achieve strong polarization squeezing. Below threshold, the self-phase-locked OPO exhibits a very rich and interesting behavior which provides a good insight into entanglement manipulation by passive operations. This opens the way to the manipulation and optimization of quantum properties in highly multimode Hilbert spaces.

References

1. N. Treps and C. Fabre, Laser Physics **15**, 187 (2005).
2. J. Laurat, L. Longchambon, T. Coudreau and C. Fabre, Opt. Lett **30**, 1177 (2005).
3. J. Laurat, T. Coudreau, and N. Treps, A. Maître, C. Fabre, Phys. Rev. Lett. **91**, 213601 (2003).

4. J. Laurat, T. Coudreau, N. Treps, A. Maître, and C. Fabre, Phys. Rev. A, **69**, 33808 (2004).
5. J. Laurat, T. Coudreau, G. Keller, N. Treps, and C. Fabre, Phys. Rev. A **70**, 042315 (2004).
6. J. Laurat, T. Coudreau, and G. Keller, N. Treps, C. Fabre, Phys. Rev. A **71**, 022313 (2005).
7. A. Heidmann, R.J. Horowicz, S. Reynaud, E. Giacobino, C. Fabre, and G. Camy, Phys. Rev. Lett. **59**, 2555 (1987); J. Mertz, T. Debuisschert, A. Heidmann, C. Fabre, and E. Giacobino, Opt. Lett. **16**, 1234 (1991); J. Gao, F. Cui, C. Xue, C. Xie, and K. Peng, Opt. Lett. **23**, 870 (1998).
8. P. Grangier, J.-M Courty, and S. Reynaud, Opt. Commun. **89**, 99 (1992).
9. L.-M. Duan, G. Giedke, J. I. Cirac, and P. Zoller, Phys. Rev. Lett. **84**, 2722 (2000).
10. G. Giedke, M.M. Wolf, O. Krüger, R.F. Werner, and J.I. Cirac, Phys. Rev. Lett **91**, 107901 (2003).
11. G. Vidal and R.F. Werner, Phys. Rev. A **65**, 032314 (2002).
12. A. Einstein, B. Podolsky, and N. Rosen, Phys. Rev. **47**, 777 (1935).
13. M. Reid and P. Drummond, Phys. Rev. Lett. **60**, 2731 (1989); M. Reid, Phys. Rev. A **40**, 913 (1989); M. Reid and P. Drummond, Phys. Rev A **41**, 3930 (1991).
14. Z.Y. Ou, S.F. Pereira, H.J. Kimble, and K.C. Peng, Phys. Rev. Lett. **68**, 3663 (1992).
15. K. McKenzie, N. Grosse, W.P. Bowen, S.E. Whitcomb, M.B. Gray, D.E. McClelland, and P.K. Lam, Phys. Rev. Lett. **93**, 161105 (2004).
16. E.J. Mason and N.C. Wong, Opt. Lett. **23**, 1733 (1998).
17. C. Fabre, E.J. Mason, and N.C Wong, Optics Communications **170**, 299 (1999).
18. A. Pikovsky, M. Rosenblum, and J. Kurths, *Synchronization*, Cambridge University Press, 2001.
19. L. Longchambon, J. Laurat, T. Coudreau, and C. Fabre, Eur. Phys. J. D **30**, 279 (2004).
20. L. Longchambon, J. Laurat, T. Coudreau, and C. Fabre, Eur. Phys. J. D **30**, 287 (2004).
21. N. Korolkova, G. Leuchs, R. Loudon, T.C. Ralph, and C. Silberhorn, Phys. Rev. A **65**, 052306 (2002).
22. W.P. Bowen, R. Schnabel, H.-A. Bachor, and P.K. Lam, Phys. Rev. Lett. **88**, 093601 (2002).
23. V. Josse, A. Dantan, L. Vernac, A. Bramati, M. Pinard, and E. Giacobino, Phys. Rev. Lett. **91**, 103601 (2003).
24. J. Hald, J.L. Sørensen, C. Schori, and E.S. Polzik, Phys. Rev. Lett. **83**, 1319 (1999).
25. M.M. Wolf, J. Eisert, and M.B. Plenio, Phys. Rev. Lett **90**, 047904 (2003).
26. G. Adesso, A. Serafini, and F. Illuminati, Phys. Rev. A **70**, 022318 (2004).

Chapter 12

Accessing the Phase Quadrature of Intense Non-Classical Light State

O. Glöckl, U. L. Andersen and G. Leuchs

Institut für Optik, Information und Photonik
Max–Planck–Forschungsgruppe
Günther–Scharowsky–Straße 1 / Bau 24, 91058 Erlangen, Germany

email: leuchs@kerr.physik.uni-erlangen.de

Phase sensitive measurements are required to characterize the quantum state of a light field in the regime of continuous variables. The standard approach to do this is to use homodyne detection. However, if the signal beam under investigation is very intense, this technique is not appropriate. We present a setup for performing sub-shot-noise measurements of the phase quadrature of intense pulsed light without the use of a separate local oscillator. A Mach Zehnder interferometer with an unbalanced arm length is used to detect the fluctuations of the phase quadrature at a single sideband frequency. Using that tool, short entangled light pulses generated in optical fibres could be applied efficiently in quantum information and quantum communication protocols. We show experimentally that the setup can be used to demonstrate the non–separability of a pair of quadrature-entangled beams.

1. Introduction

Up to now, many different experimental realizations of quadrature entanglement sources based on different non–linear effects have been carried out. Second order non–linearities are utilized in optical parametric oscillation,[1–8] third order non–linearities in atoms[9] and fibres.[10] Many protocols in the field of quantum information and quantum communication with continuous variables[11] such as dense coding[12] or quantum teleportation,[13] rely on quadrature entanglement. Moreover, for efficient implementations of the protocols it is crucial to have access to a pair of conjugate quadrature components on these entangled states. In most experimental realizations, entanglement of dim beams and vacuum beams has been achieved. The standard approach to access information about the field quadrature components in these realizations is to use a homodyne detector.[14–16] However, for intense light states homodyne detection is not appropriate. Due to the brightness of the signal beam, high power of the local oscillator is required giving rise to technical difficulties as the detectors may saturate or even be damaged. In some early quantum optical experiments on very intense non–classical light states,[17,18] the quadratures were accessed by reflecting the intense light of a single ended cavity.[19,20] However, if intense pulsed light is used, the requirements on the dispersion properties and the stability of the cavity are quite demanding. Especially for the case of entangled, short light pulses generated in optical fibres utilizing the Kerr non–linearity, no simple tool to perform phase quadrature measurements were at hand. We therefore developed an alternative approach in which the fluctuations of the phase quadrature are measured at a certain sideband frequency without the use of any local oscillator or resonator. An interferometric setup reminiscent of that used by Inoue and Yamamoto[21] to determine the longitudinal mode partition noise is shown to be capable of performing quantum-optical measurements of the phase quadrature below the shot-noise level, by introducing a phase shift between the carrier and the sidebands.[22] The setup allows for easy switching between the measurement of the phase quadrature and the amplitude quadrature and was used to fully characterize quadrature entanglement of a pair of intense pulsed beams. In this chaper, we start the discussion by introducing the optical modes which we examine in the sideband picture. Then, the Mach–Zehnder Interferometer for the phase measurements is discussed in detail: First, the principle of operation is described. Then examples of measurements of the phase quadrature on a phase modulated beam, an amplitude squeezed beam and a pair of quadrature entangled beams are presented.

2. Sideband Picture

For the description of quantum noise and squeezed states we use the sideband model, which facilitates the presentation of the phase measurement techniques. The sideband model is a simplified version of the multimode description discussed for example in the references.[23–26] The field mode $\hat{a}$ under investigation is decomposed into its bright carrier component (internal local oscillator) with the amplitude α (α is assumed to be real) at the optical frequency ω and pairs of sideband modes placed symmetrically around the carrier at the frequencies $\omega + \Omega$ and $\omega - \Omega$

$$\hat{a} = \alpha \exp(i\omega t) + \delta\hat{a}_+ \exp(i(\Omega + \omega)t) + \delta\hat{a}_- \exp(i(\Omega - \omega)t), \quad (1)$$

where $\delta\hat{a}_+$ and $\delta\hat{a}_-$ are the single mode field operators at the frequencies $\omega + \Omega$ and $\omega - \Omega$. The sideband modes contribute to the quantum noise of the mode $\hat{a}$ at the frequency Ω. If the noise contained in the field is purely white, there is an equally strong noise contribution at all other frequencies. However, for our examination it is sufficient to consider only three modes as described in (1), because we want to estimate the noise power of the fluctuations at one particular frequency Ω (around a certain bandwidth $\delta\Omega$) using a spectrum analyzer. The only significant contribution to the signal at the frequency Ω stems from the beating of these three modes. This can be seen as follows. Suppose we perform direct photodetection of mode $\hat{a}$

$$\begin{aligned}\hat{n} = \hat{a}^\dagger\hat{a} = \alpha^2 &+ \alpha\delta\hat{a}_+ \exp(i\Omega t) + \alpha\delta\hat{a}_- \exp(-i\Omega t)\\ &+ \alpha\delta\hat{a}_+^\dagger \exp(-i\Omega t) + \alpha\delta\hat{a}_-^\dagger \exp(i\Omega t).\end{aligned} \quad (2)$$

By spectral analysis of the resultant photocurrent, we obtain the fluctuations of the amplitude quadrature $\delta\hat{X}_a^\Omega$ at the sideband frequency Ω. $\delta\hat{X}_a^\Omega$ contains contributions from the lower and the upper sideband:

$$\begin{aligned}\delta\hat{X}_a^\Omega = &\ \delta\hat{a}_+ \exp(i\Omega t) + \delta\hat{a}_- \exp(-i\Omega t)\\ &+\delta\hat{a}_+^\dagger \exp(-i\Omega t) + \delta\hat{a}_-^\dagger \exp(i\Omega t).\end{aligned} \quad (3)$$

In the experiment, the spectral variance of the photocurrent which contains the quantum fluctuations of both sideband modes is determined. This means that the quadrature fluctuations are not measured directly, rather their spectral variances are determined. To characterize the quantum state of an optical system, the variances of a pair of conjugate quadrature variables must be measured. To access quadrature components different from the amplitude, a phase shift between the carrier and the sideband modes must be introduced before direct detection is performed. In particular, the phase quadrature fluctuations are obtained by a $\pi/2$ phase shift introduced in the carrier mode in Eq. (1)

$$\hat{a} = \alpha \exp(i\omega t + \pi/2) + \hat{\delta} a_+ \exp(i(\Omega + \omega)t) + \delta\hat{a}_- \exp(i(\Omega - \omega)t). \quad (4)$$

The phase quadrature $\delta\hat{Y}_a^\Omega$ noise of mode $\hat{a}$ can now be accessed in direct detection

$$\begin{aligned}\delta\hat{Y}_a^\Omega = &-i\delta\hat{a}_+ \exp(i\Omega t) - i\delta\hat{a}_- \exp(-i\Omega t)\\ &+i\delta\hat{a}_+^\dagger \exp(-i\Omega t) + i\delta\hat{a}_-^\dagger \exp(i\Omega t).\end{aligned} \quad (5)$$

For coherent states, the sideband modes are in the vacuum state, and the upper and the lower sidebands are not correlated. For squeezed light states on the other hand, quantum correlations between pairs of sidebands are introduced.[23–25,27–29] Thus, the overall noise, *i.e.* the noise of the joint quantum state of both sideband modes, is reduced for one particular quadrature component while the noise is enhanced for the conjugate quadrature component. For example, for an amplitude squeezed beam $\mathrm{Var}(\delta\hat{X}_a^\Omega) < 1$ whereas $\mathrm{Var}(\delta\hat{Y}_a^\Omega) > 1$. The uncertainties of the quadratures of the individual sideband modes on the other hand are enlarged.

As mentioned above, measurements of the amplitude quadrature can be performed easily in direct detection while for the phase quadrature measurements, a relative phase shift between the carrier and the sidebands must be introduced. There are essentially three ways by which the phase quadrature information can be accessed. The standard tool to extract the phase information is to use homodyne detection.[14–16] In such a detection protocol the signal field is superimposed with a strong external local oscillator field to probe the sideband modes of the signal field. Arbitrary quadratures could be accessed by changing the relative interference phase of the local oscillator with the signal field, however, the intensity of the local oscillator must be much larger than that of the signal field. An alternative method to the homodyne detector is the use of a phase detuned cavity:[19,20] A single-ended reflecting cavity introduces a frequency dependent phase shift in the field. If a pair of sidebands, symmetrically located around the carrier frequency, is outside the linewidth of the cavity they do not experience a phase shift whereas the carrier is phase shifted with an amount proportional to the detuning of the cavity. Thus it is possible to access all quadratures; the exact quadrature being measured depends on the detuning. A complete rotation from amplitude to phase quadrature is only possible for sidebands lying outside the bandwidth of the cavity. In this book chapter we will focus on an alternative approach to measure phase quadrature fluctuations without the need of neither local oscillators in a distinct spatial mode nor phase-shifting cavities. We use an interferometric method where the phase shift of the carrier is obtained through free space propagation and two mode interference accomplished in an unbalanced Mach-Zehnder interferometer.

In the following description of the phase interferometer, we further simplify the notation for the quantum state of the light mode $\hat{a}$ under investigation. We still consider those three modes that contribute to the quadrature signals discussed above. However, we employ a rotating frame at the optical frequency ω. The field is thus described by the bright carrier mode expressed by its classical amplitude α. Furthermore, the quantum noise arising from the sideband pair that is separated from the carrier by Ω is described by a single operator $\delta\hat{a}^{\Omega}$, which contains all classical and quantum mechanical fluctuations, the mean value $\langle\delta\hat{a}^{\Omega}\rangle$ being zero. The field mode can therefore be written as

$$\hat{a} = \alpha + \delta\hat{a}^{\Omega}. \tag{6}$$

Direct detection of this mode leads in linear approximation to a constant part proportional to the mean intensity of the optical mode $\hat{a}$ and to a fluctuating part associated with the amplitude quadrature at the frequency Ω:

$$\hat{n} = \hat{a}^{\dagger}\hat{a} = \alpha^2 + \alpha\hat{X}_a^{\Omega}. \tag{7}$$

The amplitude quadrature component $\delta\hat{X}_a^{\Omega}$ is defined in the standard way $\delta\hat{X}_a^{\Omega} = (\delta\hat{a}^{\Omega})^{\dagger} + \delta\hat{a}^{\Omega}$. However, we should stress that $\delta\hat{X}_a^{\Omega}$ contains contributions from both sideband modes at the frequencies $\omega - \Omega$ and $\omega + \Omega$ as described in equation (3). Both sidebands are detected simultaneously using direct detection.

3. Phase Measuring Interferometer — Principle of Operation

The principle of the phase measuring interferometer is depicted in the box in Fig. 1. It is a Mach–Zehnder interferometer with an arm length difference ΔL. The basic idea is that in the long arm a relative phase shift between the carrier and the sideband modes is introduced. Upon interference at the second beam splitter, an effective rotation of the carrier by $\pi/2$ is achieved such that the phase noise of the signal input becomes direct detectable. The difference of the photocurrents obtained from modes $\hat{c}$ and $\hat{d}$ contains noise contributions originating only from the phase quadrature fluctuations of the signal field, while the noise contributions from the vacuum mode cancel out as we shall see below.

The bright input mode $\hat{a} = \alpha + \delta\hat{a}^{\Omega}$ is split into two parts at a 50:50 beam splitter, hence it is mixed with vacuum $\delta\hat{v}^{\Omega}$. The two resulting modes $\hat{e}$ and $\hat{f}$ are given by

$$\hat{e} = \frac{1}{\sqrt{2}}[\alpha + \delta\hat{a}^{\Omega} + \delta\hat{v}^{\Omega}], \tag{8}$$

$$\hat{f} = \frac{1}{\sqrt{2}}[\alpha + \delta\hat{a}^{\Omega} - \delta\hat{v}^{\Omega}]. \tag{9}$$

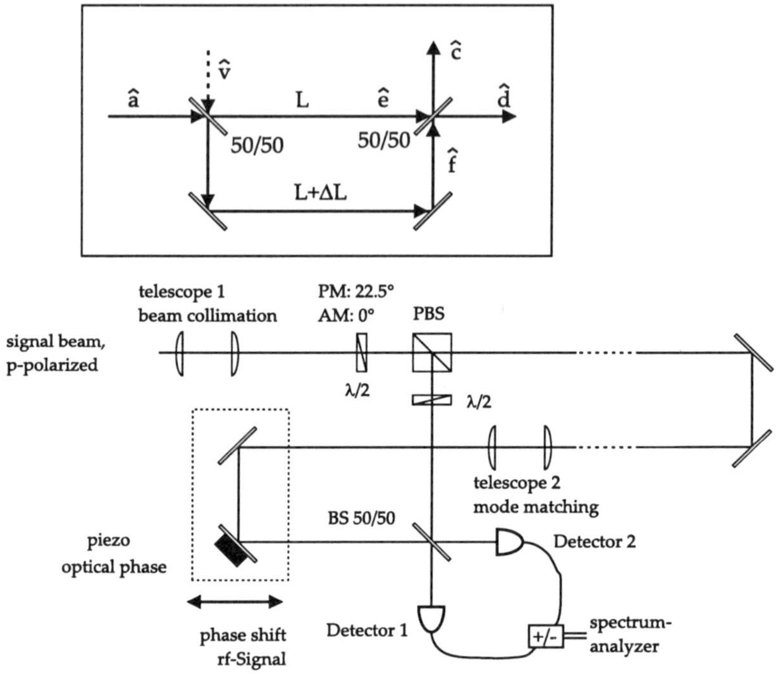

Fig. 1. Detailed illustration of the phase measuring device. At the first beam splitter, the splitting ratio could be adjusted to switch between ampitude– (AM) and phase– (PM) quadrature measurements. Telescopes are included for mode matching. PBS: Polarizing beam splitter, BS: beam splitter, $\lambda/2$: waveplates.

Mode $\hat{f}$ is delayed with respect to mode $\hat{e}$ by the time $\tau = \Delta L/c$, where c denotes the speed of light. This delay effectively introduces a relative phase factor $\exp(-i\Omega\tau) = \exp(-i\theta)$ between the sideband modes and the carrier. As the sideband frequency Ω and the frequency of the carrier ω differ by a factor $\sim 10^7$, the optical phase shift $\exp(-i\varphi)$ could be adjusted independently. Therefore, the output modes $\hat{c}$ and $\hat{d}$ after the interferometer are described by

$$\hat{c} = \frac{1}{2}[\alpha + \alpha\exp(-i\varphi) + \delta\hat{a}^{\Omega} + \delta\hat{a}^{\Omega}\exp(-i\theta)\exp(-i\varphi) + \delta\hat{v}^{\Omega} - \delta\hat{v}^{\Omega}\exp(-i\theta)\exp(-i\varphi)], \quad (10)$$

$$\hat{d} = \frac{1}{2}[-\alpha + \alpha\exp(-i\varphi) - \delta\hat{a}^{\Omega} + \delta\hat{a}^{\Omega}\exp(-i\theta)\exp(-i\varphi) - \delta\hat{v}^{\Omega} - \delta\hat{v}^{\Omega}\exp(-i\theta)\exp(-i\varphi)]. \quad (11)$$

Direct detection is performed in both output ports of the interferometer, *i.e.* the fluctuations in the photon numbers $\hat{n}_c = \hat{c}^\dagger\hat{c}$ and $\hat{n}_d = \hat{d}^\dagger\hat{d}$ are detected at the sideband frequency Ω. From the sum and the difference of the detected signals we obtain (in linear approximation) information about certain combinations of the quadrature components at the measured frequency Ω

$$\begin{aligned}\delta\hat{n}_c^\Omega + \delta\hat{n}_d^\Omega = \\ &\frac{1}{2}\alpha[\delta\hat{X}_{a,0}^\Omega + \exp(-i\theta)\delta\hat{X}_{a,0}^\Omega + \delta\hat{X}_{v,0}^\Omega - \exp(-i\theta)\delta\hat{X}_{a,0}^\Omega], \quad (12)\\ \delta\hat{n}_c^\Omega - \delta\hat{n}_d^\Omega = \\ &\frac{1}{2}\alpha[\delta\hat{X}_{a,-\varphi}^\Omega + \exp(-i\theta)\delta\hat{X}_{a,\varphi}^\Omega + \delta\hat{X}_{v,-\varphi}^\Omega - \exp(-i\theta)\delta\hat{X}_{a,\varphi}^\Omega]. \quad (13)\end{aligned}$$

Here, we introduced the short hand notation for the general quadrature component $\delta\hat{X}_{a,\varphi}^\Omega = \exp(i\varphi)(\delta\hat{a}^\Omega)^\dagger + \exp(-i\varphi)\delta\hat{a}^\Omega$. The photon number fluctuations in frequency domain are linked to the time domain fluctuations by a Fourier transformation. Due to the properties of Fourier transformation, care has to be taken when complex conjugation is applied to phase factors of the rf–signal $\exp(-i\theta)$: These factors must not be conjugated.[22] Choosing the phase shifts $\theta = \pi$ and $\varphi = \pi/2 + 2k\pi$ (k is an integer) it follows

$$\delta\hat{n}_c^\Omega + \delta\hat{n}_d^\Omega = \alpha\delta\hat{X}_{v,0}^\Omega, \quad (14)$$

$$\delta\hat{n}_c^\Omega - \delta\hat{n}_d^\Omega = \alpha\delta\hat{X}_{a,-\pi/2}^\Omega. \quad (15)$$

In the difference channel, the noise contribution from the unused input port of the interferometer cancels, the signal is direct proportional to the fluctuations of the phase quadrature at the frequency Ω. The sum signal provides a shot noise reference, as only the fluctuations originating from the vacuum mode remain.

To achieve a phase shift of π at the detected sideband frequency Ω, the required delay ΔL is given by

$$\Delta L = \frac{c\pi}{\Omega} = \frac{cT}{2} = \frac{c}{2f}, \quad (16)$$

T is the period of the signal at the frequency Ω, $f = \Omega/(2\pi)$ and c is the speed of light. When a pulsed laser system with repetition frequency $f_{\text{rep}} = 1/T_{\text{rep}}$ is used, an additional condition on ΔL is imposed as interference occurs only when two pulses overlap at the beam splitter:

$$\Delta L = ncT_{\text{rep}} = cnf_{\text{rep}}. \quad (17)$$

$T_{\rm rep}$ is the time between two successive pulses and n is an integer. Possible measurement frequencies f are therefore

$$f = \frac{f_{\rm rep}}{2n}. \tag{18}$$

At the repetition frequency of 82 MHz of our laser system, the delay must be a multiple of 3.66 m corresponding to the distance between two successive pulses. Since we measure at a frequency of 20.5 MHz, an arm–length difference of 7.32 m is required.

4. Phase Measuring Interferometer — Setup and Efficiency

The detailed setup of the Mach–Zehnder interferometer and the detection system is depicted in Fig. 1. The first beam splitter is a combination of a $\lambda/2$–waveplate and a polarizing beam splitter. It is therefore possible to switch between measurements of the amplitude and the phase quadrature. In the former case, the linearly polarized input is adjusted such that all light propagates through one arm when passing through the interferometer. The second beam splitter together with the detector pair form a balanced detection system. For phase measurements, the input power is distributed equally in the two interferometer arms. The fields from the short and the long arm have to interfere at the second beam splitter. The interference phase at the optical frequency is adjusted such that both output ports have equal intensity. The difference of the dc–signals of the photodetectors serves as error signal and is fed back onto the piezo mirror to adjust the optical phase.

For quantum optical measurements, high efficiency of the measurement devices is required as losses lead to degradation of squeezing and quantum correlations. The losses in the interferometer are governed mainly by (1) the efficiency of the photodiodes, (2) the losses at the different optical components (3) non–perfect interference contrast.

The efficiency of the photodiodes (Epitaxx ETX500) is about 90%, and the losses introduced at all the optical components (lenses, waveplates, beam splitters) is of the same size in spite of anti–reflexion coatings. Due to non–perfect mode overlap of the fields at the beam splitter, additional vacuum noise is mixed into the system in a similar way as in homodyne detection, thus reducing the non–classical properties of the light field. The losses due to imperfect mode matching $1-\eta$ are directly linked to the visibility via the relation $\mathcal{V}^2 = \eta$ where $\mathcal{V}$ denotes the visibility. An interference contrast $\mathcal{V} < 1$ is the result of inadequate temporal and spatial overlap of the pulses. The temporal overlap of two pulses and therefore the phase shift of the rf–signal is adjusted roughly by introducing the delay ΔL. Fine adjustment is achieved via small shifts of the piezo mirror. To optimise the spatial overlap and to match the phase fronts, two telescopes are inserted into the setup, one in front of the interferometer, and one inside the long arm. An interference contrast $\mathcal{V}$ of up to 95% was achieved experimentally leading to an efficiency of about 90%. Taking into account all loss mechanisms from diode inefficiency, propagation losses and non–perfect interference contrast, the overall efficiency of the phase measuring device could be estimated to be roughly $0.9 \times 0.9 \times 0.9 = 0.73$.

5. Generation of Quadrature Entanglement

Before we proceed describing some results of phase measurements with the interferometric setup on classical and non–classical beams, we will introduce our source for the generation of squeezed and entangled short light pulses. As entanglement source we use a fibre optical setup similar to the one described by Silberhorn *et al.*[10] Intense ultrashort light pulses in a regime that allows for soliton propagation are used to generate non–classical states of light in fibres.[30–32] Due to the high intensity, these pulses experience a significant Kerr non–linearity and become quadrature squeezed. Amplitude squeezing, that could be observed in direct detection, is generated using an asymmetric fibre Sagnac interferometer.[33–35] By employing a polarization maintaining fibre, two amplitude squeezed beams of orthogonal polarization could be generated simultaneously. The Sagnac loop therefore consists of an 8 m long piece of polarization maintaining fiber and an asymmetric 90:10 beam splitter. At the output of the Sagnac interferometer, between 3 and 4 dB of amplitude squeezing can be observed in direct detection. For the generation of bright continuous variable entanglement, these squeezed light fields are made to interfere at a balanced 50:50 beam splitter.[2,36] The quality of the entanglement generated with this method depends on the initial squeezing of the two input beams. In our experiment we used short pulses of about 130 fs at a center wavelength of 1530 nm and at a repetition rate of 82 MHz. The setup of the entanglement source is shown in Fig. 2. Interference of the two orthogonally polarized squeezed light fields is obtained on a polarizing beam splitter when the polarizations are rotated properly using a $\lambda/2$–waveplate. To compensate the temporal walk off of the pulses due to the birefringence of the fibre, a Michelson interferometer is placed in front of the Sagnac loop. In the following section, the non–

classical states from the Sagnac loop (squeezed and entangled) are used to check the performance of the phase measuring device that was introduced in the previous sections.

6. Different Phase Measurements

6.1. *Phase modulated laser beam*

As a first measurement to check the performance of the interferometer, the noise of a phase modulated coherent beam was examined. A phase modulation was imposed onto the beam at a frequency of 20.5 MHz which coincides with the operating frequency of the interferometer. We made sure that pure phase modulation was imposed on the beam with no residual amplitude

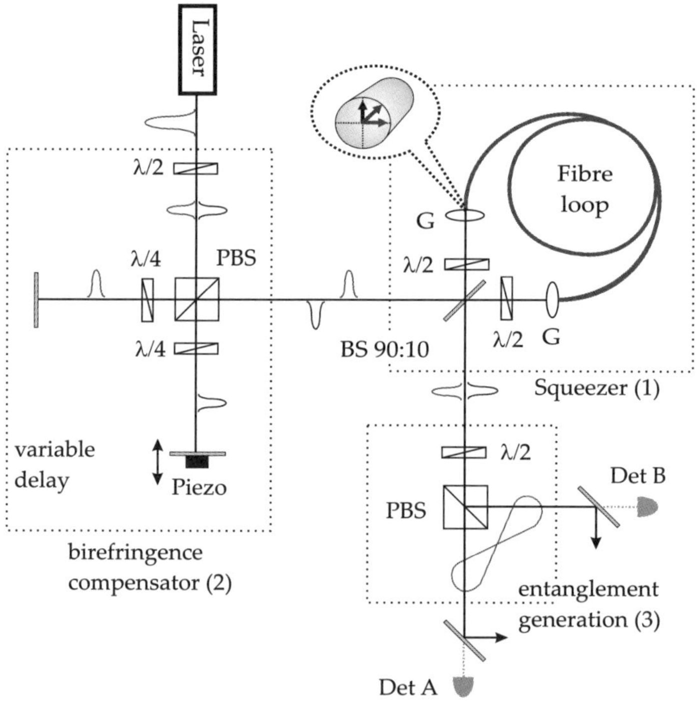

Fig. 2. Setup for entanglement generation. Two squeezed beams of orthognal polarization are generated in an asymmetric fibre Sagnac interferometer. To generate quadrature entanglement, the squeezed beam interfer at an beam splitter. PBS: polarizing beam splitter, BS: beam splitter, G: Gradient index lens, Det A and B: Detectors for phase lock, $\lambda/4$ and $\lambda/2$: retardation plates.

modulation. The graph in Fig. 3 shows two traces in the frequency interval from 20–21 MHz: (1) The result of an amplitude measurement, where no modulation is observed and (2) the result of a phase measurement where a high signal level is observed at the modulation frequency of 20.5 MHz. This is a first hint that our interferometer is indeed a phase measuring device.

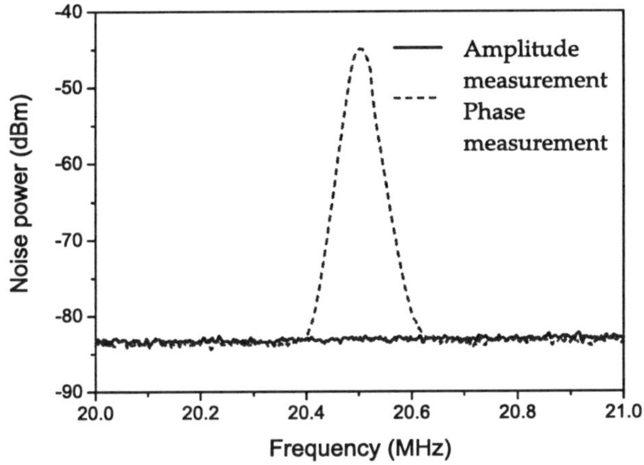

Fig. 3. Amplitude measurement (solid line) and phase measurement (dashed line) on a phase modulated beam. A large signal is only observed for phase measurements, no modulation is seen in the amplitude measurement. The resolution bandwidth was 30 kHz, the video bandwidth 100 Hz. The electronic noise level was at -93 dBm.

6.2. *Phase noise measurements of intense, short, amplitude squeezed pulses from a fibre*

In the next step, the excess phase noise generated in the fibre Sagnac interferometer through self–phase modulation and other spurious phase noise effects introduced in the light field such as scattering of photons at acoustic phonons (GAWBS)[37–39] is observed. The initial squeezing was about 3 dB below the shot noise reference. The goal of the following measurements was to check the behaviour of the phase noise (obtained from the difference channel, see Fig. 1) and the respective shot noise level (obtained from the sum channel, see Fig. 1) upon attenuation. The behaviour is compared to the theoretical prediction, especially in order to verify that the sum channel could be used as a shot noise reference. The squeezed pulsed beam was directed into the phase measuring interferometer. The noise level of the sum and the difference channel of the detector pair was measured for different values of transmission T of the input beam at a variable attenuator. The spectral noise power at the measurement frequency 20.5 MHz of the sum channel (filled squares) and the difference channel (open circles) is plotted in a logarithmic scale in Fig. 4a as a function of the transmission of the input beam. In Fig. 4b the noise level of the difference channel (phase noise

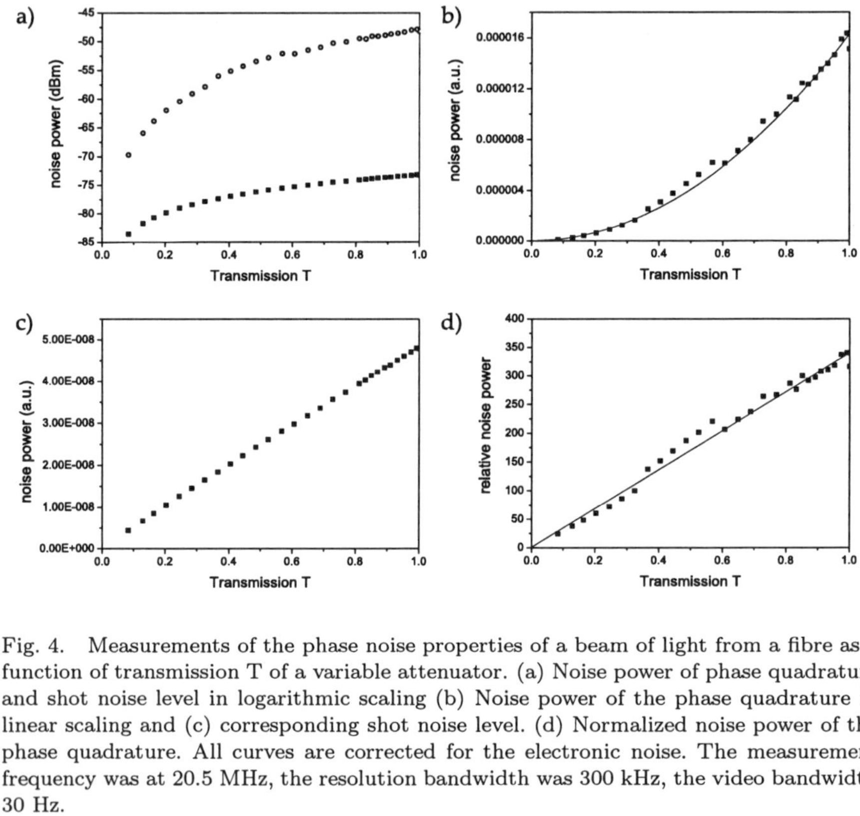

Fig. 4. Measurements of the phase noise properties of a beam of light from a fibre as a function of transmission T of a variable attenuator. (a) Noise power of phase quadrature and shot noise level in logarithmic scaling (b) Noise power of the phase quadrature in linear scaling and (c) corresponding shot noise level. (d) Normalized noise power of the phase quadrature. All curves are corrected for the electronic noise. The measurement frequency was at 20.5 MHz, the resolution bandwidth was 300 kHz, the video bandwidth 30 Hz.

of the signal beam) is plotted in a linear scale, while the noise level of the sum channel (shot noise level) is plotted in Fig. 4c. The noise power of the difference channel decreases quadratically with increasing attenuation, the expected behaviour of large noise signals. In the graph, the expected noise level of a signal which is initially 25 dB above the shot noise level is plotted, the curve agrees nicely with the measured values. On the other hand, the sum signal decreases linearly with the attenuation which is the characteristic behaviour of a shot noise limited signal. The sum channel is therefore used as a shot noise reference. In Fig. 4d the noise power of the phase signal normalized to the shot noise level is plotted. As expected, the signal decreases linearly in that case. The solid line represents the calculated behaviour of a signal with 25 dB of noise above the shot noise level. These attenuation measurements show that the signals measured with the interferometer reliably deliver the correct noise levels and agree nicely with theoretical models that describe the noise behaviour upon attenuation.

6.3. *Sub-shot noise phase quadrature measurements*

So far, we have only considered measurements on intense beams with very large signals or noise in the phase quadrature. However, for quantum optical measurements, often an accuracy better than the quantum noise level is required. One possibility to check that such sub–shot noise measurements are feasible with the interferometer would be to inject phase squeezed beams into the device. However, our fibre optical setup cannot produce phase squeezed light. To demonstrate the possibility of measurements below the shot noise level, phase measurements on a pair of quadrature entangled beams were performed simultaneously with two different interferometers and the correlations in the photocurrents were measured.

The experimental setup is depicted in Fig. 5. Two amplitude squeezed bright beams are used to generate entanglement. With the measurement devices, the noise characteristics of the squeezed input beams as well as the correlations in the amplitude and the phase quadratures of the entangled beam pair are examined. First, the noise reduction of the two squeezed input modes $\hat{a}$ and $\hat{b}$ was checked. To do that, one of these modes was blocked. The power of the other mode is distributed equally to both interferometers at the beam splitter that is usually used for the entanglement generation. The interferometers in the amplitude measurement configuration together with the beam splitter comprise a balanced detection scheme. We observed 2.5 dB and 2.7 dB of noise reduction for the two input beams. As a result of the relatively lossy measurement method, these values of noise reduction are slightly lower than those obtained in direct detection directly after the 50:50 beam splitter.

In the second step, the two squeezed beams were brought to interference to generate entanglement. First, the correlation properties of the entangled beam pair in the amplitude quadrature were determined. Both interferometers were then configured to measure the amplitude quadrature. In

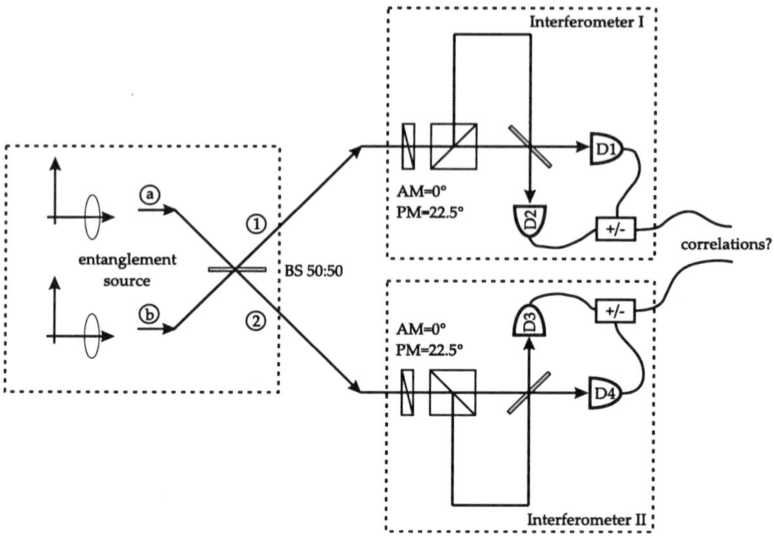

Fig. 5. Setup to measure the correlations in the phase quadratures of an entangled beam pair. Entanglement is generated by the interference of two amplitude squeezed beams a and b at a 50:50 beam splitter. The correlations in the amplitude as well as the phase quadrature are verified using two phase measuring interferometers in each entangled beam.

Fig. 6a, several traces are plotted: Trace 2 represents the quantum noise limit for the combined system whereas traces 3 show the noise level of the individual entangled beams. A high level of excess noise originating from the phase noise of the squeezed input beams is observed. However, strong anti–correlations are present as can be seen from trace 1: The sum signal $V(\delta\hat{X}_1 + \delta\hat{X}_2)$ drops below the quantum noise level by 2.6 dB, while the difference signal $V(\delta\hat{X}_1 - \delta\hat{X}_2)$ is 6 dB above the noise level of the individual modes. The squeezing variance in the amplitude quadrature, *i.e.* the variance $V(\delta\hat{X}_1 + \delta\hat{X}_2)$ normalized to the variance of the quantum noise level is $V_{\rm sq}^{+}(\delta\hat{X}) = 0.55$. The results of the correlation measurement of the phase quadrature are shown in Fig. 6b. Again, the quantum noise limit is indicated by trace 2 and the phase noise signal of the individual modes is presented by traces 3. Correlations of 1.3 dB below the shot noise level are observed in the difference variance $V(\delta\hat{Y}_1 - \delta\hat{Y}_2)$. This corresponds to a squeezing variance $V_{\rm sq}^{-}(\delta\hat{Y}) = 0.74$ in the phase quadrature. As expected, the sum signal of the phase quadrature correlation measurement is 6 dB above the noise level of the individual beams.

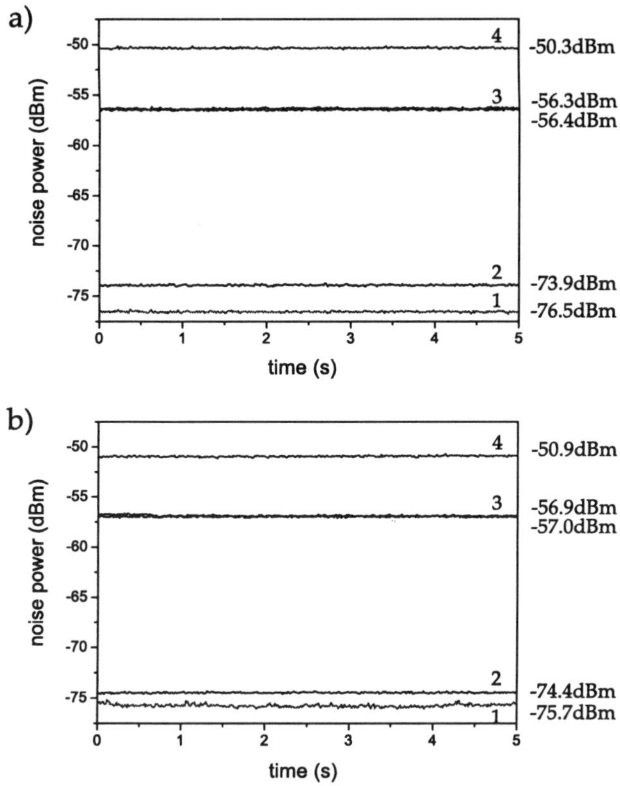

Fig. 6. Experimental results of correlation measurements of the amplitude (a) and the phase quadrature component (b) on an entangled beam pair. All traces are corrected for the electronic noise level at -83.9 dBm. The measurement frequency was 20.5 MHz, the resolution bandwidth 300 kHz and the video bandwidth 30 Hz. In each graph the noise level (traces 1) of the correlation signal is shown together with the corresponding shot-noise level (traces 2), the noise level of the individual beams (traces 3), and the signal with the anticorrelations (traces 4).

The discrepancy between the observed correlations in the amplitude and the phase quadrature has mainly two reasons. First, due to an imperfect interference contrast, the interferometer is less efficient in the phase measuring setting than in the amplitude setting, hence reduced correlations are observed. Second, slightly different correlations in the amplitude and the phase quadrature component are induced in the entanglement generation process when the splitting ratio of the beam splitter used for entanglement generation is not perfectly symmetric. As a result, the correlations in the phase quadrature are reduced compared to the correlations in the amplitude quadrature. The asymmetry in the correlation signals becomes more detrimental if very high excess noise is present in the amplitude squeezed input states.

Nevertheless, from the above measurements we can conclude that the two mode state generated in the fibre optical setup is entangled. According to the non–separability criterion for continuous variables[40,41] the state is entangled since

$$V_{\rm sq}^{+}(\delta\hat{X}) + V_{\rm sq}^{-}(\delta\hat{Y}) = 0.55 + 0.74 = 1.29 < 2. \tag{19}$$

7. Conclusions

The measurements with the Mach–Zehnder interferometer show that this device is suitable to perform phase quadrature measurements at a certain sideband frequency Ω on intense pulsed light states. It was demonstrated that by simply rotating the polarization of the input state, it is possible to switch between amplitude and phase quadrature measurements. The most important result is that even very small signals below the quantum noise limit could be resolved using that interferometric technique. The non–classical, short light pulses produced with the fibre optical setup together with the interferometric setup for phase quadrature measurements are therefore interesting for quantum optical experiments and for the implementation of quantum information and quantum communication protocols. Two or more parties can perform local measurements of the amplitude and the phase quadrature on their subsystems. The interferometric phase measurement device has already been applied successfully in an experimental demonstration of a quantum erasing experiment with continuous variables,[42] but it also paves the way to other types of quantum optical experiments. For instance using a similar setup we have recently shown that the quantum correlated sidebands can be separated into two spatial modes.[29]

References

1. Z. Ou, S. Pereira, H. Kimble, and K. Peng, Phys. Rev. Lett. **68**, 3663 (1992).
2. A. Furusawa, J. Sørensen, S. Braunstein, C. Fuchs, H. Kimble, and E. Polzik, Science **282**, 706 (1998).
3. Y. Zhang, H. Wang, X. Y. Li, J. T. Jing, C. D. Xie, and K. C. Peng, Phys. Rev. A **62**, 023813 (2000).
4. C. Schori, J. L. Sørensen, and E. S. Polzik, Phys. Rev. A **66**, 033802 (2002).

5. W. P. Bowen, R. Schnabel, P. K. Lam, and T. C. Ralph, Phys. Rev. Lett. **90**, 043601 (2003).
6. T. Aoki, N. Takei, H. Yonezawa, K. Wakui, T. Hiraoka, A. Furusawa, and P. van Loock, Phys. Rev. Lett. **91**, 080404 (2003).
7. J. Wenger, A. Ourjoumtsev, R. Tualle-Brouri, and F. Grangier, Eur. Phys. J. D **32**, 391 (2004).
8. J. Laurat, T. Coudreau, G. Keller, N. Treps, and C. Fabre, Phys. Rev. A **70**, 042315 (2004).
9. V. Josse, A. Dantan, A. Bramati, M. Pinard, and E. Giacobino, Phys. Rev. Lett. **92**, 123601 (2004).
10. C. Silberhorn, P. K. Lam, O. Weiß, F. König, N. Korolkova, and G. Leuchs, Phys. Rev. Lett. **86**, 4267 (2001).
11. S. L. Braunstein and A. K. Pati, (Eds.), *Quantum Information with Continuous Variables*, Kluwer Akademic Press, Dordrecht 2003.
12. S. L. Braunstein and H. Kimble, Phys. Rev. A **61**, 2302 (2000).
13. S. L. Braunstein and H. J. Kimble, Phys. Rev. Lett. **80**, 869 (1998).
14. H. P. Yuen and J. H. Shapiro, IEEE Trans. Inf. Theory **26**, 78 (1980).
15. H. P. Yuen and V. Chan, Opt. Lett. **8**, 177 (1983).
16. L.-A. Wu, M. Xiao, and H. J. Kimble, J. Opt. Soc. Am. B **4**, 1465 (1987).
17. R. M. Shelby, M. D. Levenson, S. H. Perlmutter, R. G. DeVoe, and D. F. Walls, Phys. Rev. Lett. **57**, 691 (1986).
18. H. A. Bachor, M. D. Levenson, D. F. Walls, S. H. Perlmutter, and R. M. Shelby, Phys. Rev. A **38**, 180 (1988).
19. M. D. Levenson, R. M. Shelby, A. Aspect, M. D. Reid, and D. F. Walls, Phys. Rev. A **32**, 1550 (1985).
20. P. Galatola, L. A. Lugiato, P. M. G., P. Tombesi, and G. Leuchs, Opt. Comm. **85**, 95 (1991).
21. S. Inoue and Y. Yamamoto, Opt. Lett. **22**, 328 (1997).
22. O. Glöckl, U. L. Andersen, S. Lorenz, C. Silberhorn, N. Korolkova, and G. Leuchs, Opt. Lett. **29**, 1936 (2004).
23. C. M. Caves, Phys. Rev. D **26**, 1817 (1982).
24. C. M. Caves and B. M. Schumaker, Phys. Rev. A **31**, 3068 (1985).
25. B. Yurke, Phys. Rev. A **32**, 300 (1985).
26. S. L. Braunstein and D. D. Crouch, Phys. Rev. A **43**, 330 (1991).
27. J. Gea-Banacloche and G. Leuchs, J. Mod. Opt. **34**, 793 (1987).
28. J. Zhang, Phys. Rev. A **67**, 054302 (2003).
29. E. H. Huntington, G. N. Milford, C. Robilliard, T. C. Ralph, O. Glöckl, U. L. Andersen, S. Lorenz, and G. Leuchs, Phys. Rev. A **71**, 041802(R) (2005).
30. P. D. Drummond and S. J. Carter, J. Opt. Soc. Am. B **4**, 1556 (1987).
31. M. Rosenbluh and R. Shelby, Phys. Rev. Lett. **66**, 153 (1991).
32. P. D. Drummond, R. M. Shelby, S. R. Friberg, and Y. Yamamoto, Nature **365**, 307 (1993).
33. M. J. Werner, Phys. Rev. Lett. **81**, 4132 (1998).
34. S. Schmitt, J. Ficker, M. Wolff, F. König, A. Sizmann, and G. Leuchs, Phys. Rev. Lett. **81**, 2446 (1998).
35. D. Krylov and K. Bergman, Opt. Lett. **23**, 1390 (1998).

36. G. Leuchs, T. Ralph, C. Silberhorn, and N. Korolkova, J. Mod. Opt. **46**, 1927 (1999).
37. R. M. Shelby, M. D. Levenson, and P. W. Bayer, Phys. Rev. Lett. **54**, 939 (1985).
38. R. M. Shelby, M. D. Levenson, and P. W. Bayer, Phys. Rev. B **31**, 5244 (1985).
39. R. M. Shelby, P. D. Drummond, and S. J. Carter, Phys. Rev. A **42**, 2966 (1990).
40. L. Duan, G. Giedke, J. Cirac, and P. Zoller, Phys. Rev. Lett. **84**, 2722 (2000).
41. R. Simon, Phys. Rev. Lett. **84**, 2726 (2000).
42. U. L. Andersen, O. Glöckl, S. Lorenz, G. Leuchs, and R. Filip, Phys. Rev. Lett. **93**, 100403 (2004).

Chapter 13

Experimental Polarization Squeezing and Continuous Variable Entanglement via the Optical Kerr Effect

V. Josse, A. Dantan, A. Bramati, M. Pinard and E. Giacobino
Laboratoire Kastler Brossel, Université Pierre et Marie Curie,
Case 74, 4 place Jussieu, 75252 Paris Cedex 05, France

email: elg@spectro.jussieu.fr

J. Heersink, U. L. Andersen, O. Glöckl and G. Leuchs
Insitut für Optik, Information und Photonik
Max-Planck Forschungsgruppe, Universität Erlangen-Nürnberg
Günther-Scharowsky-Strasse 1, Bau 24, Erlangen, 91058, Germany

email: joel.heersink@optik.uni-erlangen.de

Motivated by their easy manipulation, the quantum polarization states of light, described by the Stokes operators, are of particular interest in quantum information science. In this chapter we present two different experimental schemes which produce polarization squeezing and various forms of continuous variable entanglement. Both of these schemes, cold atomic samples and optical fibers, involve the nonlinear optical Kerr effect. We emphasize their common characteristics, as they are illustrative of the general properties satisfied by a wide variety of systems. Building upon these analogies, we propose a unifying method to find the quantum properties of a given experiment, and we interpret both experiments within this framework. While the cold atoms experiment was used to explore and develop the theoretical concepts, the particularly simple and efficient scheme employing optical fibers represents a promising resource for future developments in quantum information and communication.

1. Introduction

The polarization of light is a familiar phenomenon. However, it is less known that a mathematical description of the properties of the polarization of light was already published in 1852 by G.G. Stokes,[1] in which he defined what are today known as the Stokes parameters and introduced a description of light based on three independent parameters: S_1, S_2, S_3.[2] The classical polarization state of light can be measured directly on a simple detector, and the respective setups consist only of combinations of waveplates followed by a polarizing beam splitter. The difference in the signals at the output ports determines the value of a given Stokes parameter (Fig. 1). In any measurement configuration, the sum, referred to as the Stokes parameter S_0, is proportional to the total beam power.

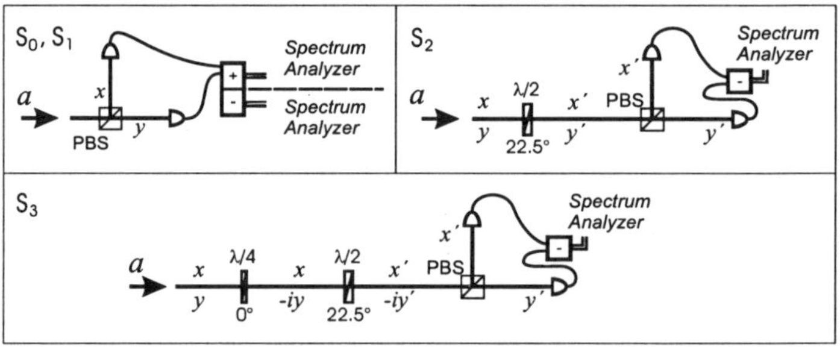

Fig. 1. Detection setups for the Stokes parameters. The unknown polarization state is split on a polarizing beam splitter and measured with two detectors. Upper left: S_0 and S_1; upper right: S_2; below: S_3.

These practical measurements are the primary reason why the quantum polarization state of light has attracted a lot of attention in continuous variable quantum optics. The first experiment to exploit the quantum properties of polarization was performed by Grangier *et al.* in 1987 in a squeezed-light-enhanced polarization interferometer.[3] First proposals to use the Stokes parameters as a resource for quantum optics were made in the early 1990s by Chirkin *et al.*[4,5] Further theoretical contributions were made by Luis *et al.*[6] The first explicit experimental demonstration was achieved by Hald *et al.* in the context of quantum memory.[7,8] Such a promising application sparked broad interest, resulting in an experiment-oriented formulation of the theory.[9] In the ensuing years polarization squeezing and entanglement have been demonstrated in a variety of nonlinear systems: Optical Parametric Oscillators[10–12] ($\chi^{(2)}$ nonlinearity), optical fibers[13–15] and cold atomic samples[16–18] (both $\chi^{(3)}$ nonlinearities). Ongoing theoretical research in many areas,[19,20] including teleportation,[21] as well as the recent demonstration of quantum memory in which the polarization state of light has been coupled to the spin state of atoms[8] will keep this field active.

The theory of the quantum polarization state of light is well documented in the literature and is discussed in detail in the chapter by N. Korolkova in this book. We therefore concentrate here on the experimental aspects of this field. In particular, we consider the interaction of bright beams in optical Kerr media, that is optical fibers and cold atomic samples, and its ability to generate polarization squeezing and entanglement. The chapter is organized as follows: the first part is dedicated to polarization squeezing, the main features are underlined and the experiments using optical fibres and cold atoms are presented. In a second part we address continuous variable entanglement: fundamental properties are first reviewed before being applied to the experimental schemes.

2. Polarization Squeezing

2.1. *Definition*

The characterization of the quantum features of a polarization state of light relies on the measurement of the quantum Stokes operators.[9] These are defined in analogy to their classical counterparts:

$$\hat{S}_0 = \hat{a}_x^\dagger \hat{a}_x + \hat{a}_y^\dagger \hat{a}_y \ , \qquad \hat{S}_1 = \hat{a}_x^\dagger \hat{a}_x - \hat{a}_y^\dagger \hat{a}_y, \tag{1}$$

$$\hat{S}_2 = \hat{a}_x^\dagger \hat{a}_y + \hat{a}_y^\dagger \hat{a}_x \ , \qquad \hat{S}_3 = i(\hat{a}_y^\dagger \hat{a}_x - \hat{a}_x^\dagger \hat{a}_y), \tag{2}$$

where $\hat{a}_x$ and $\hat{a}_y$ are two orthogonally polarized modes. Following the non-commutation of the photon annihilation and creation operators, $\hat{a}_{x/y}$ and $\hat{a}_{x/y}^\dagger$, these Stokes parameters obey the relations: $[\hat{S}_0, \hat{S}_i] = 0$ and $[\hat{S}_i, \hat{S}_j] = 2i\hat{S}_k$ with $\{i, j, k\} = 1, 2, 3$ and $i \neq j \neq k$. These relations lead to Heisenberg inequalities and therefore to the presence of intrinsic quantum fluctuations. However this fundamental noise depends on the mean polarization state and special attention has to be paid to the definition of polarization squeezing, in contrast to standard single mode squeezing.

Let us consider a situation where the light is linearly polarized along the x axis ($\langle \hat{a}_x \rangle = \alpha_x \in \Re$ and $\langle \hat{a}_y \rangle = 0$). The only non-trivial Heisenberg inequality reads $\Delta^2 \hat{S}_2 \, \Delta^2 \hat{S}_3 \geq |\langle \hat{S}_1 \rangle|^2 = \alpha_x^4$, where $\Delta^2 \hat{S}_j$ refers to the variance $\langle \hat{S}_j^2 \rangle - \langle \hat{S}_j \rangle^2$. Generally, polarization squeezing is achieved if the variance $\Delta^2 \hat{S}_\theta$, the variance of a Stokes operator rotated by θ in the $\hat{S}_2$-$\hat{S}_3$ plane where $\langle \hat{S}_\theta \rangle = 0$, is below the shot noise level (compare with Fig. 2):

$$\Delta^2 \hat{S}_\theta \leq |\langle \hat{S}_1 \rangle| = \alpha_x^2 \quad \text{where} \quad \hat{S}_\theta = \cos\theta \, \hat{S}_2 + \sin\theta \, \hat{S}_3. \tag{3}$$

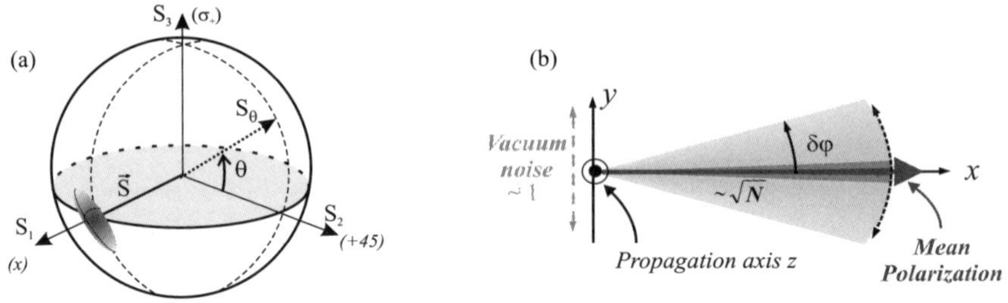

Fig. 2. (a) Representation of a polarization squeezed state on the Poincaré sphere. (b) Jitter of the mean field polarization direction due to the fluctuations of vacuum mode orthogonally polarized.

Following this example, the definition of polarization squeezing is easily extended to the general case where the classical polarization vector is aligned along a given direction $\vec{S}$ in Stokes space. Noting $\hat{S}_\perp$ a Stokes parameter belonging to the orthogonal plane, called the "dark plane" since $\langle \hat{S}_\perp \rangle = 0$, it reads:

$$\Delta^2 \hat{S}_\perp \leq |\langle \vec{S} \rangle| = \langle \hat{S}_0 \rangle. \tag{4}$$

2.2. *Connection to vacuum squeezing*

Let us further consider the case where the light is polarized along the x axis, that is $\hat{a}_x$ is the mean field and $\hat{a}_y$ is the orthogonal vacuum or dark mode. It can be seen from Eq. 2 that the relevant fluctuations of $\hat{S}_2$ and $\hat{S}_3$ are proportional to the quadrature fluctuations of $\hat{a}_y$:

$$\delta \hat{S}_2 = \alpha_x (\delta \hat{a}_y^\dagger + \delta \hat{a}_y) \equiv \alpha_x \delta \hat{X}_y(0), \tag{5}$$

$$\delta \hat{S}_3 = i\alpha_x (\delta \hat{a}_y^\dagger - \delta \hat{a}_y) \equiv \alpha_x \delta \hat{Y}_y(0), \tag{6}$$

where the quadratures, for a given angle θ, are defined in the standard fashion: $\hat{X}_\alpha(\theta) = (\hat{a}_\alpha^\dagger e^{i\theta} + \hat{a}_\alpha e^{-i\theta})$ and $\hat{Y}_\alpha(\theta) = \hat{X}_\alpha(\theta + \frac{\pi}{2})$. To simplify the equations in the following, $\hat{X}$ and $\hat{Y}$ will refer to $\hat{X}(0)$ and $\hat{Y}(0)$ respectively.

These relations provide a nice picture of the fundamental polarization fluctuations being driven by the noise of the orthogonal vacuum mode (see Fig. 2(b)). Furthermore it traces the origin of the Heisenberg inequality for the Stokes operators to the usual Heisenberg inequality satisfied by the quadratures of the electromagnetic field: $\Delta^2 \hat{X} \Delta^2 \hat{Y} \geq 1$. It then becomes clear that *polarization squeezing is equivalent to vacuum squeezing in the orthogonal polarization mode*:

$$\Delta^2 \hat{S}_\theta \leq \alpha_x^2 \quad \Leftrightarrow \quad \Delta^2 \hat{X}_y(\theta) \leq 1. \tag{7}$$

2.3. *Generation of polarization squeezing: an example*

From the above analysis, we conclude that polarization squeezing can be achieved if one mixes a bright beam with an independently produced squeezed vacuum beam.[3] On the other hand, recent experiments have shown that polarization squeezing can also be produced directly, either in optical fibers[22,15] or in cold atomic media.[16]

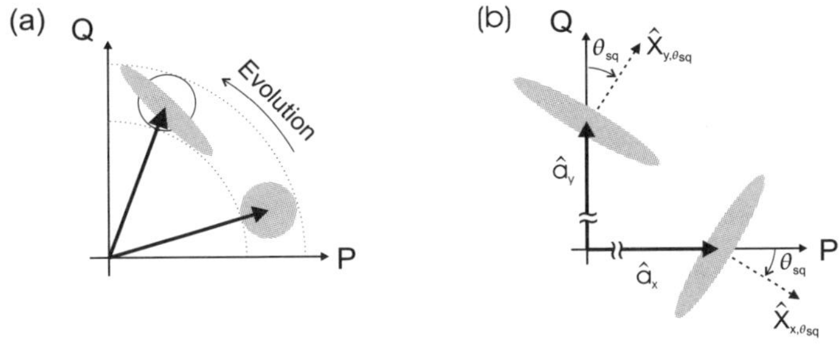

Fig. 3. (a) Representation in phase space or the Fresnel diagram of the effect of the Kerr nonlinearity on a coherent beam (bottom right). The curved arrow indicates the direction of evolution. (b) Diagram of two orthogonally polarized (x, y) squeezed states; the squeezed quadrature (defined with respect to the mean field of each mode) is the same for the two modes.

Here we will focus on a particular property that is connected to the experimental realizations presented in Sec. 3: if the system considered is able to produce two bright, symmetric and independent squeezed beams in orthogonal polarizations, then polarization squeezing is achieved. By symmetric, we mean two modes with the same amplitude and the same squeezed quadrature (with angle θ_{sq} relative to the mean field as in Fig. 3(b)).

To clarify this point we use the simple example of circularly polarized light, but a straightforward generalization can be readily made. Let us choose the two independent squeezed modes to be the $\hat{a}_x$ and $\hat{a}_y$ modes with their mean values phase shifted by $\pi/2$ (see Fig. 3(b)): $\langle \hat{a}_y \rangle = i \langle \hat{a}_x \rangle = i\alpha/\sqrt{2}$, $\alpha \in \Re$. The light is then circularly polarized with $\hat{a}_{\sigma+}$ as the mean field and $\hat{a}_{\sigma-}$ as the vacuum orthogonal mode:

$$\hat{a}_{\sigma+} = -\frac{1}{\sqrt{2}}(\hat{a}_x - i\hat{a}_y) \quad \text{with} \quad \langle \hat{a}_{\sigma+} \rangle = -\alpha, \tag{8}$$

$$\hat{a}_{\sigma-} = \frac{1}{\sqrt{2}}(\hat{a}_x + i\hat{a}_y) \quad \text{with} \quad \langle \hat{a}_{\sigma-} \rangle = 0. \tag{9}$$

In the Poincaré sphere the mean vector is along $\hat{S}_3$ and the relevant

Stokes parameters $\hat{S}_\theta$ are located in the dark plane spanned by $\hat{S}_1$ and $\hat{S}_2$ ($\hat{S}_\theta = \cos(\theta)\hat{S}_1 + \sin(\theta)\hat{S}_2$). From the previous analysis, $\hat{S}_\theta$ is related to the quadrature fluctuations of $\hat{a}_{\sigma_-}$:

$$\delta\hat{S}_\theta \approx -\alpha\delta\hat{X}_{\sigma_-}(\theta) = -\alpha(\delta\hat{X}_x(\theta) - \delta\hat{Y}_y(\theta))/\sqrt{2}. \tag{10}$$

Assuming the two quadrature noises are uncorrelated and squeezed for the same angle relative to the amplitude quadrature, polarization squeezing is found for $\hat{S}_{\theta_{sq}}$:

$$\Delta^2\hat{S}_{\theta_{sq}} = \alpha^2(\Delta^2\hat{X}_x(\theta_{sq}) + \Delta^2\hat{Y}_y(\theta_{sq}))/2 < \alpha^2. \tag{11}$$

3. Polarization Squeezing via Kerr Effect

3.1. *The optical Kerr effect*

The optical Kerr effect is a special case of four wave mixing, a nonlinear process in which four unique modes can be coupled. This phenomenon has been incorporated into a general theory of the interaction of light with optical media, namely the material polarizability. For media exhibiting inversion symmetry, *e.g.* silica glass, all even order terms in the electric field of the polarizability vanish. In the special case of degenerate four wave mixing, where the electric field components are all with the same frequency and polarization, this results in the Kerr effect (also called Self Phase Modulation) which is characterized by an intensity dependent refractive index:

$$n = n_0 + n_2 I \quad \text{with} \quad n_2 = \frac{3}{4}\frac{Re(\chi^{(3)})}{n_0^2\epsilon_0^2 c}, \tag{12}$$

where n_i is the ith order of the refractive index, $\chi^{(3)}$ is the third order electromagnetic susceptibility and I is the intensity given by $I = \frac{1}{2}n_0\epsilon_0 c\,|\mathbf{E}|^2$, where $\mathbf{E}$ is the electromagnetic field.

The ability of the Kerr effect to produce squeezed light can be intuitively understood in a simple single mode picture represented in Fig. 3(a): since different amplitudes experience different rotations in phase space, the uncertainty circle (corresponding to the shot noise) of the input field is transformed into an ellipse, corresponding to a squeezed state.

The first proposal to exploit degenerate four wave mixing in optical media for squeezing was made in 1979.[23] A number of theoretical works examined this possibility, for example[24–26] and it was recognized that optical fibers[27] as well as gaseous atomic media[28] are good candidates for squeezers. The first experimental realization of squeezed optical states using the $\chi^{(3)}$ nonlinearity was based on nondegenerate four wave mixing in sodium vapor.[29] The degenerate case, Kerr squeezing, was first seen with cw lasers in fibers[30] and then in atomic beams.[31] Some years later experiments using optical solitons in glass fibers[32] and in cold atomic samples[33] were achieved. In this chapter, we describe how the last two methods can be exploited to generate polarization squeezing and entanglement.

3.2. *Polarization squeezing with cold atoms*

3.2.1. *Nonlinear atom-light interaction in an optical cavity*

Since the 1980's atomic media were thought to be good candidates to achieve high squeezing values. Indeed, for light close to resonance with an atomic transition, high nonlinearities can be obtained. However this interaction is usually associated with spurious effects, such as absorption and other noises arising from the dynamical atom-light interaction, *e.g.* resonance fluorescence noise.[33,34] To obtain large squeezing values a thorough understanding of the interaction is required and a fine balance has to be found between nonlinearities and added noise. We thus use cold atoms in a magnetic optical trap (MOT) to avoid Doppler broadening and to facilitate precise control of the interaction parameters, *i.e.* the detuning between the atoms and light.

In the following, we focus on the nonlinear interaction arising from the saturation of an optical transition. This saturation leads to an intensity dependent refractive index and the atomic medium behaves (to a first approximation) like a Kerr medium. In the simple case of a two level system and for low excitation, the atomic susceptibility χ_{at} varies linearly with I:

$$\chi_{at} = \chi_0(\Delta)\frac{1}{1+s} \approx \chi_0(\Delta)(1-s) \quad \text{with} \quad s = \frac{I}{I_{sat}}\frac{1}{1+(\Delta/\gamma)^2}, \tag{13}$$

where χ_0 characterizes the linear response, γ is the dipole relaxation rate, Δ is the detuning and I_{sat} the saturation intensity.[35]

The atoms are placed inside an optical cavity and, as a consequence, the nonlinearities are enhanced. It is well known that a sufficiently strong Kerr interaction produces bistable behavior in the light transmitted by the cavity and that, at the turning point of the bistability curve, the quantum fluctuations of the light can be greatly reduced.[28,36]

3.2.2. *Principle of polarization squeezing generation*

In previous work the goal was to produce quadrature squeezing using cold cesium atoms.[37] The incident light was then tuned close to resonance with

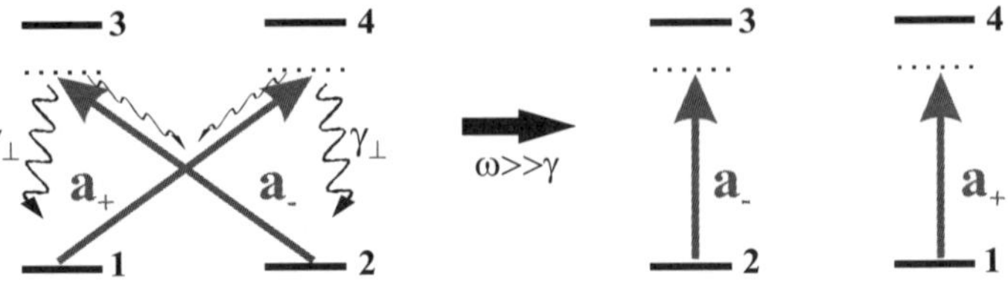

Fig. 4. Schematic energy level diagram for the X-like four-level system: $\gamma_\perp + \gamma_\parallel = \gamma$ is the optical dipole decay rate; Δ is the (large) detuning from resonance. At high frequency the X-like four-level system becomes equivalent to two independent two-level systems.

the closed transition $6S_{1/2}$, F=4 to $6P_{3/2}$, F=5, and *circularly polarized.* The optical pumping process allows the atoms to be modeled as an ensemble of two level atoms as described above. Quadrature squeezing of almost 40% was observed[37] at the output of the cavity.

In contrast, if the incident light is *linearly polarized* then the two-level model is no longer applicable. Because of optical processes, the different Zeeman sublevels of the $6S_{1/2}, F = 4$ hyperfine state are almost equally populated and an X-like four-level atomic medium is well-adapted to describe the interaction (Fig. 4). A detailed theoretical study can be found in,[38] and we summarize the main results.

First, due to the competitive optical pumping between the circular components, $\sigma_\pm$, of the light, the linear polarization becomes unstable inside the cavity above a given intensity threshold. The resonances of the cavity are strongly modified and polarization switching is observed.[16] Interestingly, this polarization switching can be interpreted as being driven by the so called "self-rotation" effect, and, in the case of linearly polarized light, vacuum squeezing on the orthogonal polarization mode was thought to be generated[39] via this process. Therefore, in the domain where the light remains linearly polarized, we would expect to observe polarization squeezing at the output of the cavity. In our experiment, self rotation is, however, related to the optical pumping process, which is very noisy, and we demonstrated that no squeezing could be expected from this source.[38]

This additional noise is, however, restricted to low frequencies, *i.e.* below the optical pumping rate, and if we consider fluctuations at higher frequencies we recover polarization squeezing due to the usual Kerr nonlinearity associated with the saturation of the optical transitions. This can be understood as follows: at high frequency we are looking at time scales smaller than the optical pumping time that couples the two transitions involving

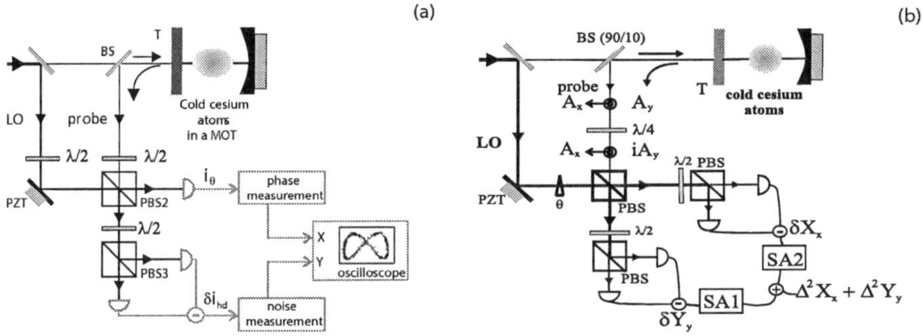

Fig. 5. Experimental setup of the cold atoms experiment. (a) Measurement scheme for the Stokes parameters. (b) Measurement scheme for the nonseparability criterion. PBS: polarizing beam splitter; BS: 90/10 beam splitter; $\lambda/2$: half-wave plate; PZT: piezo-electric ceramic.

the $\sigma_\pm$ components of the light. Then, to a good approximation the atomic medium can be modeled as two independent transitions (see Fig. 4). Below the polarization switching threshold, the light remains linearly polarized and the circular components experience the same nonlinearity. Following the analysis presented in Sec. 2.3, we find polarization squeezing at the cavity output in the "dark" $\hat{S}_2$-$\hat{S}_3$ plane.

3.2.3. *Experimental setup*

The experimental setup used is described in detail in Ref. 33 and a schematic is seen in Fig. 2(a). The cesium atoms are cooled in a standard magneto-optical trap which operates with three mutually orthogonal, circularly polarized trapping beams generated by a Ti:Sapphire laser and an inhomogeneous magnetic field. The trapping Ti:Sapphire laser is detuned by 3 times the linewidth of the upper state on the low frequency side of the $6S_{1/2}$, F=4 to $6P_{3/2}$, F=5 transition. To prevent the atoms from being optically pumped to the $6S_{1/2}$, F=3 state, we superimpose a diode laser tuned to the $6S_{1/2}$, F=3 to $6P_{3/2}$, F=4 transition onto the trapping beams. We use a 25 cm long linear cavity built around the cell. The cavity is close to the hemifocal configuration with a waist of 260 μm. The coupling mirror has a transmission of 10%; the rear mirror is highly reflecting. Hence, the cavity is close to the bad cavity limit in which the cavity linewidth ($\kappa = 5$ MHz) is larger than the atomic linewidth ($\gamma = 2.6$ MHz). We probe the atoms with a linearly polarized laser beam detuned by about 50 MHz to the red of the $6S_{1/2}$, F=4 to $6P_{3/2}$, F=5 transition. The optical power of the probe beam ranges from 5 to 15 μW. Under these conditions, the number of atoms interacting with the light beam is between 10^6 and 10^7.

Usually, measurement of the Stokes parameters can be carried out directly (see Fig. 1). In this set-up, however, the power of the probe beam is insufficient ($\sim \mu W$) and we must resort to the use of a strong local oscillator (LO). The fluctuations of the vacuum mode $\hat{a}_y$ are measured using a standard homodyne detection setup. Following Eq. 6 the photocurrent can be expressed in terms of the fluctuations of $\hat{S}_2$ and $\hat{S}_3$:

$$\delta i_{hd} \propto \cos\theta_{hd}\, \delta\hat{S}_2 + \sin\theta_{hd}\, \delta\hat{S}_3 \equiv \delta\hat{S}_{\theta_{hd}}, \tag{14}$$

where θ_{hd} is the relative phase between the LO and the mean field. As θ_{hd} is varied, we detect the fluctuations of the Stokes parameter $\hat{S}_{\theta_{hd}}$. Hence, we obtain information about the Stokes parameters simply by measuring both the relative phase θ_{hd} and the quadrature noise of the vacuum mode. This measurement is readily performed by setting the half-wave plate before PBS2 such that the $\hat{a}_y$ mode is sent to the homodyne detector; the mean field, $\hat{a}_x$, goes through the other port of the beam splitter and is detected together with a portion of the LO by a photodiode (see Fig. 2). The phase is determined via the interference signal between LO and $\hat{a}_x$ ($i_\theta \propto \cos\theta_{hd}$). The two signals i_θ and δi_{hd} are sent to an oscilloscope, giving the curves of Fig. 6.

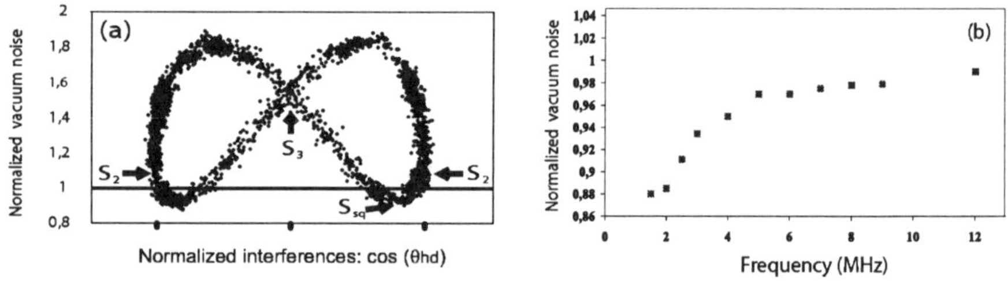

Fig. 6. (a) Measurement of the Stokes operators fluctuations versus the normalized interference signal: $\cos\theta_{hd}$. Here the Stokes operators rotated by an angle $\theta_{sq} = 35° \pm 5°$, in the $\hat{S}_2$-$\hat{S}_3$ plane, is squeezed. (b) Polarization squeezing value versus the frequency. The squeezing is limited by the cavity bandwidth of 5 Mhz.

3.2.4. *Results*

The normalized quadrature noise of $\hat{a}_y$ is plotted as a function of the relative phase between the mean field and the LO in Fig. 6(a). In agreement with

Eq. (6), it can be seen that the $\hat{S}_2$ noise is given by the extreme points $\theta_{hd} = 0, \pm\pi$ and that of $\hat{S}_3$ by the center point $\theta_{hd} = \pi/2$. Here a linear combination of $\hat{S}_2$ and $\hat{S}_3$ is squeezed. This combination depends on many experimental parameters, such as the measurement frequency; for instance at 6 MHz the $\hat{S}_3$ parameter was squeezed.[16]

In Fig. 6(b) we report the squeezing as the frequency is varied: maximum squeezing is measured at low frequencies and the squeezing decreases with increasing frequency due to cavity bandwidth effects. Since we could not measure at lower frequencies we could not observe the added noise arising from the optical pumping process, typically below 1 MHz.

3.3. *Polarization squeezing with optical fibers*

3.3.1. *Nonlinear interaction of light in a glass fiber*

Optical fibers are a natural choice for quantum communication. The interaction of light with fibers (linear and nonlinear) is well understood from many years of research and use in telecommunications. The first photon number squeezing experiment was carried out by Slusher *et al.*[29] in 1985. Squeezing in optical fibers was demonstrated a year later by Shelby *et al.*,[30] who exploited the Kerr nonlinearity in glass fibers with a continuous wave source. This experiment was limited by excess noise in the fiber arising from guided acoustic wave Brillouin scattering (GAWBS).[40] Also in this year, the use of a Mach-Zehnder interferometer exploiting the Kerr effect was proposed[27] and it served as a basis for the ensuing experiments with optical fibers. This was advantageous, as it was easily integrable in experiments attempting to avoid the damaging GAWBS by using ultrashort laser pulses. These experiments employed solitons[22,32,41] as well as zero-dispersion pulses,[42,43] both of which were successful.

The $\chi^{(3)}$ nonlinearity in fibers is weak in comparison to other media, *i.e.* cold atoms, with a value for the second order refractive index of $n_2 \approx 2 \times 10^{-16}$ cm^2/W. Using ultrashort laser pulses (down to 100 fs) and confining the light to a very small core (diameter of several μm), and a long fiber (typically 10s of meters), nonlinearities comparable to those in cold atoms are produced. However the squeezing is not generated in the amplitude (see Fig. 3) and can not be observed directly unless action is taken to rotate the squeezing ellipse in phase space.[44] Homodyning with a phase shifted local oscillator is often impractical for these intense pulses made of 10^8-10^{10} photons each. Different techniques have been implemented to observe this squeezing: phase-shifting cavities,[45] spectral filtering,[46,47] balanced interferometers[32,42,48] as well as asymmetric interferometers.[41,49,50] In a sense polarization squeezing methods provide new tools for the direct observation of squeezing.

3.3.2. *Generation of polarization squeezing*

The squeezing produced in glass fibers can be naturally employed for the generation of polarization squeezing. Both orthogonal axes of a polarization maintaining fiber are used simultaneously, as shown in Fig. 7, to produce two orthogonally polarized Kerr squeezed modes. If both incoming pulses are set to have the same amplitude, they experience the same nonlinearity in the fiber and thus exhibit the same degree of quadrature squeezing ($\Delta^2 \hat{X}_{sq}$) for the same angle θ_{sq} (relative to their mean value) as in Fig. 3(b). As they propagate separately in the fiber (due to the birefringence they overlap only at the output), their fluctuations are moreover uncorrelated: this experimental scheme is well described by the discussion of Sec. 2.3, in which polarization squeezing has been predicted.

The first experimental implementation used an asymmetric Sagnac interferometer[13] (see Fig. 7(a)) that produced two amplitude squeezed modes ($\theta_{sq} = 0$) on both horizontal, x, and vertical, y, axes. Here a zero relative phase shift was set between these modes, denoted $\hat{a}_x$ and $\hat{a}_y$, resulting in a $\hat{S}_2$ polarized beam. Applying the method of Sec. 2.3 to this case or referring to,[13] we find that $\hat{S}_1$ is squeezed with equal magnitude as the amplitude squeezing and $\hat{S}_3$ is antisqueezed.

An improvement of this setup was demonstrated recently.[15] Two Kerr squeezed beams were generated in a single pass scheme (see Fig. 7(b)), thereby avoiding the losses due to imperfect mode matching in an asymmetric Sagnac interferometer. This setup enables polarization squeezing at any input power, in contrast to the previous scheme.

In this simplified method, the amplitude fluctuations of both modes, $\hat{a}_x$ and $\hat{a}_y$, are shot noise limited and the squeezing appears for a rotated quadrature, $\theta_{sq} \neq 0$. For practical reasons related to the measurement scheme, the relative phase shift between $\hat{a}_x$ and $\hat{a}_y$ was set to $\pi/2$, leading to a $\hat{S}_3$ polarized beam. Here, we recover precisely the configuration analyzed in Sec. 2.3 (see Fig. 3(b)) and we know from Eq. 11 that polarization squeezing is visible in a Stokes parameter rotated in the $\hat{S}_1 - \hat{S}_2$ plane, $\hat{S}_{\theta_{sq}}$. This angle corresponds to the rotation of the Kerr squeezed ellipse in phase space and depends on the input power and pulse form, as well as on the fiber type and length.

3.3.3. *Experimental setup*

Two setups have been used in the production of polarization squeezing: an asymmetric Sagnac interferometer with a 93:7 splitting ratio[13] (Fig. 7(a)), and a single pass configuration[15] (Fig. 7(b)). The laser source for both experiments was a passively mode locked Cr^{4+}:YAG laser producing 130 fs pulses at a wavelength of 1500 nm and with a repetition rate of 163 MHz. These pulses were coupled with equal intensity into the two polarization axes of a birefringent fiber (FS-PM-7811 from 3M). The fiber lengths used were 14.2 m in the Sagnac interferometer and 13.3 m for the single pass configuration.

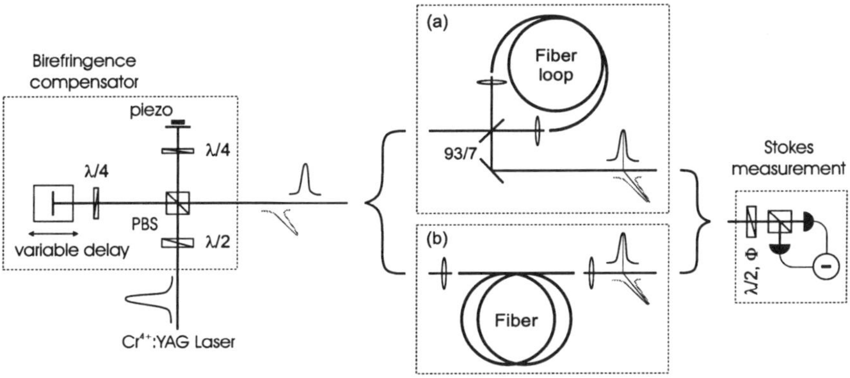

Fig. 7. Schematic of the setup used to produce polarization squeezing using (a) an asymmetric Sagnac fiber loop, and (b) a single pass of an optical fiber. The Stokes parameters of the output were measured using the setups of Fig. 1.

To ensure that the two pulses overlap optimally after the fiber, the propagation distance difference caused by the fiber birefringence was compensated before the fiber. This was achieved using a Michelson-like interferometer which splits horizontal and vertical polarizations, giving them a relative delay. Pre-compensation has the advantage of minimizing losses in the squeezed beams. The relative phase of the pulses was locked using an active feedback system to control a PZT mounted on one of the compensator's end mirrors. The detection setup consisted of a series of waveplates, depending on the parameter measured, a polarization beam splitter, and a pair of balanced AC/DC detectors based on the Epitaxx 500 photodiodes. To measure rotated Stokes parameters in the $\hat{S}_1$-$\hat{S}_2$ plane, we simply use a $\lambda/2$ waveplate, rotated by angle Φ, in front of a polarization beam splitter

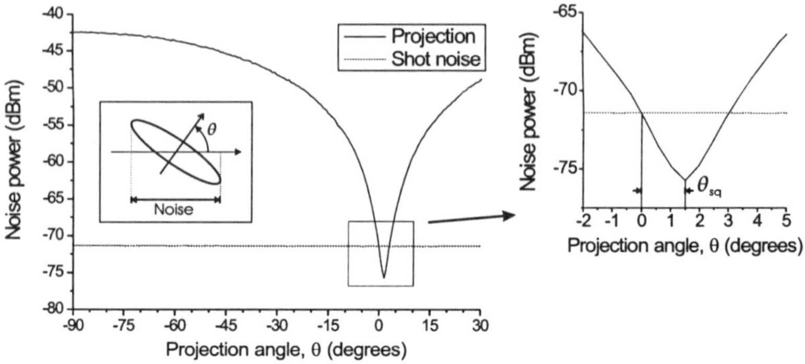

Fig. 8. Noise against phase-space rotation angle for the rotation of the measurement $\lambda/2$ waveplate for a pulse energy of 83.7 pJ using 13.3 m 3M FS-PM-7811 fiber. Inset: Schematic of the projection principle for angle θ. The measurement frequency is 17.5 MHz.

(PBS). The difference of the photocurrents of the two PBS output ports gives a measure of $\hat{S}_\theta$, with $\theta = 4\Phi$:

$$\delta i_- \propto \cos 4\Phi \, \delta\hat{S}_1 + \sin 4\Phi \, \delta\hat{S}_2 = \hat{S}_{4\Phi}. \tag{15}$$

3.3.4. *Results*

The results of the polarization squeezing experiment using two amplitude squeezed beams (Fig. 7(a)) exhibited a maximum noise reduction of 3.4±0.3 dB; in the anti-squeezed Stokes parameter 23.5±0.3 dB of excess noise was measured. These values are similar in magnitude to previous experiments generating amplitude squeezing using asymmetric fiber interferometers[41,49,50] and further details can be found in Ref. 13.

The single pass method of polarization squeezing generation (Fig. 7(b)) improves upon these results. For a total pulse energy of 83.7 pJ (soliton energy 56±4 pJ), Fig. 8 shows the measured noise as the waveplate is rotated. We find a periodic variation between very large noise and squeezing, as expected from the rotation of a squeezed state. Plotted on the x-axis is the projection angle θ, inferred from the waveplate angle Φ. For $\theta = 0$ (corresponding to an $\hat{S}_1$ measurement), we find a noise value equal to the shot noise. Rotation of the state by θ_{sq} makes the squeezing in the system observable by projecting out only the squeezed axis of the uncertainty ellipse. Further rotation brings a rapid increase in noise as the excess phase noise (composed of the anti-squeezing and spurious classical noise) becomes visible.

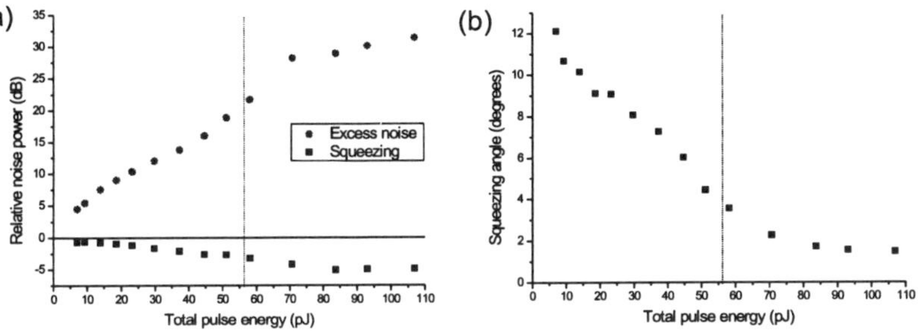

Fig. 9. Results for 13.3 m 3M FS-PM-7811 fiber as a function of total pulse energy: a) the squeezing and excess phase noise and b) the squeezing angle. The energy at which a first order soliton is generated (56±4 pJ) is shown by the dashed line. The measurement frequency is 17.5 MHz

The squeezed and anti-squeezed quadratures and the squeezing angle, θ_{sq}, of this state were investigated as a function of the total pulse energy (Fig. 9). The maximum squeezing observed is 5.1 ± 0.3 dB for an energy of 83.7 pJ. Squeezing saturation is seen at high power, most likely due to the overwhelming excess phase noise which distorts the uncertainty ellipse. The losses of the setup were found to be 20.5%: 4% from the fiber end, 7.8% from optical elements and 10% from the photodiodes. Thus we infer a maximum polarization squeezing of 8.8 ± 0.8 dB. This value agrees with theoretical predictions of Corney and Drummond.[51] Investigating the squeezing angle, θ_{sq}, we find that the rotation of the uncertainty region necessary to observe squeezing decreases with increasing power. This is expected since, despite an increasing anti-squeezing, the amplitude noise of a pure Kerr squeezed beam remains constant. At high power the curve of θ_{sq} against pulse energy tends to a small constant value as the excess noise dramatically increases, which renders difficult a clean projection of the squeezed quadrature.

4. Polarization and Quadrature Entanglement

Entanglement lies at the heart of quantum information and communication science protocols and has therefore drawn a lot of attention for discrete and continuous variables. Since the first experimental demonstration of continuous variable entanglement in 1992,[52] numerous ways of measuring and producing entanglement have been proposed and realized. Since this subject is addressed in many chapters of this book, particularly in the one by N. Korolkova, we focus here on the experimental implementation using cold atoms[18] and optical fibers,[14,53] which have both led to quadrature and polarization entanglement.

We will first discuss some general properties satisfied by bipartite quadrature entanglement and their extension to the polarization variables. From these results we propose an original method to find the maximum quadrature entanglement when no information is known *a priori* about a given two-mode system. We then apply this technique to demonstrate quadrature entanglement with cold atoms. Finally, the optical fiber experiments illustrate how the concept of quadrature entanglement can be extended to polarization entanglement.

4.1. *General properties of continuous variable entanglement*

4.1.1. *Probing and quantifying entanglement*

Let us consider the correlations between two modes $\hat{a}_a$ and $\hat{a}_b$ of the same frequency that satisfy $[\hat{a}_a, \hat{a}_b^\dagger] = \delta_{ab}$ and may correspond to either spatially separated beams or orthogonally polarized modes of the same beam. These correlations may appear between the quadratures that we previously called $\hat{X}_a(\theta)$ and $\hat{Y}_a(\theta)$ (Sec. 2.2). Since $[\hat{X}_a(\theta) + \hat{X}_b(\theta), \hat{Y}_a(\theta) - \hat{Y}_b(\theta)] = 0$, maximal correlations and anticorrelations are allowed between conjugate quadratures.

Among the different techniques to probe entanglement in continuous variables, the so-called "EPR" criterion (formulated by Reid[54] along the lines of the famous Einstein, Poldolsky, Rosen paradox[55]) and the more recent nonseparability criterion[56,57] have been mostly used so far due to their simple physical implementations. The nonseparability criterion reads:

$$\mathcal{I}_{a,b} = \min_{\theta} \frac{1}{2}[\Delta^2(\hat{X}_a - \hat{X}_b)(\theta) + \Delta^2(\hat{Y}_a + \hat{Y}_b)(\theta)] < 2. \tag{16}$$

It gives a sufficient condition for entanglement but also provides, for symmetric Gaussian states, a quantitative measure of entanglement via the "entanglement of formation".[58] As such, it will be used extensively in the following.

Historically, this criterion was derived for the quadrature variables, but it can easily be extended to polarization variables as first suggested by Korolkova *et al.*[9]. Here, we consider the Stokes parameters associated with two spatially separated beams A and B (four modes of the electromagnetic field). For instance, polarization entanglement between $\hat{S}_1^{A,B}$ and $\hat{S}_3^{A,B}$ is achieved if the following criterion is satisfied:

$$\mathcal{I}^S_{A,B} = \Delta^2(\hat{S}^A_1 - \hat{S}^B_1) + \Delta^2(\hat{S}^A_3 + \hat{S}^B_3) < 2(|\langle\hat{S}^A_2\rangle| + |\langle\hat{S}^B_2\rangle|). \qquad (17)$$

A discussion about this criterion is found in the chapter by N. Korolkova. In Sec. 2.2 we emphasized the link between the fluctuations of the "dark" Stokes parameters ($\langle\hat{S}_\perp\rangle = 0$) and the quadrature fluctuations of the orthogonal vacuum mode. When polarization entanglement between "dark" Stokes operators is considered, the criterion in Eq. 17 is therefore equivalent to the usual criterion of Eq. 16.[11,18] This link provides a clear physical understanding of polarization entanglement as a direct consequence of quadrature entanglement between orthogonal vacuum modes. This is well illustrated by the optical fiber experiment in Sec. 4.3.2.

4.1.2. *Finding maximum entanglement in a two mode system*

Before going on to the experimental descriptions, we would like to address some general properties of a two-mode system. As this discussion is found in detail elsewhere,[17] we will focus here on a more specific question: *how can one find and measure the optimal entanglement generated by a two-mode system?*

The method is of particular interest since it can be applied to various schemes in which correlations between orthogonally polarized modes may be observed (as in cold atoms (Sec. 4.2), polarization maintaining fibers[15] (Sec. 4.3), or non-degenerate optical parametric amplifiers[52,59]), but also when entanglement is created between spatially separated modes by interfering nonclassical beams on a beam splitter.[60,53,61,62]

Let us consider the latter example, which constitutes the most common method to produce entanglement: two squeezed beam (denoted $\hat{a}_u$ and $\hat{a}_v$) are produced independently – for instance with two OPOs,[60] OPAs [61], or by using the Kerr nonlinearity of optical fibers [53] – and are then combined on a 50/50 beam splitter. The relative phase shift between the beams, ϕ, is chosen to mix the squeezed and anti-squeezed quadratures. This situation is illustrated in Fig. 10(a) where the two inputs are squeezed for the same quadratures, $\hat{Y}_{u,v}$, and $\phi = \pi/2$. The correlations between the output modes $\hat{a}^*_a$ and $\hat{a}^*_b$ are found to be:

$$\left\{\begin{array}{l} \hat{a}^*_a = (\hat{a}_u \; - \; i\hat{a}_v)/\sqrt{2} \\ \hat{a}^*_b = (\hat{a}_u \; + \; i\hat{a}_v)/\sqrt{2} \end{array}\right\} \Leftrightarrow \left\{\begin{array}{l} \frac{1}{2}\Delta^2(\hat{X}_{a^*} - \hat{X}_{b^*}) = \Delta^2\hat{X}_v \; < 1 \\ \frac{1}{2}\Delta^2(\hat{Y}_{a^*} \; + \; \hat{Y}_{b^*}) = \Delta^2\hat{Y}_u \; < 1 \end{array}\right\}, \qquad (18)$$

leading to $\mathcal{I}_{a^*,b^*} < 2$. One can show that these modes are maximally entangled in the sense that $\mathcal{I}_{a^*,b^*}$ is minimal, *i.e.* $\mathcal{I}_{a^*,b^*} = \min_{a',b'} \mathcal{I}_{a',b'}$ where $\hat{a}'_a$

and $\hat{a}'_b$ refers to modes that can be obtained from $\hat{a}_u$ and $\hat{a}_v$ via any beam splitter transformation.

So far, spatially separated modes have been considered, but one should note that these results also hold in the case where correlations appear between orthogonally polarized modes. If one is able to find two uncorrelated squeezed modes which are orthogonally polarized, *i.e.* horizontally and vertically, then the maximally correlated modes are obtained via the same transformations in Eq. 18, and correspond to circularly polarized modes. They can be easily separated using a quarter wave plate and a polarizing beam splitter as depicted in Fig. 10(b). As a general fact, any beam splitter transformation can be achieved with polarization optics, meaning that the results presented here are valid for both spatially separated or orthogonally polarized modes.

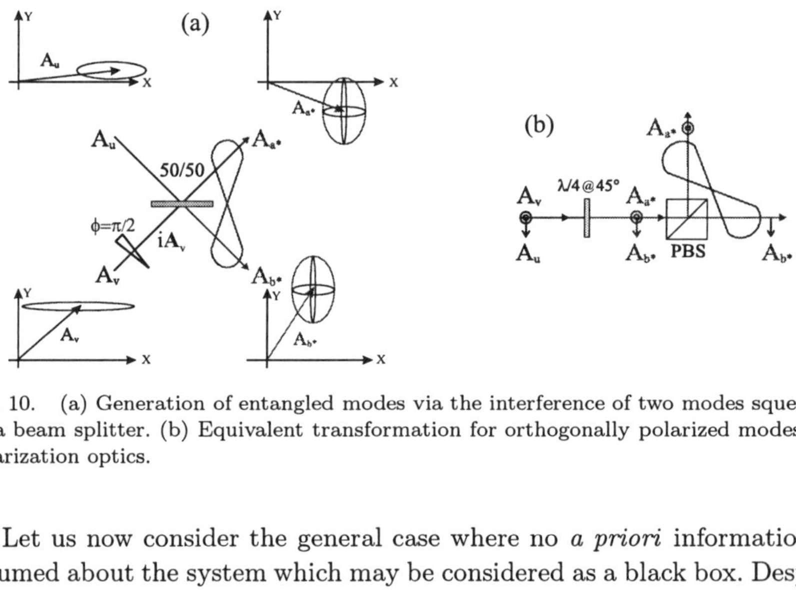

Fig. 10. (a) Generation of entangled modes via the interference of two modes squeezed on a beam splitter. (b) Equivalent transformation for orthogonally polarized modes via polarization optics.

Let us now consider the general case where no *a priori* information is assumed about the system which may be considered as a black box. Despite the apparent complexity of this problem, we begin by asking the question: *What is the optimal strategy to find the maximally correlated modes, denoted by* $\hat{a}^*_a$ *and* $\hat{a}^*_b$*, which satisfy* $\mathcal{I}_{a^*,b^*} = \min_{a',b'} \mathcal{I}_{a',b'}$*?*

Since the two input beams considered above may have been previously correlated via a unknown interaction, it seems that no general answer can be given. However, we have shown in Ref. 17 that there always exist 1) two "uncorrelated" modes $\hat{a}_u$ and $\hat{a}_v$ that satisfy $\langle \delta\hat{a}_u \delta\hat{a}_v \rangle = 0$ and which 2) exhibit minimum quadrature noise in the same quadrature, for instance

$\hat{Y}_{u,v}$ as in Fig. 10(a). These properties are of fundamental importance since, as far as the nonseparability criterion is concerned, *this system behaves as if the modes $\hat{a}_u$ and $\hat{a}_v$ are completely independent*[17,a]. It follows that we can always think of the quantum properties of two Gaussian modes, characterized here by $\mathcal{I}_{a,b}$ and their quadrature noises $\hat{X}_{u,v}(\theta)$, as being created by the mixing of two independent beams on a beam splitter. Considering the example of mixing two squeezed beams discussed above, the maximally entangled modes $\hat{a}_a^*$ and $\hat{a}_b^*$ are:

$$\hat{a}_a^* = (\hat{a}_u + i\hat{a}_v)/\sqrt{2}, \quad \hat{a}_b^* = (\hat{a}_u - i\hat{a}_v)/\sqrt{2}, \tag{19}$$

and their correlation, $\mathcal{I}_{a^*,b^*}$, is given by:

$$\mathcal{I}_{a^*,b^*} = \min_{a,b} \mathcal{I}_{a,b} = \Delta^2 \hat{X}_u^{min} + \Delta^2 \hat{X}_v^{min}. \tag{20}$$

Consequently, *the ability of the system to produce entanglement is directly related to the sum of the minimum noises of the u and v modes.* We find in particular the well-known results: at least one squeezed mode is necessary to obtain quadrature entanglement, and also that two maximally squeezed modes produce maximal entanglement.

4.1.3. *Application and representation in the Poincaré sphere*

Once the "uncorrelated" modes u and v have been found (this can be easily done experimentally, see Sec. 4.2.2), the properties sketched in the previous paragraph provide a general framework for finding the system's quantum properties. This method is of particular interest for a wide range of complex systems where correlations between two modes are produced simultaneously. It has already been successfully applied to the cold atoms experiment discussed here but also to the case where a birefringent plate has been inserted inside an optical parametric oscillator.[59]

Since experiments dealing with two Gaussian modes behave in a similar manner, we propose, as a unifying picture, to represent the quantum properties of a system in the classical Poincaré sphere. Given the special role played by the "uncorrelated" basis, we define the classical Stokes parameters from the u, v modes:

$$S_0' = A_u^* A_u + A_v^* A_v, \qquad S_1' = A_u^* A_u - A_v^* A_v,$$
$$S_2' = A_u^* A_v + A_v^* A_u, \qquad S_3' = i(A_v^* A_u - A_u^* A_v).$$

[a]However, they are not *stricto sensu* independent since $\langle \delta\hat{a}_u^\dagger \delta\hat{a}_v \rangle$ can be non zero.

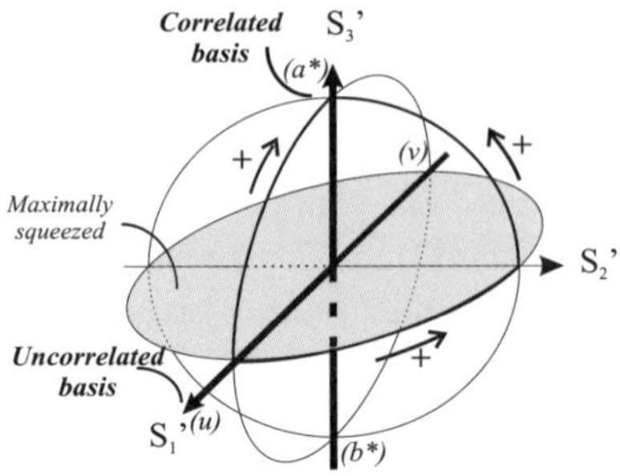

Fig. 11. Representation of the quantum properties (squeezing and correlations) in the Poincaré sphere defined from the $\hat{a}_u$ and $\hat{a}_v$ modes. The arrows with sign "+" correspond to increasing correlations, which are maximal at the poles (corresponding to the modes $\hat{a}_a^*$ and $\hat{a}_b^*$). Many properties that have not been discussed in the text can be reported in the sphere, as, for instance, the fact that the squeezing is maximal for the polarization basis located in the equatorial plane.

For a particular polarization basis (or an equivalent beam splitter transformation), represented by a diagonal axis in the Poincaré sphere, we can estimate different properties such as the squeezing and the correlation $\mathcal{I}_{a,b}$. The evolution of these parameters as we move on the sphere (*i.e.* as we change the way the modes $\hat{a}_u$ and $\hat{a}_v$ are combined) is shown in Fig. 11. We recall that maximal correlations are found for the modes defined by Eq. 19, which are at the poles of the Poincaré sphere. It is also interesting to note that maximal squeezing is observed in the equatorial plane.[17]

4.2. *Entanglement generation with cold atoms*

4.2.1. *Principle*

Coming back to the system where an x-polarized beam interacts with a cloud of cold cesium in an optical cavity (see Fig. 2(b)), we can easily find the maximally correlated modes. For high frequencies, the answer is straightforward since the system behaves as if the two transitions were independent. The modes $\sigma_\pm$ are independent and squeezed for the same quadrature (from Eq. 9 $\langle\hat{a}_{\sigma_+}\rangle = -\langle\hat{a}_{\sigma_-}\rangle$). Thus, they naturally satisfy the conditions to be the particular modes introduced in the previous section: $\hat{a}_u \equiv \hat{a}_{\sigma_+}$ and $\hat{a}_v \equiv \hat{a}_{\sigma_-}$. The maximally correlated modes are then obtained directly using the transformation in Eq. 19 and they can be shown to be the modes polarized at $\pm 45°$ (see Eq. 19):

$$\hat{a}_a^* = (\hat{a}_{\sigma_+} + i\hat{a}_{\sigma_-})/\sqrt{2} \equiv -e^{-i\frac{\pi}{4}}\hat{a}_{+45}$$
$$\hat{a}_b^* = (\hat{a}_{\sigma_+} - i\hat{a}_{\sigma_-})/\sqrt{2} \equiv e^{-i\frac{\pi}{4}}\hat{a}_{-45}.$$

Furthermore, since $\hat{a}_{\sigma_\pm}$ are completely independent, one can show that the modes $\hat{a}_x$ and $\hat{a}_y$ exhibit the same squeezing, as should all modes linearly polarized relative to the $(\hat{a}_{\sigma_-}, \hat{a}_{\sigma_+})$ basis, and are also uncorrelated. These properties are summarized in Fig. 12, where the Poincaré sphere is defined as usual ($\hat{a}_x$ and $\hat{a}_y$ being the reference modes as in Eq. 2). The vector along $\hat{S}_1$, representing the mean polarization state, is also shown. We deduce that the beam is polarization squeezed as the orthogonal vacuum mode (aligned along $-\hat{S}_1$) lies in the "squeezed plane" shaded in gray (see Sec. 2.2 and 3.2.2).

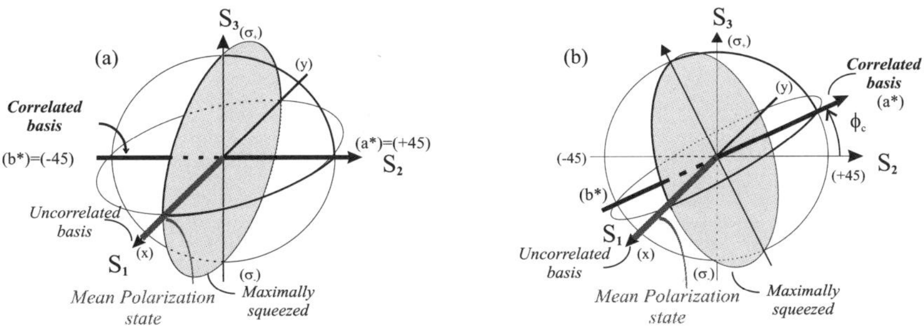

Fig. 12. Quantum properties generated by the interaction of the atoms and a linearly polarized light. a) High frequency noise analysis, where the modes located in the $\hat{S}_1$-$\hat{S}_3$ plane are uncorrelated and squeezed, whereas entanglement in found for the modes polarized at 45°. b) Low frequency analysis where correlations appear between the $\hat{a}_{\sigma_\pm}$ component of light.

If we now consider the low frequency noise, the situation is much more complex since the modes $\hat{a}_{\sigma_\pm}$ become correlated via the optical pumping process. One could then find the new $(\hat{a}_u, \hat{a}_v)$ basis using the general formulation in Ref. 17. However, the symmetry properties between the circular components of the light when the polarization is linear automatically lead to $\langle \delta\hat{a}_x \delta\hat{a}_y \rangle = 0$. The x and y modes are then always uncorrelated, regardless of frequency. To fully characterize the u and v modes, one has to look at their minimum quadrature noises.

At high frequency, it can be shown that $\hat{a}_{x,y}$ are squeezed for orthogonal quadratures; a $\pi/2$ phase shift must be introduced to retrieve squeezing in the same quadrature and the correct "uncorrelated" basis can be defined as: $\hat{a}_{u'} \equiv \hat{a}_x$ and $\hat{a}_{v'} \equiv i\hat{a}_y$. This approach naturally leads to the same conclusions as discussed above (where the u, v modes were defined using the $\hat{a}_{\sigma\pm}$ modes).

However, the behavior is different at low frequency. Indeed, the correlations between $\hat{a}_{\sigma+}$ and $\hat{a}_{\sigma-}$ introduce a rotation between the minimum quadrature noise of the x and y modes: they are no longer squeezed for orthogonal quadratures. To counteract this effect the "uncorrelated" basis is defined in that case as $\hat{a}_{u'} \equiv \hat{a}_x$ and $\hat{a}_{v'} \equiv ie^{i\phi_{cor}}\hat{a}_y$. Using Eq. 19, the maximally entangled modes are directly:

$$\hat{a}_a^* \equiv (\hat{a}_u + i\hat{a}_v)/\sqrt{2} = (\hat{a}_x \; - \; e^{i\phi_{cor}}\hat{a}_y)/\sqrt{2},$$
$$\hat{a}_b^* \equiv (\hat{a}_u - i\hat{a}_v)/\sqrt{2} = (\hat{a}_x \; + \; e^{i\phi_{cor}}\hat{a}_y)/\sqrt{2}.$$

This situation is depicted Fig. 12(b), where the correlation axis is rotated by an angle ϕ_{cor}. From Eq. 20 it follows that the maximum entanglement is still given by the minimum noise of $\hat{a}_x$ and $\hat{a}_y$:

$$\mathcal{I}_{a^*,b^*} = \Delta^2\hat{X}_x^{min} + \Delta^2\hat{X}_y^{min}. \tag{21}$$

4.2.2. *Measurement and results*

As explained in the previous section, for high frequencies the $\hat{a}_x$ and $\hat{a}_y$ modes are squeezed for orthogonal quadratures. One can set $\hat{a}_u = \hat{a}_x$ and $\hat{a}_v = i\hat{a}_y$, which are now squeezed for the same quadrature. Then the maximally entangled modes are the $\pm 45°$ modes relative to the x, y basis. This gives us the relevant quantity, $\mathcal{I}_{+45,-45}(\theta)$, which is to be measured. Using $\hat{a}_{\pm 45} = (\hat{a}_x \pm \hat{a}_y)/\sqrt{2}$, the inseparability criterion for the $\pm 45°$ modes can be expressed directly in terms of the x, y modes variances with $\hat{X}_u(\theta) = \hat{X}_x(\theta)$ and $\hat{X}_v(\theta) = \hat{Y}_y(\theta)$:

$$\mathcal{I}_{+45,-45}(\theta) = \Delta^2\hat{X}_x(\theta) + \Delta^2\hat{Y}_y(\theta). \tag{22}$$

When θ corresponds to the angle θ_{sq} of the squeezed quadrature of $\hat{a}_x$, both variances are below unity, and $\mathcal{I}_{+45,-45}(\theta_{sq}) < 2$.

In order to experimentally check the inseparability (Eq. 22), we need to simultaneously measure the fluctuations of $\hat{a}_x$ and $i\hat{a}_y$. At the output of the cavity, we insert a quarter-wave plate that rotates the noise ellipsoid of the vacuum mode $\hat{a}_y$ by $\pi/2$, the beam is mixed on a beam splitter with a local oscillator (LO), and the two resulting beams are sent into two balanced homodyne detectors [Fig. 2(b)] which allow for the simultaneous measurement of $\hat{X}_x(\theta)$ and $\hat{Y}_y(\theta)$. The sum of these two signals gives $\mathcal{I}_{+45,-45}(\theta)$, a typical measurement being plotted in Fig. 13(a). It minimal value is about 1.92 demonstrating the existence of entanglement.

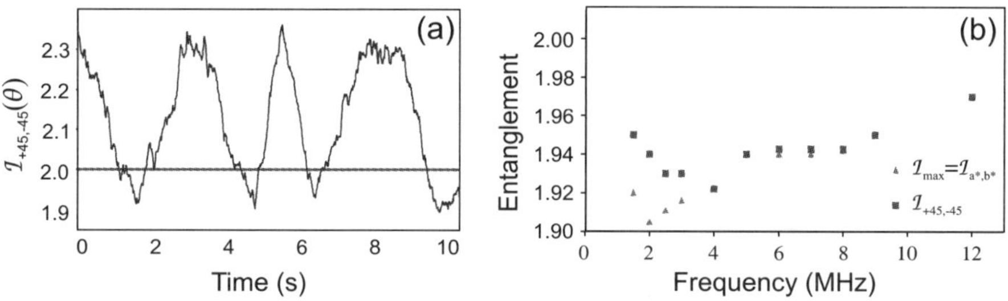

Fig. 13. (a) Direct measurement of $\mathcal{I}_{+45,-45}(\theta)$ at a frequency of 5 MHz, when the relative phase θ between the LO and the mean field mode is varied in time. (b) Entanglement for the ±45° polarized modes $\mathcal{I}_{+45,-45}$ compared to the maximal entanglement $\mathcal{I}_{max}$ versus frequency .

Note that, unlike the usual entanglement detection schemes involving two successive measurements, the present method is based on a single measurement. Provided that a proper set of wave plates are placed in front of the beam splitter shown in Fig. 2, the correlations $\mathcal{I}_{a,b}(\theta)$ between any pair of orthogonal polarization modes can be measured.[17] For instance, we checked, consistently with the general method described in Sec. 4.1.2, that the modes $\hat{a}_x$ and $i\hat{a}_y$ are indeed "uncorrelated" ($\langle \delta\hat{a}_x \delta\hat{a}_y \rangle = 0$). As shown in Ref. 17, the quantity $\mathcal{I}_{x,y}(\theta)$ should not depend on θ, leading to a clear experimental signature. This property has been verified,[18] unambiguously proving that maximal entanglement of the system is given by $\Delta^2 \hat{X}_x^{min} + \Delta^2 \hat{X}_y^{min}$.

Finally we compare in Fig. 13(b), the maximal entanglement to the entanglement of the modes ±45° as a function of the frequency. This confirms the results discussed above: for low frequencies the maximally entangled modes are no longer the ±45° modes.

4.3. *Entanglement generation from fibers*

4.3.1. *Direct generation of quadrature entanglement*

Following the ideas introduced in Sec. 4.1, we analyze here both optical fiber setups in Fig. 7. They are both two mode systems and therefore we may ask if quadrature entanglement is generated and for which modes it is maximal.

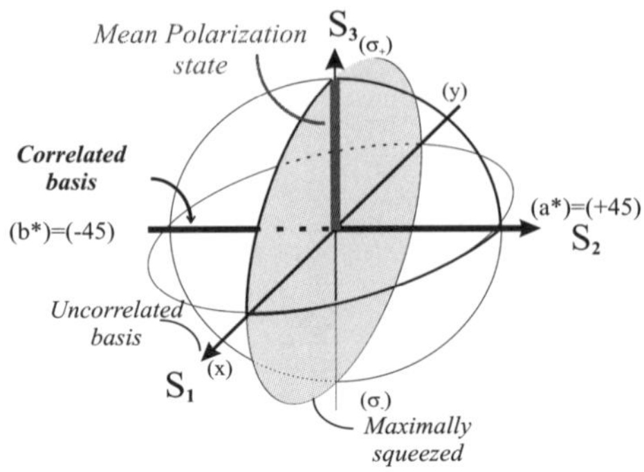

Fig. 14. Summary of the quantum properties of the light for both fiber setups depicted in Fig. 7. In both cases, if the light is circularly polarized, the modes polarized at ±45° are quadrature entangled.

Refering to Sec. 3.3.2, the answer is straightforward and is moreover the same for both setups. The two modes $\hat{a}_x$ and $\hat{a}_y$ are uncorrelated and are therefore good candidates to be the u and v modes; however, they should be phase-shifted appropriately so as to exhibit squeezing for the same quadrature. Since both modes experience the same nonlinearity, $\hat{a}_x$ and $\hat{a}_y$ are naturally squeezed for the same quadrature angle θ_{sq}, relative to their mean values as in Fig. 3(b). If we consider the situation where the birefringent compensator (in front of the fiber's input) is set to assure a $\pi/2$ relative phase shift between the horizontal and vertical axes, the output beam is circularly polarized (the mean Stokes parameter is along $\hat{S}_3$) and $\hat{a}_x$ and $\hat{a}_y$ are squeezed for *orthogonal quadratures*. Then, as in the cold atoms experiment, the modes defined by $\hat{a}_u = \hat{a}_x$, $\hat{a}_v = i\hat{a}_y$ satisfy the required conditions. Applying directly the general results given by Eq. 19 and Eq. 20, it follows that the maximally entangled modes are the modes polarized at ±45°, and their entanglement is given by:

$$\mathcal{I}_{+45,-45} = \Delta^2 \hat{X}_x^{min} + \Delta^2 \hat{X}_y^{min} = 2\Delta^2 \hat{X}_{sq}, \tag{23}$$

where $\Delta^2 \hat{X}_{sq}$ refers to the squeezing of both modes. Those properties are reported in the Poincaré sphere in Fig. 14.

It turns out that the only formal difference between these two experimental schemes lies in the relative squeezed quadrature angle, θ_{sq}, shown in Fig. 3(b): it is zero for an amplitude squeezed beam and rotated for the single pass scheme. This is of particular importance, since correlations will not appear for the same set of conjugate quadratures.

For the asymmetric Sagnac interferometer setup, amplitude squeezed modes are generated ($\theta_{sq} = 0$): the entanglement therefore appears between the phase and amplitude quadratures.[63] Despite the fact that they require involved detection schemes,[64,65] the amplitude and phase correlations can be measured: an entanglement of 0.80 has been measured in,[53,b] resulting from a squeezing of $4dB$ ($\Delta^2\hat{X}_{sq} \simeq 0.40$).

Let us now consider the single pass system of Fig. 7(b) where squeezing as high as 8.8 dB (corrected for losses) has been measured; we infer an impressive value for the quadrature entanglement: $\mathcal{I}_{+45,-45} \simeq 0.23$. However, this entanglement can not be measured as in Ref. 53, since the squeezing angle θ_{sq} is non-zero. A complete measurement would require homodyne detection techniques as described in Sec. 4.2.2, which are in this case technically impossible due to the brightness of the field.

4.3.2. *Polarization entanglement: method and results*

In Sec. 4.1.1, we underlined the link between quadrature entanglement and polarization entanglement. Building upon this discussion, it becomes clear that two configurations will lead to polarization entanglement: 1) overlapping two quadrature entangled weak beams with two strong local oscillators (that provide the mean polarizations) as in[11,18] or 2) mixing two polarization squeezed beams, in the same manner as quadrature entanglement is generated by mixing two squeezed beams.[9,14]

This last method is demonstrated using an efficient setup which requires minimal resources - one polarization squeezer, the output of which was split on a 50/50 beam splitter, seen in Fig. 15(c). The polarization squeezed beam was produced with the asymmetric Sagnac interferometer as described in Sec. 3.3.2 and the relative phase shift between the horizontal and vertical axis here was set to zero: the beam is $\hat{S}_2$ polarized ($\langle \hat{a}_{+45}\rangle = \alpha$ and $\langle \hat{a}_{-45}\rangle = 0$) whereas squeezing appears in $\hat{S}_1$ and anti-squeezing in $\hat{S}_3$. These conjugate variables are linked to the amplitude ($\Delta^2\hat{X}_{sq} < 1$) and phase ($\Delta^2\hat{Y}_{anti-sq} > 1$) quadratures of the squeezed orthogonal vacuum mode $\hat{a}_{-45}$.

A detailed theoretical explanation of the generation of polarization entanglement is derived in the chapter by N. Korolkova in this book as well as in Ref. 9, and we sketch here the main ideas. The beam splitter transfor-

[b]Spatially separated modes were considered; therefore, the modes "polarized at ±45°" must be understood as the ones obtained by mixing the $\hat{a}_x$ and $\hat{a}_y$ modes on a 50/50 beam splitter.

mation results in two independent transformations for the two orthogonally polarized $\hat{a}_{+45}$ (mean field) and $\hat{a}_{-45}$ (orthogonal vacuum mode). For the mean field this causes the output beams A and B to both be polarized along $\hat{S}_2$. For the orthogonal mode, the beam splitter mixes a vacuum squeezed mode with the vacuum input, resulting in entanglement as in the transformation of Eq. 18. This quadrature entanglement is given by $1+\Delta^2\hat{X}_{sq}$ and produces polarization entanglement between $\hat{S}_1$ and $\hat{S}_3$.

The correlations are shown in Fig. 15, together with the respective noise of the different Stokes parameters. As expected from Ref. 14, non-classical correlations are found in $\hat{S}_1$, whereas the $\hat{S}_3$ anticorrelations lie at the shot noise level. The non-separability criterion of Eq. 17 is then:

$$\frac{\Delta^2(\hat{S}_{1,A}-\hat{S}_{1,B})+\Delta^2(\hat{S}_{3,A}+\hat{S}_{3,B})}{\langle\hat{S}_{2,A}\rangle+\langle\hat{S}_{2,B}\rangle} = \Delta^2\hat{X}_{sq} + 1 = 1.52 < 2, \quad (24)$$

proving that a correlated non-separable quantum state in the Stokes variables has been generated. This entanglement could be improved using two polarization squeezed beams, however, with greater experimental cost. Further, the new efficient single pass polarization squeezer (see Fig. 7(b)) could also be implemented, the difference being that the correlations will appear between $\hat{S}_{sq}$ and $\hat{S}_{anti-sq}$, which should be rotated with respect to $\hat{S}_1$ and $\hat{S}_3$.

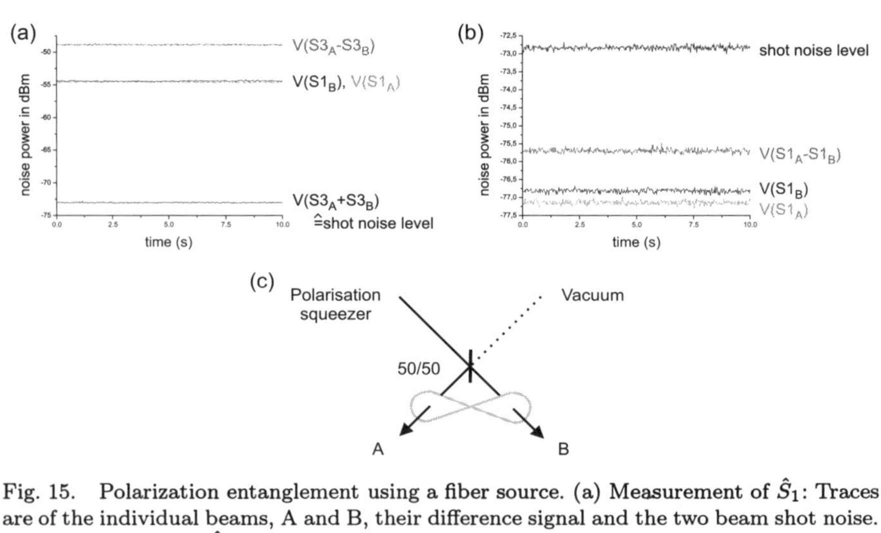

Fig. 15. Polarization entanglement using a fiber source. (a) Measurement of $\hat{S}_1$: Traces are of the individual beams, A and B, their difference signal and the two beam shot noise. (b) Measurement of $\hat{S}_3$: As in a) but with the sum signal of A and B. (c) Schematic of the setup used for resource efficient generation of polarization entanglement.

5. Conclusions and Outlook

We have presented a theoretical discussion about polarization squeezing and continuous variable entanglement, as well as their experimental realizations with cold atoms and optical fibers. Comparisons have been drawn between polarization and quadrature fluctuations. A general method to find the maximum entanglement of any given system, not only those using the optical Kerr interaction, was described and this served as a framework for the interpretation of the results for both setups.

Although polarization squeezing and entanglement can be interpreted in terms of vacuum squeezing and entanglement, it nonetheless provides a powerful new tool in the context of quantum information. In particular, we demonstrate a very simple and efficient way to produce polarization squeezing with optical fiber, leading to 5.1 dB of noise reduction. This can be further improved and could serve as a basic resource for polarization entanglement generation.

Research in the field of quantum polarization states promises to be fruitful, as the understanding of nonclassical polarization beams is now well-developed. The sources presented in this chapter have been characterized and are ready to be further optimized and to be implemented in quantum communication protocols, *e.g.* teleportation, quantum key distribution and cloning. The advantage of direct detection of the Stokes variables as well as their implications in atom-light interactions will continue to make the polarization state of light an exciting field of research in quantum science.

References

1. G.G. Stokes, *On the composition an resolution of streams of polarized light from different sources*, Trans. Cambridge Phil. Soc. **9**, 399 (1852).
2. M. Born and E. Wolf, *Principles in optics*, 7th ed., Cambridge University Press, Cambridge 1999.
3. P. Grangier, R.E. Slusher, B. Yurke, and A. LaPorta, *Squeezed-light-enhanced polarization interferometer*, Phys. Rev. Lett. **59**, 2153 (1987).
4. A.S. Chirkin, A.A. Orlov, and D.Yu. Paraschuk, *Quantum theory of two-mode interactions in optically anisotropic media with cubic nonlinearities: Generation of quadrature- and polarization-squeezed light*, Kvantovay Elektronika **20**, 999 (1993), Quantum Elec. **23**, 870-874 (1993).
5. N.V. Korolkova and A.S. Chirkin, *Formation and conversion of the polarization-squeezed light*, J. Mod. Opt. **43**, 869 (1996).
6. A. Luis and L.L. Sánchez-Soto, *Quantum phase difference, phase measurements and Stokes operators*, Prog. Opt. **41**, 421 (2000).
7. J.L. Sørensen, J. Hald, and E.S. Polzik, *Quantum noise of an atomic spin polarization measurement*, Phys. Rev. Lett **80**, 3487 (1998).

8. B. Julsgaard, J. Sherson, J.I. Cirac, J. Fiuršášek, and E.S. Polzik, *Experimental demonstration of quantum memory for light*, Nature **432**, 482 (2004).
9. N. Korolkova, G. Leuchs, R. Loudon, T.C. Ralph, and C. Silberhorn, *Polarization squeezing and continous-variable polarization entanglement*, Phys. Rev. A **65**, 052306 (2002).
10. W.P. Bowen, R. Schnabel, H.A. Bachor, and P.K. Lam, *Polarization squeezing of continuous variable Stokes parameters*, Phys. Rev. Lett. **88**, 093601 (2002).
11. W.P. Bowen, N. Treps, R. Schnabel, and P.K. Lam, *Experimental demonstration of continuous variable polarization entanglement*, Phys. Rev. Lett. **89**, 253601 (2002).
12. U.L. Andersen and P. Buchhave, *Polarization squeezing and entanglement produced by a frequency doubler*, J. Opt. B **5**, 486 (2003).
13. J. Heersink, T. Gaber, S. Lorenz, O. Glöckl, N. Korolkova, and G. Leuchs, *Polarization squeezing of intense pulses with a fiber-optic Sagnac interferometer*, Phys. Rev. A **68**, 013815 (2003).
14. O. Glöckl, J. Heersink, N. Korolkova, G. Leuchs, and S. Lorenz, *A pulsed source of continuous variable polarization entanglement*, J. Opt. B **5**, S492 (2003).
15. J. Heersink, V. Josse, G. Leuchs, and U.L. Andersen, *Efficient polarization squeezing in optical fibers*, Opt. Lett. **30**, No. 10 (2005).
16. V. Josse, A. Dantan, L. Vernac, A. Bramati, M. Pinard, and E. Giacobino, *Polarization squeezing with cold atoms*, Phys. Rev. Lett. **91**, 103601 (2003).
17. V. Josse, A. Dantan, A. Bramati, and E. Giacobino, *Entanglement and squeezing in a two-mode system: theory and experiment*, J. Opt. B **6**, 532 (2004).
18. V. Josse, A. Dantan, A. Bramati, E. Giacobino, *Continous variable entanglement using cold atoms*, Phys. Rev. Lett. **92**, 123601 (2004).
19. G.S. Agarwal and S. Chaturvedi, *Scheme to measure quantum Stokes parameters and their fluctuations and correlation*, J. Mod. Opt. **50**, 711 (2003).
20. N. Korolkova and R. Loudon, *Nonseparability and squeezing of continuous polarizaiton variables*, Phys. Rev. A **71**, 032343 (2005).
21. A. Dolińska, B.C. Buchler, W.P. Bowen, T.C. Ralph, P.K. Lam, *Teleportation of continuous-variable polarizaiton states*, Phys. Rev. A **68**, 052308 (2003).
22. M. Margalit, C.X. Yu, E.P. Ippen, and H.A. Haus, *Cross phase modulation squeezing in optical fibers*, Opt. Ex. **2**, 72 (1998).
23. H.P. Yuen and J.H. Shapiro, *Generation and detection of two-photon coherent states in degenerate four-wave mixing*, Opt. Lett. **4**, 334 (1979).
24. M.D. Reid and D.F. Walls, *Quantum statistics of degenerate four-wave mixing*, Opt. Comm. **50**, 406 (1984).
25. R.S. Bondurant, P. Kumar, J.H. Shapiro, M. Maeda, *Degenerate four-wave mixing as a possible source of squeezed-state light*, Phys. Rev. A **30**, 343 (1984).
26. P. Kumar and J.H. Shapiro, *Squeezed-state generation via forward degenerate four-wave mixing*, Phys. Rev. A **30**, 1568 (1984).

27. M. Kitagawa and Y. Yamamoto, *Number-phase minimum-uncertainty state with reduced number uncertainty in a Kerr nonlinear interferometer*, Phys. Rev. A **34**, 3974 (1986).
28. S. Reynaud, C. Fabre, E. Giacobino, and A. Heidmann, *Photon noise reduction by passive optical bistable systems*, Phys. Rev. A **40**, 1440 (1989).
29. R.E. Slusher, L.W. Hollberg, B. Yurke, J.C. Mertz, and J.F. Valley, *Observation of squeezed states generated by four wave mixing in an optical cavity*, Phys. Rev. Lett. **55**, 2409 (1985).
30. R.M. Shelby, M.D. Levenson, R.G. DeVoe, and D.F. Walls, *Broad-band parametric deamplification of quantum noise in an optical fiber*, Phys. Rev. Lett. **57**, 691 (1986).
31. M.G. Raizen, L.A. Orozco, M. Xiao, T.L. Boyd, and H.J. Kimble, *Squeezed-state generation by the normal modes of a coupled system*, Phys. Rev. Lett **59**, 198 (1987).
32. M. Rosenbluh and R.M. Shelby, *Squeezed optical solitons*, Phys. Rev. Lett. **66**, 153 (1991).
33. A. Lambrecht, E. Giacobino, and J.M. Courty, *Optical nonlinear dynamics with cold atoms in a cavity*, Opt. Comm. **115**, 199 (1995).
34. M.D. Reid, *Quantum theory of optical bistability without adiabatic elimination*, Phys. Rev. A **37**, 4792 (1988).
35. C. Cohen-Tannoudji, J. Dupont-Roc, and G. Grynberg *Atomphoton interactions: basic processes and applications* , John Wiley & Sons, Inc., New York, NY, 1992.
36. L. Hilico, C. Fabre, S. Reynaud, and E. Giacobino, *Linear input-output method for quantum fluctuations in optical bistability with two-level atoms*, Phys. Rev. A **46**, 4397 (1992).
37. A. Lambrecht, T. Coudreau, A.M. Steimberg, and E. Giacobino, *Squeezing with cold atoms*, Europhys. Lett. **36**, 93 (1996).
38. V. Josse, A. Dantan, A. Bramati, M. Pinard, and E. Giacobino, *Polarization squeezing in a four-level system*, J. Opt. B **5**, S513 (2003).
39. J. Ries, B. Berzger, and A.I. Lvovsky, *Experimental vacuum squeezing in rubidium vapor via self rotation*, Phys. Rev. A **68**, 02581 (2003).
40. R.M. Shelby, M.D. Levenson, P.W. Bayer, *Resolved forward Brillouin scattering in optical fibers*, Phys. Rev. Lett. **54**, 939 (1985).
41. S. Schmitt, J. Ficker, M. Wolff, F. König, A. Sizmann, G. Leuchs, *Photon-number squeezed solitons from an asymmetric fiber-optic Sagnac interferometer*, Phys. Rev. Lett. **81**, 2446 (1998).
42. K. Bergman and H.A. Haus, *Squeezing in fibers with optical pulses*, Opt. Lett. **16**, 663 (1991).
43. K. Bergman, H.A. Haus, M. Shirasaki, *Sub-shot-noise measurement with fiber-squeezed optical pulses*, Opt. Lett. **18**, 643 (1993).
44. A. Sizmann and G. Leuchs *The optical Kerr effect and quantum optics in fibers* Prog. Opt. **39**, 373 (1999).
45. M.D. Levenson, R.M. Shelby, S.H. Perlmutter, *Squeezing of classical noise by nondegenerate four-wave mixing in an optical fiber*, Opt. Lett. **10**, 514 (1985).

46. S.R. Friberg, S. Machida, M.J. Werner, A. Levanon, and T. Mukai, *Observation of optical soliton photon-number squeezing*, Phys. Rev. Lett. **77**, 3775 (1996).
47. S. Spälter, M. Burk, U. Strößner, A. Sizmann, and G. Leuchs, *Propagation of quantum properties of sub-picosecond solitons in a fiber*, Opt. Ex. **2**, 77 (1998).
48. C.X. Yu, H.A. Haus, and E.P. Ippen, *Soliton squeezing at the gigahertz rate in a Sagnac loop*, Opt. Lett. **26**, 669 (2001).
49. D. Krylov and K. Bergman, *Amplitude-squeezed solitons from an asymmetric fiber interferometer*, Opt. Lett. **23**, 1390 (1998).
50. M. Fiorentino, J.E. Sharping, P. Kumar, D. Levandovsky, and M. Vasilyev, *Soliton squeezing in a Mach-Zehnder fiber interferometer*, Phys. Rev. A **64**, 031801 (2001).
51. J.F. Corney and P.D. Drummond, *Quantum noise in optical fibers*, J. Opt. Soc. Am. B **18**, 139 and 153 (2001).
52. Z.Y. Ou, S.F. Pereira, H.J. Kimble, and K.C. Peng, *Realization of the Einstein-Poldosky-Rosen paradox for continuous variables*, Phys. Rev. Lett. **68**, 3663 (1992).
53. C. Silberhorn, P.K. Lam, O. Weiß, F. König, N. Korolkova, and G. Leuchs, *Generation of continuous variable Einstein-Poldosky-Rosen entanglement via the Kerr nonlinearity in an optical fibre*, Phys. Rev. Lett. **86**, 4267 (2001).
54. M.D. Reid, *Demonstration of the Einstein-Poldosky-Rosen paradox using nondegenerate parametric amplification*, Phys. Rev. A **40**, 913 (1989).
55. A. Einstein, B. Poldosky, and N. Rosen, *Can quantum-mechanical description of physical reality be considered complete?*, Phys. Rev. **47**, 777 (1935).
56. L.M. Duan, G. Giedcke, J.I. Cirac, and P. Zoller, *Inseparability criterion for continous variables systems*, Phys. Rev. Lett. **84**, 2722 (2000).
57. R. Simon, *Peres-Horodecki separability criterion for continous variable systems*, Phys. Rev. Lett. **84**, 2726 (2000).
58. G. Giedke, M.M. Wolf, O. Krüger, R.F. Werner, and J.I. Cirac, *Entanglement of formation for symmetric Gaussian states*, Phys. Rev. Lett. **91**, 107901 (2003).
59. J. Laurat, T. Coudreau, G. Keller, N. Treps, and C. Fabre, *Effects of mode coupling on the generation of quadrature Einstein-Podolsky-Rosen entanglement in a type-II optical parametric oscillator below threshold*, Phys. Rev. A **71**, 022313 (2005).
60. A. Furusawa, J.L. Sørensen, S.L. Braunstein, C.A. Fuchs, H.J. Kimble, and E.S. Polzik, *Unconditional quantum teleportation*, Science **282**, 706 (1998).
61. W.P. Bowen, N. Treps, R. Schnabel, and P.K. Lam, *Experimental demonstration of continuous variable polarization entanglement*, Phys. Rev. Lett. **89**, 253601 (2002).
62. E.H.Huntington, G.N. Milford, C. Robilliard, O. Glöckl, U.L. Andersen, S. Lorenz, and G. Leuchs, *Demonstration of the spatial separation of the entangled quantum sidebands of an optical field* Phys. Rev. A **71**, 041802(R) (2005).

63. G. Leuchs, T.C. Ralph, C. Silberhorn, N. Korolkova, *Scheme for the generation of entangled solitons for quantum communications*, J. Mod. Opt. **46**, 1927 (1999).
64. N. Korolkova, C. Silberhorn, O. Glöckl, S. Lorenz, C. Marquardt, and G. Leuchs, *Direct experimental test of nonseparability and other quantum techniques using continous variables of light*, Eur. Phys. J. D. **18**, 229 (2002).
65. O. Glöckl, U.L. Andersen, S. Lorenz, C. Silberhorn, N. Korolkova, and G. Leuchs, *Sub-shot-noise phase quadrature measurement of intense light beams*, Opt. Lett. **29**, 1936 (2004).
66. S. Huard, *Polarization of light*, John Wiley and Sons, New York 1997.
67. W.P. Bowen, R. Schnabel, P.K. Lam, and T.C. Ralph, *Experimental Investigation of Criteria for Continous Variable Entanglement*, Phys. Rev. Lett. **90**, 043601 (2003).
68. J.M. Jauch and F. Rohrlich, *The theory of photons and electrons*, Addison-Wesley, Reading, MA, 1955.
69. M.S. Kim, W. Son, V. Bužek, and P.L. Knight, *Entanglement by a beam splitter: Nonclassicality as a prerequisite for entanglement*, Phys. Rev. A **65**, 032323 (2002).

Chapter 14

High-Fidelity Quantum Teleportation and a Quantum Teleportation Network

Nobuyuki Takei, Hidehiro Yonezawa, Takao Aoki
and Akira Furusawa

*Department of Applied Physics, School of Engineering, The University of Tokyo
7-3-1 Hongo, Bunkyo-ku, Tokyo 113-8656, Japan and
CREST, Japan Science and Technology (JST) Agency
1-9-9 Yaesu, Chuo-ku, Tokyo 103-0028, Japan*

email: akiraf@ap.t.u-tokyo.ac.jp

Quantum teleportation is experimentally demonstrated using optical field modes. We describe quantum teleportation for various input quantum states: a coherent state, a squeezed state and quantum entanglement. Furthermore this protocol is extended to a multipartite protocol known as a quantum teleportation network which exploits multipartite entanglement.

1. Introduction

Quantum teleportation[1] is an essential protocol in quantum communication and quantum information processing.[2,3] This protocol enables reliable transfer of an unknown quantum state from one location to distant another one. This transfer is achieved by utilizing shared quantum entanglement and classical communication between two locations. Quantum teleportation can also be combined with other operations to construct advanced quantum circuits.[2,3] For example, it may be used for the fault-tolerant implementation of complicated quantum gates that would otherwise degrade a quantum state.[4,5] Since the teleported state will be manipulated in subsequent operations for both quantum communication and information processing, it is required to realize a high-fidelity teleporter which preserves the fragile information of a quantum state.

Quantum teleportation was originally proposed for a qubit system[1] and the experiments with qubits have been realized using the polarization state[6] and the time-bins state[7] of a single photon, and the trapped ion as well.[8,9] This protocol has been extended to a continuous-variable (CV) system[10,11] employing the Einstein–Podolsky–Rosen (EPR) correlation.[12] So far several experiments for CVs have been demonstrated for a coherent state input[13–16] and a squeezed state input[17] using quadrature-phase amplitudes of optical field modes. Teleportation of quantum entanglement, *i.e.*, entanglement swapping[18,19] has been also realized with optical field modes.[16,20] Furthermore, CV quantum teleportation has been extended to a multipartite protocol known as a quantum teleportation network,[21,22] which exploits multipartite entanglement. Such a multipartite protocol is expected to form fundamental components for larger-scale quantum communication and information processing.

In the following sections of this chapter we describe the experimental realization of CV quantum teleportation and a teleportation network. First we discuss teleportation between two parties for various quantum state inputs, and then we mention a teleportation network among three parties by utilizing tripartite entanglement.[23]

2. Quantum Teleportation

Here we describe the experiments of CV teleportation for various quantum state inputs: a coherent state, a squeezed state and (bipartite) quantum entanglement. We first briefly summarize teleportation procedure and the generation of an EPR state which is utilized as a resource of teleportation.

The quantum state to be teleported in the experiments is that of an electromagnetic field mode (frequency sidebands of a carrier beam). An electromagnetic field mode is represented by an annihilation operator $\hat{a}$ whose real and imaginary parts ($\hat{a} = \hat{x} + i\hat{p}$) correspond to quadrature-phase amplitude operators with the canonical commutation relation $[\hat{x}, \hat{p}] = i/2$ (units-free, with $\hbar = 1/2$). We use the Heisenberg picture to describe the evolution of the quantum state.

The scheme of quantum teleportation is illustrated in Fig. 1. First sender Alice and receiver Bob share EPR beams (modes A and B). EPR beams can be generated by combining two squeezed vacuum states at a half beam splitter with a $\pi/2$ phase shift as depicted in the bottom of Fig. 1, where two ellipses indicate the squeezed quadrature of each beam. Two output beams emerging from the beam splitter corresponds to a two-mode squeezed vacuum state,[24] and they show the quantum entanglement of $\hat{x}_{\mathrm{A}} - \hat{x}_{\mathrm{B}} = \sqrt{2}e^{-r}\hat{x}_1^{(0)}$ and $\hat{p}_{\mathrm{A}} + \hat{p}_{\mathrm{B}} = \sqrt{2}e^{-r}\hat{p}_2^{(0)}$, where r is the squeez-

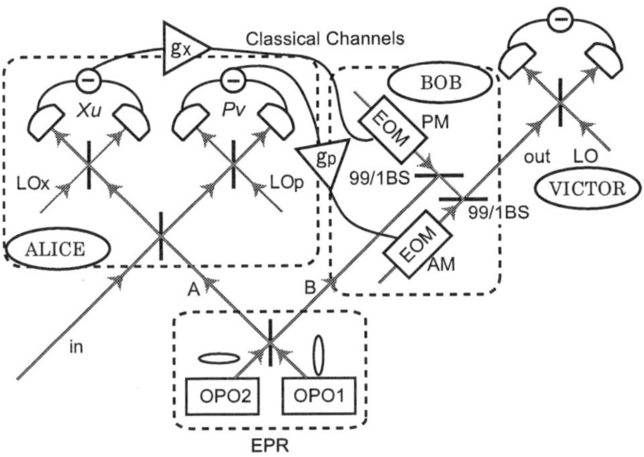

Fig. 1. Schematic set-up of the experiment for quantum teleportation.

ing parameter and a superscript (0) denotes an initial vacuum mode. The mode i (=1,2) represents the mode from a squeezer i, which is an optical parametric oscillator (OPO) driven below oscillation threshold in the experiments (the details of the OPO will be explained later). In the limit of infinite squeezing, the generated state becomes the ideal EPR state:[12] $\hat{x}_{\rm A} - \hat{x}_{\rm B} \to 0, \hat{p}_{\rm A} + \hat{p}_{\rm B} \to 0$. The existence of entanglement between these modes can be easily checked by the sufficient inseparability criterion:[25,26]

$$\Delta_{\rm A,B} \equiv \langle[\Delta(\hat{x}_{\rm A} - \hat{x}_{\rm B})]^2\rangle + \langle[\Delta(\hat{p}_{\rm A} + \hat{p}_{\rm B})]^2\rangle < 1, \quad (1)$$

where the variances of a vacuum state are $\langle(\Delta\hat{x}^{(0)})^2\rangle = \langle(\Delta\hat{p}^{(0)})^2\rangle = 1/4$. If this inequality holds, the relevant modes are entangled.

Sharing the EPR beams, Alice performs so-called "Bell measurement" on her entangled mode $(\hat{x}_{\rm A}, \hat{p}_{\rm A})$ and an unknown input mode $(\hat{x}_{in}, \hat{p}_{in})$. Note that, for the purpose of verifying the protocol, an input state is created by Victor (the "verifier") and unknown to both Alice and Bob in principle. The Bell measurement instrument consists of a half beam splitter and two optical homodyne detectors. She combines mode A and the input at the beam splitter and measures $\hat{x}_u = (\hat{x}_{in} - \hat{x}_{\rm A})/\sqrt{2}$ and $\hat{p}_v = (\hat{p}_{in} + \hat{p}_{\rm A})/\sqrt{2}$ with the detectors. These measured values x_u and p_v for $\hat{x}_u$ and $\hat{p}_v$ are sent to Bob through classical channels with normalized gains g_x and g_p, respectively.

The normalized gains are defined as $g_x = \langle\hat{x}_{out}\rangle/\langle\hat{x}_{in}\rangle$ and $g_p = \langle\hat{p}_{out}\rangle/\langle\hat{p}_{in}\rangle$, where a subscript *out* denotes a teleported mode. The gains are adjusted in the manner of Ref. 15. Typical values for the measured gains are $g_x = 1.00 \pm 0.02$ and $g_p = 0.99 \pm 0.02$, respectively, in the experiments. Therefore these gains are fixed throughout this chapter and treated as unity for simplicity.

Let us write Bob's initial mode before the measurement of Alice as: $\hat{x}_{\rm B} = \hat{x}_{in} - (\hat{x}_{\rm A} - \hat{x}_{\rm B}) - \sqrt{2}\hat{x}_u$ and $\hat{p}_{\rm B} = \hat{p}_{in} + (\hat{p}_{\rm A} + \hat{p}_{\rm B}) - \sqrt{2}\hat{p}_v$. Note that in this step Bob's mode remains unchanged. After measuring $\hat{x}_u$ and $\hat{p}_v$ at Alice, these operators collapse and reduce to certain values. Receiving her measurement results, Bob displaces his mode as $\hat{x}_{\rm B} \to \hat{x}_{out} = \hat{x}_{\rm B} + \sqrt{2}g_x x_u$, $\hat{p}_{\rm B} \to \hat{p}_{out} = \hat{p}_{\rm B} + \sqrt{2}g_p p_v$ and accomplishes the teleportation. Here we write explicitly the gains g_x and g_p to show the meaning of them, but they are treated as unity as mentioned before. In the experiments, displacement operation is performed by using electro-optical modulators (EOMs) and highly reflecting mirrors. Bob modulates two beams by using amplitude and phase modulators (AM and PM) based on the classical information from Alice, because quantum states are frequency sidebands of the carrier beam. Note that two beams are used to avoid the mixing of amplitude and phase modulations in Ref. 16. The amplitude and phase modulations correspond to the displacement of p and x quadratures, respectively. The modulated beams are combined with Bob's mode $(\hat{x}_{\rm B}, \hat{p}_{\rm B})$ at 99/1 beam splitters (99/1BSs).

The teleported output mode becomes

$$\begin{aligned}\hat{x}_{out} &= \hat{x}_{in} - (\hat{x}_{\rm A} - \hat{x}_{\rm B}) = \hat{x}_{in} - \sqrt{2}e^{-r}\hat{x}_1^{(0)},\\ \hat{p}_{out} &= \hat{p}_{in} + (\hat{p}_{\rm A} + \hat{p}_{\rm B}) = \hat{p}_{in} + \sqrt{2}e^{-r}\hat{p}_2^{(0)}.\end{aligned} \tag{2}$$

In the ideal case ($r \to \infty$), the EPR state is the state for which $\hat{x}_{\rm A} - \hat{x}_{\rm B} \to 0$ and $\hat{p}_{\rm A} + \hat{p}_{\rm B} \to 0$. Then the teleported state is identical to the input state. In real experiments, however, the teleported state has additional fluctuations. Without entanglement, *i.e.* $r = 0$, at least two units of a vacuum fluctuation are added. These additional fluctuations are called quduties which must be paid for crossing the boundary between classical and quantum domains.[11]

Finally Victor analyzes an output state from Bob's station and verifies the success of quantum teleportation. The verification is usually carried out using a fidelity which quantifies how well the teleported state resembles the original input state; the fidelity is 1 in the ideal teleportation. If an input is a pure state $|\psi_{in}\rangle$, the fidelity F becomes $F = \langle\psi_{in}|\hat{\rho}_{out}|\psi_{in}\rangle$.[27] In this case, it corresponds to an overlap between an input state $|\psi_{in}\rangle$ and an output state $\hat{\rho}_{out}$. If an input is a mixed state, the fidelity is provided as follows:[28]

$$F = \left\{ \mathrm{Tr} \left[\left(\sqrt{\hat{\rho}_{in}} \hat{\rho}_{out} \sqrt{\hat{\rho}_{in}} \right)^{1/2} \right] \right\}^2 . \quad (3)$$

Quantum teleportation succeeds when the fidelity exceeds the classical limit which is the best achievable value of the fidelity without the use of entanglement.

Before moving on to the implementation of quantum teleportation, let us summarize the experimental method of generating a squeezed vacuum state which is a common technique to all experiments described in the following sections. Each squeezed vacuum state is generated from a subthreshold OPO with a potassium niobate crystal (length 10mm). The crystal is temperature-tuned for type-I noncritical phase matching. Each OPO cavity is a bow-tie-type ring cavity which consists of two spherical mirrors (radius of curvature 50 mm) and two flat mirrors. The round trip length is about 500 mm and the waist size in the crystal is 20μm. The transmittance of the output coupler is about 12%. The output of a continuous wave Ti:Sapphire laser at 860nm is frequency doubled in an external cavity with the same configuration as the OPOs. The output beam at 430nm is divided into some beams to pump the OPOs.

2.1. *Teleportation of a coherent state*

We first describe quantum teleportation of a coherent state.[16] The experiments for such a state have been demonstrated in some groups.[13–16] Therefore the success criteria, namely the classical limit for the state has been well-understood using the fidelity F and it is given by the fidelity $F = 0.5$.[27,29,30] The meaning of value of the fidelity has been also studied very well.[31,32] The value of 2/3 is referred to as the no-cloning limit, because surpassing this limit warrants that the teleported state is the best remaining copy of the input state.[31] Note that the fidelity for a coherent state input may be used as a benchmark for the performance of the realized teleporter, being compared with other experiments.

Surpassing the no-cloning limit has another meaning. A teleporter surpassing the limit can transfer the following nonclassicality in an input quantum state. It is possible to transfer a negative part of the Wigner function of a quantum state like the Schrödinger-cat state $|\psi_{cat}\rangle \propto |\alpha\rangle \pm |-\alpha\rangle$ and a single photon state.[32] The negative part is a signature of the nonclassicality.[33] Moreover $F > 2/3$ allows one to transfer quantum entanglement which shows nonclassical correlation to the extent that $F > 2/3$ could be achieved

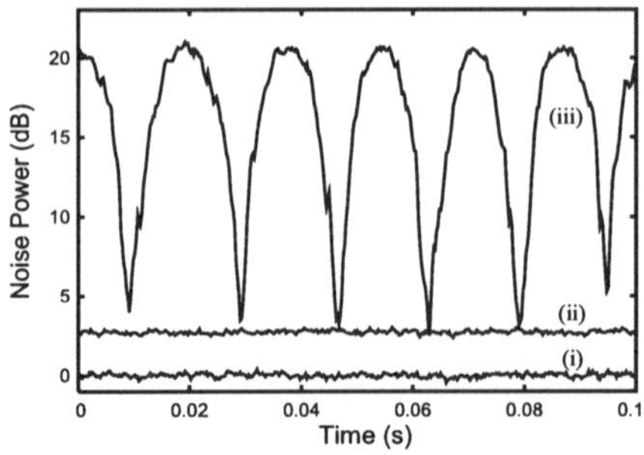

Fig. 2. The measurement results of the teleported state for a coherent state input in x quadrature. Each trace is normalized to the corresponding vacuum noise level. (i) the corresponding vacuum noise level $\langle(\Delta\hat{x}_{out}^{(0)})^2\rangle = 1/4$. (ii) the teleported state for a vacuum input. (iii) the teleported state for a coherent state input with the phase scanned. The measurement frequency is centered at 1 MHz, and the resolution and video bandwidths are 30kHz and 300 Hz, respectively. Traces i and ii are averaged 20 times.

if it is used as a resource of teleportation.[19] Since some quantum circuits may rely on the nonclassicality of the input state, it is desirable to realize a high-fidelity teleporter which preserves the nonclassicality.

The experimental realization of teleportation beyond the no-cloning limit is reported in Ref. 16, in which the fidelity of 0.70 is obtained. In the experiment, quantum states are frequency sidebands at $\pm$1MHz of an optical carrier beam. Therefore a coherent state can be generated by applying phase modulation at the frequency with EOM to the carrier beam. This modulated beam is put into the input mode.

Figure 2 shows the measurement results of the teleported mode for the coherent state. The measured displacement of the state is 20.7 ± 0.2dB compared to the corresponding vacuum noise level. Note that the variance of the teleported state for a vacuum input (trace ii) corresponds to that for a coherent state input, because a vacuum state is one of coherent states. The measured values of the variances are $\langle(\Delta\hat{x}_{out})^2\rangle = 2.82 \pm 0.09$dB and $\langle(\Delta\hat{p}_{out})^2\rangle = 2.64 \pm 0.08$ dB (not shown). In the classical case without entanglement, which is so-called classical teleportation, the teleported state has three units of a vacuum fluctuation; one unit from the input state and the other two from the quduties, and the variances of the teleported state become 4.77dB (see Eq. (2) with $r = 0$). The observed noise is smaller than the classical case, showing the success of teleportation.

We can evaluate the performance of teleportation using the fidelity. The classical limit for a coherent state input is derived by averaging the fidelity for a randomly chosen coherent state input.[27] The classical limit of the averaged fidelity F_{av} is 0.5.[30] In real experiments, however, it is impossible to take an average over the whole phase space. But if the gains of the classical channels are unity $g_x = g_p = 1$, the averaged fidelity is identical to the fidelity for a particular coherent state input ($F_{av} = F$).[27] This is because the fidelity with unity gains can be determined by only the variances of the teleported state, *i.e.*, independent of displacement of a coherent state input. The fidelity for a coherent state input at unity gains can be written as[13,29]

$$F = \frac{2}{\sqrt{(1 + 4\sigma_{out}^x)(1 + 4\sigma_{out}^p)}}, \tag{4}$$

where $\sigma_{out}^x = \langle(\Delta\hat{x}_{out})^2\rangle$ and $\sigma_{out}^p = \langle(\Delta\hat{p}_{out})^2\rangle$. The fidelity obtained from the measured variances is $F = 0.70 \pm 0.02$.[16] This result clearly shows the success of teleportation of a coherent state beyond the no-cloning limit. Moreover the correlation of the EPR beams is examined. The EPR beams show the entanglement of $\Delta_{\mathrm{A,B}} = 0.42 \pm 0.01$, from which the expected fidelity of $F = 0.70 \pm 0.01$ is calculated using Eq. (2). The experimental result is in good agreement with the calculation.

2.2. *Teleportation of a squeezed state*

We next turn to the experiment of quantum teleportation of a squeezed state. A squeezed vacuum state is defined as the state with variance smaller than the vacuum variance in one quadrature at the expense of increased variance in the other, for example, $\langle(\Delta\hat{x})^2\rangle < 1/4 < \langle(\Delta\hat{p})^2\rangle$.[24] Note that this state belongs to the class of minimum-uncertainty states. Here we describe the experiment which verify that the squeezed variance of a squeezed state is certainly transferred through a teleportation process.[17]

In any real experiments, a squeezed vacuum is degraded due to some inevitable losses and necessarily becomes a mixed state, so that it is regarded as a squeezed thermal state[34] and no longer the minimum-uncertainty state. However, as long as its squeezed variance is smaller than the vacuum variance, we call the mixed state a squeezed vacuum in this chapter. Assuming that x quadrature is squeezed, its variances are written as follows:

$$\begin{cases} \sigma_{in}^x = \langle(\Delta\hat{x}_{in})^2\rangle = e^{-2s}\coth(\beta/2)/4 \\ \sigma_{in}^p = \langle(\Delta\hat{p}_{in})^2\rangle = e^{+2s}\coth(\beta/2)/4 \end{cases}, \tag{5}$$

where s is the squeezing parameter for the input and $\coth(\beta/2)/4$ is the variance of an initial thermal state. β is the inverse temperature $1/2k_{\mathrm{B}}T$

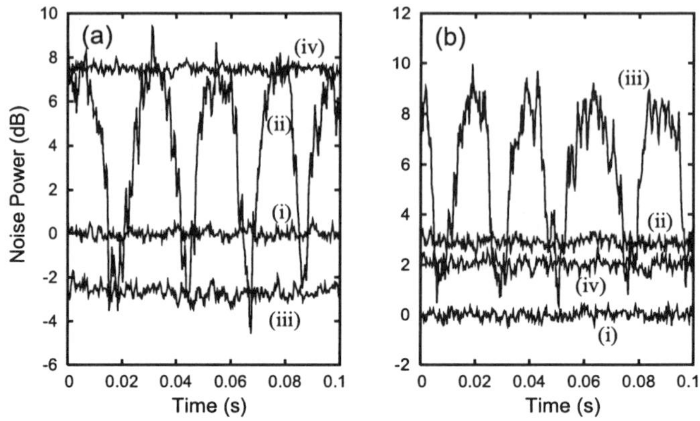

Fig. 3. (a) The measurement results on the input squeezed state. (i) the corresponding vacuum noise level; (ii) the variance of the squeezed state with the LO phase scanned; (iii) and (iv) are the minimum and the maximum noise levels with the LO phase locked. (b) The measurement results on the output states in x quadrature. (i) the corresponding shot noise level; (ii) the vacuum state input; (iii) the squeezed state input with the phase of the input state scanned; (iv) the minimum noise levels with the phase of the input state locked. The measurement conditions are the same as for Fig. 2.

where k_B is the Boltzmann constant and T is temperature. Note that general squeezed thermal states include both the squeezed vacuum states and the states with the squeezed variances larger than the vacuum variance. In the experiment, however, a squeezed vacuum state with the squeezed variance $\sigma_{in}^x < 1/4$ is teleported (see Fig. 3a).

In the experiment, both a vacuum state and a squeezed vacuum are teleported to verify that the squeezed variance of a squeezed vacuum state is transferred through a teleportation process. It is expected that the teleported squeezed state shows the smaller variance in x quadrature than that for the case of a vacuum input ($s = 0,\ \beta \to \infty$):

$$(\sigma_{out}^x)_{sq} < (\sigma_{out}^x)_{vac}. \tag{6}$$

Similarly the inequality of $(\sigma_{out}^p)_{sq} > (\sigma_{out}^p)_{vac}$ should be expected to hold.

Before performing teleportation, the input squeezed vacuum is first measured with homodyne detection. Figure 3a shows the measurement results. The squeezing and antisqueezing are -2.66 ± 0.49dB and 7.45 ± 0.17dB, respectively, compared to the vacuum noise level. In the teleportation experiment, the relative phase of the input is properly adjusted and locked so that x quadrature is squeezed.

We then proceed to the experiment of quantum teleportation of the squeezed state. Figure 3b shows the Victor's measurement results of the output state from Bob. In the case of a vacuum input (trace ii), the observed noise levels are 2.90±0.21dB for $(\sigma^x_{out})_{vac}$ and 3.01±0.19dB for $(\sigma^p_{out})_{vac}$ (not shown), respectively. The variances are smaller than the classical case (4.77dB) due to the quantum entanglement. These results indicate the success of quantum teleportation of the vacuum. Based on these variances and the measurement results σ^x_{in}, σ^p_{in} of the squeezed input, we can calculate the expected output variances $(\sigma^x_{out})_{sq}$ and $(\sigma^p_{out})_{sq}$ for the squeezed input using Eq. (2). The expected variances are calculated as 1.71±0.58dB and 8.24±0.31dB for $(\sigma^x_{\rm out})_{sq}$ and $(\sigma^p_{\rm out})_{sq}$, respectively.

The squeezed vacuum state is subsequently teleported, as shown in Fig. 3b (trace iv). The minimum noise level is 2.03±0.24dB for $(\sigma^x_{out})_{sq}$, while the maximum noise level is 8.18±0.17dB for $(\sigma^p_{out})_{sq}$ (not shown). These results are in good agreement with the expected variances. The squeezed variance of the teleported state is clearly smaller than that of the teleported vacuum state in x quadrature, and then the inequality $(\sigma^x_{out})_{sq} < (\sigma^x_{out})_{vac}$ is satisfied. Similarly the inequality $(\sigma^p_{out})_{sq} > (\sigma^p_{out})_{vac}$ holds. Therefore the squeezed variance of the squeezed vacuum input is certainly teleported.

Note that the observed squeezed variance of the teleported state is larger than Victor's shot noise level, which shows that the output state is not a nonclassical state. In order to get a nonclassical state at Bob's place, the quantum entanglement with stronger correlation is required. If three squeezed state used for the input and the EPR beams have the same degree of squeezing, more than 4.77dB squeezing is needed. It is a next challenge to generate a teleported state whose variance is below the shot noise level.

Finally we describe the evaluation of the performance of the teleporter using the fidelity. Since a vacuum is one of coherent states, the fidelity F_{vac} for the vacuum teleportation at unity gains is simply given by Eq. (4). From the measured variances, the fidelity of 0.67 ± 0.02 is obtained for quantum teleportation of a vacuum input, which exceeds the classical limit of 0.5. This result clearly shows the success of quantum teleportation of a vacuum state. But this classical limit cannot be applied to the case of the squeezed state input, since the fidelity depends on an input state like Eq. (3).

It could be considered that a (mixed) squeezed vacuum input is transformed into a squeezed thermal state through the imperfect teleportation process.[35] For these squeezed states, the fidelity F_{sq} in Eq. (3) can be written explicitly as follows:[36–38]

$$F_{sq} = \frac{2\sinh(\beta_{in}/2)\sinh(\beta_{out}/2)}{\sqrt{Y}-1}, \tag{7}$$
$$Y = \cosh^2(s_{in}-s_{out})\cosh^2[(\beta_{in}+\beta_{out})/2] - \sinh^2(s_{in}-s_{out})\cosh^2[(\beta_{in}-\beta_{out})/2].$$

Using Eq. (5), the squeezing parameter s_j and the inverse temperature β_j ($j = in,\ out$) are obtained from the measured variances. Therefore we can calculate the fidelity F_{sq}.

In the experiment, the fidelity F_{sq} is applied to the particular input state. The fidelity for the "perfect" classical teleportation could be calculated using the measurement results on the input state and Eqs. (2) and (7). The fidelity for the classical case F_{sq}^C is calculated as 0.73 ± 0.04, and this value is regarded as the classical limit for the input. In the quantum teleportation with the EPR beams, the obtained fidelity is $F_{sq}^Q = 0.85 \pm 0.05$ which is higher than the classical limit. This fact shows the success of quantum teleportation of the squeezed state.

Note that the classical limit calculated here can be applied only to the particular input state. The classical limit for a general mixed squeezed state has not investigated very much so far. The discussion has just started,[17,37], and the success criteria for such a state remains a topic for future study.

2.3. *Entanglement swapping*

Next we describe the experiment of teleportation of quantum entanglement, *i.e.* entanglement swapping, with the same high-fidelity teleporter as a coherent state input.[16] Entanglement swapping enables two parties to share quantum entanglement, where they have never directly interacted.[18,19] This protocol may be useful in making nonlocal correlations over long distances. Therefore it can play a significant role in quantum communication.[2,3]

The scheme for entanglement swapping is illustrated in Fig. 4. Two pairs of entangled beams denoted by EPR1 and EPR2 are generated by combining squeezed vacuum states at half beam splitters. The EPR1 beams consist of mode A and B, and they are utilized as a resource of teleportation. They shows the quantum entanglement of $\Delta_{\mathrm{A,B}} = 0.42 \pm 0.01$, allowing us to achieve the fidelity of 0.70 ± 0.01 for a coherent state input as mentioned before. On the other hand, the EPR2 beams consist of two sub-systems: 'reference' and 'input'. While the reference is kept during a teleportation process, the input is teleported to an output station. After the process,

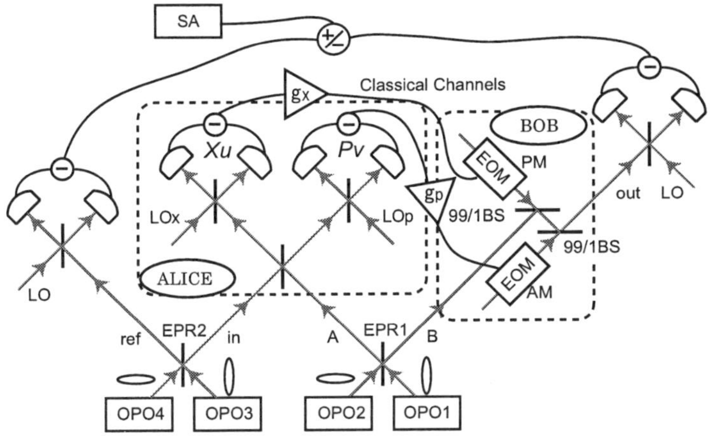

Fig. 4. The experimental set-up for teleportation of quantum entanglement.

the success of this protocol is verified by examining quantum entanglement between the reference and the output: $\Delta_{ref,out} < 1$.

Before performing the experiment of entanglement swapping, the noise power of each mode for EPR2 beams and the initial correlation between the modes are measured with homodyne detection and a spectrum analyzer (SA). For the reference mode, the noise levels of 5.23 ± 0.14dB and 4.44 ± 0.14dB are obtained for x and p quadratures, respectively (Fig. 5a). Similarly, the noise levels for the input mode are 5.19 ± 0.13dB and 4.37 ± 0.14dB for x and p quadratures (not shown). By making electrical subtraction or summation of the homodyne detection outputs, the noise levels of -3.19 ± 0.13dB and -4.19 ± 0.14dB are observed for x and p quadratures as shown in Fig. 5b. From these values, the measured variance of $\Delta_{ref,in} = 0.43 \pm 0.01 < 1$ is obtained. This result shows the existence of the quantum entanglement between the input and the reference.

We then proceed to the experiment of entanglement swapping. The state in the reference mode does not change in the process. For the output mode, the noise levels of 6.06 ± 0.12dB and 5.47 ± 0.14dB are obtained for x and p quadratures, respectively, as shown in Fig. 6a. Because of the imperfect teleportation, some noises are added to the teleported state, resulting in the larger variances than that of the reference. Figure 6b shows the results of the correlation measurement. The observed noise levels are -0.25 ± 0.13dB and -0.60 ± 0.13dB for x and p quadratures, respectively, yielding $\Delta_{ref,out} = 0.91 \pm 0.02 < 1$. This result clearly shows the existence

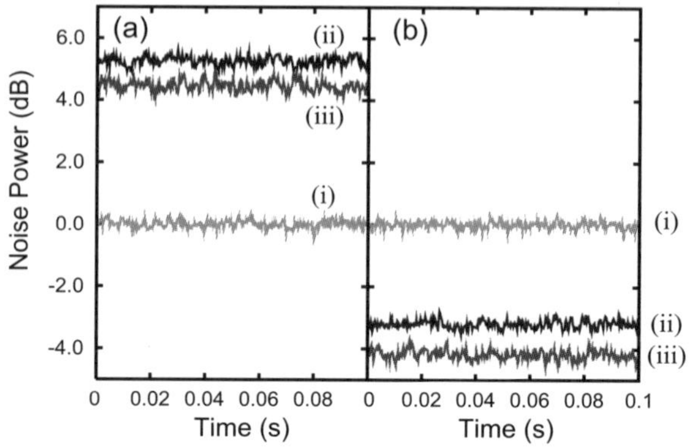

Fig. 5. Correlation measurement for EPR1 beams. (a) The measurement result of the reference mode alone. (i) $\langle(\Delta\hat{x}^{(0)}_{ref})^2\rangle = \langle(\Delta\hat{p}^{(0)}_{ref})^2\rangle = 1/4$; (ii) $\langle(\Delta\hat{x}_{ref})^2\rangle$; (iii) $\langle(\Delta\hat{p}_{ref})^2\rangle$. (b) The measurement result of the correlation between the input mode and the reference mode. (i) $\langle[\Delta(\hat{x}^{(0)}_{ref} - \hat{x}^{(0)}_{in})]^2\rangle = \langle[\Delta(\hat{p}^{(0)}_{ref} + \hat{p}^{(0)}_{in})]^2\rangle = 1/2$; (ii) $\langle[\Delta(\hat{x}_{ref} - \hat{x}_{in})]^2\rangle$; (iii) $\langle[\Delta(\hat{p}_{ref} + \hat{p}_{in})]^2\rangle$. The measurement condition is the same as that of Fig. 2.

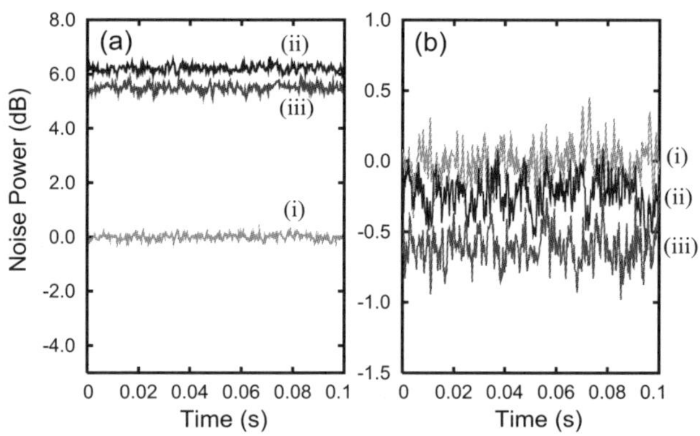

Fig. 6. Correlation measurement results of the teleportation of quantum entanglement. (a) The measurement result of the output mode alone. (i) $\langle(\Delta\hat{x}^{(0)}_{out})^2\rangle = \langle(\Delta\hat{p}^{(0)}_{out})^2\rangle = 1/4$; (ii) $\langle(\Delta\hat{x}_{out})^2\rangle$; (iii) $\langle(\Delta\hat{p}_{out})^2\rangle$. (b) The measurement result of the correlation between the output mode and the reference mode. (i) $\langle[\Delta(\hat{x}^{(0)}_{ref} - \hat{x}^{(0)}_{out})]^2\rangle = \langle[\Delta(\hat{p}^{(0)}_{ref} + \hat{p}^{(0)}_{out})]^2\rangle = 1/2$; (ii) $\langle[\Delta(\hat{x}_{ref} - \hat{x}_{out})]^2\rangle$; (iii) $\langle[\Delta(\hat{p}_{ref} + \hat{p}_{out})]^2\rangle$. The measurement condition is the same as that of Fig. 2.

of quantum entanglement between the output and the reference, and also indicates the success of entanglement swapping with unity gains.

3. Quantum Teleportation Network

Here we describe the experimental realization of a quantum teleportation network where coherent states are exchanged among three parties,[21,22] exploiting tripartite entanglement.[23] This is the demonstration of a manipulation of multipartite entanglement. The manipulation is essential for realization of more complicated quantum information processing and quantum communication among many parties.

Quantum teleportation network is a quantum communication network linked by quantum teleportation. For example, in a tripartite network, three parties (we call Alice, Bob and Claire) are connected on the network where they are spatially separated and previously share tripartite entanglement. They can only use local operations and classical channels to communicate with each other. In some respects, quantum teleportation network is similar to bipartite quantum teleportation. In both schemes the parties share quantum entanglement, and send a quantum state using local operations and the classical channels. But the properties of tripartite entanglement make it different from the bipartite teleportation in other respects.

If Alice sends a quantum state to Bob, what role does Claire play? Let us remind that the three parties are in the tripartite entangled state. The third party Claire also has a quantum correlation with the other parties. Thus Alice and Bob need Claire's information to succeed in teleportation. In other words, Claire can control the transfer of the quantum state from Alice to Bob by restricting their access to her information. This is a clear manifestation of tripartite entanglement.

Tripartite entanglement for CVs can be generated by using three squeezed vacuum states and two beam splitters,[21,23] as illustrated in the center of Fig. 7. Three ellipses indicate the squeezed quadratures for the squeezed states and these states are combined at beam splitters with transmittance/reflectivity of 1/2 and 1/1. In the limit of infinite squeezing, the state is the CV analogue[21,23] of Greenberger-Horne-Zeilinger (GHZ) state.[39] The CV GHZ state is a maximally entangled state and a simultaneous eigenstate of zero total momentum ($p_1 + p_2 + p_3 = 0$) and zero relative positions ($x_i - x_j = 0 \quad i, j = 1, 2, 3$). The entanglement properties of the CV GHZ state are very fragile under partial losses of the state. For example, if one of the three subsystems is traced out, the remaining state ($\hat{\rho}_{\rm AB}$, $\hat{\rho}_{\rm AC}$, $\hat{\rho}_{\rm BC}$) is completely unentangled.[40] Therefore without Claire's information the quantum entanglement between Alice and Bob vanishes, and quantum teleportation is no longer possible.

There is another important point to be made when bipartite quantum teleportation is developed into a tripartite quantum teleportation network. Only if we use a fully inseparable tripartite entangled state, we can succeed in teleportation between an arbitrary pair in the network. Namely, each party can play any of the three roles: a sender, a receiver and a controller. Note that if we use a partially entangled state, we may succeed in teleportation for a particular combination of the sender, the receiver and the controller, but may fail for other combinations. From this point of view, a *truly* tripartite quantum protocol is defined as a protocol that succeeds only if a fully inseparable tripartite entanglement is used. In order to verify success of a truly tripartite quantum protocol, we need to succeed in teleportation for at least two different combinations.[23,41] For example, the experiment by Jing *et al.*,[42] a controlled dense coding for a particular combination, shows only partial success and is not sufficient for the demonstration of a truly tripartite quantum protocol. In the experiment described below, quantum teleportation for three different combinations is demonstrated. This is the first demonstration of a truly tripartite quantum protocol.

Here we present the experiment of a quantum teleportation network.[22] Figure 7 shows the schematic of the experimental setup. Tripartite entangled states[23] are distributed to Alice, Bob and Claire. We represent the operators for each mode as $(\hat{x}_i, \hat{p}_i)$ $(i = \mathrm{A, B, C})$ in the Heisenberg representation. We first consider the teleportation with the combination of sender Alice, receiver Bob, and controller Claire.

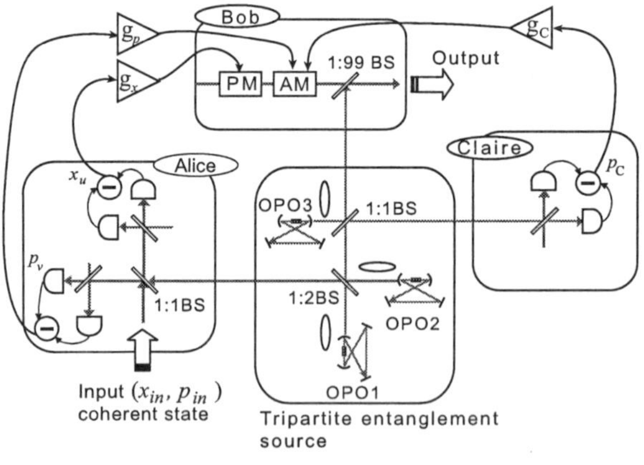

Fig. 7. The experimental set-up for a quantum teleportation network. This schematic indicates the teleportation from Alice to Bob under the control of Claire.

First Alice performs so-called "Bell measurement" on her entangled mode $(\hat{x}_\mathrm{A}, \hat{p}_\mathrm{A})$ and an unknown input mode $(\hat{x}_{in}, \hat{p}_{in})$. The input state is a coherent state and a frequency sideband of a carrier beam. The Bell measurement instrument is the same as the previous section, and it consists of a half beam splitter and two optical homodyne detectors. She measures $\hat{x}_u$ and $\hat{p}_v$ with the detectors. Figure 8a shows the measurement result of $\langle(\hat{p}_v)^2\rangle$ at Alice ($\langle(\hat{x}_u)^2\rangle$ is not shown). The variance of a vacuum input $\langle(\Delta\hat{p}_v)^2\rangle$, which is equivalent to that of a coherent state input, is 3.7dB compared to the corresponding vacuum noise level $\left\langle(\Delta\hat{p}_v^{(0)})^2\right\rangle$, while the noise level for x quadrature is 2.1 dB (not shown). Note that the measured noise levels for x and p quadratures are asymmetric. This is because tripartite entanglement is generated by two x and one p quadrature squeezed vacuum states.[23]

The measured values x_u and p_v for $\hat{x}_u$ and $\hat{p}_v$ are sent through the classical channels with unity gains $g_x = g_p = 1$. The third party Claire measures her entangled state $\hat{p}_\mathrm{C}$ itself. Note that Claire does not measure the x quadrature. Figure 8b shows her measurement result $\langle(\Delta\hat{p}_\mathrm{C})^2\rangle$, and the observed noise level is 5.7 dB. Claire sends this result to Bob through the classical channel with gain g_C.

It is convenient to write Bob's initial mode before the measurement of Alice and Claire as $\hat{x}_\mathrm{B} = \hat{x}_{in} - (\hat{x}_\mathrm{A} - \hat{x}_\mathrm{B}) - \sqrt{2}\hat{x}_u$, $\hat{p}_\mathrm{B} = \hat{p}_{in} + (\hat{p}_\mathrm{A} + \hat{p}_\mathrm{B} + g_\mathrm{C}\hat{p}_\mathrm{C}) - \sqrt{2}\hat{p}_v - g_\mathrm{C}\hat{p}_\mathrm{C}$. Note that in this step Bob's mode remains unchanged. After measuring $\hat{x}_u$, $\hat{p}_v$ and $\hat{p}_\mathrm{C}$, these operators collapse and reduce to certain values. Receiving these measurement results, Bob displaces his mode as $\hat{x}_\mathrm{B} \to \hat{x}_{out} = \hat{x}_\mathrm{B} + \sqrt{2}x_u$, $\hat{p}_\mathrm{B} \to \hat{p}_{out} = \hat{p}_\mathrm{B} + \sqrt{2}p_v + g_\mathrm{C}p_\mathrm{C}$ and accomplishes the teleportation. The displacement operation is performed by using EOMs in the same way as the previous sections.

The teleported mode becomes

$$\begin{aligned} \hat{x}_{out} &= \hat{x}_{in} - (\hat{x}_\mathrm{A} - \hat{x}_\mathrm{B}), \\ \hat{p}_{out} &= \hat{p}_{in} + (\hat{p}_\mathrm{A} + \hat{p}_\mathrm{B} + g_\mathrm{C}\hat{p}_\mathrm{C}). \end{aligned} \tag{8}$$

In the ideal case, total momentum $\hat{p}_\mathrm{A} + \hat{p}_\mathrm{B} + \hat{p}_\mathrm{C}$ and relative position $\hat{x}_\mathrm{A} - \hat{x}_\mathrm{B}$ have zero-eigenvalues $p_\mathrm{A} + p_\mathrm{B} + p_\mathrm{C} = 0$ and $x_\mathrm{A} - x_\mathrm{B} = 0$ simultaneously, and the teleported state is identical to the input state ($g_\mathrm{C} = 1$). In a real experiment, however, the teleported state has additional fluctuations. Without entanglement, at least two units of a vacuum fluctuation (quduties) are added ($g_\mathrm{C} = 0$).

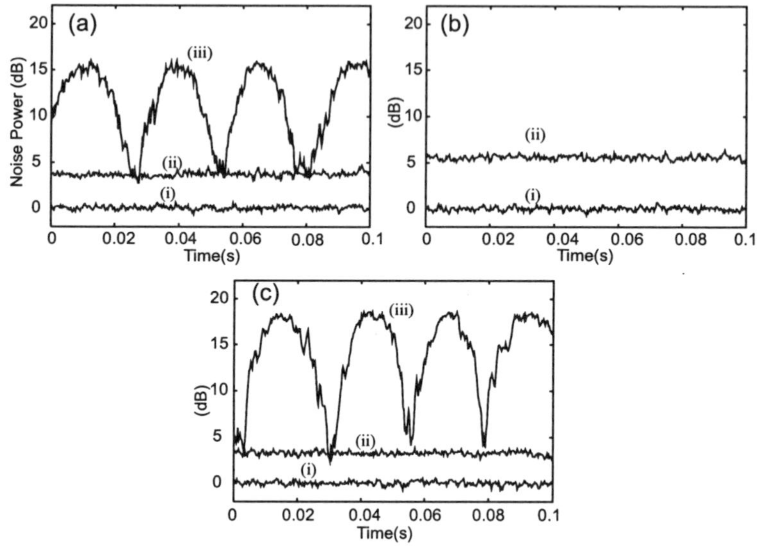

Fig. 8. Quantum teleportation from Alice to Bob under the control of Claire. (a) Alice's measurement results for p quadrature. (i) $\left\langle(\Delta\hat{p}_v^{(0)})^2\right\rangle = 1/4$; (ii) $\langle(\hat{p}_v)^2\rangle$ $(= \langle(\Delta\hat{p}_v)^2\rangle)$ for a vacuum input; (iii) $\langle(\hat{p}_v)^2\rangle$ for the coherent state input with the phase scanned. (b) Claire's measurement results for p quadrature. (i) $\left\langle(\Delta\hat{p}_{\rm C}^{(0)})^2\right\rangle$=1/4; (ii) $\langle(\hat{p}_{\rm C})^2\rangle$ $(= \langle(\Delta\hat{p}_{\rm C})^2\rangle)$. (c) The measurement results of the teleported states for p quadrature. (i) $\left\langle(\Delta\hat{p}_{out}^{(0)})^2\right\rangle = 1/4$; (ii) $\langle(\hat{p}_{out})^2\rangle$ $(= \langle(\Delta\hat{p}_{out})^2\rangle)$ for a vacuum input; (iii) $\langle(\hat{p}_{out})^2\rangle$ for the coherent state input. The measurement conditions are the same as for Fig. 2.

Figure 8c shows measurement result of the teleported mode for p quadrature with the gain $g_{\rm C} = 1.02 \pm 0.03$. The noise level of a vacuum input $\langle(\Delta\hat{p}_{out})^2\rangle$ is 3.3dB compared to the corresponding vacuum noise level for p quadrature, while the noise level for x quadrature is 3.5dB (not shown). The measured displacement of the input is about 18dB, which is 3dB higher than that of the input at Alice in Fig. 8a. It shows that the classical channel's gains are almost unity, because the displacement at Alice is 3dB lower than the actual input because of the input beam splitter.

To evaluate the performance of teleportation, the fidelity of Eq. (4) is used. Although the classical limit ($F = 0.5$) of teleportation for a coherent state is derived for the case of two parties in Refs. 27 and 30, it can be applied to the case of three parties.[21] In a classical case, three parties have no quantum correlation with each other. Thus the third party can not improve the performance of teleportation beyond the classical limit. Although the

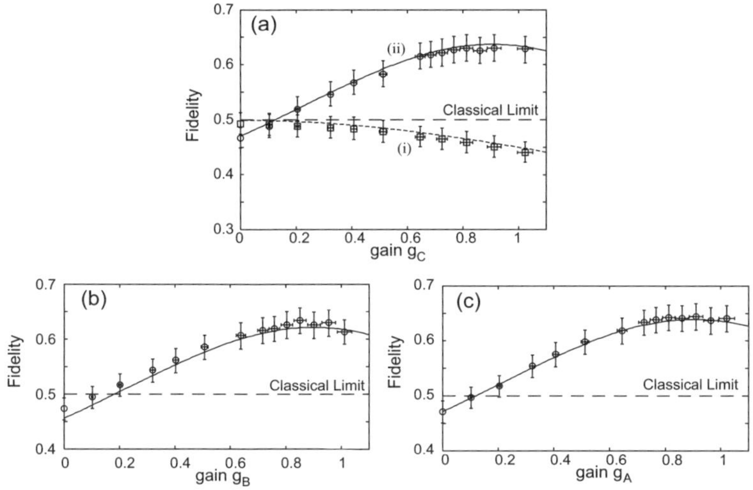

Fig. 9. The controller's gain dependences of the fidelities. (a) The fidelity of the teleportation from Alice to Bob under the control of Claire. (i) The teleportation without entanglement. (ii) The teleportation with tripartite entanglement. The best fidelity 0.63 ± 0.02 is obtained at $g_C \simeq 0.9$. The solid lines represent the theoretical curves calculated from the experimental conditions. (b) The fidelity of the teleportation from Alice to Claire under the control of Bob. The best fidelity 0.63 ± 0.02 is obtained at $g_B \simeq 0.9$. (c) The fidelity of the teleportation from Claire to Bob under the control of Alice. The best fidelity 0.64 ± 0.02 is obtained at $g_A \simeq 0.9$.

gains are set to unity: $g_x = g_p = 1$, the third party's gain g_C is changed. The best fidelity is obtained at the optimum gain g_C which is determined by the degree of the squeezing.[21]

The fidelity calculated from the variances of the teleported state is plotted as a function of g_C in Fig. 9a. Without entanglement, the fidelity is lower than 1/2. Quantum teleportation fails and optimum g_C is zero because Claire has no correlation with the other parties. With tripartite entanglement, the fidelity $F = 0.63 \pm 0.02$ ($g_C \simeq 0.9$) is obtained, which clearly shows success of quantum teleportation between Alice and Bob. At $g_C = 0$, however, quantum teleportation fails. This is because the tripartite entanglement used in this experiment behaves like the GHZ state. To succeed in teleportation, Alice and Bob need Claire's information. If Claire does not send her information to them, the fidelity becomes even lower than that without entanglement. This clearly shows that Claire controls success or failure of the teleportation.

So far the experiment for the particular combination, sender Alice, receiver Bob, and controller Claire, has been demonstrated. Note that again, to verify success of a truly tripartite quantum protocol, experiments for at least two different combinations should be performed. Now we describe experiments of teleportation for other two combinations. The one combination is sender Alice, receiver Claire, and controller Bob. The other is sender Claire, receiver Bob, and controller Alice. The configuration of the experimental setup is changed only locally, while the global configuration remains unchanged. Namely, the paths distributing the tripartite entangled states remain unchanged throughout the experiment. On the other hand, each party changes his or her setup locally according to their roles.

The gain dependence of the teleportation fidelity from Alice to Claire and from Claire to Bob are shown in Fig. 9b and 9c, respectively. Both figures show almost the same dependence as Fig. 9a. This ensures that the tripartite entanglement source utilized in the experiment has the same capability to perform teleportation for different combinations. The best fidelities are 0.63 ± 0.02 and 0.64 ± 0.02 ($g_{\mathrm{B}} \simeq g_{\mathrm{A}} \simeq 0.9$), respectively, which are greater than the classical limit $F = 1/2$ and show success of teleportation for these two combinations. In total, three different combinations are demonstrated. These results show success of a quantum teleportation network, *i.e.*, a truly tripartite quantum protocol.

4. Conclusion and Outlook

In this chapter we have presented the experimental realization of CV quantum teleportation for various input states. For a coherent state input, the fidelity of 0.70 is achieved. This value exceeds the no-cloning limit of 2/3. In the case of a squeezed state input, the squeezed variance is preserved through the teleportation process, and the performance of the process is evaluated using the fidelity for a mixed state. Teleportation of quantum entanglement, *i.e.*, entanglement swapping is realized, showing that this protocol enables two parties to share quantum entanglement where they have never directly interacted with each other. Furthermore we have described the experiment of a quantum teleportation network for CVs where coherent states are exchanged among three parties, exploiting a tripartite entanglement. This is considered to be the first demonstration of a truly tripartite quantum protocol.

The techniques used in this chapter are easily extendable to other quantum protocols, such as quantum telecloning,[43,44] a quantum teleportation network containing more than three parties[21] and error correction.[45] Improving the performance of a teleporter and realization of multipartite quantum protocols are of particular importance to realize more complicated quantum computation and quantum communication.

Acknowledgments

This work was partly supported by the MEXT and the MPHPT of Japan, and Research Foundation for Opto-Science and Technology.

References

1. C. H. Bennett *et al.*, Phys. Rev. Lett. **70**, 1895 (1993).
2. M. A. Nielsen and I. L. Chuang, *Quantum Computation and Quantum Information*, Cambridge University Press, Cambridge 2000.
3. S. L. Braunstein and A. K. Pati, *Quantum information with continuous variables*, Kluwer Academic Publishers, Dordrecht 2003.
4. D. Gottesman and I. L. Chuang, Nature (London) **402**, 390 (1999).
5. S. D. Bartlett and W. J. Munro, Phys. Rev. Lett. **90**, 117901 (2003).
6. D. Boumeester *et al.*, Nature (London) **390**, 575 (1997).
7. I. Marcikic *et al.*, Nature (London) **421**, 509 (2003).
8. M. Riebe *et al.*, Nature (London) **429**, 734 (2004).
9. M. D. Barrett *et al.*, Nature (London) **429**, 737 (2004).
10. L. Vaidman, Phys. Rev. A **49**, 1473 (1994).
11. S. L. Braunstein and H. J. Kimble, Phys. Rev. Lett. **80**, 869 (1998).
12. A. Einstein, B. Podolsky, and N. Rosen, Phys. Rev. **47**, 777 (1935).
13. A. Furusawa *et al.*, Science **282**, 706 (1998).
14. W. P. Bowen *et al.*, Phys. Rev. A **67**, 032302 (2003).
15. T. C. Zhang *et al.*, Phys. Rev. A **67**, 033802 (2003).
16. N. Takei *et al.*, Phys. Rev. Lett. **94**, 220502 (2005).
17. N. Takei *et al.*, Phys. Rev. A **72**, 042304 (2005).
18. J.-W. Pan *et al.*, Phys. Rev. Lett. **80**, 3891 (1998).
19. S. M. Tan, Phys. Rev. A **60**, 2752 (1999).
20. X. Jia *et al.*, Phys. Rev. Lett. **93**, 250503 (2004).
21. P. van Loock and S. L. Braunstein, Phys. Rev. Lett. **84**, 3482 (2000).
22. H. Yonezawa, T. Aoki, and A. Furusawa, Nature (London) **431**, 430 (2004).
23. T. Aoki *et al.*, Phys. Rev. Lett. **91**, 080404 (2003).
24. D. F. Walls and G. J. Milburn, Quantum Optics (Springer, Berlin, 1994).
25. L.-M. Duan *et al.*, Phys. Rev. Lett. **84**, 2722 (2000).
26. R. Simon, Phys. Rev. Lett. **84**, 2726 (2000).
27. S. L. Braunstein, C. A. Fuchs, and H. J. Kimble, J. Mod. Opt. **47**, 267 (2000).
28. R. Jozsa, J. Mod. Opt. **41**, 2315 (1994).
29. S. L. Braunstein *et al.*, Phys. Rev. A **64**, 022321 (2001).
30. K. Hammerer *et al.*, Phys. Rev. Lett. **94**, 150503 (2005).

31. F. Grosshans and P. Grangier, Phys. Rev. A **64**, 010301(R) (2001).
32. M. Ban, Phys. Rev. A **69**, 054304 (2004).
33. U. Leonhardt, *Measuring the quantum state of light*, Cambridge University Press, Cambridge 1997.
34. G. Adam, J. Mod. Opt. **42**, 1311 (1995).
35. M. Ban, M. Sasaki, and M. Takeoka, J. Phys. A: Math. Gen. **35**, L401 (2002).
36. J. Twamley, J. Phys. A: Math. Gen. **29**, 3723 (1996).
37. H. Jeong, T. C. Ralph, and W. P. Bowen, quant-ph/0409101 (2004).
38. M. Ban, J. Opt B, **6**, 224 (2004).
39. D. M. Greenberger *et al.*, Am. J. Phys. **58**, 1131 (1990).
40. W. Dür, G. Vidal, and J. I. Cirac, Phys. Rev. A **62**, 062314 (2000).
41. P. van Loock and A. Furusawa, Phys. Rev. A **67** 052315 (2003).
42. J. Jing *et al.*, Phys. Rev. Lett. **90**, 167903 (2003).
43. P. van Loock and S. L. Braunstein, Phys. Rev. Lett **87**, 247901 (2001)
44. S. Koike *et al.*, Phys. Rev. Lett. **96**, 060504 (2006).
45. S. L. Braunstein, Nature (London) **394**, 47 (1998).

Chapter 15

Quantum State Sharing with Continuous Variables

Tomáš Tyc

Institute of Theoretical Physics, Masaryk University, 61137 Brno, Czech Republic

Barry C. Sanders

Institute for Quantum Information Science, University of Calgary, Alberta T2N 1N4, Canada

Thomas Symul, Warwick P. Bowen, Andrew Lance and Ping Koy Lam

Quantum Optics Group, Department of Physics, Faculty of Science, Australian National University, ACT 0200, Australia

Quantum state sharing is an important quantum information protocol that deals with a secret quantum state distributed among a group of players, certain subsets of which can access the state while other subsets cannot get any information about it. Continuous-variable quantum state sharing uses light to carry the quantum information and linear optics for the state distribution and extraction, and has been realized experimentally for three players, any two of which are authorized to extract the secret state. The experimentally achieved fidelity of quantum state sharing exceeds the fidelity accessible without quantum resources, which puts the experimental realization well within the quantum domain. Quantum state sharing can be combined into complex quantum information protocols and can be expected to become an important part of full-scale quantum information processing protocols with continuous variables.

1. Introduction

Quantum communication is concerned with delivery of quantum information from a sender to one or more receivers, and the importance of quantum communication derives from its superiority over classical communication in terms of security,[1,2] information packing,[3,4] and transmitting quantum information with a network such as a distributed quantum computer. Quantum key distribution is an especially well-studied example of quantum communication; in this case Alice and Bob are on the same team and collaborate to share quantum information over a public channel to produce a secure random key for cryptographic applications. In this example Alice and Bob trust each other but no-one else. Quantum key distribution has been generalized to multiple parties who all trust each other.

Quantum state sharing builds on the principle of threshold secret sharing in information theory, which provides a secure means for sending information from a "dealer" to multiple "players" who cannot be trusted. In contrast to quantum key distribution scenarios where the the players trust each other (after authentication), secret sharing distributes information so that it can only be extracted by authorized subsets of players, and unauthorized subsets are denied any information. In threshold secret sharing, authorized subsets are those with membership not fewer than the threshold number and all subsets of players with fewer members than this threshold number are denied information. In quantum state sharing, the dealer entangles the quantum information into a state that resides in a larger Hilbert space and distributes components of this state to the players. The entanglement protocol is publicized and designed that only authorized subsets of players can collaborate to disentangle the state and extract the original quantum state whereas unauthorized subsets obtain states that are completely independent of the dealer's original state.

One important application of quantum state sharing is that a quantum state is shared by multiple, unreliable components in a quantum network; the state is thus robust against component failures as it can be reconstructed from remaining components. In this sense quantum state sharing is a special case of quantum error correction, and our theory and experiment represent significant advances in studies of quantum error correction for continuous variable systems; for further reading on quantum error correction, in continuous variable systems, see Refs. 5, 6, 7.

Quantum state sharing was first proposed under the title "How to share a quantum secret",[8] which is close to the name "Quantum secret sharing"[9] previously used to describe a protocol designed to enhance the security of classical information transmission against eavesdroppers by quantum means. To prevent misunderstandings caused by the similar nomenclature for distinct protocols, the term "quantum state sharing" (QSS) has been adopted to represent the quantum version of classical secret sharing protocols. At the same time, the term "secret" is still used for the quantum state to be shared, and we will use it in this chapter.

Whereas threshold quantum state sharing was introduced in the context of discrete quantum information theory,[8,10] it was first realized as a continuous variable quantum information protocol,[11,12] as squeezing and linear optics are the key requirements and are well developed quantum optics technologies. Furthermore the expensive part of quantum state sharing, namely the squeezers, scales nicely with an increasing number of players: the number of single-mode squeezers does not exceed the number of players by more than one.[13] A challenge in continuous variable quantum state sharing is the prevalence of gaussian states, which are non-orthogonal; hence complete extraction of quantum information by authorized subsets and complete denial of quantum information to unauthorized subsets can only be achieved in asymptotic limits of infinitely many photons. For finite systems fidelity is a useful figure of merit used to determine whether the protocol has been successful or not. The success of a continuous variable threshold quantum state sharing protocol has been reported,[11,12] where the observed fidelities exceed the upper bound that could be achieved without having an entangled resource; this criterion is analogous to the threshold fidelity for claiming success with continuous variable quantum teleportation.

In this chapter we explain the theory of quantum state sharing, describe the experiment, and provide the experimental results and corresponding analysis. The theoretical description is presented in the Heisenberg picture for the case of three players and a threshold of two players; an alternative analysis for an arbitrary number of players, which is presented in the Schrödinger picture, has been presented by Tyc and coworkers.[14,13]

2. Classical Secret Sharing

In classical information theory, secret sharing was proposed by Shamir[15] in 1979. The authorized groups that are able to access the secret are in advance chosen by the dealer and form the so-called access structure; the unauthorized groups form the adversary structure. Classical secret sharing exists for any adversary structure satisfying the natural requirement that with any group of players it contains also its subsets. An example of the access structure is the (k, n) threshold scheme in which there are total n players and any k of them can obtain the secret, while less than k cannot.

The realization of classical secret sharing is based on monotone span programs[16,10] and takes advantage of the properties of vector spaces and matrices over finite number fields. Both the encoding and decoding of the secret is achieved using linear functions over a finite field $\mathbb{F}$ and requires relatively little computational power. The simplest example of classical secret sharing is the $(2, 2)$ threshold scheme in which one player obtains a random bit r and the other player obtains the bit $r \oplus s$ where s is the secret bit and $\oplus$ means adding modulo 2. The players can then recover the secret bit by adding their shares modulo 2; at the same time, each single player has a random bit and cannot find any information about the secret on his own.

3. Quantum State Sharing with Discrete Variables

Classical secret sharing can be quite straightforwardly adapted to quantum state sharing in which the secret is quantum information carried by a pure or mixed state of a given quantum system. This system may be entangled with another system, so it is possible to share only a part of a complex quantum state. The dealer encodes the secret into an entangled state of n quantum systems by applying a suitable unitary transformation to the secret and $n - 1$ ancilla states, and distributes the n systems to the n players. If players from some authorized group wish to recover the secret, they apply another unitary transformation on their shares, and extract the secret state as a result, as well as some additional trash states. At the same time, players from an unauthorized group can undertake any action with their shares but the result will never show any kind of correlation with the original secret. The extracted secret has all the properties of the original secret, which means that possible entanglement of the secret with another system is also recovered.

In quantum state sharing, unlike classical secret sharing, not every access structure is allowed. In particular, the no-cloning theorem[17] prevents two disjoint groups of players from extracting the quantum secret because this would effectively mean cloning. For (k, n) threshold schemes this implies that only schemes with $n < 2k$ exist.[8] Also, in the extraction procedure one has to make sure that the extracted secret is disentangled from all other shares, which places some additional conditions on both the access structure and the extraction procedure. It turns out that one of these conditions is equivalent to the requirement of the no-cloning theorem. Some examples of access structures are shown in Fig. 1.

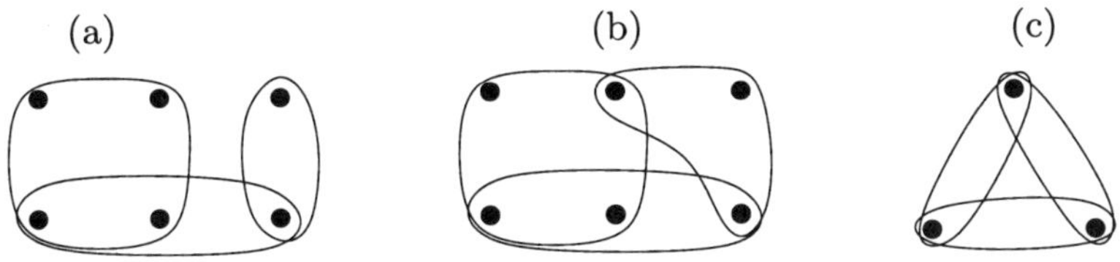

Fig. 1. Three examples of access structures; only the minimal authorized sets are shown. The access structure in (a) is allowed in classical secret sharing but not in quantum state sharing as two disjoint groups of players can access the secret; the access structure in (b) is allowed also in the quantum case, and (c) shows the access structure of the $(2, 3)$ threshold scheme.

Among quantum state sharing schemes there is an important class of the so-called self-dual access structures that enable encoding a pure secret state into a pure entangled state of the shares. A self-dual access structure has the following property: for every division of all players into two groups, exactly one group can extract the secret. It turns out that any access structure that is not self-dual can be derived from some self-dual one by discarding some shares.[10] Therefore exploring only self-dual structures is sufficient for describing quantum state sharing. As can be seen easily, a (k, n) threshold scheme is self-dual only if $n = 2k - 1$, so out of threshold schemes, only the $(k, 2k - 1)$ schemes are of interest.

Quantum state sharing can be implemented with both discrete and continuous variables. The discrete-variable version employs qudits, *i.e.*, quantum systems with discrete bases, as the carriers of the secret and the shares, and the discrete version of quantum state sharing has been developed,[8,10] in a tight analogy to classical secret sharing. For the rest of this chapter, we focus on the continuous-variable version of quantum state sharing that has been developed recently both theoretically[13,14,18] and experimentally.[11,12]

4. Quantum State Sharing with Continuous Variables

Continuous-variable (CV) version of quantum state sharing was proposed by Tyc and Sanders in 2002[14] and employs quantum systems with continuously indexed bases to carry the secret and the shares. The most important example of such systems are modes of electromagnetic field that also provide an advantageous realization of QSS employing interferometry with both passive optical elements (mirrors, beam splitters and phase shifters) and active elements (squeezers). Compared to the discrete-variable version of quantum state sharing, the CV QSS has both advantages and disadvantages. The key advantage is the possibility of its realization employing currently available optical elements. The major disadvantage of continuous-variable quantum state sharing is the necessity of infinitely squeezed states for the scheme to work perfectly. This is a general feature of CV protocols, *e.g.* of quantum teleportation.[19]

The theory of CV quantum state sharing was initially developed in the Schrödinger picture[13,14] in analogy to discrete-variable quantum state sharing. However, later the Heisenberg picture was employed,[11,12,18] which treats physical actions as transformations of Hermitian operators acting on the Hilbert space of the shares. We will use this approach in the following.

The CV system carrying the secret or its component (share) is described by the canonical position and momentum operators $\hat{x}$ and $\hat{p}$. The eigenvalue spectrum of $\hat{x}$ and $\hat{p}$ is the continuous set of real numbers $\mathbb{R}$, where also the term "continuous-variable" comes from. The operators $\hat{x}$ and $\hat{p}$ satisfy the commutation relation $[\hat{x}, \hat{p}] = \mathrm{i}$ (here we set the Planck constant $\hbar$ to unity). In optical systems, the position and momentum operators are referred to as the quadrature-phases of the field and can be measured via optical homodyne detection.[20] It is common to use scaled quadrature operators $\hat{X}^+, \hat{X}^-$ that are related to the field annihilation and creation operators $\hat{a}, \hat{a}^\dagger$ as

$$\hat{X}^+ = \hat{a} + \hat{a}^\dagger, \quad \hat{X}^- = \mathrm{i}(\hat{a}^\dagger - \hat{a}) \tag{1}$$

and satisfy the commutation relation $[\hat{X}^+, \hat{X}^-] = 2\mathrm{i}$, from which the Heisenberg uncertainty relation $\Delta\hat{X}^+\Delta\hat{X}^- \geq 1$ follows.

4.1. *Linear mode transformations*

Linear canonical (commutator-preserving) transformations of the mode quadratures are sufficient for performing a general CV quantum state sharing protocol. A transformation of this type over m modes that does not mix the positions with momenta can be written as

$$\hat{X}_i^{+\prime} = \sum_{j=1}^{m} T_{ij}\hat{X}_j^+, \qquad \hat{X}_i^{-\prime} = \sum_{j=1}^{m} S_{ij}\hat{X}_j^-, \tag{2}$$

where the matrices T and S satisfy $S = (T^{-1})^{\mathrm{T}}$ with ^{T} meaning the transposition.

Among the transformations (2), there is an important class of passive transformations for which the matrices T and S are orthogonal. Such transformations can be realized physically by passive optical elements only, that is, linear mode couplers (usually beam splitters) and phase shifters. Realizing non-orthogonal transformations of the field quadratures requires using active optical elements (such as optical parametric oscillators or feed-forward elements) and is much more challenging experimentally.

It turns out[21] that any matrix T from Eq. (2) can be decomposed as $T = O_2 D O_1$, where the matrices O_1 and O_2 are orthogonal and the matrix $D = \mathrm{diag}\,(d_1, \ldots, d_m)$ is diagonal. This way, any transformation (2) can be realized as a sequence of three steps: the first and last steps are passive transformations corresponding to O_1 and O_2, respectively. The middle step consists of m single-mode squeezing operations (phase-sensitive amplifications), each corresponding to one diagonal element of D and scaling the quadratures $\hat{X}^+$ and $\hat{X}^-$ by d_i and $1/d_i$ times, respectively. There are also other ways of realizing the transformation (2). Some of them will be discussed on the example of the $(2, 3)$ threshold scheme.

For succinctness, we shall not explain in detail a general quantum state sharing scheme. Instead, we will sketch the principle of the $(k, 2k-1)$ threshold scheme first and then explain the $(2, 3)$ scheme in detail, including its experimental realization.

5. The $(k, 2k-1)$ CV Quantum State Sharing Threshold Scheme

In the $(k, 2k-1)$ threshold scheme, there are $2k-1$ players and any k of them can extract the secret. Initially, the dealer prepares the secret state $|\psi\rangle$ and $2k-2$ ancilla states, of which $k-1$ are squeezed in the quadrature $\hat{X}^+$ and $k-1$ are squeezed in $\hat{X}^-$. To encode the secret, the dealer applies a particular transformation (2) on this initial state and distributes the resulting modes (shares) to the players. The transformation (2) has to satisfy certain conditions[14] to guarantee the access structure.

For each group of k players, there exists a transformation, again of the type (2), that enables extracting the secret. The fidelity of such extraction depends on the squeezing of the ancilla states of the dealer and for infinite squeezing, the extraction of the secret is perfect. If the squeezing is finite, which is of course always the case in experiment, then the extracted secret is partially degraded by noise from the squeezed quadratures of the dealer's ancilla states. At the same time, for infinite squeezing of the dealer's ancillas, any $k-1$ shares provide no information whatsoever about $|\psi\rangle$ due to the no-cloning theorem (as the remaining k players can fully extract the secret). If the squeezing is finite, then some information may escape to the adversaries.

It turns out that there is some freedom in the extraction procedure of the collaborating players, which can be exploited for optimizing the scheme in terms of number of active elements. It has been shown[13] that the extraction procedure can always be achieved with no more than two active single-mode optical elements (squeezers) plus passive elements (see Fig. 2). The encoding procedure of the dealer can be achieved, apart from preparation of the ancilla states, with passive elements only.

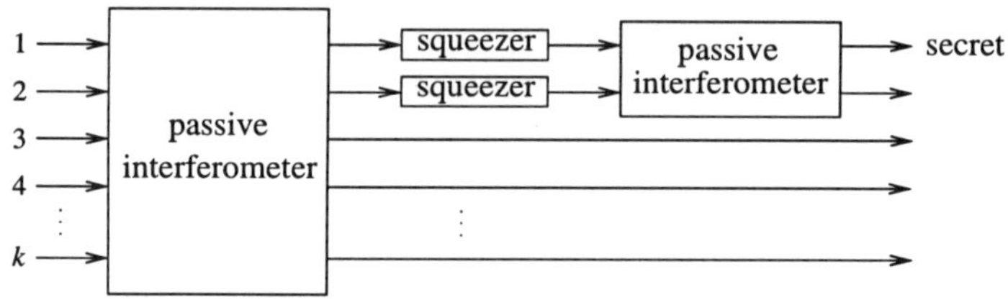

Fig. 2. The extraction procedure for the $(k, 2k-1)$ threshold scheme: the shares of the k collaborating players are first combined in a k-port passive interferometer. Two interferometer outputs are then amplified in two independent phase-sensitive amplifiers (single-mode squeezers) and the resulting beams are combined in a two-port passive interferometer (beam splitter). The extracted secret emerges from one of the outputs.

6. The $(2,3)$ Threshold Scheme

The $(2,3)$ threshold scheme is the simplest non-trivial scheme with a self-dual access structure. More importantly, it is the scheme that has been realized experimentally.[11,12] In the following we describe in detail a particular example of this scheme that has been realized experimentally.

6.1. *Encoding the secret state*

The dealer of the secret has initially the secret state and two squeezed ancillary states, the first one having reduced fluctuation of the quadrature $\hat{X}^-$ and the second one of $\hat{X}^+$. We will denote the quadratures of the three states by $X_{\rm S}^{\pm}$, $\hat{X}_{\rm sqz1}^{\pm}$ and $\hat{X}_{\rm sqz2}^{\pm}$, respectively. The squeezed states can be obtained from the vacuum states by applying the optical parametric amplification process (OPA). The dealer then performs the encoding

$$\begin{aligned} \hat{X}_{\rm player1}^{\pm} &= \hat{X}_{\rm S}^{\pm}/\sqrt{2} + (\hat{X}_{\rm sqz1}^{\pm} + \hat{X}_{\rm sqz2}^{\pm})/2 \\ \hat{X}_{\rm player2}^{\pm} &= \hat{X}_{\rm S}^{\pm}/\sqrt{2} - (\hat{X}_{\rm sqz1}^{\pm} + \hat{X}_{\rm sqz2}^{\pm})/2 \\ \hat{X}_{\rm player3}^{\pm} &= (\hat{X}_{\rm sqz1}^{\pm} - \hat{X}_{\rm sqz2}^{\pm})/\sqrt{2} \end{aligned} \qquad (3)$$

that can be achieved with a passive transformation in two steps (see Fig. 3). First, the ancillas are combined on a 1:1 beam splitter, thus forming an approximate EPR pair;[22] the quadratures $\hat{X}^+$ of the resulting beams are correlated while the quadratures $\hat{X}^-$ are anticorrelated. One of the beam splitter outputs is then combined with the secret state on another 1:1 beam splitter, which yields the first two shares; the last share is the second beam of the EPR pair. The three shares are then distributed to the players.

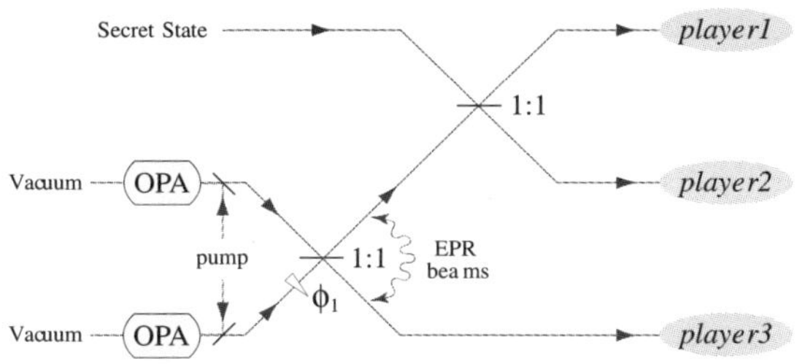

Fig. 3. The encoding of the secret in the (2, 3) threshold scheme: two ancilla states created in two optical parametric oscillators (OPA) and squeezed in the quadratures $\hat{X}^-, \hat{X}^+$, respectively, are combined on a 1:1 beam splitter, thus forming an approximate EPR pair. One of the beams is then combined with the secret state on another 1:1 beam splitter. The beam splitter outputs form two shares, the third share is formed by the second output of the first beam splitter.

The method of secret extraction depends on the collaborating pair of players who are extracting the secret. The authorized group $\{1, 2\}$ formed by players 1 and 2 can employ a passive transformation while the groups $\{1, 3\}$ and $\{2, 3\}$ need an active transformation to extract the secret.

6.2. *Extraction of the secret state by players 1 and 2*

If players 1 and 2 wish to extract the secret, they simply combine their shares on a 1:1 beam splitter (see Fig. 4 (a)). This way, a Mach-Zehnder interferometer is effectively formed of the dealer's and players' operations, and therefore we refer to this extraction scheme as to the Mach-Zehnder protocol. The quadratures of the beam splitter outputs are then

$$\hat{X}^{\pm}_{1\text{out}} = (\hat{X}^{\pm}_{\text{player1}} + \hat{X}^{\pm}_{\text{player2}})/\sqrt{2} = \hat{X}^{\pm}_{\text{S}}, \quad (4)$$
$$\hat{X}^{\pm}_{2\text{out}} = (\hat{X}^{\pm}_{\text{player1}} - \hat{X}^{\pm}_{\text{player2}})/\sqrt{2} = (\hat{X}^{\pm}_{\text{sqz1}} + \hat{X}^{\pm}_{\text{sqz2}})/\sqrt{2}.$$

We see that the quadratures of the first output exactly replicate the quadratures of the original secret, so the secret is obtained at the first beam split-

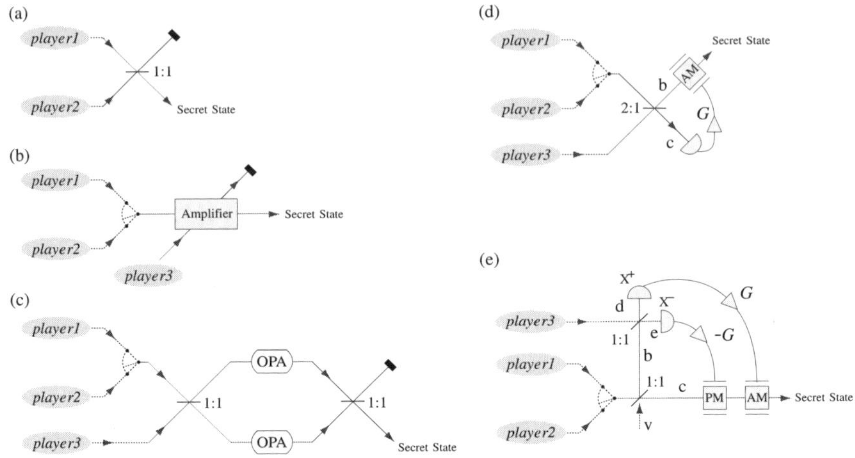

Fig. 4. The extraction of the secret in the (2, 3) threshold scheme: (a) players 1 and 2 simply combine their shares on a 1:1 beam splitter; players 1 and 3 (or 2 and 3) can employ (b) a phase-insensitive amplifier protocol, (c) two optical parametric amplifier protocol, (d) single feed-forward or (e) double feed-forward extraction protocols. The switch symbols represent the collaboration of either player 1 or player 2 with player 3 in the extraction protocols.

ter output. As the quadratures $\hat{X}^{\pm}_{\mathrm{sqz1}}$ and $\hat{X}^{\pm}_{\mathrm{sqz2}}$ do not enter the output quadratures $\hat{X}^{\pm}_{\mathrm{1out}}$, the authorized group $\{1, 2\}$ can extract the secret state to an arbitrary precision, independent of the amount of squeezing employed in the dealer protocol.

An especially simple (2, 2) threshold scheme can be naturally derived from the (2, 3) threshold scheme by keeping shares 1 and 2 and discarding share 3. Then it is even not necessary to use two squeezed ancilla states in the dealer's part of the protocol; it is enough to mix the secret state with a thermal state with a very large mean number of photons N (ideally $N \to \infty$) on a 50/50 beam splitter. Such a thermal state is equivalent to one output of the dealer's first beam splitter after tracing over the other output (share 3). The players then recombine their shares on another 50/50 beam splitter, thus again forming a Mach-Zehnder interferometer and extracting the secret state at one of the outputs.

6.3. *Extraction of the secret state by players 1 and 3*

The extraction of the secret by the authorized groups $\{1, 3\}$ and $\{2, 3\}$ is more complicated than for the group $\{1, 2\}$ because of the asymmetry of shares 1 (or 2) and 3 with respect to the content of the anti-squeezed ancillary state quadratures (see Eqs. (3)) that have to be canceled. An active operation is required for the secret extraction and can be realized experimentally in several different ways. In the following we discuss only the secret extraction by the group $\{1,3\}$ as the extraction by the group $\{2,3\}$ is almost identical.

Ideally, players 1 and 3 perform the two-mode active operation

$$\hat{X}^{\pm}_{\text{1out}} = \sqrt{2}\hat{X}^{\pm}_{\text{player1}} \mp \hat{X}^{\pm}_{\text{player3}}$$
$$\hat{X}^{\pm}_{\text{2out}} = \mp\hat{X}^{\pm}_{\text{player1}} + \sqrt{2}\,\hat{X}^{\pm}_{\text{player2}}\,. \tag{5}$$

This operation leads to the following quadratures of the first output:

$$\hat{X}^{+}_{\text{1out}} = \hat{X}^{+}_{\text{S}} + \sqrt{2}\,\hat{X}^{+}_{\text{sqz2}}, \qquad \hat{X}^{-}_{\text{1out}} = \hat{X}^{-}_{\text{S}} + \sqrt{2}\,\hat{X}^{-}_{\text{sqz1}}. \tag{6}$$

Eq. (6) shows that for infinite squeezing of the quadratures $\hat{X}^{-}_{\text{sqz1}}$ and $\hat{X}^{+}_{\text{sqz2}}$ of the dealer's ancilla states, the secret is exactly replicated at the first output.

6.3.1. *Phase insensitive amplifier protocol*

Theoretically, transformation (5) can be achieved directly by employing a phase insensitive amplifier[23] (see Fig. 4 (b)). However, experimentally it is extremely difficult to directly access the both input fields of the phase insensitive amplifier, which makes this method of secret extraction impossible at present.

6.3.2. *Two optical parametric amplifier protocol*

The second option for realizing the transformation (5) takes advantage of the decomposition of general linear mode transformations described in Sec. 4.1. The two shares are first interfered on a 1:1 beam splitter, the two resulting beams are each noiselessly amplified using phase sensitive optical parametric amplifiers with amplifying gains of $\sqrt{2}$ and $1/\sqrt{2}$, respectively, and the resulting beams are finally interfered on another 1:1 beam splitter (see Fig. 4 (c)). This way transformation (5) is completed, which yields the secret at one output. However, this extraction scheme requires significant quantum resources, namely two optical parametric amplifiers with precisely controlled amplifying gains and high non-linearity, and is susceptible to various types of losses and inefficiencies. Because of these problems, alternative extraction protocols have been developed that do not rely on optical parametric amplifiers, but instead utilize linear optics.

6.3.3. *Single feed-forward extraction protocol*

An alternative extraction protocol for players 1 and 3 employs linear optics and an electro-optic feed-forward. The experimental feasibility of this scheme is paid by the expense that the secret is not extracted in its original form, but subject to a unitary parametric (squeezing) transformation.

In this scheme, the shares 1 and 3 are first interfered on a 2:1 beam splitter, and one of the outputs is subject to balanced homodyne detection that measures its quadrature $\hat{X}^+$ (see Fig. 4 (d)). The detected signal is then imparted onto the $\hat{X}^+$ quadrature of the second beam splitter output with a well-defined electronic gain ($G = 2\sqrt{2}$) via an electro-optic feed-forward loop. The gain G as well as the beam splitter ratio 2:1 are chosen such that the anti-squeezed quadratures of the dealer's ancilla states cancel in the output quadratures. The quadratures of the output after the feed-forward are

$$\hat{X}^+_{\text{out}} = \sqrt{3}\,(\hat{X}^+_{\text{S}} + \sqrt{2}\hat{X}^+_{\text{sqz2}}), \quad \hat{X}^-_{\text{out}} = \frac{1}{\sqrt{3}}\,(\hat{X}^-_{\text{S}} + \sqrt{2}\hat{X}^-_{\text{sqz1}}). \tag{7}$$

Eqs. (7) show that in the limit of infinite squeezing in the dealer's protocol, this procedure reconstructs the quadratures of the secret state up to a scaling transformation and the extracted secret is directly related to the original secret state via a local unitary squeezing transformation. This transformation consists of scaling the quadratures $\hat{X}^+$ and $\hat{X}^-$ by the factors $\sqrt{3}$ and $1/\sqrt{3}$, respectively. The extracted secret contains all the quantum information contained in the original secret. To restore the original form of the secret, the squeezing transformation would have to be undone, which would require additional considerable quantum resources.

6.3.4. *Double feed-forward extraction protocol*

The disadvantage of the single feed-forward extraction protocol of reconstructing a squeezing-transformed secret can be avoided by employing a double feed-forward extraction protocol. In this scheme (see Fig. 4 (e)), share 1 is interfered with a vacuum state on a 1:1 beam splitter and one of the output beams is then interfered with share 3 on the second 1:1 beam splitter. The quadrature $\hat{X}^+$ of the first output of this second beam splitter is then measured as well as the quadrature $\hat{X}^-$ of its second output. The detected values of the quadratures are then transferred on the second output of the first beam splitter via an electronic feed-forward. The quadratures of the output state can be expressed as

$$\hat{X}^{+}_{\rm out} = \hat{X}^{+}_{\rm S} + \sqrt{2}\,\hat{X}^{+}_{\rm sqz2}, \quad \hat{X}^{-}_{\rm out} = \hat{X}^{-}_{\rm S} + \sqrt{2}\,\hat{X}^{-}_{\rm sqz1}, \tag{8}$$

which shows that in the case of infinite squeezing in the dealer protocol, the secret state is extracted to an arbitrary precision. This protocol has advantages over the previous protocols as it uses linear optics in the extraction procedure and the extracted state is in the same form as the original secret.

7. Characterization of the Extraction Quality

To characterize the quality of the extraction of the secret state, two measures are commonly used. The first one is the fidelity $\mathcal{F}$ which measures the overlapping of the extracted secret with the original. This measure, derived from the teleportation experiments,[24,25] presents the advantage of giving a single figure of merit to characterize the success of the quantum state sharing experiment. On the other hand, such a single figure does not indicate in general the level of quantum correlation between the shared and extracted secrets. The other option is to determine the signal transfer $\mathcal{T}$ and added noise $\mathcal{V}$ from the shared state to the extracted state. This measure is employed for quantum non-demolition (QND) experiments.[26]

7.1. *Fidelity*

Fidelity measures the overlap between the shared and extracted secrets; for pure secret states $|\psi\rangle$ it is defined as $\mathcal{F} = \langle\psi|\hat{\rho}_{\rm out}|\psi\rangle$,[27] where $\hat{\rho}_{\rm out}$ is the extracted secret density matrix. A fidelity of $\mathcal{F} = 1$ implies perfect overlap between the secret and extracted states and corresponds to state extraction with infinite precision.

We can define the classical fidelity for a given group of players as the extraction fidelity in case of the absence of quantum resources in the dealer's protocol, that is, for $\Delta\hat{X}^{\pm}_{\rm sqz1,2} = 1$. Assuming that the secret is a coherent state, the maximum achievable classical fidelities for the authorized groups can be calculated as

$$\mathcal{F}^{\rm clas}_{\{1,2\}} = 1, \qquad \mathcal{F}^{\rm clas}_{\{1,3\}} = \mathcal{F}^{\rm clas}_{\{2,3\}} = 1/2. \tag{9}$$

The average classical fidelity limit for the quantum state sharing scheme can be determined by averaging the maximum classical fidelity achievable by all the authorized groups. For the $(2,3)$ QSS scheme, the average classical fidelity is $\mathcal{F}^{\rm clas}_{\rm avg} = (\mathcal{F}_{\{1,2\}} + \mathcal{F}_{\{1,3\}} + \mathcal{F}_{\{2,3\}})/3 = 2/3$. This limit can only be exceeded using quantum resources in the dealer protocol. The average classical fidelity achievable for a general (k,n) quantum state sharing scheme with the secret in a coherent state is given by $\mathcal{F}_{\text{avg}}^{\text{clas}} = k/n$.

The maximum achievable classical fidelities can also be calculated for the adversary groups formed by the individual players:

$$\mathcal{F}_{\{1\}}^{\text{clas}} = \mathcal{F}_{\{2\}}^{\text{clas}} = 1/2, \qquad \mathcal{F}_{\{3\}}^{\text{clas}} = 0. \tag{10}$$

For increasing squeezing in the dealer protocol, the fidelities for the individual players approach zero, corresponding to decreasing overlap between the secret state and the individual shares, while the fidelities for the authorized groups approach unity, corresponding to a perfect extraction.

7.2. *Signal transfer and added noise*

In quantum state sharing, state extraction can also be characterized in terms of the signal transfer $\mathcal{T}$ to and additional noise $\mathcal{V}$ on the extracted state. These measures provide complementary information about state extraction compared with the fidelity measure.

Signal transfer is defined as $\mathcal{T} = R_{\text{out}}^{+}/R_S^{+} + R_{\text{out}}^{-}/R_S^{-}$, where $R^{\pm}$ are the quadrature signal-to-noise ratios. The additional noise on the extracted state is given by product of the quadrature conditional variances, $\mathcal{V} = V_{\text{S|out}}^{+} V_{\text{S|out}}^{-}$, that describe the amount of additional noise on the quadratures of the extracted secret state. Perfect state extraction corresponds to $\mathcal{T} = 2$ and $\mathcal{V} = 0$. Unlike fidelity that reaches its maximum possible value only if the extracted state is in the same form as the original secret, both $\mathcal{T}$ and $\mathcal{V}$ are invariant with respect to unitary transformations of the extracted state. Thus the measures $\mathcal{T}$ and $\mathcal{V}$ are more useful than $\mathcal{F}$ for describing the quality of state extraction *e.g.* for the single feed-forward protocol that extracts the secret up to a squeezing transformation (see Eqs. (7)).

The classical limit (no squeezing in the dealer protocol) of $\mathcal{T}$ and $\mathcal{V}$ for the authorized group $\{1,2\}$ can be calculated as

$$\mathcal{T}_{\{1,2\}}^{\text{clas}} = 2, \qquad \mathcal{V}_{\{1,2\}}^{\text{clas}} = 0, \tag{11}$$

which corresponds to state extraction to an arbitrary precision. For the $\{1,3\}$ and $\{2,3\}$ authorized groups, the maximum achievable signal transfer and the minimum achievable additional noise is given by

$$\mathcal{T}_{\{1,3\}}^{\text{clas}} = \mathcal{T}_{\{2,3\}}^{\text{clas}} = 1, \qquad \mathcal{V}_{\{1,3\}}^{\text{clas}} = \mathcal{V}_{\{2,3\}}^{\text{clas}} = 1/4. \tag{12}$$

For no squeezing in the dealer protocol, the $\{1\}$ and $\{2\}$ adversary groups can reach the values $\mathcal{T}_{\{1\}}^{\text{clas}} = \mathcal{T}_{\{2\}}^{\text{clas}} = 1$ and $\mathcal{V}_{\{1\}}^{\text{clas}} = \mathcal{V}_{\{2\}}^{\text{clas}} = 1/4$, so players 1 or 2 can individually reach some information about the secret. As the squeezing is increased in the dealer protocol, however, the amount of information the adversary groups obtain approaches zero. In the limit of infinite squeezing the adversary groups obtain no information about the secret state, corresponding to $\mathcal{T} = 0$ and $\mathcal{V} = \infty$, while the access structures obtain a perfect extraction.

8. Experimental Realization of the (2, 3) Threshold Scheme

The (2,3) threshold quantum state sharing scheme has been recently demonstrated experimentally.[11,12] The experimental setup is composed of a Nd:YAG laser at 1064 nm that is pumping a hemilitic $MgO : LiNbO_3$ second harmonic generator. The frequency-doubled light is used to pump two hemilitic $MgO : LiNbO_3$ optical parametric amplifiers that are producing two beams squeezed 4.5 ± 0.2 dB below the shot noise limit. The squeezed beams are used to produce a pair of entangled beams whose entanglement is estimated to be $V^{+}_{\mathrm{EPR1|EPR2}} V^{-}_{\mathrm{EPR1|EPR2}} = 0.58 \pm 0.02 < 1$ using the EPR criterion.[28]

In the experiment, the secret quantum state is a displaced coherent state at the sideband frequency of 6.12 MHz of the coherent laser field. The secret state is mixed with one beam of the EPR entangled pair and the two resulting beams, as well as the remaining beam of the EPR pair, are distributed to the three players as shown in Fig. 3. The secret is then extracted either from shares 1 and 2 using the Mach-Zehnder extraction protocol (see Fig. 4 (a)), or from share 2 and 3 (alternatively from 1 and 3) by the single feed forward extraction protocol (see Fig. 4 (d)). The shared and extracted secrets are measured using a balanced homodyne detection system via a configuration of removable mirrors.

To increase the security of the scheme, additional Gaussian noise was added onto the three shares using electro-optic modulation techniques. This noise does not degrade the secret extracted by the authorized groups while it reduces the information that can escape to adversary players if the dealer uses finite squeezing.

In order to determine the fidelity, an a posteriori symplectic transform is applied to the extracted state of access structures {2,3}. The fidelities obtained for different gains are represented in Fig. 5 (a). The best fidelities obtained are $\mathcal{F}_{\{1,2\}} = 0.95 \pm 0.05$ and $\mathcal{F}_{\{2,3\}} = 0.62 \pm 0.02$ corresponding to the best average fidelity of $\mathcal{F}_{\mathrm{avg}} = 0.73 \pm 0.02$, which exceeds the classical limit of $\mathcal{F}^{\mathrm{clas}}_{\mathrm{avg}} = 2/3$.

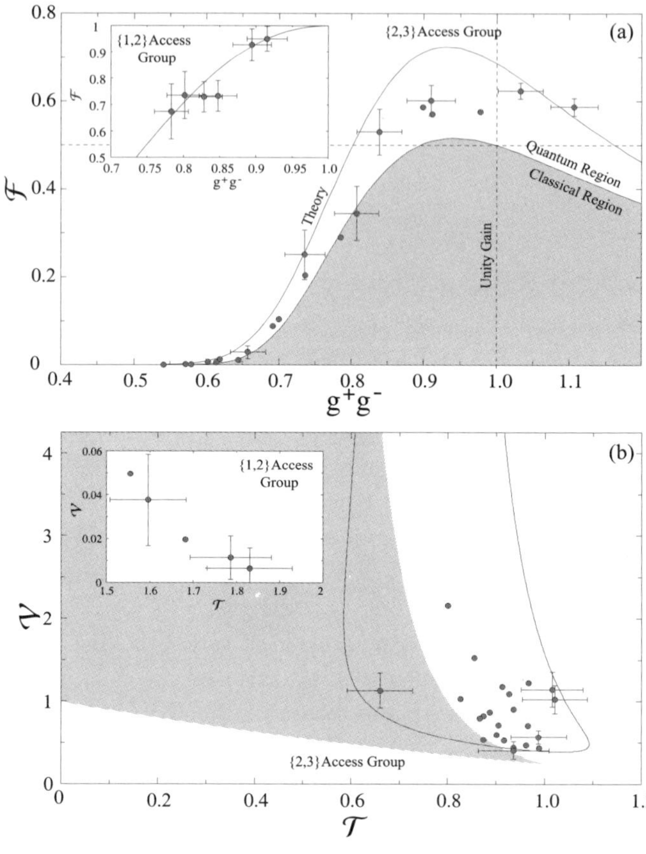

Fig. 5. (a) Experimental fidelity $\mathcal{F}$ for the $\{2,3\}$ authorized group with 4.5 dB of squeezing in the dealer protocol as a function of the optical gain product g^+g^-. Solid line is the theoretical curve with 3.5 dB of additional Gaussian noise, and a feed-forward detector efficiency of $\eta_{\mathrm{ff}} = 0.93$. Grey area marks the classical region for the authorized group. Inset shows fidelity for the {1,2} authorized group as a function of the optical gain product. (b) Experimental signal transfer $\mathcal{T}$ and added noise $\mathcal{V}$ for the authorized group $\{2,3\}$ with -4.5 dB of squeezing in the dealer protocol. Solid line is the theoretical curve, the grey area again marks the classical region for the authorized group $\{2,3\}$. Inset shows $\mathcal{T}$ and $\mathcal{V}$ for the authorized group $\{1,2\}$.

The experimental signal transfer $\mathcal{T}$ and added noise $\mathcal{V}$ are represented in Fig. 5 (b). The best values obtained are $\mathcal{T}_{\{1,2\}} = 1.83 \pm 0.10$, $\mathcal{V}_{\{1,2\}} = 0.01 \pm 0.01$ and $\mathcal{T}_{\{2,3\}} = 1.01 \pm 0.06$, $\mathcal{V}_{\{2,3\}} = 0.41 \pm 0.11$. As these results exceed the theoretical classical limits, they clearly demonstrate the quantum nature of the (2,3) quantum state sharing scheme.

9. Conclusion

In conclusion we have demonstrated quantum state sharing in a continuous variable system. Moreover this demonstration is not post-selected, which is a limitation of discrete variable quantum optical implementations of quantum information protocols: the quantum state sharing experiment is 'on demand'. Fidelities were observed to be well within the quantum domain, which corresponds to the regime for which entangled sources are required. Although the extracted secret state was not of the same form as the original secret state, it is equivalent up to a known unitary transformation with no free parameters; this state is then sufficient for anticipated quantum state sharing applications and saves the cost of one squeezer.

Further work is required to extend continuous variable quantum state sharing to more than three players, but the method is clear from our theoretical analyses, and fortunately the number of squeezers does not have to exceed the number of players by more than one.[13] Another important area for future research is partial sharing of quantum states. In this case the dealer does not distribute the entire quantum state to the players, instead providing one component of an entangled state to the players and retaining the other component. Partial sharing of quantum states would provide an excellent test of the performance of quantum state sharing experiments with respect to preserving the fragile entanglement between the two components of the dealer's state. Furthermore it is interesting to note that the average fidelity of a quantum state sharing apparatus would not inform the user as to the performance of partial state sharing, in analogy to partial continuous variable quantum teleportation.[29]

As discussed at the beginning of this chapter, quantum state sharing is a special case of quantum error correction,[8] which is an important consideration in protecting continuous variable quantum information.[6] Another interesting application would be incorporating quantum state sharing with continuous variable quantum key distribution[30] for the case that the multiple players cannot be trusted in small groups.

Quantum state sharing can be viewed within the broader framework of a quantum information network, consisting of nodes where information is stored and processed, connected by quantum channels. Many protocols have been proposed to take advantage of the quantum nature of the channels. These include dense coding[3] which allows information to be sent faster than is allowed classically; quantum state transfer and entanglement distribution;[31–33] quantum teleportation;[34] and quantum cryptography.[35]

Predominantly, these protocols focus on communication between pairs of nodes. However, complexities arise since many users, or nodes, will always be present in a realistic network. Furthermore, in some instances significant benefits can be obtained by multi-node protocols, over their two-node counterparts. The importance of protocols for multi-node communication is therefore fast becoming apparent. Quantum state sharing is one such protocol, facilitating a broad range of multi-node quantum communication objectives such as hierarchical quantum cryptography, and multi-node quantum error correction. Other related multi-node protocols include distributed quantum computation systems,[36] quantum teleportation networks,[37] quantum cryptography systems immune to the presence of many nodes,[38] distributed quantum cryptography systems,[39] and multi-node dense coding.[40,41]

These examples show how quantum state sharing can be combined into general quantum information protocols with continuous variables. Most importantly, quantum state sharing is a major addition to our quantum information toolbox that will be required for full-scale quantum information processing with continuous variables.

References

1. C. H. Bennett and G. Brassard, in *Proc. of IEEE Inter. Conf. on Computers, Systems and Signal Processing, Bangalore, India*, IEEE, New York 1984, p. 175.
2. A. K. Ekert, Phys. Rev. Lett. **67**, 661 (1991).
3. C. H. Bennett and S. J. Wiesner, Phys. Rev. Lett. **69**, 2881 (1992).
4. H. Buhrman, R. Cleve, J. Watrous and R. de Wolf, Phys. Rev. Lett. **87**, 167902 (2001).
5. S. Lloyd and J.-J. E. Slotine, Phys. Rev. Lett. **80**, 4088 (1998).
6. S. L. Braunstein, Phys. Rev. Lett. **80**, 4084 (1998).
7. R. L. Barnes, arXiv quant-ph/0405064.
8. R. Cleve, D. Gottesman, and H.-K. Lo, Phys. Rev. Lett. **83**, 648 (1999).
9. M. Hillery, V. Bužek and A. Berthiaume, Phys. Rev. A **59**, 1829 (1999).
10. A. D. Smith, arXiv quant-ph/0001087.
11. A. M. Lance, T. Symul, W. P. Bowen, B. C. Sanders, and P. K. Lam, Phys. Rev. Lett. **92**, 177903 (2004).
12. A. M. Lance, T. Symul, W. P. Bowen, B. C. Sanders, T. Tyc, T. C. Ralph, P. K. Lam, Phys. Rev. A **71**, 33814 (2005).
13. T. Tyc, D. Rowe and B. C. Sanders, J. Phys. A **36**, 7625 (2003).
14. T. Tyc and B. C. Sanders, Phys. Rev. A **65**, 42310 (2002).
15. A. Shamir, Comm. of the ACM **22**, 612 (1979).
16. M. Karchmer and A. Wigderson, In Proc. 8th Ann. Conf. Struct. Compl. Theory, San Diego, 102 (1993).

17. W. K. Wootters and W. H. Zurek, Nature **299**, 802 (1982).
18. A. M. Lance, T. Symul, W. P. Bowen, T. Tyc, B. C. Sanders and P. K. Lam, New J. Phys. **5**, 4 (2003).
19. L. Vaidman, Phys. Rev. A **49**, 1473 (1994); S. L. Braunstein and H. J. Kimble, Phys. Rev. Lett. **80**, 869 (1998); A. Furusawa *et al.*, Science **282**, 706 (1998).
20. H. P. Yuen and J. H. Shapiro, IEEE Trans. Inf. Theory **IT-25**, 179 (1979); **IT-26**, 78 (1980).
21. S. L. Braunstein, arXiv quant-ph/9904002.
22. A. Einstein, B. Podolsky and N. Rosen, Phys. Rev. **47**, 777 (1935).
23. C. M. Caves, Phys. Rev. D **26**, 1817 (1982).
24. A. Furusawa *et al.*, Science **282**, 706 (1998).
25. W. P. Bowen, N. Treps, B. C. Buchler, R. Schnabel, T. C. Ralph, H.-A. Bachor, T. Symul, and P. K. Lam, Phys. Rev. A **67**, 032302 (2003).
26. J.-Ph. Poizat, J.-F. Roch, and P. Grangier, Ann. Phys. (Paris)**19**, 265 (1994).
27. B. Schumacher, Phys. Rev. A **51**, 2738 (1995).
28. T. C. Ralph and P. K. Lam, Phys. Rev. Lett. **81**, 5668 (1998).
29. T. J. Johnson, S. D. Bartlett and B. C. Sanders, Phys. Rev. A **66**, 042326 (2002).
30. F. Grosshans and P. Grangier, Phys. Rev. Lett. **88**, 057902 (2002).
31. J. I. Cirac, P. Zoller, H. J. Kimble and H. Mabuchi, Phys. Rev. Lett. **78**, 3221 (1997).
32. L.-M. Duan, M. Lukin, J. I. Cirac, and P. Zoller, Nature **414**, 413 (2001).
33. A. Kuzmich, W. P. Bowen, A. D. Boozer, A. Boca, C. W. Chou, L.-M. Duan and H. J. Kimble, Nature **423**, 731 (2003).
34. C. H. Bennett, G. Brassard, C. Crepeau, R. Jozsa, A. Peres, and W. K. Wootters, Phys. Rev. Lett. **70**, 1895 (1993).
35. C. H. Bennett, Phys. Rev. Lett. **68**, 3121 (1992).
36. J. I. Cirac, A. K. Ekert, S. F. Huelga and C. Macchiavello, Phys. Rev. A **59**, 4249 (1999).
37. H. Yonezawa, T. Aoki, and A. Furusawa, Nature **431**, 430 (2004).
38. P. D. Townsend, Nature **385**, 47 (1997).
39. Y.-A. Chen, A.-N. Zhang, Z. Zhao, X.-Q. Zhou, C.-Y. Lu, C.-Z. Peng, T. Yang and J.-W. Pan, quant-ph/0502131 (2005).
40. J. Jing, J. Zhang, Y. Yan, F. Zhao, C. Xie and K. Peng, Phys. Rev. Lett. **90**, 167903 (2003).
41. D. Bruss, G. M. D'Ariano, M. Lewenstein, C. Macchiavello, A. Sen(De), and U. Sen, Phys. Rev. Lett. **93**, 210501 (2004).

Chapter 16

Experimental Quantum Cloning with Continuous Variables

Ulrik L. Andersen,[1] Vincent Josse,[1] Norbert Lütkenhaus[2] and Gerd Leuchs[1]

[1] *Institut für Optik, Information und Photonik,*
Max-Planck Forschungsgruppe, Universität Erlangen-Nürnberg,
Günther-Scharowsky-Str. 1, 91058 Erlangen, Germany

[2] *Institute für Theoretische Physik I,*
Institut für Optik, Information und Photonik,
Max-Planck Forschungsgruppe, Universität Erlangen-Nürnberg,
Staudtstr. 7, 91058 Erlangen, Germany

In this chapter we present a scheme for optimal Gaussian cloning of optical coherent states. Its optical realization is based entirely on simple linear optical elements and homodyne detection. This is in contrast to previous proposals where parametric processes were suggested to be used for optimal Gaussian cloning. The optimality of the presented scheme is only limited by detection inefficiencies. Experimentally we achieved a cloning fidelity of up to 65%, which almost touches the optimal value of 2/3.

1. Introduction

An electro-magnetic field of light in an unknown pure quantum state cannot be exactly copied. Lying at the heart of quantum mechanics, this property, known as the "no-cloning theorem"[1,2] has fundamental consequences in quantum information science: it is, for instance, the essence of security in quantum cryptography[3] and it also puts some limitations on the distribution of quantum information.[4] However, it is possible, using a so-called quantum cloning machine, to make approximate copies with higher quality than could be done using any classical means.[5,7,8] In this chapter we describe a recent experiment that demonstrates near optimal cloning of a coherent state by the use of such a quantum cloning machine.[9]

Although quantum information protocols were initially developed in the context of discrete variables (where the information is encoded in qubits), continuous variable (CV) quantum protocols constitute a rapidly emerging alternative and offer several distinct advantages, such as unconditional processing of information, as well as the relative ease in manipulating and preparing CV states.[10] Hence these protocols have attracted much attention and many interesting experiments have been realized in this regime.[11–18] However, quantum cloning has heretofore been restricted to the discrete variable regime where the polarization state of single photons has been conditionally copied.[22–25] Despite many theoretical proposals,[26,27,29,30] unconditional quantum cloning of coherent states was indeed still missing in the toolbox of CV protocols, probably due to the difficult implementations of the proposed schemes relying on parametric interactions. In contrast, in this chapter we show that linear optics and homodyne detection suffice.

2. Theory

First let us introduce the framework of this chapter, where the continuous variables considered are the quadrature amplitudes of the electro-magnetic field. In particular we consider the amplitude quadrature $\hat{x}$ and the phase quadrature $\hat{p}$, which are related to the photon annihilation operator via $\hat{a} = \frac{1}{2}(\hat{x}+i\hat{p})$. The non-commutativity of $\hat{x}$ and $\hat{p}$ (with the normalization $[\hat{x}, \hat{p}] = 2i$) implies the presence of quantum noise. For coherent states (which will be subject to cloning) this quadrature noise has Gaussian statistics with identical variances ($\Delta^2\hat{x}_{in} = \Delta^2\hat{p}_{in} = 1$); the state is therefore uniquely described by the expectation values $x_{in} = \langle\hat{x}_{in}\rangle$ and $p_{in} = \langle\hat{p}_{in}\rangle$ (see Fig. 1). Here we will consider a Gaussian cloning machine, which means that the input Gaussian state is mapped onto other output Gaussian states.

The efficiency of the cloning machine is typically quantified by the fidelity, which gauges the similarity between an input state $|\alpha_{in}\rangle$ and an output state described by the density operator, ρ_{out}. It is defined by $F = \langle\alpha_{in}|\rho_{out}|\alpha_{in}\rangle$.[28] This function (that gives a measure of the overlap between the signal and the clones in phase space) depends in general both on the amplitudes (via the cloning gains defined as $g_x = x_{clone}/x_{in}$ and $g_p = p_{clone}/p_{in}$) and on the output noise variances $\Delta^2\hat{x}_{clone}$ and $\Delta^2\hat{p}_{clone}$. Nevertheless, in the particular case of unity gains, the fidelity simply reads:[11]

$$F = \frac{2}{\sqrt{(1 + \Delta^2 x_{clone})(1 + \Delta^2 p_{clone})}}. \tag{1}$$

2.1. *Classical cloning*

In conventional classical information processing, information is encoded in states which can be copied at will. A standard approach to do this uses a measure-and-prepare strategy,[33,34,36] where the states are optimally measured and subsequently reconstructed based on the measurement outcomes (see upper drawing of Fig. 1): The signal is projected onto a coherent state basis by mixing it with vacuum at a 50/50 beam splitter.[35] Amplitude and phase are measured at the beam splitter outputs and the gained information (which is optimal) is used to construct the clones. This is the classical approach also referred to as classical teleportation. However, using such a procedure to clone a quantum state (for instance here a coherent state of light), two additional units of quantum noise are added to the clones: one due to the measurement (the fundamental penalty for measuring simultaneously two non commuting variables) and one due to the reconstruction procedure. Although this method enables the production of an infinite number of clones ($1 \rightarrow \infty$ cloner), the optimal fidelity is limited to 1/2,[11,33,34,36] rendering this approach as a classical probability distributor.

2.2. *Quantum cloning*

2.2.1. *Previous proposals*

What kind of transformation is needed to break the classical boundary? It has been shown that a quantum circuit for optimal $1 \rightarrow 2$ quantum cloning in all Hilbert space dimensions with no *a priori* information consists of four C-NOT transformations.[4] In particular, in the continuous variable regime, four continuous variable C-NOT operations, with appropriate prepared input ancilla states, can be used to construct the optimal coherent state cloner. This reasoning was used in the proposal of Fiurasek *et al.*,[30] where it was suggested that coherent state cloning can be performed using the off-resonant interaction between light and four atomic ensembles, each of which performs a CNOT operation. The resulting clones are stored in the quantum states of the collective atomic spins of the clouds. It was also mentioned that the realization could be simplified by replacing the last two CNOT operations by a homodyne measurement of a Stokes variable, the information of which was fed back onto the remaining two samples. Finally

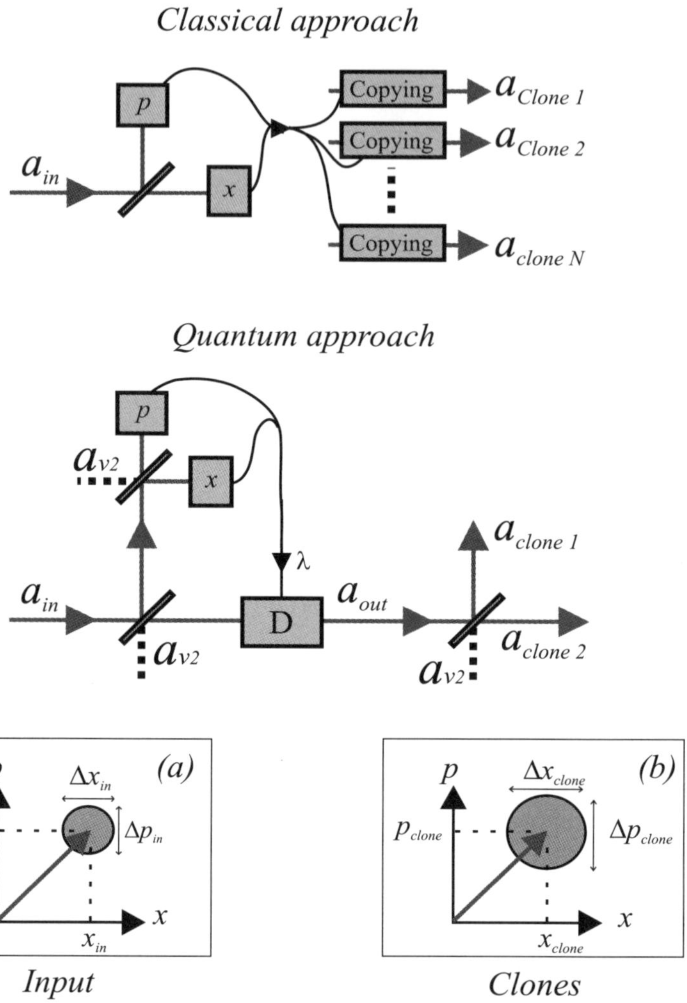

Fig. 1. Schematic drawing showing the principle of the CV cloning schemes using a classical approach and an optimal quantum approach. In the classical approach, the input coherent state $\hat{a}_{in}$ is optimally estimated (by measuring simultaneously $\hat{x}$ and $\hat{p}$) and subsequently recreated based on the measurement estimation. Contrarily, the optimum quantum approach measures only a part of the state optimally (again by measuring simultaneously $\hat{x}$ and $\hat{p}$) and uses this estimation to displace (D) the other part with a gain λ. The displaced state a_{out} is then split into two optimal clones, $\hat{a}_{clone1}$ and $\hat{a}_{clone2}$. $\hat{a}_{v1}$, $\hat{a}_{v2}$ and $\hat{a}_{v3}$ are vacuum inputs. Inserts (a) and (b) are displaying the characteristics of the input state and one of the clones in a phase space diagram.

a scheme by which one clone is stored in an atomic ensemble while the other clone is encoded into an electromagnetic field was proposed.

In Ref. 26 it was suggested to use three non-degenerate optical parametric amplifiers. However a cloning machine based on this approach is never optimal since it would require unrealistic infinite squeezing, and furthermore the experimental realization of this machine is rather complicated.

Finally, Fiurasek[29] and Braunstein *et al.*[27] proposed to use a single phase insensitive amplifier and a beam splitter to perform optimal Gaussian cloning of coherent states. In this scheme the amplifier is simply set to amplify the input state by a factor of two while the beam splitter divides the amplified state into two clones. Their scheme was generalized to encompass N→M cloning. In this case an N-splitter, collecting the N modes into one mode, was placed in front of the amplifier, which is then succeeded by an M-splitter that divides the output mode into M modes. Fiurasek[29] also suggested a scheme for asymmetric cloning by embedding the symmetric cloning scheme into a Mach-Zehnder interferometer.

2.2.2. *Our proposal*

Our approach to cloning is schematically illustrated in Fig. 1 and runs as follows. At the input side of the cloning machine the injected unknown coherent state is divided by a 50/50 beam splitter. One output beam is transformed into

$$\hat{x}_s = \frac{1}{\sqrt{2}}(\hat{x}_{in} - \hat{x}_{v1}) \tag{2}$$

$$\hat{p}_s = \frac{1}{\sqrt{2}}(\hat{p}_{in} - \hat{p}_{v1}) \tag{3}$$

for the amplitude and phase quadrature respectively, where $\hat{x}_{v1}$ and $\hat{p}_{v1}$ refer to the uncorrelated vacuum modes at the beam splitter and $\hat{x}_{in}(\hat{p}_{in})$ is the amplitude (phase) quadrature of the input signal. The other output is optimally estimated: the state is split at yet another 50/50 beam splitter and at one output the amplitude quadrature

$$\hat{x}_m = \frac{1}{2}(\hat{x}_{in} + \hat{x}_{v1}) + \frac{1}{\sqrt{2}}\hat{x}_{v2} \tag{4}$$

is measured while the phase quadrature

$$\hat{p}_m = \frac{1}{2}(\hat{p}_{in} + \hat{p}_{v1}) - \frac{1}{\sqrt{2}}\hat{p}_{v2} \tag{5}$$

is measured at the other output port. Here $\hat{x}_{v2}$ and $\hat{p}_{v2}$ represent uncorrelated vacuum modes at the beam splitter (see Fig. 1). After the measurements the quantum quadratures become classically random variables. This classical information is then used to displace the other half of the input state with an appropriate scaling factor noted λ. Let us pause at this point to have a closer look at the feedforward transformation in the Heisenberg picture.

Feedforward means that we perform an arbitrary projection measurement on one subsystem and then apply a controlled unitary operation on another subsystem, that depends on a parameter which is chosen as a function of the measurement result. In our realization we perform homodyne measurements in two different subsystems, represented by the observables $\hat{x}_m$ and $\hat{p}_m$, and the measurement outcomes are represented by their eigenstates and corresponding eigenvalues x_m and p_m. On the signal subsystem (the system which is not measured) we act with a unitary operation which depends on the measurement outcomes of the measurements, that is in general terms, we apply a unitary operation $\hat{U} = e^{-i\hat{H}(x_m,\hat{s})}$ with a Hamiltonian that depends on the measurement outcomes x_m and p_m and an operator $\hat{s}$ acting on the signal system. We can now describe the total feedforward alternatively in the following picture: We apply to the *total system* an interaction described by $\hat{U} = e^{-i\hat{H}(\hat{x}_m,\hat{s})}$ where the operator $\hat{x}_m$ replaces the measurement result x_m in $\hat{H}(x_m, \hat{s})$. This is sufficient to describe the evolution of the signal system since we trace out the "measured" subsystems after the interaction. The two pictures are illustrated in Fig. 2.

In our scheme, where $\hat{x}_m$ and $\hat{p}_m$ are measured with outcomes x_m and p_m, we displace the signal system according to the displacement operators $\hat{D}(x_m) = \exp(\lambda x_m(\hat{a}_s - \hat{a}_s^\dagger))$ and $\hat{D}(p_m) = \exp(\lambda p_m(\hat{a}_s - \hat{a}_s^\dagger))$ along the real and imaginary axis respectively, where $\hat{a}_s = \frac{1}{2}(\hat{x}_s + i\hat{p}_s)$. Using our alternative picture by the global unitary operator we get a total unitary operator: $\hat{U} = \exp(\lambda\hat{x}_m(\hat{a}_s - \hat{a}_s^\dagger)\exp(\lambda\hat{p}_m(\hat{a}_s - \hat{a}_s^\dagger)$. This simply gives rise to the transformations: $\hat{x}_s \to \hat{U}^\dagger\hat{x}_s\hat{U} = \hat{x}_s + \lambda\hat{x}_m$ and $\hat{p}_s \to \hat{U}^\dagger\hat{p}_s\hat{U} = \hat{p}_s + \lambda\hat{p}_m$. Therefore the displaced states in our scheme are

$$\hat{x}_{out} = (\frac{1}{\sqrt{2}} + \frac{\lambda}{2})\hat{x}_{in} + (\frac{\lambda}{2} - \frac{1}{\sqrt{2}})\hat{x}_{v1} + \frac{\lambda}{\sqrt{2}}\hat{x}_{v2},$$
$$\hat{p}_{out} = (\frac{1}{\sqrt{2}} + \frac{\lambda}{2})\hat{p}_{in} + (\frac{\lambda}{2} - \frac{1}{\sqrt{2}})\hat{p}_{v1} - \frac{\lambda}{\sqrt{2}}\hat{p}_{v2}. \tag{6}$$

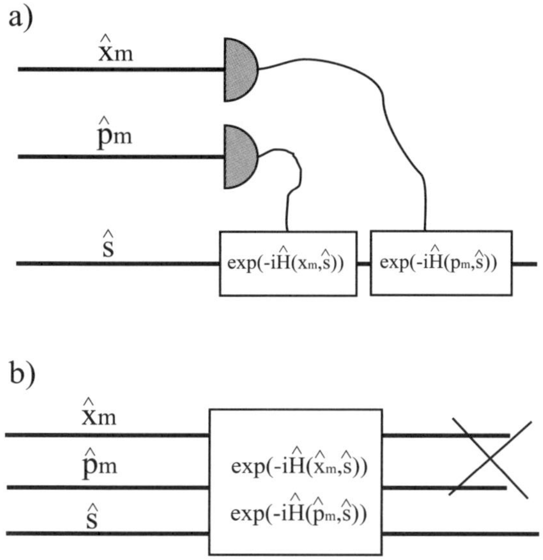

Fig. 2. Diagram of the feedforward loop using two different pictures. a) This shows the standard picture and resembles the actual experiment: The two subsystems are measured, the outcomes of which are fedforward and used to control the signal system via a unitary operation. b) Alternatively the feedforward operation can be replaced by a unitary operation followed by tracing out the controlling system.

In a final step the state is split into two beams by a 50/50 beam splitter, and for a scaling factor of $\sqrt{2}$ the machine generates two output clones:

$$\begin{aligned}
\hat{x}_{clone1} &= \hat{x}_{in} + \frac{1}{\sqrt{2}}\hat{x}_{v2} + \frac{1}{\sqrt{2}}\hat{x}_{v3}, \\
\hat{p}_{clone1} &= \hat{p}_{in} - \frac{1}{\sqrt{2}}\hat{p}_{v2} + \frac{1}{\sqrt{2}}\hat{p}_{v3}, \\
\hat{x}_{clone2} &= \hat{x}_{in} + \frac{1}{\sqrt{2}}\hat{x}_{v2} - \frac{1}{\sqrt{2}}\hat{x}_{v3}, \\
\hat{p}_{clone2} &= \hat{p}_{in} - \frac{1}{\sqrt{2}}\hat{p}_{v2} - \frac{1}{\sqrt{2}}\hat{p}_{v3},
\end{aligned} \tag{7}$$

where $\hat{x}_{v3}$ and $\hat{p}_{v3}$ are uncorrelated vaccum noise entering the last beam splitter. The transformations in Eq. (7) are known to describe an optimal Gaussian cloning machine for a flat input distribution.[27,29] In particular we see that it is invariant with respect to rotation and displacement in phase space as required by a phase covariant cloner. The variances of the clones for the amplitude and phase quadratures are

$$\Delta^2\hat{x}_{clone} = \Delta^2\hat{x}_{in} + 1, \tag{8}$$

$$\Delta^2\hat{p}_{clone} = \Delta^2\hat{p}_{in} + 1. \tag{9}$$

The fidelity is then 2/3 (Eq. (1)), which is exactly the optimal fidelity for Gaussian cloning of coherent states[31] We note that it has recently been shown that, allowing for non-Gaussian cloning transformations, the fidelity can be slightly enhanced to 0.6826.[32]

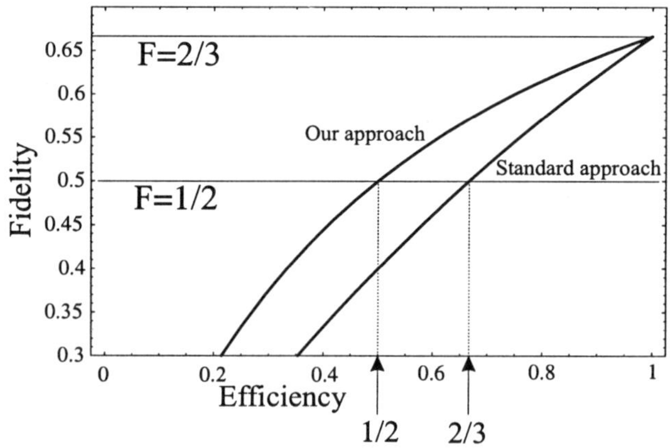

Fig. 3. The fidelity versus the efficiency of the cloner. For our approach the efficiency refers to the measurement efficiency of half of the signal, whereas for the standard approach the efficiency represents the efficiency by which the state is coupled into the amplifer. We clearly see that our approach is more robust than the standard approach.

In contrast to the classical approach where two units of vacuum are added, using the quantum approach only one unit of quantum noise is added. This single unit of vacuum noise corresponds exactly to the unavoidable price that must be paid when a joint measurement on two conjugate quadratures on a single system is performed. This demonstrates the existence of a close link between optimal measurement theory and cloning.

2.3. *Robustness*

It is easy to show that the cloning approach presented above is very robust with respect to small variations in the beamsplitting ratios as well as to losses in the measurement system. Varying the beamsplitting ratios from 40/60 to 60/40 gives rise to only a tiny variation in the added noise contributions and hence the cloning fidelity. For example the fidelity stays above 65% when the reflectivity of the first beam splitter is varied from 30% to 70%. Furthermore, the robustness of the scheme with respect to the efficiency in the detection process (which is basically the only loss introduced in the machine) is illustrated in Fig. 3. We see that the classical boundary is crossed only when the measurement efficiency is less than 50%. This is compared to the standard realization of a cloner (consisting of a standard amplifier followed by a beam splitter), in which the main part of the loss occurs at the incoupling to the amplifier. Plotting the fidelity as a function of the incoupling loss give us a curve which is decreasing more rapidly than in the previous case, hereby proving the robustness of our scheme.

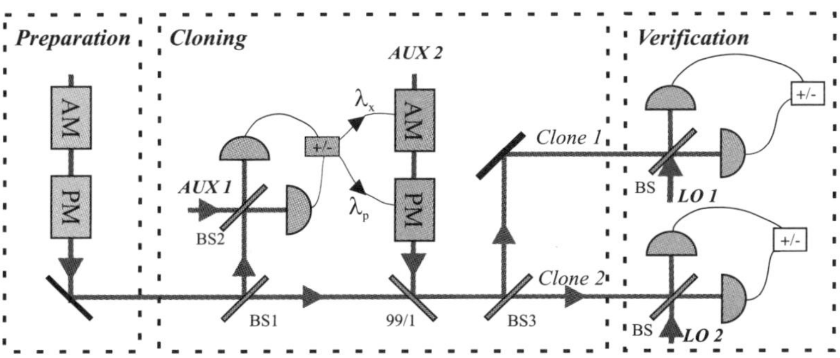

Fig. 4. Schematic of the experimental cloning setup. It is divided into three boxes defining the preparation stage (where an arbitrary input coherent state can be generated), the cloning stage (where two clones are produced) and the verification stage (where the quality of the cloning process is quantified using homodyne detection). BS: Beam splitter, λ: Electronic gain, LO: Local oscillator, AM: Amplitude modulator, PM: Phase modulator and AUX: Auxiliary beam.

3. Experiment

We now proceed to the discussion of the experimental demonstration of the quantum cloning protocol described above. First we present the experimental setup shown in Fig. 2. The laser source for our experiment was a monolithic Nd:YAG(yttrium aluminum granat) nonplanar ring laser at 1064nm delivering 500mW of power in a single transverse mode. A small part of the power was used to create an input signal to the cloning machine, whereas the rest served as local oscillator beams and auxiliary beams. The setup comprised three parts: a preparation stage, a cloning stage and finally a verification stage.

3.1. *Preparation*

Following previous realizations of CV quantum protocols,[11,14,16] we define the quantum state to be modulation sidebands at a fixed frequency (here 14.3 MHz with a bandwidth of 100 kHz) of a bright light field. At this frequency the laser was found to be shot noise limited, ensuring a pure coherent input state. An arbitrary input state is then easily generated by independently controling the modulations of the amplitude (x_{in}) and the phase (p_{in}), using two electro-optical modulators. A feedback loop was implemented to stabilize the optical power of the light emitted by the amplitude modulator in order to keep a consistent power level, thereby facilitating the evaluation procedure.

3.2. *Cloning*

The prepared state is then directed to the cloning machine where it is divided into two halves by the first beam splitter (BS1). One of the halves is combined with an auxiliary beam (AUX1) at the second beam splitter (BS2) with a $\pi/2$ relative phase shift and balanced intensities. The sum and difference of the photocurrents generated by two high quantum efficiency photodiodes then provide the simultaneous measurement of amplitude and phase.[37] (This approach is similar to the method where two independent local oscillators probe the amplitude and phase quadratures of the two beam splitter outputs as described in the previous section.) These electronic signals are amplified appropriately to ensure unity cloning gains and are subsequently used to displace the other half of the signal beam in a way that prevents losses: the feedforward modulation is first applied to an auxiliary beam (AUX2) which is combined with the signal beam on a 99/1 beam splitter.[11] In a final step, the clones are generated at the output of the third beam splitter (BS3). These states are then finally directed to two independent homodyne detectors for verification.

3.3. *Verification*

To characterize the performance of the cloning machine, the spectral noise properties of the two clones are measured by two homodyne detectors with strong local oscillator beams (LO1 and LO2). Since the statistics of the involved light fields are Gaussian we need only measure two conjugate quadratures to fully characterize the states. Therefore the homodyne detectors were set to measure stably - employing electronic servo feedback loops - either the amplitude or the phase quadrature. We note that the input state is also measured by the same homodyne detectors, to ensure a consistent comparison between the input state and the clones.

3.4. *Results*

An example of a cloning run is reported in Fig. 3. The spectral densities of the amplitude and phase quadratures are here shown over a 1 MHz frequency span for the input state (ii) and the two clones (iii). From these traces the coherent amplitude of the various fields, $x_{in,out}$ and $p_{in,out}$, are measured by the heights of the peaks at 14.3 MHz relative to the quantum noise level (i). We measure an average photon number of 62 in a 100 kHz window. As evident from the figure, the electronic gains of the feed-forward loops are adjusted such that the cloning gains are close to unity (which corresponds to a 6 dB difference between the measured input signal and the output signals, due to the degradation of the input signal by BS1 and BS3). In order to simplify the following analysis of the measurement data, we will assume unity gains and will later consider the consequences of small deviations from unity which is the case for real cloning machines. From Fig. 3 it is also evident that additional noise has been added to the clones, relative to the input state which is a result of the cloning action. In order to accurately quantify the performance of the cloning machine, we estimated precisely this amount of added noise at 14.3 MHz (in a 100 kHz bandwidth). To do so, we switched off the modulations of the input beam, and recorded the noise in a zero span measurement over 2 seconds. These results are displayed in Fig. 4, where the added noise in amplitude and phase are reported for both clones. To avoid an erroneous underestimation of the noise power, the traces are corrected to account for the detection efficiencies of the two homodyne stations (which amount to 78.5% and 77.5%). After correction the noise contributions are 3.28±0.13 dB (3.16±0.13 dB) and 3.20±0.11 dB (3.15±0.13 dB) for the amplitude quadrature and phase quadrature of clone 1(2). From these data, the fidelities of the two copies can be easily determined using Eq. 9 and are found to be 64.3±0.01 (clone 1) and 65.2±0.01 (clone 2). These values clearly demonstrate the successful operation of our cloning machine, since they significantly surpass the

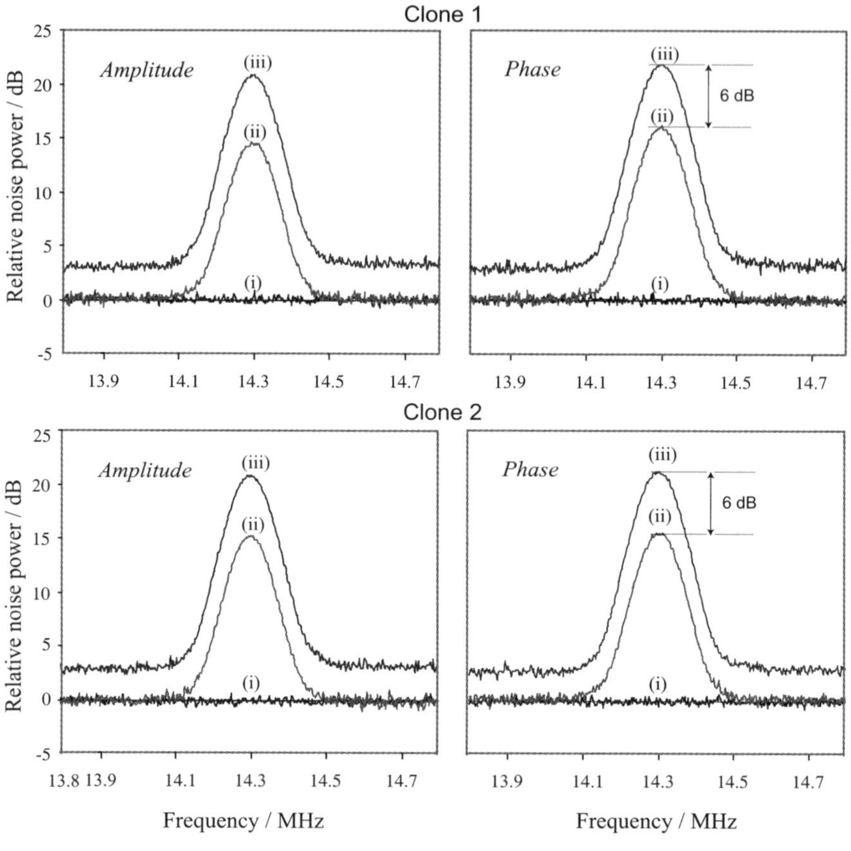

Fig. 5. Results of the quantum cloning protocol for one particular coherent state input. Spectral densities for amplitude and phase are shown (with an 1 MHz span and an 100kHz bandwidth) for the input state (traces (ii)) and both output clones 1 and 2 (traces (iii)). All the results have been normalised to the quantum noise limit (i). As can be seen, the coherent amplitudes are conserved during the cloning transformation (unity gain) but fundamental noise has been added, preventing exact replication, which is a direct consequence of the "no-cloning" theorem. The signal considered here corresponds to a bright modulation with an average photon number of n=62 per unit bandwidth per unit time (from Ref. 9).

maximum classical fidelity of 50% and approach the optimal value (within the Gaussian regime) of $2/3 \approx 66.7\%$.

The ideal cloner is phase insensitive since it adds equal and uncorrelated noise to all quadratures of the input state. To validate that our cloner was indeed phase insensitive, we recorded the spectral variances of all quadra-

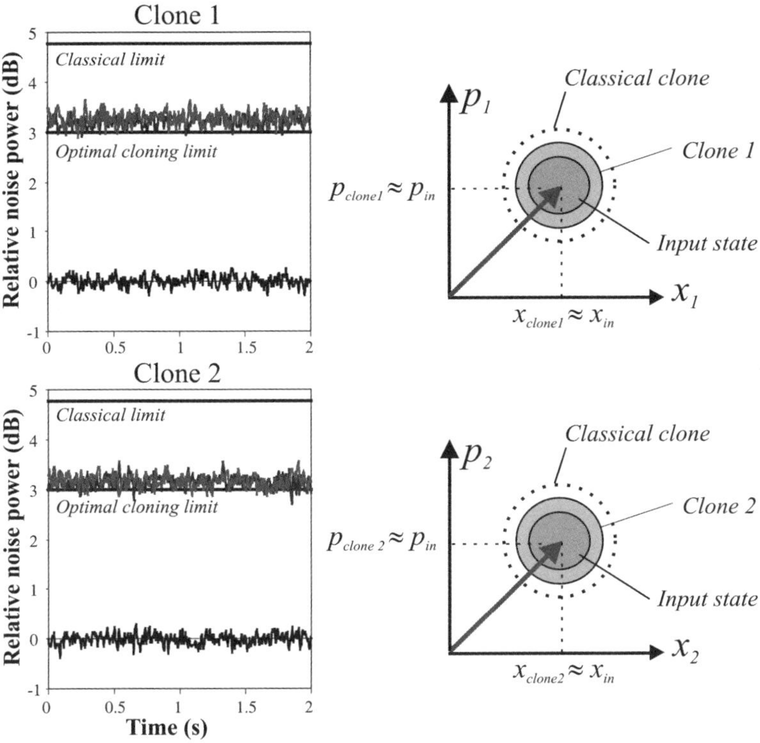

Fig. 6. Precise estimation of the added noise at 14.3 MHz in the quantum cloning protocol. The input modulation being switched off, the amplitude (red trace) and phase (blue trace) quadrature noise of both clones are recorded in a zero span measurement (over 2 seconds in a 100kHz bandwidth) and normalized to the quantum noise level (black trace). As reference levels, the optimal cloning limit (corresponding to +3dB, *i.e.* one unit of quantum noise) and the classical limit (corresponding to +4.8dB, *i.e* two units of quantum noise) are shown by solid lines. On the right hand side we schematically represent the clones in the phase space diagram where different noise sources have been artificially separated: small disc: quantum noise level of the coherent input, large disc: added noise for the experimentally achieved clones, dotted line: contour of the noise area for classical clones (from Ref. 9).

tures of the two output clones by scanning the phases of the local oscillators in the two homodyne detectors. Such measurements are depicted in Fig. 6, from which we conclude that the cloner is almost phase insensitive due to the lack of structure of the noise traces. This also demonstrates the purity of the displacements in the feedforward loop, since a cross coupling between the two quadratures will result in intra quadrature correlations, which in

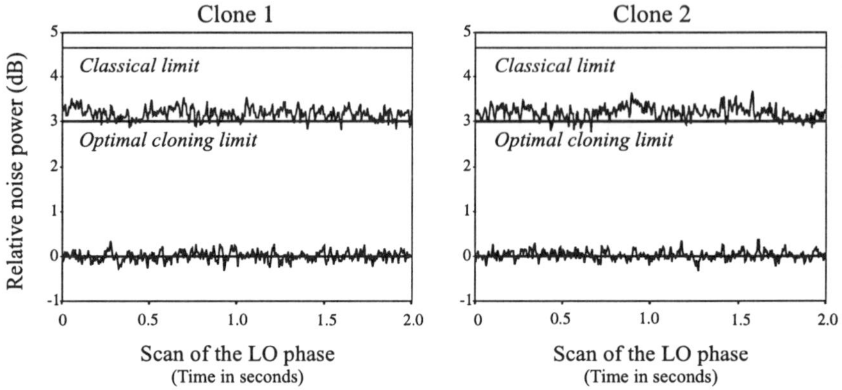

Fig. 7. Here we show a demonstration of the phase insensitivity of our quantum cloner. The spectral variances (with a band width of 100kHz) of clone 1 (left) and clone 2 (right) are measured as the local oscillator phases are scanned. This allows full determination of the output Gaussian states. Since the traces are almost structureless, we can conclude that the clones are phase insensitive.

turn will result in an oscillating structure of the noise traces while scanning the local oscillator.

The performance of our system is limited solely by imperfections in the in-line feedforward loop, which include non-unity quantum efficiency of the diodes, electronic noise of the detector circuit and non-perfect interference contrast at the beam splitter BS2 (see Fig. 2). The electronic noise was overcome by designing new detectors with ultra high sensitivity at the measurement frequency:[38] With these detectors the electronic noise was negligible and our machine was limited only by the non-optimum interference contrast, measured to 99%, and the diode quantum efficiencies of about 95%. This amounts to an overall detection efficiency of 93%. Note that the efficiency of this detector system is much higher than it is for the systems used in the verification stage. This is because we paid special attention to the selection photo diodes as well as the mode overlap in the former system, due to its importance to the efficiency of the cloning machine. As a result of the imperfect in-line detector, the expected fidelity is

$$F = \frac{2\eta}{2\eta + 1}, \tag{10}$$

where η is the efficiency of the detector. With the values stated above we calculate an expected fidelity of 65%, which is in nice agreement with the experimental results. Note that even with an imperfect detection system, the fidelity is close to the optimum of 2/3, proving the robustness of the cloning scheme.

4. Non-Unity Gain Cloning

In the above discussion on the experimental evaluation of the cloning machine we assumed unity cloning gains. However, experimental imperfections lead to a small deviation from unity. The gains were precisely determined by monitoring the spectral densities at 14.3 MHz over 2 s for the input and output states when a strong modulation corresponding to 230 photons in a 100 kHz window was applied. The accurately determined gains were found to be $g_{x1} = 0.96 \pm 0.01$ and $g_{p1} = 1.00 \pm 0.01$ for clone 1 and $g_{x2} = 1.03 \pm 0.01$ and $g_{p2} = 1.03 \pm 0.01$ for clone 2 for the amplitude and phase quadratures respectively. For non-unity cloning gains the expression for the "single-shot" fidelity is[11]

$$F_{(1,2)} = \frac{2\exp\left(-\frac{\alpha_{x,in}^2(1-g_{x(1,2)})^2}{1+\Delta^2 x_{in}} - \frac{\alpha_{p,in}^2(1-g_{p(1,2)})^2}{1+\Delta^2 p_{in}}\right)}{\sqrt{(1+\Delta^2 x_{clone(1,2)})(1+\Delta^2 p_{clone(1,2)})}}. \tag{11}$$

As a result of the deviations from unity gain, the fidelity depends on the photon number of the input coherent state (or equivalently the expectation values of the amplitude and phase quadratures). Therefore the figure of merit is defined by an average of the "single-shot" fidelities: $F_{ave(1,2)} = \int P(|\alpha_{in}\rangle)F_{(1,1)}d(\alpha_{in})$ where $P(|\alpha_{in}\rangle)$ is the probability distribution over a predefined set of input states.[33,39] Assuming this input distribution to be Gaussian with a width of Δn, the average fidelity is

$$F_{ave} = \frac{2}{\sqrt{((1+\Delta^2 x_{clone}) + 2\Delta n(1-g_x)^2)((1+\Delta^2 p_{clone}) + 2\Delta n(1-g_p)^2)}} \tag{12}$$

for both clones. As a result of the *a priori* information about the input distribution of coherent states the optimal fidelity obtained by classical means (*e.g.* by classical teleportation) is increased. The average classical fidelity is given by the expression[33,36]

$$F_{clas,ave} = \frac{\frac{1}{2} + \Delta n}{\frac{1}{2} + 2\Delta n}. \tag{13}$$

For example, considering a Gaussian distributed set of input coherent states with a spread in photon number of $\Delta n = 50$ (which is a huge number in quantum information science), the average quantum fidelities equal 61.3% and 61.5%, while the average classical fidelity is 50.3%. Based on the measured gains, the optimal fidelity with this set of states is 63%. Despite the fact that the gains are not exactly unity, the obtained fidelities are far above the classical limits and approach the optimal limits for a large set of input states, demonstrating the suitability of this cloning machine for realistic experimental quantum information tasks.

5. Other Cloning Functions

The presented scheme is easily extendable to an asymmetric cloner which produces output clones of different quality,[29] a procedure which is crucial in controlled partial information transfer between different parties in a network. Furthermore, our cloner can also be easily generalized to a N→M Gaussian cloner which takes N originals and produces M clones:[31] An N-splitter combining N identical input modes into one spatial mode is placed in front of the cloning machine. The transmittivity T of the beam splitter PBS1 as well as the feedforward gain λ should be chosen appropriately ($T = N/M$ and $\lambda = \sqrt{2(M-N)/N}$) and finally the clones are produced using an M-splitter, dividing the output from the feedforward loop into M spatial modes.[27,29,31] Hence, using the approach outlined in this chapter the optimal N→M Gaussian cloner is composed of only beam splitters and homodyne detection. There is no need for a nonlinearity to perform quantum cloning as previously proposed; the nonlinearity can be substituted with simple linear optics and homodyne measurements.

6. Conclusion

In conclusion, we have in this chapter described a quantum cloning protocol by which a coherent light state has been unconditionally cloned with an achieved fidelity around 65%, far above the classical limit of 1/2 and almost reaching the fundamental limit of 2/3. In contrast to previous theoretical proposals our scheme is based entirely on simple linear optical elements and homodyne detection. This approach enables near optimal partial transfer of quantum information between parties, which probably is an important step towards the realization of a complex quantum communication network. Furthermore, the scheme presented here allows for a malicious eavesdropper to optimally attack certain quantum cryptographic protocols.

Acknowledgments

This work has been supported by the Schwerpunktprogramm 1078 of the Deutsche Forschungsgemeinschaft, the network of competence QIP of the state of Bavaria (A8) and the EU grants COVAQIAL under project no. FP6-511004 and SECOQC under project no. IST-2002-506813. ULA also acknowledges funding from the Alexander von Humboldt Foundation.

References

1. W.K. Wootters and W.H. Zurek, Nature **299**, 802 (1982).
2. D. Dieks, Phys. Lett. A **92**, 271 (1982).
3. N. Gisin, G. Ribordy, W. Tittel, and H. Zbinden, Rev. Mod. Phys. **74**, 145 (2002).
4. S. Braunstein, V. Buzek, and M. Hillery, Phys. Rev. A **63**, 052311 (2001).
5. V. Buzek and M. Hillery, Phys. Rev. A **54**, 1844 (1996).
6. V. Buzek and M. Hillery, Phys. Rev. Lett. **81**, 5003 (1998).
7. N.J. Cerf, A. Ipe, and X. Rottenberg, Phys. Rev. Lett. **85**, 1754 (2000).
8. G. Lindblad, J. Phys. A **33**, 5059 (2000).
9. U.L. Andersen, V. Josse, and G. Leuchs, Phys. Rev. Lett. **94**, 240503 (2005).
10. S. Braunstein and P. van Loock, to appear in Rev. Mod. Phys. http://lanl.arxiv.org/abs/quant-ph/0410100
11. A. Furusawa, J.L. Sørensen, S.L. Braunstein, C.A. Fuchs, H.J. Kimble, and E.S. Polzik, Science **282**, 706 (1998).
12. X. Li, Q. Pan, J. Jing, J. Zhang, C. Xie, and K. Peng, Phys. Rev. Lett. **88**, 047904 (2002).
13. F. Grosshans, G. van Assche, J. Wenger, R. Brouri, N.J. Cerf, and Ph. Grangier, Nature **421**, 238 (2003).
14. A. M. Lance, T. Symul, W. P. Bowen, B. Sanders, and P. K. Lam, Phys. Rev. Lett. **92**, 177903 (2004).
15. U. L. Andersen, O. Glöckl, S. Lorenz, G. Leuchs, and R. Filip, Phys. Rev. Lett. **93**, 100403 (2004).
16. H. Yonezawa, T. Aoki, and A. Furusawa, Nature **431**, 430 (2004).
17. B. Julsgaard, J. Sherson, J.I. Cirac, J. Fiurasek, and E.S. Polzik, Nature **432**, 482 (2004).
18. X. Jia, X. Su, Q. Pan, J. Gao, C. Xie, and K. Peng, Phys. Rev. Lett. **93**, 250503 (2004).
19. E.F. Galvao and L. Hardy, Phys. Rev. A **62** 022301 (2000).
20. F. Grosshans and N.J. Cerf, Phys. Rev. Lett. **92**, 047905 (2004).
21. S. Braunstein, V. Buzek, and M. Hillery, Phys. Rev. A **63**, 052313 (2001).
22. A. Lamas-Linares, Ch. Simon, J.C. Howell, and D. Bouwmeester, Science **296**, 712 (2002).
23. S. Fasel, N. Gisin, G. Ribordy, V. Scarani, and H. Zbinden, Phys. Rev. Lett. **89**, 107901 (2002).
24. I.A. Khan and J.C. Howell, Phys. Rev. A **70**, 010303(R) (2004).

25. M. Ricci, F. Sciarrino, C. Sias, and F. De Martini, Phys. Rev. Lett **92**, 047901 (2004).
26. G.M. D'Ariano, F. De Martini, and M.F. Sacchi, Phys. Rev. Lett. **86**, 914 (2001).
27. S.L. Braunstein, N.J. Cerf, S. Iblisdir, P. van Loock, and S. Massar, Phys. Rev. Lett. **86**, 4938 (2001).
28. B. Schumacher, Phys. Rev. A **54**, 2614 (1996).
29. J. Fiurasek, Phys. Rev. Lett. **86**, 4942 (2001).
30. J. Fiurasek, N.J. Cerf, and E.S. Polzik, Phys. Rev. Lett. **93**, 180501 (2004).
31. N.J. Cerf and S. Iblisdir, Phys. Rev. A **62**, 040301(R) (2000).
32. N.J. Cerf, O. Krüger, P. Navez, R.F. Werner, and M.M. Wolf, Phys. Rev. Lett. **95**, 070501 (2005).
33. S.L. Braunstein, C.A. Fuchs, and H.J. Kimble, J. Mod. Opt. **47**, 267 (2000).
34. F. Grosshans and Ph. Grangier, Phys. Rev. A **64**, 010301(R) (2001).
35. E. Arthur and J.L. Kelly, Bell Syst. Technol. **44**, 725 (1965).
36. K. Hammerer, M.M. Wolf, E.S. Polzik, and J. I. Cirac, Phys. Rev. Lett. **94**, 150503 (2005).
37. G. Leuchs, T.C. Ralph, Ch. Silberhorn, and N. Korolkova, Jour. Mod. Opt. **46**, 1927 (1999).
38. B. Menegozzi, U.L. Andersen, V. Josse, and G. Leuchs, manuscript in preparation.
39. P.T. Cochrane, T.C. Ralph, and A. Dolinska, Phys. Rev. A **69**, 042313 (2004).

Chapter 17

Quantum Imaging Techniques for Improving Information Extraction from Images

Claude Fabre and Nicolas Treps

Laboratoire Kastler Brossel, Université Pierre et Marie Curie, case 74, 75252 Paris cedex 05, France

Hans A. Bachor and Ping Koy Lam

ARC Centre of Excellence for Quantum-Atom Optics, Building 38 The Australian National University, Canberra ACT 0200, Australia

1. Quantum Imaging: An Example of Multimode Quantum Optics

Optical images convey a great quantity of information in a parallel way, which can be extracted using "pixellized" detectors, like CCD cameras. They are often used to record, process and store information. The information in images is inevitably degraded by uncontrolled fluctuations arising from the quantum nature of light, in the same way as the information carried by the total intensity of a simple beam of light. Recent research, labelled under the name of "Quantum Imaging", has studied the local quantum fluctuations of light. It was discovered that one could tailor them, of course within the constraint imposed by Heisenberg inequalities, and that one could produce spatial quantum entanglement and create strong correlations between measurements performed on different pixels. These ideas can be applied to the problem of information processing in images: quantum techniques have the potential to improve the performance of information extraction from images. They can be used to detect faint changes induced by the variation of some parameters, or to detect very small details in images. A detailed account of the recent progress in the field of Quantum Imaging can be found in Ref. 1.

In this chapter we will outline some of the main features of this new field of quantum optics. In particular we will describe one special aspect of quantum imaging: the improvement of information extraction in the case where only a few parameters are known to change the transverse distribution of light.

Whereas measurements performed on a total monochromatic beam are concerned only with two parameters, namely the two quadrature components, the description of images, even monochromatic, requires the measurement of a great number of parameters, several millions when using current CCD sensors for example. From a quantum point of view, this means that the state vector describing such an image will decompose over photon states of many transverse modes of the electromagnetic field, whereas a single mode is enough to describe measurement performed on a total beam. This huge change in the size of the Hilbert in which the state evolves is one of the main features of quantum imaging: it increases the complexity of the problem, but also the possibilities of finding specific quantum states of light that can be used to improve information processing in a given situation. Another characteristics of quantum imaging is the freedom of choice of the appropriate transverse mode basis: Gaussian TEM_{pq} modes or plane waves modes $exp(i\vec{q}.\vec{r})$, but also more unusual bases, as we will see below, are likely to simplify the analysis of a given problem.

2. Quantum Imaging Using Many Modes

At any point $\vec{r}$ of the transverse plane of an image, the positive frequency part of the electric field operator (assumed to have a given polarization $\vec{\epsilon}$ and frequency ω_0) can be written as:

$$\hat{E}(\vec{r}) = \sqrt{\frac{\hbar\omega_0}{2\epsilon_0 cT}} \sum_{\ell} \hat{a}_\ell u_\ell(\vec{r}) \tag{1}$$

with $\left[\hat{E}(\vec{r}), \hat{E}^\dagger(\vec{r'})\right] = \frac{\hbar\omega_0}{2\epsilon_0 cT}\delta(\vec{r} - \vec{r'})$. Here T is the exposure time, $\{u_\ell(\vec{r})\}$ a complete set of modes -an orthonormal basis for the vector space of scalar functions defined in the 2D image plane- and $\hat{a}_\ell$ the annihilation operator of photons in mode ℓ. The measurement of the field intensity on the pixel number i of area A_i, assuming perfect quantum efficiency, is described, in power units, by the operator:

$$\hat{I}_i = 2\epsilon_0 c \int_{A_i} d^2 r \hat{E}^\dagger(\vec{r})\hat{E}(\vec{r}) = \frac{\hbar\omega_0}{T}\hat{N}_i, \tag{2}$$

where $\hat{N}_i$ is a photon number operator given by the sum $\sum_{\ell,\ell'} O^i_{\ell,\ell'} \hat{a}^\dagger_\ell \hat{a}_{\ell'}$, and where $O^i_{\ell,\ell'}$ is the overlap integral of the two modes u_ℓ and $u_{\ell'}$ over the detector area A_i:

$$O^i_{\ell,\ell'} = \int_{A_i} d^2 r u_\ell(\vec{r}) u^*_{\ell'}(\vec{r}). \tag{3}$$

It is only when the detector covers the entire image plane that, thanks to the orthonormality property of the modal functions $u_\ell(\vec{r})$, the overlap integral $O^i_{\ell,\ell'}$ reduces to a delta function, and the intensity depends only on the sum of photon numbers in the different modes $\hat{N}_{tot}$. Otherwise, all the cross terms $\hat{a}^\dagger_\ell \hat{a}_{\ell'}$ contribute, and the measurement depends in a more complicated way on the properties of the state vector $|\Psi\rangle$ of the system.

For example let us assume that $|\Psi\rangle$ is a single mode state, meaning that there exists a transverse mode basis in which $|\Psi\rangle$ can be written as $|\phi\rangle \otimes |0\rangle$, where $|\phi\rangle$ is a vector spanning the Fock states of the first mode v_0 only, while $|0\rangle$ is the vacuum state in the other modes. It can be shown[2] that the noise variance of the photon number recorded by pixel i during the exposure time T obeys:

$$\Delta N_i^2 = \langle N_i \rangle \left\{ 1 + \frac{\langle N_i \rangle}{\langle N_{tot} \rangle} \left(\frac{\Delta N_{tot}^2}{\langle N_{tot} \rangle} - 1 \right) \right\}. \tag{4}$$

This relation shows that, whatever the single mode quantum state $|\phi\rangle$ in which the system is, the local intensity fluctuations are very close to the shot noise limit $\Delta N_i^2 = \langle N_i \rangle$ if the pixel covers only a very small part of the total intensity ($\langle N_i \rangle \ll \langle N_{tot} \rangle$). This is even true if $|\phi\rangle$ is a single mode Fock state $|N\rangle \otimes |0\rangle$, for which there are no fluctuations at all on the total intensity. Equation (4) implies that quantum fluctuations in very small parts of the image plane cannot be reduced below the standard quantum noise limit by using single mode non-classical beams such as squeezed beams or sub-Poissonian beams. Instead, one needs to use multimode non-classical states of light, and the smaller the regions in which one wants to observe quantum effects, the higher the number of modes filled by non-classical states will be.

2.1. *Generation of local quantum effects*

The problem of producing highly multimode non-classical states on which local squeezing could be observed (or quantum correlations between different small areas of the image plane) has been considered by many authors. Parametric down conversion in a second-order non-linear crystal pumped by a plane wave turns out to be the best candidate to produce such light.

Because of the phase matching condition, the directions of emission of the twin photons produced in this process are correlated at the quantum level. This property has been widely used in experiments performed at the photon counting level.[3–5] It is also the basis of the quantum imaging effects at the level of continuous variables. Sokolov and Kolobov[6,7] showed that free propagation in a parametric medium, in the high parametric gain regime, is able to produce light with spatial entanglement, and local squeezing for the light in the degenerate configuration, *i.e.* at a frequency equal to half the pump frequency. Lugiato and coworkers[8–10] showed that spatial entanglement could be produced at much lower pump levels with Optical Parametric Oscillators below threshold and using plane mirrors. Lugiato and Grangier[11] showed that a frequency-degenerate OPO inserted in a confocal cavity can produce a locally squeezed vacuum, that could be observed on any pixel area A_i, provided that it is symmetric with respect to the cavity optical axis. Let us mention that other nonlinear effects, such as Kerr nonlinearities, are likely to produce local quantum effects, for example in spatial solitons,[12] but with a much lower efficiency that parametric processes.

Experimental observations of local quantum effects have recently been performed in the single-pass, high parametric gain regime with intense pulsed pump lasers: observation of pure spatial quantum correlations in a single-shot image,[13] and of noiseless amplification of images, either for local temporal fluctuations[14] and recently also for pure spatial fluctuations in a single-shot image.[15] Experiments are also in progress using cw lasers and transverse-degenerate cavities such as confocal or semi-confocal cavities. What has been obtained so far is the evidence that a multimode non-classical light is produced by the device,[16,17] but so far no strong local correlations nor local squeezing has been observed.

2.2. *Improvement of optical resolution*

Another important aspect of quantum imaging is the problem of optical resolution, *i.e.* the capacity to reconstruct from an image the smallest possible details of the corresponding object. Super-resolution techniques have been studied for a long time at the classical level with the aim of beating the Rayleigh limit of resolution, which is of the order of the wavelength. In principle, if the imaging system is perfectly known, deconvolution techniques can extract the shape of a very small object from its image even if it is completely blurred by diffraction. But the noise present in the image, and ultimately the quantum noise, prevents such a perfect reconstruction procedure. The problem of object reconstruction has recently been revisited at the quantum level.[18] It was shown that it is in principle possible to improve the performance of super-resolution techniques by injecting highly multimode non-classical light in very specific transverse modes, namely the eigenmodes of the propagation through the imaging system.

3. Quantum Imaging Using a Few Modes

In the problem of optical resolution, briefly outlined in the previous subsection, nothing is *a priori* known about the content of the image under study, and the aim is to be able to perceive its smallest possible details, whatever they are: this is a very strong requirement, and it is clear that in order to improve by quantum ways the extraction of such a significant amount of information, one needs highly non-classical states, involving a very large number of modes. But there are other cases when one one wants to extract much less information from an image, about which a lot is already known *a priori*. This is what we will envision in this section. We will see that in such a case, only a few appropriate modes must be tailored at the quantum level in order to improve the quality of the information retrieval.

3.1. *Information extraction from images*

There are numerous examples where the variation of the intensity distribution in an image depends only on a few parameters p_j that one wants to determine, while the total intensity is independent of p_j. For example one wants to know the position of a single object scattering light that is much smaller that the wavelength of light and is moving in an otherwise stationary field of view: for example a fluorophore in a biological sample.[19] In other instances one deals with a distribution of light which has a fixed transverse shape and is subject to global displacement and rotation. The problem is to measure these displacements with the highest possible accuracy. This "nano-positioning" problem is encountered in many situations, where the position of a light beam depends on physical parameters that one wants to precisely determine: among others, surface cartography by Atomic Force Microscopy,[20] or determination of ultra-weak absorption by the mirage effect.[21] In all these examples, one uses a quadrant photodiode (4-pixel detector) centered on the light beam. When the particle moves in the first example, or the beam itself in the second one, an imbalance appears between the four signals delivered by the detector which can be traced back to the variation of the parameter under consideration.

There are other examples where the extraction of information requires a more complex numerical processing of the signals delivered by the pixels. For example the determination of the Fourier component at spatial frequency $\vec{q_j}$ in the image arises from the computation of quantities like $\sum_i \cos(\vec{q_j}.\vec{r_i})N_i$, where N_i is the intensity measured on the ith pixel centered on position $\vec{r_i}$.

3.2. *Determination of the eigenmodes of the measurement*

Let us generalize the previous examples, and consider the case where the measured quantity corresponds to a quantum observable that can be written as:

$$\hat{M}(p) = \sum_i g_i \hat{N}_i, \tag{5}$$

where g_i are gain coefficients that can be either positive or negative, and are very often simply equal to ± 1. The set-up is adjusted such that there is no signal when the parameter p is zero: $\left\langle \hat{M}(p=0) \right\rangle = 0$. This is a generalized *differential measurement*, interesting from an experimental point of view because all common mode noise, for example the noise arising from the classical total intensity fluctuations of the source, do not affect this quantity.

We assume for simplicity that the detector plane is a plane of constant phase for the mean field $\left\langle \hat{E}(\vec{r}) \right\rangle$ illuminating the detector. We will assume it to be real over all the plane. This allows us to define a "illuminating mode" which has the same spatial distribution as the input field amplitude:

$$v_0(\vec{r}) = \frac{\left\langle \hat{E}(\vec{r}) \right\rangle}{\sqrt{\int d^2r \left\langle \hat{E}(\vec{r}) \right\rangle^2}} \tag{6}$$

with the integral covering the entire transverse plane. It has been shown[22] that the quantum noise affecting the measurement of this quantity *arises from a single well-defined transverse mode* $v_1(\vec{r})$, called "noise mode", defined by the condition:

$$\text{if } \vec{r} \text{ inside pixel i: } v_1(\vec{r}) = f g_i v_0(\vec{r}), \tag{7}$$

where the normalization factor f satisfies $\int d^2r \left| v_1(\vec{r}) \right|^2 = 1$. $v_0(\vec{r})$ and $v_1(\vec{r})$ are orthogonal modes because the measurement has been assumed to be differential. One can then complete the mode basis and construct an orthonormal set of modes $\{v_\ell(\vec{r})\}$ comprising v_0 and v_1 as the two first modes.

More precisely, the calculation of the noise variance on the considered signal in this mode basis leads to the following simple relation:

$$\Delta^2 M(p) = (\sum_i g_i^2 \langle N_i \rangle)\Delta^2(\hat{b}_1 + \hat{b}_1^\dagger), \tag{8}$$

where $\langle N_{tot} \rangle$ is the total number of photons measured on the whole transverse plane, and $\hat{b}_1$ the photon annihilation operator in the "noise mode" v_1. One therefore sees that the noise comes indeed from a single mode, which has a shape given by (7) and depends both on the characteristics of the mean illuminating field and of the quantity calculated from the intensity measurements performed by each pixel. More precisely it is proportional to the variance of the quadrature component in the noise mode which is in phase with the mean field (assumed here to be real).

If the light used in the experiment can be described by a single mode quantum state in the illumination mode, the noise mode is in the vacuum state. Therefore $\Delta^2(\hat{b}_1 + \hat{b}_1^\dagger) = 1$, and:

$$\Delta^2 M(p) = \sum_i g_i^2 \langle N_i \rangle . \tag{9}$$

This noise variance could also be calculated from expression (5) assuming statistically independent shot noise on each pixel. This is the standard quantum noise of this measurement, independent of the actual noise in the illuminating mode v_0. This statement is a generalization of formula (4) for the local intensity noise in a single mode state.

Equation (8) allows us to determine eigenstates of the observable $\hat{M}$ corresponding to the measurement: they are tensor products of the form $|\phi_0\rangle \otimes |\phi_1\rangle \otimes |\Phi\rangle$, where $|\phi_0\rangle$ is any state in the illuminating mode, $|\phi_1\rangle$ is a perfectly squeezed vacuum in the noise mode, and $|\Phi\rangle$ is any state of zero mean value spanning over all the other modes. In particular, $|\phi_0\rangle$ can be a coherent state and $|\Phi\rangle$ can be the vacuum. Compared to the quantum states required to obtain local squeezing or to improve optical resolution mentioned in the previous section, this latter state is much simpler as it requires the superposition of only two states, of which only one is non-classical. As we will see below it can be synthesized in a simple way from single mode states. From a practical point of view, a perfectly squeezed vacuum cannot be produced by a real experiment, but any squeezed vacuum state used in the system will provide an improvement of the measurement with respect to the standard quantum noise limit given by (9).

Let us now consider the case of P signal processing schemes, allowing us to determine P parameters p_j $(j = 1, ..., P)$ by expressions analogous to (5):

$$\hat{M}(p_j) = \sum_i g_{ij} \hat{N}_i. \tag{10}$$

Assuming an illumination mode of given mean value, we can now define with Eq. (7) one noise mode v_j per measurement. These modes are orthogonal if the P corresponding measurements are independent. The common eigenstate of all these measurements is still of the form $|\phi_0\rangle \otimes |\phi_1\rangle \otimes ... \otimes |\phi_i\rangle \otimes ...$, but now with one perfectly squeezed vacuum per noise mode, *i.e.* per measurement. We are then led to the conclusion that P noiseless measurements require the production of a tensor product of $P + 1$ single mode states, of which P are squeezed. An important feature is that the resulting multimode non-classical beam allows one to *simultaneously* measure the P parameters p_j with a perfect accuracy.

3.3. *Case of beam nano-positioning*

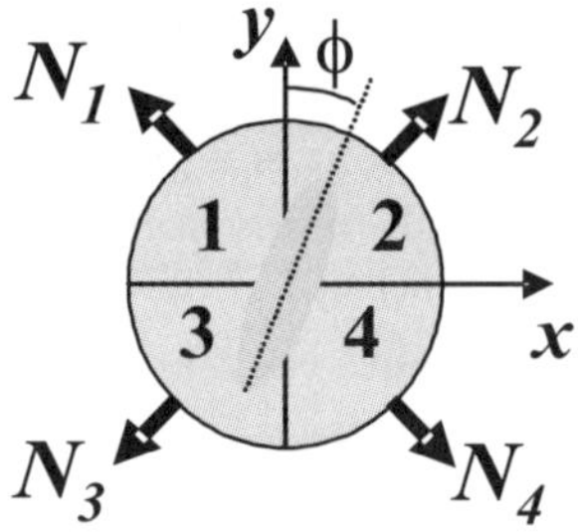

Fig. 1. Quadrant detector used to monitor the position and the rotation around its propagation axis of an asymmetric light beam (in yellow) incident on it.

Let us now come back to the problem of nano-positioning, that we have introduced at the beginning of this section. It involves a quadrant detector, *i.e.* four pixels (see Fig. 1). Displacements along the x and y direction of the beam center can be inferred respectively from the quantities $M_x = N_2 + N_4 - N_1 - N_3$ and $M_y = N_1 + N_2 - N_3 - N_4$ while rotations of a non symmetric beam around the optical axis are measured by $M_\phi = N_1 + N_4 - N_2 - N_3$. All these measurements are of the differential type, and allow us, knowing the illumination mode v_0, to define three orthogonal noise modes v_x, v_y, v_ϕ, defined from v_0 as in (7). Figure 2 gives as an example the variation along the Ox axis of the illumination mode v_0 (supposed to be the fundamental Gaussian mode TEM_{00}), and of the noise mode v_x.

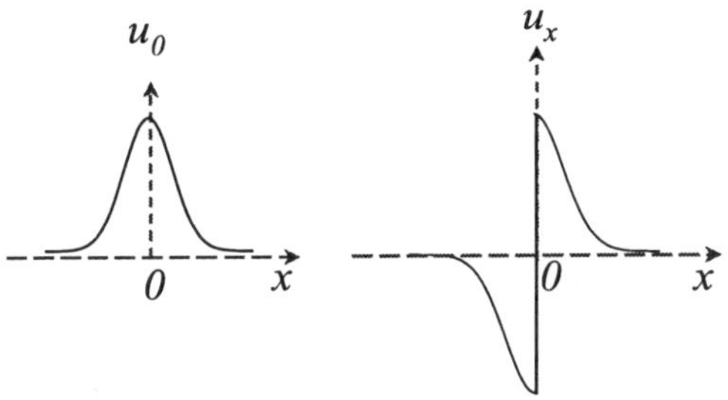

Fig. 2. shape of the illumination mode v_0 and of the noise mode v_x for the measurement of a displacement of a light beam along the Ox axis.

It is thus possible to reduce simultaneously below the standard quantum limit the fluctuations in the determination of all positioning parameters (x, y, ϕ) of an image having a shape which cannot be distorted in the transverse plane. One needs for this purpose to use a beam of light composed of a coherent state in the illuminating mode v_0, and of three squeezed vacuum states in the three noise modes and on the appropriate quadrature component. There is also an other mode basis that can be used to characterize the state: the "pixel basis" u_i $(i = 1, 2, 3, 4)$, where $u_i(\vec{r})$ is equal, within a normalization factor, to the illuminating mode $v_0(\vec{r})$ on the pixel surface A_i (namely a quadrant of the image plane), and is zero outside. It is easy to show that at the limit of perfectly squeezed vacuum states in the noise modes v_x, v_y, v_ϕ, there are *perfect quantum correlations* between the intensity fluctuations in each of the four modes: the four "pixel modes" contain what can be called "photon quadruplets", an extension of the concept of twin photons to the present four mode problem.

Let us note that a squeezing property in the u_i mode basis translates into a quantum correlation property the v_i basis. This is similar to what is obtained when squeezed beams are mixed on a beam splitter and produce entangled states at the output. This is due to the fact that the relation linking the two mode basis is actually the same as in the beam splitter case.

If one only wants to measure the transverse coordinates (x, y) of the beam center below the standard quantum noise limit, one needs, in addition to the coherent illuminating mode, only two squeezed beams in the noise modes v_x, v_y. In the pixel basis, this means that the opposite pixel modes (u_1, u_4) are filled with "twin beams", as well as the opposite pixel modes (u_2, u_3), while the intensity fluctuations on nearby quadrants (u_1, u_2) and (u_3, u_4) are uncorrelated.

4. Synthesizing a Few-Mode Quantum State for Sub-Shot Noise Beam Nano-Positioning

There are actually different ways of producing the required nonclassical multimode beams described in the previous section. The first one is to use the highly multimode beams produced by single pass parametric down conversion, which present strong quantum correlations between the intensity fluctuations measured on any two pixels with are situated at opposite points (with respect to the optical axis) in the image plane, as shown in Ref. 13. Such beams are in some respect "too non-classical", as they would increase the sensitivity of a displacement measurement performed whatever the geometry of detectors surfaces, provided that they are symmetric with respect to the optical axis. Furthermore, they require the use very intense pulse lasers as a pump. As one actually needs only two- or three-mode non-classical light, it seems more simple to "synthesize" such beams from their single mode components. We explain in the two next subsections the two techniques used so far to make such a synthesis, and the experimental results obtained with such beams.

4.1. *1D nano-positioning*

For 1D-positioning, one needs to mix a single-mode coherent beam and a single-mode vacuum squeezed mode. The simplest way to achieve that is to use a beam-splitter, with these two fields at the two input ports. The drawback of this very simple set-up is that it has two outputs, of which only one is used: it therefore introduces losses in the input channel and one of the two loss coefficient is larger than 50%, which is detrimental to squeezing properties. The best strategy is then to use a strongly unbalanced beam-splitter ($92\% - 8\%$ in the experiment described below), which will transmit most of the squeezed beam, and therefore not destroy much the quantum noise reduction effect, and a small part of the coherent beam, that remains coherent for any value of the transmission. Most of the available power is unfortunately lost in the process.

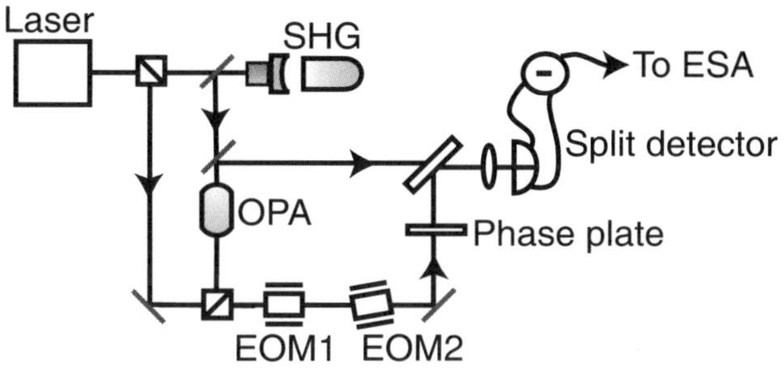

Fig. 3. Scheme of the experimental setup. SHG: second harmonic generator; OPA: optical parametric amplifier; EOM: electrooptic modulator; ESA: electronic spectrum analyzer.

The experiment has been reported in Ref. 23, and the setup is shown in Fig. 3. It comprises a squeezed light generator, consisting of a cw frequency doubled Nd:YAG laser pumping a degenerate optical parametric amplifier (OPA) that produces a stable 3.5dB squeezed beam in the TEM_{00} mode at 1064 nm. A flipped coherent mode is produced by sending the remaining part of the initial 1064 nm laser beam through an appropriate phase plate, that introduces a π phase shift between the field amplitudes of the two halves. Then the two beams are superimposed using a beam splitter as described above. The resulting two-mode beam is sent to a quadrant InGaAs detector with quantum efficiency greater than 90%. Only two of the four quadrants are used in the present experiment. The phase plate is imaged by a lens on the quadrant detector. One observes that the noise measured on the sum of the two halves, *i.e.*, on the total beam, coincides with the shot noise level, and that the noise measured on each individual half is reduced by $1.08dB$ below the quantum noise limit. As the intensity noise on each half of the beam is below the quantum noise limit while the whole beam is at shot noise, this means that there is a strong anticorrelation between the quantum fluctuations of the two halves of the squeezed beam before it is mixed with the flipped mode. The role of the π phase shift is indeed to transform these anti-correlations into correlations, as can be checked experimentally: the noise on the intensity difference between the two halves is measured at a level lying $2.34dB$ below the shot noise limit.

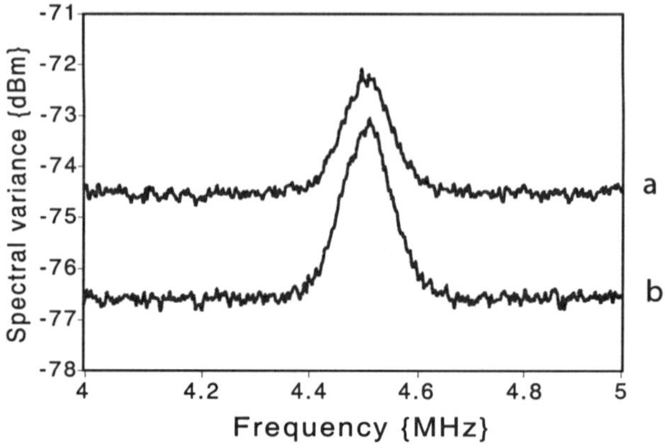

Fig. 4. Noise spectrum of the photocurrent difference in the presence of an oscillating displacement of amplitude 2.9rA and frequency 4.5 MHz.

The synthesized two-mode non-classical state has been used to improve the precision of 1D displacement measurement in the image plane. This displacement is actually a weak oscillation at 4.5 MHz of the beam position, and not a dc displacement, so that the signal is in squeezed part of the noise spectrum and also to get rid of the stray beam position fluctuations induced by mechanical or acoustic vibrations. Figure 4 shows the differential signal monitored by a spectrum analyzer when the light beam undergoes a displacement modulation with an amplitude of only 2.9rA, induced by a tilted Electro-Optic Modulator. Figure 4a shows the trace when only a single-mode coherent state is used in the measurement and gives the standard quantum limit. The peak reveals the presence of the modulated displacement. When the two-mode nonclassical beam is utilized instead (Fig. 4b), both the peak height and the noise floor are lowered, and the signal to noise (peak to floor) ratio goes from 0.68 to 1.20. The use of the non-classical beams yields therefore an improvement of the displacement measurement sensitivity by a factor of 1.7.

4.2. *2D nano-positioning*

For 2D-positioning, one needs to mix a single-mode coherent beam and two single-mode vacuum squeezed modes. The use of two beam splitters is not a good solution, since at least one of them will undergo an important loss through the mixing process, and therefore will be significantly deteriorated. One then must use a more sophisticated technique, which has the potential advantage of merging different transverse modes without any losses, in the same way as a polarizing beam splitter is able to merge onto a single beam two orthogonally polarized single mode beams: it consists in using a ring Fabry-Perot cavity designed in such a way that the different Gaussian modes TEM_{pq} resonate for different cavity lengths, and that the resonant mode is perfectly transmitted through the cavity ("impedance matched cavity"). If such a cavity is tuned on the TEM_{00} mode for example, it will transmit it perfectly, and it will reflect perfectly all the others and any linear combination of them. It will in particular reflect the flipped modes v_x and v_y, orthogonal to the TEM_{00} mode.

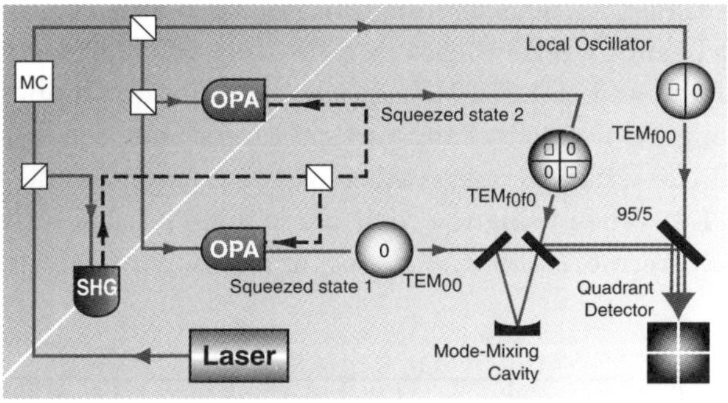

Fig. 5. Experimental setup for the detection of 2D displacement. PBS: polarising beam-splitter, SHG: second-harmonic generator, OPA: optical parametric amplifier, 4QWP: 4-quadrant wave-plate, 2QWP: 2-quadrant wave-plate, L: imaging lens, 95/5: Beam-splitter with 95% reflection and 5% transmission.

The experimental set-up[24,25] is shown in Fig. 5. In its left part, the three single mode needed for the experiment are produced: the coherent state, directly coming from the Nd:YAG laser through a mode-cleaning cavity (MC), and the two squeezed states, produces by two Optical Parametric Amplifiers (OPA) pumped by the second harmonic (SHG) of the Nd:YAG laser. These squeezed fields were observed to be 4 dB and 3 dB below the quantum noise limit, respectively. The right part of the experimental set-up is used to imprint the appropriate phase-shifts on these beams by appropriate phase plates and to mix them. For practical reasons, the illumination mode, instead of being a simple TEM_{00}, was chosen to be a horizontally flipped mode v_y. Hence, the noise mode for horizontal positioning is simply the TEM_{00} itself, and the one for vertical positioning is a "doubly flipped mode" (TEM_{f0f0} on the figure), having four quadrants consecutively phase-shifted by π. The mode mixing ring cavity, of moderate finesse ($F \simeq 30$) was tuned to resonance with the TEM_{00} mode, so that it was able to transmit with a high efficiency ($> 95\%$) the first squeezed mode and to overlap it with the second squeezed mode in the doubly flipped TEM_{f0f0} mode, reflected with an efficiency of more than 94%. The last mixing with the illumination mode was performed with the simpler method of the unbalanced $95\% - 5\%$ beam splitter. The resulting three-mode non-classical beam was finally imaged onto a quadrant detector. A simultaneous noise reduction below the standard quantum noise level was observed on both the horizontal and vertical displacement signals ($2.0dB$ squeezing for the vertical displacement, and $3.05dB$ for the vertical displacement).

An external displacement modulation was then applied to the illumination coherent beam by reflecting it on a mirror mounted on a piezo-electric transducer modulated at 4.976 MHz, giving modulation amplitudes of $1.6\text{r}A$ and $2.3\text{r}A$ along the horizontal and vertical axes. When the squeezed beams were blocked, the signal to noise ratios for the horizontal and vertical axes were 2.8 and 1.4. When using the multimode non-classical beam, the signal to noise ratios were 5.2 and 1.9, giving improvements respectively of 1.9 and 1.4.

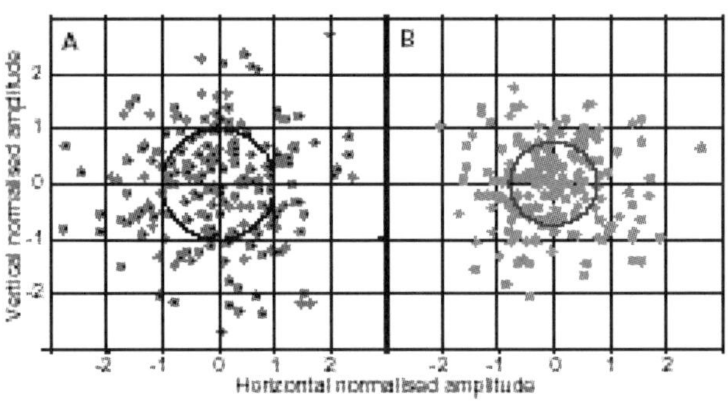

Fig. 6. Data points corresponding to successive simultaneous displacement measurements in the two directions, obtained with a coherent beam (A), and with the multimode non-classical beam (B). The mean value is indicated by the + sign at the centre of plot A. The circles indicate the standard deviations.

An important feature of the present experiment is that the measurements of the two displacements are performed simultaneously below the standard quantum noise level. Unlike the measurement of two quadratures of a single mode field, this is not forbidden by quantum mechanics because the two corresponding observables do commute. This unique feature of the displacement measurement can be exemplified on the correlation diagram of Fig. 6, for a single mode coherent beam (Fig. 6A) and for the synthesized multimode non-classical beam (Fig. 6B). Each point in these diagrams can be interpreted as an instantaneous 2D measurement of the fluctuating displacement modulation amplitude. The standard deviation of these points scattered around the mean value is displayed by the circles in the figure: the average displacement modulation fluctuation for all angles appears smaller, as expected, when using the non-classical beam than with a coherent beam.

4.3. *Optimum detection of a beam displacement*

Let us go back to the problem of measuring a displacement d along the x-axis of the TEM_{00} mode $u_{00}(\vec{r})$ having a waist w. From the point of view of the amplitude of the signal detected, the use of a split detector is not optimal: if one displaces the TEM_{00} mode, one gets the most pronounced increase of the intensity at positions where the intensity of the mode $I(x) = \int |u_{00}(x,y)dy|^2$ changes most rapidly, this is about half a beam size w away from beam centre (see Fig. 7). In contrast, we get very little change at beam

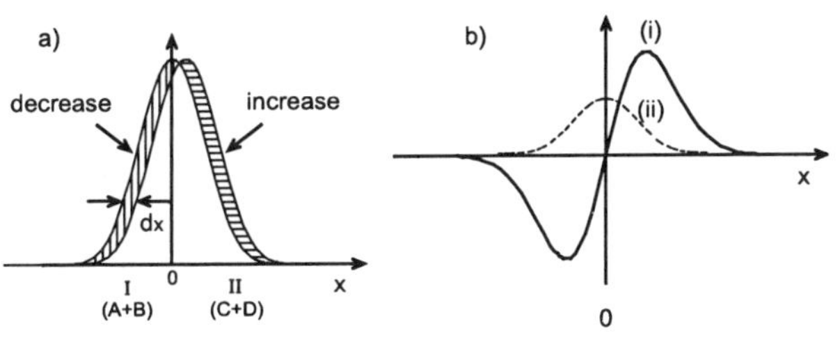

Fig. 7. a) change of intensity created by displacement dx, corresponding to the change of current in the two halves of a split detector. b) signal (solid line) and q-noise (dashed line) as a function of position x.

centre and has little intensity in the line wings. Thus one would expect that the best detection strategy is to reduce the contribution to the signal of the beam center ($x \simeq 0$), and to enhance it on the sides ($x \simeq \pm w$). This shape is quite different to what is measured by the split detector.

A first idea is to use image processing as explained in Sec. 3.2, and to make a measurement corresponding to the observable:

$$\hat{M}(d) = \sum_i x_i \hat{N}(x_i), \tag{11}$$

where $\hat{N}(x_i)$ is the intensity measured on the pixel centered at position x_i, supposed to have a very small size compared to w. This weighting factor will indeed remove the contributions from the central part of the beam, and it changes sign between the two halves, as required for a differential measurement. If one assumes that the detector total area is large enough to measure all the power of the beam, this quantity does not diverge at the edges, because the intensity recorded on the outer pixels is going to zero. Then, according to Eq. (7) the noise mode associated to this measurement is the mode $v_{noise}(\vec{r})$ defined by:

$$v_{noise}(\vec{r}) = fxu_{00}(\vec{r}) = u_{01}(\vec{r}), \tag{12}$$

where $u_{01}(\vec{r})$ is the amplitude of the TEM_{01} mode. In this case, the noise mode is a Hermite-Gauss mode, which is simple and straightforward to produce.

A second idea provides the way of determining the optimum detection strategy.([26]) Let us call $u_{00}(\vec{r}, d)$ the spatial transverse variation of a TEM_{00} mode displaced by the quantity d in the x-direction. For small displacements $d << w$, we obtain the Taylor expansion:

$$u_{00}(\vec{r}, d) = u_{00}(\vec{r}) + \frac{d}{w} u_{01}(\vec{r}), \tag{13}$$

where $u_{01}(\vec{r})$ has been defined after Eq. (12). Relation(13) means that all the information about the displacement is contained in the TEM_{01} mode, centered at the same position of the unshifted TEM_{00} mode. It is therefore the optimum detection mode, a kind of eigen-mode of the displacement measurement of TEM_{00} light beam. The best measurement strategy is to evaluate the amount of TEM_{01} mode which exists in the displaced beam, *i.e.* to measure the projection of the displaced beam on to this mode. This can be done using a cavity resonant for this mode which will transmit only the relevant projection, or more simply using a *balanced homodyne detector where the local oscillator beam is a TEM_{01} beam.* Once it is matched in size and position to the mode basis that describes the input beam we have found the optimum detection system.

If one compares the different detection schemes, the TEM_{01} mode has only about a 80% overlap with the flipped TEM_{00} mode, and consequently a split detector is only 80% efficient, as far as the signal is concerned. The remainder of the information is not detected. Let us finally note that we have to take into account the actual displacement at the location of the detector, which due to the imaging systems used and to propagation effects, will vary along the optical axis z.

It is now straightforward, at least in theory, to reduce the noise for such an optimized measurement of the displacement of a TEM_{00} beam: the intense TEM_{00} beam has to be combined with a squeezed vacuum state in the TEM_{01} mode with a squeezed quadrature in phase with the TEM_{00} field. Note that it is actually the same TEM_{01} mode which must be squeezed in order to improve the measurement defined by 12. Finally, if we wish to detect a spatial modulation, that is a periodic displacement at frequency Ω, we simply should fill the TEM_{01} mode with a squeezed vacuum having noise suppression at frequency Ω.

4.4. *Tilt and displacement measurement*

As in any homodyne measurement, two independent quantities can be measured when by varying the relative phase between the local oscillator and the input beam, which are the two quadrature components of the input field. One knows that they constitute conjugate quantities at the quantum level. In the present set-up, one of these is proportional to the displacement d of the beam. It is interesting to consider the conjugate parameter to d, measured on the orthogonal quadrature. An analysis based on a relation similar to Eq. (13) shows that the tilt of the TEM_{00} laser beam by an angle θ, at the location of the detector, gives a field distribution $u_{00}(\vec{r}, \theta)$ which has a similar simple expansion:

$$u_{00}(\vec{r}, \theta) = u_{00}(\vec{r}) + i\frac{\theta w}{\lambda} u_{01}(\vec{r}) \tag{14}$$

which holds for small amounts of tilt. This means that the information of the tilt is also encoded in the TEM_{01} mode, but on the quadrature which is out of phase with the displacement. We can therefore select a measurement of either the displacement or the tilt by setting the phase Φ_{LO} of the local oscillator beam. As in the conventional homodyne detector, the two parameters attached to the real and imaginary components, that is d and θ, represent conjugate, non-commuting variables. One consequence is that a beam with squeezed light adjusted to suppress the noise in the measurement of the displacement of the beam (real part of TEM_{01} mode) will at the same time induce excess noise in the measurement of the tilt (imaginary part of TEM_{01} mode).

We have demonstrated experimentally exactly this effect[27] (Fig. 8): we built a spatial homodyne detector with a TEM_{01} local oscillator, by selecting the TEM_{01} component with a mode cleaner cavity locked to TEM_{01}. We then combined, without losses as described in Sec. 4.2 a strong coherent beam in the TEM_{00} mode with squeezed light in the flipped mode as the input beam. This beam had a very small periodic displacement, at 5 MHz, equivalent to the quantum noise limit and about ten times larger tilt. The beam was carefully mode matched to the homodyne detector.

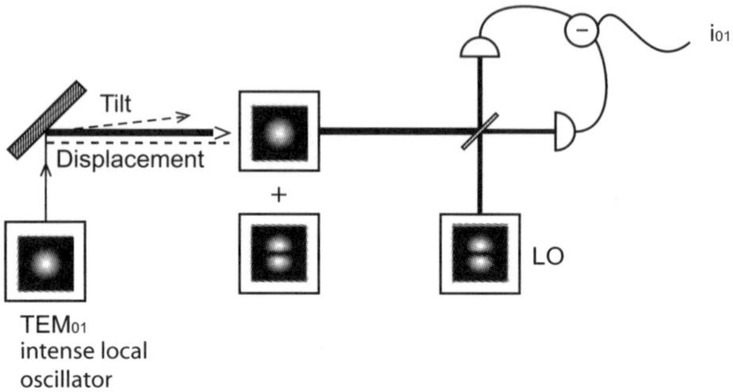

Fig. 8. Layout of a spatial homodyne detector of displacement and/or tilt.

As shown in Fig. 9, without squeezing the tilt was clearly visible while the displacement was barely detectable. In contrast, with the squeezing the noise floor was reduced, the displacement became detectable and the noise in the tilt increased, all consistent with strength of the noise reduction of -2 dB. This is a clear demonstration that we have matched the detection of both displacement and tilt to the common eigenmode of these measurements.

An important consequence of the analysis in Eqs. 13 and 14 is the fact that we can now consider *spatial entanglement.*[28] The non-commuting parameters are d and θ, and using two sources of squeezing we can produce beams with entanglement in these parameters, in complete analogy to beams which are entangled in the conventional parameters quadrature amplitude and phase. This quantum property can be used for the quantum information protocols which are based on entanglement.

5. Future Prospects

As stated in the introduction, quantum spatial effects are an expansion from the single mode quantum optics into multimode quantum optics. We have seen that it is possible to take advantage of the freedom of choice of the transverse mode basis used to describe the quantum state, which allowed us

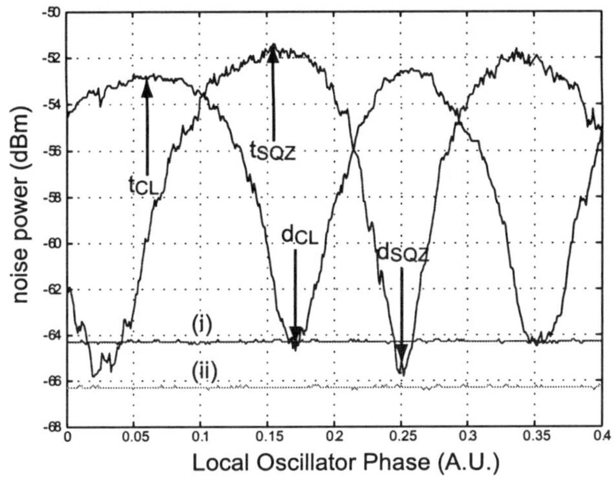

Fig. 9. Homodyne measurement using a TEM_{01} mode as a local oscillator, as a function of the local oscillator phase: the measurement cycles from displacement (close to minimum power) to tilt measurement (close to maximum power). For a coherent beam, the displacement and tilt signals are d_{CL} and t_{CL}, respectively. For a multimode squeezed beam, the displacement and tilt signals are d_{SQZ} and t_{SQZ}, respectively.(i): shot noise limit;(iv): locked signal for the displacement measurement.

to define in the specific case of displacement and tilt measurements, "noise modes", or "detection eigenmodes" of appropriate shapes. This analysis can certainly be generalized to more complex and general cases of information extraction, not only in images but also in other problems in highly multimode quantum optics.

Among the various practical examples of information extraction from images, there is one of particular importance: that of reading by optical ways bits of information stored in a medium such as a CD or a DVD disk. The technique used so far is to focus as tightly as possible a light beam on a CD surface, and monitor a change in the light reflected by a single pixel illuminated by the incident beam. This technique limits the data density to approximately one bit per λ^2. To improve it, the strategy used so far has been to use light of smaller and smaller wavelength. This "race to small wavelength" will have an end, sooner or later. To go further and match the inevitable future needs of memories with even higher storage density, one must envision other techniques allowing us to extract the digital information even when several bits are present in the focal spot: one is faced with another kind of information extraction from images, in which one has a lot a *a priori* information, as one only wants to read arrays of bits 0 and 1. A procedure similar too the one exposed in this chapter for the extraction of displacement information must then be used: first find the appropriate eigenmode that optimizes the detection, then, if quantum noise turns out to be a limiting factor, find the corresponding noise mode in which squeezed light can be injected in order to reduce the quantum fluctuations of the measured signal, and therefore the Bit Error Rate in the bit discrimination procedure. Such alley of research is currently under way in different laboratories in the world.

References

1. *Quantum Imaging, Review book*, M. Kolobov (Ed.), to be published by Springer Verlag, 2005.
2. C. Fabre, J.B. Fouet, and A. Maître, Opt. Lett. **25**, 76 (2000).
3. P. Ribeiro, S. Padua, J. Machado da Silva, and G. Barbosa, Phys. Rev. A **49**, 4176 (1994).
4. D. Strekalov, A. Sergienko, D. Klyshko, and Y. Shih, Phys. Rev. Lett. **74**, 3600 (1995).
5. T. Pittman, Y. Shih, D. Strekalov, and A. Sergienko, Phys. Rev. A **52**, R3429 (1995).
6. M. Kolobov and I. Sokolov, Phys. Lett. A **140**, 101 (1989).
7. M. Kolobov, Rev. Mod. Phys. **71**, 1539 (1999).
8. L. Lugiato and A. Gatti, Phys. Rev. Lett. **70**, 3868 (1993).
9. A. Gatti and L. Lugiato, Phys. Rev. A **52**, 1675 (1995).
10. I. Marzoli, A. Gatti, and L. Lugiato, Phys. Rev. Lett. **78**, 2092 (1997).
11. L. Lugiato and P. Grangier, J. Opt. Soc. Am. B **14**, 225 (1997).
12. E. Lantz, T. Sylvestre, H. Maillotte, N. Treps, and C. Fabre, J. Opt. B: Quantum Semiclass. Opt. **6**, S295 (2004).
13. O. Jedrkiewicz, Y.K. Jiang, E. Brambilla, A. Gatti, M. Bache, L. Lugiato, P. Di Trapani, Phys. Rev. Lett. **93**, 243601-1 (2004).
14. S.K. Choi, M. Vasilyev, and P. Kumar, Phys. Rev. Lett. **83**, 1938 (1999).
15. A. Mosset, F. Devaux, and E. Lantz, to be published in Phys. Rev. Lett. (2005).
16. M. Martinelli, N. Treps, S. Ducci, A. Maître, and C. Fabre, Phys. Rev. A **67**, 023808-1 (2003).
17. S. Gigan, L. Lopez, V. Delaubert, N. Treps, C. Fabre, and A. Maître, submitted to Phys. Rev. (2005).
18. M. Kolobov and C. Fabre, Phys. Rev. Lett. **85**, 3789 (2000).
19. C. Tischer, S. Altmann, S. Fisinger, H. Herber, E. Stelzer, and E. Florin, Appl. Phys. Lett. **79**, 3878 (1991).
20. T. Senden, Curr. Opin. Colloid Interface Sci. **6**, 95 (2001).
21. Boccara C., D. Fournier and J. Badoz, Appl. Phys. Lett. **36**, 130 (1980).
22. N. Treps, V. Delaubert, A. Maître, J.M. Courty, and C. Fabre, Phys. Rev. A **71**, 013820 (2005).

23. N. Treps, U.L. Andersen, B. Buchler, P.K. Lam, A. Maître, H. Bachor, and C. Fabre, Phys. Rev. Lett. **88**, 203601-1 (2002).
24. N. Treps, N. Grosse, W. Bowen, C. Fabre, H.A. Bachor, P.K. Lam, Science **301**, 940 (2003).
25. N. Treps, N. Grosse, W. Bowden, M.T.L. Hsu, A. Maître, C. Fabre, H.A. Bachor, and P.K. Lam, J. Opt. B, Special Issue on Fluctuations and Noise in Photonics and Quantum Optics **6**, S664 (2004).
26. M.T.L. Hsu, V. Delaubert, P.K. Lam, and W.P. Bowen,J. Opt. B: Qu. Semiclass. Opt. **6**, 495 (2004).
27. H.A. Bachor, V. Delaubert, C.C. Harb, M. Hsu, P.K. Lam, C. Fabre, and N. Treps, submitted to *J. Mod. Opt.*, (2005).
28. M.T.L. Hsu, W.P. Bowen, N. Treps, and P.K. Lam quant-phys/0501144 (2005).

Chapter 18

Squeezed Light for Gravitational Wave Detectors

Roman Schnabel
Institut für Gravitationsphysik, Leibniz Universität Hannover and Max-Planck-Institut für Gravitationsphysik (Albert-Einstein-Institut) Callinstr. 38, D-30167 Hannover, Germany

email: roman.schnabel@aei.mpg.de

1. Introduction

Gravitational waves are predicted by Albert Einstein's theory of general relativity.[1] A generally expected, indirect proof of their existence is given by the observation of the slow spiralling together of two neutron stars, caused by their loss of orbital energy to gravitational waves. The inspiral rate exactly matches the predictions of Einstein's theory.[2] This observation was made with radio telescopes because one of the neutron stars is the pulsar PSR 1913+16 and won the Nobel Prize in Physics in 1993 for the discoverers of the system, Russell Hulse and Joseph Taylor.

Gravitational waves (GWs) are emitted when huge masses are accelerated in a non-spherical event. They are distortions of space-time and quadrupole waves to first order approximation, and they possess enormous energies but hardly interact with anything. Gravitational waves carry information which is likely to revolutionize our understanding of the universe giving new insights into the nature and dynamics of gravity, space and time. Space and ground based GW detectors will be able to sense not only harmonic signals from compact binary systems, but also bursts from mergers and supernovae explosions, signals from rapidly decaying vibrational modes of new born neutron stars and background signals from the big bang. Due to their weak interaction GWs will provide information about the very early universe and will give insight deep into heavy objects.

Until now no direct observation of a gravitational wave signal has been achieved. The first attempts were made in the 1960's by Joseph Weber with resonant mass detectors. Modern versions have a typical size of a few meters and achieve impressive noise spectral densities for strain measurements of as low as $10^{-20}\,1/\sqrt{\mathrm{Hz}}$ over a bandwidth of typically 100 Hz at about one kHz. Further information about current detectors and references are provided in Ref. 3.

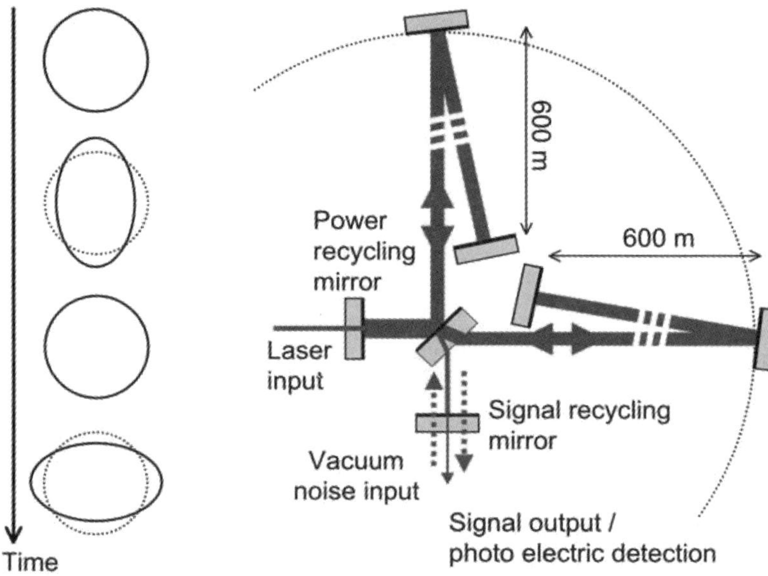

Fig. 1. **Left:** Changing distances between free falling test-masses placed on a circle caused by a gravitational wave (GW) passing in direction normal to the plane shown. **Right:** Optical layout of the GEO 600 GW detector operated close to a dark fringe. Folded arms lead to effective arm lengths of 1200 m. Power- and signal-recycling techniques are added to the simple Michelson topology to reduce quantum noise. Signal-recycling also turns the detector into a resonant device with increased sensitivity around a tunable detection frequency. Note that the relevant quantum noise enters the interferometer from its dark port and is back reflected onto the photo diode. GEO 600 technical data: up to 10 kW single mode laser power at the beam splitter, 5.6 kg mirror masses, 9.1 kg beam splitter mass.

Currently, an international array of first-generation, kilometer-scale laser interferometric gravitational-wave detectors,[3] consisting of GEO 600, LIGO, TAMA 300 and VIRGO, targeted at signals in the acoustic band from 10 Hz to 10 kHz, is going into operation. The goal is the first direct observation of gravitational waves. These first-generation detectors are all

Michelson interferometers with suspended mirrors. Injecting strong carrier light from the bright port, the anti-symmetric mode of arm-length oscillations (*e.g.* excited by a gravitational wave) yields a sideband modulation field which is detected at the dark output port. High strain sensitivity to gravitational waves is achieved by long arm lengths of 300 m up to 4 km and high circulating laser powers on the order of 10 kW which are realized due to high power build-up in the so-called *power-recycling* cavity. Most of the detectors are now close to their design sensitivity and extended periods of data taking have started in 2005. Current noise spectral densities achieved are partially lower than $10^{-22}\,1/\sqrt{\mathrm{Hz}}$ (LIGO). The second generation of GW detectors has already been designed. It will carry out searches with an amplitude sensitivity 15 times higher, which corresponds to an event rate 15^3 times higher than with first-generation interferometers. At this sensitivity, a large number and variety of gravitational wave sources is expected. The sensitivities of second-generation detectors will be limited by quantum noise and thermal noise at frequencies above 10 Hz. At lower frequencies seismic and gravity gradient noise do not allow for ground based detection. The Laser Interferometer Space Antenna space mission (LISA) is currently being designed to cover frequencies from 10^{-4} to 1 Hz. LISA is a joint ESA/NASA mission whose launch is envisaged for 2013.[3] A third generation of ground based detectors with even better sensitivity (again event rate will be increased by three orders of magnitude) is now in the early stages of discussion and conceptual design, focusing on extensive fundamental laboratory research, involving Megawatts of laser powers, diffractive optics, cooling, squeezed light, and other quantum non-demolition techniques (see below).

This contribution focuses on squeezed light injection as a *quantum non-demolition* (QND) technique for future interferometric GW detectors. The problem of detecting gravitational waves was indeed a main impetus for a detailed analysis of the ultimate sensitivity in quantum measurements. Since position and momentum are two non-commuting observables the Heisenberg uncertainty principle does not allow simultaneously vanishing noises. A well-known *standard quantum limit* (SQL) for continuous position measurement arises because the position measurement at one instant injects randomness into momentum, which becomes position noise (back-action noise) at a later time. First proposals for QND measurements were given in the late 1970's[4] but the question if the SQL was an ultimate sensitivity limit or not, remained subject to controversial discussions for some time.[5] However, intensive research during the last 25 years has revealed that the SQL can be surpassed by employing quantum correlations, which could arise naturally during the measurement process, or could be produced by injecting non-classical modes of light. Techniques that beat the SQL are now generally called QND techniques. Injection of a squeezed field into the interferometer's dark port is indeed one of the most promising QND approaches. Squeezed light injection was first proposed by Caves[6] to reduce the high laser power requirements. Later Unruh[7] and others[8–11] found and proved in different ways that squeezed light can beat the SQL of position measurement and that Michelson interferometer quantum noise can be arbitrarily reduced, limited only by the squeezing strength achieved. Recently Harms *et al.*[12] have shown that high power signal-recycled interferometers will benefit from squeezed light similarly to conventional interferometers. In third-generation detectors squeezed light might therefore be one of the new nonclassical technologies to further gain sensitivity establishing gravitational wave astronomy.

2. Quadrature Field Amplitudes in Frequency Space

The effect of gravitational waves on space time can be sensed by measuring the distance between free-falling test-masses. Distances are defined by the speed of light which makes a laser interferometer the ideal measuring device. The interferometer mirrors act as test-masses which can be made free-falling within frequencies of the detection band by appropriate suspensions. In such a measurement device the GW will introduce phase modulation sidebands on the continuous wave carrier light. The purpose of the interferometer is to transform the phase modulation into an amplitude modulation at its output port where it can be detected by a photo diode. The spectrum of the electric photocurrent then carries the information about the gravitational wave spectrum. In this section we will introduce the relevant observables of the electromagnetic laser field that contain the GW signal as well as a variety of displacement noise including quantum noise. We restrict ourselves to homodyne readout of the interferometer throughout all the sections which is chosen for second-generation detectors. First-generation detectors are operated with heterodyne readout which sets less requirements on the technical stabilization of the laser light used.

A single mode of the electromagnetic field can be described by its non-Hermitian field amplitude or annihilation operator $\hat{a}(\omega)$, obeying the commutation relation $[\hat{a}(\omega), \hat{a}^{\dagger}(\omega)] = 1$. Here, we are interested in the sideband modulation of intense carrier light which can be decomposed into upper and lower sidebands

$$\sqrt{\frac{\omega_0+\Omega}{\omega_0}}\hat{a}(\omega_0+\Omega)=\frac{\hat{a}_1(\omega_0,\Omega)+i\hat{a}_2(\omega_0,\Omega)}{\sqrt{2}}, \tag{1}$$

$$\sqrt{\frac{\omega_0-\Omega}{\omega_0}}\hat{a}^\dagger(\omega_0-\Omega)=\frac{\hat{a}_1(\omega_0,\Omega)-i\hat{a}_2(\omega_0,\Omega)}{\sqrt{2}}, \tag{2}$$

where $\hat{a}_{1,2}(\omega_0,\Omega)$ are the amplitude (sub-script 1) and phase (sub-script 2) quadrature operators of the two-photon formalism[13] acting in the space of modulation frequencies Ω of a carrier field of frequency ω_0.

The discrete commutation relation $[\hat{a}_1(\Omega),\hat{a}_2^\dagger(\Omega)]=i$ follows directly from the commutation relation of $\hat{a}(\omega)$ and $\hat{a}^\dagger(\omega)$. This relation places a fundamental limitation on how well one quadrature of an optical beam can be known, given some knowledge of the orthogonal quadrature. This can be expressed by the Heisenberg uncertainty relation

$$\Delta^2\hat{a}_1(\Omega)\,\Delta^2\hat{a}_2(\Omega)\geq\frac{1}{4}, \tag{3}$$

with the operator variances denoted by

$$\Delta^2\hat{a}_{1,2}=\frac{1}{2}\langle\hat{a}_{1,2}\hat{a}_{1,2}^\dagger+\hat{a}_{1,2}^\dagger\hat{a}_{1,2}\rangle-|\langle\hat{a}_{1,2}\rangle|^2. \tag{4}$$

For pure states minimum uncertainty is given by the equal sign in Eq. (3). The explicit frequency dependence (Ω) is often omitted for conciseness, and for $\Omega\ll\omega_0$ the following approximation to Eq. (1) is introduced

$$\hat{a}_1=\frac{a+a^\dagger}{\sqrt{2}},\quad \hat{a}_2=\frac{a-a^\dagger}{\sqrt{2}i}. \tag{5}$$

We need to keep in mind that amplitudes on the right-hand side act at different sideband frequencies; however, for all practical purposes, the quadrature amplitudes are Hermitian operators and therefore will be used as representatives of measurement results. Note that the amplitude quadrature $\hat{a}_1(\Omega)$ and the phase quadrature $\hat{a}_2(\Omega)$ are conventionally defined with respect to the intense interferometer laser field at zero frequency. Then they indeed describe amplitude and phase fluctuations (modulations) of the carrier, respectively. A laser field is best described by coherent states, and without squeezed vacuum input we will find coherent vacuum noise in both the quadratures which will be discussed in the next sections.

Let us now consider squeezing of the coherent vacuum noise at a certain frequency generated by the beat of two sidebands with the carrier. A squeezed state $|r_s,\,\theta_s,\,\alpha\rangle$ is obtained by applying the squeezing operator $S(r_s,\,\theta_s)=\exp\left[-r_s\left(\exp(-2i\theta_s)\,\hat{a}^2-\exp(2i\theta_s)\,\hat{a}^{\dagger 2}\right)\right]$ to a coherent state

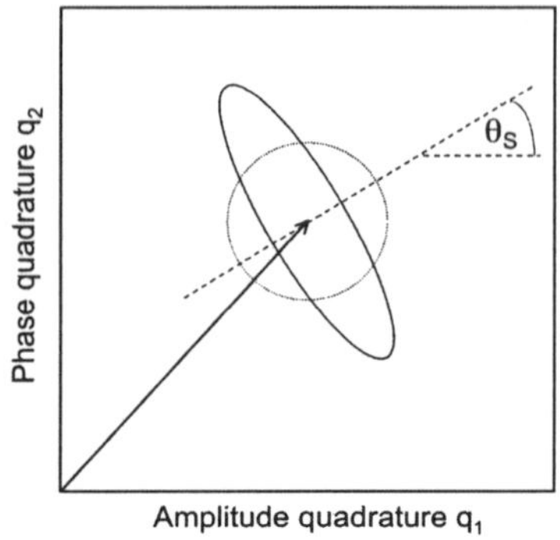

Fig. 2. Phase-space plot showing the uncertainty in a coherent state (circle) and a squeezed state (ellipse). Ellipse and circle correspond to the standard deviation of Gaussian noise, the arrow represents the field's expectation value. θ_s is the squeezing angle.

$|\alpha\rangle$.[14] The resulting squeezed state comes with the desired noise power reduction of e^{-r_s} in quadrature $\hat{a}_{\theta_s}$ compared to coherent noise. Exactly as in classical sideband modulations the phase relation of the sidebands and the carrier determines the quadrature angle, here the angle of the squeezed quadrature.

In a typical squeezing experiment optical parametric oscillation or amplification is employed. Initially the angle of the squeezed quadrature θ_s does not depend on frequency and the variance of the quadrature operator $\Delta^2\hat{a}_\theta$ is given by

$$\Delta^2\hat{a}_\theta = \cosh(2r_s) - \sinh(2r_s)\cos(2(\theta - \theta_s)). \tag{6}$$

The light beam's noise powers are therefore white in frequency space at all angles when leaving the source. If then the squeezed beam is reflected from a detuned cavity, say the signal-recycling cavity of a GW detector, quadratures get rotated by a frequency dependent angle. As a result the squeezed quadrature is not aligned with the detected quadrature for all frequencies, see Sec. 4.3.

Finally we define the time dependent normalized quadrature fields $\hat{q}_{1,2,\theta}(\Omega, t)$ whose variances are directly proportional to our measurement quantities from homodyne detection,

$$\hat{q}_{1,2,\theta}(\Omega, t) = \hat{a}_{1,2,\theta}(\Omega)e^{-i\Omega t} + \hat{a}^\dagger_{1,2,\theta}(\Omega)e^{i\Omega t}. \tag{7}$$

Note that the right-hand side of Eq. (3) is equal to 1 if the quadrature fields $\hat{q}_{1,2}(\Omega, t)$ are considered instead of its amplitudes.

3. Quantum Noise in Interferometers

All current and planned interferometric gravitational wave detectors are based on the Michelson topology with their operating point close to a dark fringe. The fact that only a small fraction of carrier light is lost towards the signal port enables power-recycling and therefore much higher laser powers. High laser power is required to reduce the shot noise (see next section) which is the quantum measurement noise of the interferometer. In Michelson interferometers a differential arm length change excites the laser field's phase quadrature (*cf.* Sec. 3.5). This corresponds to a position measurement of the test-mass mirrors. The quantum noise in this measurement is quantum noise in the phase quadrature and, for coherent quantum noise, appears as shot noise on the photo diode. As outlined in the introduction, quantum mechanics not only leads to measurement noise but additionally to back-action noise due to the Heisenberg Uncertainty Principle. The quantum back-action noise is related to the amplitude quadrature of the laser field and takes the form of radiation-pressure noise: mirror motion induced by random radiation pressure. The quantum noise in both quadratures can best be described by vacuum noise entering the interferometer through open ports. For the case of an almost optically loss free Michelson interferometer close to a dark fringe, the only relevant port is its dark port through which vacuum fluctuations enter and get back reflected onto the photo diode (Fig. 1). Consequently, in a squeezing enhanced interferometer this vacuum needs to be replaced by a squeezed vacuum, of course carefully matched to the modulation signal mode. Fig. 4 shows how spatial degeneracy between an externally generated squeezed mode and the signal mode is achieved using a Faraday rotator.

3.1. *Shot noise*

Let us consider a *simple* Michelson interferometer with neither arm cavities nor signal-recycling. The interferometer is controlled to be at its operating point using electronic servo control loops of low bandwidth. If a GW signal at a higher frequency excites the anti-symmetric mode of arm length oscillation, the formerly constant power at the signal output port changes periodically. In the topology described above the signal transfer does not depend on sideband frequency. The signal stems from phase modulation sidebands on the interferometer laser light converted to amplitude quadrature with respect to a local oscillator beam that serves to amplify the signal. The signal beats with the (spatially overlapping) vacuum mode of the same frequency giving rise to measurement quantum noise. The noise appears as counting noise of uncorrelated photons and is therefore called shot noise. Uncorrelated photons result in noise of a white spectrum. This now brings us to the conclusion that the shot noise dominated noise spectral density of a displacement measurement should also be white, *e.g.* frequency independent. The single-sided amplitude (linear) shot noise spectral density of the mirror displacement measurement in units $\mathrm{m}/\sqrt{\mathrm{Hz}}$ is given by[15]

$$\sqrt{S_{\mathrm{SN},z}} = \sqrt{\frac{\hbar c^2}{4\omega_0 P}}, \tag{8}$$

where ω_0 is the optical frequency and P the power of the carrier light. Note that the displacement measure z describes the length change of only one of the two arms. The spectral density of the measurement of a GW induced strain is given by the same expression but divided by half the interferometer's arm length L. Equation (8) shows that the smallest measurable length change (corresponding to unity signal to noise ratio) is inversely proportional to the square root of the laser power with a white spectrum for sideband frequencies much smaller than the carrier frequency, see dashed horizontal line in Fig. 3 (left).

First- and planned second-generation detectors are shot noise limited for kHz frequencies. Well stabilized high power single mode lasers are employed to reduce shot noise. Power-recycling aims at further decreases. In LIGO *arm cavities* provide an additional build-up factor and in GEO 600 *signal-recycling* further reduces shot noise. We point out that the latter two techniques introduce a frequency dependent signal transfer function and in consequence also a frequency dependent shot noise spectral density. Shot noise can also be reduced by squeezing the quantum noise in the field's phase quadrature. But before squeezing is considered, radiation pressure noise and the standard quantum limit is discussed in more detail.

3.2. *Radiation pressure noise*

The more intense the field in an interferometer, the more accurate the displacement measurement of suspended mirrors due to reduced shot noise is. Unfortunately at the same time radiation pressure noise increases which is related to (coherent) noise in the field's amplitude quadrature. Newton's law tells us that displacement due to white radiation pressure force falls off with (sideband) frequency squared. The radiation pressure noise dominated

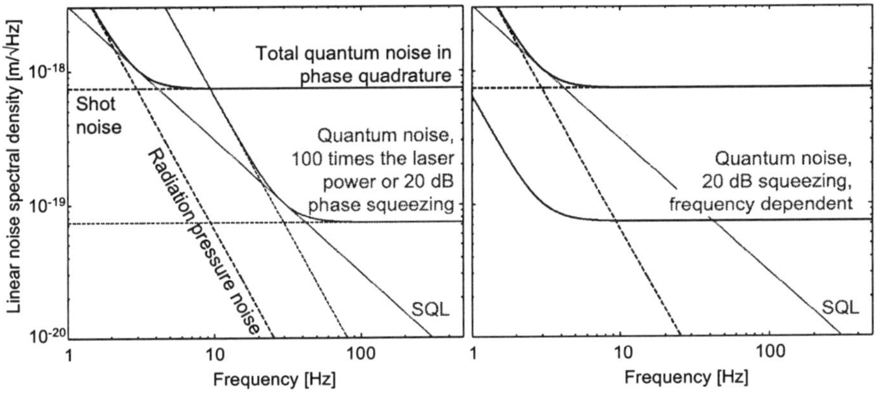

Fig. 3. **Left:** Quantum noise spectral densities of a differential arm length measurement assuming a simple Michelson interferometer with power-recycling. Detected is the phase quadrature, which contains the strongest signal. Then the interferometer's total quantum noise contains two uncorrelated contributions, shot noise and radiation pressure noise. Parameter values: 10 kW laser power at the beam splitter, 5.6 kg mirror masses and an infinite mass beam splitter. By changing the laser power the interferometer's quantum noise can be optimized for an arbitrary single detection frequency. The lower envelope of all these noise curves is the so-called standard quantum limit (SQL). Note that squeezing the phase or amplitude quadrature noise has the same effect as increasing or decreasing the laser power, respectively. Again, sub SQL performance can not be achieved. **Right:** The noise spectral density of an interferometer with injected squeezed field *can* perform better than the standard quantum limit when using frequency dependent squeezing. At the frequency where the SQL was reached without squeezing, the injected nonclassical mode should be squeezed at 45° representing quantum correlations between amplitude and phase quadrature. Simultaneously the field needs to be amplitude squeezed at very low and phase squeezed at very high frequencies to achieve the optimum noise reduction shown here.

linear spectral density of the mirrors' position measurement is then given by[15,16]

$$\sqrt{S_{\mathrm{RPN},z}} = \sqrt{\frac{\hbar\omega_0 P}{c^2 m^2 (2\pi f)^4}}, \tag{9}$$

where m is the mirror mass, ω_0 the optical frequency and P the power of the carrier light, and $f = \Omega/(2\pi)$ the sideband frequency. From Eq. (9) it is evident that radiation pressure noise is significant at low sideband frequencies where also seismic noise gives a major contribution. Generally heavier masses, lower laser power and squeezing of the amplitude quadrature can reduce back-action noise.

3.3. *Total quantum noise and the standard quantum limit*

Both, shot noise and radiation pressure noise contribute to the total quantum noise of a given interferometer. If they are not correlated, which is the case for a conventional Michelson interferometer when detecting the phase quadrature, their variances have to be added. The result is shown in the upper curve in Fig. 3 (left). It can easily be deduced from the previous sections that changing the laser power will shift the two quantum noise contributions, an example is also given in Fig. 3 (left). However, the spectral density never goes beyond the SQL (straight line with slope $1/f$). The SQL is calculated by minimizing the sum of the squares in Eqs. (8) and (9) and yields for a position measurement[15,16] (again in units $\mathrm{m}/\sqrt{\mathrm{Hz}}$)

$$\sqrt{S_{\mathrm{SQL},z}} = \sqrt{\frac{\hbar}{m(2\pi f)^2}}\,, \tag{10}$$

where m is the mirror mass and f the sideband frequency. Eq. (10) shows that the SQL is independent of laser power and falls off with sideband frequency. As we will see in the next sections the standard quantum limit *can* be surpassed if shot and radiation pressure noise are correlated. Then the total quantum noise is *not* given by the sum of the variances, *i.e.* the sum of the squares in Eqs. (8) and (9).

3.4. *Quantum non-demolition interferometers*

A measurement with sensitivity better than the standard quantum limit (SQL) is named quantum non-demolition (QND) measurement. What has to be done to surpass the SQL for a position measurement? Injected squeezing can indeed lead to sub SQL performance if the quantum measurement noise (shot noise) and the quantum back-action noise (radiation pressure noise) are correlated, see Ref. 7 to 11. Squeezing at the quadrature angle of 45° corresponds to correlations in amplitude and phase quadratures. If for all frequencies the optimum orientation of the squeezing ellipse is achieved, then for a given degree of squeezing the nonclassical noise suppression is optimal as shown in Fig. 3 (right), lower curve.

Squeezed light injection is not the only QND approach in interferometry. Especially in the last 5 years our understanding has considerably improved and more and more practical schemes for beating the SQL of a position measurement have been developed. It has been known for a long time that speed and momentum are so-called QND observables which are not subject to back-action noise. Interferometer topologies (*"speed meters"*) that mea-

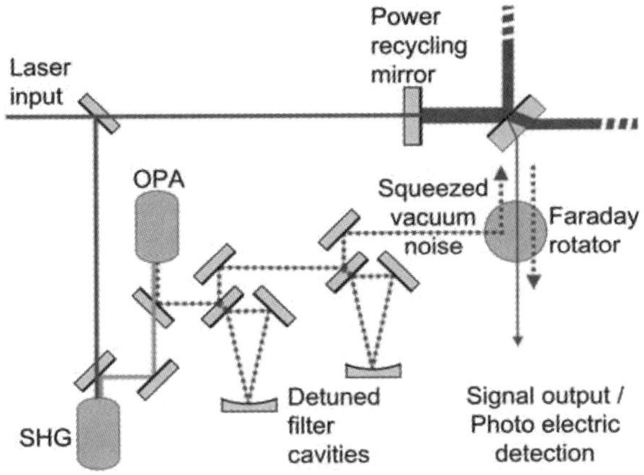

Fig. 4. A squeezed field generated in an optical parametric amplifier (OPA) might be injected into the dark port of an interferometer and matched to its signal output mode. Squeezing at detection frequencies as well as a certain frequency dependent orientation of the squeezing ellipse is required. The latter can be achieved by two detuned optical filters, as proposed by Kimble *et al.*

sure the speed instead of the position of test-masses were found. Recently it was discovered that the Sagnac interferometer is a *speed meter* by its nature.[17]

Radiation pressure leads to an intensity dependent phase shift similarly to the optical Kerr effect. An increasing intensity of the field results in an increasing phase shift on the field reflected from the mirror. This effect can couple quantum noises in amplitude and phase quadratures. An interferometer with suspended masses should therefore generate squeezed fields (ponderomotive squeezing). It has been shown that, although technically challenging, frequency dependent homodyne detection so-called *variational output*, may employ this squeezing to completely cancel the back-action noise, see Ref. 11 and references therein.

Optical bars[18] use the radiation pressure of light to generate a opto-mechanical coupling of the test-masses. The coupling between optical fields and mirror motion may modify the dynamics of the interferometer, giving rise to shifts in resonances that then beat the SQL. Recently Buoannno and Chen found that the detuned signal-recycling cavity in the GEO 600 topology leads to an *optical spring* that allows sensitivities better than the SQL[19,20] (see next section).

Common property of all these schemes is that quantum correlations are generated and employed for sub SQL performance.

3.5. *The dual-recycled Michelson interferometer*

GEO 600 is the only first-generation detector that not only uses power-recycling, but also includes the more advanced technique of (detuned) *signal-recycling*;[21] the combination of both is known as *dual-recycling.*[22] (If signal-recycling is used in combination with arm cavities it is sometimes called *resonant sideband extraction.*[23]) The idea of signal-recycling is to retro-reflect part of the signal light at the dark port back into the interferometer, establishing an additional cavity which can be set to resonate at a desired (non-zero) gravitational-wave frequency. Detuned signal-recycling transforms the interferometer into a resonant detector that leads to a well known (optical) resonance structure in the interferometer's noise curve (right minimum of curves in Fig. 5). Another effect of the detuned cavity is that the GW signal is now also found in the amplitude quadrature.

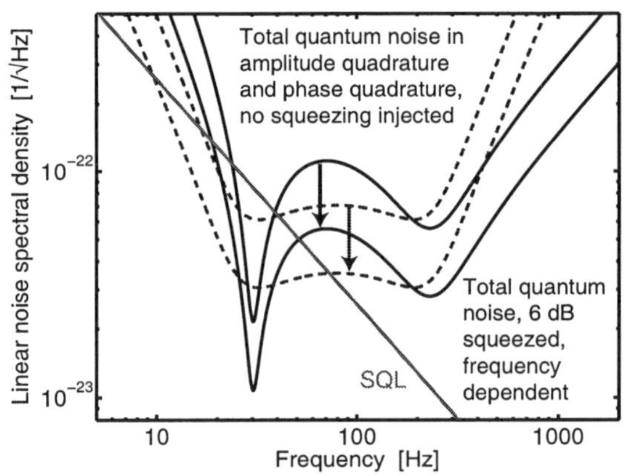

Fig. 5. Upper two curves: Expected quantum noise spectral densities of the GEO 600 design in amplitude and phase quadrature. The left minimum corresponds to the optical spring which results from opto-mechanical coupling of laser field and mirror masses. The right minimum corresponds to the optical resonance of the detuned signal-recycling cavity. If an optimally designed squeezed field is injected into the dark port a broadband sensitivity improvement is achieved, here by a factor of 2 corresponding to 6 dB of squeezing in power. This is shown by the lower traces.

Buonanno and Chen also predicted a second, opto-mechanical resonance in signal-recycled interferometers, around which the interferometer gains sensitivity. They found that an interferometer with detuned signal-recycling, like GEO 600, can already show QND performance;[19,20] neither changes to the interferometer topology nor the injection of squeezed states are necessary. As said in the previous section radiation pressure leads to a Kerr nonlinear effect. Generically a detuned signal-recycling cavity can therefore show an increased signal response, bistability and squeezing. Since radiation pressure leads to a frequency dependent phase shift the effect is not identical for all sideband frequencies but shows a resonance structure, the optical spring resonance. The shape of the resonance also depends on the field quadrature detected and corresponds to the left minimum at about 30 Hz in Fig. 5. The upper two traces show the expected quantum noise in the amplitude and phase quadratures of the GEO 600 detector. However, the sub SQL sensitivity improvement due to the optical spring will not be observed in GEO 600, because other noises, *e.g.* thermal noise limit the detector sensitivity.

It is likely that all second-generation interferometers, such as Advanced LIGO (LIGO II), will use power- as well as signal-recycling. An interesting question arises, whether such an advanced topology fully benefits from squeezed field injection, as simple interferometers do, or not. Harms *et al.*[12] recently showed that indeed the noise spectral density of a signal-recycled Michelson interferometer can be reduced over the complete detection bandwidth corresponding to the linear degree of squeezing achieved in an external source (two lower curves in Fig. 5). As first proposed by Kimble *et al.*[11] optical filter cavities (Fig. 4) can provide the required frequency dependent orientation of the squeezing ellipse. The challenging task for current experiments is therefore the demonstration of squeezing at detection frequencies and its combination with optical filter cavities.

4. Generation of Squeezed States of Light

The quantum noise of a single mode of the electric field at sideband frequency Ω can be described by the two non-commuting quadrature amplitude operators $\hat{a}_1$ and $\hat{a}_2$, see Sec. 2, Eq. (5). For coherent states the quantum noise is equally distributed. In quantum noise limited measurements of either the observable $\hat{a}_1$ or the observable $\hat{a}_2$ the signal to noise ratio can be improved by *squeezing* the noise distribution. The quantum noise in the observable of interest might be lowered while the noise budget in the orthogonal one is increased as required by the uncertainty relation, Eq. (3). Applications in GW detectors were one of the first motivations for squeezed light generation. The first demonstration of squeezed optical fields were performed in 1987 by Slusher *et al.*[24] using four wave mixing in a Sodium atomic beam. Since then several nonlinear processes such as the

Kerr effect, second harmonic generation, and optical parametric amplification have been employed to generate squeezed states. Most of the squeezing experiments have been performed in the MHz sideband spectrum. Squeezing in the acoustic detection band of planned gravitational wave detectors could not be not demonstrated due to the high technical noise levels at these frequencies. Only very recently squeezing at acoustic frequencies was demonstrated[25] as well as frequency dependent squeezing.[26] Another key issue is certainly the degree of squeezing. Up to 7 dB has been achieved.[27,28] In all those experiments optical parametric processes were employed and squeezing was generated for a carrier wavelength of 1064 nm. This wavelength is also used in all current interferometric GW detectors.

4.1. *Squeezing from optical parametric oscillation and amplification*

Fig. 6 (left) shows the schematic of a typical experiment for squeezed state generation. A well stabilized solid state laser provides about one Watt of single mode radiation. The experiments to be described here all used a monolithic Nd:YAG (Neodymium doped Yttrium Aluminum Garnet) ring oscillator. In the experiment shown in Fig. 6 (left) a mode-cleaner might be employed to reduce laser pointing and high frequency noise. A considerable amount of the laser power is frequency doubled in a second harmonic generator (SHG) built from a MgO:$LiNbO_3$ crystal inside a cavity. The SHG output is on the order of a few hundreds of milliwatts of 532 nm radiation and used to pump the optical parametric amplifier (OPA). Again a cavity built around a MgO:$LiNbO_3$ crystal is employed to increase the interaction length. In the case of an OPA a weak seed beam at fundamental frequency is also injected (in the figure shown: from top) that might be used to control the length of the OPA cavity and provides a phase reference for the squeezed field. The travelling wave at fundamental frequency emitted from the OPA (downwards) shows squeezing which is measured in a balanced homodyne detector in which an intense local oscillator is mode matched to the weak signal beam on a 50%/50% beam splitter. The LO amplifies a selected quadrature of the signal beam whereas the LO noise cancels by taking the difference current of the two photo diodes. The signal quadrature is selected by the relative phase between LO and signal beam ϕ which might be electronically locked using either a direct current (DC) or radio frequency (RF) modulation technique. A detailed description of OPA

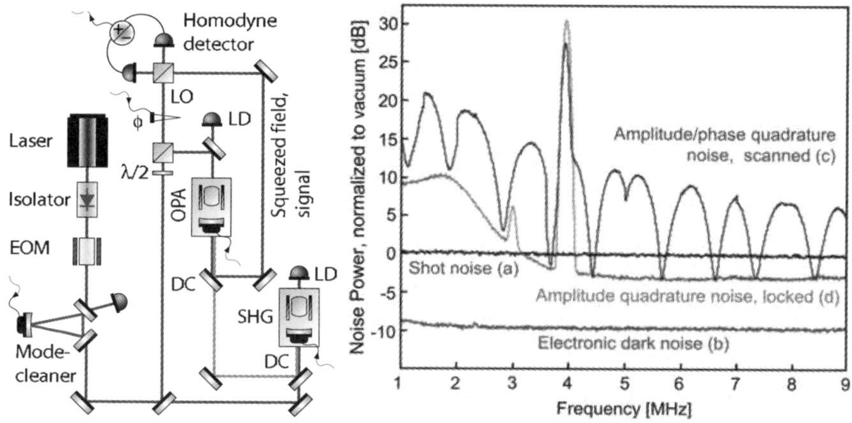

Fig. 6. **Left:** Schematic of a typical squeezing experiment using optical parametric amplification. Single mode laser radiation of up to one Watt is frequency doubled in a χ^2 nonlinear crystal. The intense output of the second harmonic generator (SHG) pumps another χ^2 nonlinear crystal with additional weak fundamental input (from top). The travelling wave mode coupled from the OPA cavity shows quadrature noise squeezing. EOM: electro-optical modulator, DC: dichroic mirror, $\lambda/2$: half wave plate, LO: local oscillator, LD: locking diode, PZT: piezoelectric transducer. **Right:** Typical spectra taken from a homodyne detector. The reference noise level is given by shot noise (a) measured with a blocked signal beam. Signal and local oscillator beams blocked result in pure electronic dark noise (b). If a squeezed signal beam from an OPA is injected and the local oscillator phase ϕ scanned the swept spectrum partially reveals squeezed and anti-squeezed quantum noise (c) as well as technical noise, which is clearly visible for example at 4 MHz. If the relative phase of LO and signal is electronically locked to the signal's amplitude quadrature, broadband squeezing of 3 dB is observed above 4 MHz (d).

experiments as well as a discussion of optical losses that limit the degree of squeezing can, for example, be found in Ref. 26.

Fig. 6 (left) shows typical spectra taken from a homodyne detector. The quantum noise reference is given by shot noise which is measured by blocking the (squeezed) signal beam, see figure caption for more detailed information.

4.2. *Squeezing at audio-band sideband frequencies*

Squeezing from OPA is predicted for arbitrarily low frequencies. Unfortunately acoustic frequencies are subject to huge technical noise which partly originates from the laser source and is present on the seed beam used. The presence of a seed beam is not a requirement, because the vacuum fluctua-

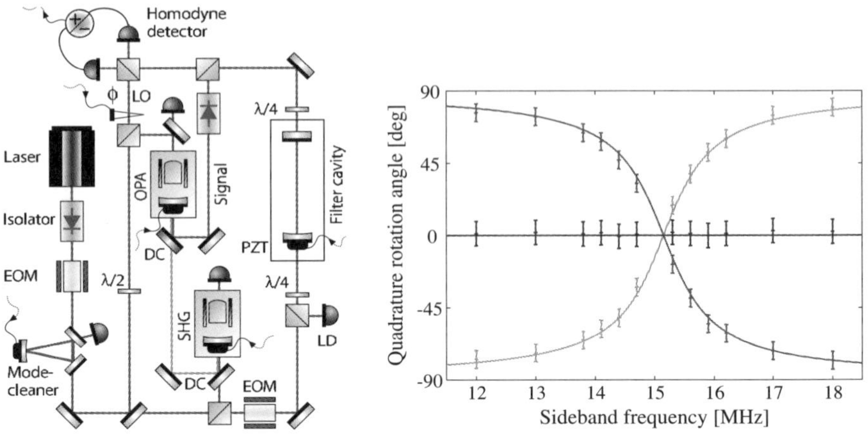

Fig. 7. **Left:** Schematic of the *frequency dependent squeezing* experiment. Amplitude squeezed light is generated in an OPA cavity of controlled length. The detuned filter cavity provides frequency dependent squeezing suitable for a broadband quantum noise reduction of a shot noise limited signal-recycled interferometer. Legend: see Fig. 6; **Right:** Measured rotations of squeezing ellipses. The two curves show the effects from +15.15 MHz (green) and from −15.15 MHz (red) detuned filter cavities. One of the cavities can be regarded as a signal-recycling cavity of a shot noise limited interferometer. After two reflections the squeezing is again observed in the initial frequency independent quadrature. The broadband squeezed quantum noise can then be adjusted to the signal carrying quadrature giving a broadband sensitivity improvement for conventional homodyne readout.

tions in the cavity mode actually initiate the parametric process. Without seeding the device is called an optical parametric oscillator (OPO). However, the seed beam is very useful to control and match the cavity length to the experiment downstream, here to the GW detector, and to provide a phase reference for the squeezed field. It therefore offers the control capability for the whole experiment. In Refs. 29 and 30 classical laser noise on an OPA squeezed beam was optically subtracted using a classically correlated ancilla laser beam. Squeezing down to 220 kHz and 80 kHz, respectively, was demonstrated with full control capability since the seed beam was kept. In Refs. 31 and 25 the devices were run as OPOs and squeezing down to 50 kHz and even down to 280 Hz were observed. In the latter experiment the phase of squeezed vacuum was controlled using a noise locking technique whereas the length of the OPO was drifting. However, all these experiments show that squeezing in the full detection band of GW detectors is feasible and will be reached by applying standard techniques for laser noise suppression.[37]

4.3. *Frequency dependent squeezing*

To fully benefit from injected squeezed states a GW detector requires *frequency dependent* manipulation and design of the orientation of the squeezing ellipse. There are two effects that come into play, firstly radiation pressure and secondly detuned cavities contribute to the final requirement. The latter was experimentally investigated in Ref. 26. The squeezed field from

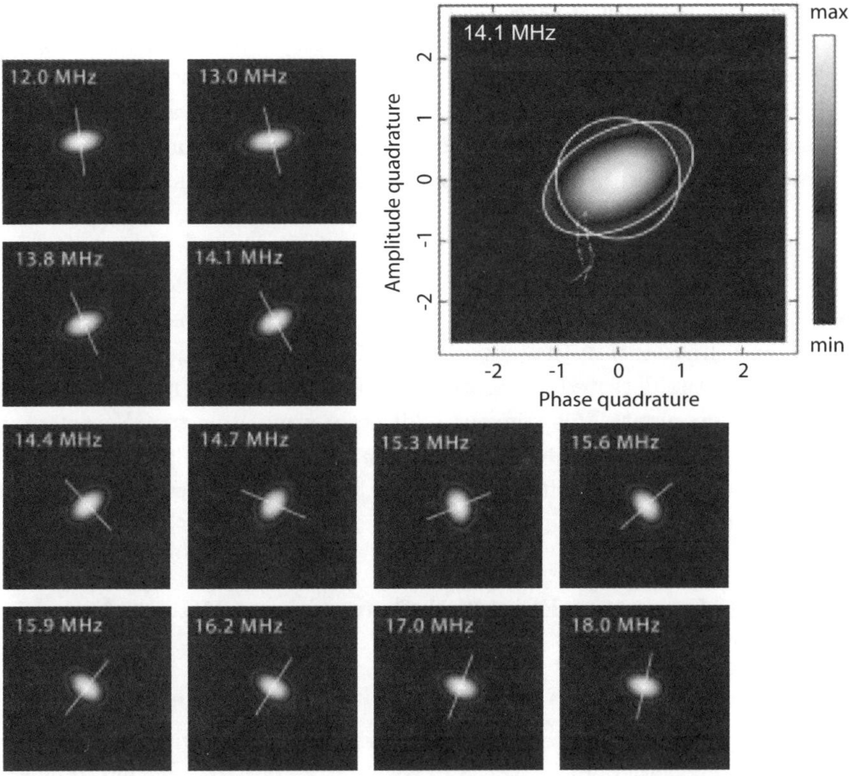

Fig. 8. **Picture top right:** From measured projections reconstructed contour plot of the Wigner function at 14.1 MHz sideband frequency revealing the so-called squeezing-ellipse. The white ellipse represents the standard deviation of the quantum noise; the white circle represents the size of the reference vacuum state. This state shows quantum correlations between phase and amplitude quadratures, *i.e.* squeezing at an angle. **Small pictures:** Measurement results on the same continuous wave laser beam at various sideband frequencies around 15 MHz. For each tomographic picture noise histograms of 100 equidistant quadrature angles were measured. In each case the laser beam was phase locked to a reference beam and the quadrature angle stably controlled and stepwise rotated. The phase reference was given by a phase modulation at 19.8 MHz.

an OPA was first reflected off a detuned cavity which was a model for a SR cavity, and then analyzed in a homodyne detector. Rotation of the squeezing ellipse around the cavity's detuning frequency was found (Fig. 7, right) and characterized by quantum state tomography. Quantum state tomography is a method to reconstruct the Wigner function from a set of measured projections of its probability densities at different quadrature angles using the inverse Radon-Transform.[32] In Fig. 8 reconstructed Wigner functions are shown which were all measured on the same laser beam but at different sideband frequencies. For these measurements the detuned filter cavity was locked to the lower sideband at $\Omega_d = -15.15$ MHz. The result clearly shows the frequency dependent orientation of the squeezing ellipse.

In the case of a interferometer the rotation is disadvantageous, since anti-squeezing rotates into the detected signal carrying quadrature. It was therefore also experimentally shown in[26] how this rotation can be compensated. The rotation can be reversed by reflecting the squeezed field beforehand from a filter cavity with opposite detuning. Fig. 7 (right) shows the rotation angles at 12 different sideband frequencies for +15.15 MHz and −15.15 MHz detuning. Adding the measured rotations resulted in the desired zero line stating that after two reflections the squeezing is always oriented with the detected quadrature giving a broadband nonclassical sensitivity improvement. This concept is directly applicable to GW detectors in their shot noise limited regime. In the case of GEO 600 a suspended cavity of the same linewidth as the signal-recycling cavity, locked to the opposite detuning, must be employed.

5. Towards Squeezing Enhanced Gravitational Wave Detectors

5.1. *Table-top experiments*

The first squeezing enhanced interferometers with sensitivities better than shot noise limit were demonstrated in the 1980's.[33,34] Here we will focus on a recent experiment bearing more resemblance to a GW detector. McKenzie *et al.*[35] employed a Michelson configuration including the technique of power-recycling and utilized a signal readout scheme compatible with an advanced gravitational wave detector, see Fig. 9 (left). The interferometer's differential mode of arm length oscillation was excited by moving one of the end mirrors at a frequency of 5.46 MHz. The differential part of the signal could be deduced from direct homodyne detection taking the sum of the photo-electric currents. The interferometer was locked slightly off the dark

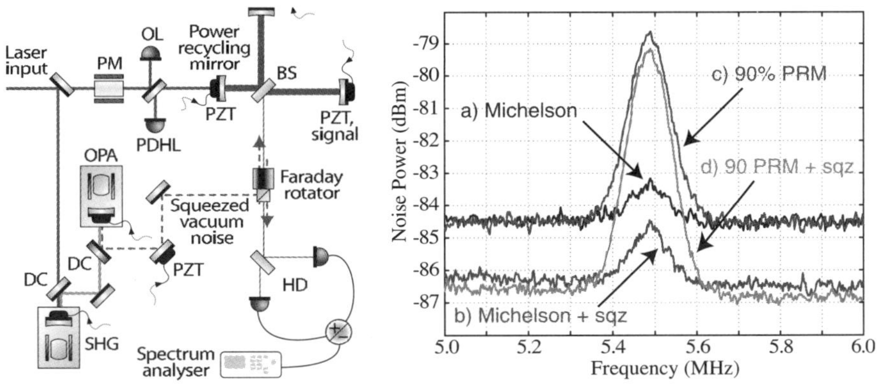

Fig. 9. **Left:** Schematic of the squeezing enhanced power-recycled Michelson interferometer (McKenzie *et al.*). Several servo control loops keep the interferometer at its operating point. PM: phase modulator, OL: offset locking detector, PDHL: Pound-Drever-Hall locking detector, PZT: piezoelectric transducer, SHG: second harmonic generator, OPA: optical parametric amplifier, BS: beam-splitter, DC: dichroic beam splitter. **Right:** (Courtesy K. McKenzie.) Results of noise floor and a signal for a) a simple Michelson, b) a simple Michelson with locked squeezing, c) a power-recycled Michelson and d) a power-recycled Michelson with locked squeezing. Electronic noise was at -93.5 dBm.

port and the carrier light that leaked through served as a local oscillator amplifying the signal. The measurement's signal to noise ratio was then improved beyond the shot noise level by replacing the ordinary coherent vacuum by a squeezed field from an optical parametric amplifier (OPA). A Faraday rotator was employed to give spatial degeneracy between the injected squeezing and the output beam, as discussed earlier. The experiment did not require frequency dependent squeezing since radiation pressure was not significant and no detuned cavities were employed in the scheme. It was therefore sufficient to control the squeezing phase with respect to the signal beam to gain broadband nonclassical noise reduction.

Fig. 9 (right) shows the nonclassical interferometer performance due to a squeezed noise floor 2.3 dB below shot noise level. The graph compares the performances of the Michelson interferometer with and without power-recycling, respectively. When power-recycling was applied not only the signal strength increased as expected due to the higher power in the arms, but also the possibility to reduce the offset from the dark port resulted in less loss and higher nonclassical noise suppression. This experiment demonstrated the compatibility of squeezed light injection, as predicted in Ref. 36, as well as a suitable control scheme.

Further investigations will also be extended to table-top squeezing enhanced dual-recycled interferometers to prepare the ground for non-classical long base line interferometers.[37]

5.2. *Outlook*

Second-generation GW detectors, like Advanced LIGO II, are designed such that they will reach the SQL at about 100 Hz. The full detection band will cover frequencies from around 10 Hz up to several kHz and major parts of the detection band will be quantum noise limited. (The first Advanced LIGO II data run is planned for after 2009.) Third-generation GW detectors are currently in their early stages of design and all nonclassical techniques described in Sec. 3.4 and also combinations of them are being considered to reach the sensitivity goal. It is very likely that the injection of squeezed states will be one of the first nonclassical techniques implemented. At detection frequencies above one kHz this technique allows the reduction of shot noise. This will be essential if higher laser power is not available or if the optics suffer from absorptive heating and thermal deformation. At lower frequencies also thermal noise needs to be reduced. The most rigorous approach is cryogenic cooling of the mirror test-masses resulting in even more stringent limitations on the maximum circulating laser power. Then squeezing will become even more valuable and sub SQL performance will be achieved.

6. Summary

The goal to detect gravitational waves has always been a significant motivation for research on the quantum measurement process, its *back-action noise*, *quantum non-demolition* techniques and squeezed light generation. More and more practical schemes for beating the SQL of a position measurement have been developed during the last years. These schemes are being discussed for future generations of gravitational wave detectors to reach the displacement/strain sensitivity goals required for gravitational wave astronomy. The quantum measurement process itself will also be subject to fundamental research since, if opto-mechanical coupling of quantum noise is significant, entanglement is created between the optical field and the suspended masses in each arm. Entanglement which is at the heart of the (microscopic) quantum world gets transferred to the macroscopic world and, for example, the study of decoherence of real macroscopic objects will be possible.

References

1. K.S. Thorne, in: *300 Years of Gravitation*, S.W. Hawking and W. Isreal (Eds.), Cambridge University Press, Cambridge, England, 1987, pp. 330–458.
2. J. M. Weisberg and J. H. Taylor, arXiv:astro-ph/0407149 (2004).
3. P. Aufmuth and K. Danzmann, New Journal of Physics **7**, 202 (2005).
4. V. B. Braginsky and F. Ya. Khalili, Rev. Mod. Phys., **68**, 1 (1996), and references therein.
5. C. M. Caves, Phys. Rev. Lett. **54**, 2465 (1985) and references therein.
6. C. M. Caves, Phys. Rev. D **23**, 1693 (1981).
7. W. G. Unruh, in: *Quantum Optics, Experimental Gravitation, and Measurement Theory*, P. Meystre and M. O. Scully (Eds.), Plenum, New York 1983, p. 647.
8. H. P. Yuen, Phys. Rev. Lett. **51**, 719 (1983).
9. M. T. Jaekel and S. Reynaud, Europhys. Lett. **13**, 301 (1990).
10. A. F. Pace, M. J. Collett, and D. F. Walls, Phys. Rev. A **47**, 3173 (1993).
11. H. J. Kimble, Y. Levin, A. B. Matsko, K. S. Thorne, and S. P. Vyatchanin, Phys. Rev. D **65**, 022002 (2001).
12. J. Harms, Y. Chen, S. Chelkowski, A. Franzen, H. Vahlbruch, K. Danzmann, and R. Schnabel, Phys. Rev. D **68**, 042001 (2003).
13. C. M. Caves and B. L. Schumaker, Phys. Rev. A, **31**, 3068 (1985); B. L. Schumaker and C. M. Caves, Phys. Rev. A **31**, 3093 (1985).
14. D. F. Walls and G. J. Milburn, *Quantum Optics*, Springer, Berlin 1995.
15. P. R. Soulson, *Interferometric gravitational wave detectors*, World Scientific, Singapore 1994.
16. Dividing by half the interferometer's arm length yields the spectral density of the measurement of a GW induced strain; and in case of arm cavities or folded arms the test-mass m has to be replaced by the reduced mass $m/2$ or $m/5$, respectively.
17. Y. Chen, Phys. Rev. D **67**, 122004 (2003).
18. V. B. Braginsky, M. L. Gorodetsky, and F. Ya. Khalili, Phys. Lett. A **232**, 340 (1997)
19. A. Buonanno and Y. Chen, Class. Quantum Grav. **18**, L95 (2001).
20. A. Buonanno and Y. Chen, Phys. Rev. D **65**, 042001 (2002).
21. B. J. Meers, Phys. Rev. D **38**, 2317 (1988).
22. G. Heinzel, K. A Strain, J. Mizuno, K. D. Skeldon, B. Willke, W. Winkler, R. Schilling, A. Rüdiger, and K. Danzmann, Phys. Rev. Lett. **81**, 5493 (1998).
23. G. Heinzel, J. Mizuno, R. Schilling, A. Rüdiger, W. Winkler, and K. Danzmann, Phys. Lett. A **217**, 305 (1996).
24. R. E. Slusher, B. Yurke, P. Grangier, A. LaPorta, D. F. Walls, and M. Reid, J. Opt. Soc. Am. B **4**, 1453 (1987).
25. K. McKenzie, N. Grosse, W. P. Bowen, S. E. Whitcomb, M. B. Gray, D. E. McClelland, and P. K. Lam, Phys. Rev. Lett. **93**, 161105 (2004).

26. S. Chelkowski, H. Vahlbruch, B. Hage, A. Franzen, N. Lastzka, K. Danzmann, and R. Schnabel, Phys. Rev. A **71**, 013806 (2005).
27. K. Schneider, M. Lang, J. Mlynek, and S. Schiller, Opt. Ex. **2**, 59 (1998).
28. P. K. Lam, T. C. Ralph, B. C. Buchler, D. E. McClelland, H.-A. Bachor, and J. Gao, J. Opt. B **1**, 469 (1999).
29. W. P. Bowen, R. Schnabel, N. Treps, H.-A. Bachor, and P. K. Lam, J. Opt. B, **4**, 421 (2002),
30. R. Schnabel, H. Vahlbruch, A. Franzen, S. Chelkowski, N. Grosse, H.-A. Bachor, W. P. Bowen, P. K. Lam, and K. Danzmann, Optics Comm. **240**, 185, (2004).
31. J. Laurat, T. Coudreau, G. Keller, N. Treps, and C. Fabre, Phys. Rev. A **70**, 042315 (2004)
32. U. Leonhardt, *Measuring the Quantum State of Light*, Cambridge University Press, 1997.
33. M. Xiao, L.-A. Wu, and H. J. Kimble, Phys. Rev. Lett. **59**, 278 (1987).
34. P. Grangier, R. E. Slusher, B. Yurke, and A. LaPorta, Phys. Rev. Lett. **59**, 2153 (1987).
35. K. McKenzie, D. A. Shaddock, D. E. McClelland, B. C. Buchler, and P. K. Lam, Phys. Rev. Lett. **88**, 231102 (2002).
36. J. Gea-Banacloche and G. Leuchs, J. Mod. Opt. **34**, 793 (1987).
37. Note added in proof: recent experimental progress in audio-band squeezing can be found in K. McKenzie, M. B. Gray, S. Goler, P. K. Lam, and D. E. McClelland, Class. Quantum Gravity **23**, S245 (2006) and in H. Vahlbruch, S. Chelkowski, B. Hage, A. Franzen, K. Danzmann, and R. Schnabel, Phys. Rev. Lett. (2006), in press; recent experimental progress in squeezed light interferometers can be found in H. Vahlbruch, S. Chelkowski, B. Hage, A. Franzen, K. Danzmann, and R. Schnabel, Phys. Rev. Lett. **95**, 211102 (2005).

Chapter 19

Continuous Variables for Single Photons

Lijian Zhang, Eran Mukamel and Ian A. Walmsley
Department of Physics, Oxford University
Clarendon Laboratory, Parks Rd. Oxford, OX1 3PU, UK
email: walmsley@physics.ox.ac.uk

Christine Silberhorn
Max-Planck-Forschungsgruppe für Optik, Information und Photonik
Günther-Scharowsky Str.1 / Bau 24, 91058 Erlangen, Germany

Alfred B. U'Ren
Department of Optics, Centro de Investigacion Cientifica y Educacion Superior de Ensenada, CICESE, Baja California, 22860 Mexico

Konrad Banaszek
Institute of Physics, Nicolaus Copernicus University, ul. Grudziadzka 5, PL-87-100 Torun, Poland

Photons have a rich structure associated with their continuous degrees of freedom, the transverse wavevector and frequency. This modal structure can play an important role in quantum information processing based on photons. In particular, information may be coded into any of the degrees of freedom, and this means that photons may represent not only qubits, but also qudits, or qunats - where the continuous degrees of freedom are involved. Coding into quantum correlations in these degrees may be usefully employed to transmit more than one bit per photon in a secure communications link. However, it may also be detrimental: for example, it can hinder the preparation of pure states via conditional detection, and thus compromise the efficacy of quantum information processing schemes based on interference. We illustrate some general criteria that are useful for source design for QIP.

1. Introduction

Continuous variable representations of the quantum state of light, based on the quadrature amplitudes of the electromagnetic field, are well known in quantum optics. The two quadratures may take on a continuum of values, but may not be determined simultaneously precisely, the imprecision being quantified by the Heisenberg Uncertainty Principle. The quantum state is usefully described in terms of quasi-probability densities, such as the Wigner function, in the phase-space of the two quadratures.

Similar phase-space representations are known in classical optics, where they yield the field distribution in position-momentum (or time-frequency) phase space. This leads, for example, to the well-known property that the direction of a light beam, and its position, are not both defined precisely. In this case the imprecision is set by Fourier's Theorem.

In the case of single photons, this kind of phase-space representation takes on a new meaning as a representation of the wavefunction of the photon. That is, the classical vector field $\vec{E}(\vec{r},t) + i\vec{B}(\vec{r},t)$, satisfying the wave equation

$$i\frac{\partial\vec{\psi}}{\partial t} = c\vec{\nabla} \times \vec{\psi} \tag{1}$$

is itself the wavefunction of the (quasi-monochromatic) particle.[1,2,a]

The parallelism implied by this connection allows one to consider operations on continuous variables associated with individual photons that cannot easily be implemented on quadrature amplitudes of fields containing many photons. Moreover, there is extrinsic utility in considering these degrees of freedom: they may be used for information processing applications. Among physical systems that can be used for practical realizations of novel information processing schemes based on intrinsically quantum phenomena, photons are ideal carriers of quantum information. Applications of photons as "flying qubits" (or qunats) include quantum cryptography, communication between distributed computational nodes in quantum networks, as well as elementary building blocks for registers in all-optical quantum computation.

For example, there are several approaches to all-optical quantum computing. These schemes generally code information in one of two ways: as

[a]The photon is a spin-one particle, and thus has a three-component wavefunction, corresponding to the components of the vector field $\vec{\psi}$. See the article by Raymer and Lvovsky in this book.

continuous information in the field amplitude, or as binary information in the polarization modes of individual photons. For both kinds of coding, schemes exist that require only linear optical elements (*i.e.* those which conserve particle number) and measurement, with feedforward. Binary coding is *e.g.* employed by Knill, LaFlamme and Milburn.[3] Their scheme uses quantum interference and photodetection to synthesize an effective nonlinear interaction between two individual photons. This effective two-qubit interaction may form the basis of a non-deterministic controlled phase gate, which scales asymptotically if additional, auxillary entangled photon multiplets are available. A second scheme uses quantum states with several photons entangled in two or more modes, in so-called cluster states and implements a program by measurement of components of these states.[4] Continuous coding is realized in another scheme: quadrature amplitudes provide a means for processing information, with the entanglement resource being found in the correlations between the fields in pairs of modes. In this model even gate operations are deterministic using linear optics. This approach achieves a non-classical information processing advantage, however, only if there are available ancillary additional elements–so-called non-Gaussian operations–that enable critical tasks such as entanglement distillation.[5]

In both approaches the appropriate multi-particle states may be generated using conditional state preparation, specifically by means of a measurement, *e.g.* photon counting, on a state consisting of the target system (and possibly an ancilla). The intrinsic structure of the photonic quantum states is crucial to the efficacy of this state preparation method, since the prepared photons must exhibit both extremely good *first-order interference*, and extremely good *second-order interference* . *First-order*, or modal, *interference* is critically sensitive to the spatio-temporal structure of the interfering photons. Great care must therefore be exercised in making sure that no distinguishing information is contained in the interfering photons that would allow one to trace their origin. For example, the Bell-state analyser proposed by Braunstein and Mann[6] is used in a number of experiments on teleportation and entanglement swapping. In order to ensure that the measurement outcome of this device is a reliable indicator of a particular Bell state, the input entangled photons should be generated optimally in single spatio-temporal modes. Good *second order interference* requires that the photons are in pure states. The quantum correlations that are present in all of the current generation of sources of photon pairs may compromise the preparation of pure states. If, for instance, single photons are generated by the detection of one photon of a pair, which event is then used as a herald for the other[7] then any entanglement in any degree of freedom between the original pair of photons means that the prepared one is in a mixed state.

A common approach to cope with these problems is the implementation of strong spatial and spectral filtering. Such a method, however, reduces the available photon sample, and achieves perfect visibility only asymptotically. Overcoming these limitations is currently one of the main challenges in the further development of quantum information processing applications in the photonic domain.

2. Space-Time Structure of Single Photons

The spatial structure of the photons, characterized by the distribution of their transverse momenta, plays an important role in multi-photon interference and in coupling efficiencies.[8,9] Therefore it is crucial to be able to properly characterize the modal structure of the photon; that is, to measure its wavefunction (or better, density matrix) directly.

2.1. *Measuring the space-time structure of photons*

Since the spatial degree of freedom of a photon provides an optical realisation of a continuous-variable quantum system, there exists an analogy between methods for measuring the photonic wavefunction and for measuring the quadrature Wigner function of a light mode.[10,11] In the latter scheme, measurements of the photon number parity operator are required. In the spatial domain, this operator has a direct equivalent in the parity of the field mode under spatial inversion. The spatial inversion operation is easily accomplished using a prism[12] or set of mirrors,[13] and therefore direct measurements of the Wigner function are possible. Indeed the apparatus is much simpler than required for the photon number parity measurement.

The technique is based on the following observation. For a complex quasi-monochromatic field $E(\boldsymbol{\xi})$ varying in the transverse plane parameterized with the two-dimensional vector $\boldsymbol{\xi}$, the Wigner function is defined according to[14,15]

$$W(\boldsymbol{x}, \boldsymbol{k}) = \frac{1}{\pi^2} \int d^2\boldsymbol{\xi} e^{2i\boldsymbol{k}\boldsymbol{\xi}} \langle E^*(\boldsymbol{x} - \boldsymbol{\xi}) E(\boldsymbol{x} + \boldsymbol{\xi}) \rangle. \tag{2}$$

Here $\boldsymbol{k}$ is the transverse spatial frequency vector and the angular brackets denote a statistical average. The above definition can be easily transformed to:

$$W(\boldsymbol{x},\boldsymbol{k}) = \frac{1}{\pi^2}\left\langle \int d^2\boldsymbol{\xi}[e^{-i\boldsymbol{k}\boldsymbol{\xi}}E(\boldsymbol{x}-\boldsymbol{\xi})]^* e^{i\boldsymbol{k}\boldsymbol{\xi}}E(\boldsymbol{x}+\boldsymbol{\xi})\right\rangle . \qquad (3)$$

This expression shows that the Wigner function at a point $(\boldsymbol{x},\boldsymbol{k})$ is given by a spatially integrated overlap of the field $e^{i\boldsymbol{k}\boldsymbol{\xi}}E(\boldsymbol{x}+\boldsymbol{\xi})$ with a complex conjugate of its replica rotated by 180^0, which corresponds to the transformation $\boldsymbol{\xi} \rightarrow -\boldsymbol{\xi}$. The field $e^{i\boldsymbol{k}\boldsymbol{\xi}}E(\boldsymbol{x}+\boldsymbol{\xi})$ can be obtained from the original input by two simple transformations: displacement inspace by $\boldsymbol{x}$ and changing the direction of propagation by $\boldsymbol{k}$, which in the paraxial approximation results in multiplying the field by a phase factor $e^{i\boldsymbol{k}\boldsymbol{\xi}}$. The above idea can be implemented to measure the Wigner function in a setup depicted in Fig. 1(a).

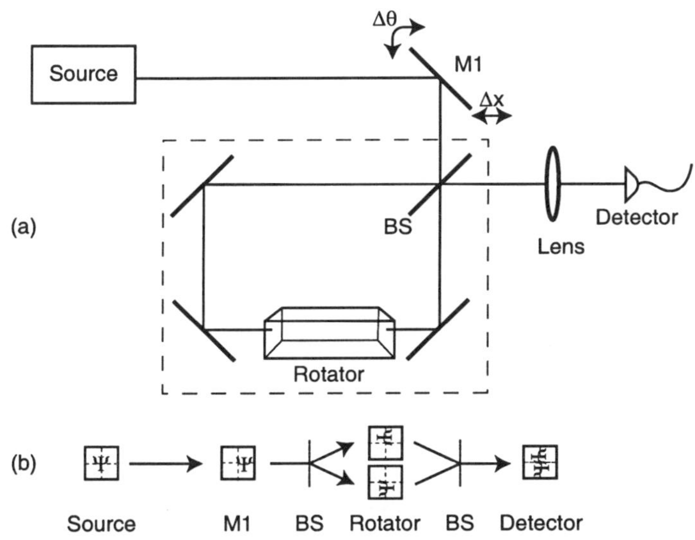

Fig. 1. Sagnac interferometer for measuring the spatial Wigner function of a single photon.

The input beam is steered using the mirror M1 into a three-mirror Sagnac interferometer. The displacement and the tilt of the mirror control the point at which the Wigner function is measured. The 50:50 beam splitter BS generates a pair of replicas of the displaced and tilted input beam travelling in opposite directions. The interferometer contains a Dove prism whose base forms 45^0 with the plane of the interferometer. Transmission through the Dove prism rotates each of the counterpropagating beams by 90^0, which adds up to the 180^0 relative rotation required to measure the

Wigner function. The operation of the setup can be most easily understood by analyzing propagation of the replicas of an exemplary image through the interferometer, as shown in Fig. 1(b). The counterpropagating replicas of the input field are recombined at the output port of the interferometer, and the emerging beam is focused on an area-integrating detector. The intensity I recorded by the detector is given by

$$I = \frac{1}{4} \int d^2\boldsymbol{\xi} |e^{-i\boldsymbol{k}\boldsymbol{\xi}} E(\boldsymbol{x} - \boldsymbol{\xi}) + e^{i(\boldsymbol{k}\boldsymbol{\xi}+\varphi)} E(\boldsymbol{x} + \boldsymbol{\xi})|^2 \tag{4}$$

and it can be decomposed into a sum of three terms, $I = I_1 + I_2 + I_{12}$. The first two of them are given by the space-integrated input intensity $|E(\boldsymbol{\xi})|^2$ and they remain constant as long as the aperture of the interferometer does not clip off any of the input field.The third term I_{12}, originating from the interference between the counterpropagating replicas is proportional to the right hand side of Eq. (3):

$$I_{12} = \frac{\pi^2 \cos\varphi}{2} W(\boldsymbol{x}, \boldsymbol{k}) \tag{5}$$

thus giving the value of the Wigner function. Here φ is the relative phase between the interfering replicas. The complete Wigner function can be therefore scanned by measuring the detector photocurrent as a function of the position and the tilt of the steering mirror, and subtracting the constant pedestal, for example, by measuring the difference photocounts between the two output ports of the interferometer.

2.2. *Measuring the joint space-time structure of photon pairs*

The analogy between quadratures and the spatial degree of freedom may be generalized to composite systems. Consider a pair of photons with perfectly correlated positions, described by a pure entangled wave function. Such a pair exemplifies the Einstein-Podolsky-Rosen state, and is entangled. This entanglement may be made manifest in several ways. First, the state violates the EPR definition of local realism. Second, it violates a Bell inequality.

2.2.1. *Test of the EPR-paradox using photon pairs*

The concept of continuous variable (CV) entanglement originates from one of the most controversial discussions about the foundations of quantum mechanics. In 1935 Einstein, Podolsky and Rosen published their famous

Gedanken experiment where they aimed to prove the incompleteness of quantum theory. The underlying argument of the EPR paradox is Einstein's principle of local reality, which proposes that physical properties exist independently from their detection and distinct particles cannot interact during their measurement process. Entangled states are characterized as two-particle states which exhibit quantum correlations for a pair of non-commuting, conjugate variables, *e.g.* position and momentum. Due to these correlations, measuring either the position or the momentum of one of the particles determines the position or the momentum, respectively, of the other particle with complete certainty. The choice of the measurement is dependent on the observer, and thus must not be incorporated into any local realistic description of the second particle. Still, the possibility of inferring both the position and the momentum of the unmeasured particle means that they constitute simultaneous realties, in violation to Heisenberg's uncertainty principle.

EPR originally studied an idealized entangled state with strict correlations given by

$$|EPR\rangle = \int_{-\infty}^{\infty} dx\, |x, x\rangle = \int_{-\infty}^{\infty} dp\, |-p, p\rangle\,,$$

however, the EPR-paradox was later generalized by Reid and Drummond[21] to non-maximally entangled states with finite correlations. Reid and Drummond derived a criterion which is applicable for measurements of conjugate field quadratures on two-mode squeezed light fields. Afterward Duan[22] *et al.*, Simon[23] and Mancini[24] *et al.* presented necessary and sufficient conditions for CV inseparability, which were all based on measurements of joint quadrature variances. The first experimental realization of the EPR paradox and CV entanglement was carried out by Ou[25] *et al.*; since then quadrature entangled fields have become an important resource for many applications in quantum information with continuous variables.

Howell[26] *et al.* reported the first realization of the EPR-paradox utilizing continuous variable degrees of freedom of single photon pairs produced by PDC. Their experimental setup utilizes avalanche detection of the single-photon state in the near and far fields, which corresponds to position and momentum or $\boldsymbol{x}$ and $\boldsymbol{k}$ measurements, respectively. By monitoring the normalized coincidence distributions Howell *et al.* demonstrate a conditional variance product (measured as joint quadrature variances) of $\Delta^2(x_A - x_B)\Delta^2(p_A + p_B) = 0.01\hbar^2$, compared with the Heisenberg limit being $0.25\hbar^2$.

In contrast to quadrature amplitude measurements, single photon detection is intrinsically implemented as direct, single-shot measurement, where no local oscillator is needed and conditioning on single events is common.

2.2.2. *Continuous-variable Bell inequality for photon pairs*

Bell inequalities provide one measure of the nonclassicality of a composite quantum system, by testing whether a local realistic model can describe the results of measurements on the system. They rely on dichotomic outcomes of measurements on each of the subsystems. Therefore it is necessary to identify a way in which a given state may yield only two possible results for the specified measurement. It is possible to identify such a way for the continuous degrees of freedom of a photon pair, by again making use of the analogy with quadrature amplitudes.

In fact, it is well known that an arbitrary system can be separated into dichotomic observables, and that these may be used to demonstrate a violation of local realism using the Bell inequalities.[16] Nonetheless, actually performing the appropriate decomposition is not always tractable.[17] An easier approach is to operate on only one two-dimensional subspace, simply discarding the photons that do not fall into this. For example, Cohen[16] shows that a Bell-state can be produced by selecting two pairs of correlated space-time modes from the continuum occupied by each EPR particle. This allows the momentum-entangled photons to be treated exactly like polarization-entangled photons. However, this procedure works by throwing away nearly all the amplitude of the EPR state, creating a spin-like state by projecting onto a dichotomic subspace of the original, continuous Hilbert space. This idea was, in fact, the basis for a practical realization by Tapster and Rarity of Bell-inequality violations using frequency and momentum entanglement of photon pairs.[18] In these experiments, pinholes were used to select just two pairs of correlated modes, thus discarding most of the correlated particles. This severely limits the signal-to-noise ratio of the experiments, and provides a potential loophole for skeptics.

The joint Wigner function of two photons in the state

$$|\psi\rangle = \int d^2\boldsymbol{k}_1 d^2\boldsymbol{k}_2 \psi(\boldsymbol{k}_1, \boldsymbol{k}_1)\hat{a}_1^\dagger \hat{a}_2^\dagger |vac\rangle, \tag{6}$$

is

$$W(\boldsymbol{x}_1, \boldsymbol{k}_1; \boldsymbol{x}_2, \boldsymbol{k}_2) =$$
$$\frac{1}{\pi^4} \int d^2\boldsymbol{\xi}_1 d^2\boldsymbol{\xi}_2 e^{2i\boldsymbol{x}_1\boldsymbol{\xi}_1} e^{2i\boldsymbol{x}_2\boldsymbol{\xi}_2} \psi^*(\boldsymbol{k}_1 - \boldsymbol{\xi}_1, \boldsymbol{k}_2 - \boldsymbol{\xi}_2)\psi(\boldsymbol{k}_1 + \boldsymbol{\xi}_1, \boldsymbol{k}_2 + \boldsymbol{\xi}_2). \quad (7)$$

A Gaussian wavefunction for the photon pair of the form:

$$\psi(\boldsymbol{k}_1, \boldsymbol{k}_2) = e^{-A(k_1^2+k_2^2)+2iB\boldsymbol{k}_1\boldsymbol{k}_2}, \quad (8)$$

then corresponds to the Wigner function:

$$W(\boldsymbol{x}_1, \boldsymbol{x}_2, \boldsymbol{k}_1, \boldsymbol{k}_2) = exp[-\frac{A}{2}(x_1^2 + x_2^2) - \frac{1}{8A}(x_1^2 + x_2^2) - \frac{B}{2A}(\boldsymbol{x}_1\boldsymbol{k}_1 + \boldsymbol{x}_2\boldsymbol{k}_2)].$$

This state violates the Bell inequality

$$-2 \leq B \leq 2 \quad (9)$$

where

$$\begin{aligned} B &= W(0,0;0,0) + W(\sqrt{J},0;0,0) + W(0,0;0,\sqrt{J}) - W(\sqrt{J},0;-\sqrt{J},0) \\ &= 1 + e^{-J/8A} + e^{-J/2A} - e^{-J\frac{5+4B}{8A}}. \end{aligned} \quad (10)$$

for $J < 1$, given $A^2 + B^2 = 1$. This is a direct spatial analog of the Bell inequality formulation for fields with Gaussian quadrature amplitude distributions,[19,20] the corresponding state in that case being a two-mode squeezed state.

Thus it is possible to violate Bell's inequality using a single set of correlations of continuous space-time variables without discarding any photons, thus closing a potentially important loophole. Rather than manipulating the input state using a projection in order to obtain a Bell-state, we instead construct a dichotomic observable that is non-zero for any biphoton state. That is, our observable does not assign any of the infinite-dimensional Hilbert space of the EPR state to the null-space of a projection. Furthermore the use of CV entanglement in photon pairs overcomes one of the main obstacles for realizations of CV entanglement in field quadratures: the degradation of the quantum correlations due to inevitable, optical losses and inefficiencies in real systems. Experiments using photon pairs are not hampered by losses, since the photon number is decoupled from the observed correlated variables such that viable results can be post-selected.

Using the Wigner distribution in real space, rather than quadrature space, also offers the advantage that the Bell inequality can be tested directly using a pair of the Sagnac interferometer described above. These can be separated by an arbitrarily large distance (Fig. 2). Multiplying the difference signals from each of the interferometers recorded at coincident times gives the joint Wigner distribution for the two points in space. By preparing a spatially entangled state with the appropriate correlations, this set of measurements will violate local realism in the sense shown above.

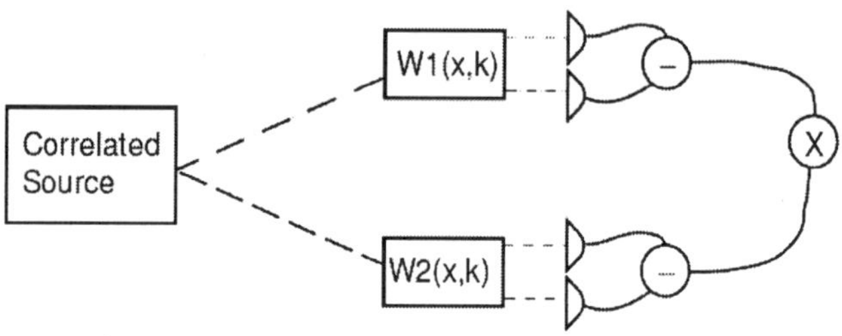

Fig. 2. Pair of Sagnac interferometers for measuring the joint spatial Wigner function of a pair of photons. This apparatus can be used to show a violation of local realism for spatially entangled photon pairs.

3. Conditional Preparation of Pure-State Single Photons

3.1. *Conditional preparation of single photons relying on PDC photon pairs*

The process of parametric downconversion (PDC) has been successfully exploited for conditional preparation of single photons.[27–30] In this technique, while photon pair emission occurs randomly (constrained to short time intervals in the case of a pulsed pump), a single "signal" photon can be heralded by the detection the conjugate "trigger" photon. This is to be contrasted with on-demand sources which emit single photons at well defined times, where imperfect matching of emission to detection modes can nevertheless result in a random selection of collected photons. The conditional preparation process generally projects the resulting heralded signal state into a mixed state[31] expressed as an incoherent sum of the available Schmidt modes.[32] What this means is that the mixedness exhibited by conditionally prepared *single photons* is determined by the level of quantum correlations present in the initial *photon pairs*; these correlations can be conveniently quantified by the cooperativity parameter calculated via a photon-pair Schmidt decomposition.[33] The conditional preparation of pure-state single photons therefore requires the ability to engineer two-photon states so as to eliminate signal-idler correlations in all degrees of freedom. In waveguided PDC,[27] the spatial correlations between signal and idler $\boldsymbol{k}$-vectors are suppressed with the result that the attainable single photon purity is limited only by spectral correlations. Thus, the combination of waveguiding technology with the engineering of spectrally uncorrelated states is expected to lead to nearly ideal conditional single photon sources.

The generation of broadband, spectrally uncorrelated PDC photon pairs necessitates the use of a pulsed pump; indeed, in the CW-pumped case, energy conservation leads to strict spectral anti-correlation between signal and idler. The spectral structure of PDC photon pairs is determined by crystal dispersion, in particular by group velocity mismatch between the pump pulse and the signal and idler photons.[34,35,37] number of methods have been put forward for controlling group velocity mismatch for the suppression of spectral correlations.[31,33,35,38,39,40] We have shown that a viable direct method is the examination of available $\chi^{(2)}$ materials and spectral ranges locating regimes where group velocity matching occurs naturally.[31,33,35] The group velocity matching condition where the reciprocal pump pulse group velocity is given as the mean of the signal and idler reciprocal group velocities[34] can lead to a *balanced* uncorrelated state where signal and idler exhibit identical bandwidths.[35] As experimentally demonstrated by Kuzucu *et al.*, this condition can also lead to frequency coincident states.[36] Unfortunately, the latter condition tends to occur, for most common $\chi^{(2)}$ materials, in the infra-red, where single photon detection is not well-developed. However, asymmetric group velocity matching, where the pump pulse is matched to either the signal or the idler, but not to both, gives an uncorrelated state in which the signal photon is broadband while the idler is nearly monochromatic at wavelengths within the detection window of well-developed silicon single photon detectors.[31]

In view of the restricted number of bulk materials, it is desirable to develop alternate tools to obtain spectral uncorrelation at arbitrary wavelengths while employing materials exhibiting large$\chi^{(2)}$-nonlinearities.[31,33,38,39,40] One possibility is the use of a segmented crystal approach, where a sequence of $\chi^{(2)}$ thin crystal segments is interspersed with linear birefringent compensators for group velocity mismatch control,[31,41] leading to an additional superlattice-like contribution to the phase matching function. A proof of principle experiment was reported in Ref. 42. This showed that, provided that the crystal/compensator materials and respective thicknesses are chosen appropriately, the maximum group velocity mismatch exhibited by such a crystal sequence can be limited to that associated with a single thin crystal segment. Furthermore, the ratio of crystal to spacer thickness controls the spectral correlation characteristics in the resulting photon pairs, with the ability to generate a spectrally uncorrelated state as a special case. The ability to control the entanglement present in the continuous variable degrees of freedom of single photons is indeed an important enabling step in the continued progress of quantum information processing in the photonic realm.

3.2. *Factorization of the wave function in signal and idler modes for bulk crystals*

The generation of pure single-photon wavepackets based on PDC photon pairs necessitates that all correlations between signal and idler modes be eliminated. This, however, does not imply the need to suppress spatio-spectral correlations for the individual signal and idler photons. For PDC in bulk crystals the spatial structure of the generated photon pairs depends not only on the crystal parameters, but also on the pump mode profile. We show in the following that the decoupling of the signal and idler modes can be achieved if spatio-spectral correlations for the individual modes are accepted. For this purpose we derive the appropriate conditions for the relation between the pump beam divergence and the crystal length.

Let us assume that the pump field is a superposition of monochromatic Gaussian modes:

$$E_p(\boldsymbol{r},t) = N_p \int d\boldsymbol{k_p}\, \alpha(\boldsymbol{k_p}) \exp\left[i(\boldsymbol{k_p}\cdot\boldsymbol{r} - \omega(\boldsymbol{k_p})t)\right] + c.c, \tag{11}$$

where N_p is a normalization constant and $\alpha(\boldsymbol{k_p})$ can be expressed as

$$\alpha(\boldsymbol{k_p}) = \int d\omega_p \alpha_\omega(\omega_p)\alpha_k(\boldsymbol{k_p},\omega_p). \tag{12}$$

The function $\alpha_k(\boldsymbol{k_p},\omega_p)$ is the angular spectrum of the monochromatic Gaussian beam with frequency ω_p. Using the standard analysis, the PDC state is then given by

$$|\Psi(t)\rangle = |vac\rangle + \mu \iint d\boldsymbol{k_s} d\boldsymbol{k_i}\, \psi(\boldsymbol{k_s};\boldsymbol{k_i})\hat{a}^\dagger(\boldsymbol{k_s})\hat{b}^\dagger(\boldsymbol{k_i})\,|vac\rangle\,. \tag{13}$$

Here the function $\psi(\boldsymbol{k_s};\boldsymbol{k_i})$ represents a weighting function for the different modes, which is determined by the pump envelope Eq. (13) and the phase matching function. Thus, $\psi(\boldsymbol{k_s};\boldsymbol{k_i})$ can be expressed in terms of pump and crystal parameters as:

$$\psi(\boldsymbol{k_s};\boldsymbol{k_i}) = \alpha_\omega(\omega_s+\omega_i)\exp\left[-\frac{w_0^2}{4}(k_x^2+k_y^2)\right]\mathrm{sinc}\left[L\left(\frac{k_x^2+k_y^2}{4k}-\frac{\Delta k_z}{2}\right)\right], \tag{14}$$

where w_0 is the pump spot size at the crystal center and L is the crystal length. The $\boldsymbol{k}$-vector is characterized by

$$k_x = k_{sx}+k_{ix} \qquad k_y = k_{sy}+k_{iy} \qquad \text{and} \qquad \Delta k_z = k - k_{sz} - k_{iz},$$

and the absolute value k is defined as $k = (\omega_s+\omega_i)n(\omega_s+\omega_i)/c$ with $n(\omega)$ denoting the refraction index of the crystal. Using the Gaussian approximation[33] $\mathrm{sinc}(x) \approx e^{-\gamma x^2}$ and noting that the transverse wavevector is $\boldsymbol{k}^{\perp 2} = k_x^2+k_y^2$, Eq. (14) yields:

$$\psi(\boldsymbol{k}^\perp,\Delta k_z) = \alpha_\omega(\omega_s+\omega_i)\exp\left[-\frac{w_0^2\boldsymbol{k}^{\perp 2}}{4}-\gamma\frac{L^2}{4}\left(\frac{\boldsymbol{k}^{\perp 2}}{2k}-\Delta k_z\right)^2\right]. \tag{15}$$

To eliminate spatial-spectral correlations between signal and idler, $\psi(\boldsymbol{k_s};\boldsymbol{k_i})$ must be factorable into individual signal and idler modes. The required factorability condition becomes apparent upon transforming from cartesian to spherical coordinates $\iiint dk_x dk_y dk_z \longrightarrow \iiint d\omega d\theta d\phi$, and expressing the phase mismatch as a Taylor expansion up to second order. We thus arrive at the conditions which ensure wavefunction factorability:

(1) $\boldsymbol{k}_s+\boldsymbol{k}_i = \boldsymbol{k}_p$ (phase matching) ,
(2) $\tan\theta = w_0/(\sqrt{\gamma}L)$, where $\theta_s=\theta_i=\theta$, and
(3) $v_s\cos\theta_s = v_i\cos\theta_i = v_p$ (group velocity matching),
where v_s, v_i and v_p are the group velocities of signal, idler and pump photons.

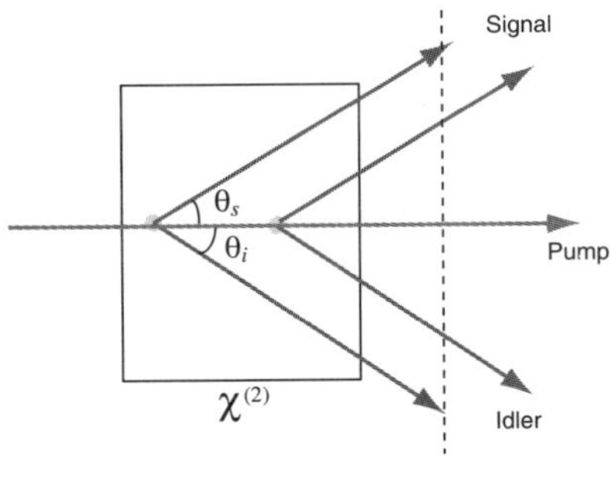

Fig. 3. Group velocity condition enables the factorability of signal and idler photon quantum states.

The first condition describes momentum conservation. The second condition gives the relation between the pump beam divergence and the crystal length. The meaning of the third condition can be understood from Fig. 3. A photon pair may be generated at any point as the pump pulse propagates through the crystal. Group velocity matching means that the z-component of the signal and idler group velocities equal the pump group velocity. Thus, at any plane normal to the z-axis, the photon pairs generated at different points will arrive at this plane at the same time. As a result, the idler detection time gives no information about the signal photon. It is this indistinguishability that guarantees the conditional single photon pure state.

4. Applications of Continuous Variables in Single Photons

The analysis of the previous section shows that source parameters control the resulting spatio-spectral correlations in PDC photon pairs. As discussed above, in the context of conditionally prepared pure-state single-photon wavepackets one aims at a factorable wavefunction. In this section we present an example where it is in fact desirable to obtain a state which is highly correlated in its continuous variable degrees of freedom ω_i, ω_s, and $\boldsymbol{k_s}$, $\boldsymbol{k_i}$. Typically, single-photon states are associated with dichotomic variables represented by their polarization or their photon number 0 or 1. By exploiting continuous variable degrees of freedom of single photons a new field for quantum information applications opens up.

4.1. *Qudit information coding*

The spatial and temporal degrees of freedom of entangled photons constitute arenas for transferring information in addition to polarization.[43] Because they are continuous variables, they have the potential to offer higher information transfer rates. Here we present the calculated results for the transmission of several bits per photon using realistic PDC sources.

The state describing PDC photon pairs is given by Eq. (13). Let us note that apart from defining the spatial properties associated with wavectors, Eq. (13) determines the spectral properties through the dispersion relation $k = \omega n(\omega)/c$. In what follows it is assumed that the state is spectrally filtered such that the signal and idler frequencies are restricted to $\omega_{s0} = \omega_{i0} = \omega_p/2$ (thus eliminating the effect of spectral correlations). The resulting joint probability amplitude is:

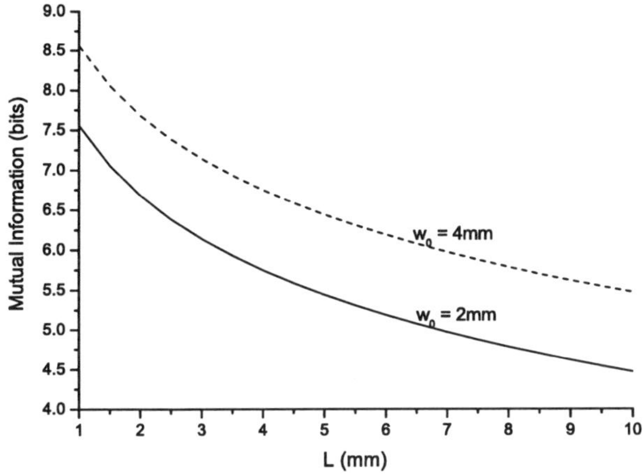

Fig. 4. Mutual information for entangled photon pairs generated by Type-I PDC in BBO crystals.

$$f(\boldsymbol{k}_s^{\perp};\boldsymbol{k}_i^{\perp}) \propto \exp\left[-\frac{w_0^2}{4}|\boldsymbol{k}_s^{\perp}+\boldsymbol{k}_i^{\perp}|^2\right] \operatorname{sinc}\left[\frac{L}{2}\left(2K-k_p-\frac{|\boldsymbol{k}_s^{\perp}-\boldsymbol{k}_i^{\perp}|^2}{4K}\right)\right], \tag{16}$$

where $K = k_s = k_i$ denotes the signal and idler wavevectors.

The mutual information $I(\boldsymbol{k}_s^{\perp};\boldsymbol{k}_i^{\perp})$ between the photons, which results from the correlations between $\boldsymbol{k}_s^{\perp}$ and $\boldsymbol{k}_i^{\perp}$, is directly determined by the joint probability distribution of the entangled photons $p(\boldsymbol{k}_s^{\perp};\boldsymbol{k}_i^{\perp}) = |f(\boldsymbol{k}_s^{\perp};\boldsymbol{k}_i^{\perp})|^2$. Similarly, the Fourier transformation of Eq. (16) determines the mutual information $I(\boldsymbol{r}_s^{\perp};\boldsymbol{r}_i^{\perp})$. Fig. 4 shows the results for the mutal information $I(\boldsymbol{k}_s^{\perp};\boldsymbol{k}_i^{\perp})$ for a realistic entangled photon source, which is generated by type-I downconversion in BBO crystals. The graph illustrates the information transfer gain if the continuous variable degrees of freedom of single photons are exploited, which has to be compared with one bit polarization coding. The amount of the shared information between PDC photons will increase with the spot size of the pump (w_0), while it decreases with the length of the crystal L. Since the mutual information indicates the correlations between the entangled photons, this result is consistent with Schmidt decomposition analysis presented by Law.[44]

The analysis in this section is based on complete continuous variable coding. Note, however, that for realistic quantum information applications, these variables must be discretisized due to limitations of the apparatus.

Thus, the information transfer rates presented here constitute an idealized upper limit; the impact of binning merits further study.

4.2. *Quantum key distribution with continuous variables and photon pairs*

The first protocol for quantum key distribution (QKD) was proposed by Bennett and Brassard[45] as an alternative to classical cryptography. This protocol, usually called BB84, and its large collection of variations, including the schemes proposed by Ekert using polarization-entangled photon pairs,[46–48] employ discrete variables to carry information.[49] There is another category of QKD protocols which employ continuous variable entangled states.[50–53] The main advantage of CV-QKD over discrete variable QKD is that it enables higher key distribution rates. In was shown in the previous section that the spatial degree of freedom can be used to transfer information between entangled photon pairs. The uncertainty relation between momentum and position suggests that these degrees of freedom can be used for QKD. A first experimental step in this direction was presented by Almeida *et al.*[54]

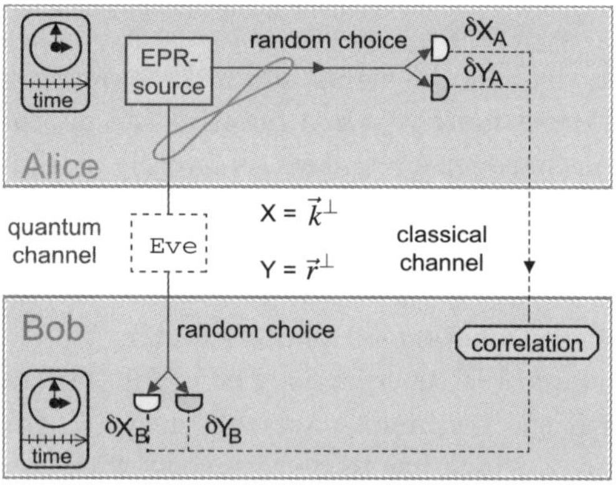

Fig. 5. A scheme for single-photon CV-QKD. The uncertainty principle $\Delta\vec{k}^{\perp}\cdot\Delta\vec{r}^{\perp} \geq 1$ prevents the simultaneous measurement of both variables.

A scheme for single-photon CV-QKD is shown in Fig. 5. The protocol, which is similar to the quadrature-entangled QKD presented in Ref. 53, is summarized as follows. Pairs of entangled photons are generated by the PDC source and transmitted to Alice (signal) and Bob (idler) separately via a quantum channel. The two receivers then measure either position ($\boldsymbol{r}^{\perp}$) or momentum ($\boldsymbol{k}^{\perp}$) of each photon they receive. The outcome registered by Alice and Bob are raw keys. They proceed to communicate via the classical channel to perform key sifting, interactive error correction and privacy amplification to extract the secure key.

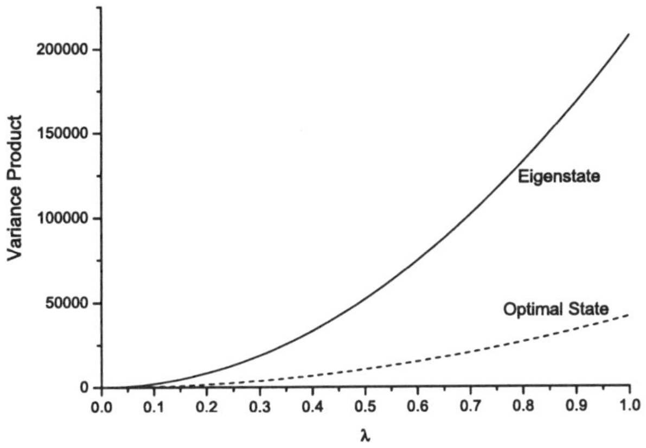

Fig. 6. Product of position-momentum variances measured by Alice and Bob, for two states sent by Eve to Bob: position eigenstate (solid line) and the optimal position-momentum squeezed state (dashed line). The initial entangled photons are generated by Type-I PDC in a 2mm BBO crystal and the spot size of the pump is 2mm.

The security of this scheme strongly depends on the quantum correlations between entangled photons. It has been demonstrated by Howell[26] *et al.* that the product of the joint quadrature variances $\Delta^2(x_A - x_B)\Delta^2(p_A + p_B)$ indicates the strength of quantum correlations between the photon pairs. For strong correlations, the product of the variance will be close to zero. Eavesdropping will decrease the correlations between Alice and Bob, thus increasing the product of the variances. Thus, by measuring the product of variances, Alice and Bob are able to detect the presence of Eve. A complete security analysis of this scheme is complicated by the wide variety of attacks that Eve can employ. In order to make a preliminary analysis, we assume a simple attack, usually called intercept-resend. In this strategy, Eve intercepts the photon sent to Bob, measures it, replaces it with a new photon based on her measurement result and sends this photon to Bob. It is the state of Eve's replaced photon which will now determine the value of the variance product which Alice and Bob ascertain. Our analysis shows that there exists an optimal state which minimizes the increase of the variance product; Fig. 6 illustrates Eve's impact on the Alice's and Bob's correlations for different states in dependence of

$$\lambda = \frac{\text{Number of photons intercepted by Eve}}{\text{Total number of photons Alice sends to Bob}}.$$

To effect classical privacy amplification, the mutual information between Alice and Bob ($I(A, B)$), and the mutual information between Alice and Eve ($I(A, E)$) are important. Fig. 7 shows the relation between $I(A, B)$, $I(A, E)$ and λ. When $I(A, B) > I(A, E)$ classical privacy amplification can be used directly to extract the secure key. This condition can be guaranteed by a measurement of the variance product.

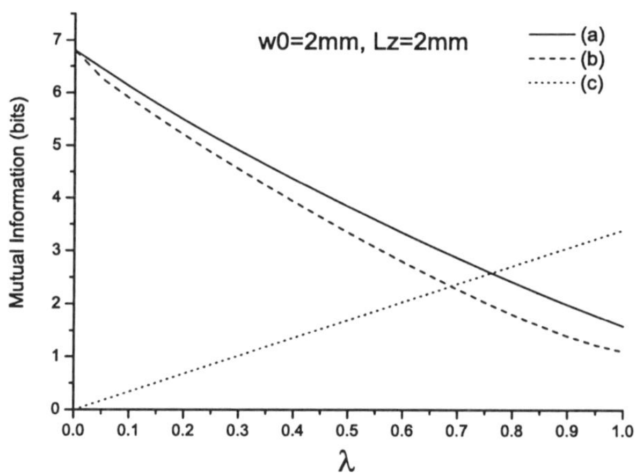

Fig. 7. (a) I_{AB} when Eve resends a position eigenstate. (b) I_{AB} when Eve resends the optimal state. (c) I_{AE}. The mutual information is independent of the state Eve sends to Bob.

5. Conclusions and Outlook

Photonic states constitute ideal carriers for quantum information, especially for quantum communication schemes where information has to be transferred between different parties. However, it is common in such applications to consider only the discrete degrees of freedom of the photonic quantum state. The intrinsic structure of the photonic states including the continuous degrees of freedom has to be taken into consideration for all quantum information processing involving interferences between photons. Uncontrolled probability amplitudes in these degrees both sets a limit to the degree to which photons from separate sources may interfere with one another, an essential ingredient in quantum information processing. On the other hand, these provide an additional extended Hilbert space into which information may be coded and manipulated. The analogy between the phase space representations of quantum states in the quadrature amplitude representation, and single photon states in the space-momentum representation shows how continuous variables of single photons can frequently be utilized in the same way which is known from quadrature-based protocols. This opens a promising route for new developments in continuous variable quantum information.

Proper control over the complete quantum state will be one of the major challenges for the further progress of the field. We have outlined both fundamental experiments and practical applications that are beginning to make use of these additional variables, and we expect that there will be many others in the future.

Acknowledgments

This work was supported by the EPSRC and by the US National Science Foundation. We are grateful for the extended interactions with M. G. Raymer that this latter source has provided, during which a number of ideas presented here were germinated. We have also benefitted from enlightening conversations with G. Leuchs, N. Lütkenhaus, T. Ralph and P. Grangier.

EM is currently with the Department of Physics, Stanford University, USA.

References

1. I. Białynicki-Birula, *Prog. in Optics XXXVI*, E. Wolf (Ed.), Elsevier, Amsterdam 1996.
2. J. E. Sipe, Phys. Rev. A **52**, 1875 (1995).
3. E. Knill, R. LaFlamme, and G. J. Milburn, Nature **409**, 46 (2001).
4. D.E. Browne, and T. Rudolph, quant-ph/0405157.
5. A. Gilchrist *et al.*, J. Opt. B **6**, S828 (2004).
6. S. L. Braunstein and A. Mann, Phys. Rev. A **51**, R1727 (1995).
7. C.K. Hong and L. Mandel, Phys. Rev. Lett. **56**, 58 (1986).
8. P. Michler, A. Kiraz, C. Becher, W. V. Schoenfeld, P. M. Petroff, L. D. Zhang, E. Hu, and A. Imamoğlu, Science **290**, 2282 (2000).
9. C. Kurtsiefer, S. Mayer, P. Zarda, and H. Weinfurter, Phys. Rev. Lett. **85**, 290 (2000).

10. S. Wallentowitz and W. Vogel, Phys. Rev. A **53**, 4528 (1996).
11. K. Banaszek and K. Wódkiewicz, Phys. Rev. Lett. **76**, 4344 (1996).
12. E.A. Mukamel, K. Banaszek, and I.A. Walmsley, Opt. Lett. **28**, 1317 (2003).
13. B. Smith and M. G. Raymer, CLEO Technical Digest, OSA Washington DC, 2004.
14. M. J. Bastiaans, Opt. Comm. **25**, 26 (1978).
15. D. Dragoman, in: *Progress in Optics XXXVII*, E. Wolf (Ed.), Elsevier, Amsterdam 1997.
16. O. Cohen, Phys. Rev. A **56**, 3489 (1997).
17. N. Gisin and A. Peres, Phys. Lett. A **162**, 15 (1997).
18. J.G. Rarity and P.R. Tapster, Phys. Rev. Lett. **64**, 2495 (1990).
19. K. Banaszek and K. Wodkiewicz, Phys. Rev. Lett. **82**, 2009 (1999); Phys. Rev. A **58**, 4345.
20. K. Banaszek and K. Wódkiewicz, Phys. Rev. A **58**, 4345 (1998).
21. M.D. Reid and P.D. Drummond, Phys. Rev. Lett. **60**, 2731 (1988).
22. L.M. Duan, G. Giedke, J.I. Cirac, and P. Zoller, Phys. Rev. Lett. **84**, 2722 (2000).
23. R. Simon Phys. Rev. Lett. **84**, 2726 (2000).
24. S. Mancini, V. Giovannetti, D. Vitali, and P. Tombesi, Phys. Rev. Lett. **88**, 120401 (2002).
25. Z.Y. Ou, S.F. Pereira, H.J. Kimble, and K.C. Peng, Phys. Rev. Lett. **86**, 3663 (1992).
26. J.C. Howell, R.S. Bennink, S.J. Bentley, and R.W. Boyd Phys. Rev. Lett. **92**, 210403 (2004).
27. A.B. U'Ren, Ch. Silberhorn, K. Banaszek, and I.A. Walmsley, Phys. Rev. Lett. **93**, 093601 (2004).
28. A.B. U'Ren, Ch. Silberhorn, J.K. Ball, K. Banaszek, and I.A. Walmsley, quant-ph/0412184 (2004).
29. O. Alibart, S. Tanzilli, D. B. Ostrowsky, and P. Baldi, quant-ph/0405075 (2004).
30. T.B. Pittman, B.C Jacobs, and J.D. Franson, Opt. Comm. **246**, 545 (2005).
31. A.B. U'Ren, Ch. Silberhorn,, K. Banaszek, I.A. Walmsley, R. Erdmann, W.P. Grice, and M.G. Raymer, Laser Physics **15**, 1 (2005).
32. C.K. Law, I.A. Walmsley, and J.H. Eberly, Phys. Rev. Lett. **84**, 5304 (2000).
33. A.B. U'Ren, K. Banaszek, and I.A. Walmsley, Quant. Inf. and Comp. **3**, 480 (2003).
34. T.E. Keller and M.H. Rubin, Phys. Rev. A **56**, 1534 (1997).
35. W.P. Grice, A.B. U'Ren, and I.A. Walmsley, Phys. Rev. A **64**, 063815 (2001).
36. O. Kuzucu, M. Fiorentino, M. A. Albota, F. N. C. Wong, and F.X. Kaertner Phys. Rev. Lett. **94**, 083601 (2005).
37. V. Giovannetti, L. Maccone, J.H. Shapiro, and F.N.C. Wong, Phys. Rev. A **66**, 043813 (2002).
38. Z.D. Walton, A.V. Sergienko, B.E. A. Saleh, and M.C. Teich, Phys. Rev. A **70**, 052317 (2004).
39. J.P. Torres, F.Maci, S.Carrasco, and L. Torner, Opt. Lett. **30**, 314 (2005).

40. J.P. Torres, M.W. Mitchell, and M.Hendrych, Phys. Rev. A **71**, 022320 (2005).
41. A.B. U'Ren, R.K. Erdmann, and I.A. Walmsley, to appear J. Mod. Opt. (2005).
42. R.K. Erdmann, A.B. U'Ren, M. de la Cruz-Gutierrez, and I.A. Walmsley, unpublished.
43. P. G. Kwiat, J. Mod. Opt. **44**, 2173 (1997).
44. C.K. Law and J.H. Eberly, Phys. Rev. Lett. **92**, 127903 (2004).
45. C.H. Bennett and G. Brassard, in: *Proc. IEEE Int. Conference on Computers, Systems and Signal Processing*, IEEE, New York 1984.
46. A.K. Ekert, Phys. Rev. Lett. **67**, 661 (1991).
47. T. Jennewein, U. Achleitner, G. Weihs, H. Weinfurther and A. Zeilinger, Phys. Rev. Lett. **84**, 4729 (2000).
48. D. Naik, C. Peterson, A. White, A. Berglund and P. Kwiat, *Phys. Rev. Lett.* **84**, 4733 (2000).
49. N. Gisin, G. Ribordy, W. Tittel, and H. Zbinden, Rev. Mod. Phys. **74**, 145 (2002).
50. T.C. Ralph, Phys. Rev. A **61**, 010303, (1999).
51. M.D. Reid, Phys. Rev. A **62**, 062308 (2000).
52. K. Bencheikh, T. Symul, A. Jankovic, and J. A. Levenson, J. Mod. Opt. **48**, 1903 (2001).
53. Ch. Silberhorn, N. Korolkova, and G. Leuchs, Phys. Rev. Lett. **88**, 167902 (2002).
54. M.P. Almeida, S.P. Walborn, and P.H. Souto Ribeiro, quant-ph/0411183 (2005).

Chapter 20

Experimental Non-Gaussian Manipulation of Continuous Variables

Jérôme Wenger, Alexei Ourjoumtsev, Julien Laurat,
Rosa Tualle-Brouri and Philippe Grangier

Laboratoire Charles Fabry de l'Institut d'Optique
CNRS UMR 8501, F-91403 Orsay, France

email: philippe.grangier@iota.u-psud.fr

Non-Gaussian operations constitute a topic of major importance, as they allow to take advantage of the whole potential of quantum continuous variables. We discuss the implementation of a non-Gaussian operation that maps individual pulses of squeezed light onto non-Gaussian states. The "de-Gaussification" effect is obtained by conditionally subtracting one photon from the squeezed vacuum thanks to a low-reflectance beam splitter followed by a photon-counter. The experimental data provides a clear evidence of phase-dependent non-Gaussian statistics, which are analyzed using standard quantum tomography techniques.

1. Introduction

Continuous variables offer an interesting alternative to discrete variables to perform quantum key distribution (QKD). Continuous variables QKD schemes using the quadrature components of amplitude and phase modulated coherent states have been recently proposed[7] and experimentally demonstrated.[2] Such coherent state protocols have first been proven to be secure against individual Gaussian attacks for any value of the line transmission.[2,3] More general proofs including non-Gaussian collective attacks have also been proposed recently.[4–7] An important practical advantage of coherent states protocols is that they can reach very high secret bit rates (greater than one million secret bits per second) in the case where the transmission of the quantum channel is high enough.[2] However, when the channel losses are higher than a few dB, the transmitted secret data gets buried into errors of various origins, that range from detectors noise to imperfect data processing. The net secret key rates is thus rapidly decreasing, and consequently the practical range of coherent states QKD appears presently limited to a few tens of kilometers. This means that coherent states QKD are not expected to do much better than photon-counting QKD[9] in terms of absolute distance.

In order to distribute quantum keys over significantly larger distances, a major challenge is to implement entanglement distillation schemes to compensate for the unavoidable perturbations induced while transfering the quantum states.[10] For finite-dimension systems, highly entangled states may indeed be distilled out of poorly entangled states at the price of a reduction of the number of total quantum states.[11] However, transposing this result to infinite-dimension systems is not simple, as fundamental limitations on the manipulation of quantum continuous variables appear. A major concern is given by Gaussian "no-go" theorems : while most (if not all) QCV transmissions so far are using light beams with Gaussian statistics, it has been shown that it is not possible to distillate entanglement from a Gaussian input to a Gaussian output by Gaussian means.[12–14] To distillate the entanglement of Gaussian states, theoretical proposals make use of non-Gaussian operations[15] or of an iterative procedure to jump "outside" the Gaussian domain and to reach it back at the end.[16] Therefore, non-Gaussian operations will play a crucial role in the future of quantum information processing to extract the whole potential of QCV.

Implementing directly non-Gaussian operations using active non-linear optical devices requires strong optical non-linearities,[17] which unfortunately are out of reach of the present technology due to the small susceptibilities of the nonlinear media. However, an alternative idea proposes to take advantage of conditional measurements (see for instance Ref. 18). If a portion of a composite entangled system is measured, the other component is *projected* according to the outcome of the measurement, following the basic postulates of quantum mechanics. The resulting operation performed on the unmeasured sub-system is then highly nonlinear. This may achieve non-Gaussian tasks using an *effective* measurement-induced nonlinearity. The rate of success of such an operation is directly given by the probability of getting a certain outcome from the measurement, which may be sufficiently high in certain cases.

For instance, conditional measurements can be performed by detecting one part of a two-mode squeezed vacuum. This setup allows to produce single photon states[19,20] or sub-Poissonian states.[21–24] Other proposals use low-reflectance beam splitters and photon-counters to conditionally add or subtract single photons. Such operations are of particular relevance to generate non-Gaussian states[25,26] and Schrödinger cats,[27–29] to increase entanglement,[30,31] teleportation fidelity[32] and squeezing,[33] and to test Bell inequalities using QCV and homodyne detections.[34–37]

In this chapter, we will discuss a non-Gaussian operation which we call "de-Gaussification", that maps short pulses of squeezed light onto non-Gaussian states. This protocol is based upon the photon subtraction from a pulsed squeezed vacuum state using a low-reflectance beam splitter and a photon-counter giving clicks when at least one photon is detected (avalanche photodiode, used in the Geiger mode). Extending this procedure to entangled EPR beams -which is fairly simple in principle- provides the first step of an entanglement distillation procedure as proposed in Ref. 16.

The experimental scheme is presented on Fig. 1. Femtosecond pulses undergo parametric amplification through a single pass in a potassium niobate nonlinear crystal, producing pulsed squeezed vacuum states with a significant squeezing of about -3 dB. Thanks to this relatively strong nonlinear interaction, "real" squeezed states are actually produced, not only parametric pairs. Therefore, high order terms beyond pair production have explicitly to be included in the analysis as they play an essential role to understand the experimental data. A small fraction ($R \ll 1$) of the squeezed vacuum beam is then taken out using a low-reflectance beam splitter, and provides trigger photons which are spatially and spectrally filtered before being detected by a silicon avalanche photodiode (APD). This photodetector has a single photon sensitivity but is unable to distinguish between the number of photons. The output of the photon-counter (= detection of at least one photon) is then used as a condition to select the transmitted pulse or to reject it. As we will show, this selection provides directly a non-Gaussian operation and maps the transmitted states onto non-Gaussian states.

To check the result of the conditioning operation, the transmitted pulse is measured by a time-resolved homodyne detection which samples one value of a chosen signal quadrature for each incoming pulse. Let us emphasize that all the processing is done in the time domain, not in the frequency domain. This allows to access to each pulse individually and to reconstruct the full statistics of the signal pulses.[38] It is thus very easy to analyse our experiment in terms of information transfers involved in quantum communication protocols.[2] The histograms presented below are obtained from these individual pulse data.

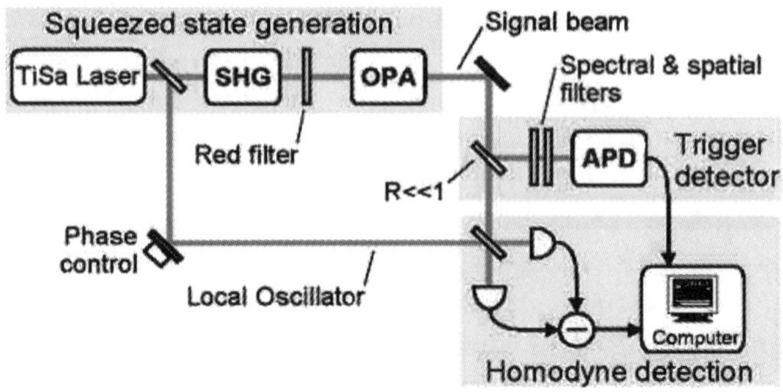

Fig. 1. Simplified experimental setup for the de-Gaussification procedure.

This chapter is organized as follows : we first discuss theoretically the proposed procedure in Sec. 2. We then describe the experimental setup in Sec. 3 before discussing the characterization of the generated non-Gaussian states in Sec. 4. Finally, we conclude in Sec. 5 by presenting some potential applications of this procedure towards QCV processing.

2. Squeezed Vacuum Degaussification: A Theoretical Approach

To begin our theoretical study, let us first recall that the (pure) squeezed vacuum state can be expressed on the Fock basis as :

$$|\Psi_s\rangle = \frac{1}{\sqrt{\cosh s}} \sum_{m=0}^{\infty} \frac{\sqrt{2m!}}{m!} \left(\frac{1}{2}\tanh s\right)^m |2m\rangle = \sum_{m=0}^{\infty} \alpha_{2m} |2m\rangle \qquad (1)$$

This expression contains only pairs of photons, which is a direct consequence of the parametric process used to generate squeezed states. Here, s stands for the compression factor (the minimum quadrature variance is then given by $V_{\min} = e^{-2s} N_0$, where N_0 is the shot noise variance). With our experimental degree of squeezing $s = 0.43$, one gets $\alpha_0 = 0.96$, $\alpha_2 = 0.27$ and $\alpha_4 = 0.10$. This shows that high order terms (beyond pair production) are clearly not negligible in our experiment.

The squeezed vacuum state $|\Psi_s\rangle$ then gets mixed with the vacuum at the beam splitter, resulting in a two-mode entangled squeezed state. Denoting as R, T the reflectivity and transmittance of the beam splitter $(R+T=1)$, the output state is :

$$|\Psi_{s,out}\rangle = \sum_{m=0}^{\infty} \sum_{k=0}^{2m} \alpha_{2m} \sqrt{\frac{2m!}{k!(2m-k)!}} \sqrt{R}^{\,k} \sqrt{T}^{\,2m-k} |k\rangle_1 |2m-k\rangle_2 \qquad (2)$$

where $|\,.\rangle_1$ is the trigger beam directed towards the photoncounter and $|\,.\rangle_2$ is the beam sent to the homodyne detection. Finally, post-triggering on the APD photocounting events resumes to applying the projector $\Pi_1 = 1 - |0\rangle_1\langle 0|_1$ to the state $|\Psi_{s,out}\rangle$ and tracing out the terms for which the number of reflected photons k is non-zero.

If the beam splitter reflectivity is small, $R \ll 1$, one may neglect terms higher than 1 on the APD beam to get a physical insight of the effects of the de-Gaussification procedure. In this limit, the state conditioned upon the photocounting events is very close to a pure state:[8]

$$|\Psi_{\text{cond}}\rangle = \frac{1}{\sqrt{\cosh^3 \zeta}} \sum_{p=0}^{\infty} \frac{\sqrt{(2p+1)!}}{p!} \left(\frac{1}{2}\tanh\zeta\right)^p \,|\,2p+1\rangle_2 \tag{3}$$

where $\tanh\zeta = (1-R)\tanh s$. This expression only contains *odd* terms on the Fock basis. Such states are well-known to be non-Gaussian states, as their Wigner function take negative values at the origin of the phase space. Intuitively, the de-Gaussification operation can be understood as the subtraction of one single photon from the squeezed vacuum, exactly as if the bosonic annihilation operator $\hat{a}$ was applied. One may also notice that this expression shows a striking similarity with the squeezing operator applied to the single Fock state $\widehat{S}(\zeta)|1\rangle$,[29] with an equivalent degree of squeezing ζ instead of s. In other words, in the limit $R \to 0$, the degaussified state $|\Psi_{\text{cond}}\rangle$ can be arbitrarily close to a squeezed photon.

In the case where the beam splitter reflectivity is non-vanishing, the conditioned state is not a pure state, but a mixed state. We have carried out a calculation for an arbitrary value of the beam splitter reflectivity taking into account the expansion of the squeezed state in a Fock state basis, including terms up to 10 photons, which is enough for our degree of squeezing. Taking into account the quantum coherence between the states for which the number of reflected photons k is the same, the probability distribution $\mathcal{P}_{\text{cond}}(x_\theta)$ for the quadrature x_θ can be written as :

$$\mathcal{P}_{\text{cond}}(x_\theta) = \sum_{k=1}^{2N} \left| \sum_{m=\lceil k/2 \rceil}^{N} \alpha_{2m} \sqrt{\frac{2m!}{k!(2m-k)!}}\, \sqrt{R}^{\,k}\, \sqrt{T}^{\,2m-k}\, \psi_{2m-k}(x_\theta) \right|^2 \tag{4}$$

where $\psi_n(x_\theta)$ is the wavefunction of the Fock state $|n\rangle$.[39]

Our complete calculation takes into account the various imperfections of the experiment: homodyne detection efficiency η, conditioning detection efficiency η_{APD}, electronic noise, dark counts and modal purity (see discussion in Sec. 4). This calculation is straightforward but tedious, and so the detailed expressions will not be written here. Figure 2 displays the results of this calculation for different values of the beam splitter reflectivity R, assuming a perfect single mode detection efficiency. The quadrature distributions show a strong non-Gaussian shape, together with a clear phase-dependence. This particularity highlights the crucial role played by high-order Fock terms in the conditioned state.

In the limit $R \to 0$, the dip in the centre of the probability distribution goes down to zero. However, when the beam splitter reflectivity is increased, Fock state terms with $n > 1$ may no longer be neglected on the APD beam, and the central dip takes a non-zero value. Strictly speaking, this is not an experimental imperfection, but an intrinsic feature of the experiment at larger R. Another interesting feature is that the non-Gaussian dip in the amplified quadrature is relatively robust to losses, as shown on Fig. 3. Therefore, it can be easily observed with our experimental parameters. The higher sensitivity of the deamplified quadrature to losses is a direct consequence of the fact that losses induce a convolution by a gaussian.[39]

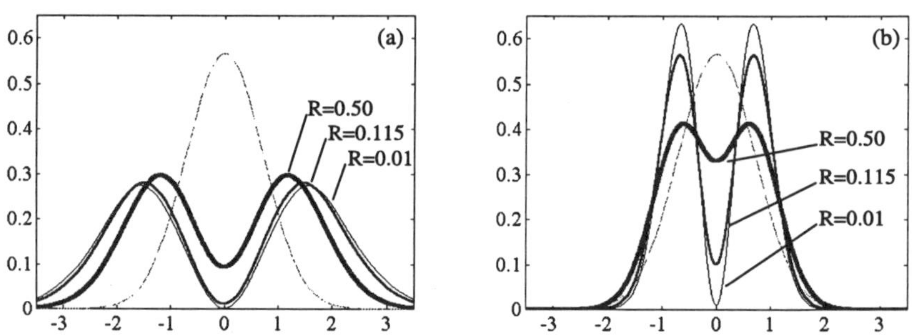

Fig. 2. Theoretical quadrature distributions of the conditioned homodyne measurements for (a) the amplified and (b) the squeezed quadrature, together with the vacuum reference (dashed line, the shot noise variance is taken equal to 1/2), for different values of the beam splitter reflectivity R. The squeezing parameter is $s = 0.43$ and perfect single mode detection is assumed.

3. Experimental Implementation

In this section, we describe the different elements of the experiment implementing the de-Gaussification procedure:[8] the pulsed squeezed vacuum source which provides the basic quantum ressource, the time-resolved ho-

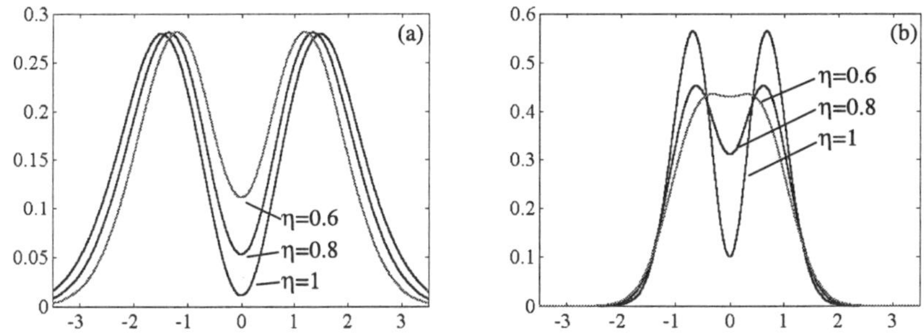

Fig. 3. Same quadrature distributions as in Fig. 2 for different values of the homodyne detection efficiency η. The squeezing parameter is $s = 0.43$ the beam splitter reflectivity is $R = 0.115$.

Fig. 4. Image of the de-Gaussification experiment: (1) indicates the homodyne detection, (2) is the conditioning photon-counting detection and (3) is the pulsed squeezed vacuum source. The photon counting channel (2) is light-shielded when the experiment is operating.

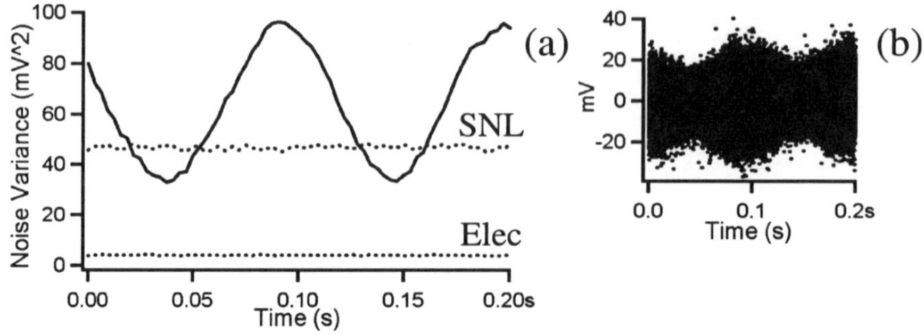

Fig. 5. Squeezed vacuum homodyne measurements in the time domain. Fig. (a) shows the noise variance (plotted in a linear scale and computed over blocks of 2,500 samples) while linearly scanning the LO phase, together with the shot noise level (SNL) and the electronic noise level. Fig.(b) displays the corresponding recorded noise pulses.

modyne detection to fully characterize the incoming quantum state and the conditioning apparatus to perform the non-Gaussian transformation.

3.1. *Pulsed squeezed vacuum generation*

The generation and characterization of squeezed states of light is a topic of major interest, since these states may be used as a direct ressource for efficient QCV protocols in quantum cryptography, entanglement generation, quantum teleportation, dense coding, or quantum computation with continuous variables. Since the landmark pulsed squeezing experiment of Slusher and coworkers,[40] ultrashort pulses have attracted lots of attention for the generation of squeezed states thanks to the availability of high peak powers and the potential for pulse shaping.

Among many nonlinear interactions producing squeezed states, the single-pass parametric amplifier appears as a relatively simple and efficient source of pulsed squeezed light.[40–45] Here, we describe a scheme for pulsed squeezed light generation using femtosecond pulses undergoing parametric deamplification through a single pass in a thin 100 μm potassium niobate ($KNbO_3$) crystal with a significant deamplification of about -3 dB. This crystal length choice allows for wide phase-matching bandwidth and avoids the conditions of large group-velocity mismatch, contrary to the previously reported use of thick $KNbO_3$ crystals.[46,47] Even for the short interaction length used here, $KNbO_3$ proved to be suitable to our applications thanks to its high nonlinear coefficient (about 12 pm/V) and non-critical phase-matching.

The experimental scheme is presented on Fig. 4. The initial pulses are obtained from a cavity-dumped titanium-sapphire laser (Tiger-CD, Time-Bandwidth Products), delivering nearly Fourier-transform limited pulses with a duration of 150 fs centered at 846 nm (FWHM 5 nm), with an energy up to 75 nJ at a pulse repetition rate of 780 kHz. These pulses are focused near the center of an a-cut 100 μm thick anti-reflection coated $KNbO_3$ crystal (FEE GmbH). The crystal is set inside a small vacuum chamber and peltier-cooled down to about -14^oC to obtain non-critical (90 degrees) type-I phase-matching for second harmonic generation (SHG) at 846 nm. The best SHG efficiency obtained was of 32% (corrected from losses), with a typical value of about 28%.

A small fraction (1%) of the fundamental beam is taken out to serve as a probe to study the classical parametric amplification occurring in a similar $KNbO_3$ crystal used in a single-pass type-I spatially degenerate configuration (DOPA). The relative phase between the probe and the pump determines the amplification or deamplification gain, which can be recorded using direct photodetection of the probe averaged power. The best classical deamplification obtained was 0.40 (-4.0 dB) with a corresponding amplification of 2.65 (+4.2 dB). However, the deamplification gain does not correspond exactly to the inverse of the amplification at high pump powers. This discrepancy can be explained by the phenomenon of gain-induced diffraction:[49] due to the Gaussian transverse dependence of the pump intensity, the portion of the probe beam closer to the propagation axis is more amplified than its wings. This non-uniform gain distorts the probe phase front, degrades phase-matching and limits deamplification. Experimentally, we have optimised the overlap between the pump and the probe beam to get the best deamplification, as a compromise has to be found between small wavefront distorsion and spatial overlap between probe and pump. The difference between the amplification and deamplification gains indicates that the general squeezed vacuum state produced by this parametric amplification is not a pure state.[50] However, as it can be seen on Fig. 6, the experimental data for both quadratures is correctly fitted by assuming a single-mode parametric gain $\exp(\pm 2s)$. Thus, since multimode effects remain reasonably small in our experimental conditions, we will use the single parameter s to describe the parametric amplification and deamplification.

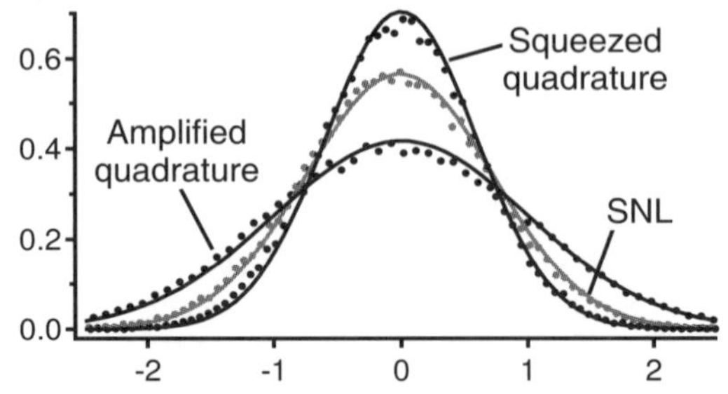

Fig. 6. Normalized probability distribution for the (unconditioned) squeezed vacuum state, obtained from the pulsed homodyne detection. The squeezed quadrature variance is 1.75 dB below SNL, while the amplified quadrature variance is 3.1 dB above. The SNL curve corresponds to the vacuum state (the shot noise variance is taken equal to 1/2).

3.2. *Time-resolved homodyne detection of pulsed squeezed vacuum*

To measure the signal quadrature components, the signal beam is made to interfere on a 50:50 beam splitter with a bright classical beam called local oscillator (LO). Each output of the beam splitter is then detected by a photodiode, the subtraction of the photocurrents is then proportional to the signal quadrature component in phase with the local oscillator LO, in the well-known balanced homodyne detection setup.[51]

A fundamental point of our experiment is that all the processing is done in the time domain and not in the frequency domain, as it is often the case even for pulsed squeezing experiments.[40] For each incoming pulse, the fast acquisition board (National Instruments PCI-6111E) samples one value of the signal quadrature in phase with the local oscillator, allowing to directly construct the histograms presented here.[2,20,45,48,53]

However, such pulsed homodyning is technically much more challenging than frequency-resolved homodyning: each arm of the detection has to be carefully balanced (with a typical rejection better than 10^{-4}) even for ultra-low frequency noises. By blocking the squeezed beam, the detection was checked to be shot-noise limited in the time domain, showing a linear dependence between LO power and the noise variance up to 2.5 10^8 photons per pulse at a repetition rate of 780 kHz and in the femtosecond regime. The electronic noise was low enough to ensure a ratio larger than 11 dB between shot noise and electronic noise variances.

With the probe beam being blocked, the amplifier generates squeezed vacuum pulses which are made to interfere with the local oscillator pulses.

Figure 5 displays the recorded quadrature variance and the corresponding noise pulses while scanning the local oscillator phase. As expected for squeezed states, the measured noise variance passes below the shot noise level (SNL) for some phase values. The measured unconditioned distributions corresponding to the squeezed and anti-squeezed quadratures, and to the vacuum noise are plotted on Fig. 6. The measured squeezing variance (with no correction) is 1.75 dB below the SNL, in good agreement with the corresponding deamplification of a probe beam (0.50 or -3 dB) and our evaluation of the overall detection efficiency $\eta_{tot} = \eta\,(1-R) = 0.66$. Here $1 - R = 0.885$ is the transmission of the conditioning beam splitter, and $\eta = 0.75$ is the homodyne detection efficiency. The value of η is obtained as $\eta = \eta_T \eta_H^2 \eta_D$, where the overall transmission $\eta_T = 0.94$, the mode-matching visibility $\eta_H = 0.92$, and the detectors efficiency $\eta_D = 0.945$ are independently measured. As one can see from the solid Gaussian curves on Fig. 6, the experimental data for both quadratures is correctly fitted by assuming a single-mode parametric gain $\exp(\pm 2s)$ with $s = 0.43$, together with the above efficiency η_{tot}. This testifies that multimode effects remain reasonably small in our experimental conditions.

3.3. *De-Gaussification apparatus*

A small fraction R of the squeezed vacuum beam is taken out from the homodyne detection channel. A crucial point is to correctly choose the conditioning beam splitter reflectivity R. As shown on Fig. 2, the dip in the quadrature distribution is deeper when R is decreased. However, the number of APD photodetection events is directly proportional to R, which means that a lower R will induce a lower success rate for the de-Gaussification procedure, and thus the whole procedure will require a longer integration time to get statistically significant results, leading to unavoidable experimental drifts. The experimental value of $R = 11.5\%$ is thus a compromise between a clear effect in the quadrature distributions and a high enough number of trigger counts per second. With $s = 0.43$ and an overall APD detection efficiency of $\eta_{\text{APD}} \approx 1.4\%$, the expected average count rate per second is $\eta_{\text{APD}}\, R \sinh^2 s\, \text{C} \approx 240$ counts/s for a pulse repetition rate of $\text{C} = 780$ kHz (we experimentally measured an average value of 280 counts/s). This value is sufficiently high compared to the APD dark counts (8 counts/s in the triggered time window).

Because the avalanche photodiode is sensitive to incoming photons in all spatio-temporal modes, tight spatial and spectral filtering are necessary to

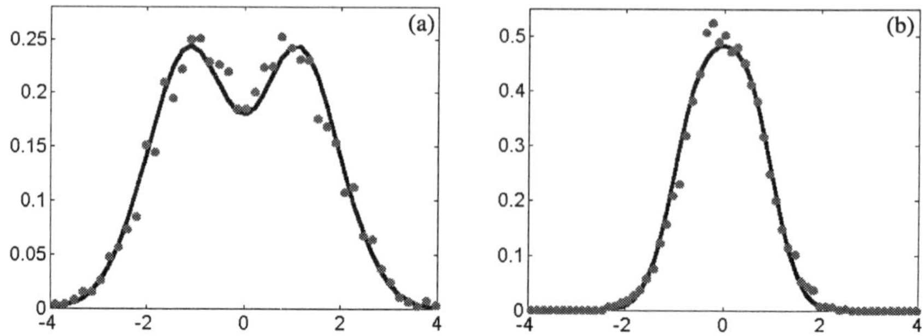

Fig. 7. Experimental (dots) and theoretical (line) quadrature distribution of the post-selected homodyne measurements for the amplified quadrature (a) and the squeezed one (b), normalized as in Fig. 6. Parameters used in the calculation are $s = 0.43$, $R = 0.115$, $\eta = 0.75$ and $\xi = 0.7$.

limit unwanted firings due to the wide parametric fluorescence emission of our system. Spatial filtering is obtained thanks to two Fourier-conjugated pinholes which allow to transmit about 17% of the relevant photons while strongly attenuating the other unwanted modes. A slit placed after a blazed grating used near the Littrow condition then selects a 3 nm spectral window centered at the laser wavelength with an efficiency of 16%. Finally, a silicon avalanche photodiode (APD) detects the trigger photons. The detection click is registered simultaneously with the homodyne signal, and can be used to post-select homodyne events to obtain non-Gaussian statistics as we will show now.

4. Characterization of the Non-Gaussian States

4.1. *Homodyne measurements and influence of experimental imperfections*

Figure 7 displays the post-selected output of the homodyne detection resulting from the de-Gaussification protocol. This figure can be compared to the recorded experimental histograms without conditioning on the APD photodetection events shown on Fig. 6. It is apparent that the curves after post-selection are no more Gaussian: the amplified quadrature exhibits a clear dip in the middle, while the deamplified quadrature is top-flattened. The experimental quadrature distribution also presents a clear phase-dependence. As seen in the analysis of Sec. 2, this may only be explained by including high order terms (beyond pair production), as these terms play an essential role to understand the experimental phase-dependence (the single photon $| 1\rangle$ is phase-independent). The theoretical curves represented on Fig. 7 are obtained from a simple single-mode model which takes into account the measured parametric gain together with various experimental imperfections (losses, imperfect mode-matching, electronic noise, dark counts and modal purity see below for details). There is a good agreement with the experimental data.

In order to characterize the experimental imperfections, let us point out that the homodyne detection and the photon-counting APD detection have quite different behaviors. Due to the interference effect with the local oscillator beam which acts as an ultra-sharp mode filter, the homodyne detection detects only the signal input that is matched with the local oscillator mode. Therefore, the homodyne detection works on a single mode basis: it is not sensitive to "real" photons that are in modes unmatched with the local oscillator, but it is quite sensitive to vacuum modes which couple into the detected mode. On the other hand, the APD detection is not sensitive to vacuum noise, but it will detect photons in any modes. Correspondingly, two experimental parameters must be used: an *homodyne efficiency* parameter η, which measures the overlap between the desired signal mode and the detected mode;[54] and a *modal purity* parameter ξ, which characterizes which fraction of the detected photons are actually in the desired signal mode.[25,55]

In the simplest approach, the homodyne efficiency can be modelized by a lossy beam splitter, taking out desired correlated photons and introducing vacuum fluctuations. However, we cannot modelize the modal purity ξ in our experiment by another lossy beam splitter, because a low value of ξ corresponds to unwanted firings of the APD, for which a squeezed vacuum is still measured at the homodyne detection port. In other words, a low value of ξ is related to a bad post-selection of the homodyne measurements, which indicates an insufficient modal filtering before the APD detection. More precisely, the measured probability distribution for a quadrature x will be taken as $\mathcal{P}(x) = \xi\ \mathcal{P}_{cond}(x) + (1-\xi)\ \mathcal{P}_{uncond}(x)$, where $\mathcal{P}_{cond}(x)$ and $\mathcal{P}_{uncond}(x)$ are respectively the conditioned (= non-Gaussian) and unconditioned (= squeezed) probability distributions, which depend on the values of s, R and η.

The procedure to measure η is well established from squeezing experiments,[38] and it can be cross-checked by comparing the classical parametric gain and the measured degree of squeezing (see Sec. 3). The procedure to measure ξ is less usual, and amounts to evaluate how many unwanted photons make their way through the spatial and spectral filters which are used on the photon counting channel. However, due to large uncertainties on the estimate of the overall APD detection efficiency, this can only yield a rough estimate on $\xi : 0.6 < \xi < 0.8$. Finally, the modal purity ξ is directly fitted to the post-selected data, and η is independantly obtained from squeezing measurements. We have plotted on Fig. 7 the conditional probability distributions, using as parameters the parametric gain $s = 0.43$ and the homodyne efficiency $\eta = 0.75$ (both evaluated from the measured squeezing displayed on Fig. 6), together with the modal purity parameter $\xi = 0.7$ fitted to the conditioned data.

4.2. *Quantum tomography of the non-Gaussian states*

In a last step, we have analysed the de-Gaussification procedure using the standard techniques of quantum tomography.[39] We have recorded an histogram with 40 bins for 6 different quadrature phase values θ between 0 and $5\pi/6$, and about 5000 points for each histogram were acquired in a 3 hours experimental run. The Wigner function displayed on Fig. 8 was then reconstructed using the Radon transform applied to the symetrized experimental data $[\mathcal{P}(x_\theta) + \mathcal{P}(-x_\theta)]/2$, without any correction for measurement efficiency (the use of the symetrized distributions allows to minimize the effects of the statistical noise on the different histograms). A clear dip is present at the origin, with a central value of $W_{\text{exp}}(0, 0) = 0.067$ while the maximum is at 0.12. The theoretical Wigner function is also displayed on Fig. 8 assuming a perfect single-mode detection $\eta = \xi = 1$ with our experimental configuration $s = 0.43$ and $R = 0.115$. The theoretical value at the origin of the phase space is clearly negative, $W_{\text{th}}(0, 0) = -0.26$, showing that the state produced by the (ideal) de-Gaussification procedure is highly non-classical.

As in quantum tomography experiments of the single photon,[19,20] the conditions to get negative values of the measured Wigner function are rather stringent. From our numerical calculations, we have observed that this requires the presence of a dip into the distribution probability associated to the squeezed quadrature. A more detailed calculation on the non-classicality of the degaussified state is given in Ref. 26 where a necessary condition on the conditioning beam splitter transmission is written given the experimental efficiencies.

Given our experimental parameters $s = 0.43$, $R = 0.115$ and $\eta = 0.75$, this condition requires a modal purity ξ better than 0.85, which was not

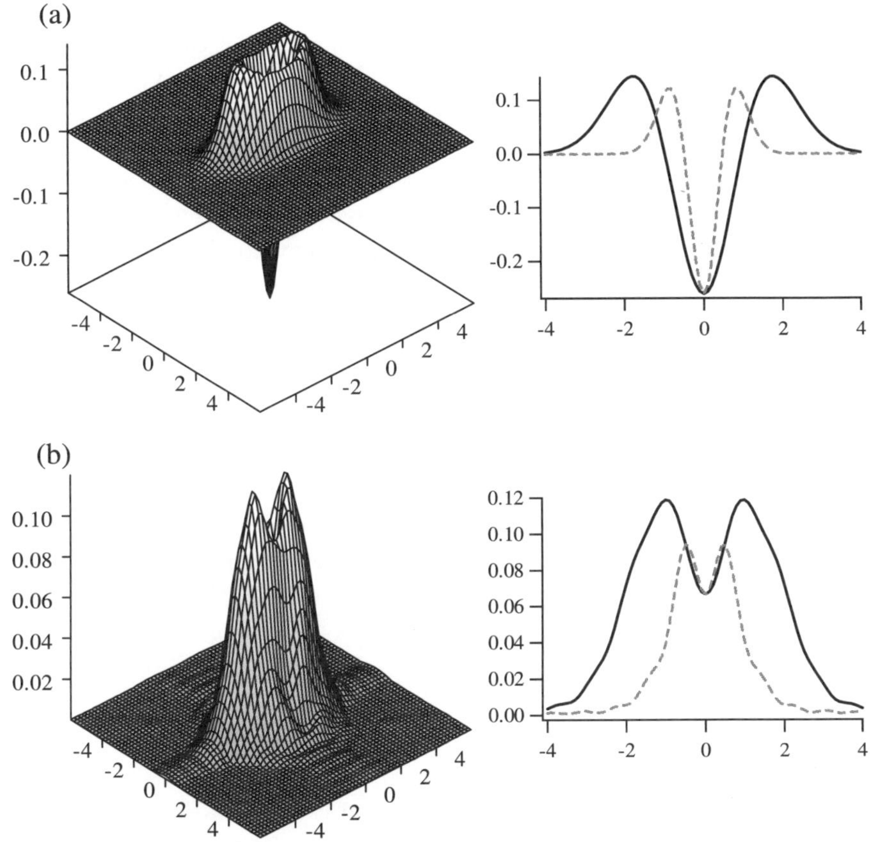

Fig. 8. (a) Theoretical Wigner function W of the output state of the de-Gaussification protocol, assuming $s = 0.43$, $R = 0.115$ and perfect detection ($\eta = \xi = 1$). (b) Reconstructed Wigner function from the experimental data ($\eta = 0.75$, $\xi = 0.7$). Cuts of the Wigner function presented on the right are made along the amplified (solid line) or the squeezed quadrature (dashed line).

experimentally attainable while keeping the APD count rate above a few tens per second. Alternatively, with the configuration $s = 0.43$, $R = 0.115$ and $\xi = 0.7$, getting a negative value of the Wigner function would require an homodyne detection efficiency η better than 85%, or in other words, a mode-matching visibility $\eta_H > 98\%$, which is difficult to reach in our present setup. Nevertheless, by playing with the whole set of experimental parameters, it should be possible to directly measure negative Wigner functions of the degaussified state in the near future.

Given our experimental parameters $s = 0.43$, $R = 0.115$ and $\xi = 0.7$, and assuming a perfect detection efficiency $\eta = 1$, the calculated Wigner function of the conditioned state theoretically takes a negative value at the origin : $W_{\mathrm{th,corr}}(0,0) = -0.06$. Everything goes as if the de-Gaussification apparatus generated a state with a negative Wigner function, but due to a limited detection efficiency, we were not able to observe this negativity. To check this from our experimental data, we have implemented an iterative expectation maximization algorithm (maximum-likelihood) developed[56,58,59] to reconstruct the quantum Wigner function of the state incoming to the homodyne detection, thus virtually compensating for the effects of detection inefficiencies (this is possible if the losses are not too important: $\eta > 50\%$ which is verified by our experiment). Unfortunately, the Wigner function calculated from our raw experimental data according to this procedure does not show negative values, but we get $W_{\mathrm{exp,corr}}(0,0) = +0.01$. The discrepancy between this result and the theoretical prediction may originate from supplementary (non-compensated) noise that affect our experimental data, such as fast phase fluctuations for instance. Let us however indicate that improvements of our setup recently allow us to reach[60] a negative value: $W_{\mathrm{exp,corr}}(0,0) = -0.13$.

5. Conclusion and Potential Applications of Non-Gaussian States

We have described the first experimental observation of a "de-Gaussification" protocol, mapping individual femtosecond pulses of squeezed light onto non-Gaussian states, by using only linear optical elements and a photocounter. To conclude, let us briefly review a few potential applications of non-Gaussian procedures.

Entanglement enhancement. Photon subtraction using low-reflectance beam splitters and photocounters is a well-know operation to enhance the quantum continuous variables entanglement ressources and improve the teleportation fidelity.[30–32] The physical idea behind this is that the single photon component (which may be fairly strong for the degaussified state) virtually bears an entanglement of 1 ebit: by sending a single photon $|1\rangle$ on a 50 : 50 beam splitter, one gets an output state $(|0,1\rangle + |1,0\rangle)/\sqrt{2}$ which contains an entanglement of exactly 1 ebit. On the contrary, the Gaussian states are dominated by the vacuum component, which does not bear any entanglement. Thus for low squeezing factors, entangled Gaussian states bear an entanglement quite near to zero. Therefore, by removing the vacuum component of Gaussian states, the de-Gaussification operation significantly improves the entanglement ressources by moving from almost zero ebit to slightly more than 1 ebit.

Entanglement distillation. Theoretical investigations have demonstrated recently the impossibility to distillate entanglement from a Gaussian input to a Gaussian output by Gaussian means.[12–14] Therefore, non-Gaussian operations have a crucial importance to fully exploit QCV quantum ressources. Extending the de-Gaussification procedure reported here to entangled EPR beams is closely related to the first step of an entanglement distillation procedure for Gaussian QCV.[16] This protocol allows to distill the entanglement ressources by jumping "outside" the Gaussian domain and reaching it back at the end, at least in an approximate way. However, implementing the whole distillation protocol is a challenging task due to the number and the complexity of the different operations required.

Schrödinger cats generation. Since Schrödinger suggested his famous cat paradox in 1935,[57] there has been great interest in generating and monitoring a quantum superposition of a macroscopic system, such as a quantum superposition of two coherent states with sufficiently large amplitudes dephased by π. The preparation of Schrödinger-cat-like states has been studied in Ref. 27 using conditional measurements on a beam splitter output. Of particular interest is the quantum superposition $|\alpha\rangle - |-\alpha\rangle$ which only contains *odd* terms in its decomposition on the Fock basis.[29] This state shows thus strong similarities with the degaussified squeezed vacuum of Eq. (3). The overlap between these two states can be measured by their fidelity $\mathcal{F} = \langle \text{cat}|\rho_{\text{cond}}|\text{cat}\rangle$. This calculation has been performed independently in Ref. 26, 8. In the limit $R \to 0$, the fidelity is very high $\mathcal{F} > 0.99$ for a cat-state with an optimized amplitude of $\alpha = 1.16$. Although this amplitude is too small to realize a "real" Schrödinger cat (this would require $\alpha > 2$), the Schrödinger kitties approximated by the degaussified squeezed vacuum can be used in the procedure described in[29] to generate with a high fidelity a cat of large coherent amplitude ($\alpha > 2$) using only beam splitters, coherent states and photocounters.

Loophole-free Bell inequality test. Recently, feasible optical setups allowing for a complete loophole-free Bell test with efficient homodyne detection have been proposed.[34–37] Since positive-definite Gaussian Wigner functions provide explicitly a hidden variables model for homodyne measurements, one has to consider non-Gaussian states to find a Bell inequality violation. In the above-cited proposals, a Non-Gaussian entangled state is generated from a two-mode squeezed vacuum by subtracting a single photon from each mode, using low-reflectance beam splitters and single-photon detectors. The maximum Bell factor achievable with such a scheme represents a Bell inequality violation of 2.3%. Taking into account the experimental constraints, it appears that there may exist an experimental window for a loophole-free Bell test with quantum continuous variables.

Acknowledgments

We thank N.J. Cerf and J. Fiurášek for very fruitful discussions and F. Grosshans for his experimental and theoretical contributions to the early steps of the experiment.

This work was supported by the European IST/FET/QIPC program (RTD projects "QUICOV" and "COVAQIAL"), and by the French programs "ACI Photonique" and "ASTRE".

References

1. F. Grosshans and Ph. Grangier, Phys. Rev. Lett. **88**, 057902 (2002).
2. F. Grosshans, G. Van Assche, J. Wenger, R. Brouri, N.J. Cerf and Ph. Grangier, Nature **421**, 238 (2003).
3. F. Grosshans, N. J. Cerf, J. Wenger, R. Tualle-Brouri, and Ph. Grangier, Quant. Inf. Comput. **3**, 535 (2003).
4. F. Grosshans and N.J. Cerf, Phys. Rev. Lett. **92**, 047905 (2004).
5. S. Iblisdir, G. Van Assche, and N.J. Cerf, Phys. Rev. Lett. **93**, 170502 (2004).
6. F. Grosshans, Phys. Rev. Lett. **94**, 020504 (2005).
7. M. Navascues and A. Acin, Phys. Rev. Lett. **94**, 020505 (2005).
8. J. Wenger, *Dispositifs impulsionnels pour la communication quantique à variables continues*, PhD Thesis, Université Paris Sud, Orsay (2004). Available online at : http://tel.ccsd.cnrs.fr/documents/archives0/00/00/69/26/
9. N. Gisin, G. Ribordy, W. Tittel, and H. Zbinden, Rev. Mod. Phys. **74**, 145 (2002).
10. H.J. Briegel, W. Dür, J.I. Cirac, and P. Zoller, Phys. Rev. Lett. **81**, 5932 (1998).
11. C.H. Bennett, D.P. DiVincenzo, J.A. Smolin, and W.K. Wooters, Phys. Rev. A **54**, 3824 (1996).
12. J. Eisert, S. Scheel, and M.B. Plenio, Phys. Rev. Lett. **89**, 137903 (2002).
13. J. Fiurášek, Phys. Rev. Lett. **89**, 137904 (2002).
14. G. Giedke and J.I. Cirac, Phys. Rev. A **66**, 032316 (2002).
15. L.M. Duan, G. Giedke, J.I. Cirac, and P. Zoller, Phys. Rev. Lett. **84**, 4002 (2000).
16. D.E. Browne, J. Eisert, S. Scheel, and M.B. Plenio, Phys. Rev. A **67**, 062320 (2003).
17. B. Yurke and D. Stoler, Phys. Rev. Lett. **57**, 13 (1986).
18. S.D. Bartlett and B.C. Sanders, Phys. Rev. A **65**, 042304 (2002).

19. A.I. Lvovsky, H. Hansen, T. Aichele, O. Benson, J. Mlynek, and S. Schiller, Phys. Rev. Lett. **87**, 050402 (2001).
20. A. Zavatta, S. Viciani, and M. Bellini, Phys. Rev. A **70**, 053821 (2004).
21. J. Fiurášek, Phys. Rev. A **64**, 053817 (2001).
22. M.G.A. Paris, M. Cola, and R. Bonifacio, Phys. Rev. A **67**, 042104 (2003).
23. J. Laurat, T. Coudreau, N. Treps, A. Maître, and C. Fabre, Phys. Rev. Lett. **91**, 213601 (2003).
24. J. Laurat, T. Coudreau, N. Treps, A. Maître, and C. Fabre, Phys. Rev. A **69**, 033808 (2004).
25. J. Wenger, R. Tualle-Brouri, and P. Grangier, Phys. Rev. Lett. **92**, 153601 (2004).
26. M.S. Kim, E. Park, P.L. Knight, and H. Jeong, arXiv quant-ph/0409218 (2004).
27. M. Dakna, J. Clausen, L. Knill, and D.G. Welsch, Phys. Rev. A **55**, 3184 (1997).
28. A. Gilchrist, K. Naemoto, W.J. Munro, T.C. Ralph, S. Glancy, S.L. Braunstein, and G.J. Milburn, J. Opt. B: Quant. Semiclass. Opt. **6**, S828 (2004).
29. A.P. Lund, H. Jeong, T.C. Ralph, and M.S. Kim, Phys. Rev. A **70**, 020101(R) (2004).
30. T. Opatrný, G. Kurizki, and D.-G. Welsch, Phys. Rev. A **61**, 032302 (2000).
31. P.T. Cochrane, T.C. Ralph, and G.J. Milburn, Phys. Rev. A **65**, 062306 (2002).
32. S. Olivares, M.G.A. Paris and R. Bonifacio, Phys. Rev. A **67**, 032314 (2003).
33. A. Kitagawa, M. Takeoka, K. Wakui, and M. Sasaki, arXiv quant-ph/0503049 (2005).
34. R. Garcia-Patron Sanchez, J. Fiurášek, N.J. Cerf, J. Wenger, R. Tualle-Brouri, and P. Grangier, Phys. Rev. Lett. **93**, 130409 (2004).
35. H. Nha and H.J. Carmichael, Phys. Rev. Lett. **93**, 020401 (2004).
36. R. Garcia-Patron Sanchez, J. Fiurášek, and N.J. Cerf, Phys. Rev. A **71**, 022105 (2005).
37. S. Olivares and M.G.A. Paris, Phys. Rev. A **70**, 032112 (2004).
38. J. Wenger, R. Tualle-Brouri, and P. Grangier, Opt. Lett. **29**, 1267 (2004).
39. U. Leonhardt, *Measuring the quantum state of light*, Cambridge University Press, Cambridge 1997.
40. R.E. Slusher, P. Grangier, A. LaPorta, B. Yurke, and M.J. Potasek, Phys. Rev. Lett. **59** 2566 (1987).
41. P. Kumar, O. Aytur, and J. Huang, Phys. Rev. Lett. **64**, 1015 (1990).
42. C. Kim and P. Kumar, Phys. Rev. Lett. **73**, 1605 (1994).
43. M.E. Anderson, M. Beck, M.G. Raymer, and J.D. Bierlein, Opt. Lett. **20**, 620 (1995).
44. E.M. Daly, A.S. Bell, E. Riis, and A.I. Ferguson, Phys. Rev. A **57**, 3127 (1998).
45. D.T. Smithey, M. Beck, M.G. Raymer, and A. Faridani, Phys. Rev. Lett. **70**, 1244 (1993).
46. A.M. Weiner, A.M. Kan'an, and D.E. Leaird, Opt. Lett. **23**, 1441 (1998).
47. D. Guzun, Y.Q. Li, and M. Xiao, Opt. Commun. **180**, 367 (2000).

48. H. Hansen, T. Aichele, C. Hettich, P. Lodahl, A.I. Lvovsky, J. Mlynek, and S. Schiller, Opt. Lett. **26**, 1714 (2001).
49. A. Laporta and R.E. Slusher, Phys. Rev. A **44** 2013 (1991).
50. J. Wenger, J. Fiurášek, R. Tualle-Brouri, N.J. Cerf, and P. Grangier, Phys. Rev. A **70**, 053812 (2004).
51. H.P. Yuen and V.W.S. Chan, Opt. Lett. **8**, 177 (1983).
52. H. Hansen, T. Aichele, C. Hettich, P. Lodahl, A.I. Lvovsky, J. Mlynek, and S. Schiller, Opt. Lett. **26**, 1714 (2001).
53. A. Zavatta, M. Bellini, P.L. Ramazza, F. Marin, and F.T. Arecchi, J. Opt. Soc. Am. B **19**, 1189 (2002).
54. F. Grosshans and P. Grangier, Eur. Phys. J. D. **14**, 119 (2001).
55. T. Aichele, A.I. Lvovsky, and S. Schiller, Eur. Phys. J. D. **18**, 237 (2002).
56. A.I. Lvovsky, J. Opt. B: Quant. and Semiclass Opt. **6** S556 (2004).
57. E. Schrödinger, Naturwissenschaften **23** 807-812; 823-828; 844-849 (1935).
58. J. Rehacek, Z. Hradil, and M. Jezek, Phys. Rev. A **63**, 040303 (2001).
59. M. Jezek, J. Fiurasek, and Z. Hradil, Phys. Rev. A **68**, 012305 (2003).
60. A. Ourjoumtsev, R. Tualle-Brouri, J. Laurat, and P. Grangier, Science **312**, 83-86 (2006).

Chapter 21

Continuous-Variable Quantum-State Tomography of Optical Fields and Photons

A. I. Lvovsky

Department of Physics and Astronomy
University of Calgary,
Calgary, Alberta T2N 1N4, Canada

email: lvov@ucalgary.ca

M. G. Raymer

Department of Physics and Oregon Center for Optics
University of Oregon
Eugene, Oregon 97403, USA

email: raymer@uoregon.edu

1. Introduction

Quantum-state tomography (QST) is the art of inferring the quantum state of an ensemble of similarly prepared physical systems by performing a set of observations. Since the first theoretical proposals[1–5] and the first experiments determining the quantum state of a light field,[6,7] the concept and practice of QST has become a standard tool in quantum-information research.[8,9] A quantum logic operation (gate device) or a quantum memory device can be fully characterized by a series of measurements of the state of a physical object (qubit) before and after interacting with the device.[8] In this chapter, we review the basis of and methods for QST of optical fields. We emphasize optical QST for the field amplitude, a continuous degree of freedom.

In quantum mechanics, it is impossible to learn the quantum state of any individual physical system.[10–12] Any series of observations on an individual system will fail to yield its initial state. Each observation, no matter how subtle, will disturb its state just enough to prevent further observations from yielding enough information for a state determination. This is the basis of quantum key distribution for cryptography.[13]

If we are provided with an ensemble of identically prepared systems, we can measure the same variable for each system, and build up a histogram of outcomes, from which a probability density can be estimated. According to the Born rule of quantum theory, this measured probability density will equal the square-modulus of the state-vector coefficients, represented in the state-space basis corresponding to the measuring apparatus. This by itself will not yet yield the full state information, since the phase of the complex state-vector coefficients will be lost. That is, this just gives us the statistics associated with the projection of the state onto a particular measurement basis.

In order to determine the *full* information about the quantum state, we must make *many* sets of measurements on many sub-ensembles, every time modifying the apparatus so that sets of projection statistics associated with many bases can be acquired. One can then combine these results to reconstruct the density matrix of the state. The data do not yield the state directly, but rather indirectly through data analysis (*i.e.*, a logical inference process). This is the basis of quantum-state tomography, or QST. For elementary review, see Refs. 9, 14.

The modern interest in QST is motivated by recent developments in measurement technologies, which allow experimenters to acquire tomographically complete information by measuring a set of observables sufficiently diverse to allow a reliable state reconstruction from the data. Such a set of complete variables was called, by Ugo Fano, a *quorum*.[15]

In Sec. 2 we review some of the reconstruction methods. The so-called deterministic method uses one of several linear integral transforms to convert the measured probability histograms into a quantitative representation of the density matrix in a chosen basis. This inverse linear transform method works well only in the limit of a very large number of data and very precise measurements, so that statistical and systematic uncertainties in the estimation of the density matrix are negligible.

In cases with smaller data sets, statistical fluctuations can lead to inaccurate, even seemingly unphysical, results for a state reconstructed by the inverse linear transform method. For example, negative values may be found on the diagonal of the reconstructed density matrix. In this case, more general methods must be used (see Ref. 8, chapters 2,3,6, and 10). These are the methods of statistical inference, the classical versions of which have been developed in traditional statistics and data analysis. These methods include the Maximum-Likelihood (MaxLik) method and the Maximum-Entropy (MaxEnt) method. The MaxEnt[16] method looks for the least biased inference of the density matrix that will generate the observed data. The MaxLik method looks for the most probable density matrix that will generate the observed data. The latter method is discussed in detail below.

It is interesting, in the context of this book, that the first experimental demonstration of optical QST using a quorum involved measurement of a set of *continuous variables* — the two quadrature-phase components (*i.e.*, real and imaginary amplitudes, Q and P) of a light wave. Using balanced homodyne detection (BHD), Smithey *et al.* measured a set of probability densities for the quadrature amplitudes of a squeezed state of light.[6,7] Each of 27 quadrature phase values was selected by setting the phase of the local oscillator field used in the BHD. The resulting probability histograms were inverted using the inverse Radon transform, familiar from medical tomographic imaging, to yield a reconstructed Wigner distribution and density matrix for a squeezed state of light. This 1993 paper introduced the term tomography into quantum optics, in the phrase "optical homodyne tomography" (OHT).

The early experiments in optical homodyne QST suffered from relatively low detection efficiency, which obscured some of the details of the squeezed state. Subsequent experiments, with higher efficiency, reaped a wealth of detail that could not have been acquired using any other known measurement technique at the time. For example, the even-odd photon-number probability oscillation predicted for squeezed vacuum states was clearly demonstrated.[17] Following this, the OHT technique was applied to single-photon wave packet states, which are achievable using parametric down conversion.[18] This experiment showed, for the first time, a quadrature-amplitude Wigner function (a phase-space quasiprobability) of a light wave having a negative value. Studies have also included OHT-like measurements for two optical modes.[19–26]

An altogether different use of QST arises when a light field is known *a priori* to contain a definite number of photons, but their distribution over geometric or polarization modes is unknown. If the number of modes is discrete, (*e.g.* in the case of polarization qubits), characterization can be done by studying relative photon number statistics in each mode and in their various linear superpositions as well as (in the case of multiple photons) photon number correlations between modes. Employing this approach, White *et al.* have implemented tomography of entangled two-qubit systems.[27] More recently, this technique has been extended to characterization of quantum dynamical processes.[28] Similar methods are used in celebrated quantum cryptography and quantum teleportation protocols. This experimental method has been recently reviewed in detail by Altepeter, James, and Kwiat[8] and is beyond the scope of this paper. However, as we discuss in Sec. 3, even in the case of discrete modes, state characterization using homodyne tomography is much more complete than by means of photon counting and can be applied to various states and protocols of discrete quantum information.[29–33]

If the distribution of light particles over electromagnetic modes is described by a continuous degree of freedom, methods of continuous-variable QST become irreplaceable. Here the question is, what quantum state describes the spatial distribution of the field? This problem is analogous to determining the spatial wave function of a massive particle. An experimental scheme based on a Sagnac interferometer has been proposed for measuring such a wave function,[34] and experimental results are partly in hand.[35] These works are outgrowths of earlier work on reconstructing the field profiles of classical light waves.[34,36–40] An alternative, holographic, technique applies when the possible spatial basis states are restricted to a small subset of the possible ones, making discrete-state QST applicable.[41,42]

The situation becomes even more interesting when the joint spatial wave function $\psi(\vec{r}_A, \vec{r}_B)$ of a pair of photons is considered.[43–45] In the case that the two photons' spatial and momentum variables are described by an entangled state, such a state measurement will provide the maximal-information characterization of the entanglement.[34,35]

2. The Principles of Homodyne Tomography

The technique of balanced homodyne detection and homodyne tomography has been extensively described in the literature, so in this section we just give a brief overview. Detailed introductions to these methods can be found in the book of Leonhardt[9] and in the book chapter by Raymer and Beck.[23]

Balanced homodyne detection (BHD) is a means to measure the amplitude of any phase component of a light mode. In BHD, the weak signal field, described by annihilation operator $\hat{a}$, and a strong coherent local oscillator (LO) field of amplitude $|\alpha_L|$ and phase θ are overlapped at a 50% reflecting beam splitter, and the two interfered fields are detected, temporally integrated, and subtracted.

Assuming perfect detection efficiency, the difference in the numbers of photoelectrons generated by the two photodiodes gives a phase-sensitive measurement of the signal field:

$$\hat{N}_- = |\alpha_L|(\hat{a}\, e^{-i\theta} + \hat{a}^\dagger\, e^{i\theta}). \tag{1}$$

As usual, the mode's annihilation operator can be expressed as a sum of Hermitian operators $\hat{a} = e^{i\,\theta}(\hat{Q}_\theta + i\hat{P}_\theta)/\sqrt{2}$, called quadrature amplitudes, with $[\hat{Q}_\theta, \hat{P}_\theta] = i$. For zero phase, $\hat{Q}_\theta$, $\hat{P}_\theta$ are denoted $\hat{Q}$, $\hat{P}$, respectively (so $\hat{Q}_\theta = \hat{Q}\cos\theta + \hat{P}\sin\theta$), and are analogous to position and momentum variables for a massive harmonic oscillator. For LO phase equal to θ, BHD measures the quadrature amplitude $Q_\theta = \hat{N}_-/(|\alpha_L|\sqrt{2})$. According to quantum mechanics, the probability density for observing the quadrature equal to Q_θ for the field in the signal mode given by the density operator $\hat{\rho}$ is

$$\mathrm{pr}(Q_\theta, \theta) = \langle Q_\theta|\hat{\rho}|Q_\theta\rangle, \tag{2}$$

where $|Q_\theta\rangle$ is the quadrature eigenstate.

The operator $\hat{a}$ in Eq. (1) should be associated with the spatiotemporal mode defined by the local oscillator. A fundamental feature of BHD is that it detects the state of the electromagnetic field precisely in this mode.[6,23,46] This allows temporal and spatial selectivity, or gating, of the signal field (not the signal intensity). This gating technique (linear-optical sampling) has application in ultrafast signal characterization.[23,47]

We emphasize that OHT normally requires pulsed, balanced homodyne detection at zero frequency (also known as DC or time-domain BHD), which results in "whole-pulse" detection. This differs from radio frequency (RF) spectral analysis of the photocurrent,[48] which yields the power in a certain intensity-modulation component. The technical details of time-domain BHD are discussed in Sec. 3.2.[23,49]

2.1. *Inverse linear transform state reconstruction*

2.1.1. *Wigner function*

When the phase is varied over a half of the complete optical cycle, quadrature amplitudes Q_θ form a quorum for QST.[5] The quantum state can be reconstructed using deterministic methods.[9,23,50,51] The homodyne detection probability (2) can be expressed in terms of a quasiprobability density called the Wigner distribution:[52]

$$W_{\hat{\rho}}(Q,P) = \frac{1}{2\pi}\int_{-\infty}^{\infty} \langle Q + \frac{1}{2}Q'|\hat{\rho}|Q - \frac{1}{2}Q'\rangle\, e^{-i\,P\,Q'}\, dQ'. \tag{3}$$

In terms of the Wigner distribution, the measurement probability is[23,46]

$$\mathrm{pr}(Q_\theta,\theta) = \int_{-\infty}^{+\infty}\int_{-\infty}^{+\infty} \delta(Q_\theta - Q\cos\theta - P\sin\theta)\, W_{\mathrm{Det}}(Q,P)dQ\,dP \qquad (4)$$
$$= \int_{-\infty}^{\infty} W_{\mathrm{Det}}(Q_\theta\cos\theta - P_\theta\sin\theta, Q_\theta\sin\theta + P_\theta\cos\theta)dP_\theta.$$

In the case of lossless photodetection, the "detected" Wigner distribution W_{Det} corresponds to the ideal Wigner distribution (3). In the case of a nonunity detector efficiency, the detected Wigner dsitribution is obtained from the former via a convolution[9,46,53,54]

$$W_{\mathrm{Det}}(Q,P) = \frac{1}{2\pi\varepsilon^2} \qquad (5)$$
$$\times \int_{-\infty}^{+\infty}\int_{-\infty}^{+\infty} \exp\left[-\frac{(Q\eta^{-1/2}-Q')^2-(P\eta^{-1/2}-P')^2}{2\varepsilon^2}\right] W_{\hat\rho}(Q,P)dQ'dP',$$

where $2\varepsilon^2 = 2\sigma^2 - 1 = 1/\eta - 1$, η being the detector's quantum efficiency.

2.1.2. *Inverse Radon transformation*

The expression (4) is known as the Radon transform,[55] and has the form of a line integral, which forms a projection of the W_{Det} function onto an axis Q_θ rotated by angle θ. It can be inverted numerically using the filtered back-projection algorithm, familiar from medical imaging[9,55] to reconstruct the phase-space density $W_{\mathrm{Det}}(Q,P)$:

$$W_{\mathrm{Det}}(Q,P) = \frac{1}{2\pi^2}\int_0^{\pi}\int_{-\infty}^{+\infty} \mathrm{pr}(Q_\theta,\theta)K(Q\cos\theta + P\sin\theta - Q_\theta)\,dQ_\theta\,d\theta, \qquad (6)$$

with the integration kernel

$$K(x) = \frac{1}{2}\int_{-\infty}^{+\infty} |\xi|\exp(i\xi x)d\xi = -\mathcal{P}\frac{1}{x^2}. \qquad (7)$$

Because the kernel is singular at $x = 0$, in numerical implementations of the inverse Radon transformation it is subjected to low pass filtering: the infinite integration limits in Eq. (7) are replaced by $\pm k_c$, with k_c chosen to reduce the numerical artifacts associated with the reconstruction while keeping the main features of the Wigner function (Fig. 1).

This was the strategy used in the first QST experiments.[6,56] In later implementations of this algorithm,[25] the summation of Eq. (6) was applied directly to acquired pairs (θ_m, Q_m):

$$W_{\text{Det}}(Q,P) = \frac{1}{2\pi^2 N}\sum_{m=1}^{N} K(Q\cos\theta_m + P\sin\theta_m - Q_m), \tag{8}$$

with phases θ_m uniformly spread over the 2π interval. This approach is simpler and more precise as it eliminates an intermediate step of binning the data and calculating individual marginal distributions associated with each phase.

From the reconstructed W_{Det}, the detected density matrix $\langle Q|\hat{\rho}_{\text{Det}}|Q'\rangle$ in the continuous Q basis can be extracted, by inverse-Fourier transforming Eq. (3). Given the reconstructed density matrix in the continuous Q basis, we can compute the matrix elements of the density operator in any other basis, for example the number basis $\rho_{nm} = \langle n|\,\hat{\rho}\,|m\rangle$. This scheme was applied to reconstruct photon-number statistics $\langle n|\,\hat{\rho}\,|n\rangle$, as well as quantum-phase statistics for squeezed and for coherent light.[7,58,59] This procedure is however unecessarily complicated and may lead to inaccurate reconstruction, especially if the detection efficiency is less than unity. The intermediate step of reconstructing the Wigner function can be sidestepped using an improved deterministic scheme known as the pattern function or quantum state sampling method introduced by D'Ariano *et al.*,[60,61] and refined several times to the present optimal form.[9,62] Figure 1(a) shows an example of calculating the density matrix using the pattern function method.

2.2. *Maximum-likelihood reconstruction*

The inverse Radon strategy, as well as other linear transform methods, suffer from several limitations. Filtered back-projection introduces a smoothing of the data, reducing the accuracy of the reconstruction.[9,55] More fundamentally, the (electromagnetic) harmonic oscillator is a quantum system of infinite dimension, and no finite amount of measurement data will constitute a quorum. There is nothing preventing statistical and systematic errors from producing an unphysical density matrix, with, for example, negative elements on the diagonal. The ripples visible in the Wigner function reconstruction in Fig. 1(a) are a direct consequence of statistical noise and are associated with unphysical high number terms in the reconstructed density matrix. Such features are typical of the inverse Radon transformation.[49,64]

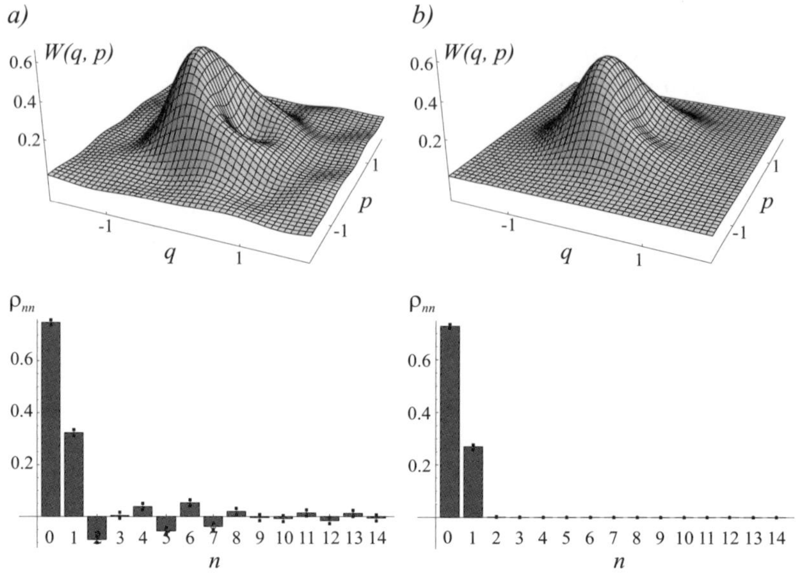

Fig. 1. Estimation of an optical ensemble from a set of 14152 experimental homodyne measurements[29] by means of the inverse Radon transformation (a) and the likelihood maximization algorithm (b). The Wigner function and the diagonal elements of the reconstructed density matrix are shown[63]. The inverse Radon transformation in (a) was performed by means of the filtered back-projection algorithm with the cutoff frequency of 6.3. The statistical uncertainties in (b) were determined by means of a Monte-Carlo simulation (see text).

These issues do not indicate an inconsistency in the method, since any negative values are within the known error bounds, which are determined from the procedure itself.[62] Still, one would like to have a data collection scheme and analysis technique that eliminates the artifacts of the inverse Radon method.

Assuming a particular density matrix $\hat{\rho}$, one can evaluate the likelihood (probability) of acquiring a particular set of measurement results that one might obtain in a quantum tomography experiment. The Ansatz of the *Maximum Likelihood* (MaxLik) estimation quantum state estimation method is to find, among the variety of all possible density matrices, the one which maximizes the probability of obtaining the given experimental data set. To date, this method has been applied to various quantum and classical problems from quantum phase estimation[65] to reconstruction of entangled optical states.[66,67]

As any statistical method, homodyne tomography is compatible with the likelihood maximization approach. The applications of MaxLik to homodyne tomography have been investigated by Banaszek, who reconstructed the photon-number distribution (the diagonal density matrix elements which correspond to a phase-randomized optical ensemble) from a Monte-Carlo simulated data set.[68,69] In a subsequent publication,[70] Banaszek *et al.* discussed the MaxLik estimation of the complete density matrix, but no explicit reconstruction algorithm was presented. Most recently, a general algorithm has been elaborated[71] and tested in homodyne reconstruction of single-mode[25] and multi-mode[26] ensembles and has proven robust and reliable.

For a given phase θ, the probability to detect a particular quadrature value q_θ is proportional to

$$\mathrm{pr}(Q_\theta, \theta) = \mathrm{Tr}[\hat{\Pi}(Q_\theta, \theta)\hat{\rho}], \tag{9}$$

where $\hat{\Pi}(Q_\theta, \theta)$ is the projector onto this quadrature eigenstate, expressed in the Fock basis as

$$\Pi_{mn}(Q_\theta, \theta) = \langle m|\hat{\Pi}(Q_\theta, \theta)|n\rangle = \langle m|Q_\theta, \theta\rangle\langle Q_\theta, \theta|n\rangle. \tag{10}$$

In the above equation the overlap between the number and quadrature eigenstates is $\langle m|Q_\theta, \theta\rangle = e^{im\theta}\psi_m(Q_\theta)$ with $\psi_m(x)$ being the Fock state wave function

$$\psi_n(x) = \langle n|x\rangle = \left(\frac{1}{\pi}\right)^{1/4} \frac{H_n(x)}{\sqrt{2^n n!}} \exp(-x^2/2). \tag{11}$$

H_n denote the Hermite polynomials.

Given a density matrix $\hat{\rho}$, the likelihood to acquire a set of experimental values (Q_j, θ_j), with a frequency f_j is

$$\mathcal{L}(\hat{\rho}) = \prod_j (\mathrm{pr}_j)^{f_j}. \tag{12}$$

A likelihood maximization algorithm[71] can be executed with the density matrix in the photon number (Fock) representation. Since the Hilbert space of optical states is of infinite dimension, the implementation of the algorithm requires its truncation so the Fock terms above a certain threshold are excluded from the analysis. This assumption conforms to many practical experimental situations in which the intensities of fields involved are *a priori* limited.

Figure 1 (b) shows that the maximum-likelihood quantum state estimation favorably compares with deterministic homodyne tomography methods. First, the unphysical ripples in the Wigner function are absent. Second, the MaxLik technique allows one to incorporate the density operator positivity and unity-trace constraints into the reconstruction procedure, thus always yielding a physically plausible ensemble.[68,69] Third, there is a possibility to incorporate the detector inefficiences by replacing the projection measurement (10) by a readily available positive operator-valued measure (POVM) of an inefficient detector.[69–71] Although the pattern-function method also can account for detector efficiency,[9] this has not been widely used due to concerns about possible numerical instability.

3. Homodyne Tomography of Discrete-Variable States

3.1. *Motivation*

In the optical implementation of the qubit, the logical value is assigned to a single photon being in one of two orthogonal modes A or B:

$$|\tilde{0}\rangle = |1_A, 0_B\rangle, \quad |\tilde{1}\rangle = |0_A, 1_B\rangle, \tag{13}$$

where the right-hand side is written in the photon number (Fock) basis for a pair of modes. Such a dual-rail optical qubit is an intuitively simple object, yet it is highly immune to decoherence and permits construction of algorithmically efficient computational gates involving only linear optical elements.[72]

Characterization of optical qubits for quantum information processing has to date been based on photon counting. A major drawback of this approach is the *a priori* assumption that the modes involved are in one of the states (13) or their linear combination. As a result, characterized are not the true quantum states of the carrier modes, but their *projections* onto a subspace spanned by the basis vectors (13). It is neglected that in actual experimental situations, most of the time the state of the modes does not belong to this subspace but is the double vacuum $|0_A, 0_B\rangle$. These overwhelming events, as well as all other events not falling into the qubit frame, are simply eliminated from the analysis.

This postselection-based approach does not provide a correct performance estimate of an experimental scheme under investigation, in particular, it does not allow one to evaluate its scalability. Application of light for quantum information processing requires a method that permits *complete* characterization of a quantum optical ensemble as an element of the full Hilbert space of optical states.

The technique of homodyne tomography (OHT) discussed in this article does not suffer from this limitation. As discussed in the previous section, this method yields *complete* characterization of any set of electromagnetic oscillators. In this section, we discuss both fundamental and technical aspects of applying continuous-variable tomography to discrete units of quantum information and review existing research results on this subject.

Technically, application of OHT in the discrete-variable domain brings both simplifications and challenges. Unlike the single-photon detectors that employ avalanche photodiodes, in homodyne detectors (HDs) ordinary, unity-gain semiconductor photodiodes are used. The latter are widely available from commercial suppliers, are cost-efficient and routinely possess high quantum efficiency, low noise, and linear response over a large dynamic range. Furthermore, single-photon detectors require thorough isolation from ambient light while in homodyne detection, the optical mode to be measured is selected by the local oscillator so the negative influence of ambient light is insignificant. On the other hand, the optical mode of the local oscillator must be perfectly matched to the mode containing the state to be measured which can be challenging if the latter comes from an independent or a semi-independent source such as a down-conversion crystal. Another challenge, of pure technical nature, is associated with the construction of a time-domain (dc) HD which possesses a sufficient sensitivity, bandwidth, and signal-to-noise ratio to implement a precise measurement of the field quadratures.

3.2. *Time-domain homodyne detection*

To date, most homodyne measurements have been performed in the frequency domain: a certain spectral component of the photocurrent difference signal (usually around 5-10 MHz where the technical noise is minimized) is used to determine the quadrature quantum noise of the optical state. While frequency-domain (ac) BHD has been successfully applied both in the continuous-wave and pulsed regimes to detect quadrature squeezing,[73] it is less useful for quantum-information related tasks as it has no capability of time resolving.

Quantum-information applications require *time-domain* homodyning. Both the local-oscillator and signal modes are pulsed; each LO pulse generates a difference photocurrent which is observed in real time and yields a single value of a field quadrature.

Time-domain BHD is more technically challenging than its frequency-domain counterpart. First, the electronics must ensure time separation of responses to individual laser pulses. The shot-noise difference charge on the scale of 10^4–10^5 photoelectons must be low-noise amplified within a bandwidth exceeding the local oscillator pulse repetition rate. Second, the measured quadrature values must not be influenced by low frequency noises. The detector must thus provide ultra-low noise, high subtraction and flat amplification profile in the *entire frequency range* from DC to at least the LO pulse repetition rate.

The first time-domain homodyne detector was implemented by Smithey *et al.*[6,74] in their original quantum tomography experiments. In those experiments, a pair of photodiodes were wired in series to subtract their output currents, and this difference signal was integrated by an Amptek A225 charge-sensitive transimpedance amplifier, allowing sub-shot-noise measurements of light pulses containing millions of photons. Since then, the DC, "whole-pulse" HD technique has been developed into a robust set of methods for measuring photon statistics and correlations on ultrafast time scales.[20,23,47,59,75,76] These studies used a separate integration channel for each photodiode, allowing the use of off-the-shelf components, and easy calibration of the shot-noise level; technical details are given in Ref. 23 An improved scheme implemented by Hansen *et al.*[49] featured a signal-to-noise ratio of 14 dB, a 85 dB subtraction efficiency, pulse repetition rate of up to 1 MHz and a 91% quantum efficiency. The detector employed a pair of Hamamatsu S3883 Si-PIN photodiodes of 300-MHz bandwidth which were mounted at a distance of only 1 cm from each other in order to minimize spurious RF interferences. The positive and negative charges produced by the optical pulses were collected and physically subtracted at a 470-pF capacitance which was much larger than the capacitances of the photodiodes (6 pF). The difference charge was then amplified using a 2SK152 FET in connection with the low-noise Amptek A250 preamplifier.

Another useful HD scheme was recently implemented by Zavatta *et al.*[77] In this detector, a broadband operational amplifier (CLC 425 from National Semiconductor[a]) was employed at the primary amplification stage which has resulted in a dramatic bandwidth increase, albeit at a reduced signal-to-noise ratio (2-5 dB over a $\sim$ 100 MHz bandwidth). This detector can be used with pulsed Ti:Sapphire lasers operating at a repetition rate of 70–80 MHz. The capabilities of both detector models[49,77] were demonstrated in

[a]This unit is no longer produced by CLC; the successor unit is LMH6624

experiments on tomographic reconstruction of the single-photon Fock state (see Sec. 3.4.1).[18,78]

3.3. *Matching the mode of the local oscillator*

One of the main challenges associated with developing the technology of optical quantum information processing is a robust and reliable method of synthesizing its primary element, the single-photon state. In order to be applicable for scalable quantum computing, the photon must be produced "on demand" with a high efficiency and in a clearly defined, highly pure spatiotemporal mode. In recent years, several experimental approaches to this problem experienced rapid development. However, none of the existing methods fully satisfy the above requirements. Existing "on demand" photon sources either possess a poor efficiency or produce a photon in a mode with poor spatiotemporal characteristics. Although applications of photons originating from such sources in simple quantum information processing protocols have been demonstrated,[79,80] they heavily rely on postselection and are thus not well scalable.

Under these circumstances, a sensible alternative is offered by heralded single-photon preparation by means of parametric down-conversion (PDC). Detection of a photon in one of the emission channels (labeled *trigger*) causes the non-local photon pair to collapse, projecting the quantum state in the remaining (*signal*) channel into a single-photon state (Fig. 2(a)). Proposed and tested experimentally in 1986 by Hong and Mandel[81] as well as Grangier, Roger and Aspect,[82] this technique has become a workhorse for many quantum optics experiments.

The biphoton produced in PDC is a complex entangled state with many parameters (spectrum, direction, polarization, etc.) of the two photons highly correlated:

$$|\Psi_{si}\rangle = \int \Psi(\omega_s, \omega_t, \vec{k}_s, \vec{k}_t)|1_{\omega_s,\vec{k}_s}\rangle|1_{\omega_t,\vec{k}_t}\rangle d\omega_s d\omega_t d\vec{k}_s d\vec{k}_t, \tag{14}$$

where ω and $\vec{k}$ denote the frequencies and wavevectors of the signal and trigger photons. If the trigger photon is measured with any uncertainty in one of these parameters, the signal photon will be prepared in a non-pure state

$$\rho_s = \mathrm{Tr}_t T(\omega_t, \vec{k}_t)\, |\Psi_{si}\rangle\langle\Psi_{si}|\,, \tag{15}$$

where $T(\omega_t, \vec{k}_t)$ is the transmission function of the filters in the trigger channel. An imperfect purity of the signal photon mode would degrade its applicability in quantum information technology where photons from independent sources are brought into interference. To our knowledge, for the first time this matter has been investigated theoretically in 1995 by Zukowski, Zeilinger, Weinfurter,[83] independently by Rarity,[84] and later confirmed in a more detailed study in 1997 by Ou,[85] These publications conclude that in order to observe strong interference effects, narrow (in comparison to the bandwidth of the pulsed pump) spectral filtering must be applied to the trigger channel of the down-converter. This implies, in particular, that high-purity photons are best obtained with sufficiently short pump pulses. Specifically in the context of OHT, theoretical treatment was given by Grosshans and Grangier[86] as well as Aichele, Lvovsky, and Schiller.[87]

One particular advantage of OHT as a method of quantum state characterization is that it measures the optical state in the mode defined by the local oscillator, thus providing information on spatiotemporal matching of the latter to the mode in question. As the pulsed classical mode of the local oscillator is usually highly coherent, it also supplies indirect information on the purity of the mode in which the measured ensemble is prepared. On the other hand, it poses a challenge of precise spatiotemporal matching between the two modes.

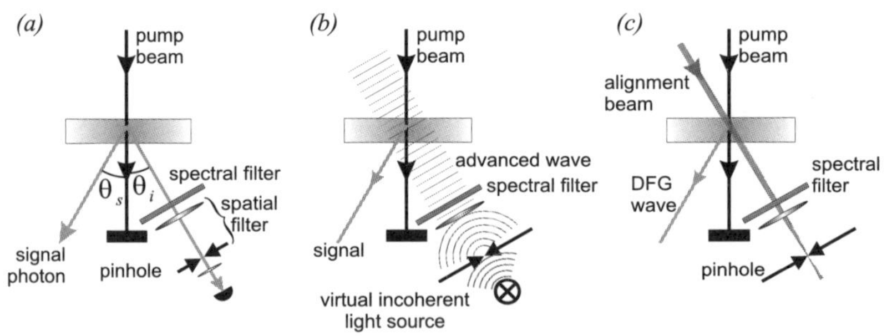

Fig. 2. (a) Preparation of single photons by conditional measurements on a biphoton state. (b) The Klyshko advanced wave model. The trigger detector is replaced with an incoherent light source, which, at the moment of detection, produces a fictitious incoherent advanced wave propagating backwards in space and time. Interaction of this wave with the pump generates a difference mode that mimics that of the conditionally prepared photon. (c) In an experiment, a laser beam, aligned for maximum transmission through all the filters, can model the advanced wave. The difference frequency (DFG) mode resembles that of the conditionally prepared photon and can be used for mode matching. From Ref. 87.

Of great help in understanding, visualizing, and implementing this mode matching experimentally is the concept of *advanced waves* proposed by Klyshko.[88–90] According to this concept, "one of the two detectors, say number 2, at the moment of registration of a photon t_2 emits back in time and space a short δ-like pulse... This pulse interacts within the excited (in a coherent state) atom[b] and the latter emits a growing wave... with a converted carrier frequency"[88] (see Fig. 2(b,c)). Moreover,[87] the classical difference-frequency wave resulting from the nonlinear interaction between the pump and Klyshko's advanced wave is *completely identical*, in its modal characteristics, to the optical mode of the single photon prepared via a conditional measurements on a down-converted pair. In other words, the advanced wave concept is not merely an informal visual tool, but a rigorous mathematical model which possesses analytic capability.

While narrow filtering of the trigger photon as the method for obtaining a high degree of spatiotemporal coherence in the signal channel is the only method so far demonstrated experimentally, it possesses a number of drawbacks, the main one being a reduced productivity. The narrower the filtering, the fewer pair creation events are registered by the trigger detector. Therefore, nowadays a strong effort is made towards circumventing this bottleneck.[91,92] One possible approach is to arrange the parametric down-conversion setup in such a way that the trigger and signal photons in the output of the down-converter are uncorrelated: the function Ψ in Eq. (14) can be written as

$$\Psi(\omega_s, \omega_t, \vec{k}_s, \vec{k}_t) = \psi_s(\omega_s, \vec{k}_s) \times \psi_t(\omega_t, \vec{k}_t). \tag{16}$$

In this case, detection of *any* photon in the trigger channel signifies that the signal photon has been emitted into a pure spatiotemporal mode defined by the function ψ_s.

The configuration of the correlation function Ψ depends primarily on the energy-conservation condition

$$\omega_s + \omega_i = \omega_p \tag{17}$$

and the phase-matching condition

$$\vec{k}_s + \vec{k}_i = \vec{k}_p. \tag{18}$$

For any pair of photons with parameters $(\omega_s, \vec{k}_s, \omega_t, \vec{k}_t)$ there must exist a pump photon $(\omega_p, \vec{k}_p)$ for which the above equations are satisfied. Because

[b]Here an atomic ensemble was assumed to play the role of the nonlinear medium.

the frequency and the wavevector are connected through dispersion relations [$d\omega = v_{gr} d|\vec{k}|$ with v_{gr} being the wave's group velocity], each of the Eqs. (17) and (18) influences both spatial and spectral characteristics of the biphoton. For example, Eq. (18) requires that, for (almost) collinear down-conversion,

$$v_{gr,s} \cos\theta_s \omega_s + v_{gr,i} \cos\theta_i \omega_i = v_{gr,p} \omega_p, \tag{19}$$

where the angles θ_s and θ_i are defined in Fig. 2(a)].

Consider just the spectral part of the biphoton correlation function $\Psi = \Psi(\omega_s, \omega_t)$. Because the pump is pulsed, its frequency ω_p has some intrinsic uncertainty and for each value of ω_i, there is a finite range of frequencies ω_s allowed by Eqs. (17) and (19), but still, $\Psi(\omega_s, \omega_t)$ is strongly anticorrelated.[93]

Grice, U'Ren, and Walmsley[93,94] proposed to eliminate these frequency correlations by exploiting an additional restriction imposed onto $\Psi(\omega_s, \omega_t)$ by the components of the signal and idler photons' momenta that are transverse to the pump, for which Eq. (18) requires

$$v_{gr,s} \sin\theta_s \omega_s - v_{gr,i} \sin\theta_i \omega_i = 0. \tag{20}$$

The signal-idler frequency pairs permitted by the above equation are highly correlated: $\omega_s \approx \omega_t$. The correlation uncertainty is determined by the geometric width of the pump laser beam inside the down-conversion crystal that can be chosen at will by the experimentalist. By manipulating this parameter as well as the configuration of the crystal, one can achieve that the correlation function defined by restrictions (17), (19), and (20) is completely separable.

The application of this method requires precise spatial filtering of the signal and idler photons, which may defeat the purpose of eliminating correlations in the biphoton spectrum. A more efficient, but also more technically challenging solution recently proposed[95] is to engineer down-conversion by integrating the nonlinear crystal into a microcavity formed by distributed Bragg reflectors created by a small, periodic modulation of the linear refractive index along the cavity axis (preferably confined by a waveguide). The cavity (a small ~ 0.1 mm gap in the center) has a much shorter length than does the nonlinear medium ($L \simeq 4$ mm). In this case, only a single cavity mode falls within the phase-matching bandwidth of the down-conversion process, leading to pure-state creation of a spectrally uncorrelated biphoton in a spatial mode defined by the cavity.

3.4. *Tomography of photons and qubits*

Once the technical challenges discussed in previous subsections have been addressed, one can apply homodyne tomography to characterizing discrete-variable quantum optical states and to solving more complex tasks of quantum-optical information technology.

3.4.1. *Single-photon Fock state*

Quantum states containing a definite number of energy quanta (Fock states) play a key role in quantum optics. They constitute the essence of the quantum nature of light and are indispensable in both theoretical description of many optical phenomena, as well as in a wide range of applications, such as quantum cryptography[96] and quantum information processing.[72,97] Another fundamental feature of the Fock states that makes their characterization interesting is that their Wigner functions take on highly nonclassical negative values. In a 2001 experiment by Lvovsky *et al.*, pulsed single photons have been prepared by conditional measurements on a biphoton state generated via parametric down-conversion and characterized by means of balanced homodyne tomography.[18]

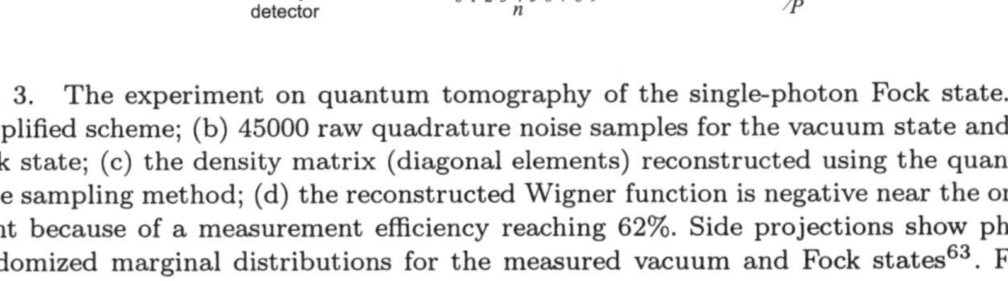

Fig. 3. The experiment on quantum tomography of the single-photon Fock state. (a) Simplified scheme; (b) 45000 raw quadrature noise samples for the vacuum state and the Fock state; (c) the density matrix (diagonal elements) reconstructed using the quantum state sampling method; (d) the reconstructed Wigner function is negative near the origin point because of a measurement efficiency reaching 62%. Side projections show phase-randomized marginal distributions for the measured vacuum and Fock states[63]. From Refs. 49,57.

The experiment employed a mode-locked Ti:sapphire laser to obtain transform-limited pulses at 790 nm with a repetition rate ~80 MHz and a pulse duration of 1.6 ps. In order to accommodate for the 1-MHz bandwidth of the available HD,[49] the laser repetition rate was reduced by transmitting the beam through an acousto-optical pulse picker that only transmitted every 100th pulse.[c] The laser beam was then directed into an optical arrangement shown in Fig. 3(a).

Most of the laser field was frequency doubled and then subjected to parametric down-conversion in a 3-mm BBO crystal, which occurred in a type-I frequency-degenerate, but spatially non-degenerate configuration. The trigger photons passed through a spatial filter and a 0.3-nm interference filter centered at the laser wavelength; the sizes of the filters were chosen to compromise between the requirement of high purity of the signal photon mode and a reasonably high pair production rate. With the electro-optical pulse picker, the trigger photon detection event rate was on the scale of 300–400 s^{-1}. In comparison with the pulse repetition rate of the laser, this number is very low, which ensured that the effect of Fock states with $n > 1$ was negligible.

A small fraction of the master laser beam — split off before the frequency-doubler — was used as the local oscillator for the homodyne system. These pulses had to be temporally and spatially mode-matched to the photons in the signal channel as described above.

In an actual experiment the reconstructed ensemble is not an ideal Fock state $|1\rangle$ but a statistical mixture whose density matrix can be approximated as

$$\rho_{\text{meas}} = \eta|1\rangle\langle 1| + (1-\eta)|0\rangle\langle 0|. \tag{21}$$

The quantity η is the measurement efficiency which depends on many factors: optical transmission of the signal photon, quantum efficiency of the HD, trigger dark counts, optical mode matching of the signal photon and the local oscillator, spatiotemporal coherence of the signal photon. The greater η, the deeper the well in the Wigner function; classically impossible negative values are obtained when $\eta > 0.5$. In this experiment, the efficiency is (62 ± 1) % so the Wigner function does become negative (Fig. 3(b-d)).[d]

Quantum state reconstruction of the state $|1\rangle$ has been reported on the vibrational degree of freedom of a trapped berillium ion[99] as well as in a cavity QED setting.[100] A special feature of the optical experiment,[18] or

[c]In a later version of the experiment,[98] this pulse picker was replaced by an electro-optical shutter placed in the local oscillator beam path. Whenever a trigger photon detection event occurred, the shutter opened to transmit a single local oscillator pulse that activated a homodyne measurement. In this way, the data collection rate was increased, in comparison to the original work,[18] by a factor of over 1000.

[d]Fig. 3 shows the result of a later experiment.[25,98] In the original work,[18] the efficiency was 55%.

of the helium-beam experiment,[14,101] is that the technique of homodyne tomography can be fully understood in the framework of classical physics. In other words, this measurement could have been conducted (and interpreted) by someone who does not believe in quantum mechanics. Yet the result obtained would appear completely absurd, incompatible with the classical picture of the world — thus providing a very strong evidence of "quantumness" of our world. It can however be shown[98] that even with *any* single-photon fraction — even below 50% — statistical mixtures of states $|0\rangle$ and $|1\rangle$ are nonclassical according to the Vogel criterion.[102]

3.4.2. *Tomography of the qubit*

Aside from its fundamental implications, the experiment described above demonstrates that all the technologies necessary for the application of OHT to discrete-variable quantum-optical states are available. One such application is the reconstruction of a dual-rail optical qubit.[26] The four-dimensional density matrix of the qubit, reconstructed using OHT, extends over the entire Hilbert space and thus reveals, for the first time, complete information about the dual-rail optical qubit as a state of the electromagnetic field. Characterizing a dual-rail qubit requires two-mode QST. A deterministic OHT scheme for state reconstruction in the Fock basis was proposed and tested numerically.[22]

A dual-rail qubit, theoretically described by the state

$$|\Psi_{\text{qubit}}\rangle = \tau|1_A, 0_B\rangle - \rho|0_A, 1_B\rangle, \tag{22}$$

is generated when a single photon state $|1\rangle$ incident upon a beam splitter with transmission and reflectivity τ^2 and ρ^2, respectively, entangles itself with the vacuum state $|0\rangle$ present in the other beam splitter input. To perform tomography measurements, a HD (associated with fictitious observers Alice and Bob) was placed into each beam splitter output channel (Fig. 4(a)). With every incoming photon, both detectors made a measurement of the field quadrature Q_A and Q_B with the local oscillators' phases set to θ_A and θ_B, respectively. The quadrature statistics collected at various phases were used to reconstruct the density matrix of the two-mode state. A detailed theoretical analysis of different aspects of such an experiment was made by Jacobs and Knight[103] as well as Grice and Walmsley.[104]

Figure 4(c,d) shows histograms of the dual-mode quadrature measurements. The two-dimensional distribution $\text{pr}_{\delta\theta}(Q_A, Q_B)$ indicates the probability of detecting a particular pair (Q_A, Q_B) of quadratures at a given

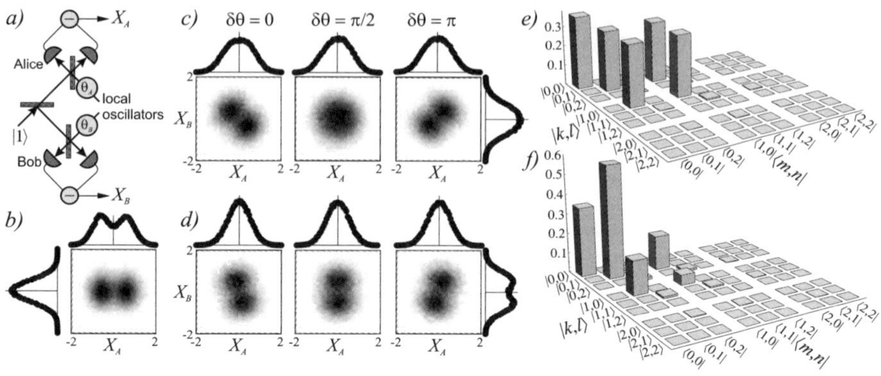

Fig. 4. (a) Scheme of the experimental setup. LOs: local oscillators. (b-d) Histograms of the experimental quadrature statistics $\mathrm{pr}_{\delta\theta}(Q_A, Q_B)$ for the "zero reflectivity" (b), symmetric (c), and highly reflective (92%, d) beam splitters. $\delta\theta = \theta_A - \theta_B$. Phase-dependent quadrature correlations are a consequence of the entangled nature of the state $|\Psi_{\mathrm{qubit}}\rangle$. Also shown are individual histograms of the data measured by Alice and Bob, which are phase-independent. Density matrix (absolute values) of the measured ensemble for the symmetric (e) and highly reflective (f) beam splitters in the photon number representation.[63] From Ref. 26.

local oscillator phase setting. These densities are the marginal distributions of the four-dimensional Wigner function of the two-mode ensemble being measured. They have been used to determine the density matrix of the two-mode state via the maximum-likelihood reconstruction technique[71] (Sec. 2.2). The positive POVM, used to describe the HDs, was modified with respect to that of an ideal HD so as to correct for the non-perfect detection efficiency associated with linear optical losses and non-unity quantum efficiency of photodiodes as well as for random variation of the input photon phase $\theta_A + \theta_B$.

The reconstructed density matrix (Fig. 4(e,f)) features a strong contribution of the double-vacuum term $|0,0\rangle\langle 0,0|$, which is a consequence of imperfect preparation of the initial single photon. The presence of this term in the measured density matrix shows that the goal of a *full* reconstruction of the qubit (including measurement of undesired terms) has been achieved.

3.4.3. *Nonlocality of the single photon and its consequences*

Whether the state (22) can be considered entangled is a widely debated issue. This controversy seems to be related to the wave-particle duality of light. If the photon is viewed as a *state* of the electromagnetic field, the notation (22) is valid and denotes an entangled object.[105] If, on the other hand, a photon is considered a particle, *i.e.* not a state but a *carrier*, *e.g.*, of a polarization state, the dual-rail qubit should be written as a superposition $|\Psi_{\text{qubit}}\rangle = \tau|A\rangle - \rho|B\rangle$ of two localizations of one photon, which may not be seen as entangled. The ideology of OHT is based on the former view.

The entangled nature of the split single photon entails its nonlocal behavior in a number of proposed experiments.[106,107,113] The experiment[26] on homodyne reconstruction of this state yields quadrature measurement statistics which converted to a dichotomic format, can be interpreted to violate, albeit with a loophole, the Bell inequality. In order to apply the Bell theorem to the experimental data, continuous results of each quadrature measurements have been converted to a dichotomic format by means of threshold discrimination. Further evidence of the entangled nature of the delocalized single photon were demonstrated is its application as a resource in quantum communication protocols such as remote state preparation[30] and teleportation.[31,108,109]

In conclusion, we notice that homodyne tomography's domain of application has experienced a rapid growth in recent years. As discussed in Sec. 3.1, homodyne tomography is irreplaceable if complete (not postselected) characterization of an ensemble is required. Among the examples are:

- ▷ quantum optical catalysis,[29] one of the first experiments demonstrating how a conditional measurement on a linear-optical system can simulate nonlinearity required for quantum information processing using optical qubits;
- ▷ an experiment on squeezing of the single photon state,[32] which combines, for the first time, the techniques of conditional preparation of single photons and pulsed squeezing, is a significant technological breakthrough in quantum optical information technology and is useful, in particular, for the distillation of quadrature-entangled two-mode squeezed states[110] as well as the preparation of the "Schrödinger cat states;[111]
- ▷ synthesis and characterization of the photon-added states[33] which paves the way towards single- and multimode quantum state engineering[112] and is interesting from the fundamental point of view: the photon-added state can be seen as a link between the particle and wave aspects of the electromagnetic field.

Traditionally, continuous-variable tomography has been applied for the measurement of quadrature-squeezed[6,64] and quadrature-entangled ensembles, whereas photon counting was employed to characterize discrete-variable states such as optical qubits and polarization-entangled photons. Quantum optics has developed along these two separate avenues that had almost no overlap. Application of continuous-variable tomography to discrete optical states allowed us, for the first time, to build a bridge between these two domains of quantum optics, and, as we demonstrated above, opened up a wide range of possibilities both in fundamental and applied aspects of the field.

References

1. R. G. Newton and B. L. Young, Ann. Phys. (New York) **49**, 393 (1968).
2. J. L. Park and W. Band, Found. Phys. **1**, 211 (1971).
3. W. Band, and J. L. Park, Am. J. Phys. **47**, 188 (1979); Found. Phys. **1**, 133 (1970); Found. Phys. 1, 339 (1971).
4. J. Bertrand and P. Bertrand, Found. Phys. **17**, 397 (1987).
5. K. Vogel and H. Risken, Phys. Rev. A **40**, 2847 (1989).
6. D. T. Smithey *et al.*, Phys. Rev. Lett. **70**, 1244 (1993).
7. D. T. Smithey *et al.*, Physica Scripta T **48**, 35 (1993).
8. *Quantum State Estimation*, Lect. Notes Phys. 649, M. Paris and J. Rehacek (Eds.), Springer, Berlin Heidelberg, 2004.
9. U. Leonhardt, *Measuring the Quantum State of Light*, Cambridge University Press, Cambridge, UK, 1997.
10. O. Alter and Y. Yamamoto, Phys. Rev. Lett. **74**, 4106 (1995).
11. O. Alter and Y. Yamamoto, *Quantum Measurement of a Single System*, Wiley, New York 2001.
12. G. M. D'Ariano and H. P. Yuen, Phys. Rev. Lett. **76**, 2832 (1996).
13. C. H. Bennett and G. Brassard, in *Proceedings of IEEE International Conference on Computers, Systems and Signal Processing, Bangalore*, IEEE, New York 1984, pp. 175-179.
14. M. G. Raymer, Contemp. Phys. **38**, 343 (1997).
15. U. Fano, Rev. Mod. Phys. **29**, 74 (1957).
16. V. Buzek and G. Drobny, J. Mod. Opt. **47**, 2823 (2000).
17. S. Schiller *et al.*, Phys. Rev. Lett. **77**, 2933 (1996).
18. A. I. Lvovsky *et al.*, Phys. Rev. Lett. **87**, 050402 (2001)
19. D. F. McAlister and M. G. Raymer, J. Mod. Opt. **44**, 2359 (1997).
20. D. F. McAlister and M. G. Raymer, Phys. Rev. A **55**, R1609 (1997).
21. E. L. Blansett *et al.*, IEEE J. Quant. Electron. **41**, 287-301 (2005).
22. M. G. Raymer, D. F. McAlister, and U. Leonhardt, Phys. Rev. A **54**, 2397 (1996).
23. M.G. Raymer and M. Beck, in: *Quantum State Estimation*, Lect. Notes Phys. 649, M. Paris and J. Rehacek (Eds.), Springer, Berlin Heidelberg, 2004.

24. M. Vasilyev, S. K. Choi, P. Kumar, and G. M. D'Ariano, Phys. Rev. Lett. **84**, 2354 (2000).
25. P. Voss *et al.*, J. Mod. Opt. **49**, 2289 (2002).
26. S. A. Babichev, J. Appel, and A. I. Lvovsky, Phys. Rev. Lett. **92**, 193601 (2004).
27. A. G. White, P. G. Kwiat, D. F. V. James, and P. H. Eberhard, Phys. Rev. Lett. **83**, 3103 (1999).
28. M.W. Mitchell *et al.*, Phys. Rev. Lett. **91**, 120402 (2003); J. B. Altepeter *et al.*, Phys. Rev. Lett. **90**, 193601 (2003).
29. A. I. Lvovsky and J. Mlynek, J *Phys. Rev. Lett.* **88** 250401 (2002).
30. S. A. Babichev, B. Brezger, and A. I. Lvovsky, Phys. Rev. Lett. **92**, 047903 (2004).
31. S. A. Babichev, J. Ries and A. I. Lvovsky, Europhys. Lett. **64**, 1 (2003).
32. R. Wenger, R. Tualle-Brouri, and P. Grangier, Phys. Rev. Lett. **92**, 153601 (2004).
33. A. Zavatta, S. Viciani, and M. Bellini, Science **306**, 660 (2004).
34. E. Mukamel *et. al.*, Opt. Lett. **28**, 1317 (2003).
35. B. Smith *et. al.*, Opt. Lett. **30**, 3365 (2005).
36. D. F. McAlister *et al.*, Opt. Lett. **20**, 1181 (1995).
37. C. Iaconis and I. A. Walmsley, Opt. Lett. **21**, 1783 (1996).
38. C.-C. Cheng and M. G. Raymer, Phys. Rev. Lett. **82**, 4807 (1999).
39. C.-C. Cheng, M. G. Raymer, and H. Heier, J. Mod. Opt. **47**, 1237 (2000).
40. K. F. Lee *et al.*, Opt. Lett. **24**, 1370 (1999).
41. R. T. Thew *et al.*, Phys. Rev. A **66**, 012303 (2002).
42. N. K. Langford *et al.*, Phys. Rev. Lett. **93**, 053601 (2004).
43. W. A. T. Nogueira *et al.*, Phys. Rev. A **66**, 053810 (2002).
44. S. P. Walborn *et al.*, Phys. Rev. A **69**, 023811 (2004).
45. T. B. Pittman *et al.*, Phys. Rev. A **53**, 2804 (1996).
46. M. G. Raymer *et al.*, J. Opt. Soc. Am. B **12**, 1801 (1995).
47. C. Dorrer *et al.*, Photon. Tech. Lett. **15**, 1746 (2003).
48. R. E. Slusher *et al.*, Phys. Rev. Lett. 55, 2409 (1985).
49. H. Hansen *et al.*, Opt. Lett. **26**, 1714 (2001).
50. G. M. D'Ariano, in: *Quantum Optics and the Spectroscopy of Solids*, T. Hakioglu and A. S. Shumovsky (Eds.), Kluwer, Dordrecht 1997, p. 139.
51. D.-G. Welsch, W. Vogel, and T. Opatrny, in *Progress in Optics*, E. Wolf (Ed.), North Holland, Amsterdam 1999, Vol. XXXIX, p. 63.
52. E. P. Wigner, Phys. Rev. A **40**, 749 (1932).
53. U. Leonhardt and H. Paul, Phys. Rev. A **48**, 4598 (1993).
54. H. Kuhn, D.-G. Welsch and W. Vogel, J. Mod. Opt. **41**, 1607 (1994).
55. G. T. Herman, *Image Reconstruction from Projections: The Fundamentals of Computerized Tomography*, Academic Press, New York 1980.
56. T. J. Dunn, I. A. Walmsley, and S. Mukamel, Phys. Rev. Lett. **74**, 884 (1995).
57. A. I. Lvovsky and S. A. Babichev, Phys. Rev. A **66**, 011801 (2002)
58. M. Beck, D. T. Smithey, and M. G. Raymer, Phys. Rev. A **48**, R890 (1993).
59. D. T. Smithey *et al.*, Phys. Rev. A **48**, 3159 (1993).

60. G. M. D'Ariano, C. Macchiavello, and M. G. A. Paris, Phys. Rev. A **50**, 4298 (1994).
61. G. M. D'Ariano, M. G. A. Paris, and M. F. Sacchi, in: *Quantum State Estimation*, Lect. Notes Phys. 649, M. Paris and J. Rehacek (Eds.), Springer, Berlin Heidelberg, 2004.
62. U. Leonhardt *et al.*, Opt. Commun. **127**, 144 (1996).
63. In this reference, the scaling convention $[X, P] = i/2$ is used instead of $[X, P] = i$ adopted in this paper. The quadrature data and Wigner function plots shown in this figure according to the convention $[X, P] = i$ can be rescaled by multiplying all axes labels by $\sqrt{2}$ while leaving the images unchanged.
64. G. Breitenbach, S. Schiller, and J. Mlynek, Nature **387**, 471 (1997).
65. J. Řeháček, Z. Hradil, M. Zawisky, S. Pascazio, H. Rauch, and J. Peřina, Phys. Rev. A **60**, 473 (1999).
66. D. F. V. James, P. G. Kwiat, W. J. Munro, and A. G. White, Phys. Rev. A **64**, 052312 (2001).
67. J. Řeháček, Z. Hradil, and M. Ježek, Phys. Rev. A **63**, 040303 (2001).
68. K. Banaszek, Phys. Rev. A **57**, 5013 (1998).
69. K. Banaszek, Acta Phys. Slov. **48**, 185 (1998).
70. K. Banaszek, G. M. D'Ariano, M. G. A. Paris, and M. F. Sacchi, Phys. Rev. A **61**, 010304 (1999).
71. A. I. Lvovsky, J. Opt. B: Q. Semiclass. Opt. **6**, (2004) S556-S559
72. E. Knill, R. Laflamme, and G. J. Milburn, Nature **409**, 46 (2001); J. L. O'Brien *et al.*, Nature **426**, 264 (2003).
73. R.E. Slusher *et al.*, Phys. Rev. Lett. **55**, 2409 (1985); L. A. Wu, H. J. Kimble, J. L. Hall, H. Wu, Phys. Rev. Lett. **57**, 2520 (1986).
74. D. T. Smithey, M. Beck, M. Belsley, and M. G. Raymer, Phys. Rev. Lett. **69**, 2650 (1992).
75. M. Munroe *et al.*, Phys. Rev. A **52**, R924 (1995).
76. E. L. Blansett *et al.*, Opt. Express **9**, 312 (2001).
77. A. Zavatta *et al.*, J. Opt. Soc. Am. B **19**, 1189 (2002).
78. A. Zavatta, S. Viciani, and M. Bellini, Phys. Rev. A **70**, 053821 (2004).
79. D. Fattal *et al.*, Phys. Rev. Lett. **92**, 037904 (2004).
80. C. Santori *et al.*, Nature **419**, 594 (2002).
81. C. K. Hong and L. Mandel, Phys. Rev. Lett. **56**, 58 (1986).
82. P. Grangier, G. Roger, and A. Aspect, Europhys. Lett. **1**, 173 (1986)
83. M. Zukowski, A. Zeilinger, and H. Weinfurter, Ann. NY Acad. Sci. **755**, 91 (1995).
84. J. G. Rarity, Ann. NY Acad. Sci. **755**, 624 (1995).
85. Z. Y. Ou, J. Opt. B: Qu. Semiclass. Opt. **9**, 599 (1997).
86. F. Grosshans and P. Grangier, Eur. Phys. J. D **14**, 119 (2001).
87. T. Aichele, A. I. Lvovsky, and S. Schiller, Eur. Phys. J. D **18**, 237 (2002).
88. D. N. Klyshko, Phys. Lett. A **128**, 133 (1988).
89. D. N. Klyshko, Phys. Lett. A **132**, 299 (1988).
90. D. N. Klyshko, Sov. Phys. Usp. **31** 74, (1988).
91. I. Walmsley and M. Raymer, Science **307**, 1733, 2005).

92. T. E. Keller and M. H. Rubin, Phys. Rev. A **56**, 1534 (1997).
93. W. P. Grice, A. B. URen, and I. A. Walmsley, Phys. Rev. A **64**, 063815 (2001).
94. A. B. U'Ren *et al.*, Laser Phys. **15**, 146 (2005).
95. M. G. Raymer, J. Noh, K. Banaszek, and I.A. Walmsley, Phys. Rev. A **72**, 023825 (2005).
96. For a review, see W. Tittel, G. Ribordy, and N. Gisin, Physics World, March 1998, page 41.
97. M. Koashi, T. Yamamoto, and N. Imoto, Phys. Rev. A **63**, 030301 (2001).
98. A. I. Lvovsky and J. H. Shapiro, Phys. Rev. **A 65**, 033830 (2002).
99. D. Leibfried *et al.* Phys. Rev. Lett. **77**, 4281 (1996).
100. G. Nogues *et al.*, Phys. Rev. **A 62**, 054101 (2000).
101. C. Kurtsiefer, T. Pfau, and J. Mlynek, Nature **386**, 150 (1997).
102. W. Vogel, Phys. Rev. Lett. **84**, 1849 (2000); L. Diósi, Phys. Rev. Lett. **85**, 2841 (2000); Th. Richter and W. Vogel, Phys. Rev. Lett. **89**, 283601 (2002).
103. K. Jacobs and P. L. Knight, Phys. Rev. A **54**, 3738 (1996).
104. W. Grice and I. A. Walmsley, J. Mod. Opt. **43**, 795 (1996).
105. S. J. van Enk, Phys. Rev. A **67**, 022303 (2003).
106. S. M. Tan, D. F. Walls, and M. J. Collett, Phys. Rev. Lett. **66**, 252 (1991); K. Banaszek and K. Wodkiewicz, *ibid.* **82**, 2009 (1999); K. Jacobs and P. L. Knight, Phys. Rev. A **54**, 3738 (1996).
107. B. Hessmo, P. Usachev, H. Heydari, and G. Björk, Phys. Rev. Lett. **92**, 180401 (2004).
108. D. T. Pegg, L. S. Phillips, and S. M. Barnett, Phys. Rev. Lett. **81**, 1604 (1998).
109. S. K. Ozdemir *et al.*, Phys. Rev. A **66**, 053809 (2002).
110. D.E. Browne *et al.*, Phys. Rev. A **67**, 062320 (2003).
111. A. P. Lund *et al.*, Phys. Rev. A **70**, 020101 (2004).
112. J. Clausen *et al.*, quant-ph/0007050.
113. B. J. Oliver and C. R. Stroud, Phys. Lett. A **135**, 407 (1989).

Part III: Atomic Continuous Variables

Chapter 22

Gaussian Description of Continuous Measurements on Continuous Variable Quantum Systems

L. B. Madsen[1] and K. Mølmer[1,2]

[1]*Department of Physics and Astronomy, University of Aarhus, 8000 rArhus C, Denmark*

[2]*Danish National Research Foundation Center for Quantum Optics*

The Gaussian state description of continuous variables is adapted to describe the quantum interaction between macroscopic atomic samples and continuous-wave light beams. The formalism is very efficient: a non-linear differential equation for the covariance matrix of the atomic system explicitly accounts for both the unitary evolution, the dissipation and noise due to the atom-light interaction, and the back-action due to homodyne optical detection on the beam after its interaction with the atoms. Applications to atomic spin squeezing and estimation of unknown classical parameters are presented, and extensions beyond the Gaussian states are discussed.

1. Introduction

Pulses of light, large atomic ensembles, and collections of more than, say, hundred trapped ions, are quantum systems where the behavior of various collective degrees of freedom is well described by quantities which have continuous spectra, *i.e.*, the systems may be described by collective effective position and momentum variables. The demonstration of quantum control of these systems varies from studies of squeezing and entanglement, over storage and retrieval of optical information in gases to high precision probing of classical properties in atomic magnetometry, atomic clocks and inertial sensors. Control is exercised via tunable interactions, by state reduction due to measurements on the systems, and by feed-back schemes in connection with measurements.

In quantum optics, the quantum properties of a continuous beam of light are normally described in the Heisenberg picture, where field operators are

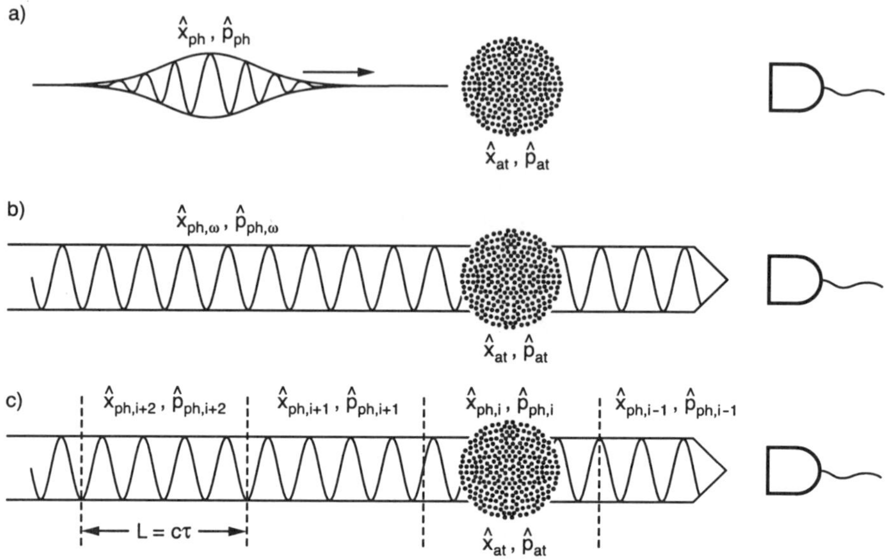

Fig. 1. Atom-light interaction. In the figure, we display a cloud of atoms described by collective continuous variables $\hat{x}_{\rm at}, \hat{p}_{\rm at}$, interacting with (a) a pulse of light and (b)-(c) a continuous wave of light. In (a) the assumption of an accurate description of the light pulse in terms of just a single mode $(\hat{x}_{\rm ph}, \hat{p}_{\rm ph})$ leads to a simple and natural description in the time domain of both the interaction of the pulse with the atoms as well as the detection process. In quantum optics, the continuous beam of light in (b) is normally described in the frequency domain, say, by canonical operators $\hat{x}_{\rm ph,\omega}, \hat{p}_{\rm ph,\omega}$. The interaction with the atoms and the measurement process, on the other hand, is more readily described in the time domain, and as discussed in detail in the text, it is technically difficult to pass from the frequency domain for the light operators to the time domain for the description of interaction and measurement process. To circumvent this problem, we introduce an effective description of the integral system in the time domain as indicated in (c). Here the beam is divided into segments of duration τ and length $L = c\tau$ each of which is assumed to be short enough to be accurately described by a single mode $\hat{x}_{{\rm ph},i}, \hat{p}_{{\rm ph},i}$, and the interaction with the atoms and the measurement is described by a succession of interactions with the individual beam segments.

expressed (often in the frequency domain) in terms of incoming vacuum fields with standard correlation functions. This input–output formalism leads, *e.g.*, to the noise spectrum of a squeezed light beam.[1] This approach accounts for the results one obtains if measurements are carried out directly on the beam, but it has been technically very difficult to describe the situation where the light beam is made subject to interaction with another quantum system and is subsequently measured. The measurement record is stochastic, and a real-time description of the measurement back-action on the probed quantum system is normally referred to quantum trajectory or Monte Carlo wave function treatments in the Schrödinger picture, which are incompatible with the frequency domain Heisenberg representation of the optical beam.

In Fig. 1, we display the interaction between light and atoms. A Gaussian state analysis was introduced recently to deal, in general terms, with the quantum properties of these systems, and tools were developed to handle interactions and measurements which preserve the Gaussian state character.[2,3] As we shall illustrate below, the Gaussian description is useful because (i) it handles the interaction between atoms and a quantized continuous-wave (cw) beam of light and (ii) it allows a description of measurement induced back–action in real time. This description thus provides a useful approach to a long standing problem in quantum optics, and it presents a theoretical treatment of physical systems and interactions of high current interest. The description is restricted to Gaussian states. A cw laser beam described by a coherent state and squeezed and quantum correlated optical beams created by down conversion are Gaussian in the field canonical variables and hence readily incorporated in our treatment. Turning now to the atoms, our approach does not describe the interaction with a single ion or atom, but a very accurate mapping exists between macroscopically spin-polarized atomic samples and a single harmonic oscillator.

In this work, we describe the practical application of the Gaussian state formalism to continuous variable systems, allowing full account of back-action due to measurement, noise, losses and inhomogeneities of the systems. The formalism is illustrated by a discussion of explicit examples concerning spin squeezing, magnetometry and entanglement. A whole tool-box can be created, describing the effect of frequency filters, finite band-width sources and detectors, finite efficiency detection, and dark counts, simply by adding extra reservoir modes. In practice, the Gaussian state for a system of n quantum harmonic oscillators, representing a number of optical beams and atomic components, is described by $2n$ mean values for the quadrature components and by a $2n \times 2n$ covariance matrix. While the evolution during measurements of mean values is stochastic, the covariance matrix is propagated in time in a deterministic way (see Sec. 2). It is a remarkable advantage of the Gaussian state description that extra physical systems and reservoir modes can be included at only little expense (two extra rows and columns in the covariance matrix per mode). In the last paragraphs of this work, it will be discussed how to develop a theory for continuous variable systems where the Gaussian description breaks down, either because of the interactions involved, because of the measurement schemes, or because of coupling of a small discrete system to collective continuous degrees of freedom.

2. Time Evolution of Gaussian States, General Theory

In this section, we introduce the Gaussian description in a general setting using existing results.[2–4] Sections 2.1 and 2.2 deal with the evolution of continuous variable systems due to a bilinear Hamiltonian and linear losses. This evolution can be solved by an affine transformation in time of the canonical operators, and all system properties are given by their mean values and their covariance matrix for which an exact treatment is provided. Section 2.3 deals with the effect of measurements on the system. The update of the system state vector or density operator conditioned on a measurement outcome is non-trivial in the most general case, but as we shall see, Gaussian states transform into other Gaussian states in a well described manner under homodyne detection on part of the system, and in this case the mean values and the covariance matrix still provide all properties of the system.

2.1. *Time evolution due to a bilinear Hamiltonian*

Let $\hat{\boldsymbol{y}} = (\hat{x}_1, \hat{p}_1, \hat{x}_2, \hat{p}_2, \ldots, \hat{x}_n, \hat{p}_n)^T$ denote the column vector of $2n$ variables with canonical commutators $[\hat{x}_i, \hat{p}_j] = i\delta_{ij}$, and let $\hat{H} = \hat{H}(\hat{x}_1, \hat{p}_1, \hat{x}_2, \hat{p}_2, \ldots, \hat{x}_n, \hat{p}_n)$ denote the Hamiltonian of the system. We shall assume that $\hat{H}$ is bilinear in the canonical variables. Heisenberg's equations of motion during time τ are then solved by a linear transformation of the operators by the matrix $\mathbf{S}_\tau$

$$\hat{\boldsymbol{y}}(t+\tau) = \mathbf{S}_\tau \hat{\boldsymbol{y}}(t). \tag{1}$$

The same transformation applies to the vector of mean values $\mathbf{m} \equiv \langle \hat{\boldsymbol{y}} \rangle$, $\mathbf{m}(t+\tau) = \mathbf{S}_\tau \mathbf{m}(t)$. From Eq. (1) and the definition of the covariance matrix $\gamma_{ij} \equiv 2\mathrm{Re}\,\langle (\hat{y}_i - \langle \hat{y}_i \rangle)(\hat{y}_j - \langle \hat{y}_j \rangle) \rangle$, we directly verify that $\boldsymbol{\gamma}$ transforms as

$$\boldsymbol{\gamma}(t+\tau) = \mathbf{S}_\tau \boldsymbol{\gamma}(t) \mathbf{S}_\tau^T \tag{2}$$

under the interaction.

2.2. *Time evolution due to dissipation and noise*

In the absence of dissipation Eq. (2) determines the evolution of the covariance matrix. In realistic situations, however, there will be sources of dissipation and noise. Dissipation leads to a reduction in the mean values of the canonical variables, and as is known from the quantum theory of damping and the fluctuation-dissipation theorem of statistical mechanics, such a reduction must be accompanied by fluctuations. In the quantum domain we must, *e.g.*, fulfill the Heisenberg uncertainty relations, also when the mean values are reduced. The generalization of Eq. (2) to the noisy case reads for small τ

$$\gamma(t+\tau) = \mathbf{L}_\tau \mathbf{S}_\tau \gamma(t) \mathbf{S}_\tau^T \mathbf{L}_\tau + \mathbf{N}_\tau, \tag{3}$$

where $\mathbf{L}_\tau$ describes the reduction of the mean values, $\mathbf{m}(t+\tau) = \mathbf{L}_\tau \mathbf{S}_\tau \mathbf{m}(t)$, and where $\mathbf{N}_\tau$ is the associated noise. In examples below, we shall give explicit forms of these matrices.

If the state of the system is initially a Gaussian state, *i.e.*, its Wigner function for the canonical variables is a Gaussian function, the evolution due to a bilinear Hamiltonian preserves the Gaussian character. The same is true for linear damping of an optical field mode, and as validated by a calculation and more detailed discussion[5] it also holds to an excellent approximation for atomic decay models.

2.3. *Time evolution due to a homodyne measurement event*

The above arguments were based on the Heisenberg picture evolution of the canonical operators, but the evolution due to measurements is more conveniently described as state reduction in a Schrödinger picture representation of the system state vector or density operator. A general representation of the state, pure or mixed, of a collection of harmonic oscillators is provided by the Wigner function $\mathcal{W}(\xi)$ with $\xi \equiv (\xi_1, ...\xi_{2n}) \in \mathbb{R}^{2n}$. This function is connected with the density matrix in position or momentum representations by a Fourier-transformation, and it provides a good intuitive picture of the phase space distribution of the system. In fact, the expectation value of any symmetrically ordered function $F_{\text{sym}}(\hat{x}_1, \hat{p}_1, ...\hat{x}_n, \hat{p}_n)$ (F_{sym} is the average of all the ways of ordering the operators defining $F(\hat{x}_1, \hat{p}_1, ...\hat{x}_n, \hat{p}_n)$), is given by the pseudo-classical expression:

$$\langle F_{\text{sym}}(\hat{x}_1, \hat{p}_1, ...\hat{x}_n, \hat{p}_n)\rangle = \int d^{2n}\xi \ \mathcal{W}(\xi) F(\xi). \tag{4}$$

We recall that we aim at a description of the state of an atomic sample subject to interaction with an optical beam which is being probed after the interaction. We hence address what happens to the quantum state of the remaining system when one of the sub-systems (with a conjugate pair of observables $\hat{x}_n, \hat{p}_n$) is subject to a measurement. Examples of measurements are positive operator valued measures with coherent state outcomes, homodyne detection which projects the measured sub-system onto a position or momentum eigenstate (equivalent to the limit of a strongly quadrature squeezed state), and number state detection. Such measurements project the $(\hat{x}_n, \hat{p}_n)$ sub-system onto a particular state which we can also describe by a Wigner function $\mathcal{W}_{\text{meas}}(\xi_{2n-1}, \xi_{2n})$. The state of the remaining system conditioned on the outcome leading to this particular state is

$$\mathcal{W}_{\text{cond}}(\xi_1, \dots, \xi_{2n-2}) = \frac{\int d\xi_{2n-1} d\xi_{2n}, \mathcal{W}(\xi_1, \dots, \xi_{2n}) \mathcal{W}_{\text{meas}}(\xi_{2n-1}, \xi_{2n})}{P_{\text{meas}}}, \tag{5}$$

with $P_{\text{meas}} = \int d\xi_1 \dots d\xi_{2n} \mathcal{W}(\xi_1, \dots, \xi_{2n}) \mathcal{W}_{\text{meas}}(\xi_{2n-1}, \xi_{2n})$.

Now turning to the Gaussian states, a series of simplifications occur. For example, the Wigner functions $\mathcal{W}_{\text{meas}}(\xi_{2n-1}, \xi_{2n})$ for coherent and squeezed states are Gaussian functions of the variables. This implies, that if the initial Wigner function is a Gaussian function of the variables, this property is maintained by the homodyne detection process. Generally, the Wigner function for a Gaussian state is fully parameterized by the mean values $\mathbf{m}$ and the covariance matrix γ:

$$\mathcal{W}_{\text{Gauss}}(\xi) = \frac{1}{\pi^n} \frac{1}{\sqrt{\det \gamma}} \exp\left(-(\xi - \mathbf{m})^T \gamma^{-1} (\xi - \mathbf{m})\right). \tag{6}$$

As the Gaussian character is also maintained by the bilinear Hamiltonian and the linear decay processes, we conclude that to describe the time evolution of a system which starts in a Gaussian state, it suffices to provide the time dependent $\mathbf{m}$ and γ.

Since part of the system is being measured upon, and hence disappears from our quantum state, cf. Eq. (5), it makes sense to write the covariance matrix in the form

$$\gamma = \begin{pmatrix} \mathbf{A} & \mathbf{C} \\ \mathbf{C}^T & \mathbf{B} \end{pmatrix}, \tag{7}$$

where the $(2n-2) \times (2n-2)$ sub-matrix $\mathbf{A}$ is the covariance matrix for the variables $\hat{\boldsymbol{y}}_1 = (\hat{x}_i, \hat{p}_1, \dots, \hat{x}_{n-1}, \hat{p}_{n-1})^T$ which are not subject to measurement, $\mathbf{B}$ is the 2×2 covariance matrix for the sub-system subject to measurement $\hat{\boldsymbol{y}}_2 = (\hat{x}_n, \hat{p}_n)^T$, and $\mathbf{C}$ is the $(2n-2) \times 2$ correlation matrix between the elements of $\hat{\boldsymbol{y}}_1$ and $\hat{\boldsymbol{y}}_2$. According to the above expressions (5)-(6), a measurement of $\hat{x}_n$ transforms $\mathbf{A}$ as[2-4]

$$\mathbf{A} \mapsto \mathbf{A}' = \mathbf{A} - \mathbf{C}(\pi \mathbf{B} \pi)^{-} \mathbf{C}^T, \tag{8}$$

where $\pi = \text{diag}(1, 0)$, and where $(\ )^-$ denotes the Moore-Penrose pseudoinverse: $(\pi\mathbf{B}\pi)^- = \text{diag}(B(1,1)^{-1}, 0)$. If we associate with the precise measurement of $\hat{x}_n$ an infinite variance of $\hat{p}_n$ and hence a total loss of correlations between $\hat{p}_n$ and the other observables, this result is equivalent to the Bayesian update of a classical Gaussian probability distribution.[6] We recognize the Moore-Penrose pseudoinverse as the normal inverse of the corresponding covariance matrix, $(\pi\mathbf{B}\pi)^- = \text{diag}(B(1,1), \infty)^{-1}$.

Unlike the covariance matrix update, which is independent of the value measured, the vector $\mathbf{m} = \langle \hat{\boldsymbol{y}} \rangle$ of expectation values will change in a stochastic manner depending on the actual outcome of the measurement. The outcome of the measurement of $\hat{x}_n$ is random, and the measurement changes the expectation value of all other observables due to the correlations represented by the covariance matrix. Let χ denote the difference between the measurement outcome and the expectation value of $\hat{x}_n$, *i.e.*, a Gaussian random variable with mean value zero and variance given by half of the appropriate covariance matrix element $B(1,1)$. It follows again from Eqs. (5)-(6), (and from the corresponding classical theory of multi-variate Gaussian distributions,) that the change of $\mathbf{m}_1 = \langle \boldsymbol{y}_1 \rangle$ due to the measurement is given by:

$$\mathbf{m}_1 \mapsto \mathbf{m}_1' = \mathbf{m}_1 + \mathbf{C}_\gamma (\pi\mathbf{B}\pi)^- (\chi, \cdot)^T, \tag{9}$$

where we use that $(\pi B \pi)^- = \text{diag}(B(1,1)^{-1}, 0)$, and hence the second entrance in the vector $(\chi, \cdot)$ need not be specified.

2.4. *Time evolution due to continuous homodyne measurements*

In the continuous interaction between a cw light beam and a cloud of atoms one faces a situation where a single system (the atoms) is continuously indirectly monitored, *e.g.*, by a homodyne detection of the light field. This raises the problems, mentioned in the introduction, of dealing simultaneously with a continuous beam and measurement induced back-action (see also Fig. 1). We have recently solved this problem for Gaussian states[7] by quantizing the light beam in short segments of duration τ and corresponding length $L = c\tau$. These beam segments are chosen so short that the field in a single segment can be treated as a single mode and such that the state of the atoms interacting with the field does not change appreciably during time τ. The evolution of the atomic system with the entire beam of light is obtained by sequential interaction with subsequent light segments.

The generic multi-mode character of the cw beam of light is treated in the Schrödinger picture in the time domain rather than in the Heisenberg picture in the time domain (cf. Fig. 1(c)).

The simplest example of continuous light-atom interaction is the one of a coherent monochromatic beam of light, corresponding to a product state of coherent states in each segment along the beam axis. In this case, the problem simplifies significantly because all segments are in the same trivial state prior to the interaction with the atoms. The segments need not be included formally in the update of the covariance matrix until it is their turn to interact with the atoms. Segments which have already interacted with the atoms may be detected instantly after the interaction, and in practice they are if the detector is placed within meters from the interaction volume. The detected segments then disappear from the formal description of the system. Prior to the interaction with the beam, we thus consider only the atomic covariance matrix $\mathbf{A}$, and in the absence of any correlation with the incident beam segment, the block-off-diagonal matrices in Eq. (7) vanish

$$\mathbf{C} = \mathbb{O}_{2\times(2n-2)} \tag{10}$$

while the field state of the incident segment is characterized by the normal noise properties of the coherent state

$$\mathbf{B} = \mathbb{I}_{2\times 2}. \tag{11}$$

The full covariance matrix is now propagated according to Eq. (3), and the matrix changes to describe the state of the atoms and the optical segment after interaction. To describe the effect on the atoms of the measurement on the field segment, we apply the measurement update formula (8) for the atomic part, and since the field segment has been observed and reduced to classical information, we are ready to turn to the interaction with the next light segment, which conveniently fits into the covariance matrix (7) in the same locations as the previous segment according to Eqs. (10)-(11). This evolution is repeated to describe in real time the interaction with a beam for any extended period of time, and the expectation value and our uncertainty about any variable of the system at the end of the interaction is readily found from the appropriate entrances in the vector $\mathbf{m}$ and the matrix γ.

In the limit of small τ the changes in γ and $\mathbf{m}$ expressed by the update formulae (3), and (8)-(9), are infinitesimally small. In this, suitably defined, continuous limit, the update formulae translate into differential equations. After application of Eq. (3), the sub-matrix $\mathbf{C}$ depends linearly on the elements of **A** and as shown in Eq. (21) and the ensuing discussion below, its elements are proportional to $\sqrt{\tau}$. **B** is essentially unchanged for short τ, and **A** changes linearly with τ. In the limit of infinitesimally small time increments, the update formula may therefore be written as a closed non-linear equation of motion for **A**:

$$\dot{\mathbf{A}} = \lim_{\tau \to 0^+} \frac{\mathbf{A}' - \mathbf{A}}{\tau} \equiv \mathbf{G} - \mathbf{DA} - \mathbf{AE} - \mathbf{AFA}, \tag{12}$$

with suitably defined matrices $\mathbf{G}, \mathbf{D}, \mathbf{E}, \mathbf{F}$. This equation is an example of a so-called matrix Ricatti equation,[8] and by the decomposition $\mathbf{A} = \mathbf{WU}^{-1}$, it can be rewritten in terms of two coupled linear equations $\dot{\mathbf{W}} = -\mathbf{DW} + \mathbf{GU}$, and $\dot{\mathbf{U}} = \mathbf{FW} + \mathbf{EU}$. Below, we shall see examples of analytical solutions to the problem based on these equations.

3. Application of the Gaussian Formalism to Atom-Light Interaction

The Gaussian formalism can be applied to describe the interaction between atomic samples and optical beams. In our examples, we consider optical Faraday rotation, which probes the collective spin ground state of a gas of atoms. To introduce the transition to an effective Hamiltonian expressed in terms of canonical variables, we discuss in some detail the interaction of an atomic ensemble with a pulse, or segment, of light.

3.1. *Stokes vector and canonical conjugate variables for light*

To make the discussion simple, at first only a single atomic sample and a single pulse or segment of a light beam will be considered. In Faraday rotation experiments, one uses light, which is linearly polarized along the, say, x-axis. The interesting quantum degree of freedom of the light pulse is not the field amplitude itself, but the intensity difference between the linearly polarized components along 45 and 135 degree directions in the xy plane, and between the two circularly polarized components with respect to the z-axis. These components are equally populated on average, but as every single x-polarized photon can be expanded as a superposition of single photon states of either pair of polarizations, their populations will fluctuate according to a binomial distribution. For a pulse with a definite number N_{ph} of photons, one may represent these populations conveniently by the components of the Stokes vector, where the x, y and z-components represent the populations difference of x and y polarizations, 45 and 135 degree polarizations and σ^+ and σ^--polarizations, respectively, *i.e.*,

$$\hat{S}_x = \frac{\hbar}{2}\left(\hat{a}_x^\dagger \hat{a}_x - \hat{a}_y^\dagger \hat{a}_y\right) = -\frac{\hbar}{2}\left(\hat{a}_+^\dagger \hat{a}_- + \hat{a}_-^\dagger \hat{a}_+\right), \tag{13}$$

$$\hat{S}_z = \frac{\hbar}{2}\left(\hat{a}_x^\dagger \hat{a}_y + \hat{a}_y^\dagger \hat{a}_x\right) = -\frac{\hbar}{2i}\left(\hat{a}_+^\dagger \hat{a}_- - \hat{a}_-^\dagger \hat{a}_+\right), \tag{14}$$

$$\hat{S}_z = \frac{\hbar}{2i}\left(\hat{a}_x^\dagger \hat{a}_y - \hat{a}_y^\dagger \hat{a}_x\right) = \frac{\hbar}{2}\left(\hat{a}_+^\dagger \hat{a}_+ - \hat{a}_-^\dagger \hat{a}_-\right). \tag{15}$$

Since the light is assumed to be linearly polarized along the x axis, $\hat{S}_x$ may be treated classically and from Eq. (13), $\hat{S}_x/\hbar = S_x/\hbar = N_{\rm ph}/2$. The Stokes vector components obey the commutator relations of a fictitious spin, and the variance of the binomial distributions are in precise correspondence with the quantum mechanical uncertainty on $\hat{S}_y$ and $\hat{S}_z$, achieving the Heisenberg limit $\mathrm{Var}(\hat{S}_y)\mathrm{Var}(\hat{S}_z) = |\langle \hbar \hat{S}_x\rangle|^2/4$.

We assume that S_x remains large and essentially unchanged during the interaction with the atomic gas, and we can then introduce the effective position and momentum operators

$$(\hat{x}_{\rm ph}, \hat{p}_{\rm ph}) = \left(\hat{S}_y/\sqrt{|\langle \hbar S_x\rangle|}, \hat{S}_z/\sqrt{|\langle \hbar S_x\rangle|}\right), \tag{16}$$

which fulfill the standard commutator relation $[\hat{x}_{\rm ph}, \hat{p}_{\rm ph}] = i$ and resulting uncertainty relation. These are the canonical conjugate variables that we wish to describe by the formalism outlined in the previous section. The initial binomial distributions of $\hat{S}_y, \hat{S}_z$ approach Gaussian distributions in the limit of large photon numbers. Moreover, the fact that the uncertainty relation is minimized in the initial state implies that this state is a Gaussian state, *i.e.*, the Wigner function for the field is a Gaussian function.[9]

3.2. *Atom-light interaction*

The physical system of interest consists of one or more macroscopic ensembles of trapped atoms interacting off-resonantly with one or more laser beams. We consider the usual electric dipole interaction between the atoms and the quantized field. First, the off-resonant coupling of the atoms with the light field is expanded in transition operators between the ground ($|FM\rangle$) and excited ($|F'M'\rangle$) hyperfine states (several excited states with different F' may be coupled to the ground state). Then, the atomic coherences pertaining to the excited states are expressed by the light fields and ground state coherences by adiabatic elimination using Heisenberg's equations of motion for the slowly varying operators. This procedure generally allows us to derive a dispersive effective Hamiltonian,[10,11] which for the N_{at} atoms reads[12]

$$\hat{H}_{\text{int},\tau} = \sum_{j=1}^{N_{\text{at}}} \sum_{M=-F}^{F} [\Big(c_{+,M}(\Delta)\hat{a}_+^\dagger \hat{a}_+ + c_{-,M}(\Delta)\hat{a}_-^\dagger \hat{a}_-\Big) |FM\rangle_j\langle FM|$$

$$+ b_M(\Delta)\Big(\hat{a}_-^\dagger \hat{a}_+|FM+1\rangle_j\langle FM-1| + \hat{a}_+^\dagger \hat{a}_-|FM-1\rangle_j\langle FM+1|\Big)], \quad (17)$$

where field creation and annihilation operators for σ^+ and σ^--polarized photons have been introduced. The first two terms describe the ac Stark shift of the ground state $|FM\rangle$ caused by the coupling to the excited $|F'M \pm 1\rangle$ states by the two field components. The coupling coefficients are given by $c_{\pm,M}(\Delta) = -2\hbar\sum_{F'}(g^{\pm}_{FM;F'M'})^2/\Delta_{F'}$ where $\Delta_{F'}$ is the detuning of the laser frequency from the upper level, and where the coupling constants $g^{\pm}_{FM;F'M'}$ are the electric dipole coupling matrix elements, $g^{\pm}_{FM;F'M'} = \sqrt{\omega_0/2\hbar\epsilon_0 Ac\tau}\, d^{\pm}_{FM;F'M'}$. These matrix elements contain the 'electric field per photon' for a plane wave field with transverse area A and length $c\tau$, and they involve the spherical tensor components of the dipole operator $\hat{\boldsymbol{d}} = -e\hat{\boldsymbol{r}}$ of the electron, $d^{\pm}_{FM;F'M'} = \langle FM|\hat{d}_{\pm}|F'M'\rangle$ and $\hat{d}_+ = -(\hat{d}_x + i\hat{d}_y)/\sqrt{2}$, $\hat{d}_- = (\hat{d}_x - i\hat{d}_y)/\sqrt{2}$. The terms in Eq. (17) proportional to $b_M(\Delta) = -2\hbar\sum_{F'} g^{+}_{FM-1;F'M} g^{-}_{FM+1;F'M}/\Delta_{F'}$ describe $\Delta M = \pm 2$ Raman transitions involving absorption and stimulated emission of a pair of photons with different polarization.

3.2.1. *Spin 1/2-case*

For much of the discussion in the rest of this work, we shall restrict ourselves to the case of atoms with only one ground and one excited level which both have total angular momenta $F = F' = 1/2$. The above Hamiltonian simplifies in this case, and noting further that the dipole matrix elements are related to the total spontaneous decay rate Γ of the upper state, $c_{\pm,\mp 1/2} = -3\hbar\Gamma\sigma/(2\tau\Delta A)$, with $\sigma = \lambda^2/(2\pi)$ the resonant photon absorption cross section, Eq. (17) then reduces to

$$\hat{H}_{\text{int},\tau} = -\sum_j \frac{3\hbar\Gamma\sigma}{2\tau\Delta A}\left(\hat{a}_+^\dagger\hat{a}_+\left|-\frac{1}{2}\right\rangle_j\left\langle-\frac{1}{2}\right| + \hat{a}_-^\dagger\hat{a}_-\left|\frac{1}{2}\right\rangle_j\left\langle\frac{1}{2}\right|\right), \quad (18)$$

where the $F = F' = 1/2$ index has been suppressed.

The atomic ensemble is initially prepared with all $N_{\rm at}$ atoms in a superposition $(|-1/2\rangle + |1/2\rangle)/\sqrt{2}$ of the two ground states with respect to the quantization axis z, *i.e.*, the total state of the atoms is initially given by $\left((|-1/2\rangle + |1/2\rangle)/\sqrt{2}\right)^{N_{\rm at}}$. In this state, the system of two-level atoms is described by a collective spin, $\hat{\boldsymbol{J}} = \frac{\hbar}{2}\sum_j \hat{\boldsymbol{\sigma}}_j$, with a component along the x-direction which attains the macroscopic value $\langle \hat{J}_x \rangle = \hbar N_{\rm at}/2$, and with a component along the z-axis, $\hat{J}_z$, which represents the population difference between the $|\pm 1/2\rangle$ states. Similarly, we may use Eq. (15) and represent the operators of the photon field in terms of the collective Stokes vector operator, $\hat{S}_z$. The Hamiltonian can, hence, be rewritten in terms of the collective spin variables for photons and atoms

$$\hat{H}_{{\rm int},\tau} = -\frac{3\Gamma\sigma}{\tau\Delta A}\hat{S}_z\hat{J}_z, \tag{19}$$

where an overall energy-shift proportional to the number of photons in the pulse segment has been neglected.

As for the photons it is convenient to introduce effective atomic position and momentum coordinates

$$(\hat{x}_{\rm at}, \hat{p}_{\rm at}) = \left(\hat{J}_y/\sqrt{|\langle \hbar \hat{J}_x \rangle|}, \hat{J}_z/\sqrt{|\langle \hbar \hat{J}_x \rangle|}\right), \tag{20}$$

for which the initial state is a minimum uncertainty Gaussian state. The last step of this analysis is then to rewrite Eq. (19) in terms of canonical conjugate variables,

$$\hat{H}_{{\rm int},\tau} = \hbar\kappa_\tau \hat{p}_{\rm at}\hat{p}_{\rm ph}, \tag{21}$$

where

$$\kappa_\tau = -\frac{3\Gamma\sigma}{\tau\Delta A}\sqrt{|\langle \hat{S}_x \rangle|}\sqrt{|\langle \hat{J}_x \rangle|}. \tag{22}$$

The Hamiltonian (21) correlates the atoms and the light fields and is bilinear in the canonical variables. Hence the theoretical formalism of Sec. 2 applies. The coupling constant κ_τ is small for realistic parameters, and a coarse grained description, where the atoms interact with one segment of light after the other, will be perfectly valid even for the macroscopic number of photons $N_{\rm ph}$ in each segment required by our Gaussian treatment. Note that $\langle \hat{S}_x \rangle$ is proportional to the number of photons in the beam segment, *i.e.*, to τ, and it follows that $\hat{H}_{{\rm int},\tau}\tau$ is proportional to $\sqrt{\tau}$ yielding a well-defined differential limit in Eq. (12).

We have emphasized the convenience of using Gaussian states, because their Schrödinger picture representation is very efficient and compact. Now, given that every segment of the optical beam becomes correlated with the atomic sample, as a function of time, the joint state of the atom and field has to be specified by a larger and larger number of mean values and second order moments. If no further interactions take place between the atoms and the light after the interaction, there is no need to keep track of the state of the total system. In practice, either the transmitted light may simply disappear or it may be registered in a detection process. In the former case, the relevant description of the remaining system is obtained by a partial trace over the field state, which produces a new Gaussian state of the atoms, which is simply given by removing the photonic lines and columns of the covariance matrix immediately after the interaction update (3). The measurement of the small Faraday rotation of the linearly polarized probe is done by a measurement of the intensity difference between the 45 and 135 degree polarization components, *i.e.*, by a measurement of the $\hat{S}_y \propto \hat{x}_{ph}$ observable, which is precisely the "homodyne" measurement described in Sec. 2.4. The atomic state is thus described by the corresponding update formula of Eq. (12).

4. Spin Squeezing in the Gaussian Description

With spin squeezed atomic ensembles, *i.e.*, samples where the variance of one of the angular momentum (spin) components is reduced compared with the coherent state value, one has the possibility to measure certain atomic and/or classical parameters beyond the precision set by the standard quantum noise.

The theory of squeezing of the collective atomic spin variable was dealt with in a series of papers,[10,11,13] and extended to include investigations of quantum non-demolition feedback schemes,[14,15] and inhomogeneous light-atom coupling.[16,17] In related works,[7,18–20] spin-squeezing of continuous variable quantum systems has been investigated in the approximation where the atomic and photonic degrees of freedom are described by a Gaussian state.

We are interested in the case, where the polarization rotation of the light field is registered, *i.e.*, the observable $\hat{x}_{\rm ph}$ is measured. The effect of measuring one of the components in a multi-variable Gaussian state is effectively to produce a new Gaussian state of the remaining variables as discussed in detail in Sec. 2. The column vector of the variables for the gas and the photon field reads $\hat{\boldsymbol{y}} = (\hat{x}_{\rm at}, \hat{p}_{\rm at}, \hat{x}_{\rm ph}, \hat{p}_{\rm ph})^T$ and the S-matrix in Eq. (1) is

$$\mathbf{S}_\tau = \begin{pmatrix} 1 & 0 & 0 & \kappa_\tau \\ 0 & 1 & 0 & 0 \\ 0 & \kappa_\tau & 1 & 0 \\ 0 & 0 & 0 & 1 \end{pmatrix}. \tag{23}$$

4.1. *Dissipation and noise*

In the probing process there is a small probability that the excited atomic levels which were adiabatically eliminated from the interaction Hamiltonian of Eq. (21) will be populated. If this happens, the subsequent decay to one of the two $M_z = \pm 1/2$ ground states occurs with the rate $\eta = \Phi \frac{\sigma}{A} \left(\frac{\Gamma^2/4}{\Gamma^2/4+\Delta^2} \right)$, where Φ is the photon flux and where the remaining parameters were defined in Sec. 3.2.1. The consequence of the decay is a loss of spin polarization since a detection of the fluorescence photons in principle can tell to which ground state the atom decayed. If every atom has a probability $\eta_\tau = \eta\tau$ to decay in time τ with equal probability into the two ground states, the collective mean spin vector is reduced by the corresponding factor $\langle \boldsymbol{J} \rangle \to \langle \boldsymbol{J} \rangle (1 - \eta_\tau)$.

Simultaneously, every photon on its way through the atomic gas has a probability for being absorbed[19] $\epsilon = N_{\rm at} \frac{\sigma}{A} \left(\frac{\Gamma^2/4}{\Gamma^2/4+\Delta^2} \right)$ (see Sec. 3.2.1 for definition of parameters). The effect of these noise contributions were discussed in detail elsewhere,[5,20] and the result for the reduction and noise matrices of the update formula of Eq. (3) reads $\mathbf{L}_\tau = {\rm diag}(\sqrt{1-\eta_\tau}, \sqrt{1-\eta_\tau}, \sqrt{1-\epsilon}, \sqrt{1-\epsilon})$, and $\mathbf{N}_\tau = {\rm diag}(\frac{\hbar N_{\rm at}}{\langle J_x(t) \rangle}\eta_\tau, \frac{\hbar N_{\rm at}}{\langle J_x(t) \rangle}\eta_\tau, \frac{\hbar N_{\rm ph}}{2\langle S_x(t) \rangle}\epsilon, \frac{\hbar N_{\rm ph}}{2\langle S_x(t) \rangle}\epsilon)$ for $\eta_\tau, \epsilon \ll 1$. The factor $\hbar N_{\rm at}/\langle \hat{J}_x(t) \rangle$ initially attains the value 2, and increases by the factor $(1-\eta_\tau)^{-1}$ in each time step τ. The factor $\hbar N_{\rm ph}/(2\langle \hat{S}_x(t) \rangle)$ is initially unity, and is approximately constant in time since the light field is continuously renewed by new segments of the light beam interacting with the atoms.

We note that when the classical x-component of the atomic spin is reduced this leads to a reduction with time of the coupling strength $\kappa_\tau \mapsto \kappa_\tau \sqrt{1-\eta_\tau}$ (see Eq. (22)).

4.2. *Solution of Ricatti equation*

We now have explicit forms for the matrices needed for our update of the Gaussian states. In the Gaussian description, the problem of spin squeezing

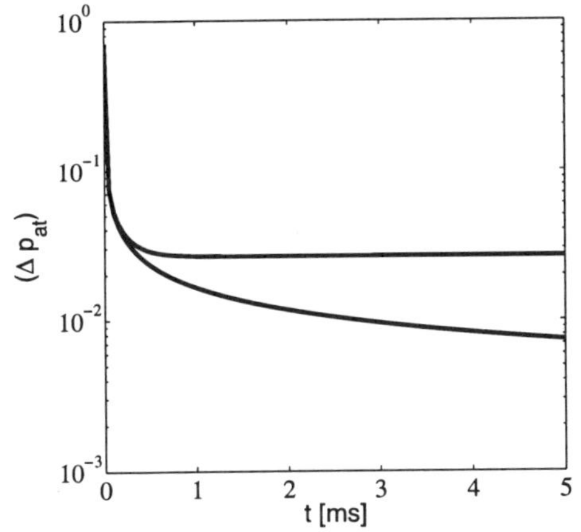

Fig. 2. Uncertainty of $p_{\rm at}$ (20) as function of time during optical Faraday rotation experiment. The effective coupling is $\kappa^2 = 1.83 \times 10^6\,{\rm s}^{-1}$. The lower curve is without inclusion of atomic decay, and the upper curve includes atomic decay with a rate $\eta = 1.7577\,{\rm s}^{-1}$ and photon absorption with a probability $\epsilon = 0.028$. These values correspond, for example, to a 2 mm^2 interaction area, 2×10^{12} atoms, 5×10^{14} photons s^{-1}, 10 GHz detuning, and 852 nm light, appropriate for the $^{133}{\rm Cs}(6S_{1/2}(F=4) - 6P_{1/2}(F=5))$ transition. Factors of order unity related to the coupling matrix elements among different states of the actual Zeeman substructure of Cs are omitted.

may be solved either by the discrete update formulae or analytically from the matrix Ricatti equation. In the latter case, we note that the covariance matrix after n iterations in the noise-less case is

$$\gamma_n = \begin{pmatrix} 2{\rm Var}(\hat{x}_{\rm at}) & 0 & 0 & 0 \\ 0 & 2{\rm Var}(\hat{p}_{\rm at}) & 0 & 0 \\ 0 & 0 & 1 & 0 \\ 0 & 0 & 0 & 1 \end{pmatrix}. \tag{24}$$

We then apply the S-matrix from Eq. (23) and find

$$\mathbf{S}_\tau \gamma_n \mathbf{S}_\tau^\dagger = \begin{pmatrix} 2{\rm Var}(\hat{x}_{\rm at})+\kappa_\tau^2 & 0 & 0 & \kappa_\tau \\ 0 & 2{\rm Var}(\hat{p}_{\rm at}) & 2\kappa_\tau {\rm Var}(\hat{p}_{\rm at}) & 0 \\ 0 & 2\kappa_\tau {\rm Var}(\hat{p}_{\rm at}) & 1+2\kappa_\tau^2 {\rm Var}(\hat{p}_{\rm at}) & 0 \\ \kappa_\tau & 0 & 0 & 1 \end{pmatrix}. \tag{25}$$

From this matrix, we determine, to lowest order in τ, $\mathbf{C}(\pi\mathbf{B}\pi)^-\mathbf{C}^T = \kappa_\tau^2 \begin{pmatrix} 0 & 0 \\ 0 & (2{\rm Var}(\hat{p}_{\rm at}))^2 \end{pmatrix}$, insert into Eq. (8), take the continuous limit and use $\kappa^2 = \kappa_\tau^2/\tau$. This procedure leads to the following differential equation for the variance of $\hat{p}_{\rm at}(\propto \hat{J}_z)$: $\frac{d}{dt}{\rm Var}(\hat{p}_{\rm at}) = -2\kappa^2 \left({\rm Var}(\hat{p}_{\rm at})\right)^2$, which is readily solved by separating the variables

$${\rm Var}(\hat{p}_{\rm at}) = \frac{1}{2\kappa^2 t + 1/{\rm Var}(\hat{p}_{\rm at,0})}, \tag{26}$$

where $\mathrm{Var}(\hat{p}_{\mathrm{at},0}) = 1/2$ is the variance of the initial minimum uncertainty state. Note that the solution to the variance of the conjugate atomic variable is $\mathrm{Var}(\hat{x}_{\mathrm{at}}) = \kappa^2 t/2 + \mathrm{Var}(\hat{x}_{\mathrm{at},0})$ with $\mathrm{Var}(\hat{x}_{\mathrm{at},0}) = 1/2$. Hence, while $\hat{p}_{\mathrm{at}}$ is squeezed, $\hat{x}_{\mathrm{at}}$ is antisqueezed to maintain the equal sign in Heisenberg's uncertainty relation.

When dissipation and noise is included the problem may still be solved analytically.[20] The expressions for the variances are quite complicated and will not be given here. Figure 2 shows the spin squeezing as a function of probing time. When atomic decay is not included, the uncertainty in $\hat{p}_{\mathrm{at}}$ is a monotonically decreasing function with time. When decay and noise is included, a minimum at t_{min} is reached whereafter the degree of squeezing starts to decrease. On the time scale of the figure, which is chosen to reflect realistic experimental time scales, the increase in $\mathrm{Var}(\hat{p}_{\mathrm{at}})$ is hardly visible.

4.3. *Inhomogeneous coupling*

One of the virtues of the Gaussian description of spin squeezing is that it is straightforwardly generalized to handle situations which are hard to approach by standard means. For example, a variation in the intensity of the light beam across the atomic sample and a large photon absorption probability both lead to an inhomogeneous atom-light coupling.[17,20] To treat such a case, the atomic gas is divided into n slices each with local light-atom coupling strength κ_i. The $2n+2$ dimensional vector of gaussian variables describing the $2n$ collective canonical position and momentum variables for the atoms, and the two collective position and momentum variables for the photon field then reads $\hat{\boldsymbol{y}} = (\hat{x}_{\mathrm{at},1}, \hat{p}_{\mathrm{at},1}, \ldots, \hat{x}_{\mathrm{at},n}, \hat{p}_{\mathrm{at},n}, \hat{x}_{\mathrm{ph}}, \hat{p}_{\mathrm{ph}})^T$, and the generalization of Eq. (21) to this case is

$$\hat{H}_{\mathrm{int},\tau} = \hbar \left(\sum_{i=1}^{n} \kappa_{\tau,i} \hat{p}_{\mathrm{at},i} \right) \hat{p}_{\mathrm{ph}}, \tag{27}$$

where the summation index covers the different slices of atoms. With this Hamiltonian and the atomic decay and photon absorption loss mechanisms, the appropriate $\mathbf{S}_\tau$, $\mathbf{L}_\tau$, and $\mathbf{N}_\tau$ matrices are readily found, and the update formulae of Sec. 2 (or a slightly modified version thereof for the optically thick gas[20]) may be applied for the determination of the covariance matrix and the mean value vector for the Gaussian variables in $\boldsymbol{y}$. The result of this calculation is a $2n \times 2n$ atomic covariance matrix, with only minor squeezing in each slice, as the quantum correlations are distributed over the entire sample. One readily obtains the noise properties of the total atomic spin components, but it is more interesting to find the smallest eigenvalue of the covariance matrix, corresponding to a specific spatial mode of the atoms which is maximally squeezed. This mode, indeed, is the one that couples most efficiently to the radiation, and it is hence this smallest eigenvalue that determines the precision with which one can estimate, *e.g.*, the Larmor rotation rate of the collective spin.[20]

5. Magnetometry in the Gaussian Description

Precision atomic magnetometry relies on the measurement of the Larmor precession of a spin-polarized atomic sample in a magnetic field.[21–23] From standard counting statistics arguments, one might expect the uncertainty in such measurements to decrease with the interaction time t and with the number of atoms $N_{\rm at}$ as $1/\sqrt{N_{\rm at}t}$. If, on the other hand, the monitoring of the atomic sample, necessary for the read-out of the estimate of the magnetic field, squeezes the atomic spin, the above limit may be surpassed. In a theoretical analysis[24] it was suggested to estimate a scalar B field by a polarization rotation measurement of a far off-resonant light beam passing through a trapped cloud of spin-1/2 atoms. By quantum trajectory theory[25] combined with the classical theory of Kalman filters,[24,26] the uncertainty in the classical field strength was found[24] to decrease as $1/(N_{\rm at}t^{3/2})$. This proposal was implemented experimentally, and indeed sub-shot-noise sensitivity was found.[27] In our analysis of the experiment,[7,28] we advocated treating all variables, including the magnetic field, as quantum variables, and to assume a Gaussian probability distribution for the classical variable, so that the entire system can be described by the covariance matrix formulation.

In the case of a scalar field directed along the y direction, the effective Hamiltonian of the system is given by

$$\hat{H}_{{\rm int},\tau} = \hbar(\kappa_\tau \hat{p}_{\rm at}\hat{p}_{\rm ph} + \mu_\tau \hat{x}_{\rm at}\hat{B}), \tag{28}$$

where $\mu_\tau = (\tau/\hbar)\beta\sqrt{|\langle\hat{J}_x\rangle|/\hbar}$ is given by the magnetic moment β, and where the B field causes a Larmor rotation of the atomic spin towards the z axis. Figure 3 shows the setup. It is the coupling of the B field to the spin-squeezed variable $\hat{p}_{\rm at}$ that makes an improved precision measurement of the magnetic field possible.[29]

The vector of variables in the case of a scalar magnetic field is $\hat{\boldsymbol{y}} = (\hat{B}, \hat{x}_{\rm at}, \hat{p}_{\rm at}, \hat{x}_{\rm ph}, \hat{p}_{\rm ph})$, and with the Hamiltonian of Eq. (28), the S-matrix is found to be[7,28]

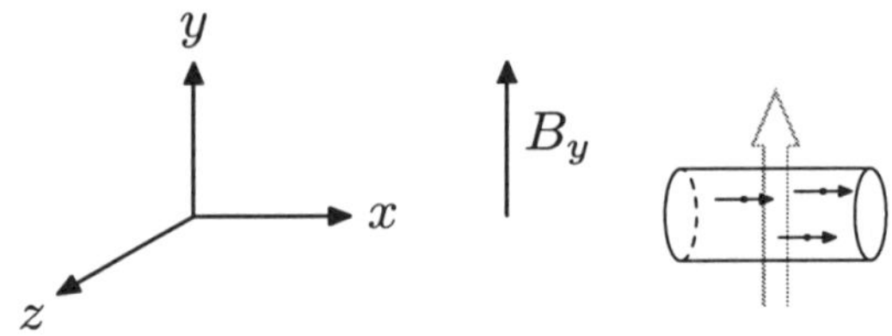

Fig. 3. Setup for measuring the y-coordinate of a magnetic field. This is done by measuring the Farady rotation of a linearly polarized optical beam propagating through the atomic gas.

$$\mathbf{S}_\tau = \begin{pmatrix} 1 & 0 & 0 & 0 & 0 \\ 0 & 1 & 0 & 0 & \kappa_\tau \\ -\mu_\tau & 0 & 1 & 0 & 0 \\ 0 & 0 & \kappa_\tau & 1 & 0 \\ 0 & 0 & 0 & 0 & 1 \end{pmatrix}. \tag{29}$$

As B_y only causes rotation perpendicular to its direction, the variable $\hat{x}_{\mathrm{at}} \propto \hat{J}_y$ does not couple to $(B_y, \hat{p}_{\mathrm{at}})$ and, hence, we only need to consider a 2×2 system with $\mathbf{y} = (B_y, \hat{p}_{\mathrm{at}})^T$. In the noise-less case, the system may now be propagated in time with the discrete update formula of Sec. 2. Alternatively we may consider the continuous limit and derive the differential equation for the covariance matrix $\mathbf{A}$ matrix of Eqs. (8)-(12) pertaining to $\mathbf{y} = (B_y, \hat{p}_{\mathrm{at}})^T$. The differential equation is on the matrix Ricatti form[8]

$$\dot{\mathbf{A}}(t) = \mathbf{G} - \mathbf{D}\mathbf{A}(t) - \mathbf{A}(t)\mathbf{E} - \mathbf{A}(t)\mathbf{F}\mathbf{A}(t), \tag{30}$$

with $\mathbf{G} = 0$, $\mathbf{D} = \left(\begin{smallmatrix} 0 & 0 \\ \mu & 0 \end{smallmatrix}\right)$, $\mathbf{E} = \mathbf{D}^T$, and $\mathbf{F} = \left(\begin{smallmatrix} 0 & 0 \\ 0 & \kappa^2 \end{smallmatrix}\right)$ where $\kappa^2 = \kappa_\tau^2/\tau$ and $\mu = \mu_\tau/\tau$. As may be checked by insertion, the solution to Eq. (30) is $\mathbf{A}_\gamma = \mathbf{W}\mathbf{U}^{-1}$, where $\dot{\mathbf{W}} = -\mathbf{D}\mathbf{W} + \mathbf{G}\mathbf{U}$ and $\dot{\mathbf{U}} = \mathbf{F}\mathbf{W} + \mathbf{E}\mathbf{U}$. The resulting solution for the variance of the B field reads:

$$\begin{aligned} \mathrm{Var}(\hat{B}(t)) &= \frac{\mathrm{Var}(\hat{B}_0)(\kappa^2 t + 1)}{\frac{1}{6}\kappa^4\mu^2\mathrm{Var}(\hat{B}_0)t^4 + \frac{2}{3}\kappa^2\mu^2\mathrm{Var}(\hat{B}_0)t^3 + \kappa^2 t + 1} \\ &\to_{t\to\infty} \frac{6}{\kappa^2\mu^2 t^3} \propto \frac{1}{N_{\mathrm{at}}^2 \Phi t^3}, \end{aligned} \tag{31}$$

The presence of noise[28] reduces the asymptotic decrease in the uncertainty with time from $1/t^3$ to $1/t$. Figure 4 shows the decrease in the uncertainty of the B field with time in a calculation with physically realizable parameters.

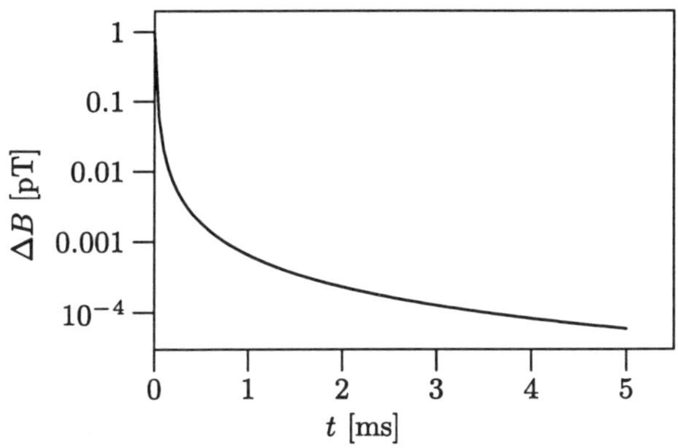

Fig. 4. Uncertainty of B field as a function of time. The value at $t = 5\,\text{ms}$ is $\Delta B_y = 5.814 \times 10^{-5}$ pT. We have chosen a segment duration $\tau = 10^{-8}$ s and corresponding field parameters $\kappa_\tau^2 = 0.0183$ and $\mu_\tau = 8.8 \times 10^{-4}$.

The new concept introduced in estimating the value of the classical B field is to treat the field itself as a quantum variable. Such an approach is not incompatible with the assumption that it is a classical parameter. We may imagine a canonically conjugate variable to B having an uncertainty much larger than required by Heisenberg's uncertainty relation and/or additional physical systems, entangled with the B-variable, in which cases the B-distribution is indeed incoherent and "classical". Also, one may argue that all classical variables are quantum mechanical variables for which a classical description suffices, and hence our theory provides the correct estimator according to the quantum theory of measurements: quantum mechanics dictates that the quantum state provides all the available knowledge about a system, and any estimator providing a tighter bound hence represents additional knowledge equivalent to a local hidden variable, and this is excluded by quantum theory. It is of course crucial that our measurement scheme corresponds to a quantum non-demolition (QND) measurement, *i.e.*, we assume that there is not a free evolution of the B-field induced by its conjugate variable which may thus remain unspecified. It is also this QND property of the measurement scheme that implies a monotonic reduction of the uncertainty of B which is consistent with the classical parameter estimation (we can not unlearn what we have already learnt about B), unlike, *e.g.*, the uncertainty of the atomic $\hat{x}_{\text{at}}$ variable which must increase when $\text{Var}(\hat{p}_{\text{at}})$ is reduced and when the atoms undergo spontaneous decay.

6. Entanglement in the Gaussian Description

The theoretical proposal[30,31] and the subsequent experimental demonstration[32] that by using only coherent light, entanglement can be generated between distant free space atomic ensembles, has attracted much attention in the quantum information community. The primary reason being that entanglement in such macroscopic ensembles of atoms is robust and easy to make because of the effective and tunable atom-light coupling. Theoretical analyses of entanglement with continuous variables has been performed in the Heisenberg picture,[30] and the state vector dynamics for a few tens of atoms has been considered by quantum trajectory methods.[33,34] Also the Gaussian description has been successfully applied to describe the entanglement generation between two macroscopic atomic ensembles due to continuous probing of collective spin variables by optical Faraday rotation.[5]

Generally, in the entanglement setup, the two gasses are polarized along opposite directions, say the positive and negative x axis. This means that the classical x components of the collective spin vectors are given by $J_{x,1} = \hbar N_{\text{at},1}/2 \equiv J_x$, and $J_{x,2} = -J_x$, and the Gaussian description is applicable with the following vector of canonical quantum variables $\hat{\boldsymbol{y}} = (\hat{x}_{\text{at},1}, \hat{p}_{\text{at},1}, \hat{x}_{\text{at},2}, \hat{p}_{\text{at},2}, \hat{x}_{\text{ph}}, \hat{p}_{\text{ph}})^T = (\frac{\hat{J}_{y,1}}{\sqrt{\hbar|J_x|}}, \frac{\hat{J}_{z,1}}{\sqrt{\hbar|J_x|}}, -\frac{\hat{J}_{y,2}}{\sqrt{\hbar|J_x|}}, \frac{\hat{J}_{z,2}}{\sqrt{\hbar|J_x|}}, \frac{\hat{S}_y}{\sqrt{\hbar|S_x|}}, \frac{\hat{S}_z}{\sqrt{\hbar|S_x|}})^T$. The Hamiltonian for either sample is given by Eq. (21). To model the light-atom interaction, the light beam is divided into segments as discussed in Sec. 2.4. The S-matrices $\mathbf{S}_{\tau,1}$ and $\mathbf{S}_{\tau,1}$ for the two gasses are readily found from Heisenberg's equation of motion for the variables in $\boldsymbol{y}$, and combined to $\mathbf{S}_\tau = \mathbf{S}_{\tau,1}\mathbf{S}_{\tau,2}$ for the full matrix. Additional evolution matrices may be defined that describe the rotation of the atomic variables of the samples and the effect of the homodyne detection.[5] The update of the system then proceeds as outlined in Sec. 2. The theory incorporates the interaction between the atoms and the optical field, atomic decay, and the measurement induced transformation of the atomic state. The reduction of the full quantum state description to a simple Gaussian state fully represented by a set of mean values and a covariance matrix makes the system straightforward to deal with numerically, and analytical results can be obtained in several important cases.

While the general problem of a measure for the entanglement between two mixed states remains unsolved, the entanglement between the atomic ensembles obtained by the continuous probing may be quantified by the Gaussian entanglement of formation[35] or the logarithmic negativity.[36] The

Gaussian description may also be used to identify the optimal performance of the entanglement scheme in the presence of atomic decay.[5]

6.1. *Entanglement and vector magnetometry*

The possibility to entangle more atomic gasses was also considered in magnetometry[28] in connection with the problem of measuring two or three components of the B field using entangled gasses. In the case of two components, say B_y and B_z, the atomic sample is split in two and one gas is polarized along x and the other along $-x$. These polarizations assure that the two observables $(\hat{J}_{y_1} + \hat{J}_{y_2})$ and $(\hat{J}_{z_1} + \hat{J}_{z_2})$, and equivalently $\hat{x}_{\mathrm{at}_1} - \hat{x}_{\mathrm{at}_2}$ and $\hat{p}_{\mathrm{at}_1} + \hat{p}_{\mathrm{at}_2}$ commute. (Note that a different sign convention for the $\hat{x}_{\mathrm{at}_2}$ variable was applied in our previous work.[28]) The interaction between the magnetic fields and the two samples is described by the following effective Hamiltonian

$$\hat{H}^m_{\mathrm{int},\tau} = \mu_\tau \hat{B}_y(\hat{x}_{\mathrm{at}_1} - \hat{x}_{\mathrm{at}_2}) + \mu_\tau \hat{B}_z(\hat{p}_{\mathrm{at}_1} + \hat{p}_{\mathrm{at}_2}). \tag{32}$$

This interaction causes changes in the atomic observables $(\hat{p}_{\mathrm{at}_1} - \hat{p}_{\mathrm{at}_2})$ and $(\hat{x}_{\mathrm{at}_1} + \hat{x}_{\mathrm{at}_2})$ proportional with B_y and B_z, respectively. To probe these changes we introduce the effective light-atom interaction

$$\hat{H}^l_{\mathrm{int},\tau} = \kappa_\tau(\hat{p}_{\mathrm{at}_1} - \hat{p}_{\mathrm{at}_2})\hat{p}_{\mathrm{ph}_1} + \kappa_\tau(\hat{x}_{\mathrm{at}_1} + \hat{x}_{\mathrm{at}_2})\hat{x}_{\mathrm{ph}_2}, \tag{33}$$

where the appropriate relative sign between the atomic variables of the two gasses can be implemented by adjusting the sign on κ_τ after the probe beams have passed through the first gas.[28] The gasses are probed by the simultaneous action of the Hamiltonian from Eqs. (32)-(33), $\hat{H}_{\mathrm{int},\tau} = \hat{H}^m_{\mathrm{int},\tau} + \hat{H}^l_{\mathrm{int},\tau}$. The vector of quantum variables is $\hat{\mathbf{y}} = (\hat{B}_z, \hat{B}_y, \hat{x}_{\mathrm{at}_1}, \hat{p}_{\mathrm{at}_1}, \hat{x}_{\mathrm{at}_2}, \hat{p}_{\mathrm{at}_2}, \hat{x}_{\mathrm{ph}_1}, \hat{p}_{\mathrm{ph}_1}, \hat{x}_{\mathrm{ph}_2}, \hat{p}_{\mathrm{ph}_2})^T$. With this state vector and the above Hamiltonian, the formalism of Sec. 2 can be directly applied and the final uncertainty of the B fields can indeed be lowered compared to the case with individual probe beams by letting the probe beams pass through both gasses and thereby entangling the two. An extension to full three dimensional vector magnetometry using three probing beams and six atomic samples can also be shown to have superior resolution in comparison with measurements on separable systems.[28]

7. Extensions of the Theory

In this section, we outline some topics which are subject to studies within the Gaussian description at the time of writing, and we discuss how to go beyond the Gaussian approximation.

7.1. *Non spin–1/2 systems*

The theory presented in Sec. 3 explicitly used the representation of the collective angular momentum variable in terms of Pauli spin matrices $\hat{\boldsymbol{J}} = \frac{\hbar}{2}\sum_j \hat{\boldsymbol{\sigma}}$. This representation was crucial for the reduction of the Hamiltonian of Eq. (19) to the form of Eq. (21) which is expressed directly in terms of the canonical $(\hat{x}, \hat{p})$ Gaussian variables. For the more general problem of excited ($|F'M'\rangle$) and ground ($|FM\rangle$) states with $F > 1/2$ and $F' > 1/2$, it is still possible to obtain an approximate solution within the Gaussian description. This more complicated problem is of both fundamental interest and of practical importance since ongoing experiments based on the Faraday-rotation scheme are carried out with such real multilevel atoms.[37,38]

Equation (17) represents the interaction with atoms with arbitrary level structure, and to deal with such atomic samples within a Gaussian description, we suggest to introduce a second quantized formalism for the atoms in which bosonic atomic field operators $\hat{\Psi}^\dagger_M$, $\hat{\Psi}_{M'}$ create and destroy atoms with the given magnetic quantum number. Note that the bosonic character merely reflects the symmetry under permutations of the atoms: The theory works for both fermionic and bosonic atoms. We can then write the collective atomic operators in the Hamiltonian in terms of the atomic field operators, $\sum_j |FM\rangle_j\langle FM'| = \hat{\Psi}^\dagger_M \hat{\Psi}_{M'}$ to obtain

$$\hat{H}_{\text{int},\tau} = \sum_{M=-F}^{F} [\left(c_{+,M}(\Delta)\hat{a}^\dagger_+\hat{a}_+ + c_{-,M}(\Delta)\hat{a}^\dagger_-\hat{a}_-\right)\hat{\Psi}^\dagger_M\hat{\Psi}_M \\ +b_M(\Delta)\left(\hat{a}^\dagger_-\hat{a}_+\hat{\Psi}^\dagger_{M+1}\hat{\Psi}_{M-1} + \hat{a}^\dagger_+\hat{a}_-\hat{\Psi}^\dagger_{M-1}\hat{\Psi}_{M+1}\right)]. \tag{34}$$

At this point we make a mean field approximation, and we expand the field operators for the light fields

$$\hat{a}_\pm \to \alpha_\pm + \delta\hat{a}_\pm, \tag{35}$$

and the atom fields

$$\hat{\Psi}_M \to \Phi_M + \delta\hat{\Psi}_M, \tag{36}$$

with *c*-numbers $\alpha_\pm$ and Φ_M, and "small" operators $\delta\hat{a}_\pm$ and $\delta\hat{\Psi}_M$. We insert Eqs. (35)-(36) into Eq. (34) and expand to second order in the operator terms. This procedure leads to a rather lengthy expression which is conveniently split into terms which are of zeroth, first and second order in the quantum fields. The classical fields are explicitly time-dependent and their dynamics is given by replacing all operators by their c-number parts in Heisenberg's equations of motion. Since we neglect operator terms above second order, the quantum part of the Hamiltonian is at most bilinear (with classical time-dependent coefficients). The linear terms cause mean drifts of the mean value of the operator components, which we can absorb in the *c*-number components. For the new quantum operators, we may then maintain $\langle \delta\hat{a}_\pm \rangle = 0$ and $\langle \delta\hat{\Psi}_M \rangle = 0$. In the resulting bilinear Hamiltonian involving the operator terms, we now make the transition to the Gaussian state description by forming a vector of variables $\hat{\boldsymbol{y}} = (\hat{x}_{-F}, \hat{p}_{-F}, \ldots, \hat{x}_F, \hat{p}_F, \hat{x}_+, \hat{p}_+, \hat{x}_-, \hat{p}_-)^T$, with $x_\pm = \sqrt{\frac{\hbar}{2}}(\delta\hat{a}_\pm + \delta\hat{a}_\pm^\dagger)$, $p_\pm = -i\sqrt{\frac{\hbar}{2}}(\delta\hat{a}_\pm - \delta\hat{a}_\pm^\dagger)$, $x_M = \sqrt{\frac{\hbar}{2}}(\delta\hat{\Psi}_M + \delta\hat{\Psi}_M^\dagger)$, $x_M = -i\sqrt{\frac{\hbar}{2}}(\delta\hat{\Psi}_M - \delta\hat{\Psi}_M^\dagger)$, and adopting the formalism of Sec. 2.

We note that the expansion of noise terms around classical mean values has been used as a standard tool in quantum optics, *e.g.*, to deal with the optical Kerr-effect, and if only unitary dynamics and losses are considered, the present approach does not offer any new insights. It is important to remember, however, that we are also able to treat the dynamics conditioned on measurements on the system.

7.2. *Quantum correlated light beams*

So far, we have treated the case of a coherent, monochromatic beam of light incident on the atomic samples. For high precision probing, atomic spin squeezing and entanglement, it has been proposed to use squeezed beams of light and twin beams, and this is indeed also possible within the Gaussian formalism. To model in a simple manner the coupling to squeezed light beams, one may simply alter the covariance matrix elements for the field operators in each beam segment prior to the interaction with the atomic sample, so that rather than the 2×2 identity matrix with equal variances of the two field quadratures in Eq. (11), we assume the form

$$\mathbf{B} = \mathrm{diag}(1/r, r), \tag{37}$$

where r is the squeezing parameter. Carrying out the calculations as described in the previous sections, we observe,[7,28] that the magnetometer resolution is improved by this parameter.

As pointed out in our analysis,[7] however, a squeezed beam of light carries correlations between the the field operators evaluated at different times. This implies, that if one observes the beam for only a very short time, one will not be able to detect the squeezing. It is well-known from the standard quantum optical analysis of the optical parametric oscillator

(OPO) (in the Heisenberg picture in the frequency domain) that there is a certain frequency band-width, Γ, of squeezing associated with the field decay rate of the cavity. Only the accumulated signal over sufficiently long times ($t \gtrsim 1/\Gamma$) will be able to extract the frequency components for which the field is squeezed. One might think, that one should therefore consider a longer string of light segments with appropriate covariance matrix elements, and carry out the update on all segments (and the atoms) every time a single optical segment is detected. In fact, there is an easier approach. The squeezed beam is produced by continuous leakage of the field inside the OPO cavity, and the temporal correlations are due to this joint source of the radiation. The most economical way to describe the interaction of atoms with a squeezed beam is therefore to incorporate the single-mode field inside the cavity in the Gaussian state formalism, and to consider again only one optical beam segment at a time, from its creation out of the cavity, interaction with the atoms, and final detection, and hence causing an update of the joint atom and cavity covariance matrix.

We have implemented such a model[39] and verified that it reproduces the known noise properties for the signal integrated over both short and long times. In addition, we have applied the model to magnetometry, and observed that for segments shorter than the inverse band-width of squeezed light, the resolution is not improved with the use of squeezed light, but after many segments and a long total interaction time, the results asymptotically approach the factor $1/r$ improvement of the simple model as one might have expected.[39]

In closing this section, we note that the Gaussian description is not restricted to the examples and extensions discussed above. Extra physical systems can be included straightforwardly by adding appropriate rows and columns to the covariance matrix. In this way, one may, *e.g.*, describe the effects of imperfect detectors and filters, and loss in optical fibres.

7.3. *Beyond the Gaussian approximation*

Looking back on the development of the theory in this work, we note that the interaction and the dissipation can be treated exactly without recourse to a Gaussian ansatz for the quantum state. In fact, Eq. (5) is a general update formula for the Wigner function under an arbitrary measurement, but in the general case this expression may be difficult to evaluate, and in particular to use as input in the next step of the continuous probing on the system. In a recent experiment,[40] a Gaussian squeezed state was mixed with the vacuum field at a beam splitter to produce an entangled two-mode field state part of which was sent to an avalanche photodiode and part of which was monitored by homodyne detection. The state of the second component conditioned on a photodiode counting event is a non-Gaussian state, as verified by a double-peaked homodyne detection signal. The process was modeled[41] by assuming that prior to the registration of a single photon, the state of the field is described by the Gaussian Wigner function $\mathcal{W}_{\text{Gauss}}(\gamma, \delta)$ with γ and δ denoting pairs of real variables of the two modes. The detection of a single photon, $|1\rangle$, corresponds to application of the Wigner function $\mathcal{W}_{meas}(\delta) = \mathcal{W}_{|1\rangle\langle 1|}(\delta) = \frac{2}{\pi}\exp(-2|\delta|^2)(-1+4|\delta|^2)$ in Eq. (5), and the Wigner function for the second beam conditioned on this state, $\mathcal{W}_c(\gamma) = \int \mathcal{W}(\gamma, \delta)\, \mathcal{W}_{|1\rangle\langle 1|}(\delta)\, d^2\delta$ is readily shown not to be a Gaussian. Another example where one 'jumps' out of the Gaussian states is in a recent proposal[42] where a setup of two beam splitters with carefully chosen properties and photodetectors allows one to produce with high fidelity a single-photon state from Gaussian squeezed vacuum input beams. We believe that a combination of the theory of Gaussian state updates and inclusion via Eq. (5) of one or a few non-Gaussian preserving measurements may be a useful approach to these problems.

As it is necessary to leave the Gaussian states to perform some quantum information tasks such as distillation of entangled states,[2,4,43] it is in general important to have tools to handle the interface between Gaussian and non-Gaussian states.

8. Conclusions and Outlook

We have presented a general formalism to treat the dynamics of $2n$ canonical variables within a Gaussian description. For a probing light beam, the concept of "segment quantization" where the beam is quantized in small fragments of duration τ and length $L = c\tau$ allows an efficient description of not only the evolution of the system subject to the Hamiltonian, but also to the measurement process through simple update formulae for the mean value vector and the covariance matrix, which fully characterize the Gaussian state. Of particular current interest in the field of quantum information and quantum communication,[10,30,32] and in precision magnetometry,[24] is the off-resonant probing of ensembles of atoms leading to a dispersive Faraday effect. This interaction is bilinear in the effective canonical variables of the system, and a description within the Gaussian framework of, *e.g.* spin squeezing,[20] magnetometry,[7,28] and entanglement[5] is straightforward.

In Sec. 7, we have outlined some possible extensions to the Gaussian description. In the future it will be interesting to develop further theory for continuous variable system which leave the Gaussian description, either because of the interactions involved, because of the measurement schemes, or because of coupling of a small discrete system to the collective continuous degrees of freedom (examples: single photons can be stored and emitted on demand by macroscopic atomic samples, trapped ions can be entangled by continuous probing with classical laser fields). Such approaches hold the potential to form Schrödinger Cat states, which may have favorable properties in high precision detection, and they may be used to implement distillation and purification protocols from quantum information theory, which are known not to work for Gaussian states and operations. The theoretical task is to identify processes that can be implemented experimentally and which break the Gaussian character, and to establish a theoretical description of the resulting states, which will invariably be much more complicated to deal with than the Gaussian states.

Acknowledgments

We thank Vivi Petersen, Jacob Sherson and Uffe V. Poulsen for useful discussions. L.B.M. was supported by the Danish Natural Science Research Council (Grant No. 21-03-0163).

References

1. D. F. Walls and G. J. Milburn, *Quantum optics*, Springer-Verlag, Berlin 1994.
2. J. Eisert and M. B. Plenio, Int. J. Quant. Inf. **1**, 479 (2003).
3. G. Giedke and J. I. Cirac, Phys. Rev. A **66**, 032316 (2002).
4. J. Fiurášek, Phys. Rev. Lett. **89**, 137904 (2002).
5. J. Sherson and K. Mølmer, Phys. Rev. A **71**, 033813 (2005).
6. P. S. Maybeck, *Stochastic Models, Estimation and Control*, Vol. 1, Academic Press, New York 1979.
7. K. Mølmer and L. B. Madsen, Phys. Rev. A **70**, 052102 (2004).
8. J. K. Stockton, J. M. Geremia, A. C. Doherty, and H. Mabuchi, Phys. Rev. A **69**, 032109 (2004).
9. E. Merzbacher, *Quantum Mechanics*, 3rd ed., Wiley, New York 1998.
10. A. Kuzmich, N. P. Bigelow, and L. Mandel, Europhys. Lett. **42**, 481 (1998).
11. Y. Takahashi *et al.*, Phys. Rev. A **60**, 4974 (1999).
12. B. Julsgaard, C. Schori, J. L. Sørensen, and E. Polzik, Quantum Information and Computation **3**, 518 (2003).
13. J. H. Müller, P. G. Petrov, D. Oblak, C. L. Garrido Alzar, S. R. Echaniz, and E. S. Polzik, Phys. Rev. A **71**, 033803 (2005).

14. L. K. Thomsen, S. Mancini, and H. M. Wiseman, J. Phys. B: At. Mol. Opt. Phys. **35**, 4937 (2002).
15. J. M. Geremia, J. K. Stockton ,and H. Mabuchi, Science **304**, 270 (2004).
16. I. Bouchoule and K. Mølmer, Phys. Rev. A **66**, 043811 (2002).
17. A. Kuzmich and T. A. B. Kennedy, Phys. Rev. Lett. **92**, 030407 (2004).
18. B. Kraus, K. Hammerer, G. Giedke, and J. I. Cirac, Phys. Rev. A **67**, 042314 (2003).
19. K. Hammerer, K. Mølmer, E. S. Polzik, and J. I. Cirac, Phys. Rev. A **70**, 044304 (2004).
20. L. B. Madsen and K. Mølmer, Phys. Rev. A **70**, 052324 (2004).
21. D. Budker *et al.*, Rev. Mod. Phys. **74**, 1153 (2002).
22. I. K. Kominis, T. W. Kornack, J. C. Allred, and M. V. Romalis, Nature (London) **422**, 596 (2003).
23. M. Auzinsh *et al.*, Phys. Rev. Lett. **93**, 173002 (2004).
24. J. M. Geramia, J. K. Stockton, A. C. Doherty, and H. Mabuchi, Phys. Rev. Lett. **91**, 250801 (2003).
25. H. Carmichael, *An Open Systems Approach to Quantum Optics*, Springer-Verlag, Berlin, Heidelberg, 1993.
26. J. K. Stockton, J. M. Geremia, A. C. Doherty, and H. Mabuchi, Phys. Rev. A **69**, 032109 (2004).
27. J. M. Geremia, J. K. Stockton, and H. Mabuchi, Phys. Rev. Lett. **94**, 203002 (2005).
28. V. Petersen, L. B. Madsen, and K. Mølmer, Phys. Rev A **71**, 012312 (2005).
29. J. M. Geremia, J. K. Stockton, A. C. Doherty, and H. Mabuchi, Phys. Rev. Lett. **91**, 250801 (2003).
30. L. M. Duan, J. I. Cirac, P. Zoller, and E. S. Polzik, Phys. Rev. Lett. **85**, 5643 (2000).
31. A. Kuzmich, L. Mandel, and N. P. Bigelow, Phys. Rev. Lett. **85**, 1594 (2000).
32. B. Julsgaard, A. Kozhekin, and E. S. Polzik, Nature (London) **413**, 400 (2001).
33. A. D. Lisi and K. Mølmer, Phys. Rev. A **66**, 052303 (2002).
34. A. D. Lisi, S. D. Siena, and F. Illuminati, Phys. Rev. A **70**, 012301 (2004).
35. G. Giedke, M. M. Wolf, O. Krüger, R. F. Werner, and J. I. Cirac, Phys. Rev. Lett. **91**, 107901 (2003).
36. K. Audenaert, J. Eisert, M. B. Plenio, and R. F. Werner, Phys. Rev. A **66**, 042327 (2002).
37. G. A. Smith, S. Chaudhury, A. Silberfarb, I. H. Deutsch, and P. S. Jessen, Phys. Rev. Lett. **03**, 163602 (2004).
38. D. Kupriyanov, O. Mishina, I. Sokolov, B. Julsgaard, and E.S.Polzik, Phys. Rev. A **71**, 032348 (2005).
39. V. Petersen, L. B. Madsen, and K. Mølmer, Phys. Rev. A **72**, 053812 (2005).
40. J. Wenger, R. Tualle-Brouri, and P. Grangier, Phys. Rev. Lett. **92**, 153601 (2004).
41. J. Zieliński and K. Wódkiewicz, private communication.

42. C. R. Myers, M. Ericsson, and R. Laflamme, e-print quant-ph/0408194 (2004).
43. D. E. Browne, J. Eisert, S. Scheel, and M. B. Plenio, Phys. Rev. A **67**, 062320 (2003).

Chapter 23

Quantum State Preparation of Spin Ensembles by Continuous Measurement and Feedback

Ramon van Handel, John K. Stockton and Hideo Mabuchi
Department of Physics and Control & Dynamical Systems
California Institute of Technology 266-33, Pasadena, CA 91125, USA

Howard M. Wiseman
Centre for Quantum Computer Technology and Centre for Quantum Dynamics School of Science, Griffith University, Brisbane, QLD 4111, Australia

Beginning with a microscopic model for the interaction between atoms and the electromagnetic field, we describe how stochastic equations for the state of the atomic ensemble conditioned upon homodyne measurement of the field can be derived. We investigate the evolution of the stochastic state both in the short time scale, where a description in terms of continuous variables (spin quadratures) is possible, and in the longer time scale where this description breaks down. Next we discuss general principles of quantum feedback control, and apply them to quantum state preparation in the atomic ensemble. We consider three cases: producing spin-squeezed states, preparing Dicke states (the long-time limit), and producing twin-ensemble squeezing. Where appropriate, we emphasize relations with experiments.

1. Introduction

State preparation is a ubiquitous concept in physics. In most experiments, the system state is initialized in a way that maximizes the utility of the subsequent procedure. Cooling, trapping, pumping, polarizing, etc. are all descriptors of state preparation procedures that reduce the entropy of a system for some benefit in performance. Increasingly, physicists are becoming able to prepare systems in uniquely quantum states. Much attention has been placed on controlling the state of single quantum objects, and incredible progress has been made in, for example, the cooling and trapping of individual atoms and ions. Yet, a great deal of attention has also been directed towards the experimental creation of entangled states in multipartite systems.[1,3–5,2] This interest has been hastened by the realization that the correlations implied by entanglement can be leveraged to enhance practical applications ranging from quantum communication[6] to precision metrology.[1,7,9,11,10]

The process of creating a particular entangled state is often less intuitive than the preparation of separable states, where only a single particle picture of the process is typically used. As a consequence, the problem of entangled state preparation has become highlighted in contrast to the often trivial separable counterpart. There are many different ways to create entanglement, including direct interaction between the constituent particles,[4,5,12,13] but in this chapter we shall focus on the creation of entanglement in spin ensembles through measurement and feedback. By driving a system with a probe field, and destructively measuring that probe, entangled states of the system can be created conditioned on the measurement record. The type of entanglement produced then reflects the degree of distinguishability of the system components during the measurement. In addition, by using intelligent feedback control to map the measurement results onto a Hamiltonian control parameter, one can potentially reduce the randomness of the final state while retaining the desired correlations induced by the conditioning.[14]

These principles have been demonstrated in recent years in several experiments using atomic ensembles, *i.e.* a large number of atomic spins, as the quantum object of interest. In one class of experiments, two spatially separated ensembles have been entangled through measurement,[1,15] creating states with applications in quantum communication protocols.[6] In another class, a single atomic spin ensemble has been probed to create entanglement within the ensemble in the form of a conditional spin-squeezed state (SSS)[3,16,2] which has applications in a variety of metrology tasks, including magnetometers[1,10] and atomic clocks.[7,9,11]

In this chapter, we model the conditional generation of entanglement within a single atomic ensemble via the continuous measurement of an optical probe interacting dispersively with the ensemble.[3,16,17] Within any given atomic ensemble, there are so many atoms that evolving the full (exponentially large) system state in time is a computational impossibility for the observer. Not only are approximate descriptions needed for economy of thought, but they are also needed by any feedback control law that aims to effectively manipulate the ensemble state in real-time. Fortunately, in the experiments we discuss, each ensemble can be aptly described by a small set of quantum continuous variables (QCV) and linear dynamics[1,18] at small times. More specifically, the evolution of the full system state can be reduced to only the expected means and variances of the collective angular momentum operators describing the ensemble. Furthermore, it is possible to connect the values of these moments to certain measures of entanglement using, for example, the definition of the spin-squeezing parameter.[13] However, like all reduced descriptions, the QCV approach is an approximation that will eventually fail, requiring a more complete description.

The body of this chapter is divided into four sections. Section 2 describes the derivation from first principles, using a simplified field-theoretical model, of stochastic dynamical equations for the atomic ensemble and the probe field. In Sec. 3 we describe how to condition the ensemble state upon the results from continuous optical measurements in the field. Section 4 presents general principles of feedback control. In Sec. 5 we demonstrates how these principles can be applied to enable deterministic quantum state preparation in atomic ensembles. This procedure is discussed in both the short time limit, where QCV approach is valid, and in the long time limit, where a more complete description is required.[19] As we proceed, we attempt to review the literature concerning measurement and feedback control of atomic ensembles, while also putting into context related, but more mathematical, works concerning estimation and control.

2. The Physical Model: From QED to Stochastic Equations

In this section we will describe a microscopic model for the class of systems we consider. The model consists of an atomic ensemble coupled weakly to an external electromagnetic field which is ultimately detected.

2.1. *System model from quantum electrodynamics*

It is well known from quantum electrodynamics[21,20] that the observable for the free electric field is given by

$$\mathbf{E}(\mathbf{r},t) = \sqrt{\frac{\hbar}{(2\pi)^3\varepsilon_0}} \sum_s \int \sqrt{\frac{\omega}{2}} \left(i a_{\mathbf{k},s} \boldsymbol{\varepsilon}_{\mathbf{k},s} e^{i(\mathbf{k}\cdot\mathbf{r}-\omega t)} + \text{h.c.}\right) d^3k, \qquad (1)$$

where $\omega = c|\mathbf{k}|$, $\boldsymbol{\varepsilon}_{\mathbf{k},s}$ are polarization vectors and $a_{\mathbf{k},s}$ are plane wave (Fourier) mode annihilation operators that satisfy the commutation relations $[a_{\mathbf{k},s}, a^\dagger_{\mathbf{k}',s'}] = \delta^3(\mathbf{k}-\mathbf{k}')\delta_{ss'}$. We will assume that the atomic ensemble (centered at the origin) interacts with the field predominantly through its

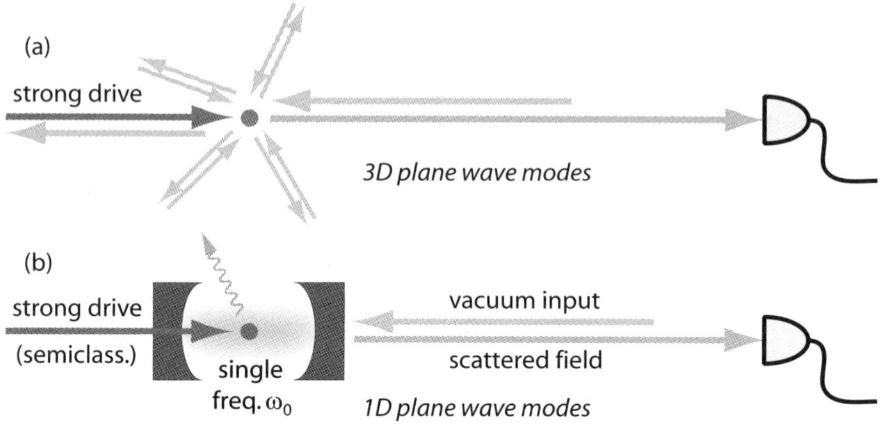

Fig. 1. (a) Schematic of the full interaction Hamiltonian, where all plane wave modes in three dimensions interact with an atomic ensemble. One of the incoming modes is coherently excited with frequency ω_0 and large amplitude; the coherent excitation scatters mainly in the forward direction. The remaining incoming modes are in the vacuum state and drive spontaneous emission. (b) Simplified one-dimensional model, where plane waves are scattered off a single-mode cavity in one direction only. The strong driving field is treated semiclassically and coherently excites the cavity mode, which has frequency ω_0. Spontaneous emission can be added to the model in a phenomenological manner.

collective dipole moment; *i.e.*, the interaction Hamiltonian will be of the form $H_I(t) = -\mathbf{d}(t) \cdot \mathbf{E}(0,t)$ where $\mathbf{d}(t)$ is the ensemble dipole operator. In practice there will be some ultraviolet cutoff, which we can obtain *e.g.* by averaging the electric field over the volume of the cloud of atoms instead of evaluating it at the origin. We will write

$$\mathbf{E}^{(+)}(\mathbf{r},t) = [\mathbf{E}^{(-)}(\mathbf{r},t)]^\dagger = \sum_s \int g(\mathbf{k})\, a_{\mathbf{k},s} \boldsymbol{\varepsilon}_{\mathbf{k},s} e^{i(\mathbf{k}\cdot\mathbf{r}-\omega t)}\, d^3k, \tag{2}$$

where $\mathbf{E} = i(\mathbf{E}^{(+)} - \mathbf{E}^{(-)})$ and $g(\mathbf{k})$ is the mode function, *e.g.* $g(\mathbf{k}) \propto \sqrt{\omega}\, e^{-d\omega^2}$ if we average $\mathbf{E}$ over a spatial Gaussian distribution.

The full interaction is sketched in cartoon form in Fig. 1a. The atomic ensemble interacts through its dipole moment with all plane wave modes in three dimensions. A strong, focused laser beam at frequency ω_0 is modeled by bringing the corresponding modes into a large-amplitude coherent state. The drive is scattered predominantly in the forward direction, and is ultimately detected. The remaining modes are in the vacuum state and drive spontaneous emission of the ensemble in all directions. This essentially complete description of the interaction embodies all the physics of the problem, and thus allows one to predict quantities such as the spontaneous emission rate. The full picture is also very complicated, however, as it requires a detailed analysis of the atomic structure, a partitioning of the field into observed and unobserved modes, etc. Such an analysis is beyond the scope of this chapter and will be presented elsewhere.

Instead, we will investigate a highly simplified model (Fig. 1b) that is widely used in quantum optics.[22–24] To justify such a model, we claim that most of the interesting physics occurs in the direction of the driving laser, as most of the light is scattered forward and observed in this direction. Hence we can approximate the system by a one-dimensional model where only the forward modes are treated exactly and the strong drive is treated semiclassically. Spontaneous emission into the eliminated modes is unobserved, and hence we could include it phenomenologically by adding decoherence. Finally, to simplify the interaction with the ensemble, we place the ensemble into a leaky single mode cavity. This allows us to treat the interaction between the ensemble and the field for a single frequency only, that of the cavity mode, which is chosen to be at the laser driving frequency ω_0. The cavity dynamics is then adiabatically eliminated to give an effective interaction between the ensemble and the external field.

Let us systematically work out this simplified model. We begin by treating the one-dimensional external field that is ultimately detected. We can obtain an expression for the field by integrating Eq. (1) over a transverse area,[24] or alternatively by directly quantizing the wave equation in one dimension.[22] We obtain

$$E(z,t) = \sqrt{\frac{\hbar}{2\pi\varepsilon_0 c}} \int_0^\infty \sqrt{\frac{\omega}{2}}\,(ia_\omega e^{-i\omega(t-z/c)} + \text{h.c.})\,d\omega, \tag{3}$$

for the electric field intensity in a single polarization state (we will assume polarized light), where $[a_\omega, a^\dagger_{\omega'}] = \delta(\omega-\omega')$. The annihilators a_ω correspond to plane wave modes in the z-direction, $\mathbf{k} = (\omega/c)\hat{\mathbf{z}}$, where positive z is defined to be on the left in Fig. 1. Thus the field for $z < 0$ is propagating towards the cavity mirror in time, whereas $z > 0$ propagates away from the mirror. In practice the cavity mirror will reverse the propagation direction, so we can reinterpret $z > 0$ as the component of the field propagating toward the detector, whereas $z < 0$ is the incident part of the field.

We now introduce the cavity mode with annihilation operator $b(t) = be^{-i\omega_0 t}$. The interaction Hamiltonian between the cavity mode and the external field is given by

$$H_{CF} = i\hbar \int_0^\infty \kappa(\omega)(a_\omega^\dagger b e^{i(\omega-\omega_0)t} - b^\dagger a_\omega e^{-i(\omega-\omega_0)t}) d\omega$$
$$= i\hbar(E^{(-)}(0,t)b(t) - b^\dagger(t)E^{(+)}(0,t)) \quad (4)$$

in the rotating wave approximation, where we have used

$$E^{(+)}(z,t) = [E^{(-)}(z,t)]^\dagger = \int_0^\infty \kappa(\omega) a_\omega e^{-i\omega(t-z/c)} d\omega. \quad (5)$$

Here $\kappa(\omega)$ does not only depend on the external field but also on the frequency-dependent transmission of the cavity mirror, and is unitless. The rotating wave approximation will be good when $\kappa(\omega)$ is very small near the resonant frequency $\omega \approx \omega_0$, *i.e.* in the weak coupling limit.

It remains to describe the driving field H_D, the atomic Hamiltonian H_A, the ensemble-cavity mode interaction H_{AC} and the spontaneous emission $H_\perp$. The interaction Hamiltonian of the resonant drive is given by

$$H_D = \hbar\mathcal{E}(b + b^\dagger), \quad (6)$$

where $\mathcal{E}$ is the drive amplitude. The remaining Hamiltonians are more variable, as they depend on the structure of the atoms in the ensemble. In particular, we get drastically different behavior when the atoms have a transition that is resonant with the cavity mode than in the far detuned case. We will consider a specific example in Sec. 2.2.

We can consider Eq. (5) as the Fourier transform of the operator distribution $\kappa(\omega)\theta(\omega)a_\omega$, where θ is the step function. We will always take the incoming field to be in the vacuum state as in Fig. 1. Each a_ω can be thought of as an independent quantum "complex Gaussian" random variable, in the sense that its "real" and "imaginary" parts $a_\omega + a_\omega^\dagger$ and $ia_\omega^\dagger - ia_\omega$ are precisely Gaussian random variables. Hence $E^{(+)}(0,t)$ will be some sort of quantum complex Gaussian noise. Note that the two quadratures $E^{(+)} + E^{(-)}$ and $iE^{(-)} - iE^{(+)}$ do not commute, so we cannot interpret $E^{(+)}(0,t)$ as a classical complex noise. Nonetheless we would like to think of the interaction picture Heisenberg equation

$$\dot{X}_t = \frac{i}{\hbar}[H_\perp + H_A + H_{AC} + H_D + H_{CF}(t), X_t] \quad (7)$$

as being driven by the noises $E^{(+)}$, $E^{(-)}$; then the statistical inference step can be formulated as finding the best estimate of the noisy time evolution of atomic observables given noisy observations of the field.

Similar problems have been studied in classical probability for about a century, and the main lessons learned there appear to carry over to the quantum case. In particular, statistical inference of continuous-time processes is essentially intractable unless we approximate the noise process by a *white noise.* In this case, the time evolution of the system is Markovian[25] (*i.e.* the distribution of future system states depends only on the present state and not on past history) and statistical inference is described by the elegant theory of Markov nonlinear filtering. In the quantum case, this results in quantum Langevin equations, or quantum stochastic differential equations. A rigorous treatment of quantum Itô theory was introduced by Hudson and Parthasarathy;[26] a heuristic treatment can be found in Gardiner and Collett.[27]

2.2. *Example: spins with dispersive coupling*

After making a Markovian approximation to the atom-probe interaction, the slow dynamics can be added back in simply by adding the corresponding Hamiltonians. We use a highly simplified model of an atomic ensemble interacting with an electromagnetic field.[14,28] Consider an atomic ensemble consisting of a set of N atoms with a degenerate two-level ground state. We will assume that all atomic transitions are far detuned from the cavity resonance, so the interaction between the atoms and the cavity is well described by the dispersive Hamiltonian $H_{AC} = \hbar\chi F_z b^\dagger b$ where F_z is the collective dipole moment of the ensemble, *i.e.* it is a spin-$N/2$ angular momentum operator, and χ determines the coupling strength. Such a Hamiltonian can be obtained, for example, by considering the full dipole coupling and then adiabatically eliminating all the excited states.

We will furthermore neglect spontaneous emission by setting $\gamma_\perp = 0$, and add a magnetic field driving term $H_A = \hbar\, h(t) F_y$ where $h(t)$ is the magnetic field strength. This term will allow us to apply feedback to the system by varying the magnetic field. The final result is that the solution to the Heisenberg equation (7) can be approximated by

$$j_t(X) \equiv U_t^\dagger X U_t, \tag{8}$$

where

$$\begin{aligned} dU_t = \Big[&\sqrt{\gamma'}\, b\, dA_t^\dagger - \sqrt{\gamma'}\, b^\dagger\, dA_t \\ &- \tfrac{1}{2}\gamma' b^\dagger b\, dt - i(h(t)F_y + \chi F_z b^\dagger b + \mathcal{E}(b + b^\dagger))\, dt\Big]\, U_t. \end{aligned} \tag{9}$$

We now adiabatically eliminate the cavity,[29–31] assuming that γ' and $\mathcal{E}$ are sufficiently large so this is a good approximation. This yields

$$dU_t = \Big[L\,dA_t^\dagger - L^\dagger\,dA_t - \tfrac{1}{2}MF_z^2\,dt - ih(t)F_y\,dt\Big]\,U_t, \tag{10}$$

where $L = \sqrt{M}F_z + 2i\mathcal{E}/\sqrt{\gamma'}$ and $M = 16\chi^2\mathcal{E}^2/(\gamma')^3$ is the effective interaction strength. We will use this highly simplified model as an example throughout this chapter.

3. Conditioning

In the previous section we considered in detail the physical interactions between an atomic ensemble and the electromagnetic field, which, after many simplifications, were condensed into Eq. (10). This expression contains all the physical dynamics of our model. We now start the second step in our program, in which we perform statistical inference of the atomic dynamics based on an observation of the field.

3.1. *Optical detection*

Before we can derive a filtering equation we must specify what measurement is performed. We will consider the case of (balanced) homodyne detection, which measures a quadrature of the outgoing field. The principles of this method are discussed in many textbooks[32,33] and a continuous time description in terms of quantum stochastic calculus can be found in Ref. 34. Homodyne detection has the advantage that it gives rise to a continuous, Wiener process-type integrated photocurrent, which is particularly convenient for continuous time feedback control.

Heuristically, consider Eq. (10) as being driven by the white noise a_t, the "derivative" of A_t. An ideal wide-band homodyne detector will measure the field observable $a_t + a_t^\dagger$ after the field has interacted with the ensemble; *i.e.*, we observe the photocurrent $I(t) = j_t(a_t + a_t^\dagger)$. As usual mathematically rigorous results are much more easily obtained in integrated form; hence we define as our observation the *integrated photocurrent*

$$Y_t = j_t(A_t + A_t^\dagger), \tag{11}$$

where $I(t)$ can be considered the "derivative" of Y_t. For a rigorous treatment directly from the quantum stochastic description we refer to Ref. 34.

Finding an explicit expression for Y_t is a straightforward exercise in the use of the quantum Itô rules. From Eqs. (10) and (11) we directly obtain

$$dY_t = 2\sqrt{M}\,j_t(F_z)\,dt + dA_t + dA_t^\dagger. \tag{12}$$

Thus clearly homodyne detection of the field provides a measurement of the system observable $F_z(t)$ corrupted by the incident field noise.

We will extend our observation model a little further. We assumed in the above analysis that the detection efficiency is perfect. In practice there will always be some technical noise added to the signal, either due to the intrinsic loss mechanisms in the photodetectors or due to noise in the detection electronics (*e.g.* amplifier noise). We will model these effects by the addition of an uncorrelated white noise term dW'_t to the observation current; *i.e.*

$$dY_t = 2\sqrt{M\eta}\, j_t(F_z)\, dt + \sqrt{\eta}\,(dA_t + dA_t^\dagger) + \sqrt{1-\eta}\, dW'_t, \qquad (13)$$

where $\eta \in (0,1]$ determines the relative strength of the technical noise ($\eta = 1$ is perfect detection.) Note that we have rescaled the current $I(t)$ so that the total corrupting noise has unit variance, *i.e.* $dY_t^2 = dt$; this gives a convenient normalization of the photocurrent. Experimentally the observed current will have some arbitrary amplification. Note also that the process Y_t is a *classical* stochastic process;[35] it commutes with itself at different times and it commutes with all system operators in the future.

3.2. *The quantum filter*

Using Eq. (10) and the quantum Itô rules we easily obtain

$$dj_t(X) = j_t(\mathcal{L}[X])\, dt + \sqrt{M}\, j_t([X, F_z])\,(dA_t^\dagger - dA_t), \qquad (14)$$

where $\mathcal{L}[X] = ih(t)[F_y, X] + MF_zXF_z - \frac{1}{2}M(F_z^2X + XF_z^2)$. We have already established the observation equation

$$dY_t = 2\sqrt{M\eta}\, j_t(F_z)\, dt + \sqrt{\eta}\,(dA_t + dA_t^\dagger) + \sqrt{1-\eta}\, dW'_t. \qquad (15)$$

Together, Eqs. (14) and (15) form the system-observation pair of our model. Equation (14) describes the time evolution of any system observable, whereas Eq. (15) describes the observed current. The goal of the filtering problem is to find an expression for $\pi_t(X) = \mathbb{E}[j_t(X)|Y_{s\le t}]$, the best estimate of the observable X at time t, given the prior observations $Y_{s\le t}$.

There are many approaches to this goal, with different levels of rigor and different perspectives. We note that there is one approach,[35,36,38,39] that rigorous and that allows the conditional expectation $\pi_t(X)$ to be derived using classical arguments of Bayesian conditioning. Space does not permit us to present this approach here, but the result is that

$$\begin{aligned} d\pi_t(X) = \pi_t(\mathcal{L}[X])\, dt + \sqrt{M\eta}\,(\pi_t(F_zX + XF_z) \\ - 2\pi_t(F_z)\pi_t(X))(dY_t - 2\sqrt{M\eta}\,\pi_t(F_z)\, dt), \end{aligned} \qquad (16)$$

which is the quantum filtering equation for our model. It is more useful to represent the filter in its adjoint form (which is the primary form derived in other approaches.[22]) To this end, we define the conditional atomic density matrix ρ_t as the matrix that satisfies $\pi_t(X) = \mathrm{Tr}[\rho_t X]$ for all X. Eq. (16) gives

$$d\rho_t = -ih(t)[F_y, \rho_t]\, dt + M\,\mathcal{D}[F_z]\rho_t\, dt + \sqrt{M\eta}\,\mathcal{H}[F_z]\rho_t\, dW_t, \tag{17}$$

where we have used the notation[37]

$$\mathcal{D}[c]\rho \equiv c\rho c^\dagger - (c^\dagger c\rho + \rho c^\dagger c)/2 \tag{18}$$

$$\mathcal{H}[c]\rho \equiv c\rho + \rho c^\dagger - \mathrm{Tr}[(c + c^\dagger)\rho]\rho \tag{19}$$

and we have defined the *innovations process*

$$dW_t = dY_t - 2\sqrt{M\eta}\,\mathrm{Tr}[\rho_t F_z]\, dt. \tag{20}$$

An important result in filtering theory is that the innovations process W_t is in fact a Wiener process.[38,39] Though we have not introduced sufficient technical machinery to prove this fact, we can can give a simple interpretation. We can write W_t in the form

$$dW_t = 2\sqrt{M\eta}\,(j_t(F_z) - \pi_t(F_z))\, dt + \sqrt{\eta}\,(dA_t + dA_t^\dagger) + \sqrt{1-\eta}\, dW_t'. \tag{21}$$

This expression consists of two parts: the last two terms are white noise terms, whereas the first term is the difference between an atomic observable and our best estimate of that observable. If our estimator is truly optimal, then the difference between the true observable and its estimate cannot be biased, *i.e.* it must be white noise. Hence the innovations process is integrated white noise, which is a Wiener process.

3.3. *Conditional spin dynamics*

Before we add control to the picture it is interesting to take a look at the open-loop properties of the filtering Eq. (17), *i.e.* without feedback, by setting $h(t) = 0$. The equation propagates a density matrix, defined as the adjoint of a set of classical conditional expectations, which carries the interpretation of the "statistically inferred" density matrix of the ensemble given the observations in the probe field. One might wonder how such a picture is related to the traditional picture of quantum measurements.

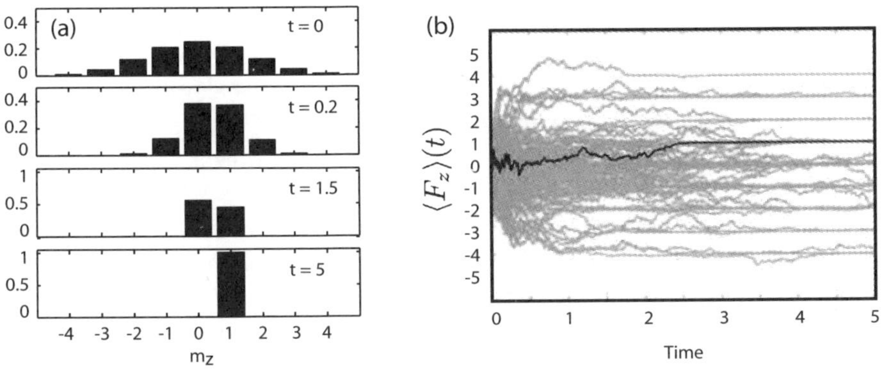

Fig. 2. (a) Time evolution of the conditional state from a coherent spin state at $t = 0$ to an eigenstate at long times. The graph shows the population of each F_z eigenstate. (b) 100 sample paths of $\pi_t(F_z)$, with $M = \eta = 1$. The dark line is the sample path shown in (a), resulting in $m_z = 1$. Von Neumann projection is clearly visible at long times.

To illustrate the filtering process we reproduce simulations of Eq. (17) for a spin $F = 5$ ensemble (*e.g.*, 10 two-level atoms).[19] Such simulations are highly simplified by the fact that the innovations process is a Wiener process. This means that we do not have to simulate the full quantum-mechanical model, Eqs. (14) and (15), to obtain a photocurrent Y_t to drive (17). Instead, we just plug in a Wiener process for the innovations, for which straightforward numerical methods are available. The results are shown in Fig. 2.

At long times the conditional state is clearly driven to one of the eigenstates of F_z, *i.e.* Dicke states,[40] just as predicted by the Von Neumann projection postulate. In fact, it can be rigorously proved that the $t \to \infty$ limit of Eq. (17) is *exactly* identical to the projection postulate, *i.e.* the probability of collapse onto each eigenstate is predicted correctly by the filtering equation.[19,36,41]

At intermediate times, $t < \infty$, the conditional state gradually collapses onto the F_z eigenstates. This process, for a single sample path, is shown in Fig. 2a. Whereas a Von Neumann measurement would take the state discontinuously from the initial state to the final collapsed state, the filtering process continuously narrows the distribution over the eigenstates until only one remains. Aside from giving a more realistic description of continuous optical measurements, this description creates an opportunity that has no analog with projective measurements: we can interfere with the collapse process while it is occuring by applying real-time feedback.

4. Quantum Feedback Control

The intrinsic randomness of quantum measurement should not dissuade the capable observer from trying to control the dynamics of a system. In fact, it should do just the opposite. The inherent uncertainty in observation is the inspiration for the use of feedback control, and promotes it to the status of fundamental.

Although the physical constraints imposed by quantum mechanics are performance limiting, quantum feedback control problems are well defined and worth pursuing for all of the same reasons engineers use control on classical systems. Furthermore, quantum feedback control, while technically difficult, is simply a branch of traditional control and amenable to the techniques developed therein.[36,42,43] Far from introducing an entirely new kind of problem, the challenges presented here highlight and motivate the extension of mathematical methods already in development elsewhere.

Note that the term "quantum feedback control" as used in this chapter refers to a particular class of problems that should be distinguished from other types of control with quantum systems. The class we consider involves the measurement of a quantum system by interaction with a quantum field. The field is destructively measured resulting in a classical measurement record. That measurement record is then processed and fed back to Hamiltonian parameters affecting the same system.

In this section we begin by discussing the types of problems and structure encountered in a typical quantum feedback control scenario, building upon the formalism developed above. Here we use language from classical control theory, and discuss the possible application of optimal and robust control theories to the quantum setting. We also emphasize experimental constraints which motivate simplifications of desired controls through model reduction.

4.1. *Separation structure*

Generally speaking, the control problem consists of finding a mapping of the measurement record onto the actuation variables such that some pre-defined task is achieved. When stated in this way the problem is very difficult to solve; after all, when we allow any functional from the photocurrent history to the control variables, it is hard to know where to start.

Fortunately we can simplify the problem description considerably using what is sometimes referred to as the *separation principle* or the *information state approach*, originally introduced in classical control theory by

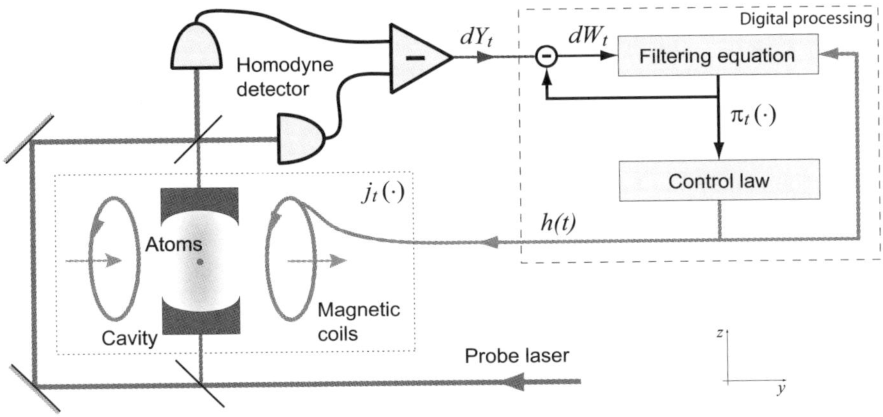

Fig. 3. Schematic of the entire feedback control problem for an atomic ensemble. On the experimental level an ensemble interacts with a probe field, as described by Eq. (14). Homodyne detection gives rise to the photocurrent (15), which is processed by a digital controller. A magnetic field is used for feedback. On the controller level, the photocurrent drives the quantum filter (16) which updates recursively the best estimate of the atomic state. The control law is a functional of the current conditional state. The innovations structure (20) allows the control design to be based directly on the filtering dynamics.

Mortensen.[44] The basic idea behind this approach is that we can never control the system more precisely than the precision with which the system state can be inferred from the observations. In many ways this is a statement of the obvious: for example if we know that the system is controlled to within some bound, then clearly we can infer that the system state is within that bound. As a consequence, the best we can do is to control the best estimate of the system state, *i.e.* the conditional state.

The advantage of this approach is that we have converted the *output feedback* control problem into a *state feedback* control problem for the filter. Operationally, we then consider the filtering equation (17) as our new "effective" dynamical equation to be controlled, where the feedback $h(t)$ can now be taken to be a function of the conditional state ρ_t as opposed to the measurement record. Control design is further simplified by the fact that the innovation, Eq. (20), is white. This means we can consider Eq. (17) as an ordinary Itô equation to be controlled, without separately modeling the statistics of the photocurrent driving noise.

The structure of the entire control setup, in the context of the model discussed in the previous sections, is shown in Fig. 3. The atomic ensemble and its interaction with the optical probe field and the magnetic control field was modeled in Sec. 2. Homodyne detection was the subject of Sec. 3.1. The photocurrent is processed by a digital control circuit which produces the feedback signal. Inside the controller, the "whitened" photocurrent drives the quantum filter, as described in Sec. 3.2. The control law is a function of the best estimate of the system state. To design the control law, however, we only need to consider the "internal" feedback loop inside the digital circuit. From the controller's effective perspective, the only role of the physical experiment is to provide the innovation dW_t, which is white by construction.

4.2. *Defining an objective*

We have separated the control design into an estimation problem, which was the subject of Sec. 3.2, and a control problem. The control problem is undefined, however, until we state a goal that our controller should achieve.

As an example, an experimentalist may want to minimize some functional of the state and control variables, *e.g.*,

$$C[h(t)] = \mathbb{E} \int_0^T \left(\pi_t(F_z^2) + \mu h(t)^2 \right) dt, \tag{22}$$

where μ is a parameter that limits the degree that the control input is applied. In principle, the minimization of this "cost function" using the dynamics of the filtering equation would produce a control law which is the optimal time-dependent mapping of the conditional state onto the control parameters. This type of problem is known as "optimal control", and is one of the primary modes of thought in classical control theory.[45–47] As is apparent from Eq. (17), the general form of the filtering equation is non-linear in the state and, as an unfortunate result, the optimal control solution is extremely difficult to find. Although both non-linear and stochastic control theories are well developed fields classically, there is still much work to be done in their intersection.

Fortunately, there are alternative methods for gaining ground on the quantum feedback control problem. First, in some instances, it is possible to linearize the dynamics of the filtering equation via moment expansions. In this case, one can readily adopt "LQG" techniques from classical control,[46,47] for linear systems (L) a cost function quadratic in linear observables and control variables as above (Q), and Gaussian dynamics (G), to solve the problem completely.[1,29,42,43,48] In any given example, the needed linearization may only work for particular initial states and limited periods of time, but the LQG results can still be remarkably far reaching.

Second, we can choose to be less demanding of our controller, and instead formulate a non-optimal goal. For instance, in the context of state preparation, we could try to find a controller that made the target state a global (stochastically) stable state, thus ensuring its eventual preparation. Although the state might not be prepared as fast as is physically possible, it is an accomplishment to know that it will eventually be prepared with unit probability. Here there is much work to be done on constructively generating controllers and methods for proving the stability, but progress has been made for some simple problems.[36]

4.3. *Robustness and model reduction*

If given the choice between a controller that works optimally under one set of ideal circumstances and a controller that works sub-optimally, but adequately, over a wide set of possible conditions, the wise experimentalist would always choose the latter. Due to unexpected modeling uncertainties and exogenous noise sources, the optimal control approach has the potential to fail catastrophically in realistic environments, a possibility that has motivated the development of "robust control" for many years.[49] One could say the reason experiments are performed at all is to test the robustness of our model and control design.

The concept of robust control has been extensively studied in the classical deterministic setting, but the same logic holds true for quantum applications. Even as quantum technology reaches its limits, there will always be some degree of non-intrinsic system uncertainty to which the system should be robust. Not surprisingly, quantum feedback techniques have been shown to enable robustness to model uncertainty in metrology applications.[1] Of course one need not draw the line too sharply between optimal and robust control, as there exist types of risk-sensitive optimal quantum control that inherently consider certain kinds of robustness.[50,51]

Another practical reason why optimal control may not be ultimately relevant is that real-time information processing takes time. Even if the modeling is perfect and there are no excess noise sources, an optimal controller may not work due to processing constraints. If the processing delay of the actual controller is large compared to the relevant time-scale of the filtering dynamics, then another approach will be needed. Despite the improving performance of programmable logic devices that might best implement the optimal control, there are few experiments with slow enough time-scales that modern electronics can be optimally effective at real-time estimation and control.[52]

5. Feedback in Atomic Ensembles

When considering systems with the potential for interesting applications related to quantum information processing, there exists a natural tendency within many physicists to consider conceptually simplified systems, *e.g.*, a single atom or ion. While much progress has been made in trapping, measuring, and controlling single particles, it has also been realized for some time that the use of atomic ensembles does not preclude the observation of uniquely quantum effects nor a simple description. As compared to alternative systems, ensembles are experimentally convenient and, by the sheer number of participants, sufficient signal can be generated to make them powerful in quantum applications, with atomic clocks being just one prominent example.

Here we consider those experiments where continuous measurement and feedback have been used to generate entanglement either within or between atomic ensembles. We begin by discussing the use of dispersive measurement to produce a spin-squeezed state in a single ensemble, and how feedback can be used to make this process deterministic. We then focus on a particular theoretical limit where the continuous variable treatment fails, but still highly entangled eigenstates of the measured F_z can be prepared by using the more complete filtering equation and an intuitive feedback law. Finally, we briefly discuss experiments and proposals involving the creation of entanglement between two ensembles with and without feedback.

5.1. *Spin squeezing in one ensemble*

For state preparation with atomic ensembles, spin-squeezed states (SSS)[12] are natural target states. These collective spin states are internally entangled, simply characterized by measured moments of the spin-operators, and useful in metrology tasks.[1,11,10] For an ensemble with N spin-f particles and collective angular momentum operators F_i, a state is defined as spin-squeezed, and entangled, if

$$\frac{2fN\langle \Delta F_z^2\rangle}{\langle F_x\rangle^2} < 1, \tag{23}$$

where the spin-state is pointing along x so that $\langle F_x\rangle = F = Nf$ and $\langle F_y\rangle = \langle F_z\rangle = 0$.[13] Methods to produce these states typically begin with an unentangled coherent spin state (CSS) with all spins exactly polarized along the x direction and realizing the equality of the uncertainty relation

$$\langle \Delta F_y^2\rangle\langle \Delta F_z^2\rangle \geq \frac{\hbar^2\langle F_x\rangle^2}{4}. \tag{24}$$

For the SSS, the equality is roughly maintained with one component $\langle \Delta F_z^2 \rangle$ squeezed smaller than the CSS value and the other $\langle \Delta F_y^2 \rangle$ anti-squeezed to compensate.

There are many ways one can imagine producing the spin-correlations within the ensemble needed for the collective state to be squeezed. Examples include using direct Hamiltonian interactions[13] and also transferring correlations from an auxiliary system, *e.g.* squeezed states of light.[54–56] We shall focus on the production of spin squeezed states via dispersive measurement, the effects of which were originally discussed and demonstrated in Refs. 2, 57. Subsequently, Thomsen, *et al.*[14,28] proposed a feedback procedure, discussed below, that used a measurement based field rotation to remove the randomness of the measurement while retaining the desired squeezing effect. A similar result can be achieved using feedback to control an optical pumping beam,[56] although this is not strictly spin-squeezing since there is no mean spin in this proposal. A procedure similar to that of[28] has recently been experimentally demonstrated, deterministically producing spin-squeezed states in cold atomic samples.[3,16] Much work continues in this direction, in particular towards creating squeezed states with the Cesium clock transition, which would considerably improve current atomic clock performance.[9]

To understand the conditional preparation of spin-squeezed states by dispersive measurement, consider the apparatus in Fig. 3. As shown above, the filtering equation is given by Eq. (17). This equation is only applicable at long times $t \gg 1/M$ if a sufficiently strong cavity is used to suppress the spontaneous emission to an insignificant level. Given existing experimental technology this is currently unrealistic; nevertheless we consider the long time dynamics for purposes of demonstration.

The filtering equation was derived using a simplified one-dimensional model of the interaction. Although this model is often an adequate description of free-space experiments where a distribution of atoms interacts with a spatially extended probe beam, there is much interest in making the model more accurate by extending it to three dimensions. A complete model would consider the scattering process where all free-space field modes interact with the atomic distribution. Some of those channels would then be measured, and the results used to condition the atomic state. In this picture, the conditional entanglement results from the indistinguishability of the atoms in the measurement and "spontaneous emission" is a term used to describe the effect of the remaining unobserved channels. A considerable amount of work remains to be done in describing collective scattering in terms of measurement theory, but much progress has been made.[58–61]

Returning to the one-dimensional model, we can extract the conditional evolution equations for the moments of any operator from the filtering equation. Under the approximation that there are many atoms and the initial collective state is nearly polarized along the x-direction, we can derive the closed set of equations

$$d\pi_t(F_z) \approx F\exp[-Mt/2]\,h(t)\,dt + 2\sqrt{M\eta}\,\pi_t(\Delta F_z^2)\,dW_t \tag{25}$$

$$d\pi_t(\Delta F_z^2) \approx -4M\eta\,\pi_t(\Delta F_z^2)^2\,dt. \tag{26}$$

These equations are obtained by truncating the exact coupled expressions for $\pi_t(F_z^n)$, calculated from Eq. (16), at $n = 2$.[1,10] This reduced description is equivalent to a classical Kalman filter[46,47] and corresponds to a local linearization of the spin dynamics, *i.e.*, to a QCV approximation.

5.2. *Dicke state preparation in one ensemble*

Eqations (25) and (26) are valid only in the short time limit $t \ll 1/M$, past which the full filtering equation is needed. At longer times terms neglected in the approximation grow to the point that the variance becomes stochastic,[1] and the moment truncation is no longer a good description. This process can be seen in Fig. 4a, where at small times the variance is deterministic, but then becomes random at longer times. Clearly, in this example the QCV approach is applicable only in the short time limit.

The deterministically shrinking variance of Eq. (26) at short times signifies that a spin squeezed state is prepared with a random offset given by Eq. (25). The idea of[28] was to choose $h(t) \propto \dot{Y}_t$ with an intelligently chosen gain such that the first term effectively cancels the second term in Eq. (25), preparing the same SSS on every trial. Although this exact procedure cannot be implemented in practice due to the infinite detector and actuator bandwidths implied by the control law, it was essentially this proportional feedback, with a broad-band filter, that was used in the experiment.[3] Because of the linearity of the dynamics in the short-time limit, the simple current-based feedback law does not perform significantly worse than a law that changes $h(t)$ more optimally according to the state $\pi_t(F_z)$.[1,18]

Given these dynamics, another control strategy would be to separate the measurement and control in time: simply measuring for a finite amount of time, turning off the probe, and using the measurement result to rotate the spin-squeezed state to the desired location. However, as pointed out

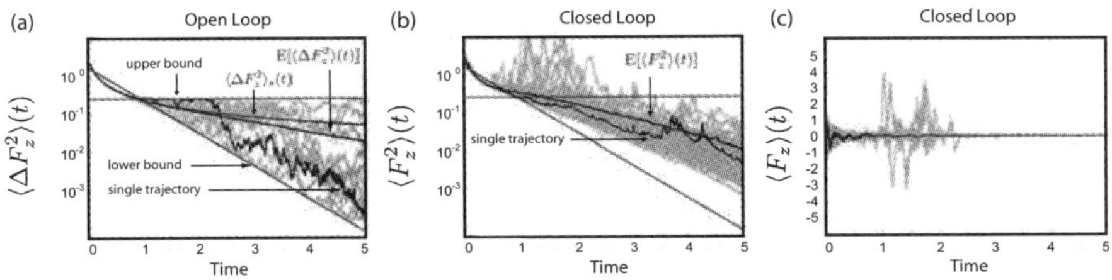

Fig. 4. (a) $\langle \Delta F_z^2 \rangle(t) = \pi_t(\Delta F_z^2)$ in open loop $h(t) = 0$. $\langle \Delta F_z^2 \rangle_s$ is the approximate variance of Eq. (26). (b) $\langle F_z^2 \rangle(t) = \pi_t(F_z^2)$ and (c) $\langle F_z \rangle(t) = \pi_t(F_z)$ with the control law (27) and $\lambda = 10$. Note that $\mathbb{E}\pi_t(F_z^2)$, the cost for preparation of the $m_z = 0$ eigenstate, decreases monotonically. All plots show 100 sample paths and $M = \eta = 1$.

in,[28] the continuous feedback approach is more robust than this procedure to *e.g.* uncertainty in the total atom number which is necessary to compute the size of the correcting rotation.

To further demonstrate the utility of continuous measurement and feedback, we now consider the long time behavior of the filtering equation, past the point in time $t > 1/M$ when the linearized description fails. As discussed in Sec. 3.3, the filtering equation stochastically prepares a random eigenstate of F_z asymptotically in time.[a] The performance of particular controllers at producing one F_z eigenstate deterministically (*i.e.* the same one on every trial) was investigated numerically in.[19]

Here it is critical to point out that, unlike with the Gaussian spin-squeezed states, a post-measurement rotation strategy will not work in this regime. If the wrong eigenstate is randomly prepared in one measurement, it cannot be transformed into the correct eigenstate by a rotation alone. Furthermore, despite the adequacy of the direct current feedback law at short times, such a controller is less useful at longer times. As opposed to state-based control, this type of control will feed noise into the system even if the target state is reached, unless the gain is turned to zero. Although certain gain-tailored schemes can be made to optimize the feedback at small times,[14,28] it is not at all obvious how such a procedure could be generalized to the long time case.

In contrast, if the control variable $h(t)$ is made a function of the conditional state, then it will naturally know when the goal has been achieved and no longer disturb the state unnecessarily. Numerically it was demonstrated[1] that with an initial x-polarized state, the control law

[a]There are other schemes that produce superpositions of F_z eigenstates conditionally but without control, based on single photon detection of an ensemble in a cavity.[62]

$$h(t) = -\lambda\, \pi_t(F_z) \tag{27}$$

appeared to deterministically prepare the highly entangled state $m_z = 0$ on every trial, as seen in Fig. 4. Thus, continuous feedback, in addition to being robust, is also capable of preparing states on every trial that would be impossible to generate deterministically with measurement and control pulses separated in time.

5.3. *Spin squeezing across two ensembles*

The creation of a collective entanglement within a single atomic ensemble can be motivated with, for example, the need for noise reduction in metrology tasks, where the system is used as a relatively localized probe of some parameter of interest. In other practical applications, like quantum communication, it is desirable to have an entangled quantum state, but with constituents separated substantially in space.[6] Indeed it has been experimentally demonstrated that by detecting a single probe beam after it passes through two spatially separate atomic ensembles, the two ensembles can be made conditionally entangled.[1]

Just as single-mode spin squeezing can be quantified with the collective variables for the one ensemble, here the "two-mode" squeezing can be quantified with the joint collective operators describing both ensembles. Furthermore, in analogy to the work of,[28] the random offset observed in the measurement process can in principle be eliminated with a suitable feedback law to deterministically produce the same two-mode spin squeezed state on every trial.[63] This work contained a careful optimization of the current-based feedback of Ref. 28, showing that this simple feedback can, in principle, prepare spin squeezed states in one and two ensembles very close to the Heisenberg limit.

In a related context, it has recently been experimentally demonstrated that the two-ensemble system may serve as an effective quantum memory for states of light.[15] This procedure differs from the deterministic state preparation discussed previously in that the state of light to be mapped onto the ensembles is not known beforehand. However, the procedure described in Ref. 15 is similar in that it does use feedback to rotate the Gaussian ensemble state in a way that maps one measured quadrature of the optical state onto the atoms, while the other unmeasured quadrature is mapped unconditionally by the interaction alone. Clearly, this process shares many of the same properties as the applications discussed previously and can similarly benefit from analyses with technical notions of robustness and optimality. Finally, this procedure becomes even more efficient if the input atomic state is a two-mode squeezed state, which highlights yet another practical application of deterministic entangled quantum state preparation.

6. Conclusion

In this chapter we have attempted to give a unified picture of a quantum feedback control setup. Starting from elementary physical interactions, as described by a field-theoretic model, we first performed statistical inference on this model, and then used this framework to develop feedback control strategies for state preparation in atomic ensembles. The latter is directly related to recent experimental work which we briefly summarized. It is our hope that such a unified picture will help linking the basic physics and experimental reality to a high-level, control-theoretic point of view.

Many open problems remain on both ends of the spectrum. On the physics side much work remains to be done on the realistic modeling of laboratory experiments. Ultimately a full three-dimensional field-theoretic model will be invaluable for quantitative comparison of theory and experiments. On the control-theoretic side many of the techniques that have been used are still heuristic in nature. Systematic, constructive design methods for nonlinear stochastic controllers, the incorporation of realistic robustness criteria, and efficient model reduction techniques with controllable approximation errors are some of the major outstanding issues. We believe that a fruitful interaction between the physics and mathematical control theory communities will open the road to significant advances in these directions.

Acknowledgments

The authors would like to thank Luc Bouten and JM Geremia for many fruitful discussions. This work was supported by the Australian Research Council, the ARO and the Caltech MURI Center for Quantum Networks. JKS acknowledges a Hertz fellowship. Some of this chapter was originally published (in extended form) in Ref. 64. These parts are reproduced here with permission of Institute of Physics Publishing Ltd.

References

1. B. Julsgaard, A. Kozhekin, and E. S. Polzik, Nature **413**, 400 (2001).
2. A. Kuzmich, L. Mandel, and N. P. Bigelow, Phys. Rev. Lett. **85**, 1594-1597 2000.

3. J. M. Geremia, J. K. Stockton, and H. Mabuchi, Science **304**, 270 (2004).
4. C. A. Sackett, D. Kielpinski, B. E. King, C. Langer, V. Meyer, C. J. Myatt, M. Rowe, Q. A. Turchette, W. M. Itano, D. J. Wineland, and C. Monroe, Nature **404**, 256 (2000).
5. O. Mandel, M. Greiner, A. Widera, T. Rom, T.W. Hänsch, and I. Bloch, Nature **425**, 937-940 (2003).
6. L.-M. Duan, M. D. Lukin, J. I. Cirac. and P. Zoller, Nature **414**, 413-418 (2001).
7. D. Leibfried, M. D. Barrett, T. Schaetz, J. Britton, J. Chiaverini, W. M. Itano, J. D. Jost, C. Langer, and D. J. Wineland, Science **304**, 1476 (2004).
8. D. J. Wineland, J. J. Bollinger, W. M. Itano, and D. J. Heinzen, Phys. Rev. A **50**, 67-88 (1994).
9. D. Oblak, J.K. Mikkelsen, W. Tittel, A.K. Vershovski, J.L. Sorensen, P.G. Petrov, C.L. Garrido Alzar, and E.S. Polzik, Phys. Rev. A **71**, 043807 (2005).
10. J. M. Geremia, J. K. Stockton, A.C. Doherty, and H. Mabuchi, Phys. Rev. Lett. **91**, 250801 (2003).
11. J.K. Stockton, JM Geremia, A.C. Doherty, and H. Mabuchi, Phys. Rev. A **69**, 032109 (2004).
12. K. Mølmer and A. Sørensen, Phys. Rev. Lett. **82**, 1838 (1999).
13. A. Sorensen, L. -M. Duan, J. I. Cirac, and P. Zoller, Nature **409**,63 (2001).
14. L. K. Thomsen, S. Mancini, and H. M. Wiseman, J. Phys. B: At. Mol. Opt. Phys. **35**,4937 (2002).
15. B. Julsgaard, J. Sherson, J.I. Cirac, J. Fiurasek, and E.S. Polzik, Nature **432**, 482 (2004).
16. JM Geremia, J.K. Stockton, and H. Mabuchi, quant-ph/0501033 (2005).
17. Greg A. Smith, S. Chaudhury, and P.S. Jessen, J. Opt. B: Quant. Semiclass. Opt. **5** 323 (2003).
18. H. M. Wiseman and A. C. Doherty, Optimal Quantum Feedback Control for Linear Systems, unpublished.
19. J.K. Stockton, R. Van Handel and H. Mabuchi, Phys. Rev. A **70**,022106 (2004).
20. C. Cohen-Tannoudji, J. Dupont-Roc, and G. Grynberg, in *Photons and Atoms: introduction to quantum electrodynamics*, Wiley 1989.
21. L. Mandel and E. Wolf, in: *Optical Coherence and Quantum Optics*, Cambridge University Press, Cambridge, United Kingdom, 1997.
22. H. M. Wiseman, *Quantum Trajectories and Feedback*, Ph.D. thesis, Univ. of Queensland 1994.
23. H. P. Yuen and J. H. Shapiro, IEEE Trans. Inf. Th. **24**, 657-668 (1978).
24. C. W. Gardiner and P. Zoller, in: *Quantum Noise*, 3rd ed., Springer, 2004.
25. B. Øksendal, in: (*Stochastic Differential Equations*, 5th ed., Springer, 1998.
26. R. L. Hudson and K. R. Parthasarathy, Commun. Math. Phys. **93** 301-323 (1984).
27. C. W. Gardiner and M. J. Collett, Phys. Rev. A **31**, 3761-3774 (1985).
28. L. K. Thomsen, S. Mancini, and H. M. Wiseman, Phys. Rev. A **65**, 061801(R) (2002).
29. A. C. Doherty and K. Jacobs, Phys. Rev. A **60**, 2700 (1999).

30. J. A. Dunningham and H. M. Wiseman, and D. F. Walls, Phys. Rev. A **55**, 1398 (1997).
31. P. Warszawski and H. M. Wiseman, Phys. Rev. A **63**, 013803 (2000).
32. M. O. Scully and M. S. Zubairy, in: *Quantum Optics*, Cambridge University Press, 1997.
33. D. F. Walls and G. J. Milburn, in: *Quantum Optics*, Springer 1994.
34. A. Barchielli, *Continual Measurements in Quantum Mechanics*, Lecture notes of the Summer School on Quantum Open Systems, Institut Fourier, Grenoble 2003.
35. V. P. Belavkin, Commun. Math. Phys. **146**, 611-635 (1992).
36. R. Van Handel, J.K. Stockton, and H. Mabuchi, IEEE Trans. Automat. Control **50**, (6) 768-780 (2004).
37. H. M. Wiseman and G. J. Milburn, Phys. Rev. A **49**, 4110 (1994).
38. V. P. Belavkin, J. Multivariate Anal. **42**, 171-202 (1992).
39. L. Bouten, M. Guţă, and H. Maassen, J. Phys. A **37**, 3189-3209 (2004).
40. R. H. Dicke, Phys. Rev. **93**, 99 (1954).
41. S. L. Adler, D. C. Brody, T. A. Brun, and L. P. Hughston, J. Phys. A **34**, 8795-8820 (2001).
42. Andrew C. Doherty, Salman Habib, Kurt Jacobs, Hideo Mabuchi, and S.M. Tan, Phys. Rev. A **62**, 012105 (2000).
43. V. P. Belavkin, in: *Proceedings, Bellman Continuum*, Sophia-Antipolis 1988, Springer-Verlag, 1988, serie: *Lecture Notes in Control and Information Sciences* 121, pp. 245-265.
44. R. E. Mortensen, Int. J. Control **4**, 455-464 (1966).
45. J. Doyle, B. Francis, and A. Tannenbaum, in: *Feedback Control Theory*, Macmillan Publishing Co., 1990.
46. O. L. R. Jacobs, in: *Introduction to Control Theory*, 2nd ed., Oxford University Press, New York 1996.
47. A. Bensoussan, in: *Stochastic Control of Partially Observable Systems*, Cambridge University Press, 1992.
48. H. M. Wiseman and A. C. Doherty, Phys. Rev. Lett. **94**, 070405 (2005).
49. K. Zhou and J. C. Doyle, in: *Essentials of Robust Control*, 1st ed., Prentice-Hall, Inc., New Jersey 1997.
50. M. R. James, Phys. Rev. A **69**, 032108 (2004).
51. M. R. James, J. Opt. B: Quantum Semiclass. Opt. **7**, S198-S207 (2005).
52. J. Stockton, M. Armen and H. Mabuchi, J. Opt. Soc. Am. B **19**, 3019 (2002).
53. M. Kitagawa and M. Ueda, Phys. Rev. A **47**, 5138 (1993).
54. A. Kuzmich, K. Mølmer, and E.S. Polzik, Phys. Rev. Lett. **79**, 4782 (1997).
55. J. Hald, J. L. Sørensen, C. Schori and E. S. Polzik, Phys. Rev. Lett **83**, 1319-1322 (1999).
56. K. Mølmer, Eur. Phys. J. D **5**, 301-305 (1999).
57. A. Kuzmich, L. Mandel, J. Janis, Y. E. Young, R. Ejnisman, and N. P. Bigelow, Phys. Rev. A **60**, 2346 (1999).
58. L. -M. Duan, J. I. Cirac, and P. Zoller, Phys. Rev. A **66**, 023818 (2002).
59. I. Bouchoule and K. Mølmer, Phys. Rev. A **66**, 043811 (2002).

60. J. H. Mueller, P. Petrov, D. Oblak, C. L. Garrido Alzar, S. R. de Echaniz, and E. S. Polzik, Phys. Rev. A **71**, 033803 (2005).
61. D.V. Kupriyanov, O.S. Mishina, I.M. Sokolov, B. Julsgaard, and E.S.Polzik, Phys. Rev. A **71**, 032348 (2004).
62. L. -M. Duan and H. J. Kimble, Phys. Rev. Lett. **90**, 253601 (2003).
63. D. W. Berry and Barry C. Sanders, Phys. Rev. A **66**, 012313 (2002).
64. R. van Handel, J.K. Stockton, and H. Mabucchi, J. Opt. B: Quantum Semiclass. Opt. **7**, S179-S197 (2005).

Chapter 24

Real-Time Quantum Feedback Control with Cold Alkali Atoms

J. M. Geremia

Physics and Control & Dynamical Systems, California Institute of Technology
Department of Physics & Astronomy, University of New Mexico

This chapter describes recent experiments involving continuous measurement and quantum feedback control of collective spin in a cloud of cold alkali atoms. While measurement by itself can generate entanglement between different atoms in the sample by virtue of conditional spin-squeezing, the squeezed state produced in any individual instance of the measurement is randomly distributed. Incorporating real-time feedback control into the observation process suppresses the statistical dispersion of the measurement outcome so that the same *a posteriori* state is achieved in every instance of the feedback-stabilized measurement.

1. Introduction

Feedback control plays a prominent role in many areas of science and technology because it enables a physical system to achieve stable, predictable behavior despite uncertainty in both its state and its dynamics. But, this motivation for employing feedback assumes fundamental importance in quantum mechanics where the world is inherently probabilistic. It is a basic fact of nature that— in the absence of external mediation — even ideal quantum measurements are unpredictable due to Heisenberg uncertainty.

However, we now know that it is frequently possible to embed an otherwise random quantum measurement within a stabilizing feedback loop designed to guide the measurement toward a reliable outcome. Since the underlying measurement is probabilistic, the control process that stabilizes it is necessarily stochastic. This property requires us to consider real-time procedures in which the quantum system is directed toward the target outcome by a time-dependent control Hamiltonian, $\hat{H}_c(t)$, applied in response to an evolving continuous measurement, *as it is performed.*

Of course, adding feedback usually makes an experiment more difficult to implement, so this prompts us to ask what we can gain by reducing measurement indeterminism. In this chapter, we focus on improving our capability for quantum state preparation.[1,2] It is well known that a measurement performed on a composite system can produce highly entangled *a posteriori* states. As a grossly simplified example, consider measuring the total z-component of the spin of two qubits. There are several possible outcomes: when $J_\mathrm{z} = 1$, the post-measurement state is given by, $|\psi_1\rangle = |\uparrow\uparrow\rangle$, and when $J_\mathrm{z} = -1$, the resulting state is $|\psi_{-1}\rangle = |\downarrow\downarrow\rangle$. Neither of these is entangled; however, when $J_\mathrm{z} = 0$, we obtain post-measurement states, $|\Psi^\pm\rangle = (|\uparrow\downarrow\rangle \pm |\downarrow\uparrow\rangle)/\sqrt{2}$, that are substantially more valuable for quantum information theoretic tasks as they are maximally entangled.

But left to chance, the desired $J_\mathrm{z} = 0$ outcome occurs only a fraction of the time. This is an endemic problem when using a probabilistic measurement for state preparation— the result is *conditioned* on the outcome. Being greedy, we would much prefer to keep all the entangling power of the measurement but discard its unwanted statistical uncertainty. That would require us to direct the outcome to $J_\mathrm{z} = 0$ in every instance. It is this topic of deterministic state preparation by a feedback-stabilized measurement that we address in this chapter. Our aim is to familiarize the reader with a recent experiment in which real-time quantum feedback control is used to produce deterministic spin-squeezed states of a cold many-atom system.[3]

We should note that the chapter by van Handel, *et al.* develops the theoretical underpinnings of continuous measurement in the quantum Markov limit. It describes quantum filtering theory,[4,5] a subject that has significantly shaped our interpretation of the experiment presented here. This chapter is designed to stand alone, but we occasionally forgo mathematical rigor in favor of physical descriptions. Our hope is that this will allow readers of both chapters to draw connections between quantum filtering theory and real experiments, for example by relating the occasionally vague parameters encountered in the stochastic filtering equations to routinely characterized numbers in modern atomic physics experiments.

Finally, it is crucial to comment that experiments which aim to manipulate quantum mechanical uncertainty must be capable of distinguishing these intrinsic fluctuations from classical noise. We refer to an experiment as being *quantum noise-limited* when non-fundamental sources of uncertainty have been reduced to the point that quantum effects dominate. In general, this is the most challenging aspect of such experiments; but indeed, it makes little sense to speak of *quantum feedback control* in the absence of quantum-limited performance.

2. The Atomic Spin System

The physical system in the experiment we describe is composed of neutral Cesium-133 atoms. Like all ground state alkalis, Cs contains an unpaired valence electron and thus exhibits intrinsic spin angular momentum (a magnetic moment) as a result of its electronic, $\hat{\mathbf{s}}$, and nuclear, $\hat{\mathbf{i}}$, degrees of freedom. While there is no net orbital angular momentum, $\hat{\mathbf{l}}$, for the $6^2\mathrm{S}_{1/2}$ Cs ground state, this is not necessarily true of excited electronic states, and we use some of these excited states to implement a continuous measurement of the atomic angular momentum by coupling the atoms to a laser field.

As such, the operator of interest is the total angular momentum,

$$\hat{\mathbf{f}} = \hat{\mathbf{s}} \otimes \hat{\mathbb{1}}_{l\otimes i} + \hat{\mathbb{1}}_s \otimes \hat{\mathbf{l}} \otimes \mathbb{1}_i + \hat{\mathbb{1}}_{s\otimes l}\hat{\mathbf{i}}. \tag{1}$$

Eigenstates, $|f, m_f\rangle$, of the atomic spin are defined in the usual manner,

$$\hat{\mathbf{f}}^2|f,m\rangle = \hbar^2 f(f+1)|f,m_f\rangle \tag{2}$$

$$\hat{f}_z|f,m_f\rangle = \hbar m_f|f,m_f\rangle, \tag{3}$$

where f and m_f are, respectively, the quantum numbers corresponding to the magnitude of $\hat{\mathbf{f}}$ and its z-axis cartesian projection, $\hat{f}_z$. For ^{133}Cs, where $s = 1/2$ and $i = 7/2$, the total spin quantum number, f, can take on the values of either $f = 3$ or $f = 4$. The atomic spin we consider in this chapter involves only the $f = 4$ hyperfine ground state manifold composed of $(2f + 1) = 9$ Zeeman degenerate sublevels indexed by $m_f = -4, -3, \ldots, 3, 4$.

The spin system in the experiment is comprised of $N \gg 1$ (it's between 10^{10} and 10^{11}) of these spin-4 atoms. However, the individual particles are indistinguishable as the optical and magnetic fields which implement our continuous measurement and feedback control address the atoms non-selectively.[3] As a result, the quantum state of the collective atomic system exhibits permutation symmetry with respect to particle exchange; swapping particles introduces no difference in the coupling between the collective system and the measurement or control fields.[6] Thus, the atomic degrees of freedom can be well-approximated as a pseudo-particle with N times the spin of a single atom. That is,

$$\hat{\mathbf{F}} = \sum_{n=1}^{N} \hat{\mathbf{f}}^{(n)}. \tag{4}$$

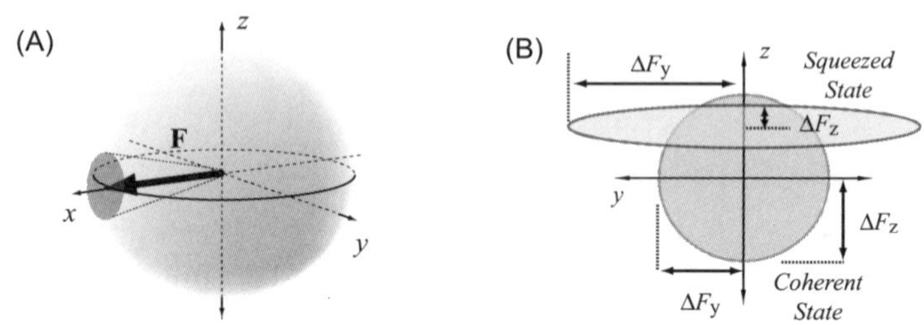

Fig. 1. Spin polarized collective atomic states can be represented by their net magnetization vector, $\mathbf{F} = \langle[\hat{F}_{\rm x}, \hat{F}_{\rm y}, \hat{F}_{\rm z}]\rangle$ along with the transverse uncertainties, $\Delta\hat{F}_{\rm y}$ and $\Delta\hat{F}_{\rm z}$.

It is readily shown that permutation symmetry restricts us to the maximum total angular momentum manifold where the quantum number of $\hat{\mathbf{F}}^2$ is $F = Nf$.[6] Under this symmetry, the cartesian components of $\hat{\mathbf{F}}$ are simple sums of the constituent particles, $\hat{F}_i = \sum_n \hat{f}_i^{(n)}$.[a]

At the beginning of the experiment, the atomic system is prepared into its initial state by independently polarizing the magnetic moments of the N atoms along the laboratory x-axis. Ideally, this polarization step drives all of the atoms into their $m_f = 4$ Zeeman sublevel (in a basis quantized along the x-axis) to produce what is known as a *coherent spin state*,

$$\hat{\rho}_{\rm CSS} = \bigotimes_{n=1}^{N} |f, f\rangle_{\rm x}\langle f, f|. \tag{5}$$

Experimentally, this preparation step is accomplished with reasonable fidelity by optically pumping along the x-axis with a circularly polarized laser field.[9,10] In the laboratory, pumping produces an approximate coherent state, $\hat{\rho}(0) \approx \hat{\rho}_{\rm CSS}$, where $\hat{\rho}(0)$ is the density operator corresponding to the quantum state of the atomic sample as it is actually prepared.

For this initial state, the expectation values of the transverse spin components, $\langle\hat{F}_{\rm y}\rangle$ and $\langle\hat{F}_{\rm z}\rangle$, are both zero. However, the operator variances, $\langle\Delta\hat{F}_{\rm y}^2\rangle$ and $\langle\Delta\hat{F}_{\rm z}^2\rangle$, are not, such that the Heisenberg-Robertson relation,

$$\langle\Delta\hat{F}_{\rm y}^2\rangle\langle\Delta\hat{F}_{\rm z}^2\rangle \geq \frac{1}{4}\hbar^2|\langle\hat{F}_{\rm x}\rangle|^2, \tag{6}$$

is satisfied. Quantum mechanically, these variances quantify the statistical dispersion of transverse spin-component measurements performed on

[a]Terms that are nonlinear in the spin operators cannot be expressed as a simple sum over particles, *i.e.*, $\hat{F}_i^q \neq \sum_n (\hat{f}_i^{(n)})^q$ for $q > 1$, but we will not encounter any of these.[7,8]

an ensemble of identically prepared atomic systems and this measurement uncertainty is generally termed *spin projection noise.*[11] It can be extremely convenient to visualize the spin-polarized quantum system as though it were a classical magnetic moment, $\mathbf{F} = [F_x, F_z, F_z]$, with transverse quantum uncertainties, $\langle \Delta \hat{F}_y \rangle$ and $\langle \Delta \hat{F}_z \rangle$, as suggested by Fig. 1(A).

The coherent spin state is a minimum uncertainty state in the sense that any reduction in $\langle \Delta \hat{F}_z^2 \rangle$ below its initial value of $F\hbar^2/2$ must be accompanied by a corresponding increase in $\langle \Delta \hat{F}_y^2 \rangle$, and vice versa. States in which one of $\Delta \hat{F}_y^2$ or $\Delta \hat{F}_z^2$ is reduced below its coherent state value of $F\hbar^2/2$ are referred to as *spin-squeezed states.*[12] A schematic of this distinction is illustrated in Fig. 1(B), where transverse uncertainty in the atomic magnetization is denoted by an ellipse with dimensions $\Delta \hat{F}_y^2$ and $\Delta \hat{F}_z^2$. It can be shown that spin-squeezing implies entanglement between the different atoms in the atomic system.[13–15] Squeezing is therefore of interest in quantum information science, particularly for reducing quantum noise limitations to the resolution of various precision measurements.[11,16,17]

2.1. *Generating spin-squeezing using measurement*

Spin-squeezing experiments based on continuous measurement operate on the principle that it is possible to weakly measure[18,19] the z-component of the collective spin.[20,21] As information about $\hat{F}_z$ is acquired, its uncertainty, $\Delta \hat{F}_z$, decreases below the coherent state value of $\hbar F/2$.[3] To compensate for this reduction, uncertainty in the orthogonal component, $\Delta \hat{F}_y$, increases (known as *anti-squeezing*) to accommodate Eq. (6).[12]

Unfortunately the squeezing produced by the raw measurement is conditional, as illustrated by the schematic in Fig. 2. The mean value around which the spin-squeezing is achieved is randomly distributed (the ellipses are not centered on the origin) in an ensemble of $\hat{F}_z$ measurements.[11,20,16] Since one cannot predict this mean z-component offset in advance of any individual measurement trajectory, the dispersion of these offsets must be considered as part of the statistical outcome of the measurement. It can be shown that the shot-to-shot distribution of z-component offsets has a variance equal to that of the initial (ideally coherent) state, $\Delta \hat{F}_z^2 \approx \hbar^2 F/2$.[16] Therefore, there is no uncertainty reduction without conditioning on the particular offset observed in a given measurement trajectory.

This is where feedback enters the picture.[22] Spin-squeezing would be deterministic if the z-component offset assumed the same value (for instance $F_z \sim 0$) in every measurement, and we use feedback to accomplish this

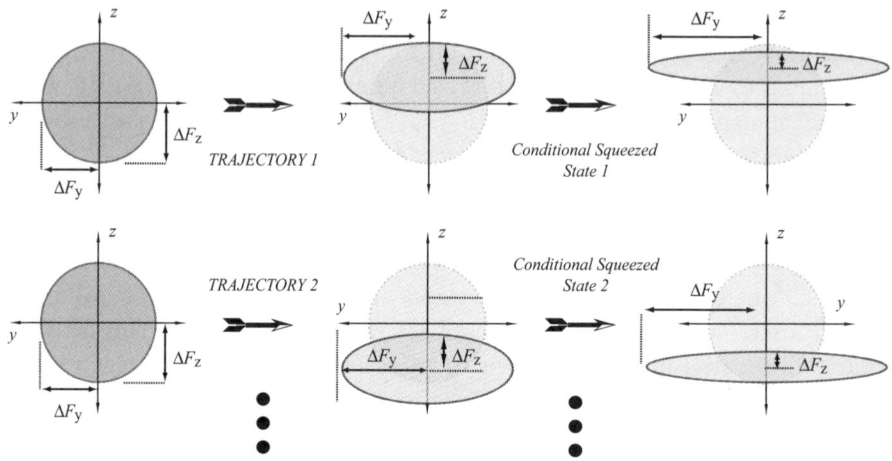

Fig. 2. Schematic of conditional spin squeezing by continuous quantum measurement. In any given measurement trajectory, uncertainty in the z-component of the collective atomic spin angular momentum is reduced below that of a coherent state. However, the orientation of the atomic magnetization vector is random in each trajectory.

objective by stabilizing the orientation of the atomic magnetization vector. This is done by driving a time-dependent y-axis magnetic field in response to the real-time measurement to lock the magnetization vector to the x-axis by Larmor precession. This idea of counteracting the random spin offset with a control Hamiltonian of the form, $\hat{H}_c = \hbar\gamma B_y(t)\hat{F}_y$ was first suggested by Thomsen, Mancini and Wiseman.[23]

3. Continuous Measurement of Spin Angular Momentum

The continuous measurement in our quantum feedback control experiment utilizes an optical probe of the collective atomic spin.[24,25] A schematic of the major components of the measurement apparatus are illustrated in Fig. 3.[7,21,24,26] Specifically, a (near) quantum-limited laser is tuned to be off-resonant with respect to the Cs $6^2S_{1/2}\rightarrow 6^2P_{3/2}$ transition at $\lambda \sim 852$ nm, the so-called D_2 line. Prior to interacting with the atomic sample, the probe laser is linearly polarized along the x-axis with a high-quality polarizer such that the atomic magnetization and probe polarization are (nearly) aligned. Following this interaction, the forward-scattered probe field is detected using a balanced quantum noise-limited polarimeter.

Qualitatively, the continuous measurement occurs as a result of spin-dependent birefringence in the atomic sample. That is, the left (−) and

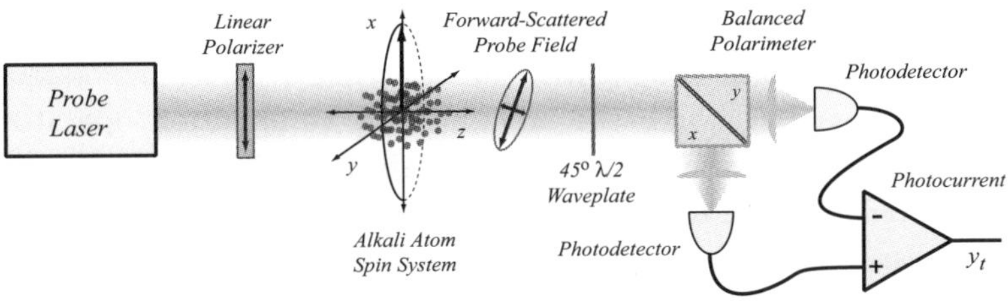

Fig. 3. Schematic of a quantum nondemolition measurement of collective spin in an Alkali atom sample. Faraday rotation of an off-resonant optical probe is detected by a balanced polarimeter. Conditioning the quantum state of the atoms on polarimeter photocurrent produces spin-squeezing.

right (+) circular polarization components of the linearly polarized probe laser experience a different index of refraction, $\delta n = (n_+ - n_-)$, that is proportional to the z-component of the atomic magnetization vector. This dispersive interaction causes the probe polarization to acquire a Faraday rotation proportional to F_z.[24,25] Therefore, the photocurrent produced by the polarimeter provides information about the quantum observable, $\hat{F}_z$.

Since the probe laser is detuned from the atomic resonance, little absorption occurs, but some probe light does scatter into electromagnetic field modes not imaged by the polarimeter detectors. This non-paraxial scattering leads to atomic decoherence from information loss.[27] In this chapter we consider the short-measurement limit where it is possible to neglect this source of decoherence without incurring significant error. Although it is not possible to achieve the Heisenberg squeezing limit, significant uncertainty reduction in the z-component of the collective spin is still possible.[23,3,7]

3.1. *Continuous measurement as a scattering process*

Continuous measurement of the atomic system is performed by detecting the forward scattered probe field after it has interacted coherently with the atoms.[7,23] In the Markov limit (similar to that derived in the chapter by van Handel, *et. al* using techniques from quantum stochastic calculus), we obtain the following mode operators for the scattered field,

$$\hat{\mathbf{E}}_t^{(+)} = \sqrt{\frac{\hbar\omega}{2\epsilon_0 V}} \left(\hat{a}_{-,t}\vec{\mathbf{e}}_- + \hat{a}_{+,t}\vec{\mathbf{e}}_+ \right), \tag{7}$$

and

$$\hat{\mathbf{E}}_t^{(-)} = \sqrt{\frac{\hbar\omega}{2\epsilon_0 V}} \left(\hat{a}^\dagger_{-,t}\vec{\mathbf{e}}^*_- + \hat{a}^\dagger_{+,t}\vec{\mathbf{e}}^*_+ \right). \tag{8}$$

Here, $\hat{a}^\dagger_{-,t}$ and $\hat{a}_{-,t}$ are the Heisenberg-picture creation and annihilation operators for the z-axis propagating mode with left circular polarization. Similarly, $\hat{a}^\dagger_{+,t}$ and $\hat{a}_{+,t}$ are the creation and annihilation operators for right circular polarization. The (complex) spherical basis vectors, $\vec{\mathbf{e}}_-$ and $\vec{\mathbf{e}}_+$, denote polarization modes with left and right helicity.

In the dispersive regime, the probe field interacts with the atomic system according to a scattering Hamiltonian of the form,

$$\hat{H}_t = \hat{\mathbf{E}}_t^{(-)} \cdot \frac{\hat{\mathbf{d}}\hat{\mathbf{d}}^\dagger}{\hbar\Delta} \cdot \hat{\mathbf{E}}_t^{(+)}, \tag{9}$$

where $\hat{\mathbf{d}}^\dagger$ and $\hat{\mathbf{d}}$ are vector dipole atomic raising and lowering operators. This Hamiltonian is readily interpreted as a scattering interaction: the atom is first brought from its ground state to a virtual excited state via the raising operator, $\hat{\mathbf{d}}^\dagger$, by annihilating a photon from the probe field through $\hat{\mathbf{E}}_t^{(+)}$. Then, the temporarily excited atom returns to a (potentially different) ground state by emitting a photon into a (potentially different) scattered probe mode via $\hat{\mathbf{d}}$ and $\hat{\mathbf{E}}_t^{(-)}$.

3.1.1. *Physical interpretation*

We can roughly interpret the measurement process as the continuous limit of the following discrete-time picture. First, the full measurement period is divided into segments of duration, δt, delimited by the points, $\{t_1, t_2, t_3, \ldots\}$. This discretization prompts us to decompose the probe field Hilbert space, $\mathscr{H}_{\mathrm{P}}$, in a similar manner,

$$\mathscr{H}_{\mathrm{P}} \equiv \bigotimes_{i=1} \mathscr{H}_{\mathrm{P}}^{t_i} = \mathscr{H}_{\mathrm{P}}^{t_1} \otimes \mathscr{H}_{\mathrm{P}}^{t_2} \otimes \mathscr{H}_{\mathrm{P}}^{t_3} \otimes \cdots, \tag{10}$$

such that the state of the field mode for the ith measurement time-slice is described by the reduced density operator, $\hat{\pi}^{t_i}(t) \in \mathscr{H}_{\mathrm{P}}^{t_i}$. We will always take the input field to be in an optical coherent state,

$$\hat{\pi}^{t_i}(t_i) = |\beta\rangle_+\langle\beta| \otimes |\beta\rangle_-\langle\beta|, \tag{11}$$

linearly polarized along the x-axis with complex amplitude, β, in both its left (−) and right (+) circular polarization components.

At the beginning of each segment, the joint state of the atoms (with reduced density operator $\hat{\rho}(t) \in \mathscr{H}_{\mathrm{A}}$) and the probe is separable,

$$\hat{\chi}(t_i) = \hat{\rho}(t_i) \otimes \hat{\pi}^{t_i}(t_i) \in \mathscr{H}_{\mathrm{A}} \otimes \mathscr{H}_{\mathrm{P}}^{t_i} \tag{12}$$

as the two have not yet interacted. But, the joint system then evolves from t_i to $t_i + \delta t$ according to the Hamiltonian in Eq. (9) and this entangles the atomic spin and probe polarization. In the Schrödinger picture, one treats the scattering process by propagating the system's state, $\hat{\chi}(t_i + \delta_t) = \hat{U}(\hat{\rho}(t_i) \otimes \hat{\pi}^{t_i})\hat{U}^\dagger$, with the time-evolution operator, $\hat{U}$ (described below).

The interaction ceases at $t_i + \delta t$ and the scattered field is subsequently measured. It turns out that the continuous limit of this detection process is best treated in the Heisenberg picture where states are time-independent, $\hat{\chi}(t_i + \delta t) = \hat{\rho}(t_i) \otimes \hat{\pi}^{t_i}(t_i)$, and one instead considers the time-evolution of the measurement operator,

$$y_t = \hat{U}_t^\dagger \left(\hat{\mathbb{1}}_{\mathrm{A}} \otimes \hat{\Upsilon}_{\mathrm{P}} \right) \hat{U}_t. \tag{13}$$

Here, $\hat{\Upsilon}_{\mathrm{P}}$ is the operator corresponding to the action of the polarimeter detectors and $\hat{\mathbb{1}}_{\mathrm{A}}$ is the identity operator on the atomic space $\mathscr{H}_{\mathrm{A}}$ (the atoms are not directly measured). This expression is referred to as the *photocurrent* since it is obtained by measuring the scattered optical field.

3.1.2. *Irreducible representation of the scattering Hamiltonian*

In order to use Eq. 13 to obtain an explicit expression for the photocurrent in terms of the magnetization, $\mathbf{F}$, we must address the details of the scattering Hamiltonian in Eq. (9) to obtain an expression for the time-evolution, $\hat{U}_t$. We begin by focusing on the operator, $\hat{\boldsymbol{\alpha}} \equiv \hat{\mathbf{d}}\hat{\mathbf{d}}^\dagger$, at the center of the interaction Hamiltonian, the *atomic polarizability tensor*.[9,28,29] It is a dyad of vector operators and thus a rank-2 spherical tensor that can be decomposed into a direct sum of irreducible components,

$$\hat{\boldsymbol{\alpha}} \equiv \hat{\mathbf{d}}\hat{\mathbf{d}}^\dagger = \hat{\boldsymbol{\alpha}}^{(0)} \oplus \hat{\boldsymbol{\alpha}}^{(1)} \oplus \hat{\boldsymbol{\alpha}}^{(2)}. \tag{14}$$

The scattering Hamiltonian therefore decomposes into irreducible spherical tensor operators,

$$\hat{H}_t = \hat{H}_t^{(0)} + \hat{H}_t^{(1)} + \hat{H}_t^{(2)}, \tag{15}$$

where $\hat{H}_t^{(0)}$ is a scalar contribution, $\hat{H}_t^{(1)}$ transforms as a vector, and $\hat{H}_t^{(2)}$ transforms as a rank-2 symmetric tensor. It can be shown that the scalar Hamiltonian is independent of the atomic state; it is the AC-Stark shift of the atomic energy levels due to the probe laser. Less trivially, the rank-2 scattering interaction vanishes when both the atomic and optical polarizations are aligned. In the experiment, both polarizations are nominally oriented along the laboratory x-axis so $\hat{H}_t^{(2)}$ vanishes to first order.

The only remaining term is the vector Hamiltonian,

$$\hat{H}_t^{(1)} = \hat{\mathbf{E}}_t^{(-)} \cdot \frac{\hat{\boldsymbol{\alpha}}^{(1)}}{\hbar\Delta} \cdot \hat{\mathbf{E}}_t^{(+)} = \sum_{f'} \frac{\omega\alpha_0\alpha_{f,f'}^{(1)}}{\epsilon_0\sqrt{2}V\Delta_{f,f'}} \hat{S}_{z,t}\hat{F}_z, \tag{16}$$

where $\hat{S}_{z,t}$ is the z-component of the probe field's Stokes (Schwinger boson) operator,

$$\hat{S}_{z,t} = \frac{1}{2}\left(\hat{a}_{+,t}^{\dagger}\hat{a}_{+,t} - \hat{a}_{-,t}^{\dagger}\hat{a}_{-,t}\right), \tag{17}$$

expressed in the Heisenberg picture. In these expressions, V is the volume of the atom cloud and $\Delta_{f,f'} = \omega_{f,f'} - \omega$ is the probe detuning with respect to the hyperfine transition between the $6^2S_{1/2}$ ground state with $f = 4$ and the excited $6^2P_{3/2}$ states with $f' = \{3, 4, 5\}$. The atom-probe coupling strength is determined by the characteristic atomic polarizability,

$$\alpha_0 = \frac{3\epsilon_0\hbar\Gamma\lambda^3}{8\pi^2}, \tag{18}$$

and angular momentum terms,[7,28,29]

$$\alpha_{f,f'}^{(1)} = \left[\frac{2f-1}{f}\alpha_f^{f-1}\delta_f^{f'+1} + \frac{2f+1}{f(f+1)}\alpha_f^f\delta_f^{f'} - \frac{2f+3}{f+1}\alpha_f^{f+1}\delta_f^{f'-1}\right], \tag{19}$$

where $\delta_f^{f'}$ is the Kronecker delta and

$$\alpha_f^{f'} = (2f+1)\frac{2j'+1}{2j+1}\left|\begin{Bmatrix} 1 & j & j' \\ i & f' & f \end{Bmatrix}\right|^2. \tag{20}$$

In these expressions, $j = 1/2$ and $j' = 3/2$ are the ground and excited state fine-structure quantum numbers for the Cs D_2 transition, $\Gamma \sim 2\pi \times 5.1$ MHz is the atomic spontaneous emission rate and $\lambda \sim 852.36$ nm is the transition wavelength.

3.1.3. *Scattering time-evolution operator*

Following its interaction with the atoms, the polarization of the scattered probe field is rotated by 45° (refer to Fig. 3) by a half-waveplate, described by the unitary operator, $\hat{U}_{\pi/2} = \hat{\mathbb{1}}_A \otimes \exp(-i\pi\hat{S}_z/2)$. It is now straightforward to obtain the time-evolution operator, $\hat{U}_t$, that describes the scattering interaction in each measurement time-slice by combining the effect of the vector Hamiltonian in Eq. (16) and the output waveplate,

$$\hat{U}_t = \hat{U}_{\pi/2}\exp\left(-i\gamma\hat{S}_z\hat{F}_z\right), \quad \gamma = \frac{3\Gamma\lambda^2}{4\pi^2 r^2}\sum_{f'}\frac{\alpha_{f,f'}^{(1)}}{\sqrt{2}\Delta_{f,f'}}. \tag{21}$$

Note that in arriving at the expression for γ we have used Eqs. (16 - 20) and have taken the interaction time to be $t_{\rm int} = 4r/3c$, the probe transit-time through the (spherical) atom cloud with radius, r.[7]

3.2. *The continuous photocurrent*

The optical measurement of the scattered probe field (refer to Fig. 3) detects the power difference in the output ports of a polarizing beam-splitter which distinguishes between x- and y- axis polarizations. It therefore implements a measurement of $\hat{Y}_{\rm P} = \hat{a}^\dagger_{{\rm y},t}\hat{a}_{{\rm y},t} - \hat{a}^\dagger_{{\rm x},t}\hat{a}_{{\rm x},t}$, which is proportional to the x-component of the probe Stokes operator in the circular basis, $\hat{\Upsilon}_{\rm P} = \hat{a}^\dagger_{+,t}\hat{a}_{-,t} + \hat{a}^\dagger_{-,t}\hat{a}_{+,t} = 2\hat{S}_{{\rm x},t}$. After evaluating Eq. (13) using the time-evolution operator derived just above we obtain

$$y_t = \hat{U}^\dagger_t \left(\hat{\mathbb{1}}_{\rm A} \otimes \hat{S}_{{\rm x},t}\right) \hat{U}_t = \sqrt{S}\hat{U}^\dagger_t F_{\rm z} \hat{U}_t + (\hat{\mathbf{E}}^{(+)}_t + \hat{\mathbf{E}}^{(-)}_t), \tag{22}$$

where S is a constant of proportionality,

$$S = \frac{1}{\hbar^2}\left[\frac{3\Gamma\lambda^2 P}{4\pi^2 r^2}\sum_{f'}\frac{\alpha^{(1)}_{f,f'}}{\Delta_{f,f'}}\right]^2 \tag{23}$$

that we call the *scattering strength* and $P = \hbar\omega|\beta|^2$ is the optical power of the probe laser [refer to Eq. (11)].

Taking a Markov limit using techniques from quantum stochastic calculus developed in the chapter by van Handel, *et. al*, we identify the input field, $\hat{\mathbf{E}}^{(+)}_t + \hat{\mathbf{E}}^{(-)}_t$, is quantum white noise. Noting that $\hat{U}_t$ and $F_{\rm z}$ commute leads us to the central result, the continuous photocurrent,

$$y_t = \eta\sqrt{S}F_{\rm z} + \sqrt{\eta}\zeta_t, \tag{24}$$

where ζ_t are Gaussian white noise increments and we have introduced the quantum efficiency, η, of the photodetectors used to perform the measurement. The variance of the noise increments is readily obtained by recognizing them as the vacuum fluctuations (shotnoise) in the probe field,

$$\Delta\zeta^2 = 2\hbar\omega P. \tag{25}$$

Finally, it is remarkably beneficial to pause for a brief discussion on the units of the quantities we just introduced— dimensional analysis (in the author's experience) can help circumnavigate much confusion when manipulating continuous measurement equations. The scattering strength, S, has units of power squared per $\hbar^2$. This makes sense since the atomic spin $F_{\rm z}$ has units of $\hbar$, thus $\sqrt{S}F_{\rm z}$ has units of power. Indeed, the polarimeter measures an imbalance in optical power between its two ports. The shot-noise variance, $\Delta\zeta^2$, has units of power squared over frequency. And for completeness, we should relate S and $\Delta\zeta^2$ to the *measurement strength*, $M = S/\Delta\zeta^2$, as it is normally defined in quantum trajectory theory.[30] M has units of frequency while the stochastic terms are in root frequency.

3.3. *Physical interpretation of the photocurrent*

Equation (24) suggests that we should actually view the continuous quantum measurement as sampling a classical random variable, F_z, from a distribution $p_0(F_z)$ corresponding to the projection noise of the initial atomic state. Of course, the sampled value of F_z in any given realization of the measurement is obscured by quantum white noise, ζ_t, which prevents the observer from knowing the value of F_z with perfect precision. The problem of discerning the value of a parameter (in this case the measurement outcome, F_z) from a noisy signal invariably leads to filtering theory; here is no exception even if the noise results from intrinsic quantum fluctuations in the scattered optical field.

In the short-measurement regime, we can just average the photocurrent,

$$\bar{y}_{[0,\tau)} = \frac{1}{\tau}\int_0^\tau y_t\, dt = \eta\sqrt{S}F_z + \sqrt{\eta}\,\bar{\zeta}_{[0,\tau)}, \tag{26}$$

over an interval that lasts from $t = 0$ to $t = \tau$. Averaging reduces the detrimental effects of the white noise since $\mathbb{E}[\zeta_t] = 0$. Unfortunately, integrating over a finite period ($\tau < \infty$) leaves a residual random increment, $\bar{\zeta}_{[0,\tau)}$, that

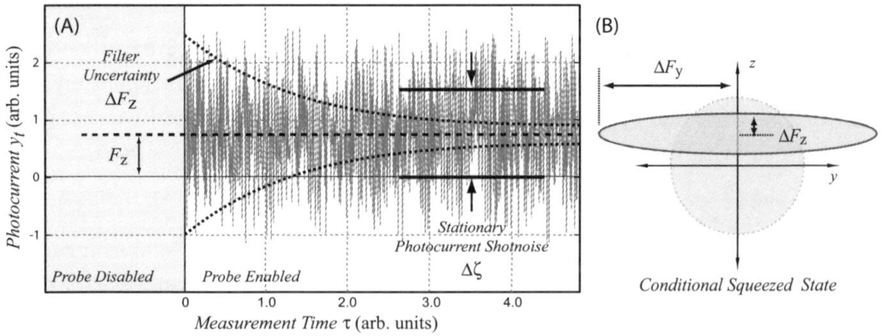

Fig. 4. This simulated measurement trajectory (A), generated according to Eq. (24) ($\eta = 1$), demonstrates many aspects of filtering a noisy photocurrent to produce conditional spin-squeezing (B).

reflects the quantum noise which survives the incomplete averaging period, $0 \leq t < \tau$. The statistics of $\bar{\zeta}_{[0,\tau)}$ are also Gaussian,

$$p_\zeta(\bar{\zeta}_{[0,\tau)}) = \frac{1}{\sqrt{2\pi\Delta\zeta_\tau^2}} \exp\left(-\frac{\bar{\zeta}^2_{[0,\tau)}}{2\Delta\zeta_\tau^2}\right), \tag{27}$$

with a variance, $\Delta\zeta_\tau^2 = \Delta\zeta^2/\tau$, that scales inversely with the measurement duration. As expected, longer averaging better suppresses the white noise.

These points are illustrated by the simulated measurement trajectory in Fig. 4. The plot begins with the probe laser turned off, during which all necessary state preparation of the atomic system such as atom trapping, cooling and optical pumping into an x-polarized coherent spin state is performed. Once the probe light is enabled at $t = 0$, the photocurrent acquires a mean offset, $\eta\sqrt{S}F_z$, proportional to the random value of F_z for that trajectory, but this mean value is masked by photocurrent noise. At short times, the signal is overwhelmed by local statistical fluctuations; however, averaging the photocurrent over a period of time suppresses the uncertainty (dotted lines) in the mean by integrating away the white noise.

4. Spin Squeezing

Reducing uncertainty in the mean value of the photocurrent by averaging is equivalent to gradually learning the value of F_z. *This process corresponds to conditional spin-squeezing*, the amount of which is determined by how well the average polarimeter photocurrent, $\bar{y}_{[0,\tau)}$, can be distinguished from the residual noise, $\bar{\zeta}_{[0,\tau)}$.[7,16] At the beginning of the measurement trajectory, before any information has been acquired about the atomic spin, the statistical distribution of possible measurement outcomes is determined entirely by $\hat{\rho}(0)$, the initial quantum state of the atomic system.

For a coherent spin state polarized along the x-axis, $\hat{\rho}(0) = |F,F\rangle_x\langle F,F|$, this distribution is Gaussian,[16,23]

$$p_0(F_z) = \frac{1}{\sqrt{\pi F\hbar^2}} \exp\left(-\frac{F_z^2}{F\hbar^2}\right), \tag{28}$$

where F_z represents the classical measurement outcome. As expected, the variance of this distribution,

$$\langle \Delta F_z^2 \rangle = \mathrm{tr}\left[\hat{F}_z^2 \hat{\rho}(0)\right] - \mathrm{tr}\left[\hat{F}_z \hat{\rho}(0)\right]^2 = \frac{1}{2}F\hbar^2, \tag{29}$$

is the coherent state projection noise. Quantum mechanically, this distribution has the interpretation that the classical outcome, F_z, will be obtained with probability $p(F_z)$ in an ensemble of $\hat{F}_z$ measurements performed on a large collection of similarly prepared coherent spin states.

4.1. *Filtering and the degree of squeezing*

Once the measurement process begins, the prior distribution $p_0(F_z)$ must be updated to remain consistent with the information gained from the measurement photocurrent. This updating process occurs by conditioning the initial distribution, $p\left(F_z|\bar{y}_{[0,\tau)}\right)$, of the measurement outcome, $\bar{y}_{[0,\tau)}$,

$$p\left(F_z|\bar{y}_{[0,\tau)}\right) = \frac{p\left(\bar{y}_{[0,\tau)}|F_z\right)p_0\left(F_z\right)}{p\left(\bar{y}_{[0,\tau)}\right)}, \tag{30}$$

using Bayes' rule. Here, $p(\bar{y}_{[0,\tau)}|F_z)$ is the probability of observing the average photocurrent, $\bar{y}_{[0,\tau)}$, given that the measurement outcome is F_z, and $p(\bar{y}_{[0,\tau)})$ is the unconditional probability of observing the photocurrent with average value $\bar{y}_{[0,\tau)}$.

The conditional distribution, $p\left(\bar{y}_{[0,\tau)}|F_z\right)$, is given by the probability for observing the stochastic increment $\bar{\zeta}_{[0,\tau)} = (\bar{y}_{[0,\tau)} - \eta\sqrt{S}F_z)/\sqrt{\eta}$,

$$p\left(\bar{y}_{[0,\tau)}|F_z\right) = p_\zeta\left((\bar{y}_{[0,\tau)} - \eta\sqrt{S}F_z)/\sqrt{\eta}\right). \tag{31}$$

Thus, the unconditional probability for the average photocurrent, $\bar{y}_{[0,\tau)}$, is obtained by integrating Eq. (31) with respect to the prior distribution,

$$\begin{aligned} p(\bar{y}_{[0,\tau)}) &= \int p_\zeta\left((\bar{y}_{[0,\tau)} - \eta\sqrt{S}F_z)/\sqrt{\eta}\right) p(F_z)\, dF_z \\ &= \frac{1}{\sqrt{\pi(\eta S F\hbar^2 + 2\Delta\zeta_\tau^2)}} \exp\left(\frac{-\bar{y}^2_{[0,\tau)}}{\eta^2 S F\hbar^2 + 2\eta\Delta\zeta_\tau^2}\right). \end{aligned} \tag{32}$$

It is now possible to evaluate the Bayesian update rule in Eq. (30),

$$p\left(F_z|\bar{y}_{[0,\tau)}\right) = \frac{p_\zeta\left((\bar{y}_{[0,\tau)} - \eta\sqrt{S}F_z)/\sqrt{\eta}\right) p_0(F_z)}{p(\bar{y}_{[0,\tau)})}, \tag{33}$$

and compute the variance of conditional measurement distribution,

$$\langle\Delta F_z^2\rangle_\tau = \int \left(F_z - \langle F_z\rangle_\tau^2\right) p(F_z|\bar{y}_{[0,\tau)})\, dF_z = \frac{F\hbar^2\Delta\zeta_\tau^2}{\eta S F\hbar^2 + 2\Delta\zeta_\tau^2}. \tag{34}$$

The degree of this squeezing, W, can be quantified via the fractional reduction in the variance of the measurement outcome with respect to the coherent state variance $\langle\Delta\hat{F}_z^2\rangle_{\mathrm{CSS}} = \hbar^2 F/2$,

$$W = \frac{\langle\Delta F_z^2\rangle_\tau}{\langle\Delta F_z^2\rangle_{\mathrm{CSS}}} = \frac{1}{1+\mathrm{SNR}^2}, \tag{35}$$

where we have defined $\mathrm{SNR} = \hbar\sqrt{2\eta S/F\Delta\zeta_\tau^2}$ as the polarimeter photocurrent signal to noise ratio relative to the initial spin coherent state. This measure of the signal to noise has the interpretation that 3 dB of spin-squeezing (in variance) is produced, $W(1) = \frac{1}{2}$, when $\mathrm{SNR} = 1$. Furthermore, some degree of squeezing is produced by any finite signal to noise ratio, provided that the initial spin state is in fact a minimum uncertainty state.

4.2. *Real-time feedback control*

Feedback has played little role in the spin-squeezing discussion so far. This is no cause for alarm; conditional quantum state reduction— decreasing $\langle \Delta F_\mathrm{z}^2 \rangle$ around a known value of F_z— occurs from filtering the continuous measurement, not from feedback. In the small-squeezing limit, feedback simply acts on top of the photocurrent filtering process derived in the previous section; it drives the value of $F_\mathrm{z} \propto \bar{y}_{[0,t)}$ to zero. This actuation is accomplished by applying a time-dependent transverse magnetic field, $\mathbf{B}_c(t) = b(t)\hat{\mathbf{y}}$, to implement a control Hamiltonian,

$$\hat{H}_c(t) = \hbar\gamma\, b(t)\hat{F}_\mathrm{y}, \tag{36}$$

where $\gamma \sim 2\pi \times 0.35$ MHz/Gauss is the $6^2\mathrm{S}_{1/2}(4)$ gyromagnetic ratio.

The feedback process can be visualized using the following not-too-inaccurate picture: if filtering the photocurrent in the initial stage of the measurement begins to reveal that its average value, $\bar{y}_{[0,t)}$, is positive (for example), the feedback controller imposes a small Hamiltonian proportional to $\hat{F}_\mathrm{y}$ to rotate the atomic magnetization around the y-axis toward the xy-plane. As F_z relaxes back to zero, the controller turns off the applied magnetic field to prevent the atoms from over-rotating. Of course, had the initial photocurrent developed a negative mean value, the controller would have imposed a Hamiltonian proportional to $-\hat{F}_\mathrm{y}$. More specifically,

$$b(t) = \int_0^t G(s) y_s\, ds, \tag{37}$$

where In our case, the kernel $G(s)$ corresponds to a single-pole analog low pass filter with a corner frequency of $f_c = 850$ kHz. This is a type of integral control since changing $b(t)$ affects the *rate* of the Larmor precession.

Finally, it should be noted that, strictly speaking, it is not technically correct to separate completely the stabilizing feedback rotation from the photocurrent statistics. Applying the Hamiltonian in Eq. (36) prevents F_z and $\hat{U}_t$ from commuting in the photocurrent derivation leading to Eq. (24). We avoid this complication when analyzing our data (described in the next

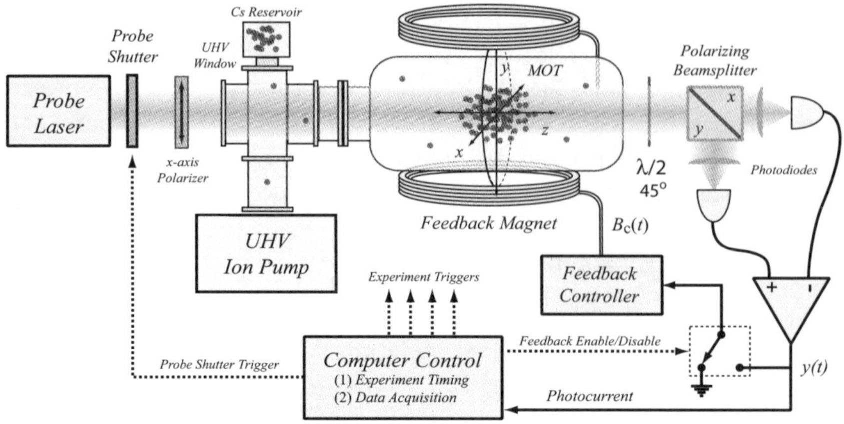

Fig. 5. Schematic of the experimental apparatus used to perform deterministic spin squeezing using continuous measurement of collective spin angular momentum of a cloud of laser cooled Cs atoms and real-time quantum feedback control. Components not shown include lasers used for atom trapping, cooling and optical pumping as well as trim coils used to zero ambient magnetic fields and their gradients.

section) by computing photocurrent statistics only after the feedback has been successful such that $b(t) \to 0$. This procedure would not be capable of achieving the Heisenberg spin-squeezing limit even if atomic decoherence due to scattering probe light into unobserved (non-paraxial) electromagnetic field modes did not occur, such as if the atoms were strongly coupled to a high-finesse optical cavity. A feedback control law capable of achieving the Heisenberg limit was proposed by Thomsen, Mancini and Wiseman.[23]

5. Deterministic Spin-Squeezing Experiment

We now have the theoretical tools necessary to address the experiment. A simplified schematic of the apparatus used to acquire the results below is provided in Fig. 5.[3,7] Roughly, it consists of three main components: (1) an ultra-high vacuum (UHV) environment containing a reservoir of ^{133}Cs atoms, (2) a shotnoise-limited, DC-balanced polarimeter for measuring the z-component of the atomic angular momentum, and (3) feedback control circuitry for applying time-dependent magnetic fields along the laboratory y-axis based on the polarimeter photocurrent.

While details can be found in the references, we give a quick overview of the atom preparation procedure. Each run of the experiment begins by collecting between 10^{10} – 10^{11} Cs atoms in a magneto-optical trap (MOT) and then these atoms to a dark spontaneous-force optical trap (dark-SPOT) to increase their density.[31] This produces a cold Cs cloud with a radius of about $r = 2$ mm at a temperature of $T \sim 10$ μK and a density of nearly 10^{12} atoms/cm^3. Following the collection and cooling stage, which requires about 4 seconds to perform, an x-polarized spin state is prepared by optically pumping the atoms using a circularly polarized 100 μW laser (pulsed for 2-4 ms) oriented along the x-axis. The pumping laser is tuned to the $6^2S_{1/2}(4)\rightarrow 6^2P_{3/2}(4)$ Cs hyperfine transition.

The continuous F_z measurement is implemented with a probe laser blue detuned from the $6^2S_{1/2}(4)\rightarrow 6^2P_{3/2}(5)$ Cs transition by $\Delta \sim 500$ MHz and is linearly polarized by a high extinction Glan-Thompson prism prior to passing through the cold atom cloud. Forward scattered probe light is detected with a polarimeter constructed from a Glan-Thompson polarizing beam splitter and a DC-balanced photodetector with $\sim$1 MHz measurement bandwidth. Magnetic fields with magnitudes up to ~ 0.5 G can be applied in arbitrary (time-dependent) directions by driving 3 pairs of computer-controlled Helmholtz coils, oriented along the laboratory x-, y-, and z-axes, with a bandwidth of ~ 1 MHz. With feedback enabled, the polarimeter photocurrent programs the y-axis coil by driving a high-bandwidth low-noise current source.

A computer controls the experiment timing and records the polarimeter output as well as diagnostic information including the atom number (measured by fluorescence imaging). The computer can enable/disable the measurement by controlling a shutter on the probe laser and it can open/close the feedback control loop via a digitally programmed switch. Ambient magnetic fields are cancelled via a combination of large external trim coils as well as smaller computer controlled balancing coils. Each atom preparation (trapping, cooling and optical pumping) and measurement cycle is synchronized with respect to the 60-Hz building power lines to suppress the effects of fast-scale field fluctuations.

5.1. *Experimental characterization of spin-squeezing*

We analyze spin-squeezing by acquiring statistical ensembles of photocurrents,

$$\mathscr{Y}_t^{(\mathrm{expt})} \equiv \left\{ y_t^{(1)}, y_t^{(2)}, \ldots, y_t^{(Q)} \right\}, \quad 0 \le t < \tau, \tag{38}$$

composed of Q distinct measurement trajectories. Within each ensemble, all experimental parameters are constant such that the photocurrents in $\mathscr{Y}_t$

assume the (ideal) form in Eq. (24) with a consistent scattering strength, S, and white-noise variance, $\Delta\zeta^2$. To analyze the conditional statistics of the measurement we consider two different averaging periods for each photocurrent in the ensemble, $y_t^{(q)} \in \mathscr{Y}_t$.

▷ **Measurement Period 1** of each trajectory is associated with the photocurrent $y_t^{(q)}$ during the time interval, $t_1 \leq t < \tau_1$. The statistical ensemble of measurement outcomes for this measurement window corresponds to the mean values,

$$\bar{\mathscr{Y}}_{[t_1,\tau_1)} \equiv \left\{ \bar{y}^{(1)}_{[t_1,\tau_1)}, \bar{y}^{(2)}_{[t_1,\tau_1)}, \ldots \bar{y}^{(Q)}_{[t_1,\tau_1)} \right\}.$$

▷ **Measurement Period 2** of each trajectory is associated with the photocurrent $y_t^{(q)}$ during the time interval, $t_2 \leq t < \tau_2$. The statistical ensemble of measurement outcomes for this measurement window corresponds to the mean values,

$$\bar{\mathscr{Y}}_{[t_2,\tau_2)} \equiv \left\{ \bar{y}^{(1)}_{[t_2,\tau_2)}, \bar{y}^{(2)}_{[t_2,\tau_2)}, \ldots \bar{y}^{(Q)}_{[t_2,\tau_2)} \right\}.$$

This division allows us to assess the degree of spin-squeezing from the variance of the conditional photocurrent distribution,

$$p(\bar{y}_{[t_2,\tau_2)} | \bar{y}_{[t_1,\tau_1)}) \sim p(\bar{y}_{[t_2,\tau_2)} - \bar{y}_{[t_1,\tau_1)}), \tag{39}$$

interpreted as the uncertainty in how well one can predict the second outcome, $\bar{y}_{[t_2,\tau_2)}$, given that the outcome of the first measurement period yielded $\bar{y}_{[t_1,\tau_1)}$. The variance of this distribution can be obtained from the statistics of the measured photocurrents,

$$\langle \Delta p^2(\bar{y}_{[t_2,\tau_2)} | \bar{y}_{[t_1,\tau_1)}) \rangle = \frac{1}{2}\left(1 + \frac{\mathrm{SNR}^2}{1+\mathrm{SNR}^2}\right) \mathrm{var}\left[\bar{\mathscr{Y}}^{(\mathrm{expt})}_{[t_2,\tau_2)} - \bar{\mathscr{Y}}^{(\mathrm{expt})}_{[t_1,\tau_1)}\right] \tag{40}$$

using the the following identity to obtain the signal to noise ratio,[7]

$$\mathrm{SNR}^2 = \frac{1}{2}\left(\frac{\mathrm{var}\left[\bar{\mathscr{Y}}^{(\mathrm{expt})}_{[t_1,\tau_1)} + \bar{\mathscr{Y}}^{(\mathrm{expt})}_{[t_2,\tau_2)}\right]}{\mathrm{var}\left[\bar{\mathscr{Y}}^{(\mathrm{expt})}_{[t_2,\tau_2)} - \bar{\mathscr{Y}}^{(\mathrm{expt})}_{[t_1,\tau_1)}\right]} - 1 \right). \tag{41}$$

5.2. *Squeezing data*

We began the experiment by operating with feedback disabled to characterize the conditional spin-squeezing produced by the measurement. Figure 6 shows the DC-coupled polarimeter photocurrent for one open-loop (feedback disabled) measurement trajectory. Time $t = 0$ corresponds to the

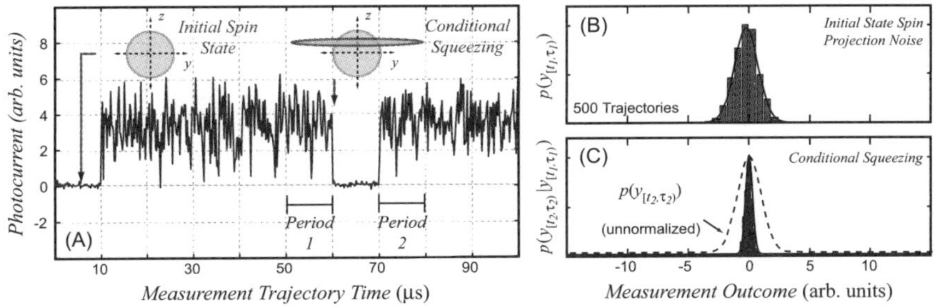

Fig. 6. An example open-loop continuous measurement record acquired with feedback disabled. The histogram (B) depicts the measurement outcome distribution $p(\bar{y}_{[t_1,\tau_1)})$ for the first measurement period. (C) depicts the conditional distribution $p(\bar{y}_{t_2,\tau_2)}|\bar{y}_{[t_1,\tau_1)})$.

beginning of the measurement record following the atom sample preparation, including trapping, cooling and optical pumping. The measurement is initiated at $t = 10$ μs by opening the probe laser shutter and the random value of F_z for that trajectory appears as an offset in the photocurrent.

We compute the outcome for the first measurement period from the mean photocurrent, $\bar{y}_{[t_1,\tau_1)}$, in the window with $t_1 = 50$ μs and $\tau_1 = 60$ μs. Immediately following this period, the measurement is disabled for 10 μs by closing the probe shutter. It is then re-enabled for the final 30 μs and the outcome for the second measurement period, $\bar{y}_{[t_2,\tau_2)}$ is obtained from the window with $t_2 = 70$ μs and $\tau_2 = 80$ μs. On a shot-to-shot basis, $\bar{y}_{[t_1,\tau_1)}$ randomly assumes different Gaussian-distributed values and the probability distribution, $p(\bar{y}_{[t_1,\tau_1)})$, of observing $\bar{y}_{[t_1,\tau_1)}$ is depicted in Fig. 6(B). This histogram was obtained from 500 continuous measurement records.

As described above, uncertainty in F_z in any individual trajectory is associated with statistical fluctuations in the average value of the photocurrent due to the presence of detection noise. To see that this uncertainty is reduced by the averaging process and that spin-squeezing is obtained, we consider the combined statistics of the two measurement periods. The outcome of the second measurement, $\bar{y}_{[t_2,\tau_2)}$, can only be predicted with improved precision provided that $\bar{y}_{[t_1,\tau_1)}$ is already known— the squeezing is conditioned on the value of $\bar{y}_{[t_1,\tau_1)}$, as expected. This can be seen from the probability distribution, $p(\bar{y}_{[t_2,\tau_2)}|\bar{y}_{[t_1,\tau_1)})$ in Fig. 6(C), which displays one tenth the variance of the initial spin projection noise. However, the average atomic state over many trajectories is not squeezed with respect to any fiducial value, such as zero, as indicated by the unconditioned distribution for the second measurement, $p(\bar{y}_{[t_2,\tau_2)})$ [dashed line in Fig. 6(C)].

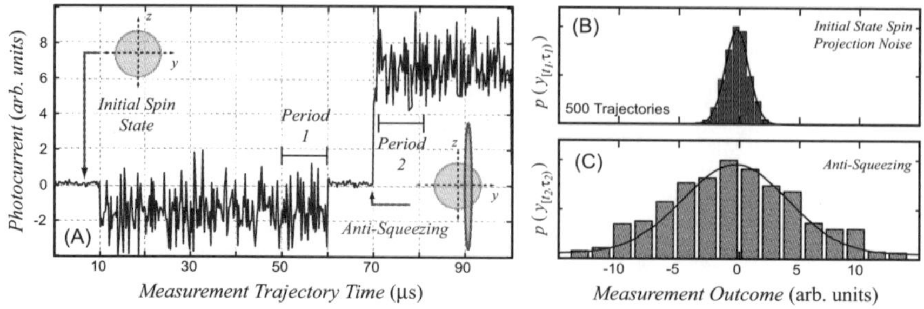

Fig. 7. (A) An example open-loop continuous measurement record acquired with feedback disabled. The histogram (B) depicts the measurement outcome distribution $p(\bar{y}_{[t_1,\tau_1)})$ for the first measiurement period. (C) depicts the distribution $p(\bar{y}_{t_2,\tau_2)})$ following a $\pi/2$ longitudinal Larmor pulse applied to reveal anti-squeezing.

It is also possible to observe anti-squeezing, the increase in uncertainty in the orthogonal transverse component, $\hat{F}_\mathrm{y}$. In Fig. 7, a precision pulsed longitudinal magnetic field is applied during the 10 μs dark period between $t = 60$ and $t = 70$ μs. This field produces a $\pi/2$ rotation around the x-axis and exchanges the transverse uncertainties, $\Delta\hat{F}_\mathrm{z}$ and $\Delta\hat{F}_\mathrm{y}$. Consequently, the anti-squeezed component is observed during the second measurement period. As expected, the distribution of measurement outcomes, $p(\bar{y}_{[t_1,\tau_1)})$, for the first period depicted in Fig. 7(B) exhibits a variance comparable to that in Fig. 6(B). However, the histogram in Fig. 7(C) reveals the substantial uncertainty increase in F_y due to anti-squeezing.

At this point, feedback stabilization of the continuous F_z measurement was enabled. As seen in Fig. 8(A), the measurement reveals an initial offset in the photocurrent that corresponds to the random value of F_z for that

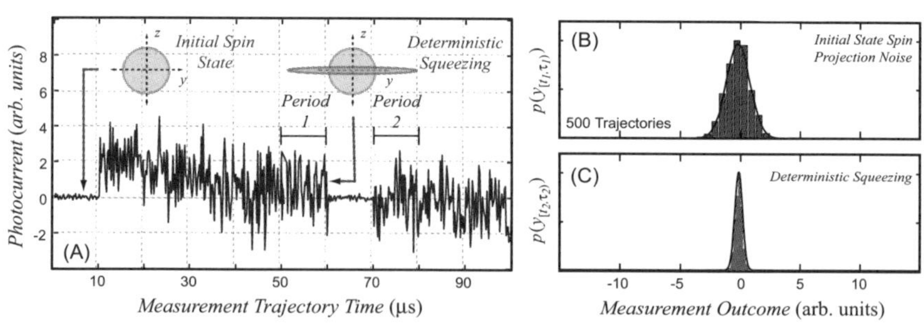

Fig. 8. (A) An example feedback-stabilized continuous measurement photocurrent. The histogram (B) depicts the measurement outcome distribution $p(\bar{y}_{[t_2,\tau_2)})$ in the absence of feedback. (C) depicts the distribution $p(\bar{y}_{t_2,\tau_2)})$ with feedback enabled to achieve deterministic spin-squeezing.

trajectory. However, this non-zero photocurrent drives a y-axis feedback control field to rotate the atomic magnetization until $F_z \sim 0$. In sharp contrast to the open-loop measurement, feedback locks the orientation of the atomic magnetization onto the x-axis despite its initial offset. The measurement $\bar{y}_{[t_2,\tau_2)}$ in the second period demonstrates that the spin state is unconditionally squeezed— the feedback-stabilized value of $\bar{y}_{[t_2,\tau_2)} \sim 0$ can be predicted with greater confidence than the initial spin projection variance. This is confirmed by the histogram of the unconditioned distribution $p(\bar{y}_{[t_2,\tau_2)})$ in Fig. 8(C). The variance of $p(\bar{y}_{[t_2,\tau_2)})$ is also about one tenth that of the initial state spin projection noise $p(\bar{y}_{[t_1,\tau_1)})$ computed in the open loop measurements [plotted in Fig. 8(B) for reference].

5.3. *Absolute spin-squeezing calibration*

Strictly speaking, the photocurrent statistics presented above do not guarantee quantum-mechanical squeezing and further analysis is necessary to do so. Many classical noise sources, including background magnetic field fluctuations and imperfect preparation of a minimum uncertainty state, are possible; the variance of the measured distribution, $p(\bar{y}_{[t_1,\tau_1)})$, could be much larger than that of a true coherent spin state. If that initial variance were sufficiently large, even the factor of 10 uncertainty reduction observed in the previous section might not be enough to bring $\langle \Delta \hat{F}_z^2 \rangle$ below $\hbar^2 F/2$.

The degree of squeezing should be evaluated by comparing the statistics of the measured conditional photocurrent distributions to an independent calibration of the coherent state-equivalent variance. This helps avoid accidentally attributing quantum squeezing to what is really a reduction in residual classical uncertainty. Fortunately, an independent calibration of the coherent state-equivalent variance is readily obtained from Eq. (24),

$$\langle \Delta p^2(\bar{y}) \rangle_{\mathrm{CSS}} \equiv \frac{SF}{2} = \frac{F}{2} \left[\frac{3\Gamma\lambda^2 P}{4\pi^2 r^2} \sum_{f'} \frac{\alpha^{(1)}_{f,f'}}{\Delta_{f,f'}} \right]^2 . \tag{42}$$

The value of the scattering strength, S, needed to evaluate $\langle \Delta p^2(\bar{y}) \rangle_{\mathrm{CSS}}$ can be obtained using readily-characterized experimental parameters such as the trap volume and probe power and detuning.[3,7] The value of S obtained in this manner can be further verified by considering full amplitude atomic Larmor precession, as depicted in Fig. 9(A). The maximum amplitude of this photocurrent is given by $y_L = \eta\sqrt{S}\hbar F$, and this calibration can be used to verify that the uncertainty of the initial optically-pumped atomic state, measured by $\langle \Delta p(\bar{y}_{[t_1,\tau_1)}) \rangle$, scales as the square root of the Larmor amplitude, y_L, when F is varied. This scaling is depicted by Fig. 9(B), measured by loading the atom trap for different lengths of time to vary the total atom number, N.

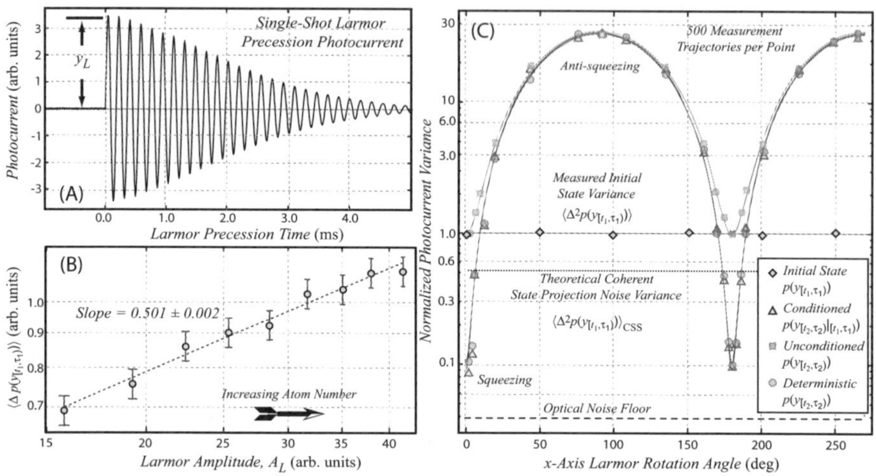

Fig. 9. Measurement statistics (A) as a function of the x-axis rotation angle applied between the two measurement periods, (B) full amplitude atom Larmor precession used to calibrate the scattering strength, S, from the Larmor height, y_L, and (C) scaling of the initial state variance.

Finally, Fig. 9(C) compares the open-loop and feedback statistics, $p(\bar{y}_{[t_2,\tau_2)})$ and $p(\bar{y}_{[t_2,\tau_2)}|\bar{y}_{[t_1,\tau_1)})$, measured in the previous section against the calibrated variance in Eq. (42). All variances in this plot are normalized with respect to that of the initial state, $p(\bar{y}_{[t_1,\tau_1)})$. The plotted curves display the expected sinusoidal form characteristic of an elliptical noise distribution[32] as a function of the x-axis rotation angle applied during the dark period between the two measurement (above we considered only $\pi/2$ pulses). We observe that the squeezing reduction in the z-component uncertainty drops below the coherent state-equivalent level, $\langle\Delta^2 p(\bar{y}_{[t_1,\tau_1)})\rangle_{\mathrm{CSS}}$ (dotted line) suggesting that quantum squeezing has been achieved. This is true in both the conditional open-loop (triangles) and deterministic feedback (circles) procedures. The unconditioned open-loop variance, $\langle\Delta^2 p(\bar{y}_{[t_2,\tau_2)})\rangle$, (squares) saturates to that of the initial state, as should be the case.

6. Conclusion

In this chapter we adopted concepts from quantum filtering theory to analyze recent experiments on continuous measurement and real-time quantum feedback control. Specifically, we considered deterministic spin-squeezing generated by embedding a continuous measurement of collective atomic spin within a feedback loop that stabilizes the outcome of the quantum measurement to a predetermined value. This required that we address the atom-probe scattering theory underlying the continuous measurement and develop a model for filtering the continuous photocurrent by averaging. Hopefully our approach provides some intuition for the mathematics encountered in quantum filtering theory by relating it to a real experiment.

Acknowledgments

The experiment described here involves work by JMG, John K. Stockton and Hideo Mabuchi. Ramon van Handel has played an instrumental role in the theoretical analysis of the results, and we thank Howard Wiseman, Andrew Doherty, Poul Jessen, Ivan Deutsch and Andrew Silberfarb for many indispensable discussions over the past several years.

References

1. J.K. Stockton, R. van Handel, and H. Mabuchi, *Deterministic dicke state preparation with continuous measurement and control*, Phys. Rev. A **70**, 22106 (2004); quant-ph/0402137.
2. R. van Handel, J.K. Stockton, and H. Mabuchi, *Feedback control of quantum state reduction*, IEEE T. Automat. Contr., to appear, 2004, quant-ph/0402093.
3. JM Geremia, J.K. Stockton, and H. Mabuchi, *Real-time quantum feedback control of atomic spin squeezing*, Science **304**, 270 (2004).
4. V.P. Belavkin, *Measurement, filtering and control in quantum open dynamical systems*, Rep. on Math. Phys. **43**, 405 (1999).
5. H. Maassen, *Quantum probability applied to the damped harmonic oscillator*, in: S. Attal and J.M. Lindsay, (Eds.), *Quantum Probability Proceedings XII*, chapter: *Quantum probability applied to the damped harmonic oscillator*, pp. 23–58, World Scientific, 2003.
6. J. K. Stockton, JM Geremia, A. C. Doherty, and H. Mabuchi, *Characterizing the entanglement of symmetric multi-particle spin-1/2 systems*, Phys. Rev. A **67**, 022112 (2003).
7. JM Geremia, J.K. Stockton, and H. Mabuchi, *Continuous measurement and conditional spin-squeezing in alkali atoms*, Phys. Rev. A, accepted, 2005, quant-ph/0501033.
8. Greg A. Smith, S. Chaudhury, A. Silberfarb, I.H. Deutsch, and P.S. Jessen, *Continuous weak measurement and nonlinear dynamics in a cold spin ensemble*, Phys. Rev. Lett. **93**, 163602 (2004).

9. W. Happer, *Optical pumping*, Rev. Mod. Phys. **44**, 169 (1972).
10. B. Julsgaard, J. Sherson, J.L. Sorensen, and E.S. Polzik, *Characterizing the spin state of an atomic ensemble using the magneto-optic resonance method*, J. Opt. B: Quantum Semiclass. Opt. **6**, 5 (2004).
11. D. J. Wineland, J. J. Bollinger, W. M. Itano, and D. J. Heinzen, *Squeezed atomic states and projection noise in spectroscopy*, Phys. Rev. A, **50**, 67 (1994).
12. M. Kitagawa and M. Ueda, *Squeezed spin states*, Phys. Rev. A, **47**, 5138 (1993).
13. A. Sørensen, L.-M. Duan, J.I. Cirac, and P. Zoller, *Many-particle entanglement with bose-einstein condensates*, Nature, **409**, 63 (2001).
14. A. Sørensen and K. Mølmer, *Entanglement and extreme spin squeezing*, Phys. Rev. Lett. **86**, 4431 (2001).
15. B. Julsgaard, A. Kozhekin, and E. S. Polzik, *Experimental long-lived entanglement of two macroscopic objects*, Nature **413**, 400 (2001).
16. JM Geremia, J.K. Stockton, A.C. Doherty, and H. Mabuchi, *Quantum kalman filtering and the heisenberg limit in atomic magnetometry*, Phys. Rev. Lett. **91**, 250801 (2003).
17. D. Oblak, J.K. Mikkelsen, W. Tittel, A.K. Vershovski, J.L. Sorensen, P.G. Petrov, C.L. Garrido Alzar, and E.S. Polzik, *Quantum noise limited interferometric measurement of atomic noise: towards spin squeezing on the cs clock transition*, Phys. Rev. A, accepted, (2003), quant-ph/0312165.
18. V. B. Braginski and F. Khalili, *Quantum Measurements*, Cambridge University Press, 1992.
19. G. Nogues, A. Rauschenbeutel, S. Osnaghi, M. Brune, J.M. Raimond, and S. Haroche, *Seeing a single photon without destroying it*, Nature **400**, 239 (1999).
20. Y. Takahashi, K. Honda, N. Tanaka, K. Toyoda, K. Ishikawa, and T. Yabuzaki, *Quantum nondemolition measurement of the spin via the paramagnetic faraday rotation*, Phys. Rev. A **60**, 4974 (1999).
21. A. Kuzmich, L. Mandel, and N. P. Bigelow, *Generation of spin squeezing via continuous quantum nondemolition measurement*, Phys. Rev. Lett. **85**, 1594–1597 (2000).
22. H. Wiseman and G. Milburn, *Quantum theory of continuous feedback*, Phys. Rev. A **49**, 2133 (1994).
23. L. K. Thomsen, S. Mancini, and H.M. Wiseman, *Spin squeezing via quantum feedback*, Phys. Rev. A **65**, 061801–1–4(R) (2002).
24. G.A. Smith, S. Chaudhury, and P.S. Jessen, *Faraday spectroscopy in an optical lattice: a continuous probe of atom dynamics*, J. Opt. B: Quant. Semiclass. Opt. **5**, 323 2003.
25. A. Silberfarb and I. Deutsch, *Continous measurement with travelling wave probes*, Phys. Rev. A **68**, 013817 (2003).
26. J. Hald, J.L. Sørensen, C. Schori, and E.S. Polzik, *Spin squeezed atoms: A macroscopic entangled ensemble created by light*, Phys. Rev. Lett. **83**, 1319 (1999).

27. H. Nha and H.J. Carmichael, *Decoherence of a two-state atom driven by coherent light*, Phys. Rev. A **71**, 013805 (2005); quant-ph/0411007.
28. I.H. Deutsch and P. Jessen, *Quantum state control in optical lattices*, Phys. Rev. A **57**, 1972 (1998).
29. C. Cohen-Tannoudji, J. Dupont-Roc, and G. Grynberg, *Atom Photon Interactions*, Wiley-Interscience, New York 1992.
30. H.M. Wiseman and G.J. Milburn, *Quantum theory of field-quadrature measurements*, Phys. Rev. A **47**, 642 (1993).
31. W. Ketterle, K. Davis, M. Joffe, A. Martin, and D.E. Pritchard, *High-densities of cold atoms in a dark spontaneous-force optical trap*, Phys. Rev. Lett. **70**, 2253 (1993).
32. L.A. Wu, H.J. Kimble, J.L. Hall, and H. Wu, *Generation of squeezed states by parametric down conversion*, Phys. Rev. Lett. **57**, 2520 (1986).

Chapter 25

Deterministic Quantum Interface between Light and Atomic Ensembles

K. Hammerer,[1] J. Sherson,[2] B. Julsgaard,[3] J. I. Cirac[1] and E. S. Polzik[2]

[1] *Max-Planck-Institut für Quantenoptik, Hans-Kopfermann-Strasse, D-85748 Garching, Germany*

[2] *QUANTOP, Danish Research Foundation Center for Quantum Optics, Niels Bohr Institute, DK-2100 Copenhagen, Denmark*

[3] *Lund Institute of Technology, Box 118, S-221 00 Lund, Sweden*

1. Introduction

Efficient quantum interface between light pulses carrying quantum information and atomic processors and storage units is an important ingredient in quantum information processing. In this chapter we describe such an interface based on the off-resonant interaction of light pulses with spin polarized atomic ensembles. The dipole interaction of light and atoms leads to two well understood phenomena: polarization rotation (phase shift) of light which depends on the quantum state of atoms and Stark shift of atomic ground states (rotation of collective atomic orientation) which depends on the quantum state of light. In addition an atomic collective orientation (we shall refer to it as "spin") of a well polarized atomic ensemble ground state is nicely described by canonical variables, just as is the state of a well polarized light pulse. As described in this chapter, such settings provide an efficient and experimentally feasible interface which has already been used to demonstrate entanglement of distant atomic objects[1] and deterministic quantum memory for light[2] described in detail later in this chapter. The range of possibilities of the interface is extended by applying quantum feedback onto light or atoms, and by the use of more than a single passage of light through atoms. The same type of interface combined with the feedback can be used for generation of spin squeezing, as described in the chapter by JM Geremia (Chapter 24). Theory for continuous measurement directly relevant to this type of interface is given in the chapters by L.B. Madsen, K. Mølmer (Chapter 22) and by R. van Handel *et al.* (Chapter 23).

It is worth emphasizing that although the interface is described here in the language of canonical continuous variables, it can be used just as well for qubit-type input states of light. Indeed, for example, when a perfect fidelity of quantum memory is predicted later in the chapter, it applies to an arbitrary single mode quantum state of light, including a qubit state. In this sense the interface provides a natural bridge between manifestly collective continuous variables of atomic ensembles and possibly discrete quantum variables for light.

An alternative approach which has been theoretically proposed as a candidate for deterministic quantum memory for light is described in the chapter by M. Fleischhauer and M. Lukin (Chapter 27).

2. Off-Resonant Interaction of Pulsed Laser Light with Spin Polarized Atomic Vapor

We consider an ensemble of spin polarized Alkali atoms whose ground state spins are probed by an off resonant pulse of weak laser light propagating in free space. We assume that we interact with a single hyperfine manifold F of the ground state. The light is blue detuned from the D_2 line, $F \rightarrow F' = F-1, F, F+1$, by $\Delta \succeq \omega_{\text{hfs}}^{\text{ex}}$, but with $\Delta \ll \omega_{\text{hfs}}^{\text{gr}}$. Here $\omega_{\text{hfs}}^{\text{ex}}$ and $\omega_{\text{hfs}}^{\text{gr}}$ are the hyperfine splittings of the excited and ground state manifolds, respectively. The effective Hamiltonian of the system is given by $H = H_{\text{at}} + H_{\text{li}} + V$ where the atomic part $H_{\text{at}} = \hbar\Omega \sum_i F_x^{(i)}$ ($F_x^{(i)}$ the x-component of the ground state spin of the i-th atom) accounts for Zeeman splitting of $\hbar\Omega$ in case an external magnetic field is applied along x. H_{li} is the free space Hamiltonian for light and the interaction term V is the level shift operator[3,4]

$$V = \int \mathrm{d}^3 r \vec{E}^{(-)}(\vec{r}) \overleftrightarrow{\alpha}(\vec{r}) \vec{E}^{(+)}(\vec{r}), \tag{1}$$

which appropriately describes the interaction of off-resonant light with atoms. We assume here implicitly that the electric field contains only frequency components within a bandwidth b around the carrier frequency ω_c of the off-resonant coherent probe pulse satisfying $\Omega \ll b \ll |\Delta|$ where Δ is the detuning from the relevant transition.

The atomic polarizability density tensor introduced in Eq. (1) is

$$\overleftrightarrow{\alpha}(\vec{r}) = \sum_i \overleftrightarrow{\alpha}^{(i)} \delta(\vec{r} - \vec{r}^{(i)}),$$

where $\vec{r}^{(i)}$ is the position of the i'th atom and $\overleftrightarrow{\alpha}^{(i)}$ is the single atom ground state polarizability $\overleftrightarrow{\alpha}$ defined by

$$\overleftrightarrow{\alpha} = \sum_{F'=F-1}^{F+1} \frac{1}{\hbar\Delta_{F'}} \vec{d}^{-}_{F,F'} \otimes \vec{d}^{+}_{F',F}.$$

Effects of thermal motion are discussed in Sec. 5.2. In the above $\vec{d}^{+}_{F',F}$ is the step up component of the dipole operator for the $F \to F'$ transition and $\Delta_{F'} = \omega_{F'} - \omega_c$ is the respective detuning. It is well known that the direct vector product in this equation can be decomposed into a scalar, vector and tensor part,

$$\overleftrightarrow{\alpha} = \frac{d^2}{\hbar(\Delta - i\Gamma/2)} \left(a_0 \mathbf{1} + a_1 \vec{F} \times + a_2 \overleftrightarrow{T} \right), \tag{2}$$

where $d = -e\langle J = 1/2 | r | J' = 3/2 \rangle$ is the relevant reduced dipole matrix element of the probed transition, Γ the excited states' decay rate, $\Delta \equiv \Delta_{F'=F+1}$ the detuning from the uppermost level, $\mathbf{1}$ is the 3×3 identity matrix and $\vec{F}\times$ has to be understood to give the vector cross product of $\vec{F}$ with the vector to the right. Now, each of the coefficients a_j is a sum of contributions from transitions to all excited states manifolds F' and depends itself on the detuning Δ. If the detuning is larger than the typical excited states' hyperfine splitting one finds that the coefficients tend to the asymptotic values

$$a_k = -\frac{(-)^{2J+F+J'+I+k}}{\sqrt{3}} \sqrt{2F+1}(2k+1) \begin{Bmatrix} J & I & F \\ F & k & J \end{Bmatrix} \begin{Bmatrix} J & J & k \\ 1 & 1 & J' \end{Bmatrix}$$
$$\times \begin{cases} 1 & \\ -\frac{1}{\sqrt{2F(F+1)}} & \text{for} \quad k = 1 \\ \frac{3}{\sqrt{10F(F+1)(2F-1)(2F+3)}} & \end{cases} \begin{matrix} 0 \\ \\ 2 \end{matrix} \tag{3}$$

where the expressions in the curly brackets are $6j$-symbols. For the D_2 line, $J = 1/2$, $J' = 3/2$, the prefactor of the tensor part, a_2, vanishes since the triple $\{J, J, k\} = \{1/2, 1/2, 2\}$ does not satisfy the triangle inequality. For Cesium, $I = 7/2$, the asymptotic values for the different coefficients are

$$a_0 = 1/6,\ a_1 = -1/48,\ a_2 = 0, \qquad \Delta \to \infty. \tag{4}$$

In the following we will assume that the detuning is large enough to neglect the effects of the tensor polarizability.

The non-hermitian part of the resulting Hamilton operator describes the effect of light absorption and loss of ground state population due to depumping in the course of interaction. In the following we will focus on the coherent interaction and, for the time being, take into account only the hermitian component. The effects of light absorption and subsequent atomic depumping are treated in Sec. 4.

Since scattering of light occurs predominantly in the forward direction[5] it is legitimate to adopt a one dimensional model such that the (negative frequency component of the) electric field propagating along z is given by

$$\begin{aligned}\vec{E}^{(-)}(z,t) &= E^{(-)}(z)\vec{e}_y + \mathcal{E}^{(-)}(z,t)\vec{e}_x\\ E^{(-)}(z) &= \rho(\omega_c)\int_b \mathrm{d}\omega a^\dagger(\omega)e^{-ikz}\\ \mathcal{E}^{(-)}(z,t) &= \rho(\omega_c)\sqrt{2\pi N_{\mathrm{ph}}/T}e^{-i(k_c z-\omega_c t)},\end{aligned}$$

where $\rho(\omega)=\sqrt{\hbar\omega/4\pi\epsilon_0 Ac}$ and A denotes the pulse's cross sectional area, N_{ph} the overall number of photons in the pulse and T its duration. We restrict the field in the x polarization to the classical probe pulse, since only the coupling of atoms to the y polarization is enhanced by the coherent probe. Furthermore we implicitly assume for the classical pulse a slowly varying envelope such that it arrives at $z=0$ at $t=0$ and is then constant for a time T. Combining this expression for the field with expression (2) for the atomic polarizability in Eq. (1) yields

$$V=-\frac{i\hbar\kappa}{\sqrt{4\pi JT}}\int_b \mathrm{d}\omega\int \mathrm{d}z j(z)\left(a(\omega)e^{-i[(k_c-k)z-\omega_c t]}-h.c.\right),$$

where we defined an atomic spin density $j_z(z)=\sum_i F_z^{(i)}\delta(z-z^{(i)})$ and a dimensionless coupling constant

$$\kappa=\sqrt{\frac{N_{\mathrm{ph}}J}{2}\frac{\omega_c a_1 d^2}{\hbar\epsilon_0 cA\Delta}}. \tag{5}$$

In the expression for the interaction V we skipped terms proportional to a_0 which will give rise only to a global phase shift (corresponding to the index of refraction of the atomic medium) and included for convenience the square root factor with $J=N_{\mathrm{at}}F$ where N_{at} is the number of atoms.

We now define slowly varying field quadratures for spatially localized modes[6,7] as

$$\begin{aligned}x(z) &= \frac{1}{\sqrt{4\pi}}\int_b \mathrm{d}\omega\left(a(\omega)e^{-i[(k_c-k)z-\omega_c t]}+h.c.\right),\\ p(z) &= -\frac{i}{\sqrt{4\pi}}\int_b \mathrm{d}\omega\left(a(\omega)e^{-i[(k_c-k)z-\omega_c t]}-h.c.\right)\end{aligned}$$

with commutation relations $[x(z), p(z')] = ic\delta(z - z')$ where the delta function has to be understood to have a width on the order of c/b. Since we assumed that $\Omega \ll b$, the time it takes for such a fraction of the pulse to cross the ensemble is much smaller than the Larmor period $1/\Omega$. During the interaction with one of these spatially localized modes the atomic state does not change appreciable and we can simplify the interaction operator to

$$V = \hbar\kappa(JT)^{-1/2} J_z p(0), \tag{6}$$

where $J_z = \sum_i F_z^{(i)}$ and we assumed that the ensemble is located at $z = 0$.

A last approximation concerns the description of the atomic spin state. Initially the sample is prepared in a coherent spin state with maximal polarization along x, *i.e.* in the eigenstate of J_x with maximal eigenvalue J. We can thus make use of the Holstein-Primakoff approximation[8] which allows us to describe the spin state as a Gaussian state of a single harmonic oscillator. The first step is to express collective step up/down operators (along x), $J_\pm = J_y \pm iJ_z$, in terms of bosonic creation and annihilation operators, $[b, b^\dagger] = \mathbf{1}$, as

$$J_+ = \sqrt{2J}\sqrt{\mathbf{1} - b^\dagger b/2J}\, b, \quad J_- = \sqrt{2J} b^\dagger \sqrt{\mathbf{1} - b^\dagger b/2J}.$$

It is easily checked that these operators satisfy the correct commutation relations $[J_+, J_-] = 2J_x$ if one identifies $J_x = J - b^\dagger b$. The fully polarized initial state thus corresponds to the ground state of the harmonic oscillator. Note that this mapping is exact. Under the condition that $\langle b^\dagger b \rangle \ll J$ one can approximate $J_+ \simeq \sqrt{2J}b$, $J_- \simeq \sqrt{2J}b^\dagger$ and therefore $J_z \simeq -i\sqrt{J/2}(b - b^\dagger)$. Introducing atomic quadratures $X = (b + b^\dagger)/\sqrt{2}$ and $P = -i(b - b^\dagger)/\sqrt{2}$ finally yields the desired expression for the interaction $V = \hbar\kappa T^{-1/2} Pp(0)$. In terms of atomic quadratures the free Hamiltonian for atoms is $H = \hbar\Omega/2(X^2 + P^2)$.

The hermitian part of the full Hamiltonian is thus given by

$$\begin{aligned} H &= H_{\text{at}} + H_{\text{li}} + V, \\ H_{\text{at}} &= \frac{\hbar\Omega}{2}(X^2 + P^2), \\ V &= \frac{\hbar\kappa}{\sqrt{T}} Pp(0). \end{aligned} \tag{7}$$

Note that in the frame rotating at the carrier frequency the action of H_{li} on the light quadratures $x(z), p(z)$ is simply $i/\hbar[H_{\text{li}}, x(z)] = -c\frac{\partial}{\partial z}x(z)$ and likewise for $p(z)$.

3. Equations of Motion

We will now solve the resulting Heisenberg equations of motion for three configurations which have been used in the experiments described in the following sections: In Sec. 3.1 we treat the most simple case of a single sample with no external field applied, a situation as for example in the experiments on spin squeezing.[9] In Sec. 3.2 we address with the setup[1,2] in which two samples were placed close to each other in two oppositely oriented magnetic fields. Finally, in Sec. 3.3 we treat the case of a single sample with an external magnetic field applied.

3.1. *Single sample without magnetic field*

In this case $\Omega \equiv 0$ in the basic Hamiltonian (7) and the Heisenberg equations of motion are simply

$$\tfrac{\partial}{\partial t}X(t) = \frac{\kappa}{\sqrt{T}}p(0,t), \qquad \tfrac{\partial}{\partial t}P(t) = 0, \tag{8a}$$

$$\left(\tfrac{\partial}{\partial t} + c\tfrac{\partial}{\partial z}\right)x(z,t) = \frac{\kappa c}{\sqrt{T}}P(t)\delta(z), \qquad \left(\tfrac{\partial}{\partial t} + c\tfrac{\partial}{\partial z}\right)p(z,t) = 0.$$

To solve this set of coupled equations it is convenient to introduce a new position variable $\xi = ct - z$ to eliminate the z dependence. New light quadratures defined by $\bar{x}(\xi,t) = x(ct-\xi,t)$, $\bar{p}(\xi,t) = p(ct-\xi,t)$ also have a simple interpretation: ξ labels the slices of the pulse moving in and out of the ensemble one after the other, starting with $\xi = 0$ and terminating at $\xi = cT$. The Maxwell equations now read

$$\tfrac{\partial}{\partial t}\bar{p}(\xi,t) = 0, \quad \tfrac{\partial}{\partial t}\bar{x}(\xi,t) = \frac{\kappa c}{\sqrt{T}}P(t)\delta(ct-\xi). \tag{8b}$$

Equations (8) have to be integrated form t=0 at which the classical pulse is assumed to enter the sample up to time T when the pulse terminates. Integrating the Maxwell-Bloch equation causes no problems and the well known solutions are given by

$$\hat{X}^{+}out = \hat{X}^{+}in + \kappa p^{\,\mathrm{in}}, \quad P^{\,\mathrm{out}} = P^{\,\mathrm{in}}, \tag{9a}$$

$$x^{\,\mathrm{out}} = x^{\,\mathrm{in}} + \kappa P^{\,\mathrm{in}}, \quad p^{\,\mathrm{out}} = p^{\,\mathrm{in}}. \tag{9b}$$

where we defined input/output operators as $\hat{X}^{+}in[out] = X(0)[X(T)]$ and $x^{\,\mathrm{in[out]}} = \frac{1}{\sqrt{T}}\int_0^T \mathrm{d}\tau\,\bar{x}(c\tau, 0[T])$ and the same for P and p.

3.2. *Two samples in oppositely oriented magnetic fields*

If two samples, described by two pairs of canonical conjugate operators $[X_k, P_k] = i\delta_{kl}$, $k, l = 1, 2$, are placed in two oppositely oriented magnetic fields the full Hamiltonian of the system is given by

$$\begin{aligned} H &= H_{\rm at} + H_{\rm li} + V, \\ H_{\rm at} &= \frac{\hbar\Omega}{2}(X_1^2 + P_1^2) - \frac{\hbar\Omega}{2}(X_2^2 + P_2^2), \\ V &= \frac{\hbar\kappa}{\sqrt{T}}(P_1 + P_2)p(0). \end{aligned} \tag{10}$$

In V we have neglected the distance between the two samples and placed them formally both at $z = 0$. Since the light modes we are dealing with are spatially localized only within c/b this is well justified as long as the distance between the samples is much smaller than this length. Note that this assumption can easily be dropped and the results below still hold in principle.

It is convenient to change in the atomic subsystem to EPR modes defined by $X_\pm = (X_1 \pm X_2)/\sqrt{2}$, $P_\pm = (P_1 \pm P_2)/\sqrt{2}$ in terms of which the free atomic Hamiltonian becomes $H_{\rm at} = \Omega(X_+X_- + P_+P_-)$. Changing to an interaction picture with respect to $H_{\rm at}$ and evaluating the Heisenberg equations yields the Maxwell-Bloch equations

$$\begin{aligned} &\tfrac{\partial}{\partial t}X_+(t) = \kappa\sqrt{\frac{2}{T}}\cos(\Omega t)p(0,t), \quad \tfrac{\partial}{\partial t}P_+(t) = 0, \\ &\tfrac{\partial}{\partial t}P_-(t) = -\kappa\sqrt{\frac{2}{T}}\sin(\Omega t)p(0,t), \quad \tfrac{\partial}{\partial t}X_-(t) = 0, \\ &\tfrac{\partial}{\partial t}\bar{x}(\xi,t) = \kappa c\sqrt{\frac{2}{T}}\left[\cos(\Omega t)P_+(t) + \sin(\Omega t)X_-(t)\right]\delta(ct-\xi), \\ &\tfrac{\partial}{\partial t}\bar{p}(\xi,t) = 0. \end{aligned}$$

The integration is again straight forward and one arrives at the input/output relations

$$\hat{X}^{+}+out = \hat{X}^{+}+in + \kappa p_{\rm c}^{\rm in} \qquad x_{\rm c}^{\rm out} = x_{\rm c}^{\rm in} + \kappa P_+^{\rm in} \tag{11a}$$

$$P_+^{\rm out} = P_+^{\rm in} \qquad p_{\rm c}^{\rm out} = p_{\rm c}^{\rm in} \tag{11b}$$

$$\hat{X}^{+}-out = \hat{X}^{+}-in \qquad x_{\rm s}^{\rm out} = x_{\rm s}^{\rm in} + \kappa\hat{X}^{+}-in \tag{11c}$$

$$P_-^{\rm out} = P_-^{\rm in} - \kappa p_{\rm s}^{\rm in} \qquad p_{\rm s}^{\rm out} = p_{\rm s}^{\rm in} \tag{11d}$$

where we defined input output operators of a cosine modulation mode as

$$p_{\rm c}^{\rm in[out]} = \sqrt{\frac{2}{T}} \int_0^T {\rm d}\tau \cos(\Omega\tau)\bar{p}(c\tau, 0[T]), \tag{12a}$$

$$x_{\rm c}^{\rm in[out]} = \sqrt{\frac{2}{T}} \int_0^T {\rm d}\tau \cos(\Omega\tau)\bar{x}(c\tau, 0[T]) \tag{12b}$$

and the same for the pairs $x_{\rm s}^{\rm in[out]}$, $p_{\rm s}^{\rm in[out]}$ with $\cos(\Omega\tau)$ replaced by $\sin(\Omega\tau)$. In deriving the second equations in (11a) and (11c) we used $\int_0^T {\rm d}\tau \cos(\Omega\tau)\sin(\Omega\tau) = \mathcal{O}(n_0^{-1}) \simeq 0$. In frequency space these modes consist of spectral components at sidebands $\omega_c \pm \Omega$ and are in fact closely related to the sideband modulation modes introduced in[10,11] for the description of two photon processes. It is easily checked that these modes are asymptotically canonical, $[x_{\rm c}^{\rm in}, p_{\rm c}^{\rm in}] = [x_{\rm s}^{\rm in}, p_{\rm s}^{\rm in}] = i[1 + \mathcal{O}(n_0^{-1})] \simeq i$, and independent, $[x_{\rm c}^{\rm in}, p_{\rm s}^{\rm in}] = \mathcal{O}(n_0^{-1}) \simeq 0$, if we assume $n_0 \gg 1$ for $n_0 = \Omega T$, the pulse length measured in periods of Larmor precession.

Note that the solution (11) falls naturally into two groups (11a),(11b) and (11c),(11d) which have each, up to a phase difference, the same structure as the solution (9) in the last section.

3.3. *Single sample in magnetic field*

Next we are dealing with the case of a single atom placed in an external magnetic field. The Hamiltonian for this system is now given by (7) with $\Omega \neq 0$. We change again to an interaction picture with respect to the Zeeman term $H_{\rm at}$. In principle the operators in the interaction picture should be denoted by X_I, P_I but in the following we will suppress the index. The resulting Maxwell-Bloch equations are

$$\tfrac{\partial}{\partial t} X(t) = \frac{\kappa}{\sqrt{T}} \cos(\Omega t) p(0,t), \tag{13a}$$

$$\tfrac{\partial}{\partial t} P(t) = \frac{\kappa}{\sqrt{T}} \sin(\Omega t) p(0,t), \tag{13b}$$

$$\tfrac{\partial}{\partial t} \bar{p}(\xi,t) = 0, \tag{13c}$$

$$\tfrac{\partial}{\partial t} \bar{x}(\xi,t) = \frac{\kappa c}{\sqrt{T}} \left[\cos(\Omega t)P(t) - \sin(\Omega t)X(t)\right] \delta(ct-\xi). \tag{13d}$$

Equations (13a), (13b) and (13c) are again integrated without difficulty and the solution in terms of cosine and sine modulation modes as defined in (12) is

$$\hat{X}^{+}out = \hat{X}^{+}in + \frac{\kappa}{\sqrt{2}} p_{\mathrm{c}}^{\mathrm{in}}, \tag{14a}$$

$$P^{\mathrm{out}} = P^{\mathrm{in}} + \frac{\kappa}{\sqrt{2}} p_{\mathrm{s}}^{\mathrm{in}}, \tag{14b}$$

$$p_{\mathrm{c}}^{\mathrm{out}} = p_{\mathrm{c}}^{\mathrm{in}}, \tag{14c}$$

$$p_{\mathrm{s}}^{\mathrm{out}} = p_{\mathrm{s}}^{\mathrm{in}}. \tag{14d}$$

Deriving the corresponding expressions for the cosine and sine components of the field in phase, $x_{\mathrm{c}}^{\mathrm{out}}$, $x_{\mathrm{s}}^{\mathrm{out}}$, raises some difficulties connected to the back action of light onto itself. Let us first integrate (13d) formally as

$$\bar{x}(\xi, t) = \bar{x}(\xi, 0) + \frac{\kappa}{\sqrt{T}} \left[\cos(\Omega\xi/c) P(\xi/c) - \sin(\Omega\xi/c) X(\xi/c).\right]$$

Using the solutions of (13a) and (13b) for intermediate times $0 \leq t \leq T$ and again the approximate orthogonality of cosine and sine over the interval $[0, T]$ we find for example for the cosine modulation mode

$$\begin{aligned} x_{\mathrm{c}}^{\mathrm{out}} &= \sqrt{\frac{2}{T}} \int_0^T \mathrm{d}\tau \cos(\Omega\tau) \bar{x}(c\tau, T) \\ &= x_{\mathrm{c}}^{\mathrm{in}} + \frac{\kappa}{\sqrt{2}} P^{\mathrm{in}} + \frac{\sqrt{2}\kappa^2}{T^{3/2}} \int_0^T \mathrm{d}\tau \int_0^\tau \mathrm{d}\tau' [\cos(\Omega\tau)^2 \sin(\Omega\tau') \bar{p}(c\tau', 0) \\ &\quad - \cos(\Omega\tau) \sin(\Omega\tau) \cos(\Omega\tau') \bar{p}(c\tau', 0)]. \end{aligned}$$

After interchanging the order of integration, $\int_0^T \mathrm{d}\tau \int_0^\tau \mathrm{d}\tau' \to \int_0^T \mathrm{d}\tau' \int_{\tau'}^T \mathrm{d}\tau$ one can perform the integration over τ. Neglecting all terms of order n_0^{-1} or less where $n_0 = \Omega T \gg 1$ one finds

$$x_{\mathrm{c}}^{\mathrm{out}} = x_{\mathrm{c}}^{\mathrm{in}} + \frac{\kappa}{\sqrt{2}} P^{\mathrm{in}} + \frac{\sqrt{2}\kappa^2}{T^{3/2}} \int_0^T \mathrm{d}\tau \frac{T - \tau}{2} \sin(\Omega\tau) \bar{p}(c\tau, 0).$$

The last term represents back action of light onto itself. It can be expressed as a sum of two terms, one proportional to $p_{\mathrm{s}}^{\mathrm{in}}$ and another one proportional to

$$p_{\mathrm{s},1}^{\mathrm{in}} = \sqrt{3} \left(\frac{2}{T}\right)^{3/2} \int_0^T \mathrm{d}\tau \left(\frac{T}{2} - \tau\right) \sin(\Omega\tau) \bar{p}(c\tau, 0).$$

It is easily verified that the back action mode defined by this equation and the corresponding expression for $x_{\mathrm{s}}^{\mathrm{back}}$ is canonical $[x_{\mathrm{s},1}^{\mathrm{in}}, p_{\mathrm{s},1}^{\mathrm{in}}] = i[1 - \mathcal{O}(n_0^{-2})] \simeq i$ and independent from all the other modes introduced so far, f.e. $[x_{\mathrm{s}}^{\mathrm{in}}, p_{\mathrm{s},1}^{\mathrm{in}}] = \mathcal{O}(n_0^{-2}) \simeq 0$. Repeating the calculation for $x_{\mathrm{s}}^{\mathrm{out}}$

with appropriate replacements and a definition of $p_{\mathrm{c},1}^{\mathrm{in}}$ analogous to the last equation finally yields

$$x_{\mathrm{c}}^{\mathrm{out}} = x_{\mathrm{c}}^{\mathrm{in}} + \frac{\kappa}{\sqrt{2}} P^{\mathrm{in}} + \left(\frac{\kappa}{2}\right)^2 p_{\mathrm{s}}^{\mathrm{in}} + \frac{1}{\sqrt{3}} \left(\frac{\kappa}{2}\right)^2 p_{\mathrm{s},1}^{\mathrm{in}}, \tag{14e}$$

$$x_{\mathrm{s}}^{\mathrm{out}} = x_{\mathrm{s}}^{\mathrm{in}} - \frac{\kappa}{\sqrt{2}} \hat{X}^{+}in - \left(\frac{\kappa}{2}\right)^2 p_{\mathrm{c}}^{\mathrm{in}} - \frac{1}{\sqrt{3}} \left(\frac{\kappa}{2}\right)^2 p_{\mathrm{c},1}^{\mathrm{in}}. \tag{14f}$$

Equation (14) describe the final state of atoms and scattered light after the pulse has passed the atomic ensemble.

4. The Role of Dissipation

In order to treat the role of dissipation let us come back to the basis of the interaction under consideration, the single atom polarizability as given in (2). The anti hermitian part of this operator describes depumping of ground state population and photon absorption. In the limit of $\Delta \gg \Gamma$ it is given by

$$\boldsymbol{\alpha}_h = \frac{id^2\Gamma}{2\hbar\Delta^2}\left(a_0\mathbf{1} + ia_1\vec{F}\times\right).$$

For a single atom the dominating loss term stems from the scalar part of the polarizability and is given by $V_{\mathrm{loss}} = i\hbar\eta\mathbf{1}/4T$ where we defined the dimensionless parameter $\eta = N_{\mathrm{ph}}a_0\omega_c\Gamma d^2/2\hbar\Delta^2\epsilon_0 Ac = N_{\mathrm{ph}}a_0\sigma\Gamma^2/4A\Delta^2$. In the Maxwell-Bloch equations this term will cause damping of the transverse spin components. For example in the case of a single sample without magnetic field as in Sec. 3.1 we get

$$\tfrac{\partial}{\partial t}X(t) = -\frac{\eta}{T}X(t) + \frac{\kappa}{\sqrt{T}}p(0,t) + \sqrt{\frac{\eta}{T}}V_X(t), \tag{15a}$$

$$\tfrac{\partial}{\partial t}P(t) = -\frac{\eta}{T}P(t) + \sqrt{\frac{\eta}{T}}V_P(t). \tag{15b}$$

We included here delta-correlated Langevin noise terms, $\langle V_a(t)V_b(t')\rangle = V_{\mathrm{at}}^2\delta(t-t')\delta_{ab}$ which in principle can be derived from microscopic considerations as it was done in Ref. 12. In practice, however, there are additional sources of decoherence of the atomic spin state. For a thermal sample of atoms contained in a glass cell light induced collisions and collisions with walls will contribute also to relaxation of transverse spin components. Without a microscopic model for the various decay processes the parameters β and V_{at}^2 have to be determined experimentally. See Sec. 5.

For weak damping the decay can be treated linearly such that the solutions to Eq. (15) can be approximated by

$$\hat{X}^{+}out = \sqrt{\beta}(\hat{X}^{+}in + \kappa p^{\text{in}}) + \sqrt{1-\beta}V_X \tag{16a}$$
$$P^{\text{out}} = \sqrt{\beta}P^{\text{in}} + \sqrt{1-\beta}V_P, \tag{16b}$$

where now $\langle V_X^2\rangle = \langle V_P^2\rangle = V_{\text{at}}^2$.

Photon absorption on the other hand, *i.e.* the scattering of light into other modes than the laser mode, can be treated similarly such that

$$x^{\text{out}} = \sqrt{\zeta}(x^{\text{in}} + \kappa P^{\text{in}}) + \sqrt{1-\zeta}V_x \tag{16c}$$
$$p^{\text{out}} = \sqrt{\zeta}p^{\text{in}} + \sqrt{1-\zeta}V_P \tag{16d}$$

again with $\langle V_x\rangle = \langle V_p\rangle = V_{\text{li}}^2$. The overall loss of photons, $\zeta = \epsilon + r$ is due to absorption during the interaction characterized by $\epsilon = N_{\text{at}}\sigma\Gamma^2/A\Delta^2$ and reflection losses, r, when light for example crosses surfaces of glass cells.

Input/output equations analogous to (16) hold also for the solutions given in Secs. 3.2 and 3.3 for the case of one or two samples placed in a magnetic field.

A central quantity in this system is the optical density on resonance $\alpha_0 = N_{\text{at}}\sigma/A$ which gives the probability for a single photon to get elastically scattered and which can be related to the other parameters as $\epsilon = \alpha_0(\Gamma/\Delta)^2$ and $\kappa^2 = \eta\alpha_0$. Note that there is an apparent trade off between having a large coupling and at the same time low atomic depumping. For a given optical density one can treat ϵ and η in a certain range as independent parameters tailoring the first by means of the detuning and the last by means of N_{ph}. This implies in particular that, if we are interested only in a certain quantity, as for example entanglement generated in the interaction or spin squeezing as it can be achieved from the state after the interaction, there will always be optimal values of ϵ and η maximizing these quantities for fixed optical density α. See in particular Sec. 9.

5. Experimental Implementations

The experimental implementation of the light-atoms quantum interface has been performed using room temperature Cesium gas contained in a glass cell. A special paraffin coating inside the cell protects atoms against collisional decoherence providing the ground state coherence time of up to 40 msec. It should be noted here that the general approach to the light-atoms interface described in this chapter is applicable to a broad range of materials, including ensembles of hot and cold atoms in the gas phase and solid state media. The medium should just fulfill the following basic requirements: it should have a long lived effectively two-level ground state, optical pumping of the ground state should be possible, the ground state should be coupled to some excited state via an optical transition, and the medium should have significant optical depth for the light resonant with this transition.

The convenient possibility of using gas at room temperature is due to the fact that the interaction is off-resonant, and hence Doppler broadening is irrelevant. The motion of atoms to the first approximation is not an obstacle if the optical beam cross section covers most of the atomic ensemble and the pulse duration is much longer than the transient time of an atom through the beam. However, even under these conditions small corrections due to atomic motion are necessary, as described in this section.

A very important experimental task is to measure the projection noise level, *i.e.* the noise of uncorrelated atoms which originates from the Heisenberg uncertainty principle in the minimum uncertainty state. First of all, this is necessary as a starting point for all quantum information protocols with continuous variables - one has to be able to initialize the system in the minimal uncertainty state. Secondly, the projection noise of atoms compared to the shot noise of light is a very convenient measure of the coupling strength between light and matter. We discuss the projection noise level in Sec. 5.1.

5.1. *Determination of the projection noise level*

In Sec. 3 the reduced variables x, p, X, and P (see *e.g.* Eqs. (9)) were defined. These are convenient since we then have a very well known commutation relation $[x, p] = i$. From this follows the Heisenberg uncertainty relation $\mathrm{Var}(x)\mathrm{Var}(p) \geq 1/4$, and for the minimum uncertainty state (symmetric in x and p) we have $\mathrm{Var}(x) = \mathrm{Var}(p) = 1/2$. For atoms these minimum uncertainty states are the coherent spin states (CSS).

If we now prepare the CSS for a single atomic sample in the absence of a magnetic field or for two oppositely oriented samples in a magnetic field and send a pulse of light through, the interaction is described by Eqs. (9) and (11), respectively. In any case, the variance of the outgoing light pulse is

$$\mathrm{Var}(x^{\mathrm{out}}) = \mathrm{Var}(x^{\mathrm{in}}) + \kappa^2 \mathrm{Var}(P^{\mathrm{in}}) = \frac{1 + \kappa^2}{2}. \tag{17}$$

The ratio of atomic to light noise is κ^2 which is conveniently found experimentally. However, it would be nice to have some theoretical prediction of the projection noise level based on independent measurements. Consider the definition of κ in Eq. 5 of Sec. 2. We have

$$\kappa^2 = \left(\frac{\omega_c a_1 d^2}{\hbar \epsilon_0 c A \Delta}\right)^2 \frac{N_{\text{ph}} J}{2}. \tag{18}$$

From this we see that the projection noise scales linearly with the macroscopic spin size J. A convenient way to measure J is to exploit the same light/matter interaction as discussed in Sec. 2 by sending a linearly polarized beam of light *along* the macroscopic spin J_x. The spin will cause a polarization rotation of this light of a magnitude depending on J.

To see this we will reuse our formalism of Sec. 2 by directing the macroscopic spin along z, $J_z = J$. In particular expression (6) shows that the x-quadrature of the outgoing, y-polarized field will be $x(z)_{\text{out}} = x(z)_{\text{in}} + \frac{\kappa}{\sqrt{JT}} J_z$. Given that the slowly varying field amplitude is connected to the quadrature by $E(z) = \rho(\omega_c)\sqrt{4\pi}x(z)$ the change in the amplitude will be $\Delta E(z) = \rho(\omega_c)\sqrt{4\pi}[x(z)_{\text{out}} - x(z)_{\text{in}}] = \rho(\omega_c)\kappa\sqrt{4\pi J/T}$. Now, if we assume that the classical pulse of light is only rotated by a small angle θ away from the x-axis such that $\Delta E(z) = \tan\theta\mathcal{E}(z) \simeq \theta\mathcal{E}(z)$ and take into account that the slowly varying envelope of the classical pulse was $\mathcal{E}(z) = 2\rho(\omega_c)\sqrt{2\pi N_{\text{ph}}/T}$ we get

$$\theta = \frac{\kappa}{2}\sqrt{\frac{2J}{N_{\text{ph}}}} = \frac{J\omega_c a_1 d^2}{2\hbar\epsilon_0 c A \Delta}. \tag{19}$$

We see that J can be found as the polarization rotation of the light pulse together with fundamental constants. Note, we wish to know the J which corresponds to *all* atoms in the cubic vapor cell. If the beam only fills the area A we only probe the fraction $A \cdot l/V$ of atoms where l is the length of the sample and V is the vapor cell volume. By replacing A with $A_{\text{eff}} \equiv V/l$ in the above equation the value of J will exactly correspond to all atoms and we need not to worry about the focusing of the probing beam for the measurement of macroscopic spin magnitude J. The polarization rotation depends on atomic density only.

5.2. *The effect of atomic motion*

When probing along a direction with no mean spin the motion of the atoms in and out of the beam will be important for the measured statistics. In our experiments the atoms have time to move across the beam several times during a pulse which will lead to an averaging effect, all atoms spend some time in the beam. In this section we discuss the implications of atomic motion on the measured projection noise level. The results are related to the work of Ref. 13.

We consider Eq. (9) for a pulse propagating through atoms which we now assume to be moving. For this reason we write $P^{\text{in}} = J_z/\sqrt{J} = \sum_i p_i F_z^{(i)}/\sqrt{J}$ where p_i is the probability of finding the i'th atom inside the light beam and $F_z^{(i)}$ is an individual atomic spin operator. This definition leads to the variance

$$\begin{aligned}\text{Var}(x^{\text{out}}) &= \text{Var}(x^{\text{in}}) + \frac{\kappa^2}{J}\left\langle\left(\sum_{i=1}^{N_{\text{at}}} p_i F_z^{(i)}\right)^2\right\rangle \\ &= \frac{1}{2} + \frac{\kappa^2}{2}\left(\frac{1}{N}\sum_{i=1}^{N}\langle p_i^2\rangle\right).\end{aligned} \tag{20}$$

In the second step we assumed the coherent spin state where all atoms are independent with $\text{Var}(F_z^{(i)}) = F/2$. The last parenthesis can be expressed in terms of the mean and variance of p_i since $\text{Var}(p_i) = \langle p_i^2\rangle - \langle p_i\rangle^2$. Let us define $p = \langle p_i\rangle$ and $\text{Var}(p_i) = \sigma^2 \cdot p^2$. With this definition σ is the relative standard deviation of p. Then we get

$$\text{Var}(x^{\text{out}}) = \frac{1 + \kappa^2 \cdot p^2(1+\sigma^2)}{2}. \tag{21}$$

The average fraction of time p each atom spends inside the beam is clearly $p = A/A_{\text{eff}}$ and we may absorb this p into κ by replacing A with A_{eff} in the estimate of κ^2 in Eq. (18). This is again equivalent to letting J denote the macroscopic spin size of *all* atoms in the sample. In addition we see there will be a relative increase in the measured projection noise of σ^2 which depends only on the mean and variance of the time each atom spends inside the beam.

Let us now discuss the scaling of σ^2 with simple physical parameters. The fact that the variance may be non-zero arises from the finite time available for the averaging process carried out by the atomic motion. A typical traversing time across the vapor cell is $\tau = L/v_0$ where L is the cell dimension and v_0 is *e.g.* the one-dimensional rms speed of the atoms. We may think of this atomic motion as n independent journeys across the vapor cell volume, where $n \approx T/\tau = Tv_0/L$. We then model the motion through the beam with mean occupancy p by assuming in each walk across the cell volume that either (1) the atom spends all the time τ inside the beam. This should happen with probability p, or (2) the atom spends all the time τ outside the beam which should happen with probability $1-p$.

We then count the number of times n_{inside} that an atom was inside the beam out of the possible n journeys. In this simple model n_{inside} is a stochastic variable which is binomially distributed with mean np and variance $np(1-p)$. We are interested in the fraction of time ($\approx n_{\text{inside}}/n$) spent inside the beam. It follows $\langle n_{\text{inside}}/n\rangle = p$ and $\sigma^2 = \text{Var}([n_{\text{inside}}/n]/p) = (1-p)/np$. Hence the simple model will predict

$$p = \frac{A}{A_{\text{eff}}} \quad \text{and} \quad \sigma^2 = \frac{(A_{\text{eff}} - A)L}{ATv_0}, \tag{22}$$

where A_{eff} is the effective transverse area of the vapor cell which is equal to L^2 for a cubic cell. Note the characteristic scaling with T^{-1} and with the area ($A_{\text{eff}} - A$) *not* covered by the light beam (when A is close to its maximum value A_{eff}). We note that due to the simplicity of the above model the absolute numbers should only hold as an order of magnitude estimate. Numerical simulations performed for a cubic cell have shown that the relative variance σ^2 is roughly four times smaller than the estimate above. Also, due to the Doppler broadening, the effective detuning will differ from atom to atom and cause an increase in σ^2. We shall not discuss these issues theoretically anymore but just state that putting this discussion together with Eqs. (18) and (19) we end up with a convenient way to predict the projection noise level as we discuss next.

5.3. *Predicting the projection noise level*

Let us now connect Eqs. (18), (19), and (21) in order to predict the measured atomic to shot noise ratio which we shall denote by κ^2_{th}. In Eqs. (18) and (19) we replace A by A_{eff} in order that J denotes the total spin size of all atoms in the sample. We assume that we can use the same A_{eff} for both of these equations. This is valid for a cubic cell or an irregular cell with the same sample length in both the z and x directions. We find

$$\kappa^2_{\text{th}} = \frac{(1+\sigma^2)\Gamma\lambda^3 PT\theta}{32\pi^2 A_{\text{eff}}\Delta\hbar c} = \frac{56.4 \cdot P[\text{mW}] \cdot T[\text{ms}] \cdot \theta[\text{deg}] \cdot (1+\sigma^2)}{A_{\text{eff}}[\text{cm}^2] \cdot \Delta[\text{MHz}]}. \tag{23}$$

To reach this equation we related the number of photons N_{ph} to power and pulse duration by $P \cdot T = \hbar\omega_c N_{\text{ph}}$ and the dipole moment d can be shown[14] to be related to the FWHM line width Γ of the optical transition by $d^2 = 12\pi\epsilon_0\hbar c^3\Gamma/\omega^3$. In the second step we insert $\Gamma = 5.21\text{MHz}$, $\lambda = 852.3\text{nm}$. We express the remaining physical quantities in convenient units from an

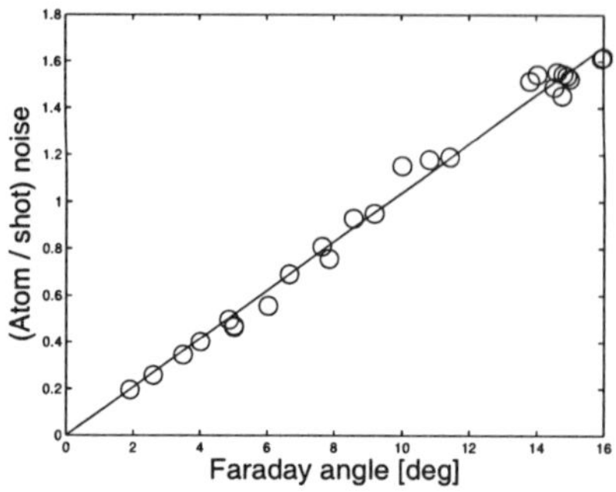

Fig. 1. Measured atomic noise relative to shot noise of light. The linearity is a clear signature of the projection noise limitation. The slope $\kappa^2 = 0.104(2) \cdot \theta$ should be compared to the theoretical value of $\kappa^2_{\text{theory}} = 0.115 \cdot \theta$ from Eq. (23) with $\sigma^2 = 0$. There is a discrepancy of 11% which will be even worse if we include a non-zero σ^2. ($T_{\text{probe}} = 2.0$ms, $\Delta = 700$MHz, $P = 4.5$mW).

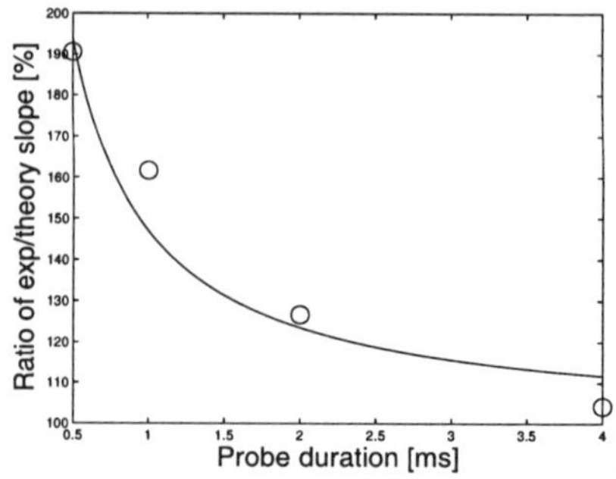

Fig. 2. Slope of measured κ^2 vs. θ normalized to the experimentally predicted level (without the factor $1+\sigma^2$) vs. T_{probe}. The fit gives $\kappa^2_{\text{exp}}/\kappa^2_{\text{th}} = 1 + 0.47(13)/T$[ms].

experimental point of view. In case of light losses between atoms and the detector, it can be shown that the correct prediction is found if the *detected* power P is inserted above.

Turning to experimental investigation, in Fig. 1 we see an example where we plot the measured noise relative to the shot noise of light. The data are clearly linear. With $\Delta = 700$MHz, $T = 2.0$ms, $P = 4.5$mW, and $\sigma^2 = 0$ for the moment, we predict a linear slope of 0.115 which is somewhat higher than the measured one. Including the σ^2 from atomic motion will only make this discrepancy worse but we definitely have the right order of magnitude.

To test the scaling properties predicted in the atomic motion calculations, we fix the power P, detuning Δ and macroscopic spin size J but vary the probe duration T. The measured noise is plotted in Fig. 2 relative to the prediction (23) with $\sigma^2 = 0$. We see that as T is increased we do see a lower and lower noise level which corresponds to the decreasing σ^2. The solid line in the figure represents a fit where $\sigma^2 = (0.47 \pm 0.13)/T$[ms]. To compare this to the simple model (22) we estimate our beam diameter to be 1.6cm which gives, $A \approx 2.0$cm, we have $L = 3.0$cm, $v_0 = 13.7$cm/ms (cesium at room temperature). For $T = 1$ms we get the prediction $\sigma^2 = 0.44$. This is in very good agreement with the measured data but this must be taken as purely coincidental. As mentioned before, numerical simulations of atomic

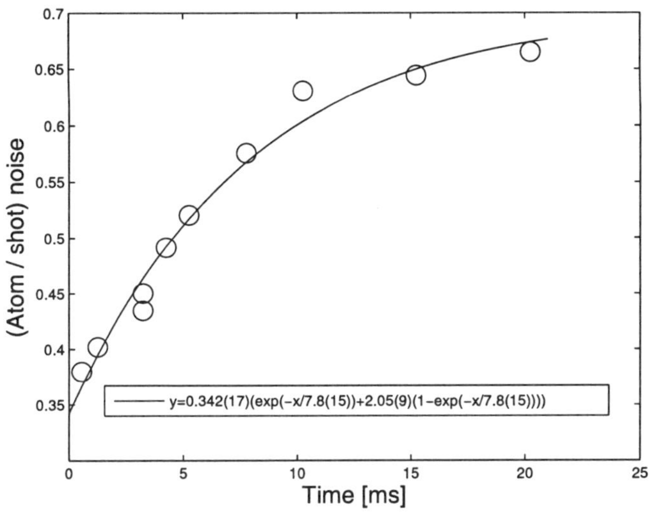

Fig. 3. Coherent state noise compared to the completely unpolarized spin noise. The data is taken with a vapor cell in which the spin life time is very short. The noise level increases on a time scale of roughly 8ms to the thermal equilibrium level. The increase in noise is consistent with predictions for the coherent and unpolarized spin states.

motion have shown that the variance estimate (22) is almost four times too high. The high experimental value must be attributed to the additional Doppler broadening effect. We also note the relatively high uncertainty of 0.13. But all together we have a qualitative understanding of the physics and a quantitative agreement within a few tens of percent.

5.4. *Thermal spin noise*

Another issue concerning the projection noise level is the question of thermal spin noise. For the establishment of the correct noise level we must be in the CSS with high precision. For the CSS the spin is completely polarized along the x-direction and $\mathrm{Var}(j_y) = \mathrm{Var}(j_z) = F/2 = 2$ for the $F = 4$ ground state. As a very different example we may consider a completely unpolarized sample. We then have by symmetry $\mathrm{Var}(j_x) = \mathrm{Var}(j_y) = \mathrm{Var}(j_z) = (j_x^2 + j_y^2 + j_z^2)/3 = F(F+1)/3 = 20/3$. This is a factor of 10/3 higher and even for a fairly good polarization the thermal noise may be significant. In our experiments with quantum information protocols we exceed a spin polarization of 99% which means that the thermal noise must be very small compared to the true projection noise. The degree of spin polarization has been measured independently with methods similar to those in Ref. [Julsgaard *et al.* on magnetometers].

A nice illustration of the fact that we get lower noise for the CSS than in the unpolarized case is given in Fig. 3. Experimentally, we perform measurements on very poor vapor cells where the macroscopic spin life time is small. We do optical pumping as well as possible and wait for some variable delay time before probing the spin noise. For long times the spins will reach thermal equilibrium, where the noise of each atom in $F = 4$ contributes 20/3. The fraction of atoms in $F = 4$ is 9/16, the remaining 7/16 are in the $F = 3$ state and do not contribute because of the large detuning. Initially, all atoms are in $F = 4$ in the CSS and they each contribute the value 2 to the noise. Hence the measured noise must be on the form

$$\text{Measured noise} \propto 2 \cdot \exp(-\Gamma t) + \frac{20}{3} \cdot \frac{9}{16}(1 - \exp(-\Gamma t)). \tag{24}$$

The predicted ratio of final to initial noise is thus $15/8 \approx 1.88$. Experimentally we find the ratio 2.05 ± 0.09 which is consistent. To sum up, there is strong evidence that we really do create the CSS with the correct minimum uncertainty noise.

5.5. *Quantumness of the noise*

After the discussion in Sec. 5.2 of the extra contribution to the measured quantum projection noise from atomic motion one may speculate whether we really see true quantum noise and whether the extra contribution gives rise to an unjustified additional coupling strength between light and atoms.

To answer the first question, the atomic motion leads to the re-definition of the collective spin operators $J_y \to \sum p_i F_y^{(i)}$ and $J_z \to \sum p_i F_z^{(i)}$. We then do not have a real angular momentum, but there is still sense in these quantum variables. The commutator is

$$\left[\sum_{i=1}^{N_{\text{at}}} p_i F_y^{(i)}, \sum_{i=1}^{N_{\text{at}}} p_i F_z^{(i)}\right] = \sum_{i=1}^{N_{\text{at}}} p_i^2 \left[F_y^{(i)}, F_z^{(i)}\right] = i \sum_{i=1}^{N_{\text{at}}} p_i^2 F_x^{(i)}. \tag{25}$$

This leads to the Heisenberg uncertainty relation (for a highly polarized sample with $F_x \approx F$)

$$\text{Var}\left(\sum_{i=1}^{N} p_i F_y^{(i)}\right) \text{Var}\left(\sum_{i=1}^{N} p_i F_z^{(i)}\right) \geq \left(\frac{J}{2} p^2 (1 + \sigma^2)\right)^2. \tag{26}$$

Since we for the coherent spin state may calculate

$$\text{Var}\left(\sum_{i=1}^{N} p_i F_z^{(i)}\right) = \frac{J}{2} p^2 (1 + \sigma^2) = \text{Var(CSS)}, \tag{27}$$

we see that this corresponds to the minimum uncertainty state. The measured noise is then limited by the Heisenberg uncertainty principle and we confidently call this projection noise. To maintain the correct commutation relation $[X, P] = i$ we experimentally normalize the atomic operators to the *measured* projection noise, *i.e.* instead of defining $X = J_y/\sqrt{J}$ we effectively have $X = \sum_i p_i F_y^{(i)}/\sqrt{Jp^2(1+\sigma^2)}$.

This brings us to the second question of this section. Is the extra contribution parametrized by σ^2 a cheap way to reach the quantum limit and therefore an unjustified advantage? The answer is no, the atomic motion is a dis-advantage after a closer look. To see this, imagine that we perform some manipulations of atoms by one laser pulse and subsequently probe these manipulations by another laser pulse. Since atoms move during interactions the probed quantum operator changes in time. Comparing the operator at the 1st and 2nd times we get

$$\begin{aligned}\mathrm{Var}\left(\sum_{i=1}^{N} p_{i,\mathrm{2nd}} F_{z,\mathrm{2nd}}^{(i)} - \sum_{i=1}^{N} p_{i,\mathrm{1st}} F_{z,\mathrm{1st}}^{(i)}\right) &= \sum_{i=1}^{N} \mathrm{Var}(F_{z,\mathrm{1st}}^{(i)}) \left\langle (p_{i,\mathrm{2nd}} - p_{i,\mathrm{1st}})^2 \right\rangle \\ = \frac{J}{2} \cdot 2p^2\sigma^2 = 2\mathrm{Var}(\mathrm{CSS})(1-\sqrt{\beta}) \quad \text{with} \quad \sqrt{\beta} &= \frac{1}{1+\sigma^2}.\end{aligned} \tag{28}$$

We assumed $p_{i,\mathrm{1st}}$ and $p_{i,\mathrm{2nd}}$ to be uncorrelated, when an atom hits the cell wall its new direction is completely random. Also, we took $F_{z,\mathrm{1st}}^{(i)} = F_{z,\mathrm{2nd}}^{(i)}$. This corresponds to having no decoherence at all apart from the effect of atomic motion which is the only effect studied in this calculation. Furthermore we take $\mathrm{Var}(F_{y,\mathrm{2nd}}^{(i)}) = \mathrm{Var}(F_{z,\mathrm{2nd}}^{(i)}) = F/2$ which is always a good approximation for highly polarized samples. To interpret the above calculations we consider a standard decoherence calculation. Consider a true spin operator J_z subject to decoherence parametrized by the number β such that

$$J_{z,\mathrm{1st}} \longrightarrow J_{z,\mathrm{2nd}} = \sqrt{\beta} J_{z,\mathrm{1st}} + \sqrt{1-\beta} J_{\mathrm{vac}} \text{ with } \mathrm{Var}(J_{\mathrm{vac}}) = \frac{J}{2} = \mathrm{Var}(\mathrm{CSS}). \tag{29}$$

Then the operator changes by an amount characterized by the variance

$$\begin{aligned}\mathrm{Var}\left(J_z^{\mathrm{2nd}} - J_z^{\mathrm{1st}}\right) &= \mathrm{Var}\left(J_z^{\mathrm{1st}}(1-\sqrt{\beta}) - \sqrt{1-\beta} J_{\mathrm{vac}}\right) \\ &= J(1-\sqrt{\beta}) = 2\mathrm{Var}(\mathrm{CSS})(1-\sqrt{\beta})\end{aligned} \tag{30}$$

which is exactly the same as in (28). We must interpret the result such that motion of atoms inevitably gives rise to an effective decoherence. So to answer the initial question, with a high σ^2 it may seem easier to reach the quantum limited regime in actual measurements but this limit becomes decreasingly useful for higher σ^2. Higher σ^2 will not lead to higher fidelity protocols.

6. Entanglement Generation and Verification

Two atomic samples described by the modes (X_1, P_1) and (X_2, P_2) are entangled in the usual EPR-sense when the following criterion is fulfilled:

$$\mathrm{Var}(X_-) + \mathrm{Var}(P_+) = \frac{1}{2}\left[\mathrm{Var}(X_1 - X_2) + \mathrm{Var}(P_1 + P_2)\right] < 1. \tag{31}$$

This statement can be proved rigorously[15] and it can be understood intuitively by noting that in the case the samples are independent and in the minimum uncertainty state with $\mathrm{Var}(X_1) = \mathrm{Var}(X_2) = \mathrm{Var}(P_1) = \mathrm{Var}(P_2) = 1/2$ we have the equality fulfilled which is the best we can obtain classically. For entanglement generation we thus need to reduce the variance below this level.

Experimentally we perform the following steps: First the atoms are prepared in the oppositely oriented coherent states corresponding to creating the vacuum states of the atomic modes (X_+, P_+) and (X_-, P_-). Next a pulse of light called the *entangling pulse* is sent through atoms and we measure the two operators x_s^{out} and x_c^{out} with outcomes A_1 and B_1, respectively. These results bear information about the atomic operators X_-^{in} and P_+^{in}, respectively, see Eq. (11). The knowledge gained from the measurement reduces variances $\mathrm{Var}(X_-)$ and $\mathrm{Var}(P_+)$. To prove we have an entangled state we must confirm that the variances of X_- and P_+ fulfill the criterion (31). That is we need to know the mean values of X_- and P_+ with a total precision better than unity. For this demonstration we send a second *verifying pulse* through the atomic samples again measuring x_s^{out} and x_c^{out} with outcomes A_2 and B_2. Now it is a matter of comparing A_1 with A_2 and B_1 with B_2. If the results are sufficiently close the state created by the first pulse was entangled.

Now let us be more quantitative. The interaction (11) mapping the atomic operators X_- and P_+ on light is very useful for a strong κ and useless if $\kappa \ll 1$. We will describe in detail the role of κ for all values. A convenient way of deducing κ experimentally is given in Sec. 5.1. This κ will be used below to predict theoretically the amount on entanglement observed. Experimentally, we need to deduce the statistical properties of the state created by the *entangling pulse*. Based on the measurement results

A_1 and B_1 of this pulse we must predict the mean value of the second measurement outcome. If $\kappa \to \infty$ we ought to trust the first measurement completely since the initial noise of $x_{s,c}^{\text{in}}$ is negligible, *i.e.* $\langle A_2 \rangle = A_1$ and $\langle B_2 \rangle = B_1$. On the other hand, if $\kappa = 0$ we know that atoms must still be in the vacuum state such that $\langle A_2 \rangle = \langle B_2 \rangle = 0$. It is natural to take in general $\langle A_2 \rangle = \alpha A_1$ and $\langle B_2 \rangle = \alpha B_1$. We need not know a theoretical value for α to proceed. The actual experimental value can be deduced from the data. If we repeat the measurement cycle N times with outcomes $A_1^{(i)}$, $B_1^{(i)}$, $A_2^{(i)}$, and $B_2^{(i)}$, the correct α is found by minimizing the conditional variance

$$\begin{aligned} &\text{Var}(A_2|A_1) + \text{Var}(B_2|B_1) = \\ &\qquad \min_{\alpha} \frac{1}{N-1} \sum_{i}^{N} \left((A_2^{(i)} - \alpha A_1^{(i)})^2 + (B_2^{(i)} - \alpha B_1^{(i)})^2 \right). \end{aligned} \tag{32}$$

In order to deduce whether we fulfill the entanglement criterion (31) we compare the above to our expectation from (11). For the verifying pulse we get

$$\begin{aligned} \left\langle \left(x_s^{\text{out}} - \langle x_s^{\text{out}} \rangle\right)^2 \right\rangle &= \left\langle \left(x_s^{\text{in,2nd}} + \kappa \left[X_-^{\text{ent}} - \langle X_-^{\text{ent}} \rangle\right]\right)^2 \right\rangle \\ &= \frac{1}{2} + \kappa^2 \text{Var}(X_-^{\text{ent}}), \end{aligned} \tag{33}$$

and similarly

$$\left\langle \left(x_c^{\text{out}} - \langle x_c^{\text{out}} \rangle\right)^2 \right\rangle = \frac{1}{2} + \kappa^2 \text{Var}(P_+^{\text{ent}}), \tag{34}$$

where $x_{s,c}^{\text{in,2nd}}$ refer to the incoming light of the *verifying pulse* which has zero mean. X_-^{ent} and P_+^{ent} refer to the atoms after being entangled. We see that the practical entanglement criterion becomes

$$\begin{aligned} \text{Var}(A_2|A_1) + \text{Var}(B_2|B_1) &= 1 + \kappa^2 \left(\text{Var}(X_-^{\text{ent}}) + \text{Var}(P_+^{\text{ent}})\right) \\ &< 1 + \kappa^2 = \text{Var}(A_1) + \text{Var}(B_1). \end{aligned} \tag{35}$$

In plain English, we must predict the outcomes A_2 and B_2 with a precision better than the statistical spreading of the outcomes A_1 and B_1 with the additional constraint that A_1 and B_1 are outcomes of quantum noise limited measurements.

6.1. *Theoretical entanglement modeling*

Above we described the experimental procedure for generating and verifying the entangled states. Here we present a simple way to derive what we expect for the mean values (*i.e.* the α-parameter) and for the variances $\mathrm{Var}(X_-^{\mathrm{ent}})$ and $\mathrm{Var}(P_+^{\mathrm{ent}})$.

We calculate directly the expected conditional variance of A_2 based on A_1:

$$\begin{aligned}&\left\langle\left(x_s^{\mathrm{out,2nd}}-\alpha x_s^{\mathrm{out,1st}}\right)^2\right\rangle\\=&\left\langle\left(x_s^{\mathrm{in,2nd}}-\alpha x_s^{\mathrm{in,1st}}+\kappa\left[X_-^{\mathrm{ent}}-\alpha X_-^{\mathrm{in}}\right]\right)^2\right\rangle\\=&\frac{1}{2}(1+\alpha^2+\kappa^2(1-\alpha)^2).\end{aligned} \tag{36}$$

In the second step we assumed that the measurement is perfectly QND and without any decoherence, *i.e.* $X_-^{\mathrm{ent}}=X_-^{\mathrm{in}}$. By taking the derivative with respect to α we obtain the theoretical minimum

$$\begin{aligned}\mathrm{Var}(A_2|A_1)+\mathrm{Var}(B_2|B_1)&=1+\frac{\kappa^2}{1+\kappa^2}\\\Rightarrow \mathrm{Var}(X_-^{\mathrm{ent}})+\mathrm{Var}(P_+^{\mathrm{ent}})&=\frac{1}{1+\kappa^2}\end{aligned} \tag{37}$$

obtained with the α-parameter

$$\alpha=\frac{\kappa^2}{1+\kappa^2}. \tag{38}$$

We see that in principle any value of κ will lead to creation of entanglement. The reason for this is our prior knowledge to the entangling pulse. Here the atoms are in the coherent state which is as well defined in terms of variances as possible for separable states. We only need an "infinitesimal" extra knowledge about the spin state to go into the entangled regime.

It is interesting to see what happens to the conjugate variables X_+ and P_- in the entangling process. This is governed by Eq. (11). We do not perform measurements of the light operator $p_{s,c}^{\mathrm{in}}$ so all we know is that both X_+^{in}, P_-^{in} and $p_{s,c}^{\mathrm{in}}$ are in the vacuum state. Hence $\mathrm{Var}(X_+^{\mathrm{ent}})=\mathrm{Var}(P_-^{\mathrm{ent}})=(1+\kappa^2)/2$ and we preserve the minimum uncertainty relation $\mathrm{Var}(X_+^{\mathrm{ent}})\mathrm{Var}(P_+^{\mathrm{ent}})=\mathrm{Var}(X_-^{\mathrm{ent}})\mathrm{Var}(P_-^{\mathrm{ent}})=1/4$.

6.2. *Entanglement model with decoherence*

Practically our spin states decohere between the light pulses and also in the presence of the light. We model this decoherence naively by putting the entire effect between the two pulses, *i.e.* we assume there is no decoherence in presence of the light but a larger decoherence between the pulses. We may then perform an analysis in complete analogy with the above with the only difference that $X_-^{\text{ent}} = \sqrt{\beta}X_-^{\text{in}} + \sqrt{1-\beta}\hat{V}_x$ where $\hat{V}_x$ is a vacuum operator admixed such that $\beta = 0$ corresponds to a complete decay to the vacuum state and $\beta = 1$ corresponds to no decoherence. Completing the analysis we find the theoretical conditional variances

$$\begin{aligned} \text{Var}(A_2|A_1) + \text{Var}(B_2|B_1) &= 1 + \kappa^2 \frac{1 + (1-\beta)\kappa^2}{1+\kappa^2} \\ \Rightarrow \text{Var}(X_-^{\text{ent}}) + \text{Var}(P_+^{\text{ent}}) &= \frac{1 + (1-\beta)\kappa^2}{1+\kappa^2} \end{aligned} \tag{39}$$

obtained with α-parameter

$$\alpha = \frac{\sqrt{\beta}\kappa^2}{1+\kappa^2}. \tag{40}$$

In the limit $\beta \to 1$ these results agree with (37) and (38). For $\beta \to 0$ we have $\alpha \to 0$ (outcomes A_1 and B_1 are useless) and the variance approaches that of the vacuum state which is a separable state.

6.3. *Experimental entanglement results*

For the experimental demonstration of entanglement generation we need to employ the criterion (35). To this end we perform several measurements of $x_{s,c}$ as a function of the macroscopic spin size in order to find the projection noise level. This has already been discussed in Sec. 5.1. In Fig. 4 we show both the noise of the first and second pulses (with squares and circles). These have the same noise level since the measurement is a QND-type. We plot only the atomic part of the noise by subtracting the shot and electronics noise of a single light pulse.

Next we compare the results of the first and second pulse measurements as discussed above. In Fig. 4 the tip down triangles show the conditional variance $\text{Var}(A_2|A_1) + \text{Var}(B_2|B_1)$ normalized to shot noise and with shot and electronics noise subtracted. According to (35) we thus plot $\kappa^2(\text{Var}(X_-^{\text{ent}}) + \text{Var}(P_+^{\text{ent}}))$. The fact that the points are lower than the straight line (κ^2) is a direct indication that the entanglement criterion (31) is fulfilled. For the higher densities the reduction is 25% but we note that entanglement is also observed for smaller densities with $\kappa^2 < 1$. The corresponding α-parameters from the minimization procedure (32) are plotted in Fig. 4 with tip up triangles.

The expected entangled noise level in the ideal case is given by (37). This is drawn as the dash-dotted curve (κ^2 times $1/(1+\kappa^2)$ in order to normalize to the straight line). We see the conditional variance lies higher than this

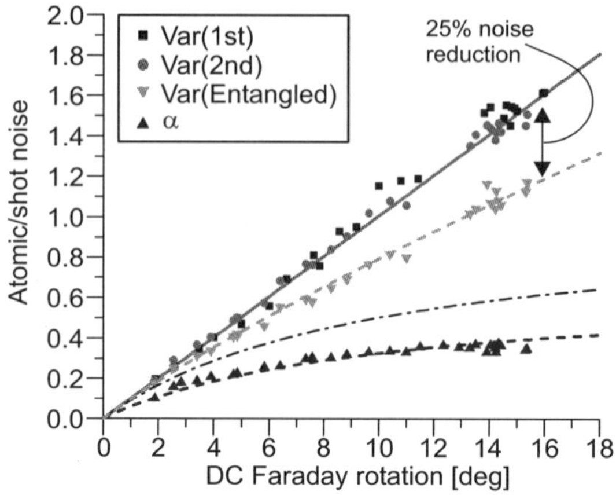

Fig. 4. Atomic noise in units of shot noise as a function of the macroscopic spin size (measured by DC Faraday rotation). Squares show the 1st pulse noise which are identical to the data in Fig. 1. The circles show 2nd pulse noise. Tip down triangles show the noise of the entangled states as estimated by (32) with shot and electronics noise subtracted. Tip up triangles show the weight factor α. The two dashed curves trough triangles is the model described in Eqs. (39) and (40) with $\sqrt{\beta} = 0.65$. The dash-dotted curve is the theoretically best for the triangles ($\beta = 1$). The states created by the first pulse measurements are really entangled states (according to the criterion (35)) since the noise is clearly below the straight line fit. We observe up to 25% noise reduction. Note, entanglement is observed for low densities also with $\kappa^2 < 1$.

curve and hence the entanglement is worse than expected. According to (38) we also would expect the α-parameters to lie on the same dash-dotted curve in the ideal case. It is clearly not the case, the experimental α-parameters are lower which indicates that the results A_1 and B_1 can not be trusted to as high a degree as expected.

Let us try to apply the simple decoherence model given by Eqs. (39) and (40). Taking the decoherence parameter $\sqrt{\beta} = 0.65$ we get the dashed lines in the figure. These match nicely the experimental data. We conclude that the simple decoherence model has some truth in it and we must accept that the entangled state created can only be verified to be around "65% as good" as expected in an ideal world.

7. Quantum Memory

In Eqs. (9) and (11) describing the interaction of light with a single sample without a magnetic field and two oppositely oriented samples in a homogeneous magnetic field respectively, one of the light variables is mapped onto one of the atomic variables. Since the two expressions are equivalent we will stick to the generic formalism of Eq. (9) in the following. This represents a natural starting point for a quantum memory protocol in which the entire light mode described by the two non-commuting variables x^{in} and p^{in} is faithfully stored. In the so-called "direct mapping protocol" of[2] the mapping is completed by measuring the remaining light quadrature $x^{\text{out}} = x^{\text{in}} + \kappa P^{\text{in}}$ and feeding the result back into the atomic X with a gain of g:

$$X^{\text{out}} = X^{\text{in}} + \kappa p^{\text{in}} \tag{41a}$$

$$P^{\text{out}\,\prime} = P^{\text{out}} - g x^{\text{out}} = P^{\text{in}}(1 - \kappa g) - g x^{\text{in}}. \tag{41b}$$

If $\kappa = g = 1$ and the initial atomic state is assumed to be a coherent state with zero mean value as discussed in Sec. 5 the mean values of both light variables will be stored faithfully in the atoms. Although the initial atomic state has zero mean it is a quantum mechanical fluctuating state and any uncanceled atomic part will increase the noise of the final state and thus degrade the quantum mapping performance. In the following we will deal with storage of coherent states of light, *i.e.* vacuum states which are displaced by an unknown amount in phase space. For the storage of an arbitrary coherent light state the remaining X^{in} contribution limits the storage fidelity to 82%. This can be remedied by initially squeezing the atomic state, in which case 100 % fidelity can be reached in the limit of infinite squeezing.

For a quantum memory we require 1) that the light state to be stored is supplied by a third party in an unknown state and 2) that this state is mapped onto an atomic state with a fidelity higher than the best classical fidelity. For coherent states it was recently shown in Ref. 16 that the optimum classical strategy is to split the unknown light pulse in two using a beam splitter and measure separate light quadratures in each of the two parts. The measurement results are then fed back onto the atomic sample initially in the vacuum state. If the unknown light stated is taken from a Gaussian distribution with mean photon number, $\bar{n}$, the best classical fidelity was shown to be:

$$F_{\text{class}} = \frac{1+\bar{n}}{1+2\bar{n}} \to \frac{1}{2}, \qquad \bar{n} \to \infty. \tag{42}$$

This means that F_{class} decreases monotonously from unity for the vacuum state to 1/2 for the an arbitrary coherent state. For experimental reasons discussed below we will map with non-unity gain. In this case we will have

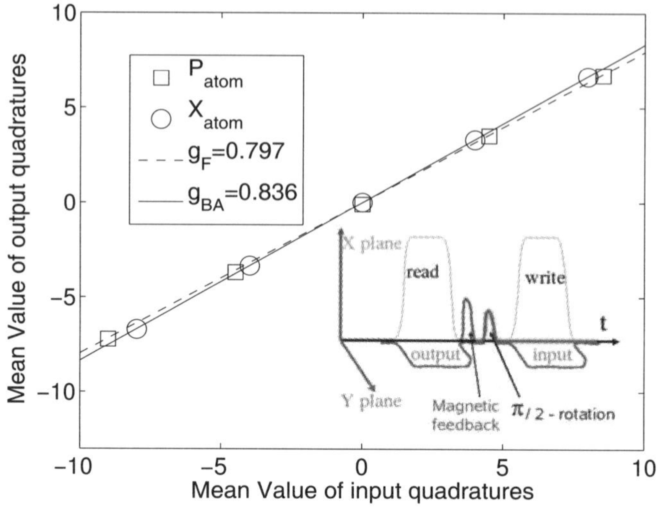

Fig. 5. Mean value of the read out pulse as a function of the mean values of the input light variables, x^{in} and p^{in}, to be stored. Inset: the strong classical and the weak quantum pulses in opposite polarizations. Between the input and the output pulses are the feedback pulse and the optional $\pi/2$ pulse.

to restrict the range of input states and the achieved fidelity will have to be compared with Eq. (42) with an appropriate $\bar{n}$.

7.1. *Experimental verification of quantum memory*

A storage cycle like the one described above involves 1) preparation of the initial atomic state in a coherent state via optical pumping, 2) mapping of one of the light quadratures through the off-resonant Faraday interaction, Eq. 9, and 3) storage of the second light quadrature by a direct measurement of this quadrature and subsequent feedback into the atoms. In a complete quantum memory one would now apply a procedure to retrieve the stored state. As discussed in Sec. 7.3 until very recently no protocol for quantum memory retrieval had been proposed which was experimentally feasible for the current implementation so instead we have determined the quality of mapping by performing a destructive reconstruction of the stored state statistics.This is done by waiting for a time τ and then sending a readout light pulse through the atomic sample. This will then give information of the atomic P^{out} or X^{out} if a $\pi/2$ rotation in the atomic XP space is performed prior to the readout pulse. Repeating this 10,000 times, the statistics for the atomic variables after the storage procedure can be reconstructed. The first

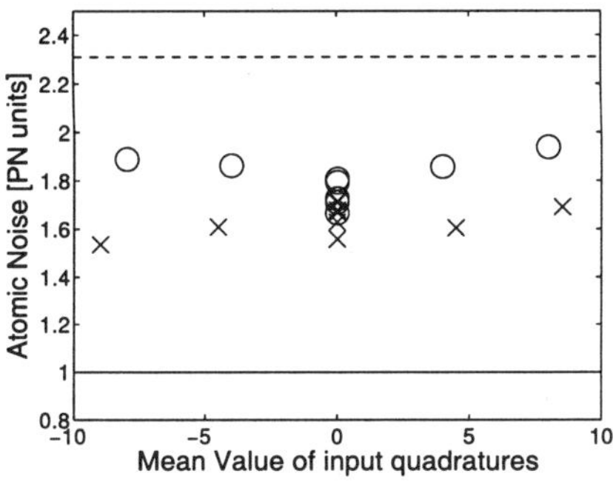

Fig. 6. Variance of the readout quadratures with the intrinsic shot noise of the readout pulse subtracted as a function of the mean values of the input light variables. Full drawn curve: the variance for a perfect quantum storage. Dotted curve: classical limit on the variances for $n_o = 4$. Crosses: Var(P^{atom}), circles: Var(X^{atom}).

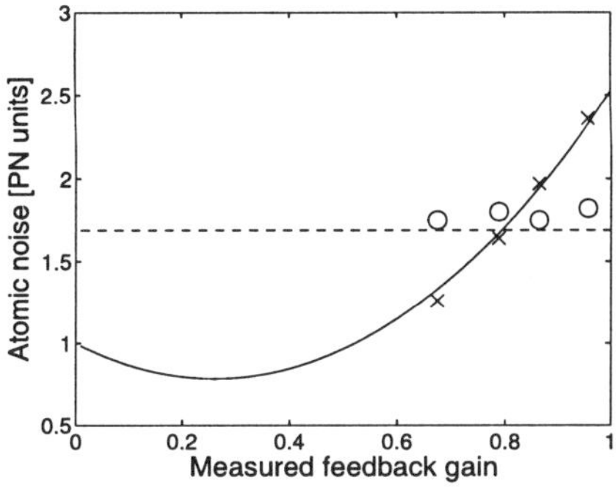

Fig. 7. Atomic variances as a function of feedback gain with realistic noise parameters included. Crosses: Var(P^{atom}), circles: Var(X^{atom}).

thing to check is that the mean value of the stored state depends linearly in the mean value of the input light state. This is shown in Fig. 5. First of all we note that the linear dependence is clear for both quadratures. This completes the proof of classical memory performance. The next thing to note is that the slope is not unity, which means that the stored state has a different mean value than the input state. The reason for this will be discussed further below. For the quadrature mapped straight from the back action of the light onto the atoms we have the gain of $g_{\text{BA}} = 0.836$ and for the quadrature mapped via the feedback we have $g_{\text{F}} = 0.797$.

In order to verify quantum storage we also need to consider the shot-to-shot fluctuations in the stored state, which for a Gaussian state are fully characterized by the variance of the state. The experimentally reconstructed variances of the atomic quadratures, that is the variance of the readout pulse with one unit of shot noise subtracted, are shown in Fig. 6. Also shown is the ideal quantum limit on the variance for a perfect mapping and the classical limit for $n_o = 4$. As can be seen the variance is more or less independent of the mean value of the input light quadratures. The fidelity of the stored state for a Gaussian distribution of input states with mean photon number n_o can be calculated given the measured gains and variances, σ_x and σ_p, according to:

$$F = \frac{2}{\sqrt{(2n_o(1-g_{\text{BA}})^2 + 1 + \sigma_x)(2n_o(1-g_{\text{F}})^2 + 1 + \sigma_p)}}. \tag{43}$$

With the experimentally measured values we get $F = (66.7 \pm 1.7)\%$ for $\bar{n} = 4$ and $F = (70.0 \pm 2.0)\%$ for $\bar{n} = 2$ with the corresponding classical boundaries of 55.4% and 59.6% respectively. This verifies that the storage of the light state in fact constitutes a quantum memory according to the criteria stated above. The results shown were obtained for a storage time, $\tau = 0.5$, but the memory has been shown to work for up to 4 ms. Note that we have chosen to calculate the fidelity as the average of the squared overlap between the stored state and the ideally stored state. For a non-unity gain this decreases very rapidly although one could argue that a storage with an arbitrary but known gain constitutes just as useful a memory as the unity gain memory. If analyzed solely based on the added noise, our memory would therefore perform better than the previously stated results, which can therefore be viewed as a lower bound on memory capability.

7.2. *Decoherence*

The gain lower than unity is caused by decoherence of the mapped atomic state and reflection losses of light. As discussed in Sec. 4 both of these can be described by an admixture of vacuum components on. With atomic and light dampings of β and ζ respectively Eq. (41) transforms to:

$$X^{\text{out}} \to \sqrt{\beta}(X^{\text{in}} + \kappa p^{\text{in}}) + \sqrt{1-\beta}V_X \tag{44a}$$

$$P^{\text{out}} \to (\sqrt{\beta} - g\kappa\sqrt{\zeta})P^{\text{in}} - g\sqrt{\zeta}x^{\text{in}} + \sqrt{1-\beta}V_P - g\sqrt{1-\zeta}V_x. \tag{44b}$$

We see that p^{in} and x^{in} are mapped with gains $g'_{\text{BA}} = \sqrt{\beta}\kappa$ and $g'_{\text{F}} = g\sqrt{\zeta}$ respectively. The variances can easily be calculated to be:

$$\text{Var}(X^{\text{out}}) = 1 + g'^2_{\text{BA}} \tag{45a}$$

$$\text{Var}(P^{\text{out}}) = 1 + \frac{g'^2_{\text{F}}}{\zeta} + \frac{g'^2_{\text{F}} g'^2_{\text{BA}}}{\beta} - 2g'_{\text{F}} g'_{\text{BA}}. \tag{45b}$$

In Fig. 7 we show the atomic variances as a function of the feedback gain with the experimentally determined decoherence values of $\beta = 0.37$ and $\zeta = 0.75$. From this it is clear that because of the decoherence and light loss increasing the feedback gain towards unity the noise will grow dramatically. With this, the fidelity quantified by Eq. (43) can be optimized with respect to the feedback gain. As can be seen from the values of β and ζ the light loss and atomic decoherence is significant. The high light loss is due to the fact that the glass cells containing the atomic vapor were not anti-reflection coated. Therefore each glass-air interface contributes about 4% loss. The main source of atomic decoherence is light assisted collisions, which change the phase of the atoms without affecting the spin projection. The atoms will thus be driven towards the coherent state, which justifies the use of the simple model of beam splitter admixture of vacuum.

7.3. *Quantum memory retrieval*

As mentioned in Ref. 2 the stored state can in principle be retrieved by inverting the roles of light and atoms in the direct mapping protocol. This would involve first an interaction between a read-out light beam and the atomic sample acting as a storage medium. According to Eq. (9) this would map P_A onto the light. Next X_A has to be measured and feedback applied to the read-out beam according to the result of the measurement. Since, however, the atomic measurement requires a certain time during which the read-out pulse propagates at the speed of light, the feedback is only possible for pulse durations of the order of up to some tens of nanoseconds. In the experiments of Ref. 2 pulses of millisecond duration ($\sim$ 300km) are required in order to obtain a sufficiently high interaction strength, and the inverse direct mapping protocol is thus infeasible for this experimental realization.

Several years ago a retrieval scheme, which did not involve measurements, but instead two orthogonal passages of the read-out pulse was proposed.[17] In each passage one of the atomic quadratures are mapped onto the light pulse, and in this way retrieval fidelities of up to 82% can be achieved without squeezing and 100% with squeezing. Unfortunately in order to preserve the QND nature of each of the two interactions the light has to pass entirely through the atomic medium before proceeding to the second passage. This again renders the protocol inapplicable to all setups requiring "long" pulses.

Recently this problem was eliminated by solving the complex dynamics arising from simultaneous passage of the light beam through the atomic medium.[18] The two-pass protocol mentioned above will thus work for any pulse duration. If the light is reflected back after the second passage, thus completing four passages, and a time dependent interaction strength is applied perfect retrieval can be achieved without requiring squeezed initial states.

8. Quantum Teleportation

8.1. *Basic protocol*

In this section we will show how the multi mode entanglement between light and atoms generated in the scattering process on a single cell placed in a magnetic field (see Sec. 3.3) can be employed for the teleportation of a coherent state from light onto atoms. The teleportation protocol is a simple generalization of the standard protocol for continuous variable teleportation using EPR-type entangled states.[19,20] A detailed description can be found in Ref. 21. Figure 8 depicts the basic scheme which, as usually, consists of a Bell measurement and a feedback operation.

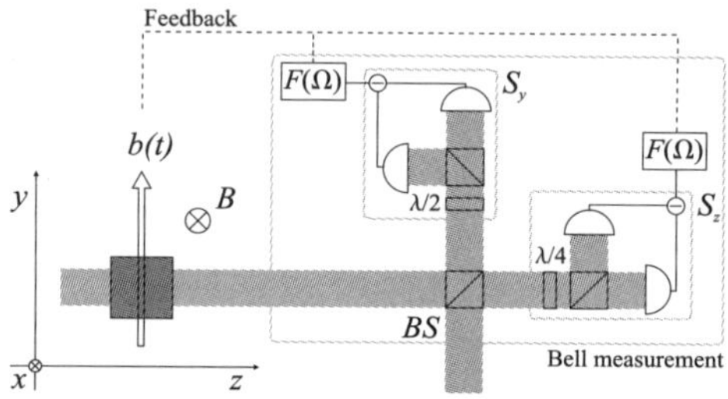

Fig. 8. Scheme for teleportation of light onto atoms: Classical pulse and scattered light (linearly polarized along y) are overlapped with a with a coherent pulse (linearly polarized along z) at beam splitter *BS*. By means of standard polarization measurements Stokes vector components S_y and S_z are measured at one and the other port respectively, realizing the Bell measurement. Conditional displacement of the atomic spin which can be achieved by applying a properly timed transverse magnetic field $b(t)$.

Input The coherent state to be teleported is encoded in a pulse which is linearly polarized orthogonal to the classical driving pulse and whose carrier frequency lies at the upper sideband, *i.e.* at $\omega_c+\Omega$. The pulse envelope has to match the one of the classical pulse. Canonical operators y, q with $[y, q] = i$ describing this mode can conveniently be expressed in terms of cosine and sine modulation modes, analogous to Eq. (12), defined with respect to the carrier frequency. One finds

$$y = \frac{1}{\sqrt{2}} \left(y_{\rm s} + q_{\rm c} \right), \quad q = -\frac{1}{\sqrt{2}} \left(y_{\rm c} - q_{\rm s} \right). \tag{46}$$

A coherent input amounts to having initially $\mathrm{Var}(y) = \mathrm{Var}(q) = 1/2$ and an amplitude $\langle y \rangle$, $\langle q \rangle$ with mean photon number $n_{\rm ph} = (\langle y \rangle^2 + \langle q \rangle^2)/2$.

Bell measurement This input is combined at a beam splitter with the classical pulse and the scattered light. At the ports of the beam splitter

Stokes vector components S_y and S_z are measured by means of standard polarization measurements. Given the classical pulse in x polarization this amounts to a homodyne detection of in- and out-of-phase fields of the orthogonal polarization component. The resulting photo-currents are numerically demodulated to extract the relevant sine and cosine components at the Larmor frequency. Thus one effectively measures the commuting observables

$$\tilde{x}_{\rm c} = \frac{1}{\sqrt{2}}\left(x_{\rm c}^{\rm out} + y_{\rm c}\right), \qquad \tilde{x}_{\rm s} = \frac{1}{\sqrt{2}}\left(x_{\rm s}^{\rm out} + y_{\rm s}\right), \tag{47}$$
$$\tilde{q}_{\rm c} = \frac{1}{\sqrt{2}}\left(p_{\rm c}^{\rm out} - q_{\rm c}\right), \qquad \tilde{q}_{\rm s} = \frac{1}{\sqrt{2}}\left(p_{\rm s}^{\rm out} - q_{\rm s}\right).$$

Let the respective measurement results be given by $\tilde{X}_{\rm c}$, $\tilde{X}_{\rm s}$, $\tilde{Q}_{\rm c}$ and $\tilde{Q}_{\rm s}$.

Feedback Conditioned on these results the atomic state is then displaced by an amount $\tilde{X}_{\rm s} - \tilde{Q}_{\rm c}$ in X and $-\tilde{X}_{\rm c} - \tilde{Q}_{\rm s}$ in P. This can be achieved by means of two fast radio-frequency magnetic pulses separated by a quarter of a Larmor period. In the ensemble average the final state of atoms is simply given by

$$\hat{X}^+ fin = \hat{X}^+ out + \tilde{x}_{\rm s} - \tilde{q}_{\rm c}, \quad P^{\rm fin} = P^{\rm out} - \tilde{x}_{\rm c} - \tilde{q}_{\rm s}. \tag{48}$$

Relating these expressions to input operators we find by means of Eqs. (14), (46) and (47)

$$\hat{X}^+ fin = \left(1 - \frac{\kappa}{2}\right)\hat{X}^+ in - \frac{1}{\sqrt{2}}\left(1 - \frac{\kappa}{2}\right)^2 p_{\rm c}^{\rm in} + \frac{1}{\sqrt{2}} x_{\rm s}^{\rm in} - \frac{1}{\sqrt{6}}\left(\frac{\kappa}{2}\right)^2 p_{\rm c,1}^{\rm in} + y,$$
$$P^{\rm fin} = \left(1 - \frac{\kappa}{2}\right)P^{\rm in} - \frac{1}{\sqrt{2}}\left(1 - \frac{\kappa}{2}\right)^2 p_{\rm s}^{\rm in} - \frac{1}{\sqrt{2}} x_{\rm c}^{\rm in} - \frac{1}{\sqrt{6}}\left(\frac{\kappa}{2}\right)^2 p_{\rm s,1}^{\rm in} + q.$$

Teleportation fidelity Taking the mean of the last equations with respect to the initial state all contributions due to input operators and back action modes vanish such that $\langle \hat{X}^+ fin\rangle = \langle y\rangle$ and $\langle P^{\rm fin}\rangle = \langle q\rangle$. Thus the amplitude of the coherent input light pulse is mapped on atomic spin quadratures as desired. In order to proof faithful teleportation also the variances have to be conserved. It is evident that the final atomic spin variances will be increased as compared to the coherent input. These additional terms describe unwanted excess noise and have to be minimized by a proper choice of the coupling κ. As a figure of merit for the teleportation

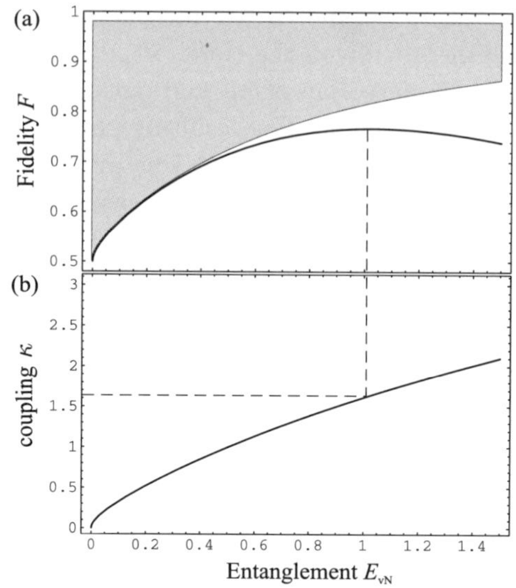

Fig. 9. (a) Theoretical limit on the achievable fidelity F versus entanglement between atoms and light measured by the von Neumann entropy E_{vN} of the reduced state of atoms. The gray area is unphysical. For moderate amounts of entanglement our protocol is close to optimal. (b) Coupling strength κ versus entanglement. The dashed lines indicate the maximal fidelity of $F = .77$ which is achieved for $\kappa = 1.64$.

protocol we use the fidelity, *i.e.* squared overlap, of input and final state. Given that the means are transmitted correctly the fidelity is found to be $F = 2\left[(1+2\mathrm{Var}(\hat{X}^{+}fin))(1+2\mathrm{Var}(P^{\mathrm{fin}}))\right]^{-1/2}$. In Fig. 9 we take advantage of the fact that the amount of entanglement between light and atoms is a monotonously increasing function of κ such that we can plot the fidelity versus the entanglement. This has the advantage that we can compare the performance of our teleportation protocol with the canonical one[19,20] which uses a two-mode squeezed state of the same entanglement as a resource and therefore maximizes the teleportation fidelity for the given amount of entanglement. The theoretical fidelity achievable in our protocol is maximized for $\kappa \simeq 1.64$ corresponding to $F \simeq .77$. But also for experimentally more feasible values of $\kappa \simeq 1$ can the fidelity well exceed the classical limit[22,16] of $1/2$ and, moreover, comparison with the values achievable with a two-mode squeezed state shows that our protocol is close to optimal.

Under realistic conditions the teleportation fidelity will be degraded by noise effects like decoherence of the atomic spin state, light absorption and reflection losses and also because the coupling constant κ is experimentally limited to values $\kappa \simeq 1$. On the other hand the classical fidelity bound to be beaten will be somewhat higher than 1/2 since the coherent input states will necessarily be drawn according to a distribution with a finite width in the mean photon number $\bar{n}$. Under this conditions the feedback has in general be performed with non unit gain factors for the four measurement outcomes. For symmetry reasons it is enough to deal with two, g_x, g_q, for the measurement outcomes of sine and cosine components of x and q respectively. Treating decay of atoms and absorption of light as indicated in Sec. 4 one derives the fidelity in its dependence on κ, the decay parameters β and ϵ for atoms and light respectively and the gain factors g_x, g_q, see Ref. 21 for details. Optimization with respect to the latter parameters yields the results shown in Fig. 10. We assume a realistic value $\kappa = 0.96$ for the coupling constant and a mean number of photons $\bar{n} = 4$ for the distribution of the coherent input. For feasible values of β, $\epsilon \precsim 0.2$ the average fidelity is still well above the classical bound on the fidelity.[16,22] This proves that the proposed protocol is robust against the dominating noise effects in this system.

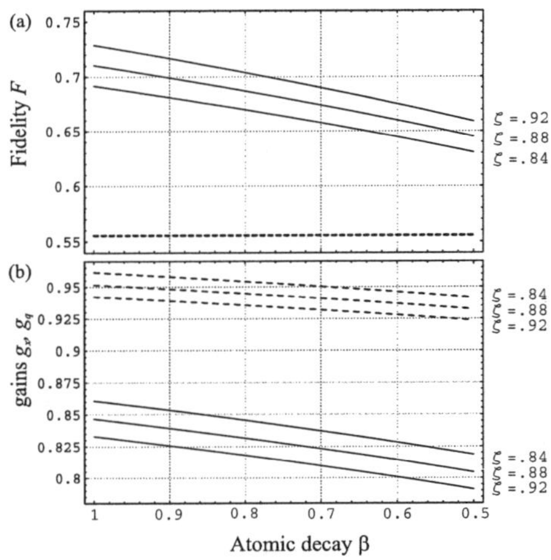

Fig. 10. (a) Average fidelity achievable in the presence of atomic decay β, reflection and light absorption losses $\epsilon = 8\%$, .12%, .16%, coupling $\kappa = .96$ and Gaussian distributed input states with mean photon number $\bar{n} = 4$. The fidelity benchmark is in this case 5/9 (dashed line). (b) Respective optimal values for gains g_x (solid lines) and g_q (dashed lines).

9. Multipass Interface

In this section we refer to the interaction of a single sample with light without magnetic field applied, *i.e.* the situation which was analyzed in Sec. 3.1. We examine how multiple applications of the interaction interspersed by simple polarization rotations can significantly enhance the creation of relevant quantities such as entanglement, the EPR uncertainty or squeezing.[23] We will show also that these protocols for multiple passes are robust against noise.

In the following it will be convenient to describe the Gaussian state of light and atoms in the Schrödinger picture by means of the correlation matrix which is defined by $\gamma_{i,j} = \mathrm{tr}\{\rho[(R_i - \langle R_i\rangle), (R_j - \langle R_j\rangle)]_+\}$, $i, j = 1, \ldots, 4$, where $\vec{R} = (X, P, x, p)$ and $[.,.]_+$ denotes the anti commutator. The correlation matrix of a coherent state of atoms and light is then just the 4×4 identity matrix.

The state after a single pass of a pulse of light through the atomic ensemble is described in terms of input-output relations as

$$\gamma_{\mathrm{out}} = \bar{D}(\eta,\epsilon)S(\kappa)\gamma_{\mathrm{in}}S(\kappa)^T\bar{D}(\eta,\epsilon) + D(\eta,\epsilon)\gamma_{\mathrm{noise}}, \tag{49}$$

where the scattering matrix

$$S(\kappa) = \begin{pmatrix} 1 & 0 & 0 & \kappa \\ 0 & 1 & 0 & 0 \\ 0 & \kappa & 1 & 0 \\ 0 & 0 & 0 & 1 \end{pmatrix} \tag{50}$$

and $D(\eta,\epsilon) = \mathrm{diag}(\eta,\eta,\epsilon,\epsilon)$, $\bar{D}(\eta,\epsilon) = \sqrt{\mathbf{1} - D(\eta,\epsilon)}$, $\gamma_{\mathrm{noise}} = \mathrm{diag}(2,2,1,1)$. The output state is a weighted sum of a coherent contribution and a noise component γ_{noise} whose form is due to the fact that the field decay is accompanied by a vacuum noise contribution and the atomic decay both contributes to noise due to the breaking of correlations among the atoms and due to the atoms once decayed being still present in the sample, explaining the factor of 2 in the atomic component of γ_{noise}. In principle, the noise introduced in atoms increases with the decay of the mean polarization, but this effect is negligible for the example presented (see Ref. 7 for a refined model for this interaction using the same formalism).

The state created after several passes can be calculated by iterating the map defined by Eq. (49). Note, however, that the coupling strength κ depends on the polarizations along x and that these classical variables will decay from pass to pass as $\langle J_x\rangle_{\mathrm{out}} = (1-\eta)\langle J_x\rangle_{\mathrm{in}}$, $\langle S_x\rangle_{\mathrm{out}} = (1-\epsilon)\langle S_x\rangle_{\mathrm{in}}$.

For the n-th step the remaining coupling strength is hence reduced $\kappa_n = [(1-\eta)(1-\epsilon)]^{n/2}\kappa$. Reflection losses can be taken into account by replacing ϵ by $\zeta = \epsilon + r$ where r is the overall reflectivity of mirrors, cell etc. Equation (49) provides then readily a recursion relation

$$\gamma_n = \bar{D}(\eta,\zeta)S(\kappa_n)\gamma_{n-1}S(\kappa_n)^T\bar{D}(\eta,\zeta) + D(\eta,\zeta)\gamma_{\text{noise}} \tag{51}$$

for the state after n passes which can be solved exactly.

The effect of n consecutive passes is comparable to that of a single pass performed with an n times increased optical density. This is clear from the meaning of α_0 and becomes manifest in the group property $S(\kappa)S(\lambda) = S(\kappa+\lambda)$ of the scattering matrix (50). This indicates that the strategy of multiple passes is especially interesting for low optical densities.

We are here especially interested in three quantities characterizing the quantum properties of the state generated: (a) the Gaussian Entanglement of Formation (GEOF),[24] the only available physical Entanglement measure for mixed Gaussian bipartite states, (b) the closely related[25] EPR uncertainty of the combined atom+field system, which indicates how close the state is to a maximally entangled EPR state, given for the present states by $\Delta EPR = \frac{1}{2}[\Delta^2(x_{\text{at}} - p_{\text{ph}}) + \Delta^2(p_{\text{at}} - x_{\text{ph}})]$, and finally (c) the atomic (and light) squeezing achievable either by a QND measurement (homodyne detection of light) or by means of a particular disentangling operation at the end of the multi pass protocol. In the following we will optimize these quantities for a given optical density α_0, photon loss ζ, and number of steps n with respect to the atomic depumping η. As explained in Sec. 4 this strategy relies on the fundamental connection $\kappa^2 = \alpha_0\eta$.

The dependence of the GEOF and the EPR variance on the number of passes is shown in Fig. 11. In general it can be shown under the assumption of vanishing reflection losses ($r = 0$) that for given optical density and number of steps n there exist optimal choices for η and ϵ such that, taking formally $n \to \infty$, the GEOF tends to infinity. The EPR-variance is limited by 0.5, or 3 dB of squeezing, which is also evident in Fig. 11.

The multipass scheme is capable of improving these features significantly. In Ref. 26 it was shown in a pure state analysis that entanglement and squeezing is created at a maximal rate if one switches from $H \propto p_{\text{at}}p_{\text{ph}}$ to an interaction $H \propto -x_{\text{at}}x_{\text{ph}}$ in every second step which can be done by applying a a proper unitary operation and its adjoint before and after an interaction. The effect of the switching becomes clear if one approximates $\exp(ix_{\text{at}}x_{\text{ph}}\kappa)\exp(-ip_{\text{at}}p_{\text{ph}}\kappa) \simeq \exp[-i(p_{\text{at}}p_{\text{ph}} - x_{\text{at}}x_{\text{ph}})\kappa + o(\kappa^2)]$. To first order this interaction creates a two-mode squeezed state. In particular the

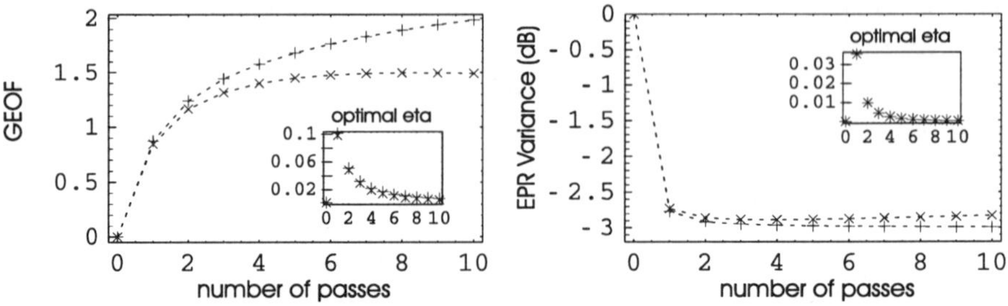

Fig. 11. GEOF and EPR variance vs. number of passes: For given n both quantities are maximized with respect to η and ζ. The optimal values for η are shown in the inserts. It is always best to have $\zeta = r$ corresponding to $\epsilon \ll \eta$. $+'$es refer to the case $r = 0$, $\times'$es to $r = 2\%$. The optical density is $\alpha_0 = 25$.

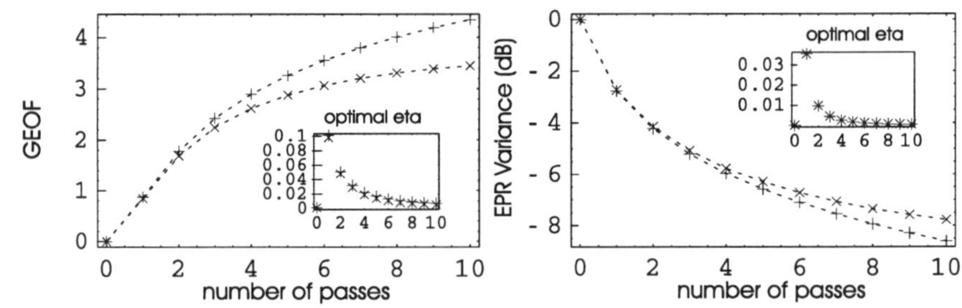

Fig. 12. GEOF and EPR variance vs. number of passes including polarization rotations: $+'$es refer to the case $r = 0$, $\times'$es to $r = 2\%$. Optical density $\alpha_0 = 25$.

growth is linear in n and thus provides an exponential improvement as compared to the scheme without switching. The final state after n passes follows from Eq. (51) by taking the scattering matrix to be $S(\kappa)^T$ - corresponding to an interaction $H \propto -x_{\mathrm{at}}x_{\mathrm{ph}}$ - in every second step. Figure 12 shows how the quantities of interest develop. In comparison with the unswitched case, the GEOF is roughly doubled and the EPR squeezing is no longer limited to 3 dB. In the limit of $n \to \infty$ the resulting state approximates a maximally entangled EPR state which can as well be used as a resource for continuous variable teleportation.

After multiple passes (with or without switching of polarizations) neither light nor atomic quadratures are squeezed separately. In order to obtain such local squeezing an additional operation has to be carried out. One possibility is to perform a destructive homodyne detection of light, which - in the unswitched scheme - amounts to a QND measurement of the atomic

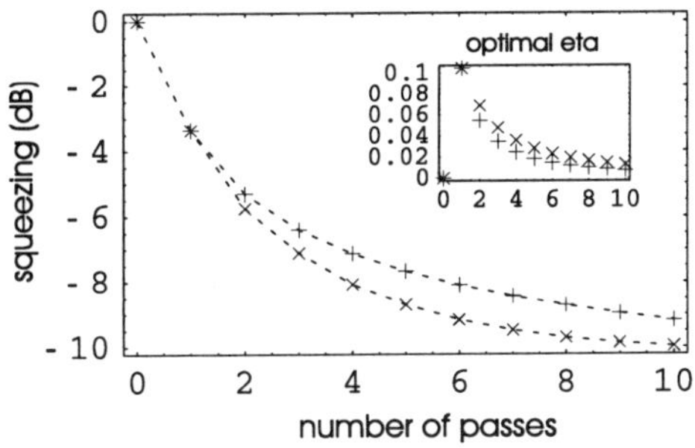

Fig. 13. Atomic squeezing after homodyne detection of light: Unswitched scheme (QND measurement) " + " and switched scheme " × ". $\zeta = r = 2\%$, $\alpha_0 = 25$.

p-quadrature and yields a squeezed state of atoms while the light is lost. Performing the same measurement on one half of an EPR state - as it arises in the switched scheme - also leaves the other system in a squeezed state. Figure 13 displays the atomic squeezing after a homodyne detection of light for both schemes. The switching provides a small advantage even though the actual interaction has lost its QND character. The trade off between squeezing and spontaneous emission noise has also been discussed in Ref. 27 for a different type of interaction.

10. Prospects

The off-resonant interaction of light with spin-polarized atomic ensembles described in this chapter has turned out to be an extremely useful resource in quantum information processing. Two distinct avenues for using such interaction have emerged: combining it with a quantum measurement on transmitted light and using two- and multi-pass interaction of light with atoms. Future experiments already proposed theoretically and outlined in this chapter include teleportation of atomic states, entanglement swapping of atomic states, complete deterministic quantum memory for light, including the retrieval of the atomic memory state back onto light, and generation of entangled and squeezed states of light via interaction with atoms. An interesting perspective is the exploration of the multimode features of atomic ensembles. Spatially extended atomic samples can, in principle, store a number of spatial light modes, opening the way towards multimode atomic memories with high capacity. Another promising direction is the extension of this approach to discrete variables still using collective atomic states. This extension may require another kind of measurement, for example, photon counting techniques, which proved so useful in the proposal for a quantum repeater.

Acknowledgments

B. Julsgaard is supported by the Carlsberg Foundation. We acknowledge funding from EU project COVAQIAL and the Danish National Research Foundation.

References

1. B. Julsgaard, A. Kozhekin and E.S. Polzik, Nature **413**, 400 (2001).
2. B. Julsgaard, J. Sherson, J.I. Cirac, J. Fiurasek, and E.S. Polzik, Nature **432**, 482 (2004).
3. W. Happer, Rev. Mod. Phys. **44**, 169 (1972).
4. D.V. Kupriyanov, O.S. Mishina, I.M. Sokolov, B. Julsgaard, and E.S. Polzik, *Multimode entanglement of light and atomic ensembles via off-resonant coherent forward scattering*, Phys. Rev. A **71**, 032348 (2005).
5. J.H. Müller, P. Petrov, D. Oblak, C.L.G. Alzar, S.R. de Echaniz, and E.S. Polzik, *Diffraction effects on light-atom ensemble quantum interface*, Phys. Rev. A **71**, 033803 (2005).
6. A. Silberfarb and I.H. Deutsch, Phys. Rev. A **68**, 13817(2003).
7. L.B. Madsen and K. Mølmer, Phys. Rev. A **70**, 052324 (2004).
8. C. Kittel, *Quantum Theory of Solids*, Wiley 1987, New York.
9. J.M. Geremia and J.K. Stockton, and H. Mabuchi, Science **304**, 270 (2004).
10. C.M. Caves and B.L. Schumaker, Phys. Rev. A **31**, 3068 (1085).
11. B.L. Schumaker and C.M. Caves, Phys. Rev. A **31**, 3093 (1985).
12. L-M. Duan, J.I. Cirac, P. Zoller, and E.S. Polzik, Phys. Rev. Lett. **85**, 5643 (2000).
13. A. Kuzmich and T.A.B. Kennedy, Phys. Rev. Lett. **92**, 030407 (2004).
14. B. Julsgaard, *Entanglement and Quantum Interactions with Macroscopic Gas Samples*, University of Aarhus, 2003, available at `http://www.phys.au.dk/main/publications/PhD/`.
15. L.-M. Duan, G. Giedke, J.I. Cirac and P. Zoller,*Inseparability Criterion for Continuous Variable Systems*, Phys. Rev. Lett. **84**, 2722 (2000).
16. K. Hammerer, M.M. Wolf, E.S. Polzik, and J.I. Cirac, Phys. Rev. Lett. **94**, 150503 (2005).
17. A. Kuzmich and E.S. Polzik, *Quantum Infomation with Continuous Variables*, S. L. Braunstein and A. K. Pati (Eds.), Klüwer, Dordrecht, pp. 231-265 (2003).
18. J. Sherson, A.S. Sørensen, J. Fiurášek, K. Mølmer, and E.S. Polzik, *Light Qubit Storage and Retrieval using Macroscopic Atomic Ensembles*, quant-ph/0505170 (2005).
19. L. Vaidman, Phys. Rev. A **49**, 1473 [1994).
20. S.L. Braunstein and H.J. Kimble, Phys. Rev. Lett. **80**, 869 (1998).

21. K. Hammerer, E.S. Polzik and J.I. Cirac, in preparation.
22. S.L. Braunstein and H.J. Kimble and C.A. Fuchs, J. Mod. Opt. **47**, 267 (2000).
23. K. Hammerer, K. Mølmer, E. S. Polzik and J.I. Cirac, Phys. Rev. A **70**, 044304 (2004).
24. M.M. Wolf, G. Giedke, O. Krüger, R.F. Werner, and J.I. Cirac, Phys. Rev. A **69**, 052320 (2004).
25. G. Giedke, M.M. Wolf, O. Krüger, R.F. Werner, and J.I. Cirac, Phys. Rev. Lett. **91**, 107901 (2003).
26. B. Kraus, K. Hammerer, G. Giedke, and J.I. Cirac, Phys. Rev. A **67**, 042314 (2003).
27. A. Andre and M. D. Lukin, Phys. Rev. A **65**, 053819 (2002).

Chapter 26

Long Distance Quantum Communication with Atomic Ensembles

C. W. Chou, S. V. Polyakov, D. Felinto, H. de Riedmatten,
S. J. van Enk† and H. J. Kimble

Norman Bridge Laboratory of Physics
California Institute of Technology, 12-33
Pasadena, California 91125, USA
†Bell Labs, Lucent Technologies
600-700 Mountain Ave, Murray Hill, NJ 07974, USA

1. Introduction

The ability to distribute quantum resources over long distances is a key element for the development of several practical applications in the field of quantum information.[1] In this way, major efforts have been made in the past decade to extend the maximum distances for the observation of entanglement or for the realization of quantum cryptography. Using optical fibers as the quantum channel, entanglement has already been achieved between photons separated by 50 km of fiber,[2] and quantum cryptography is now a commercial technology[3] that allows quantum key distributon over distances up to around 100 km.[4–6] Progress has also been made towards entanglement distribution in free space, with the achieved distances of about 10 km on ground atmosphere[7,8] being a substantial step toward the development of satellite-based quantum communication.[9–11]

However, in spite of all these developments, the quantum communication protocols that have been tested up to now suffer from some fundamental limitations on the maximum achievable distance.[12] Quantum key distribution setups, for example, seem to be approaching their limits. The problem is the absorption of photons in the quantum channel, which grows exponentially with the length of the channel. In classical communication, the attenuation problem is solved in a straightforward way by amplification of the signal. It is not possible, however, to generate perfect copies of a quantum state,[13] which prevents the amplification techniques to be used for quantum communication.

A first step to solve this problem was the development of techniques for entanglement purification.[14,15] These techniques were inspired by quantum algorithms for error correction, and are designed to obtain a set of transmited states of high fidelity (*i.e.*, high overlap with a maximally entangled state) out of a larger set with lower fidelity, if this fidelity is still above a certain value F_{min}. Based on these ideas, a solution for the long-distance-problem in quantum communication was then formulated with the development of the concept of quantum repeater.[16,17] The idea is to divide the quantum channel in several segments with lengths that allow the transmission of information with fidelity $F > F_{min}$, and then apply the entanglement purification algorithms in a hierarchical way, beginning in the smaller portions and going up to larger regions until a high fidelity is obtained for the entanglement of pairs of particles located on the extremes of the communication channel. The connection between the different segments is done by entanglement swapping.[18,19] The essential advantage of the quantum repeater protocol is that the number of trials to entangle a pair of particles in this case grows only polynomially with the distance, and not exponentially as in the protocols usually used for quantum communication.

The development of quantum repeaters, however, introduces new challenges which need to be addressed. It is necessary to entangle distant qubits, store them over operational time scales, and perform collective local operations on several of these qubits. Quantum memory is essential, since all purification protocols have a probabilistic nature. If the system had no memory, the purification of all segments would have to succeed at the same time for the algorithm to work, which is improbable. The requirement of quantum memory implies, in practice, that the local qubits need to be stored in atomic internal states. Since the communication between distant sites still needs to be done by photons, a quantum repeater should rely on systems that present strong atom-photon coupling. Usually this strong coupling is obtained with the aid of high finesse cavities, in the field known as Cavity Quantum Electrodynamics (CQED). However, in spite of important recent advances,[20–22] these techniques still require a reasonably complicated experimental setup, which could prevent the application on a larger scale of quantum communication and error correction protocols for CQED, as the ones described in Refs. 23 and 24, respectively.

This was the context in which Duan, Lukin, Cirac and Zoller (DLCZ) proposed an alternative protocol[25] for the implementation of quantum repeaters with atomic ensembles, which would require a much simpler experimental apparatus. The DLCZ protocol introduces novel ideas in several levels, from the physical process that couples light to atoms, to the way entanglement is achieved between distant atomic ensembles, and also on the details of the entanglement purification process. The main objective of the present chapter is to give an overview of this protocol and of the developments made towards its experimental implementation. Emphasis will be given on the experimental investigation carried out by our group at Caltech,[26–28] but different developments by other groups will also be discussed alongside.[29–32]

Section 2 provides a description of the basic elements of the protocol, while the rest of the chapter is dedicated to the experimental progress. Sections 3 through 5 describe then the realization of the first step in the implementation of the DLCZ protocol, *i.e.*, the control of the coupling between photons and atomic ensemble. Section 3 describes the first observations of such coupling as required by the protocol, through spontaneous Raman scattering on the single-photon level. In Sec. 4, it is shown that this process can also be used as a conditional source of single photons, a new application that was not in the original DLCZ proposal. Section 5 describes an in-depth analysis of the temporal structure of the two-photon wavepacket emitted by the sample. This gives detailed information on readout and decoherence timescales, which are important to the characterization of the quantum memory in our experimental setup.

Sections 6 and 7 discuss further developments. In Sec. 6, we report recent improvements in the coherence time of our system, and discuss the prospects for further increases. In Sec. 7, we discuss the experimental strategies to take the next step in the implementation of the DLCZ protocol: the generation and characterization of entanglement between atomic ensembles by the detection of a single photon. Finally, in Sec. 8 we draw our conclusions, and perspectives for future developments.

2. DLCZ Protocol for Quantum Repeaters

The building block of the DLCZ protocol is an ensemble of N_a identical atoms with lambda-type energy level configuration as shown in Fig. 1. The metastable lower states $|g\rangle$ and $|s\rangle$ can be, *e.g.*, hyperfine or Zeeman sublevels of the electronic ground state of alkali-metal atoms, thus ensuring a long coherence lifetime. First, all atoms are prepared in the state $|g\rangle$.

By sending in a weak, off-resonant laser pulse, one atom of the ensemble might be transfered from $|g\rangle$ to $|s\rangle$, thus emiting a photon at a frequency or polarization different from the original exciting field. A key element of the protocol is the collective enhancement of this spontaneous Raman scattering in a forward direction, which is determined by the spatial mode of the laser pulse and the geometry of the excitation region.[33] If the laser intensity is low enough so that two excitations are very unlikely, the detection of the photon generated in this process is a signature that the ensemble was excited to a symmetrical collective state,[25,33] which can be explicitly written as

$$|1_a\rangle = \frac{1}{\sqrt{N_a}} \sum_{i=1}^{N_a} |g\rangle_1 \cdots |s\rangle_i \cdots |g\rangle_{N_a}, \tag{1}$$

where the sum goes over all atoms addressed by the laser pulse, and $|1_a\rangle$ indicates the state of the atomic ensemble with just one excitation. This is the "writing" step of the protocol (Fig. 1a).

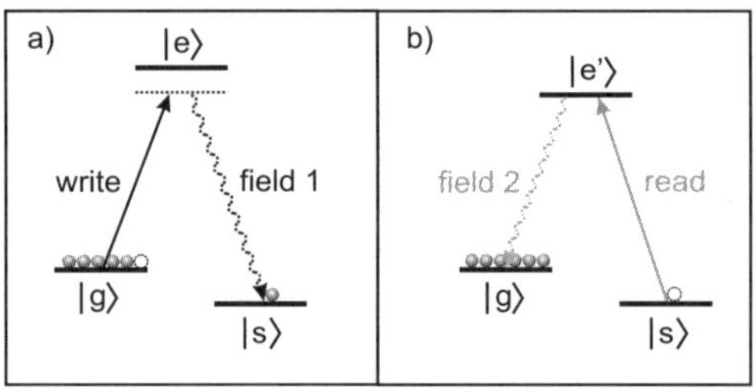

Fig. 1. Relevant level structure of the atoms in the ensemble for (a) writing and (b) reading processes, with $|g\rangle$ the initial ground state and $|s\rangle$ the ground state for storing a qubit. $|e\rangle$ and $|e'\rangle$ are excited states. The transition $|g\rangle \to |e\rangle$ is initially coupled by a classical laser pulse (write beam) detuned from resonance, and the forward-scattered Stokes light (field 1) comes from the transition $|e\rangle \to |s\rangle$, which has different polarization or frequency to the write light. A classical read pulse then couples the transition $|s\rangle \to |e'\rangle$, leading to the emission of forward-scattered anti-Stokes light (field 2) from the transition $|e'\rangle \to |g\rangle$.

Since the excitation probability χ is very small, the whole state of the system consisting of atoms and forward-scattered mode of light is in the following form:

$$|\phi\rangle = |0_a\rangle|0_1\rangle + e^{i\beta}\sqrt{\chi}\,|1_a\rangle|1_1\rangle + O(\chi), \tag{2}$$

where $\chi << 1$, $|n_1\rangle$ stands for the state of the forward-propagating light with n excitations, β is an arbitrary phase, and $|0_a\rangle \equiv \bigotimes_i^{N_a} |g\rangle_i$. $O(\chi)$

represents all the other possible excitation processes, which in the ideal case occur with probabilities of order χ^2. The system remains in this state for a time on the order of the lifetime of the ground states. The state of Eq. (2) shows already a certain amount of entanglement (small, because χ is small) in the excitation-number basis between the atomic ensemble and the forward-scattered light field. One can thus manipulate the state of the atoms by acting on the state of the emitted field. Note, however, that the state of Eq. (2) and in the following equations can be considered entangled only if one knows the phase β.

By sending in a second ("read") pulse resonant with the $|s\rangle \rightarrow |e'\rangle$ transition, the state of the atomic ensemble can be transferred (read out) to another forward-propagating light field 2 at the $|e'\rangle \rightarrow |g\rangle$ transition (see Fig. 1b). In this way, it is possible to access the quantum state of the atoms. This reading process is then closely related to low-light-level Electromagnetically Induced Transparency.[34–36] The case in which $|e\rangle = |e'\rangle$ is called, in the following, a three-level scheme of excitation, while at four-level excitation schemes $|e\rangle \neq |e'\rangle$. Note also that the resultant state of the two forward-scattered modes (1,2) can be written, in the ideal case, as

$$|\phi_{1,2}\rangle = |0_1\rangle|0_2\rangle + \sqrt{\chi}\,|1_1\rangle|1_2\rangle + \chi\,|2_1\rangle|2_2\rangle + O(\chi^{3/2}), \tag{3}$$

i.e., the photon numbers in the two modes are correlated, precisely as for parametric down conversion.[37]

The scheme to create an entangled pair of ensembles is shown in Fig. 2. Two write pulses are sent into two ensembles, L and R. At the output of the ensembles, the scattered fields and the ensembles are in the state:

$$\begin{aligned} |\Phi_{LR}\rangle &= |\Phi_L\rangle \otimes |\Phi_R\rangle \\ &= \left[|0_a\rangle_L|0_1\rangle_L + e^{i\beta_L}\sqrt{\chi}\,|1_a\rangle_L|1_1\rangle_L + O_L(\chi)\right] \\ &\quad \otimes \left[|0_a\rangle_R|0_1\rangle_R + e^{i\beta_R}\sqrt{\chi}\,|1_a\rangle_R|1_1\rangle_R + O_R(\chi)\right]. \end{aligned} \tag{4}$$

The scattered fields are later combined on a 50-50 beam splitter, with outputs directed to two photodetectors. In the ideal case, neglecting terms of order χ describing two or more excitation, detection of a photon in either detector then projects the state of the ensembles to the entangled state

$$|\phi_{L,R}\rangle = \frac{1}{\sqrt{2}}\left(|0_a\rangle_L|1_a\rangle_R \pm e^{i\eta_1}|1_a\rangle_L|0_a\rangle_R\right), \tag{5}$$

where η_1 relates to the difference of phase shifts in the two channels connecting the ensembles to the beam splitter (see Sec. 8), and the initial phase difference $\beta_L - \beta_R$. The sign $\pm$ depends on which detector records the event.

Note that the presence of certain sources of noise, especially dark noise on the detectors, modifies the above pure state to

$$\hat{\rho}_{LR} = \frac{1}{c_0 + 1}\left(c_0 |0_a\rangle_L |0_a\rangle_R \langle 0_a|_R \langle 0_a|_L + |\phi_{L,R}\rangle\langle\phi_{L,R}|\right), \tag{6}$$

where c_0^2 gives the probability for a detection not related to excitations on the ensembles. This state is called an effective maximally entangled (EME) state.[25] The DLCZ protocol is designed to be resilient to this important kind of noise, since it always requires further detection events to proceed. The detection events without excitations in the relevant atomic modes have then a high probability of being discarded right after they occur. This is the main origin of the "build-in entanglement purification" characteristic of the protocol.[25] Note that in writing (6), we have neglected higher-order terms involving pairs of excitations (one in each ensemble) and two excitations in either ensemble, which are intrinsic to the protocol of DLCZ, as well as diverse imperfections relevant to actual experimental implementations.

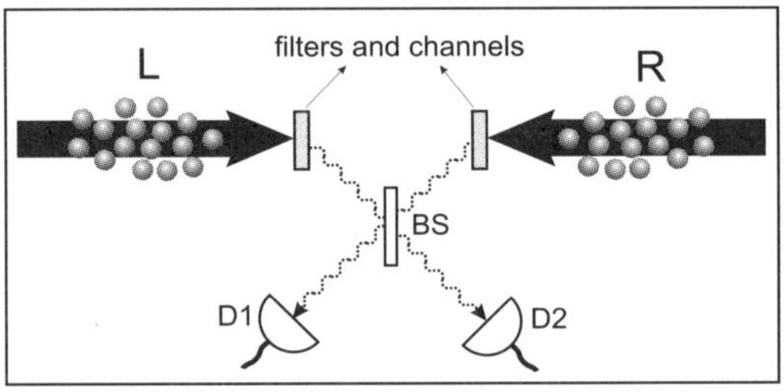

Fig. 2. Setup for generating entanglement between two atomic ensembles L and R. The ensembles are pencil-shaped, and excited by synchronized writing pulses. After filtering, the Stokes pulses are collected and coupled to optical channels. The pulses after the transmission channels interfere at a 50%-50% beam splitter BS, with outputs directed towards two single-photon detectors D_1 and D_2. Ideally, if D_1 or D_2 records a detection event, the process is finished and entanglement is successfully generated. Otherwise, the system is restored to its initial state and the process is repeated until D_1 or D_2 records a click.

The maximum distance between the above entangled ensembles is limited by the attenuation length of the communication channels of field 1. To extend the distance between entangled ensembles, one applies the entanglement swapping scheme, Fig. 3. In this way, two pairs of ensembles need first to be prepared in the entangled state described by Eq. (5). Note that, since the entanglement process is probabilistic, the two pairs of ensembles do not necessarily become entangled at the same time. Due to the system memory, however, once a pair is entangled, one can hold the entangled state and wait for the other pair to attain entanglement. This is an important advantage of the quantum repeater idea, which is responsible for the polynomial growth in number of trials with the distance between the final entangled pair. Other quantum communication schemes, like the quantum relay[38] that require all parts to be entangled at the same time, present an exponential growth with distance.

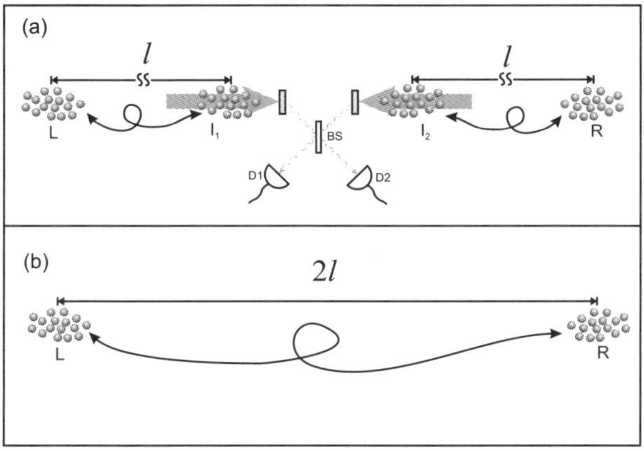

Fig. 3. Setup for extending the distance between entangled ensembles (entanglement swapping).[25] Two pairs of ensembles - L and I_1, and I_2 and R - are initially distributed at three sites L, I, and R. Each of the pairs is prepaired in an EME state in the form of Eq. (6). (a) The stored atomic excitations of two nearby ensembles I_1 and I_2 are converted into light by two simultaneous read pulses. The emitted anti-Stokes fields interfere at a 50%-50% beam splitter, and then are detected by the single photon detectors D_1 and D_2. (b) If either D_1 or D_2 clicks, the protocol is successful and an EME state is established between the ensembles L and R with twice the communication distance. Otherwise, the process fails, and the previous entanglement generation and swapping need to be repeated until a click is recorded in D_1 or D_2.

Once the two entangled pairs are obtained, the distance of entanglement is then doubled through the entanglement connection scheme shown in Fig. 3. Two read pulses are sent into ensembles I_1 and I_2 located in the middle, and the scattered fields are combined on a 50-50 beam splitter, Fig. 3a. A single click in either the detectors prepares then ensembles L and R in an EME state like the one of Eq. (6) (Fig. 3b). The "vacuum" coefficient c_0 now also includes the probability of reading the two excitations from the pairs (I_1,I_2) at the same time, leaving no remaining excitation in L or R.

The EME states created between distant ensembles can then be used to implement several entanglement-based communication schemes, such as quantum cryptography, teleportation, and Bell-inequality measurements. We refer the reader to the original DLCZ paper[25] for the details on how to implement each of these schemes. In order to appreciate the broad applicability of the DLCZ protocol, however, it is important to note that a pair of entangled ensembles can be used to represent a qubit, with the states $\{|0_a\rangle_L|1_a\rangle_R, |1_a\rangle_L|0_a\rangle_R\}$ mapping to the usual qubit basis, $\{|0\rangle, |1\rangle\}$, in the ideal case. An experiment in this context has been recently reported by Matsukevich and Kuzmich.[32]

The probabilistic character of the DLCZ scheme, however, imposes stringent requirements for the coherence times for the quantum memories at the various sites in Fig. 3. Namely, the coherence time τ_{memory} for each memory is required to be much longer than the time Δt_c taken to establish an entanglement connection between two nodes in the network. Since Δt_c is given approximately by the inverse of the rate R_s of detection events for the photons 1 generated by the write pulse, we require roughly

$$\tau_{memory} \gg (P_1 R_1)^{-1}. \quad (7)$$

Here P_1 gives the probability of detection of a scattered photon from a write pulse in each trial of the experiment and R_1 gives the rate at which successive trials can be carried out. The maximum repetition rate is given by the inverse of the time light takes to travel from one site to the other, so that $R_1 < c/l$. Since $|\chi|^2 \ll 1$ and the propagation and detection efficiencies from an atomic ensemble to D_1, D_2 are also small, $P_1 \sim 10^{-2}$, which with $l \gtrsim 10$ km implies $\tau_{memory} \gg 10^{-3}$ seconds. Notice that τ_{memory} results, in practice, in a new limit for the maximum distance for communication with this protocol. The search for larger memory times is then one of the major goals in the actual implementation of the protocol, as discussed in more detail in Sec. 6.

3. Nonclassical Photon Pairs from an Atomic Ensemble

As the first step to realize the DLCZ protocol, our group investigated the correlation between the fields 1 and 2 at the single photon level.[26] Since photon 2 is generated by reading out the stored collective excitation, these correlation measurements probe directly the nature of the quantum memory in our experiments. We were particularly interested in testing the quantum character of the field correlations in the single-photon regime required by the protocol.

The experimental setup for this investigation is illustrated in Fig. 4. The atomic sample consists of a magneto-optical trap (MOT) of cesium atoms, where the Cs hyperfine manifolds $\{|6S_{1/2}, F = 4\rangle; |6S_{1/2}, F = 3\rangle; |6P_{3/2}F' = 4\rangle\}$ correspond to levels $\{|g\rangle; |s\rangle; |e\rangle\}$, respectively. The period for each trial is 4 μs, in which the trap light is switched off during 1μs. In this "dark" period the write and read pulses are sent into the MOT to generate the two forward scattered fields 1 and 2. In the first experiment[26] these fields were then directed onto two single-photon detectors D1 and D2. In the following experiments[27,28] four detectors were used, as illustrated in Fig. 4. The write pulse is made sufficiently weak so that the probability χ to scatter one Raman photon into field 1 is much less than unity for each pulse. The read pulse is about 100 times more intense than the write pulse to optimize the readout efficiency of the atomic excitation to field 2.

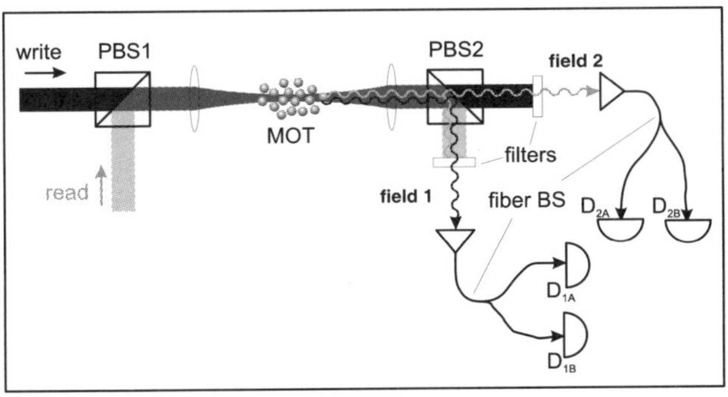

Fig. 4. Experimental setup. Write and read pulses propagate sequentially into a cloud of cold Cs atoms (MOT), generating pairs of correlated output photons 1 and 2. The write and read pulses have orthogonal polarizations, are combined into a single input at polarizing beam splitter PBS1, and are then focused in the Cs MOT with a waist of approximately 30 μm. The output fields are split by PBS2, which also serves as a first stage of filtering the (write, read) beams from the (1,2) fields. For example, field 2 is transmitted by PBS2 to be subsequently registered by detector D_2 while the read pulse itself is reflected by 90° at PBS2. Further filtering is achieved by passing each of the outputs from PBS2 through separate frequency filters. SM stands for single mode fibers.

A challenging aspect of the experiment is to separate the classical pulses from the weak nonclassical fields, since they are temporally and spatially overlapped, and their frequencies are only 9 GHz apart. In the first experiment[26] the filtering had three stages. First, field 1 (2) was separated from the write (read) pulse in a polarizing beam splitter (PBS) right after the MOT chamber. Later, the leakage of the excitation pulses that still escapes the PBS in the wrong direction was spectrally filtered by optically-pumped vapor cells.[39] Finally, field 1(2) was distinguished from the read (write) pulse by temporal gating of the detection. A further improvement was to use a four-level scheme of excitation,[27] in which write and read pulses are 42 nm apart. This allows a fourth filtering stage by narrow-bandwith optical filters, and the study of correlations with temporally overlapped write and read pulses.[27,28]

In order to investigate the quantum nature of the correlations, we use the fact that there exists a well-defined border between the classical and quantum domains for fields 1 and 2 that can be operationally accessed via coincidence detection, as was first demonstrated in the pioneering work by Clauser.[40] In this way, we measure the joint detection probability $p_{1,2}$ for both detectors (D1 and D2) to record an event in the same trial, and the singles probabilities p_1 and p_2 to register an event in detectors D1 and D2, respectively. By splitting field i with a 50-50 beam splitter and directing the output to the two detectors, the joint probabilities $p_{i,i}$ are also measured, where $i = 1$ or 2. Fields for which the Glauber-Sudarshan phase-space function is well-behaved (*i.e.*, classical fields) are constrained by a Cauchy-Schwarz inequality for the various probabilities,[40–42] namely:

$$R = \frac{[\tilde{g}_{1,2}(t)]^2}{\tilde{g}_{1,1}\,\tilde{g}_{2,2}} \leq 1, \tag{8}$$

where $\tilde{g}_{1,1} \equiv p_{1,1}/p_1^2$, $\tilde{g}_{2,2} \equiv p_{2,2}/p_2^2$, $\tilde{g}_{1,2}(t) \equiv p_{1,2}/(p_1 p_2)$, and t denotes the time separation between the detection of photons 1 and 2.

In our first experiment,[26] we measured $\tilde{g}_{1,1} = 1.739 \pm 0.020$ and $\tilde{g}_{2,2} = 1.710 \pm 0.015$. This is in correspondence to the expectation that fields 1 and 2 should each exhibit Gaussian statistics with $\tilde{g}_{1,1} = \tilde{g}_{2,2} = 2$ for the DLCZ protocol in the ideal case, which is here degraded by diverse sources of background counts.[41] By contrast, for the cross-correlations of fields 1 and 2, we record $\tilde{g}_{1,2}(\delta t) = 2.335 \pm 0.014$, with $\delta t = 415$ ns giving the time difference between the rising edges of write and read pulses. Hence the inequality (8) for classical fields is strongly violated, namely $R = 1.84 \pm 0.06 \nleq 1$, where all errors indicate the statistical uncertainties. This violation of the Cauchy-Schwarz inequality (8) clearly demonstrates the nonclassical character of the correlations between photons 1 and 2. Moreover, the measured coincidence rates in this experiment explicitly documented the cooperative nature of the emission process.[41]

Several developments on these measurements of non-classical correlations followed from this first work. Jiang *et al.*,[30] at University of Science and Technology of China, used a magnetically-shielded vapor cell instead of the MOT of the previous work. In this configuration, they were able to violate the non-classical inequality with a delay between write and read pulses of 2 μs, for pulse durations of about 1 μs, explicitly obtaining $R = 1.34 \pm 0.05$. In a setup in our group, the use of a 4-level scheme of excitation improved the filtering, and allowed measurements of large violations of the inequality (8) by decreasing the delay between the excitation pulses.[27] Further increase in the violation was observed in the analysis of the temporal structure of the correlations between photons 1 and 2 (see Sec. 5 and Ref. 28). Table 1 summarizes the measured values of R and $\tilde{g}_{1,2}$ for these several situations.

Table 1. Measurements of the violation of the classical inequality $R \leq 1$ and associated cross-correlation function $\tilde{g}_{1,2}$ for photons 1 and 2. δt denotes here the time between the rising edges of write and read pulses, and T_p gives their duration. τ is the temporal window for detection.

Experimental conditions	$\tilde{g}_{1,2}$	R
Cs MOT, 3-level scheme, $\delta t = 405$ ns, $T_p \approx$ 40 ns, $\tau = 60$ ns, Ref. 26	2.335 ± 0.014	1.84 ± 0.06
Rb vapor cell, 3-level scheme, $\delta t = 2$ μs, $T_p \approx 1$ μs, $\tau = 1$ μs, Ref. 30	2.043 ± 0.031	1.34 ± 0.05
Cs MOT, 4-level scheme, $\delta t = 200$ ns, $T_p \approx$ 140 ns, $\tau = 200$ ns, Ref. 27	7.30 ± 0.04	23 ± 1
Cs MOT, 4-level scheme, $\delta t = 50$ ns, $T_p \approx$ 140 ns, $\tau = 200$ ns, Ref. 27	10.20 ± 0.06	53 ± 4
Cs MOT, 4-level scheme, $\delta t = 200$ ns, $T_p \approx 140$ ns, Optimum-windows analysis with $\tau = 30$ ns, Ref. 28	21.4 ± 0.5	202 ± 60
Cs MOT, 4-level scheme, $\delta t = 50$ ns, $T_p \approx 140$ ns, Optimum-windows analysis with $\tau = 30$ ns, Ref. 28	23.0 ± 0.5	292 ± 57

Another closely related set of correlation measurements have been reported in Refs. 29 and 31 by the group of M. Lukin at Harvard University. In Ref. 29, strong intensity correlations were observed between macroscopic fields generated in the spontaneous Raman process, and their quantum nature was demonstrated by an analysis of the fluctuation spectral density with respect to the shot-noise, or vacuum-state limit. This experiment was carried out in the regime of large photon number for the $(1, 2)$ fields with $N \approx 10^3 - 10^4$ and the associated fluctuations on the order of $N^{1/2} \gg 1$. In Ref. 31, nonclassical correlations were observed in the regime $N \approx 1$, and the sub-Poissonian nature of the anti-Stokes field generated by the read pulse was demonstrated, as discussed in more detail in the following section.

4. Atomic Ensemble as Conditional Source of Single Photons

The large degree of quantum correlation between fields 1 and 2 observed in Ref. 27 (see Table 1) supports the presence of field correlations for single photons that are required by the DLCZ protocol. This observation, however, allows applications that go beyond this protocol. Most notably, it indicates that this system might work as a good conditional source of single photons.

Efforts to generate single-photon wave packets can be broadly divided into techniques that provide photons "on demand" (*e.g.* quantum dots[43–45] or single atoms[46] coupled to microcavities) and those that produce photons as a result of conditional measurement on a correlated quantum system. For conditional generation, the detection of one photon from a correlated pair results in a one-photon state for the second field, as was first achieved using "twin" photons from atomic cascades[40,47] and parametric down-conversion.[49] The photon pair production discussed so far in this chapter introduces then a new avenue for producing single photons via conditional measurement, *i.e.*, using optically thick atomic ensembles with quantum memory. The system memory adds the important capability of controlling the delay between the trigger detection and the single photon.

In order to investigate directly the single photon character of photon 2, we have measured the suppression of the two-photon component in the state of photon 2 conditioned on the detection of photon 1. This two-photon component is observed through the correlation function $w(1_2, 1_2|1_1)$ as: $w(1_2, 1_2|1_1) \equiv p^{(c)}(1_2, 1_2|1_1)/[p^{(c)}(1_2|1_1)]^2$. Here, $p^{(c)}(1_2, 1_2|1_1)$ is the conditional probability for detecting two photons from field 2 once an initial photon from field 1 was detected, and $p^{(c)}(1_2|1_1)$ is the probability for detection of one photon 1_2 given a detection event 1_1. Bayes' theorem allows the conditional probabilities to be written in terms of single and joint probabilities, so that[47,48]

$$w \equiv w(1_2, 1_2|1_1) = \frac{p_1\, p_{1,2,2}}{p_{1,2}^2}, \tag{9}$$

where $p_{1,2,2}$ gives the joint probability of detecting one photon in field 1 and two photons in field 2. Classical fields must satisfy the inequality $w \geq 1$. For independent coherent states, we should find $w = 1$, while for thermal beams, $w = 2$. However, for the state $|\phi_{12}\rangle$ of Eq. (3) we should have $w = 4\chi$, which is much smaller than 1 for small χ and approaches the ideal case $w \to 0$ for a "twin" Fock state $|1_1 1_2\rangle$. For the state represented by $\tilde{g}_{1,2}$,

$\tilde{g}_{1,1}$ and $\tilde{g}_{2,2}$ the value of the correlation function w can also be expressed as: $w = \tilde{g}_{1,1}\tilde{g}_{2,2}/\tilde{g}_{1,2}$.

To measure w directly, the record of photodetection events on all four detectors of Fig. 4 was used to calculate the probabilities appearing in Eq. 9, with the results of this analysis shown in Fig. 5. Significantly, as the degree of cross-correlation expressed by $\tilde{g}_{1,2}$ increases (*i.e.*, decreasing χ), w drops below the classical level of unity, indicative of the sub-Poissonian character of the conditional state of field 2. With $\delta t = 200$ ns, $w = 0.34 \pm 0.06$ for $\tilde{g}_{1,2} = 7.3$, while with $\delta t = 50$ ns, $w = 0.24 \pm 0.05$ for $\tilde{g}_{1,2} = 10.2$. The solid line in Fig. 5 is given by $w = A/\tilde{g}_{1,2}$, where $A = \langle\tilde{g}_{1,1}\rangle\langle\tilde{g}_{2,2}\rangle$ is a constant determined from the measurements of $\tilde{g}_{1,1}$ and $\tilde{g}_{2,2}$. No corrections for dark counts or other backgrounds have been applied to the data in Fig. 5. We conclude that the observations shown in Fig. 5 represent a sizable nonclassical effect in support of the conditional generation of single photons for field 2.

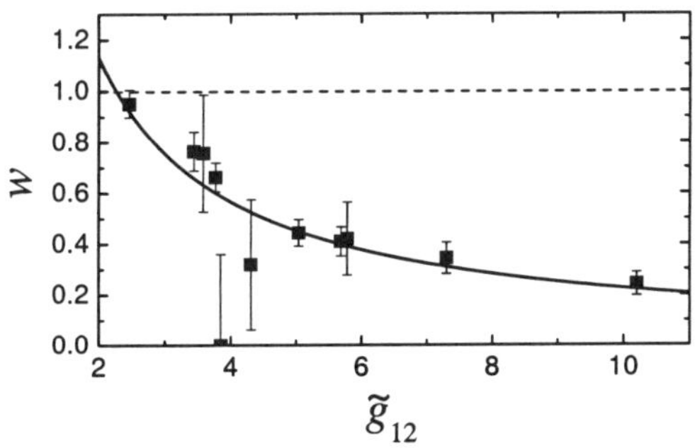

Fig. 5. Threefold correlation function w for a detection event 1_1 for field 1 followed by two events $(1_2, 1_2)$ for field 2 versus the normalized cross correlation $\tilde{g}_{1,2}$. $w < 1$ for sub-Poissonian fields in support of the single-photon character of field 2. Statistical uncertainties are indicated by the error bars. The full curve is given by $w = A/\tilde{g}_{1,2}$, where $A = \langle\tilde{g}_{1,1}\rangle\langle\tilde{g}_{2,2}\rangle$ is a constant independently determined from measurements of $\tilde{g}_{1,1}$ and $\tilde{g}_{2,2}$.

Another observation of the single photon character of field 2 was reported by Eisaman *et al.*[31] Measurements of the second order intensity correlation function $g^{(2)}_{nS}(AS)$ of field 2 as a function of the mean photon number in field 1 revealed a parameter region where $g^{(2)}_{nS}(AS) < 1$ (see Fig. 6), indicating the nonclassical character of the anti-Stokes photon states. The Mandel Q parameter[37] for field 2 was calculated from the results of Fig. 6 to be $Q = -0.09 \pm 0.03$ for $n_S = 2$ ($Q \geq 0$ for classical states and $Q = -1$ for Fock states).

5. Temporal Structure of the Nonclassical Correlations

As discussed so far in this chapter, central to the DLCZ protocol is the ability to write and read collective spin excitations into and out of an atomic ensemble, with efficient conversion of discrete spin excitations to single-photon wave packets. A critical aspect of such wave packets is that they are emitted into well defined spatiotemporal modes to enable quantum interference between emissions from separate ensembles (*e.g.*, for entanglement based quantum cryptography.[25])

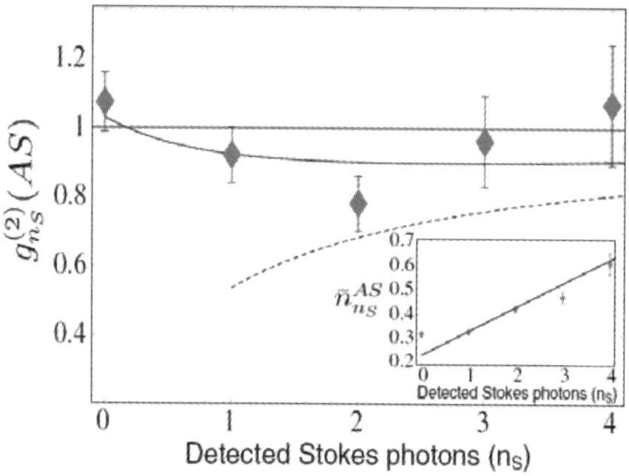

Fig. 6. Conditional nonclassical state generation.[31] Diamonds show experimentally measured values of the second-order intensity correlation function $g^{(2)}_{n_S}(AS)$ for the anti-Stokes pulse conditioned on the detection of n_S photons in the Stokes channel. The inset shows the mean number of photons in the anti-Stokes pulse $\bar{n}^{AS}_{n_S}$ as a function of n_S. For classical light, $g^{(2)} \geq 1$. The solid and dashed lines come from theoretical models explained in Ref. 31.

Following initial measurements of pulse shapes for the fields (1,2) (Fig. 2 of Ref. 29), Eisaman *et al.* have reported a detailed investigation of the manipulation of the temporal profile for the Stokes and anti-Stokes fields[31] in the few photon regime. As shown in Fig. 7, these authors achieved good correspondence between their measurements and the theoretically calculated photon fluxes. The observed dynamics provide evidence for the collective nature of the process, and indicates possible ways to control the shape of the single-photon wave packets.

The high efficiencies achieved in the work of Ref. 27 enabled us to investigate in detail the temporal properties of the nonclassical correlations between emitted photon pairs,[28] providing a direct look at different im-

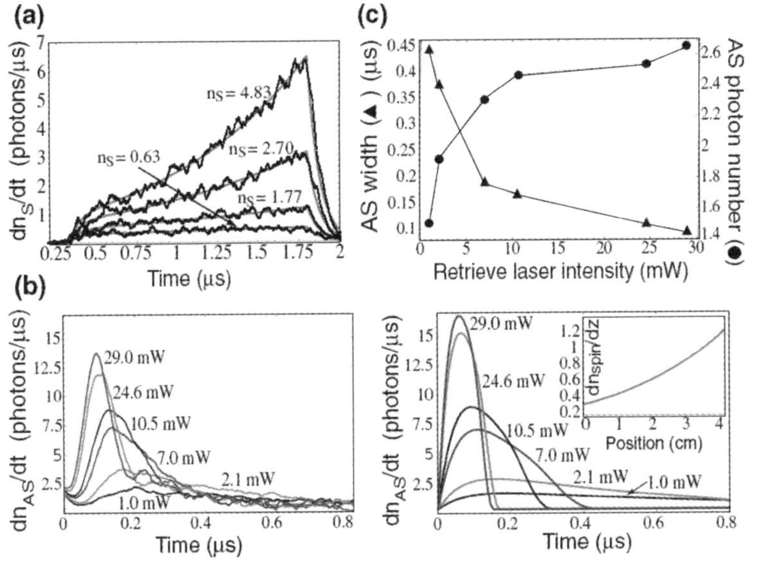

Fig. 7. Stokes and anti-Stokes pulse shapes, from Ref. 31. (a) Experimentally measured and theoretically calculated values of the Stokes photon flux dn_S/dt. For each plot, n_S represents the total number of photons emitted from the cell. Write laser power was varied from 25 μW to 100 μW. (b) Experimentally measured (left) and theoretically calculated (right) values of the anti-Stokes photon flux $dn_{AS} = dt$. The experimental pulse shapes correspond to a Stokes pulse with $n_S \approx 3$ photons, and the theoretical curves assume an initial spin-wave with $n_{spin} = 3$ excitations and an optical depth of ≈ 20. Each curve is labeled with the power of the retrieve laser. (b and inset) Theoretical calculation of the number of flipped spins per unit length $dn_{spin} = dz$ (cm^{-1}) for $n_{spin} = 3$. (c) Measured anti-Stokes pulse width (full-width at half-max) and total photon number as a function of the retrieve laser intensity.

portant features of the complete two-photon wave packet (field 1 + field 2) generated by the system. To address the photon statistics, we use four avalanche photodetectors, a pair for each field, as in the previous section. In the following analysis, our main quantity of interest is $p_\tau(t_1, t_2)$, the joint probability for photoelectric detection of photon 1 at time t_1 and photon 2 at time t_2 within a time window given by τ. The times for this quantity are counted starting from the beginning of the write pulse. Two other quantities necessary to measure R with temporal resolution are $p_\tau(t_1, t_1)$ and $p_\tau(t_2, t_2)$, the joint probabilities for two-photon events in fields 1 and 2, respectively. These quantities are determined from the record of time-stamped detections on all four photodetectors, allowing us to measure autocorrelations and cross correlations simultaneously. The detectors have a time resolution of 2 ns (minimum bin size), but usually we need to consider larger bins to acquire enough events for the statistics. The specific expression that we use to calculate R from these quantities is

$$R_\tau(t_1,t_2) \equiv \frac{[p_\tau(t_1,t_2)]^2}{p_\tau(t_1,t_1)p_\tau(t_2,t_2)}. \tag{10}$$

Another important quantity to measure is the joint probability $q_\tau(t_1,t_2)$ for detections from fields 1 and 2 coming from different trials, *i.e.*, uncorrelated. In this case, t_1 and t_2 have different time origins, they begin on the rising edge of write pulses from different trials. This quantity is expected to be proportional to the product of intensities of the fields 1 and 2, in reasonable correspondence to the form shown in Figs. 8(b) and 8(d) for our roughly rectangular write and read pulses, but distinctively different from $p_\tau(t_1,t_2)$ in Figs. 8(a) and 8(c). The comparison of $p_\tau(t_1,t_2)$ and $q_\tau(t_1,t_2)$ quantifies the degree of two-photon correlation for the system.

In our experiment, we focus on two cases: (I) nearly simultaneous application of write and read pulses with offset $\delta t = 50$ ns less than the duration of either pulse, and (II) consecutive application of write and read pulses with $\delta t = 200$ ns. Results for $p_\tau(t_1,t_2)$ and $q_\tau(t_1,t_2)$ are presented in Fig. 8. For both $\delta t = 50$ and 200 ns, $p_\tau(t_1,t_2) \gg q_\tau(t_1,t_2)$, indicating the strong correlation between fields 1 and 2, with the maximal ratio $\tilde{g}_{1,2} = p_\tau(t_1,t_2)/q_\tau(t_1,t_2) \gtrsim 30$. In Fig. 8, $\tau = 4$ ns, leading to statistical errors of about 8% for the largest values shown.

In case (I), Fig. 8a shows that $p_\tau(t_1,t_2)$ peaks along the line $t_2 - t_1 = \delta t_{12} \simeq 50$ ns with a width $\Delta t_{12} \simeq 60$ ns, in correspondence to the delay δt_{12} and duration Δt_{12} for read-out associated with the transition $|b\rangle \to |e'\rangle \to |a\rangle$ given an initial transition $|a\rangle \to |e\rangle \to |b\rangle$.[29] In case ($II$) with the read pulse launched 200 ns after the write pulse, excitation is "stored" in the atomic ensemble until the readout. The production of correlated photon pairs should now be distributed along $t_2 \simeq \delta t + \delta t_{12}$ with width $\simeq \Delta t_{12}$. Instead, as shown in Fig. 8c, $p_\tau(t_1,t_2)$ peaks towards the end of the write pulse (*i.e.*, $t_1 \gtrsim 100$ ns), and near the beginning of the read pulse (*i.e.*, $200 \lesssim t_2 \lesssim 300$ ns). Early events for field 1 lead to fewer correlated events for field 2, as $p_\tau(t_1,t_2)$ decays rapidly beyond the line $t_2 - t_1 = \tau_d \simeq 175$ ns. The marked contrast between $p_\tau(t_1,t_2)$ for $\delta t = 50$ and 200 ns results in a diminished ability for the conditional generation of single photons from excitation stored within the atomic ensemble[27] and, more generally, for the implementation of the DLCZ protocol for increasing Δt. The underlying mechanism is decoherence within the ensemble, as will be discussed in the next section.

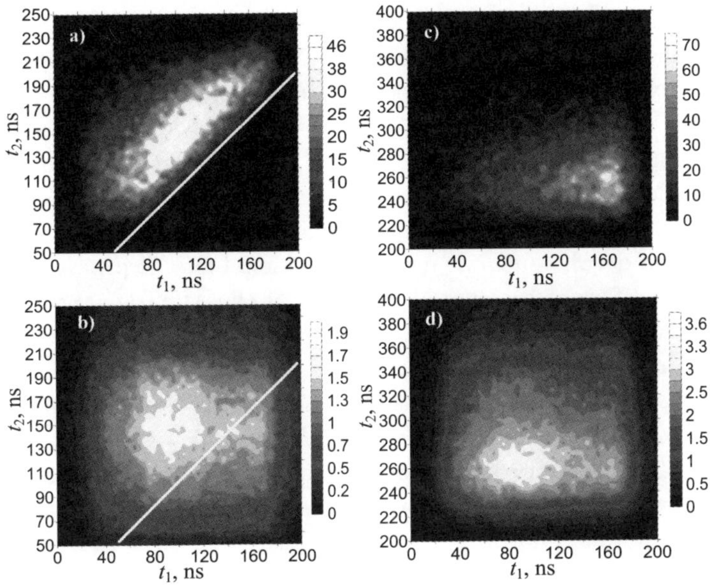

Fig. 8. Probability of joint detection scaled by 10^9 from fields (1,2) at times (t_1,t_2) with $t_1 = t_2 = 0$ at the beginning of the write pulse. $p_\mathcal{T}(t_1, t_2)$ comes from joint detections within the same trial. $q_\mathcal{T}(t_1, t_2)$ gives the probability of joint detection from different trials, *i.e.*, uncorrelated. (a) $p_\mathcal{T}$ and (b) $q_\mathcal{T}$ for overlapped write and read pulses, $\delta t =$ 50 ns. The solid line corresponds to $t_2 = t_1$. (c) $p_\mathcal{T}$ and (d) $q_\mathcal{T}$ for consecutive non-overlapping write and read pulses, $\delta t = 200$ ns. Note that $p_\mathcal{T} \gg q_\mathcal{T}$ in both situations, indicating the strong correlation between fields 1 and 2.

The temporal dependence of R closely follows $p_\mathcal{T}(t_1, t_2)$.[28] To deduce R from Eq. 10, we acquired the joint detection probabilities $p_\mathcal{T}(t_1, t_1)$ for field 1 and $p_\mathcal{T}(t_2, t_2)$ for field 2 from the same record of photoelectric events as for Figs. 8a and 8c. The maximal observed ratio R in a short temporal window of 30 ns is $R^{\max} = 292 \pm 57$ for $\Delta t = 50$ ns and $R^{\max} = 202 \pm 60$ for $\delta t = 200$ ns ($R = 198 \pm 33$ in the neighboring bin), which strongly violate the classical inequality $R \leq 1$. The relatively large errors in R arise predominantly from the uncertainties in $p_\mathcal{T}(t_1, t_1)$ and $p_\mathcal{T}(t_2, t_2)$, as the rate of detection of two photons in the same field is much lower than that of the correlated detection.

6. Decoherence in the Atomic Ensemble

As discussed in the previous section, the temporal structure of the two-photon wave packet emitted by the sample is strongly affected by decoherence effects, once the read and write pulses are delayed such that they do not overlap. This decoherence results in a strong decrease of the correlations between photons 1 and 2. The timescale for this decrease is on the order of 100 ns, very short for any process that relies only on atomic ground-state coherence.

This decoherence timescale, however, is consistent with the inhomogenous broadening of the ground state caused by the quadrupole magnetic field of the trap. In order to confirm that this is indeed the origin of the observed fast decoherence, we developed a model for the process taking into account the effect of the spatial variation of the MOT magnetic field and the resulting splitting of the atomic Zeeman states, and compared these results with measurements of $\tilde{g}_{1,2}$ as a function of the delay δt between read and write pulses. Figure 9a shows the result of this comparison.

In Fig. 9a, the normalized correlation function $\tilde{g}_{1,2}$ was determined from the ratio of integrated coincidence counts to singles counts over the entire detection window of 200 ns. The initial growth of $\tilde{g}_{1,2}$ for small δt comes from the fact that photon 2 can only be detected after the generation of a photon 1, since we initially optically pump all atoms to the $F = 4$ hyperfine state. In this way, if the read pulse arrives prior to the write pulse ($\delta t = -T_p$) we should observe no coincidence counts, and the number of coincidences should grow as the read pulse is delayed toward positive values. If the decoherence timescale is much longer than the pulse duration and the read pulse is weak, the integral number of coincidences should reach a maximum when the pulses are exactly back to back ($\delta t = T_p$), and then stay constant at this maximum for further delays. For strong read pulses, the maximum can be achieved at a shorter timescale, on the order of $\delta t \approx \delta t_{12}$ (see comments to Fig. 8a), staying constant later on. For the experiments described in the previous sections, the decoherence time is on the order of the pulse duration and the read pulse is strong, resulting in a maximum at $\delta t < T_p$ followed by a fast decay, as described by the open circles in Fig. 9a.

Since this strong decoherence seems to be related to the MOT quadrupole field, an immediate way to improve the coherence time is then to turn off this field. This needs to be done, however, as fast as possible in order to keep the high optical density of the atomic medium together with a high repetition rate for interrogation trials. The measurements of $\tilde{g}_{1,2}(\delta t)$ in this situation are shown in Fig. 9a (squares). Note that the coherence time improves significantly, but there is still a decay related to residual magnetic fields ($\approx 30\,\mathrm{mG}$) in the MOT region. When comparing the curves with and without magnetic field, it is important to have in mind that they were taken at different coupling conditions and atomic densities, so that the change in their maxima should not be attributed only to a change in the magnetic field. The modification in the rate of loss of coherence, however, can only be observed by turning off the field.

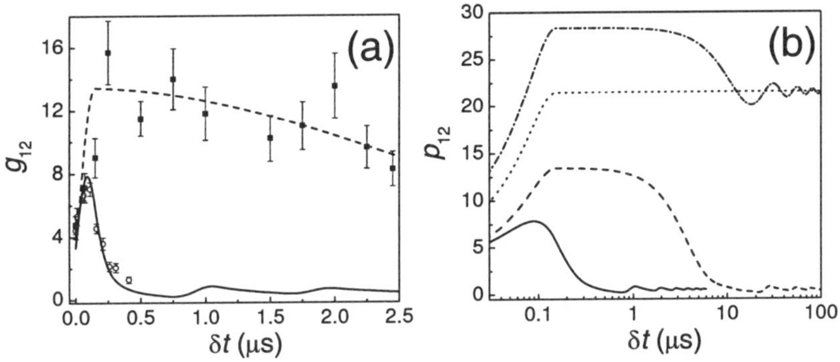

Fig. 9. (a) Variation of $\tilde{g}_{1,2}$ with the delay δt between write and read pulses. The open circles (filled squares) give the experimental results in the presence (absence) of the MOT magnetic field. The solid (dashed) curve is the theoretical result for $K = 1.1$ MHz ($K = 40$ kHz). (b) Variation of $\tilde{p}_{1,2}$ with the delay δt between write and read pulses for (solid curve) $K = 1.1$ MHz and an unpolarized sample, (dashed curve) $K = 40$ kHz and an unpolarized sample, and (dash-dotted curve) $K = 40$ kHz and an initially spin polarized sample with all atoms in $|F = 4, m_F = 0\rangle$. The dotted curve corresponds to an initially spin polarized sample classically excited by fields with polarizations such that only a magnetic insensitive transition is allowed, see text for details. In (b), the same arbitrary scaling factor was used for all curves.

In order to understand in detail the shape of the experimental curves in Fig. 9a, we extended the theoretical treatment of Duan *et al.*[33] to include the reading process as well as the full set of Zeeman states for the $F = 3, 4$ hyperfine levels. The sample of Cs atoms is assumed to be initially unpolarized and distributed over the same range of magnetic fields as for the MOT. With write and read pulses that approximate those used in our experiment and separated by δt, we calculate the joint probability $p_{1,2}^{th}(\delta t)$ to generate a pair of photons in the fields $(1, 2)$. We compare the quantity $\tilde{p}_{1,2} \equiv \xi p_{1,2}^{th}$ to the measured $\tilde{g}_{1,2}$ by way of a single overall scaling parameter ξ for all δt, as the rate of single counts in fields $(1, 2)$ is measured not to depend on δt (to within 20%). The form of $\tilde{p}_{1,2}$ strongly depends upon the inhomogeneity of Zeeman splitting across the MOT, which is described by the parameter $K = \mu_B g_{F_g} Lb/h$, where L is the MOT diameter, b is the gradient of the magnetic field for the MOT, and g_{F_g} is the Landé factor. The solid curve in Fig. 9a is the theoretical result for an initially unpolarized sample with $K = 1.1$ MHz, as for the experiment with the quadrupole field on (where $L \approx 3.6$ mm and $b \approx 8.4$ G/cm). The fast decay of the correlations is then adequately described by the model.

The dashed curve in Fig. 9a gives the decay expected for a magnetic-field gradient such that $K = 40$ kHz. This gives a reasonable approximation to the behavior of $\tilde{g}_{1,2}$ under the action of the residual magnetic field, even though the spatial dependence of this field can be more complicated than a simple linear gradient. The change in K from 1.1 MHz to 40 kHz is consistent with the reduction of the ground state linewidth between the two cases, as measured directly by Raman spectroscopy[50] in our setup.

The theory developed to explain the data in Fig. 9a, on the other hand, can also be used to devise new ways to improve the system. The inclusion of the Zeeman structure in the theory, for example, allows the study of different polarization schemes for both classical excitation and photon detection. It also allows the investigation of the role of the atomic initial state on the measured correlations. In Fig. 9b we give two examples of possible ways to improve the system. The solid and dashed lines in the figure represent the two experimental conditions of Fig. 9a (initially unpolarized samples with $K = 1.1$ MHz and $K = 40$ kHz), but now with the same scaling factor. The dash-dotted curve shows how the $K = 40$ kHz curve changes if the system is initially spin polarized, with all atoms in the $|F = 4, m_F = 0\rangle$ state. Note that in this case the value of $\tilde{g}_{1,2}$ considerably increases, and the system develops a plateau coming from the predominant transition $|F = 4, m_F = 0\rangle \rightarrow |F = 3, m_F = 0\rangle \rightarrow |F = 4, m_F = 0\rangle$, which is magnetic-field insensitive. Furthermore, it is possible to devise a polarization scheme of excitation that allows only this specific transition for any δt, as *e.g.* when the write pulse and field-1 detection are σ^+ polarized, and the read pulse and field-2 detection are σ^-. This is the case for the dotted curve in Fig. 9b.

The idealized improvements described by the dotted and dash-dotted curves of Fig. 9b, however, will probably be limited by two effects which are not taken into account by the theory. First, in our experimental setup we should see a decay with a timescale on the order of 100 μs due to the average time the cold atoms take to cross the 30 μm beam waist of the classical write and read pulses. Second, the theory assumes the presence of a magnetic field predominantly in the z direction, which defines the quantization axis. This can be obtained by applying an extra DC magnetic field along that direction,[51,52] but any residual transverse field should lead to some decay of the plateau. In spite of these restrictions, however, we believe that such improvements could lead to an increase of perhaps two orders of magnitude over the largest experimental decoherence time of Fig. 9a. It is also clear that there is a benefit in the careful preparation of the initial state for the magnitude of the measured correlations. This is an important point that should also be taken into account when considering the implementation of the DLCZ protocol in vapor cells.

It is essential to have in mind, however, that all coherence-time values discussed above are still far from fulfilling the requirement imposed by Eq. (7) on τ_{memory}. The largest memory times that can be obtained in vapor cells or MOTs (of the order of milliseconds) should be enough to provide proof of principle demonstrations of all of the critical elements of the DLCZ protocol. However, in order for this protocol to become really an alternative to long distance quantum communication, new technologies must be introduced. An association of purely optical traps (*e.g.*, a FORT[53]) with good magnetic shielding may eventually provide the required duration for the coherence time.

7. Prospect for Entanglement between Distant Ensembles

All previous sections discussed experiments related to the control of the photon generation process that is the basis of the DLCZ protocol. As explained in Sec. 2, however, this is just the first step. Entangling two spatially separated atomic ensembles is now the next essential step towards the realization of the protocol. In this section, we discuss the requirements to realize experimentally such entanglement and how to verify it.

As already mentioned in Sec. 2, to create entanglement between two spatially separated ensemble, L and R, we start by simultaneously exciting the two ensembles with a weak write pulse. At that point, the state of the system is given by Eq. (4). The photons emitted in the forward scattered-mode in each ensemble are then directed towards the input modes of a 50-50 beam splitter, as shown in Fig. 2. If the excitation probability is small enough and provided that the two fields entering the beam splitter are indistinguishable, the information concerning the origin of those fields is erased, and a detection event in the photodetector projects the two ensembles onto an entangled state. In an idealized setting, the resulting state is

$$|\Psi_{LR}\rangle = \left[\epsilon_L|1_a\rangle_L|0_a\rangle_R \pm e^{i\eta_1}\epsilon_R|0_a\rangle_L|1_a\rangle_R\right], \tag{11}$$

where $\epsilon_L(\epsilon_R)$ is the normalized amplitude of photon generation from ensemble $L(R)$, with $\epsilon_L^2 + \epsilon_R^2 = 1$. The phase η_1 is the sum of two contributions:

$\eta_1 = \Delta\beta + \Delta\gamma$, where $\Delta\beta$ is the phase difference of the classical write beams at the two ensembles L and R, and $\Delta\gamma$ is the difference of phase shift acquired by the field 1 during the propagation from the ensembles to the beam splitter. Note that this phase η_1 should be well-defined operationally and remain constant, in order to create a measurable entangled state.

To verify that an entangled state of the two ensembles has been created, the first step is to map the delocalized atomic excitation into a field state, by applying simultaneously a strong read beam at the two ensembles. Assuming that the mapping succeeds with a probability unity in an idealized case, the conditional overall state of the system becomes:

$$|\Psi_{LR}\rangle = \left[\epsilon_L|1_2\rangle_L|0_2\rangle_R \pm e^{i(\eta_1+\eta_2)}\epsilon_R|0_2\rangle_L|1_2\rangle_R\right], \tag{12}$$

where the phase η_2 is, similarly to η_1, the sum of two contributions, taking into account the phase difference of the read beams at the ensembles and the propagation phase of field 2. In this idealized setting, the atomic entangled state has now been mapped onto a photonic state entangled in the photon number basis (one photon with the vacuum), in two different spatial modes L and R. To show entanglement, there are two possibilities. The first one is to use the scheme proposed by Tan *et al.*[54] to demonstrate non-locality of a single photon (Fig. 10a). The second one is to reconstruct the density matrix of the entangled state, a procedure known as quantum tomography (Fig. 10b). The former method has the advantage that it detects entanglement irrespective of assumptions about multiple (and zero) excitations. on the other hand, it does not determine how much entanglement one created. Tomography does give a quantitative answer but requires one to explicitly take into account multiple- and zero-excitation probabilities.

Of course, the states in Eqs. (3)-(6) and (11)-(12) are idealizations relative to an actual experiment. Specifically, Eq. (6) is not realistic in that (a) terms with 2 or more excitations in either ensemble are neglected, (b) it is assumed that all excitations are in the correct "modes", both for the signal mode and the atomic ensembles, (c) it is assumed that each excitation of the atomic medium leads to a signal photon, and (d) diverse sources of uncorrelated background light are omitted. Hence any verification protocol must be expanded beyond the idealizations of Eqs. (11),(12) to provide a robust, model independent determination of entanglement.[55]

An initial attempt to create entanglement between two ensembles located within the same MOT has been reported by Matsukevich and Kuzmich.[32] Unfortunately, due to a verification protocol that does not address many of the system's imperfections listed above, the interpretation of their results is strongly model dependent.[55] Although the experiment demonstrated a substantial amount of coherence between the two ensembles, appropriate analysis and measurements are missing to determine conclusively the creation of entanglement. Particularly, we understand that the measurements reported in Ref. 32 can be reproduced by states of the two ensembles that are unentangled.[55]

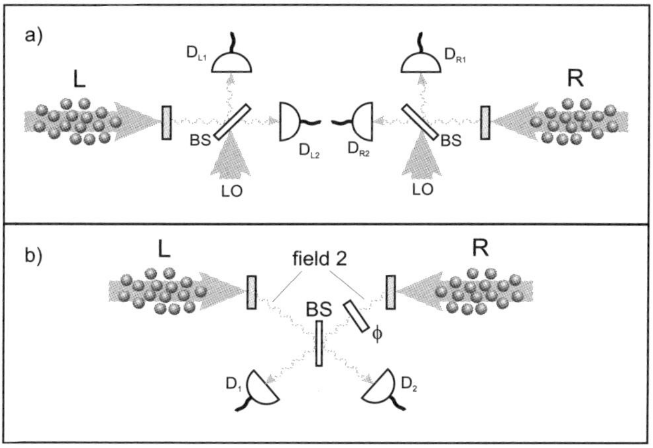

Fig. 10. Schematic setup for verification of entanglement between two atomic ensembles L and R, for entanglement generated as suggested by the DLCZ protocol. Conditioned on the detection of a first photon 1, the single excitation stored in the pair of ensembles is retrieved by read pulses that simultaneously excite each ensemble. (a) In the single photon non-locality scheme, the two output channels from the ensembles are directed to separate homodyne-detection setups. Correlations between the two detection regions are then used to verify if a Bell's inequality is violated, in which case entanglement between the two output channels is demonstrated. LO stands for local oscillator. (b) In the quantum tomography scheme, the two output channels from the ensembles are directed to the same beam splitter, and a phase ϕ can be introduced in one of the channels in a controlled way. The overall detection probabilities for single and double events are then obtained, together with the visibility of fringes observed with the variation of ϕ. These quantities are then used to reconstruct the density matrix ρ of the system, from which a measure of entanglement can be calculated.

7.1. *Single photon non-locality*

The scheme of Tan, Walls, and Collett[54] to demonstrate single photon non-locality is a double homodyne measurement, in which the signals from the two modes L and R are mixed with a local oscillator (LO) at a 50-50 beam splitter, as shown in Fig. 10a. The intensity of the LO must be very weak, such that when a photodetection occurs after the beam splitter, there is in principle no way of telling if the photon came from the signal or from the LO. If the coincidence count rate between two detectors placed after the beam splitters is recorded as a function of the phase shift between the two LOs, an interference fringe is expected with a visibility:[54]

$$V = \frac{1}{(\alpha^2 + 1)}, \tag{13}$$

where α is the amplitude of the LO. If α is very small, the visibility tends towards one. If V is higher than $1/\sqrt{2}$, a Bell inequality can be violated when four detectors are used.

This experiment requires the detection of triple coincidences, since the entanglement is always conditioned on the detection of a first photon. To date, only one experiment has demonstrated this scheme, using conditional single photons generated from parametric down conversion.[56] However, the background due to coincidences from either two signal photons or two photons from the LO had to be subtracted, in order to obtain enough visibility to violate a Bell inequality. Moreover, the standard analysis needs to be extended beyond the ideal state of Eq. (12) to include the various imperfections of any implementation of the DLCZ protocol.[25]

7.2. *Quantum tomography*

Another possibility to demonstrate entanglement is to reconstruct the density matrix $\rho_{L,R}$ of the photonic state in the two modes L and R and to calculate the amount of entanglement using *e.g.* Wootters' formula.[57] As previously discussed, the state given by Eq. 12 is an ideal state. In a real experiment, the state will be a mixed state containing contaminations from vacuum and higher order photon states, due to inherent atomic excitation statistics and various background light sources. If one neglects the events where two excitations are created in the same ensemble,[58] the total photonic state is an entangled state of two qubits living in a Hilbert space spanned by: $|1_2\rangle_L|0_2\rangle_R$, $|0_2\rangle_L|1_2\rangle_R$, $|1_2\rangle_L|1_2\rangle_R$ and $|0_2\rangle_L|0_2\rangle_R$. The density matrix can be written (for simplicity, we ignore all coherence terms between states with different numbers of excitations):

$$\rho_{L,R} = \begin{pmatrix} p_{00} & 0 & 0 & 0 \\ 0 & p_{01} & D & 0 \\ 0 & D^* & p_{10} & 0 \\ 0 & 0 & 0 & p_{11} \end{pmatrix}, \tag{14}$$

where D satisfies $|D|^2 \leq p_{01}p_{10}$. The diagonal elements, which are the probabilities p_{ij} to find i photons in mode L and j photon in mode R, can be determined by directly counting the number of photons in each mode. The coherence term D can be determined by adding a phase shift in one mode and then mixing the resulting state on a 50/50 beam splitter (see Fig. 10b). By recording the count rate after the beam splitter as function of the phase shift, one can measure an interference fringe with a visibility V. It can be shown that $V = 2|D|/(p_{10} + p_{01})$. Once the density matrix has been reconstructed, the amount of entanglement can be determined using Wootters' formula. The result for the concurrence C is:

$$C = max(2|D| - 2\sqrt{(p_{00}p_{11})}, 0). \tag{15}$$

The entanglement is given by:

$$E = h\left(\frac{1 + \sqrt{1 - C^2}}{2}\right) \tag{16}$$

with $h(p) = -p\log_2 p - (1-p)\log_2(1-p)$. We see that the probabilities p_{00} and p_{11} to have no excitation and one excitation in each mode, respectively, are essential in order to determine the amount of entanglement. Advantages of this method are that it requires no LO and that the coherence terms (which requires interferometric stability) can be determined with the record of double coincidences.

Finally, it is important to have in mind that probabilities such as p_{00} and p_{11} must be deduced from corresponding photodetection probabilities, *e.g.*, Q_{00} and Q_{11}. The actual experimental values (p_{00}^{exp}, p_{11}^{exp}, ...) to be used as inputs for the previous expressions to evaluate the amount of entanglement in the system must always be obtained by an inversion procedure of the direct measurements of photodetection probabilities. This inversion procedure takes into account, *e.g.*, the propagation and detection efficiencies in the experimental setup.

8. Concluding Remarks

The DLCZ protocol represents today one of the most promising schemes to solve the problem of long distance quantum communication. It suggests new ways to store information in atomic ensembles, to manipulate it, and to generate entanglement between remote systems. In this chapter, we have presented a discussion of initial experimental steps towards the realization of this protocol[25] that are being pursued by groups worldwide.[26–32] These experiments have already made important progress to generate and store single quanta in atomic ensembles.

However, much work remains to be done in order to demonstrate an elementary version of a quantum repeater using the DLCZ protocol. In particular, many experimental challenges must be solved in order to connect distant atomic ensembles. Nevertheless, the results presented in this chapter already indicate that the implementation of this protocol will lead to the development of many important tools for various tasks in quantum information science and quantum control.

Acknowledgments

The authors gratefully acknowledge fruitfull discussions with W. Bowen, A. Boca, J. McKeever, and D. Boozer. This work is supported by ARDA, by the Caltech MURI Center for Quantum Networks, and by the NSF. D.F. is supported by a fellowship from CNPq (Brazilian agency). H.d.R. is supported by a fellowhip from the Swiss National Science Foundation.

References

1. *The Physics of Quantum Information*, D. Bouwmeester, A. Ekert, and A. Zeilinger, Springer-Verlag, Berlin, Germany, 2001.
2. I. Marcikic, H. de Riedmatten, W. Tittel, H. Zbinden, M. Legré, and N. Gisin, Phys. Rev. Lett. **93**, 180502 (2004).
3. See the website of the companies *MagiQ* (www.magiqtech.com) and *id Quantique* (www.idquantique.com).
4. N. Gisin, G. Ribordy, W. Tittel, and H. Zbinden, Rev. Mod. Phys. **74**, 145 (2002).
5. D. Stucki, N. Gisin, O. Guinnard, G. Ribordy, and H. Zbinden, New J. Phys. **4**, 41 (2002).
6. C. Gobby, Z.L. Yuan, and A.J. Shields, Appl. Phys. Lett. **84**, 3762 (2004).
7. K.J. Resch, M. Lindenthal, B. Blauensteiner, H.R. Bhm, A. Fedrizzi, C. Kurtsiefer, A. Poppe, T. Schmitt-Manderbach, M. Taraba, R. Ursin, P. Walther, H. Weier, H. Weinfurter, and A. Zeilinger, Opt. Exp. **13**, 202 (2005).
8. C.-Z. Peng, T. Yang, X.-H. Bao, J.-Zhang, X.-M. Jin, F.-Y. Feng, B. Yang, J. Yang, J. Yin, Q. Zhang, N. Li, B.-L. Tian, and J.-W. Pan, quant-ph/0412218 (2004).
9. B.C. Jacobs and J.D. Franson, Opt. Lett. **21**, 1854 (1996).
10. R.J. Hughes, J.E. Nordholt, D. Derkacs, and C.G. Peterson, New J. Phys. **4**, 43 (2002).
11. C. Kurtsiefer, P. Zarda, M. Halder, H. Weinfurter, P.M. Gorman, P.R. Tapster, and J.G. Rarity, Nature **419**, 450 (2002).
12. G. Brassard, N. Lütkenhaus, T. Mor, and B.C. Sanders, Phys. Rev. Lett. **85**, 1330 (2000).
13. W.K. Wootters, and W.H. Zurek, Nature **299**, 802 (1982).

14. C.H. Bennett, G. Brassard, S. Popescu, B. Schumacher, J.A. Smolin, and W.K. Wootters, Phys. Rev. Lett. **76**, 722 (1996).
15. D. Deutsch, A. Ekert, R. Jozsa, C. Macchiavello, S. Popescu, and A. Sanpera, Phys. Rev. Lett. **77**, 2818 (1996).
16. H.-J. Briegel, W. Dür, J.I. Cirac, and P. Zoller, Phys. Rev. Lett. **81**, 5932 (1998).
17. W. Dür, H.-J. Briegel, J.I. Cirac, and P. Zoller, Phys. Rev. A **59**, 169 (1999).
18. C.H. Bennett, G. Brassard, C. Crépeau, R. Jozsa, A. Peres, and W.K. Wootters, Phys. Rev. Lett. **70**, 1895 (1993).
19. M. Zukowski, A. Zeilinger, M.A. Horne, and A.K. Ekert, Phys. Rev. Lett. **71**, 4287 (1993).
20. *Cavity Quantum Electrodynamics*, P. Berman (Eds.), Academic Press, San Diego 1994.
21. H.J. Kimble, Phys. Scr., T **76**, 127 (1998).
22. H. Mabuchi and A.C. Doherty, Science **298**, 1372 (2002).
23. J.I. Cirac, P. Zoller, H.J. Kimble, and H. Mabuchi, Phys. Rev. Lett. **78**, 3221 (1997).
24. S.J. van Enk, J.I. Cirac, and P. Zoller, Science **279**, 205 (1998).
25. L.-M. Duan, M.D. Lukin, J.I. Cirac, and P. Zoller, Nature **414**, 413 (2001).
26. A. Kuzmich, W.P. Bowen, A.D. Boozer, A. Boca, C.W. Chou, L.-M. Duan, and H.J. Kimble, Nature **423**, 731 (2003).
27. C.W. Chou, S.V. Polyakov, A. Kuzmich, and H.J. Kimble, Phys. Rev. Lett. **92**, 213601 (2004).
28. S.V. Polyakov, C.W. Chou, D. Felinto, and H.J. Kimble, Phys. Rev. Lett. **93**, 263601 (2004).
29. C.H. van der Wal, M.D. Eisaman, A. Andr, R.L. Walsworth, D.F. Phillips, A.S. Zibrov, and M.D. Lukin, Science **301**, 196 (2003).
30. W. Jiang, C. Han, P. Xue, L.-M. Duan, and G.-C. Guo, Phys. Rev. A **69**, 043819 (2004).
31. M.D. Eisaman, L. Childress, A. Andr, F. Massou, A.S. Zibrov, and M.D. Lukin, Phys. Rev. Lett. **93**, 233602 (2004).
32. D. N. Matsukevich and A. Kuzmich, Science **306**, 663 (2004).
33. L.-M. Duan, J.I. Cirac, and P. Zoller, Phys. Rev. A **66**, 023818 (2002).
34. M.D. Lukin, Rev. Mod. Phys. **75**, 457 (2003).
35. D.A. Braje, V. Balić, G.Y. Yin, and S.E. Harris, Phys. Rev. A **68**, 041801 (2003).
36. D.A. Braje, V. Balić, S. Goda, G.Y. Yin, and S.E. Harris, Phys. Rev. Lett. **93**, 183601 (2004).
37. L. Mandel and E. Wolf, *Optical Coherence and Quantum Optics*, Cambridge Univ. Press, Cambridge, UK, 1995.
38. B.C. Jacobs, T.B. Pittman, and J.D. Franson, Phys. Rev. A **66**, 052307 (2002).
39. E.B. Alexandrov, M.V. Baladas, D. Budker, D. English, D.F. Kimball, C.-H. Li, and V.V. Yashchuk, Phys. Rev. A **66**, 042903 (2002); Phys. Rev. A **70**, 049902 (2004).
40. J.F. Clauser, Phys. Rev. D **9**, 853 (1974).

41. Supplementary Information accompanying Ref. 26 at http://www.nature.com/nature/journal/v423/n6941/suppinfo/nature01714.html.
42. *Optical Coherence and Quantum Optics*, L. Mandel and E. Wolf, Cambridge Univ. Press, Cambridge, UK, 1995.
43. P. Michler, A. Kiraz, C. Becher, W. V. Schoenfeld, P. M. Petroff, Lidong Zhang, E. Hu, and A. Imamoglul, Science **290**, 2282 (2000).
44. E. Moreau, I. Robert, J.M. Gérard, I. Abram, L. Manin, and V. Thierry-Mieg, Appl. Phys. Lett. **79**, 2865 (2001).
45. M. Pelton, C. Santori, J. Vuckovic, B. Zhang, G.S. Solomon, J. Plant, and Y. Yamamoto, Phys. Rev. Lett. **89**, 233602 (2002).
46. J. McKeever, A. Boca, A. D. Boozer, R. Miller, J. R. Buck, A. Kuzmich, and H. J. Kimble, Science **303**, 1992 (2004).
47. P. Grangier, G. Roger, and A. Aspect, Europhys. Lett. **1**, 173 (1986).
48. P.G. Kwiat and R.Y. Chiao, Phys. Rev. Lett. **66**, 588 (1991).
49. C.K. Hong and L. Mandel, Phys. Rev. Lett. **56**, 58 (1986).
50. J. Ringot, P. Szriftgiser, and J.C. Garreau, Phys. Rev. A **65**, 013403 (2001).
51. G. Avila, V. Giordano, V. Candelier, E. de Clercq, G. Theobald, and P. Cerez, Phys. Rev. A **36**, 3719 (1987).
52. S.-I. Ohshima, Y. Nakadan, and Y. Koga, IEEE Trans. Instrum. Meas. **37**, 409 (1988).
53. J.D. Miller, R.A. Cline, and D.J. Heinzen, Phys. Rev. A **47**, 4567 (1993).
54. S.M.Tan, D.F.Walls, and M.J.Collett, Phys.Rev.Lett. **66**, 252 (1991).
55. S.J. van Enk and H.J. Kimble, submitted to publication.
56. B. Hessmo, P. Usachev, H. Heydari, and G. Björk, Phys.Rev.Lett. **92**, 180401 (2004).
57. W.K. Wootters, Phys.Rev.Lett 80, 2245 (1998).
58. This approximation is applied just to explain the basic elements in the idea of quantum tomography. In the actual experiment, components with larger photon number have necessarily to be taken into account. In particular, the probabilities p_{02} and p_{20} arise in the same order as does p_{11}. In this more general case, one cannot apply Wootters' formula, but instead one could calculate the negativity, as defined by G. Vidal and R.F. Werner [Phys. Rev. A **65**, 032314 (2002)], to determine the amount of entanglement.

Chapter 27

Decoherence and Decoherence Suppression in Ensemble-Based Quantum Memories for Photons

Michael Fleischhauer[1] and Claudia Mewes[1,2]

[1] *Fachbereich Physik, Technische Universität Kaiserslautern, D-67653 Kaiserslautern, Germany*
email: mfleisch@physik.uni-kl.de

[2] *Department of Physics, Ohio State University, Columbus, Ohio 43210, USA*

We discuss the effects of decoherence processes on the fidelity of ensemble-based quantum memories for photons and potential ways to suppress them. We analyse individual decoherence processes, where each atom is coupled to an independent reservoir as well as collective reservoir interactions. It is shown that despite the large entanglement of the storage states in the atomic ensemble, the sensitivity to decoherence does in general not increase with the number of atoms. This is due to the existence of equivalence classes of storage states characterised by the same projection onto the relevant quasi-particle modes, which contain the information about the stored photon state. It is shown furthermore that it is possible to construct a collective quantum memory which possesses a two-dimensional decoherence free subspace, provided all atoms are coupled to individual and independent reservoirs. This subspace consists of a pair of special collective states which have a large effective distance in state space. The probability of mutual transitions between these states due to single-atom errors scales like $1/N$ with N being the total number of atoms. Providing a sufficiently large energy gap to all other states suppresses transitions out of the subspace. In this way qubits can be protected from decoherence due to both, spin flips and dephasing processes.

1. Introduction

An important element of quantum information processing with photons[1,2] is a reliable quantum memory capable of a faithful storage of the quantum state of photons. Such a memory plays a key role in long-distance, secure quantum communication and quantum teleportation[3–7] as well as in network quantum computing.[8] Photons are one of the best suited information carriers, and atoms are reliable, and in the case of spin components of the electronic ground-state, also long lived storage units. A controllable and decoherence insensitive way of coupling between light and atoms is provided by Raman transitions. Although the conceptually simplest and for processing purposes best suited storage system for photonic qubits are *individual* atoms they do not allow a faithful transfer of quantum states from and to the radiation field, unless strongly coupling resonators are used. The technically difficult requirements of strong-coupling cavity QED can be alleviated however, if large ensemble of atoms are used. Here coherent and reversible transfer techniques for individual photon wave-packets[9–16] and cw light fields[17–20] have been proposed and in part experimentally implemented.

The advantage of a light-matter interface with atomic ensembles is the enhanced coupling between collective many-atom states and the radiation field. However, the corresponding excitations of the ensemble are highly entangled many-particle states if non-classical states of light or single photons are stored. Entangled states are known to be very sensitive to decoherence and one could naively expect that the lifetime of quantum correlations decreases with the number of atoms involved. On the other hand the storage states corresponding to a single photonic qubit are of the so-called W-class, which is known to preserve entanglement as long as the number of single-particle errors is small. It is therefore important to analyse the effect of unwanted environmental influences on the fidelity of the quantum memory.

Decoherence is usually modelled by coupling the system to a large reservoir of *e.g.* harmonic oscillators. Here two principle types of couplings can be distinguished which are indicated in Fig. 1. These are individual interactions (Fig. 1a) where each particle couples to its own independent reservoir and collective interactions (Fig. 1b) where all particles couple to the same bath. Decoherence processes are most generally described by completely positive maps of the density operator W of the system

$$W \longrightarrow W' = \sum_j E_j \, W \, E_j^\dagger, \qquad \sum_j E_j^\dagger E_j = \mathbf{1} \tag{1}$$

where E_j are the so-called Kraus operators. For the case of individual reservoir interactions the E_j are functionals of single-atom operators

$$E_j^{\rm ind} = E_j^{\rm ind}[\sigma_j, \sigma_j^\dagger]. \tag{2}$$

In the other case of a collective reservoir coupling the E_j correspond to specific collective modes described by operators

$$\Sigma_k = \sum_{j=1}^{N} \eta_{kj}\sigma_j, \quad \text{where} \quad \sum_{j=1}^{N} \eta_{jk}^* \eta_{jk'} = \delta_{kk'}. \tag{3}$$

$$E_k^{\text{coll}} = E_k^{\text{coll}}[\Sigma_k, \Sigma_k^\dagger]. \tag{4}$$

An example for the first type of coupling are collisions with atoms of a different species, an example for the second is a fluctuating magnetic field leading to random but correlated Zeeman shifts of all atoms. In reality the character of reservoir interactions will be in between the two limiting cases, *i.e.* the map (1) is neither diagonal in the individual-atom nor in the collective-mode representation.

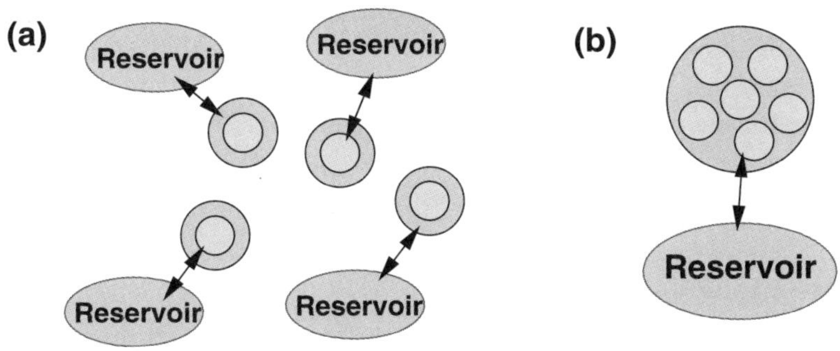

Fig. 1. Individual (a) vs. collective reservoir coupling (b).

We will show that for the storage and retrieval of a field state in a collective quantum memory only excitations in certain quasi-particle modes, the dark-state polaritons, are of importance. Thus all storage states with the same projection to this mode can be considered equivalent. Due to the existence of these equivalence classes there is no enhanced sensitivity to decoherence processes in an N atom system as compared to a single atom system, if they result from individual reservoir couplings. The same holds for collective decoherence processes if they affect all collective modes with comparable strength. This applies for *e.g.* to external magnetic fields with isotropic and homogeneous fluctuations.

Quantum error correction requires a fidelity for all individual storage and gate operations better than one part in 10^4.[22,23] Thus it is important to develop efficient techniques to suppress the influence of decoherence. Among the currently discussed strategies are geometric quantum computation[24] and the use of decoherence-free subspaces (DFS).[25] Although a DFS of dimension one is easy to find, the interesting case $d \geq 2$ requires in general special symmetries of the system–reservoir interaction, see Fig. 1b. An example for this is Dicke subradiance, where a tightly confined ensemble of two level atoms couples to the radiation vacuum.[21] The identical coupling of all atoms to the same vacuum modes generates a large DFS, the sub-radiant states. In many experimentally relevant circumstances, as for example in the case of a dilute atomic vapour, such symmetries do not exist, however, and the coupling to the environment is more accurately described by a model of individual and uncorrelated reservoirs, as in Fig. 1a.

As suggested by Kitaev for Majorana fermions[26,27] and by Dorner *et al.* for collective atom states in a 1-D lattice[28] qubits can effectively be protected from decoherence without requiring a highly symmetric reservoir coupling, if they are stored in special collective states of interacting multi-particle systems. We here show that such special states can also be found in the collective photon memories. Storing photonic qubits in these states and providing for an additional nonlinear interaction a *decoherence free subspace* of dimension 2 emerges.

2. Two-Mode Quantum Memory

In this section we introduce and discuss a model of a collective quantum memory based on the adiabatic rotation of dark-state polaritons.[10,11,14] Much of the following discussion applies however also for ensemble-based quantum memories which employ different storage and retrieval mechanisms.[16,17–20]

In order to simplify the discussion we will restrict ourselves to a quantum memory for a two-mode radiation field, realized for example in a weak-coupling resonator allowing for two orthogonal polarisations of the same frequency described by annihilation and creation operators $a_\pm, a_\pm^\dagger$.[10] First we reexamine the adiabatic transfer scheme[10,11] for this case. We consider an ensemble of N 5-level atoms with internal states $|a_\pm\rangle$, $|b\rangle$ and $|c_\pm\rangle$ resonantly coupled to the two quantum modes $a_\pm$ and classical control field of Rabi-frequencies $\Omega_\pm$ as shown in Fig. 2.

The dynamics of this system is described by the Hamiltonian ($E_b = \hbar\omega_b = 0$):

$$H = H_0 + H_I \ , \tag{5}$$

with

$$H_0 = \hbar\omega\left(a_+^\dagger a_+ + a_-^\dagger a_-\right) + \hbar\omega_a \sum_{j=1}^{N}\left(\sigma_{a_+a_+}^j + \sigma_{a_-a_-}^j\right)$$
$$+\hbar\omega_c \sum_{j=1}^{N}\left(\sigma_{c_+c_+}^j + \sigma_{c_-c_-}^j\right), \tag{6}$$

and

$$H_I = \hbar g \sum_{j=1}^{N}\left(a_+\sigma_{a_+b}^j + a_-\sigma_{a_-b}^j\right) + \hbar\Omega \mathrm{e}^{-i\nu t}\sum_{j=1}^{N}\left(\sigma_{a_+c}^j + \sigma_{a_-c}^j\right) + \text{h.c.}\,. \tag{7}$$

Here $\sigma_{\mu\nu}^j = |\mu\rangle_{jj}\langle\nu|$ is the flip operator of the jth atom and the vacuum Rabi-frequency is assumed to be equal for all atoms. All spatial phases have been absorbed in the definition of the atomic states.

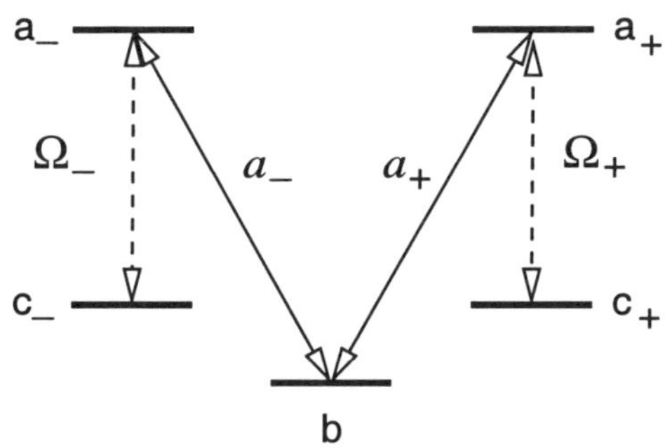

Fig. 2. Scheme for parallel storage of two modes $a_\pm$ with orthogonal polarisations. All atoms are initially prepared in state $|b\rangle$. $\Omega_\pm$ represent the Rabi-frequencies of coherent control fields.

When all atoms are initially prepared in level $|b\rangle$ the only states coupled by the interaction are the totally symmetric Dicke-states[21]

$$|\mathbf{b}\rangle = |b_1, b_2 \ldots b_N\rangle, \tag{8}$$

$$|\mathbf{a}_\pm^1\rangle = \frac{1}{\sqrt{N}}\sum_{j=1}^{N}|b_1 \ldots a_{\pm j} \ldots b\rangle, \tag{9}$$

$$|\mathbf{c}_\pm^1\rangle = \frac{1}{\sqrt{N}}\sum_{j=1}^{N}|b_1 \ldots c_{\pm j} \ldots b\rangle, \tag{10}$$

etc..

The couplings within the sub-systems consisting of the atomic states and the two radiation modes corresponding to a single excitation are shown in Fig. 2. The set of collective states can be separated into groups with specific excitation numbers $\{n_+, n_-\}$ in the two polarisation modes. Due to the symmetry of the interaction there is no coupling between these classes.

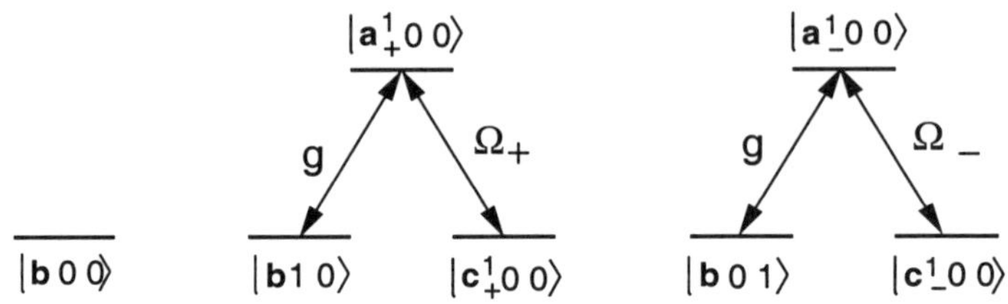

Fig. 3. Coupling of bare eigenstates of atom plus cavity system for at most one photon. $|\mathbf{b}, 1, 0\rangle$ denotes collective atomic state $|\mathbf{b}\rangle$, one photon in mode a_+ and zero photons in mode a_-.

In the following we will restrict ourselves to resonance conditions on all transitions. In this case the interaction of the N-atom system with the quantised radiation modes has a double-family of dark-states, *i.e.* adiabatic eigenstates with vanishing component of the excited states $|a_\pm\rangle_j$

$$|D, n, m\rangle = \sum_{k=0}^{n}\sum_{l=0}^{m} \xi_{nk}\xi_{ml}(-\sin\theta_+)^k(\cos\theta_+)^{n-k} \times \\ \times(-\sin\theta_-)^l(\cos\theta_-)^{m-l}|\mathbf{c}_+^k\mathbf{c}_-^l, n-k, m-l\rangle, \quad (11)$$

with $\xi_{nk} \equiv \sqrt{n!/(k!(n-k)!)}$. The mixing angles $\theta_\pm$ are defined as

$$\tan\theta_\pm(t) \equiv \frac{g\sqrt{N}}{\Omega_\pm(t)}, \quad (12)$$

and can be controlled by the external fields $\Omega_\pm(t)$. In (11) n and m denote the number of photons in the modes a_+ and a_- respectively. Although the dark states are degenerate, there are no transition between them even if non-adiabatic corrections are taken into account due to the symmetry of the interaction Hamiltonian.

The dark states of the N-atom system can be identified as quasi-particle excitations of the so-called dark-state polaritons $\Psi_\pm$ in the space of atoms and cavity mode[11]

$$\Psi_\pm = \cos\theta_\pm(t)\, a_\pm - \sin\theta_\pm(t)\, \frac{1}{\sqrt{N}} \sum_{j=1}^{N} \sigma^j_{bc_\pm}. \tag{13}$$

They are superpositions of the resonator mode and the collective spin corresponding to the ground-state transition $|b\rangle \leftrightarrow |c_\pm\rangle$.

$$|D,n,m\rangle_N = \frac{1}{\sqrt{n!m!}} \left(\Psi_+^\dagger\right)^n \left(\Psi_-^\dagger\right)^m |\mathbf{b},0,0\rangle. \tag{14}$$

Associated with the dark polaritons are the orthogonal superpositons

$$\Phi_\pm = \sin\theta_\pm(t)\, a_\pm + \cos\theta_\pm(t)\, \frac{1}{\sqrt{N}} \sum_{j=1}^{N} \sigma^j_{bc_\pm}. \tag{15}$$

which are called bright-polaritons. To obtain a complete set of operators in the relevant state space of the cavity photons and N atoms in internal states $|b\rangle$ and $|c_\pm\rangle$ we need in addition the operators

$$\Phi_{\pm l} = \frac{1}{\sqrt{N}} \sum_{j=1}^{N} \sigma^j_{bc_\pm} \exp\left\{2\pi i \frac{lj}{N}\right\}, \quad l = 1\ldots N-1, \tag{16}$$

which will also be referred to as bright polaritons. In the limit of small atomic excitations the polariton operators obey in first order of $1/N$ bosonic commutation relations.[14]

Adiabatically rotating the mixing angles $\theta_\pm$ from 0 to $\pi/2$ leads to a complete and reversible transfer of all photonic states to a collective atomic excitation if the maximum number $n_{\max}$ of photons in both modes $n+m$ is less than the number of atoms N. If the initial quantum state of the single-mode light field is described by the density matrix $\hat\rho_f = \sum_{n,m}^{n_{\max}} \sum_{k,l}^{n_{\max}} \rho^{kl}_{nm}\, |n,k\rangle\langle m,l|$, where the numbers n,m correspond to mode a_+ and k,l to mode a_-, the transfer process generates a quantum state of collective excitations according to

$$\begin{gathered}
\sum_{n,m}^{n_{\max}} \sum_{k,l}^{n_{\max}} \rho^{k,l}_{nm}\, |n,k\rangle\langle m,l| \otimes |\mathbf{b}\rangle\langle\mathbf{b}| \\
\downarrow \\
\sum_{n,m}^{n_{\max}} \sum_{k,l}^{n_{\max}} \rho^{k,l}_{nm}\, |D,n,k\rangle\langle D,m,l| \\
\downarrow \\
|0\rangle\langle 0| \otimes \sum_{n,m}^{n_{\max}} \sum_{k,l}^{n_{\max}} \rho^{k,l}_{nm}\, |\mathbf{c}_+^n \mathbf{c}_-^k\rangle\langle \mathbf{c}_+^m \mathbf{c}_-^l|\ .
\end{gathered} \tag{17}$$

In order to understand the adiabatic dynamics of the write and read-out processes, we will now take into account spontaneous decay from the two excited states. To this end we have to introduce a super-operator $\mathcal{L}$ acting on the N-atom density operator ρ:

$$\mathcal{L}\rho = -\frac{i}{\hbar}[H,\rho] - \frac{\gamma}{2}\sum_{j=1}^{N}\Big(\sigma^j_{a_+a_+}\rho + \rho\sigma^j_{a_+a_+}\Big) + \Big(''+'' \to ''-''\Big). \quad (18)$$

Here H is the Hamiltonian (5). Adiabatically eliminating the excited states, assuming two-photon resonance, and expressing $\mathcal{L}$ in terms of the polariton operators yields in a rotating frame

$$\mathcal{L}\rho = -\frac{i}{\hbar}\big[H_{\text{eff}},\rho\big] \quad (19)$$
$$-\frac{\Omega_+^2(t)}{2\gamma}\sum_{l=1}^{N-1}\Big(\Phi^\dagger_{+,l}\Phi_{+,l}\rho + \rho\Phi^\dagger_{+,l}\Phi_{+,l} - 2\Phi_{+,l}\rho\Phi^\dagger_{+,l}\Big) + \Big(''+'' \to'' -''\Big),$$

$$H_{\text{eff}} = \hbar\omega_c\Big\{\Psi^\dagger_+\Psi_+ + \Psi^\dagger_-\Psi_- + \sum_{l=0}^{N-1}\Big(\Phi^\dagger_{+,l}\Phi_{+,l} + \Phi^\dagger_{-,l}\Phi_{-,l}\Big)\Big\}. \quad (20)$$

We here see two important points: First of all, the adiabatic dynamics does not couple different polariton modes. Secondly, all bright polariton excitations $\Phi_{\pm,l}$, $l = 1,2,\ldots,N-1$ decay by optical pumping, *i.e.* by excitation to the excited state and successive spontaneous emission if $\Omega_\pm \neq 0$, while the dark polaritons $\Psi_\pm$ as well as the bright polaritons $\Phi_\pm$ are immune to spontaneous emission.

In the following we are interested in the storage of a single photonic qubit spanned by the two polarisation states $\{|1,0\rangle, |0,1\rangle\}$, *i.e.* the initial state

$$|\psi_0\rangle = \Big(\alpha|1,0\rangle + \beta|0,1\rangle\Big)\otimes|\mathbf{b}\rangle \quad (21)$$

is mapped to the collective state $|\psi_1\rangle = |0,0\rangle \otimes |\phi_1\rangle$ where

$$|\phi_1\rangle = \frac{\alpha}{\sqrt{N}}\sum_j |b_1,\ldots,c_{+j},\ldots,b_N\rangle$$
$$+\frac{\beta}{\sqrt{N}}\sum_j |b_1,\ldots,c_{-j},\ldots,b_N\rangle \quad (22)$$

through adiabatic rotation of the dark state

$$|D\rangle = \alpha|D,1,0\rangle + \beta|D,0,1\rangle,$$
$$= \Big(\alpha\Psi^\dagger_+ + \beta\Psi^\dagger_-\Big)|\mathbf{b},0,0\rangle \qquad \text{with} \quad \theta_+ = \theta_-. \quad (23)$$

3. Equivalence Classes of Storage States and Sensitivity to Decoherence

As can be seen from Eq. (22) the storage state of the photonic qubit is a maximally entangled N atom state of the W-type. These states are quite sensitive to certain decoherence processes. For example if a single atom undergoes a spin flip from internal state $|b\rangle$ to $|c_+\rangle$ or $|c_-\rangle$ the resulting state is almost orthogonal to (22). If p denotes the probability that one atom undergoes such a transition the total probability P_{error} to end up in an (almost) orthogonal state scales as $P_{\text{error}} \sim 1-(1-p)^N \sim pN$. Thus one might naively expect that for the storage of a single photon the collective quantum memory will have an N times enhanced sensitivity to decoherence as compared to a single-atom device. We will show in the following that this is in general not the case.

Inverting Eqs. (13) and (15) one finds

$$a_\pm = \cos\theta_\pm(t)\,\Psi_\pm + \sin\theta_\pm(t)\,\Phi_\pm. \tag{24}$$

Thus in the read-out process of the quantum memory, *i.e.* when rotating $\theta_\pm$ from $\pi/2$ to 0, only the dark polariton excitations $\Psi_\pm$ are transferred to the cavity modes. All other polariton excitations decay in the read-out process by optical pumping. Thus only the modes $\Psi_\pm$ are relevant for the stored quantum state. For example if W denotes the total density operator of the combined atom-cavity system after the writing process, all states W' that are identical to W when tracing out the bright polariton modes $\Phi_\pm$ and $\Phi_{\pm l}$, *i.e.* for which holds

$$\text{Tr}_\Phi\Big\{W'\Big\} = \text{Tr}_\Phi\Big\{W\Big\} \tag{25}$$

reproduce the same state of the radiation field. States with the property (25) define an *equivalence class*. The importance of these equivalence classes stems from the fact that transitions among the states of the same class due to unwanted environmental interactions do not affect the fidelity of the quantum memory.

3.1. *Individual reservoir interactions*

On the level of individual atoms the storage occurs within the three-state system consisting of $|b\rangle$, $|c_+\rangle$ and $|c_-\rangle$. We may safely neglect decoherence processes involving the excitation of other states. Then decoherence caused by individual and independent reservoir interactions can be described by the action of the two-level Pauli operators

$$X_\pm^j = \sigma_{bc_\pm}^j + \sigma_{c_\pm b}^j, \quad Z_\pm^j = \left[\sigma_{bc_\pm}^j, \sigma_{c_\pm b}^j\right], \quad Y_\pm^j = i\sigma_{bc_\pm}^j - i\sigma_{c_\pm b}^j. \tag{26}$$

$X_\pm^j$ describes a symmetric spin flip of the jth atom between states $|b\rangle$ and $|c_\pm\rangle$, $Z_\pm^j$ a phase flip, and $Y_\pm^j$ a combination of both. In addition there are corresponding processes involving states $|c_+\rangle$ and $|c_-\rangle$:

$$X_0^j = \sigma_{c_+c_-}^j + \sigma_{c_-c_+}^j, \quad Z_0^j = \left[\sigma_{c_+c_-}^j, \sigma_{c_-c_+}^j\right], \quad Y_0^j = i\sigma_{c_+c_-}^j - i\sigma_{c_-c_+}^j. \tag{27}$$

Every single-atom error can be expressed in terms of the above operators by a completely positive map of the total density operator

$$W_0 \longrightarrow W_1 = \sum_\mu p_\mu \frac{\Theta_\mu W_0 \Theta_\mu}{\mathrm{Tr}\left\{\Theta_\mu W_0 \Theta_\mu\right\}}, \tag{28}$$

where the $p_\mu \geq 0$ are the probabilities for the map Θ_μ, where $\Theta_\mu \in \{\mathbf{1}, X_\pm^j, Y_\pm^j, Z_\pm^j, X_0^j, Y_0^j, Z_0^j\}$ and $\sum_\mu p_\mu = 1$. We here have introduced a normalisation denominator although $\mathrm{Tr}\left\{\Theta_\mu W_0 \Theta_\mu\right\} = 1$ for all Θ_μ. This is because after expressing the Θ_μ in terms of polariton operators, we want to make explicit use of their Bose character, which is however only approximate.

In order to see the effect of the different decoherence processes after tracing out the bright polariton modes, we now express the operators Θ_μ in terms of polaritons in the limit $\theta_\pm = \pi/2$. This yields:

$$X_\pm^j = \frac{1}{\sqrt{N}} \left\{\sum_l \left(\eta_{jl}\Phi_{\pm l}^\dagger + \eta_{jl}^*\Phi_{\pm l}\right) - \Psi_\pm - \Psi_\pm^\dagger\right\}, \tag{29}$$

$$Y_\pm^j = \frac{i}{\sqrt{N}} \left\{\sum_l \left(\eta_{jl}\Phi_{\pm l}^\dagger - \eta_{jl}^*\Phi_{\pm l}\right) - \Psi_\pm + \Psi_\pm^\dagger\right\}, \tag{30}$$

$$Z_\pm^j = \frac{1}{N} \left[\sum_l \eta_{jl}^*\Phi_{\pm l} - \Psi_\pm, \sum_m \eta_{jm}\Phi_{\pm m}^\dagger - \Psi_\pm^\dagger\right], \tag{31}$$

and

$$X_0^j = \frac{1}{N}\Big(\sum_l \eta_{jl}\Phi_{+l}^\dagger - \Psi_+^\dagger\Big)\Big(\sum_m \eta_{jm}^*\Phi_{-m} - \Psi_-\Big) + h.a., \tag{32}$$

$$Y_0^j = \frac{i}{N}\Big(\sum_l \eta_{jl}\Phi_{+l}^\dagger - \Psi_+^\dagger\Big)\Big(\sum_m \eta_{jm}^*\Phi_{-m} - \Psi_-\Big) - h.a., \tag{33}$$

$$Z_0^j = \frac{1}{N}\Big(\sum_l \eta_{jl}\Phi_{+l}^\dagger - \Psi_+^\dagger\Big)\Big(\sum_l \eta_{jl}^*\Phi_{+l} - \Psi_+\Big) - \Big(\text{terms with } + \longleftrightarrow -\Big). \tag{34}$$

Here $\eta_{jl} = \exp\{-2\pi lj/N\}$. Since the $\Phi's$ and $\Psi's$ obey bosonic commutation relations up to corrections $\mathcal{O}(1/N)$, one sees from (31) that $Z_\pm^j = 0 + \mathcal{O}(1/N)$. For example within the bosonic quasi-particle approximations the action of a phase flip $Z_\pm^j$ cannot be calculated. However one can draw the conclusion that a single-atom phase error only contributes in first order of $1/N$. From the other equations one recognises an important property: If we assume that the initial state W_0 is an ideal storage state, *i.e.* without bright polariton excitations, we find that after tracing out the bright polariton states only decoherence contributions of order $\mathcal{O}(1/N)$ survive. For example:

$$\mathrm{Tr}_\Phi\Big(X_\pm^j W_0 X_\pm^j\Big) = \Big(1 - \frac{1}{N}\Big)\rho_0 + \frac{1}{N}(\Psi_\pm^\dagger + \Psi_\pm)\rho_0(\Psi_\pm^\dagger + \Psi_\pm). \tag{35}$$

where $\rho_0 = \mathrm{Tr}_\Phi(W_0)$. The factor $1/N$ in front of the second term exactly compensates for the fact that each atom can independently undergo a spin flip, which would yield an enhancement factor of N. Thus we see that due to the existence of equivalence classes, the collective quantum memory does not show an enhanced sensitivity with respect to single-atom errors.

3.2. *Collective reservoir interactions*

In the previous sub-section we considered individual and uncorrelated reservoir interactions of the atoms. We now discuss collective reservoir interactions, restrict ourselves however to Markovian decoherence processes. In this case the dynamics of the N-atom density operator can be described by a Lindblad equation of the type:

$$\frac{\mathrm{d}}{\mathrm{d}t}\rho\Big|_{\text{decoh}} = \mathcal{L}\rho = \sum_{k=1}^N \frac{\gamma_k}{2}\Big(2\Sigma_k\,\rho\,\Sigma_k^\dagger - \rho\,\Sigma_k^\dagger\Sigma_k - \Sigma_k^\dagger\Sigma_k\,\rho\Big) \tag{36}$$

where the $\Sigma_k, \Sigma_k^\dagger$ are the collective operators introduced in (3). In order to compare the effect of collective decoherence processes on the ensemble with that on an individual atom we express the collective operators in terms of single-atom variables. This yields

$$\frac{\mathrm{d}}{\mathrm{d}t}\rho\Big|_{\mathrm{decoh}} = \sum_{i,j=1}^{N} \frac{C_{ij}}{2}\Big(2\sigma_j\,\rho\,\sigma_i^\dagger - \rho\,\sigma_i^\dagger\sigma_j - \sigma_i^\dagger\sigma_j\,\rho\Big) \tag{37}$$

where $C_{ij} = \sum_{k=1}^N \eta_{ik}^*\eta_{jk}\gamma_k$. Let $\gamma_{\min}$ be the smallest collective decay rate. Then one easily verifies

$$\frac{\mathrm{d}}{\mathrm{d}t}\rho\Big|_{\mathrm{decoh}} \geq \gamma_{\min}\sum_{j=1}^{N}\Big(2\sigma_j\,\rho\,\sigma_j^\dagger - \rho\,\sigma_j^\dagger\sigma_j - \sigma_j^\dagger\sigma_j\,\rho\Big). \tag{38}$$

A quantum state stored in an individual atom would decay with a rate faster than the minimum collective decay rate. Thus if the collective decoherence rates $\gamma_\pm$ of the relevant dark-state polariton modes are comparable to the minimum collective rate $\gamma_{\min}$, the decoherence time of the collective storage state is again not much different from that of the quantum state stored in an individual atom.

4. Decoherence Suppression and Decoherence-Free Subspaces

In the following section we will show that the two-mode storage scheme shown in Fig. 2 can be used to create a (quasi) decoherence-free subspace of dimension 2. The two collective states

$$|\mathbf{c}_\pm^\mathbf{1}\rangle = \frac{1}{\sqrt{N}}\Big[|c_\pm bb\ldots b\rangle + \cdots + |bbb\ldots c_\pm\rangle\Big] \tag{39}$$

corresponding to a stored single photon in the two modes $a_\pm$ have an important property: Calculating the transition probability amplitude from one state to the other upon a spin flip of any one atom from $|c_+\rangle$ to $|c_-\rangle$ or vice versa, one finds that it scales as $1/N$. This can be seen most easily by expressing the spin-flip operator $X_j = |c_+\rangle_{jj}\langle c_-| + h.a. = \sigma^j_{c_+b}\sigma^j_{bc_-} + h.a.$ in terms of polaritons

$$\begin{aligned} X_j &= \frac{1}{N}\Big(\sum_l \eta_{jl}\Phi_{l+}^\dagger - \Psi_+^\dagger\Big)\Big(\sum_m \eta_{jm}^*\Phi_{m-} - \Psi_-\Big) + h.a. \\ &= \frac{1}{N}\Big(\Psi_+^\dagger\Psi_- + \Psi_-^\dagger\Psi_+\Big) + \cdots\,, \end{aligned} \tag{40}$$

where the omitted terms involve excitations or de-excitations of bright-polariton modes and $\eta_{jl} = \exp\{-2\pi lj/N\}$. One recognises that a single-atom spin flip causes a transitions between $|\mathbf{c}_+^\mathbf{1}\rangle$ and $|\mathbf{c}_-^\mathbf{1}\rangle$ with amplitude $1/N$. The corresponding transition probability scales as $1/N^2$. Thus even after taking into account that anyone of the N atoms can undergo a spin flip independently of all others, which leads to another factor of N, the total transition probability scales only as $1/N$. Consequently in a sufficiently large ensemble real transitions between the two states induced by single-atom spin-flips can practically be neglected.

If in addition a collective interaction between the atoms of the ensemble is provided corresponding to an effective Hamiltonian of the form

$$H_{\text{gap}} = -\hbar\omega_g \Big[|\mathbf{c}_+^\mathbf{1}\rangle\langle\mathbf{c}_+^\mathbf{1}| + |\mathbf{c}_-^\mathbf{1}\rangle\langle\mathbf{c}_-^\mathbf{1}|\Big], \tag{41}$$

the subspace of storage states $\mathcal{M} \equiv \{|\mathbf{c}_+^\mathbf{1}\rangle, |\mathbf{c}_-^\mathbf{1}\rangle\}$ is energetically shifted by an energy gap $\hbar\omega_g > 0$ below all other states. If the gap is sufficiently large, *e.g.* larger than the thermal energy of the reservoirs, transitions from $\mathcal{M}$ into all other states are exponentially suppressed. Hence $\mathcal{M}$ should be to a very good approximation a decoherence-free subspace with respect to individual spin-flips.

In many quantum system it is not difficult to suppress the effect of one specific type of environmental interaction. A truly decoherence-free subspace should however be insensitive to all relevant decoherence processes. In the present scheme the latter consist of spin-flips, phase flips and combinations of the two. Thus the question arises how phase-flips affect the subspace $\mathcal{M}$. A phase flip of the jth atom can be described by the action of the operator $Z_j = \sigma^j_{c_+c_+} - \sigma^j_{c_-c_-} = \sigma^j_{c_+b}\sigma^j_{bc_+} - \sigma^j_{c_-b}\sigma^j_{bc_-} =:$

$$\begin{aligned} Z_j &= \frac{1}{N}\Big(\sum_l \eta_{jl}\Phi^\dagger_{l+} - \Psi^\dagger_+\Big)\Big(\sum_l \eta^*_{jl}\Phi_{l+} - \Psi_+\Big) \\ &\quad - \Big(\text{terms with } + \longleftrightarrow -\Big) \\ &= \frac{1}{N}\Big(\Psi^\dagger_+\Psi_+ - \Psi^\dagger_-\Psi_-\Big) + \cdots, \end{aligned} \tag{42}$$

where the omitted terms again include excitations or de-excitations of bright polaritons. One recognises that a dephasing also leads to contributions within the subspace $\mathcal{M}$ only with amplitude $1/N$. All other terms that involve the excitation or de-excitation of bright polaritons are exponentially suppressed by the energy gap (41). We thus conclude that the subspace $\mathcal{M}$ is a quasi decoherence free subspace with respect to both, spin flips and dephasing processes. It is remarkable that the QDFS exists in the present system despite the fact that we have assumed no special symmetry in the reservoir coupling but rather individual uncorrelated bath interactions of the atoms. This is because the qubit states are delocalized many-particle states. If a collective rather than an individual reservoir interaction would have been assumed, the subspace $\mathcal{M}$ would not show the reduced sensitivity to decoherence. The assumption of uncorrelated individual reservoirs is however very well justified in a dilute system like a gas or a vapour.

To support the above arguments about the existence of a quasi decoherence-free subspace we proceed by analysing a particular model for the reservoir interaction of the atoms describing both spin-flip and dephasing. The model is indicated in Fig. 4. The atomic system is limited to the three relevant states $|b\rangle$, $|c_+\rangle$, and $|c_-\rangle$. It is important that dissipation should not be added on a phenomenological level here. The strong collective interaction (41) needs to be taken into account first before including perturbative reservoir interactions and applying the standard Born-Markov approximations. We thus start from a full Hamiltonian approach of the reservoir interaction and eliminate the bath degrees of freedom later.

Consider individual bosonic reservoirs $A_j, \ldots, F_j$ with finite thermal energy $k_B T$ and with mode operators $A_{jk}, A_{jk}^\dagger$ etc. Environmentally induced spin flips can then be described by the interaction Hamiltonian

$$
\begin{aligned}
H_{\text{spin}} &= \hbar \sum_{j=1}^{N} \sum_{k} \Big[g_{c_+c_-}^{k}\, A_{jk} + h.a. \Big] \Big[\sigma_{c_+c_-}^{j} + h.a. \Big] \\
&+ \hbar \sum_{j=1}^{N} \sum_{k} \Big[g_{c_+b}^{k}\, B_{jk} + h.a. \Big] \Big[\sigma_{c_+b}^{j} + h.a. \Big] \\
&+ \hbar \sum_{j=1}^{N} \sum_{k} \Big[g_{c_-b}^{k}\, C_{jk} + h.a. \Big] \Big[\sigma_{c_-b}^{j} + h.a. \Big],
\end{aligned} \tag{43}
$$

where the $g_{\mu\nu}^{j}$ denotes the coupling strength of atom j. It should be noted that the use of a rotating-wave approximation is in general not possible here due to the degeneracy of the involved states. In a similar way dephasing of the three states can be modelled by second order processes in the bosonic reservoir modes $\{D_k, E_k, F_k\}$

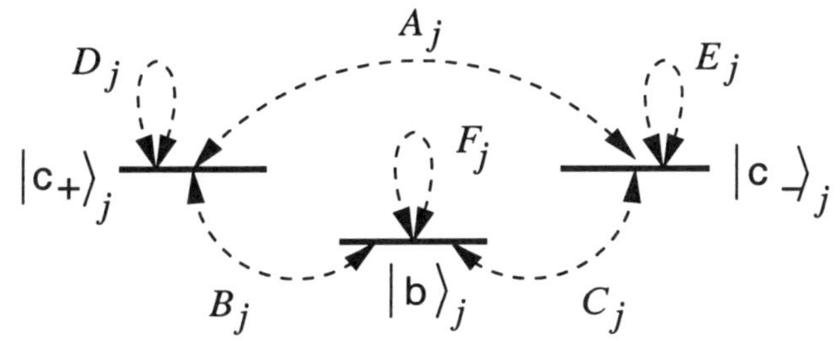

Fig. 4. Model for interactions of an atom with individual and independent bosonic reservoirs decribing spin-flips (A_j, B_j, C_j) and dephasing (D_j, E_j, F_j).

$$H_{\text{deph}} = \hbar \sum_{j=1}^{N} \sum_{k,s} \Big[\kappa^j_{c_+ ks}\, D^\dagger_{jk} D_{js}\, \sigma^j_{c_+c_+} + \kappa^j_{c_- ks}\, E^\dagger_{jk} E_{js}\, \sigma^j_{c_-c_-} + \kappa^j_{bks}\, F^\dagger_{jk} F_{js}\, \sigma^j_{bb} \Big]. \tag{44}$$

The reservoir operators $D^\dagger_{jk} D_{jk}$ etc. in (44) have a non-vanishing mean value in the thermal state. These mean values can be absorbed in the free Hamiltonian of the atoms and will not be considered further. Starting from the gap Hamiltonian (41) and the interactions with the reservoirs (43) and (44), eliminating the reservoir degrees of freedom within the standard Born-Markov approach we can derive a master equation for the reduced density operator of the atomic ensemble. The calculation is rather lengthy but straight forward. Here only the main results will be presented. Disregarding level shifts caused by the bath interaction we find for the populations in states $|\mathbf{c}^1_\pm\rangle$ the following density matrix equations in the interaction picture:

$$\begin{aligned}\frac{\mathrm{d}}{\mathrm{d}t}\rho_{\mathbf{c}_+\mathbf{c}_+} = &-\Big(1-\frac{1}{N}\Big)\, f_A(\omega_g)\, \overline{n}(\omega_g)\, \rho_{\mathbf{c}_+\mathbf{c}_+} \\ &-\frac{1}{N}\, f_A(0)\, \big[2\,\overline{n}(0)+1\big] \Big(\rho_{\mathbf{c}_+\mathbf{c}_+} - \rho_{\mathbf{c}_-\mathbf{c}_-}\Big) \\ &-2\, f_B(\omega_g)\, \overline{n}(\omega_g)\, \rho_{\mathbf{c}_+\mathbf{c}_+} \\ &-2\Big(1-\frac{1}{N}\Big)\Big(\alpha_D(\omega_g) + \alpha_F(\omega_g)\Big)\rho_{\mathbf{c}_+\mathbf{c}_+} \\ &+ \text{decay from outside of } \mathcal{M},\end{aligned} \tag{45}$$

and a similar equation for the matrix element $\rho_{\mathbf{c}_-\mathbf{c}_-}$. Here

$$\overline{n}(\omega) = \frac{\mathrm{e}^{-\beta\hbar\omega_k}}{1-\mathrm{e}^{-\beta\hbar\omega_k}} \tag{46}$$

is the thermal occupation number of the reservoirs at temperature $\beta = 1/k_B T$ and energy $\hbar\omega$.

$$f_A(\omega) = 2\pi h_A(\omega)|g_{c_+c_-}(\omega)|^2, \tag{47}$$

$$\alpha_D(\omega) = \pi \int_0^\infty d\omega' h_D(\omega')h_D(\omega+\omega')|\kappa_D(\omega+\omega',\omega)|^2 \times \overline{n}(\omega+\omega')\Big[\overline{n}(\omega')+1\Big] \tag{48}$$

etc. with $h_i(\omega)$ being the density of oscillator states at frequency ω. The terms in Eq. (45) can easily be interpreted. The first term describes the decay out of the symmetric (dark-polariton) state $|\mathbf{c}_+^1\rangle$ into non-symmetric (bright-polariton) states with the same number of excitations in internal states $|c_\pm\rangle$. This process is exponentially suppressed by the term $\overline{n}(\omega_g)$ due to the presence of the energy gap. The terms on the second line correspond to an excitation exchange between the collective states $|\mathbf{c}_+^1\rangle$ and $|\mathbf{c}_-^1\rangle$. As argued before this process is suppressed by a factor $1/N$. The third term describes spin flips to the collective ground state $|\mathbf{b}\rangle$. Also this process is exponentially suppressed by the factor $\overline{n}(\omega_g)$ due to the energy gap. Finally the fourth term is a decay of population in the symmetric (dark-polariton) collective states into non-symmetric (bright-polariton) collective states caused by dephasing of the single-atom states $|c_+\rangle$ and $|b\rangle$. As can be seen from Eq. (48) this contribution is also exponentially suppressed since it contains the term $\overline{n}(\omega+\omega_g) \le \overline{n}(\omega_g)$.

In a similar way one finds an equation for the off-diagonal density-matrix element $\rho_{\mathbf{c}_+\mathbf{c}_-}$:

$$\begin{aligned}
\frac{d}{dt}\rho_{\mathbf{c}_+\mathbf{c}_-} = &-\Big(1-\frac{1}{N}\Big)\, f_A(\omega_g)\,\overline{n}(\omega_g)\,\rho_{\mathbf{c}_+\mathbf{c}_-} \\
&-\frac{1}{N}\, f_A(0)\,\big[2\,\overline{n}(0)+1\big]\Big(\rho_{\mathbf{c}_+\mathbf{c}_-}-\rho_{\mathbf{c}_-\mathbf{c}_+}\Big) \\
&-3\,\Big(f_B(\omega_g)+f_C(\omega_g)\Big)\,\overline{n}(\omega_g)\,\rho_{\mathbf{c}_+\mathbf{c}_-} \\
&-\Big(1-\frac{1}{N}\Big)\Big(\alpha_D(\omega_g)+\alpha_E(\omega_g)+2\alpha_F(\omega_g)\Big)\rho_{\mathbf{c}_+\mathbf{c}_-} \\
&-\frac{1}{N}\Big(\alpha_D(0)+\alpha_E(0)\Big)\rho_{\mathbf{c}_+\mathbf{c}_-} \\
&+\text{ decay from outside of } \mathcal{M}.
\end{aligned} \tag{49}$$

The first four lines of Eq. (49) are analogous to corresponding terms in Eq. (45) and can be understood in a similar way. They are all either exponentially small or suppressed by a factor $1/N$. The fifth line is a new contribution corresponding to collective dephasing. It is suppressed by a factor $1/N$ as expected from previous discussion, see Eq. (42).

Thus the decoherence model of Fig. 4 exactly reproduces the anticipated behaviour. All decoherence processes resulting from individual and uncorrelated reservoir interactions of the atoms are either suppressed by the energy gap (41) or are proportional to $1/N$. The latter is due to the large effective distance of the collective states in state space. In this way a quasi decoherence free subspace of dimension two is generated which allows to protect a stored photonic qubit from decoherence much more efficiently than possible in quantum memories based on single particles. A remarkable property of the present scheme is that the presence of this QDFS does not require a symmetric coupling of the atoms to a common reservoir as often assumed in this context.[25] Rather the opposite limit of individual, uncorrelated reservoirs is considered, which is the natural situation for dilute ensembles.

5. Summary

In the present paper we have studied the influence of individual and certain collective decoherence processes on the fidelity of a quantum memory for photons based on ensembles of atoms. Despite the fact that the atomic storage states corresponding to non-classical states of the radiation field are highly entangled, the system shows no enhanced sensitivity to decoherence as compared *e.g.* to single-atom storage systems. This is due to the existence of equivalence classes of storage states corresponding to the excitations of one specific eigenmode of the system. It was shown that all states which have the same reduced density operator after tracing out the N bright-polariton modes, reproduce the same quantum state of light in the read-out process. For similar reasons no stringent requirements for preparation of the atomic system before the storage exist. It is sufficient that the number of atoms remaining in the storage level $|c\rangle$ after preparation of the ensemble is small compared to the total number of atoms, which can easily be achieved by optical pumping.

The present paper proves that atomic ensembles are suitable systems for the storage of quantum states of the radiation field even in the presence of decoherence processes. It should be noted that this conclusion does however not apply to the quantum gate between stored photonic qubits based on dipole blockade[29] nor to the photon detection scheme based on the mapping to ensembles.[30]

Acknowledgments

The authors would like to thank M. D. Lukin and R. Unanyan for stimulating discussions. The support of the DFG through the SPP 1078 "Quanteninformationsverarbeitung" and the EU network QUACS is gratefully acknlowedged. C. M. also thanks the Studienstiftung des Deutschen Volkes for financial support.

References

1. D.P. DiVincenzo, Fortschr. Physik, **48**, 771 (2000).
2. P. Zoller, J. I. Cirac, Luming Duan, J. J. Garcia-Ripoll, quant-ph/0405025.
3. L.-M. Duan, J. I. Cirac, P. Zoller, and E. S. Polzik, Phys. Rev. Lett. **85**, 5643 (2000).
4. L.-M. Duan, M. D. Lukin, J.I. Cirac, P. Zoller, Nature **414**, 413 (2001).
5. B. Julsgaard, A. Kozhekin, and E.S. Polzik, Nature **413**, 400 (2001).
6. A. Kuzmich, W. P. Bowen, A. D. Boozer, A. Boca, C. W. Chou, L.-M. Duan, and H. J. Kimble, Nature **423**, 731 (2003).
7. C. H. van der Wal, M. D. Eisaman, A. André, R. L. Walsworth, D. F. Phillips, A. S. Zibrov, and M. D. Lukin, Science 10859461 (2003).
8. J. I. Cirac, P. Zoller, H. Mabuchi, and H. J. Kimble, Phys.Rev.Lett. **78**, 3221 (1997).
9. J.R. Csesznegi and R. Grobe, Phys. Rev. Lett. **79**, 3162 (1997).
10. M.D. Lukin, S.F. Yelin, and M. Fleischhauer, Phys. Rev. Lett. **84**, 4232 (2000).
11. M. Fleischhauer, and M.D. Lukin, Phys. Rev. Lett. **84**, 5094 (2000).
12. D.F. Phillips, A. Fleischhauer, A. Mair, R.L. Walsworth, and M.D. Lukin, Phys. Rev. Lett. **86**, 783 (2001).
13. C. Liu, Z. Dutton, C.H. Behroozi, and L.V. Hau, Nature **409**, 490 (2001).
14. M. Fleischhauer and M.D. Lukin, Phys. Rev. A **65**, 022314 (2002).
15. M.D. Lukin, Rev. Mod. Phys. (2003).
16. B. Julsgaard, J. Sherson, J.I. Cirac, J. Fiurasek, and E. S. Polzik, Nature **432**, 482 (2004).
17. A. Kuzmich, K. Mølmer, and E.S. Polzik, Phys. Rev. Lett. **79**, 4782 (1997),
18. J. Hald, J.L. Sørensen, C. Schori, and E.S. Polzik, Phys. Rev. Lett. **83**, 1319 (1999).
19. A. Kuzmich and E.S. Polzik, Phys. Rev. Lett. **85**, 5639 (2000).
20. C. Schori, B. Julsgaard, J L. Sørensen, and E.S. Polzik, Phys. Rev. Lett. **89**, 057903 (2002).
21. R.H. Dicke, Phys. Rev. **93**, 99 (1954).
22. E. Knill, R. Laflame, and W. H. Zurek, Proc. R. Soc. London Ser. A **454**, 365 (1998).
23. J. Preskill Proc. R. Soc. London Ser. A **454**, 469 (1998).
24. see *e.g.*: V. Vedral, Int. J. Quantum Inf. **1**, 1 (2003) an references.
25. see *e.g.*: D.A. Lidar and K.B. Whaley in: *Irreversible Quantum Dynamics*,

F. Benatti and R. Floreanini (Eds.), pp. 83-120, Springer, Lecture Notes in Physics, vol. 622, Berlin 2003.

26. A. Kitaev, cond-mat/0010440
27. L.S. Levitov, T.P. Orlando, J.B. Majer, and J.E. Mooij, cond-mat/0108266
28. U. Dorner U.P. Fedichev, D. Jaksch, M. Lewenstein, and P. Zoller, Phys. Rev. Lett. **97**, 073601 (2003).
29. M.D. Lukin, M. Fleischhauer, R. Cote, L.M. Duan, D. Jaksch, J.I. Cirac, and P. Zoller, Phys. Rev. Lett. **87**, 037901 (2001).
30. A. Imamoğlu, Phys. Rev. Lett. **89**, 163602 (2002).

Index